ILLUSTRATED POWER S[illegible] DIAGNOSTICS

(6.0L - 2006-07)

2008 EDITION
FIRST PRINTING

HERITAGE INFORMATION SYSTEMS

Published By Heritage Information Systems
P.O. Box 141
Cochranville, PA 19330

ISBN: 978-0-9725784-1-7

To purchase an electronic version of this manual and all of our other products, visit us on the web at:
www.etruckfacts.com

HOW TO USE THIS MANUAL

This manual uses illustrated diagnostic steps to aid **QUALIFIED** technicians. Every effort has been made to include all of the needed data and information within each diagnostic test.

Below is a list of the chapters contained in this manual along with a brief description of the chapter.

Chapter 1: DESCRIPTION & OPERATION: This chapter provides a brief but concise description of the Powertrain Control System and its components.

Chapter 2: DIAGNOSTIC METHODS: This chapter contains test procedures that may be used within the Diagnostic Tests (Chapter 5).

Chapter 3: SYMPTOM CHARTS: This chapter provides technicians with a Diagnostic process when there are no Diagnostic Trouble Codes (DTC's)

Chapter 4: DIAGNOSTIC SUBROUTINES: This chapter provides Hard Start or No Start and Performance Diagnostic procedures. Both the Symptom Charts (Chapter 3), and the Diagnostic Tests (Chapter 5) will direct technicians to certain Diagnostic Subroutines to aid in the diagnostic process.

Chapter 5: Diagnostic Trouble Code (DTC) Index & Diagnostic Tests: The DTC index will tell you what Diagnostic Test and which test frame to enter to begin the diagnostic process. The first page of every Diagnostic Test contains a DTC chart as well as valuable reference information and connector identification.

Chapter 6: Reference Values: This chapter contains data that is called for in the throughout this entire manual. Also in this chapter is a chart of Diagnostic Trouble Codes (DTC's) and possible causes.

Chapter 7: Acronyms & Definitions: This chapter contains acronyms and their definitions that are used throughout this entire manual. This chapter also contains a Legend of commonly used illustrations.

TABLE OF CONTENTS

Where Is The Best Place To Start In This Manual?:

Chapter 5: DTC Index & Diagnostic Tests is usually the best place to start. After pulling DTC's, use the DTC index to direct you to the appropriate Diagnostic Test.

If No DTC's are present, perform the Diagnostic Test QT on page 5-293.

DESCRIPTION & OPERATION
INDEX

NOTES

Diesel Vehicle Emission Control Information

Decal
Each vehicle has a decal containing emission control information that applies specifically to the vehicle and engine.

The emission decal shows the model year, engine displacement, rated horsepower, fuel rate and idle speed.

Diesel On Board Diagnostics System

Overview
The California Air Resources Board (CARB) began regulation of On Board Diagnostic (OBD) for diesel vehicles sold in California beginning with the 1997 model year. OBD requires monitoring of emission-related components. The malfunction indicator lamp (MIL) is required to light and alert the driver of a malfunction and the need for service of the emission control system. A diagnostic trouble code (DTC) is associated with the MIL identifying the specific area of the fault.

Note: The OBD system is used on vehicles under 14,000 GVW.

The OBD system meets government regulations by monitoring the emission control system. When a system or component exceeds emission thresholds or a component operates outside of tolerance, a DTC will be stored and the MIL will be illuminated.

The OBD monitors detect system faults and initiates DTC setting and MIL activation. Fault detection strategy and MIL operation are associated with drive cycles. An OBD pending DTC is stored in the PCM keep alive random access memory (RAM) when a fault is first detected. The MIL is turned on after two consecutive drive cycles with faults. The DTC is cleared after 40 engine warm-up cycles without the fault being detected once the MIL is turned off. Once a monitor turns on the MIL, it will require three consecutive drive cycles without a fault for the MIL to turn off. The operation of each of the OBD monitors is discussed in detail within this section.

The On Board diagnostic computer program in the electronic powertrain control module (PCM) coordinates the OBD self-monitoring system. This program controls all the monitors and interactions, DTC and MIL operation, freeze frame data and scan tool interface.

Freeze frame data describes stored engine conditions such as state of the engine RPM and load at the point the first fault is detected. This data is accessible with the scan tool to assist in repairing the vehicle.

OBD inspection maintenance (IM) readiness DTC P1000 indicates that not all of the OBD monitors have been completed since the PCMs keep alive random access memory was last cleared. In some states, it may not be possible to obtain vehicle registration if P1000 is detected during inspection. To erase DTC P1000 from the PCM, operate the vehicle until the DTC is cleared using the manufacturer's specified drive cycle.

The On Board Diagnostic System is comprised of the comprehensive component monitor, glow plug monitor, misfire detection monitor and exhaust gas recirculation (EGR) monitor.

Diesel On Board Diagnostics System (Continued)

Diesel On Board Diagnostics Monitors

This section provides a general description of each OBD monitor. In these descriptions, the monitor strategy, hardware, testing requirements and methods are presented together to provide an overall understanding of each monitor operation.

Each illustration depicts the PCM as the main focus with the primary inputs and outputs for each monitor. The numbers to the left of the PCM represent the inputs used by each of the monitor strategies to enable or activate the monitor. The components and subsystems to the right of the PCM represent the hardware and signals used while performing the tests and the systems being tested. The comprehensive component monitor (CCM) illustration has numerous components and signals involved and is shown generically. When referring to the illustrations, match the numbers to the corresponding numbers in the monitor descriptions for a better comprehension of the monitor and associated DTCs.

These monitor descriptions are intended as general information only.

Deviations From Standard Gasoline Implementation of OBD

1. Readiness (i.e.: all monitors complete) is based on diagnostics for the following 6.0L diesel engine systems:
    - Comprehensive Component Monitor (CCM)
    - Misfire Detection Monitor
    - EGR Monitor

   The glow plug monitor is part of the comprehensive component monitor.

   Readiness is based on every OBD (component) having run sufficiently to have found a fault without regard to whether or not a fault exists.
2. The drive cycle provided in the Diagnostic Method section is used to clear a P1000 code. The diesel system differs substantially from the gasoline system.
3. The command to clear DTCs will ONLY clear P1000 if all drive cycle testing has been satisfied. All other detected DTCs will be cleared with a CLEAR code command from the scan tool if the fault that caused the DTC is no longer present.

Comprehensive Component Monitor

The comprehensive component monitor (CCM) is an on-board strategy designed to monitor a fault in any electronic component or circuit that provides input or output signal to the PCM and is not exclusively monitored by another monitor system. Inputs and outputs are considered inoperative when a failure exists due to a lack of circuit continuity, out-of-range value, or a failed rationality check.

The CCM covers many components and their circuits. The tests vary depending on the hardware, function, and type of signal. For example, analog inputs are typically checked for opens, shorts, out of range values and rationality. This type of monitoring is performed continuously. These tests may require the monitoring of several components and can only be performed under the appropriate test conditions. Some outputs are also monitored for the correct function by observing the reaction of the control system to a given change in the output command. An example of this would be the injection control system.

In general, the CCM covers a broad range of individual component and circuit checks and testing is performed under various conditions. The CCM is enabled after the ignition switch is turned on for three MIL is activated if the fault detected affects emissions. All of the CCM tests are also performed during on demand self-test.

The following is an example of some of the input and output components monitored by the CCM. The components monitored may belong to the engine, transmission or any other PCM supported subsystem.

Diesel On Board Diagnostics System (Continued)

1. Inputs:
   - engine oil temperature (EOT), accelerator pedal position (APP), camshaft position (CMP)
2. Outputs:
   - injection pressure regulator (IPR), exhaust gas recirculation (EGR) valve
3. The MIL is activated after a fault is detected, if the fault detected affects emissions.

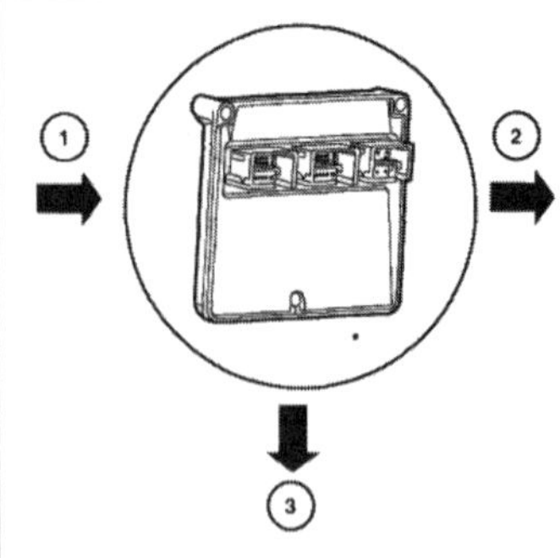

Comprehensive Component Monitor

Misfire Detection Monitor

The misfire detection monitor is an on-board strategy designed to monitor engine misfire and identify the specific cylinder in which the misfire has occurred. Misfire is defined as lack of combustion in a cylinder due to poor compression, fuel delivery or mechanical engine failure. The misfire detection monitor will be enabled only when certain base engine conditions are first satisfied. Input from the following sensors is required to enable the monitor: engine oil temperature (EOT), crankshaft position (CKP), mass fuel desired (MFDES), exhaust pressure (EP), intake air temperature (IAT), fuel level (FLI) and injector control pressure (ICP).

1. The CKP signal generated is the main input used in determining cylinder misfire.
2. The input signal generated by the CKP sensor is derived by sensing the passage of the teeth from the crankshaft position wheel, which is mounted on the crankshaft.
3. The input signal to the PCM is then used to calculate the time between CKP edges and also crankshaft rotational velocity and acceleration. By comparing the accelerations of each cylinder event, the power loss of each cylinder is determined. When the power loss of a particular cylinder is sufficiently less than a calibrated value and other criteria is met, the suspect cylinder is determined to have misfired.

Misfire Type B:

Upon detection of a misfire type B (1,000 revolutions), which will exceed the emissions threshold or cause a vehicle to fail an inspection and maintenance tailpipe emissions test, the MIL will illuminate and a DTC will be stored.

The DTC associated with multiple cylinder misfire is DTC P0300.

The DTCs associated with cylinder misfire are P0301, P0302, P0303, P0304, P0305, P0306, P0307 and P0308.

DTC	Description	Possible Causes	Diagnostic Aides
P0300 Random Misfire	The random misfire DTC indicates multiple cylinders are misfiring or the PCM cannot identify which cylinder is misfiring.	Oil/fuel aeration Base engine	
P0301 — P0308 Misfire Detection Monitor	The misfire detection monitor is designed to monitor engine misfire and identify the specific cylinder effected due to poor compression or any other cause.	Base engine Injector Circuitry	If the MIL is on steady state due to a misfire, this will indicate the threshold for emissions was exceeded.

Glow Plug Monitor

The 6.0L diesel engine utilizes a glow plug monitor (GPM) system designed to detect failed glow plugs or failed wiring in the glow plug system. DTCs indicate which cylinder has failed glow plugs or failed glow plug wiring.

Diesel On Board Diagnostics System (Continued)

The glow plug system is composed of a solid state glow plug control module (GPCM), glow plugs and the associated wiring harness. The glow plug on-time is controlled by the PCM and is a function of oil temperature, barometric pressure and battery voltage. The PCM enables the GPCM which powers the individual glow plugs. Glow plug on-time, varies between 1 to 120 seconds. In addition to the PCM control, the GPCM limits the glow plug operation to 180 seconds regardless of PCM commanded on-time. The power to the glow plugs is provided through the GPCM solid state drivers directly from the vehicle battery. The GPCM monitors and detects individual glow plug functionality, and the control and communication links to the PCM. The failures detected by the GPCM are passed to the PCM using a serial communication signal on the diagnostic line.

Note: Wait-to-start lamp on time is controlled by the PCM and is independent from GPCM on-time.

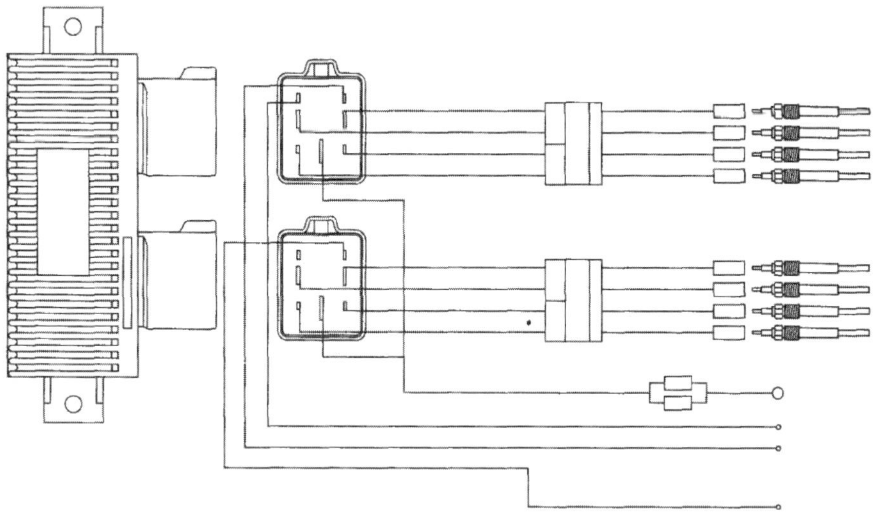

Glow Plug System

The key on engine off (KOEO) test is performed in order to test the GPCM control circuit for failure. Glow plugs are not operated during this test.

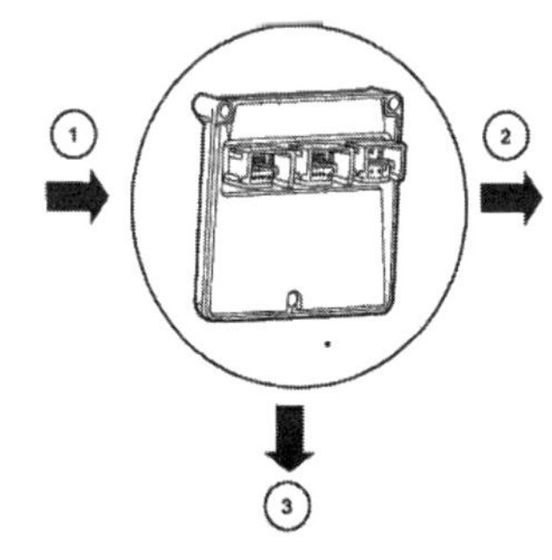

PCM Glow Plug Monitor

The glow plug monitor self-test is a functional KOER test of the PCM performed on demand with the engine running and the A/C off. The PCM will activate the GPCM which monitors the glow plugs. The pedal may be used to increase the engine speed to increase voltage if needed. A fault must be present at the time of testing for the test to detect a fault. The DTCs will be sent to the PCM on the diagnostic line and output to the scan tool.

The following is an example of some of the input and output components monitored by the PCM. The components monitored belong to the engine system.

1. Inputs:
   • engine oil temperature (EOT), barometric pressure sensor (BARO)
2. Outputs:
   • GPCM
3. The MIL is activated after a fault, if the fault detected affects emissions.

Malfunction Indicator Lamp (MIL)

The malfunction indicator lamp (MIL) informs the driver that the PCM has detected an OBD emission-related component or system fault. When this occurs on a California calibrated engine, an OBD DTC will be set. On a federal calibrated engine and transmission, certain faults will also illuminate the MIL.

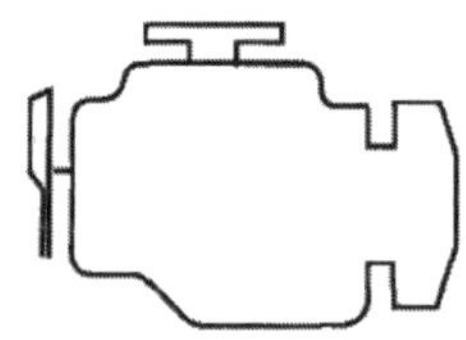

Malfunction Indicator Lamp (MIL)

- On all vehicles, the MIL is located in the instrument cluster.
- The indicator is illuminated by a PCM message sent through the standard corporate protocol (SCP) communications network to the instrument cluster.
- An engine or transmission with federal calibration operating in the failure mode effects management (FMEM) will cause the MIL to be illuminated.
- To turn off the MIL after a repair, a clear DTCs command from the scan tool must be sent to the PCM.
- For any MIL concern, go to, Symptom Charts.

Diesel Electronic Engine Control (EEC) System

Overview

The PCM contains both engine and transmission microprocessors. Operating information, as well as fault information, is communicated between the two processors via controller area network (CAN) communications. Both can be programmed individually, however replacement of the PCM is as an assembly.

Fuel control is accomplished by the PCM and the fuel injection control module (FICM). The two modules communicate operational information via CAN communications.

The electronic control (EC) system provides optimum control of the engine and transmission through the enhanced capability of the PCM. The EC system also has an on-board diagnostics monitoring system (on-board diagnostics) with features and functions to meet federal regulations on exhaust emissions.

The EC system has two major divisions: hardware and software. The hardware includes the PCM, FICM, sensors, switches, actuators, solenoids, and interconnecting terminals. The software in the PCM provides the strategy control for outputs (engine and transmission hardware) based on the values of the inputs to the PCM. EC hardware and software are discussed in this section.

The PCM receives information from a variety of sensor and switch inputs. Based on the strategy and calibration stored within the memory chip, the PCM generates the appropriate output. The system is designed to minimize emissions and optimize fuel economy and driveability. The software strategy controls the basic operation of the engine and transmission, provides the OBD strategy, controls the MIL, communicates to the scan tool via the data link connector (DLC), allows for flash electrically erasable programmable read only memory (FEEPROM), and controls failure mode effects management (FMEM).

Modifications to OBD Vehicles

Modifications or additions to the vehicle may cause incorrect operation of the OBD system. Performance modifications that cause a Ford part to fail may not be covered by the Ford New Vehicle Limited Warranty. Burglar alarms, cellular telephones and CB radios must be carefully installed. Do not install these devices by tapping into or running wires close to powertrain control system wires or components.

Diesel Powertrain Control Software

Engine RPM Limiter
The PCM limits engine RPM by cutting off fuel whenever the engine RPM limit is detected. RPM limits are as follows:

- F-Series 250/550 & E-Series — 4000 RPM

The purpose of the engine RPM limiter is to prevent damage to the powertrain.

Flash Electrically Erasable Programmable Read Only Memory
The flash electrically erasable programmable read only memory (FEEPROM) is an integrated circuit (IC) within the PCM. This integrated circuit contains the software code required by the PCM to control the powertrain. One feature of the FEEPROM is that it can be electrically erased and then reprogrammed without removing the PCM from the vehicle. If a software change is required to the PCM, the module no longer needs to be replaced, but can be reprogrammed at the dealership. The reprogramming is done through the DLC.

Failure Mode Effects Management
Failure mode effects management (FMEM) is an alternate system strategy in the PCM designed to maintain vehicle operation if one or more critical sensor inputs fail.

When a sensor input is perceived to be out-of-limits by the PCM, an alternative strategy is initiated. The PCM substitutes a fixed value and continues to monitor the incorrect sensor input. If the suspect sensor operates within limits, the PCM returns to the normal engine running strategy.

FMEM operation will result in continuous memory DTCs during normal engine operation and when performing key on engine running (KOER) self-test mode.

High Speed-Controller Area Network (HS-CAN)
HS-CAN is based on SAE J2284, ISO-11898 and is a serial communication language protocol used to transfer messages (signals) between electronic control modules or nodes. Two or more signals can be sent over one CAN network circuit allowing two or more electronic modules or nodes to communicate with each other. This communication or multiplexing network operates at 500k/sec (kilobytes per second) and allows the electronic modules to share their information messages.

Included in these messages is diagnostic data sent over the CAN High (+) and CAN Low (-) lines to the DLC. PCM connection to the DLC is typically done with a two wire, twisted pair cable used for network interconnection. The diagnostic data such as self-test or PIDs can be accessed with a scan tool. Information on scan tool equipment is described in Diagnostic Methods.

Keep Alive Random Access Memory (RAM)
The PCM stores information in keep alive random access memory (RAM) (a memory integrated circuit chip) about vehicle operating conditions, and then uses this information to compensate for component variability. Keep Alive RAM remains powered when the vehicle key is off so that this information is not lost.

Multiplexing
The increased number of modules on the vehicle dictate a more efficient method of communication. Multiplexing is the process of communicating several messages over the same signal path. This process allows multiple modules to communicate with each other through the signal path (CAN2H/CAN2L). Modules communicate with the PCM using SCP or CAN communications, which determines the priority in which the signals are sent. Refer to SCP or High Speed-Controller Area Network (HS-CAN) for more information.

Diesel Powertrain Control Software (Continued)

Power and Ground Signals

Vehicle Power

When the key is turned to the start or run position, battery positive voltage (B+) is applied to the coil of the PCM power relay. Since the other end of the coil is wired to ground, this energizes the coil and closes the contacts of the PCM power relay. Vehicle power (VPWR) is now sent to the PCM and the EC system as VPWR.

Vehicle Reference Voltage

The vehicle reference voltage (VREF) is a positive voltage (about 5.0 volts) that is an output by the PCM. This is a consistent voltage that is used by the three-wire sensors.

Signal Return

The signal return (SIG RTN) is a dedicated ground circuit used by most electronic control (EC) sensors and some other inputs.

Power Ground

Power ground (PWR GND) is an electric current path return for VPWR voltage circuit. The purpose of the PWR GND is to maintain sufficient voltage at the PCM.

Gold-Plated Pins

Some engine control hardware components have gold-plated pins on the connectors and mating harness connectors to improve electrical stability for low draw current circuits and to enhance corrosion resistance. The EC components equipped with gold terminals will vary by vehicle application.

Note: Damaged gold terminals should only be replaced with new gold terminals.

Powertrain Control Module

The center of the electronic control (EC) system is the powertrain control module (PCM). The PCM contains both engine and transmission microprocessors. Operating information, as well as fault information, is communicated between the two processors via controller area network (CAN) communications. Both can be programmed individually, however replacement of the PCM is as an assembly. The PCM has three electrical connectors (122 pins total). The PCM receives input from sensors and other electronic components (switches, relays, etc.) and places this information into random access memory (RAM) or keep alive RAM. Based on information programmed into its read-only memory (ROM), the PCM generates output signals to control various relays, solenoids and actuators. For vehicles equipped with manual transmissions, only two electrical connectors are used.

The TCM controls the power monitor strategy. The power monitor monitors engine speed at idle. If an abnormal engine speed increase is detected, the power monitor will command the fuel injection control module (FICM) off in an effort to maintain a normal idle speed. For manual transmission vehicles, the power monitor is the only active TCM function. DTCs will be stored if the power monitor is active.

Standard Corporate Protocol

The standard corporate protocol (SCP) is a communication language used by Ford Motor Company for exchanging bi-directional messages (signals) between stand-alone modules and devices. Two or more signals can be sent over one circuit.

Included in these messages is diagnostic data that is output over the BUS (+) and BUS (-) lines to the data link connector (DLC). This information is accessible with a scan tool. Information on this equipment is described in Diagnostic Methods.

Vehicle Speed Limiter

Note: Maximum speed may vary with load and axle ratio.

For F-250/350, Excursion and E-Series applications, a diesel engine equipped vehicle is limited to a maximum speed of approximately 150 km/h (95 mph). For F-450/550 applications, a diesel engine equipped vehicle is limited to a maximum speed of approximately 130 km/h (81 mph).

Diesel PCM Inputs

Accelerator Pedal Position (APP) Sensor

The accelerator pedal position (APP) sensor is a three-track potentiometer that is used to calculate driver demand for fuel quantity. The sensor receives VREF voltage from the PCM and provides a variable voltage signal directly proportional to the accelerator pedal position. A PCM detected fault of one of the three sensor track signals will permit normal operation. A fault with two sensor signals will only allow the engine to operate at idle.

Air Conditioning Pressure Switch

The A/C pressure switch (ACPSW) is used for additional A/C system pressure control. The ACPSW is also referred to as the refrigerant containment switch.

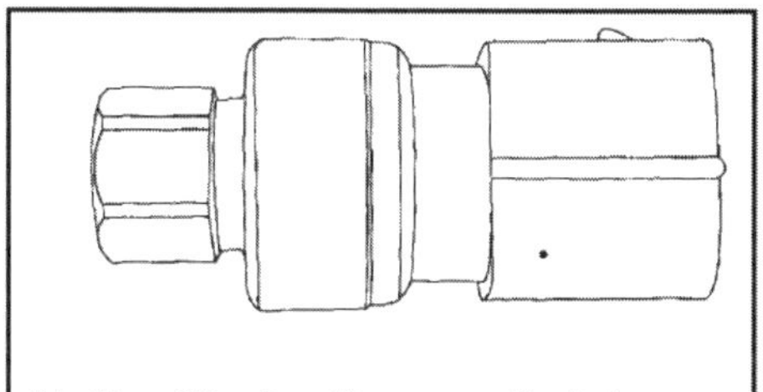

Air Conditioning Pressure Switch

For refrigerant containment control, the normally closed high pressure contacts open at a predetermined A/C head pressure. This turns off the A/C by opening the A/C demand circuit, preventing the A/C pressure from rising to a level that would open the A/C high pressure relief valve.

An Excursion may be equipped with the optional electronic automatic temperature control (EATC).

Analog Manifold Absolute Pressure Sensor

The analog manifold absolute pressure (MAP) sensor is a variable capacitor sensor that is supplied a 5-volt reference signal by the PCM and returns a voltage signal to the PCM relative to intake manifold pressure. The sensor voltage increases as pressure increases. The MAP sensor allows the PCM to determine engine load to calculate fuel quantity. In addition, the MAP signal is used to control smoke by limiting fuel quantity during acceleration until a specified boost pressure is obtained and is utilized by the PCM for EGR system calculations and control.

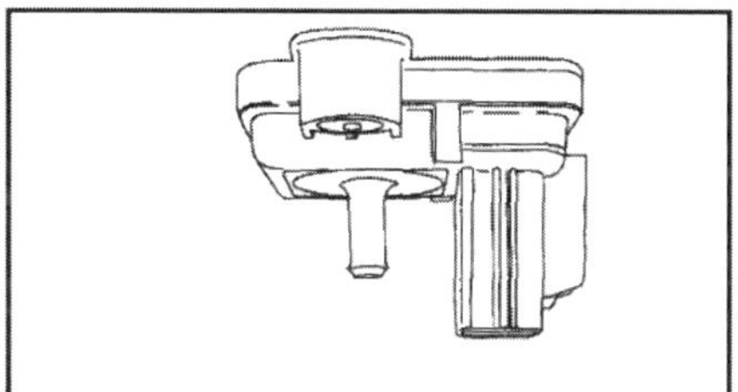

Analog Manifold Absolute Pressure (MAP) Sensor

A MAP signal fault detected by the PCM will cause the PCM to calculate an estimated manifold pressure based on known engine conditions.

Barometric Pressure Sensor

The barometric pressure (BARO) sensor is a variable capacitor sensor that processes a signal indicating atmospheric pressure. This allows the PCM to compensate for altitude. The PCM uses this information to calculate injection timing and glow plug control. The BARO sensor is located behind the lower steering column opening finish panel.

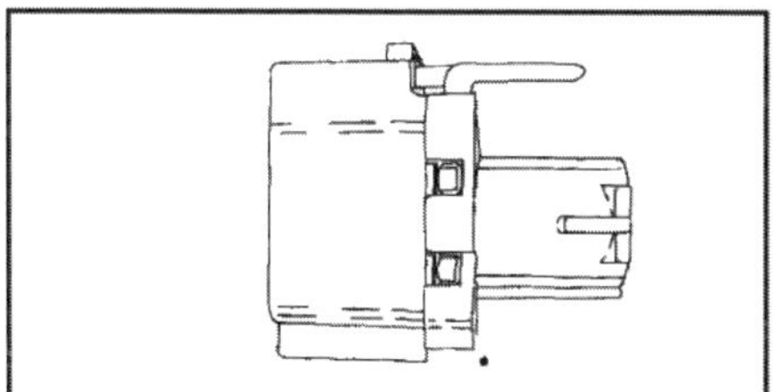

Barometric Pressure (BARO) Sensor

A BARO sensor fault will result in an out-of-range signal to the PCM.

Diesel PCM Inputs (Continued)

Brake Pedal Position (BPP) Switch

The brake pedal position (BPP) switch signals the PCM with a battery positive voltage (B+) signal whenever the vehicle brake pedal is applied.

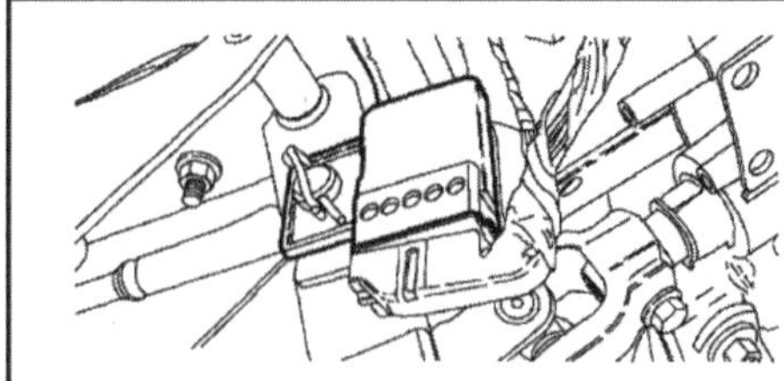

Brake Lamp Switch

The signal informs the PCM to disengage the torque converter clutch, speed control and auxiliary idle control (if equipped).

If all the stoplamp bulbs are burned out (open), a high voltage is present at the PCM due to a pull-up resistor in the PCM. This provides fail-safe operation in the event the circuit to the BPP switch has failed.

Brake Pressure Applied Switch

Note: The brake pressure applied switch is present on vehicles equipped with speed control.

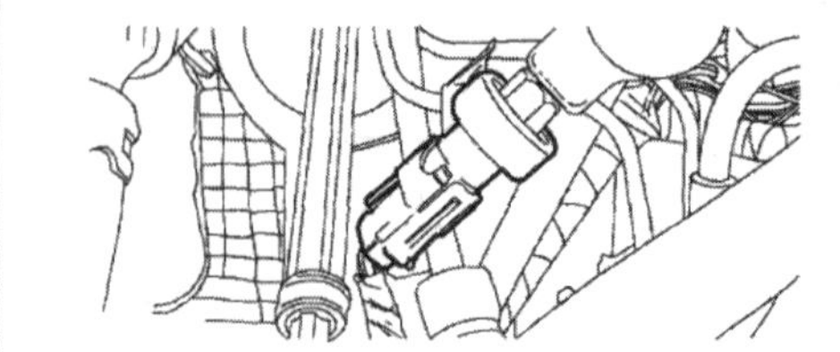

Brake Pressure Applied Switch

All vehicles have a single brake pressure applied (BPA) switch. A BPA switch provides a backup for the brake pedal position (BPP) switch. Normally, a brakes-applied signal from the BPP switch will disengage the speed control. If the BPP switch signal is lost, the BPA switch will then supply the brakes-applied signal to the speed control system.

Camshaft Position Sensor

The camshaft position (CMP) sensor is a variable reluctance sensor, which will respond to a rotating trigger protruding from the camshaft. The trigger is a single 9.525 mm diameter peg approximately 18 camshaft degrees wide, projecting 3-5 mm from the camshaft. The sensor will produce a sine wave in response to the peg as it passes the sensor. The sensor output is required to determine camshaft position.

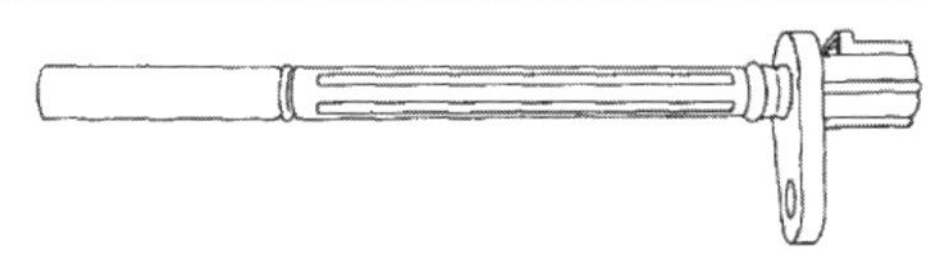

Camshaft Position (CMP) Sensor

Clutch Pedal Position Switch

The clutch pedal position (CPP) switch is an input to the PCM indicating the clutch pedal position. The CPP sends battery voltage to the PCM when the clutch is engaged (foot off of pedal) and zero voltage when the clutch is disengaged (pedal applied).

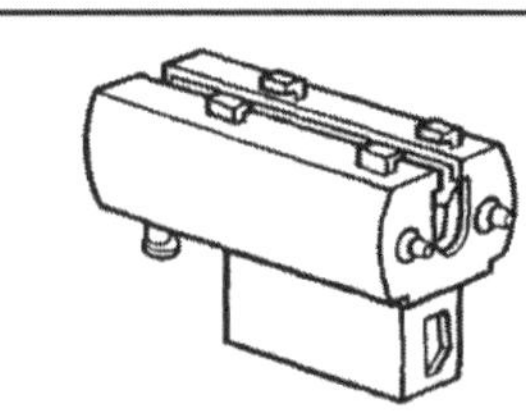

Clutch Pedal Position (CPP) Switch

Cold Idle Kicker

Cold idle kicker strategy provides an increase in idle speed during cold engine warm up, of up to 1100 RPM (normally 725 RPM for manual, 650 RPM for auto), for a faster warm-up to operating temperature during extended idle conditions. This is accomplished by the PCM, which monitors the EOT sensor input and adjusts the RPM accordingly, to a maximum of 1100 RPM.

The idle speed is increased proportionately when the engine oil temperature is below 70°C (158°F) and the engine has been at idle for more than 2 minutes. Applying the brake pedal, clutch or accelerator pedal will deactivate the cold idle kicker strategy and return the idle speed to 650 RPM.

Crankshaft Position Sensor

The crankshaft position (CKP) sensor is a variable reluctance sensor which will respond to a rotating actuator positioned on the crankshaft. The actuator is a 60-2 tooth steel disk with 58 evenly spaced teeth and a minus 2 tooth slot. The sensor will produce a sine wave for each tooth edge of the actuator. The sensor's output is required to determine crankshaft speed, position and acceleration.

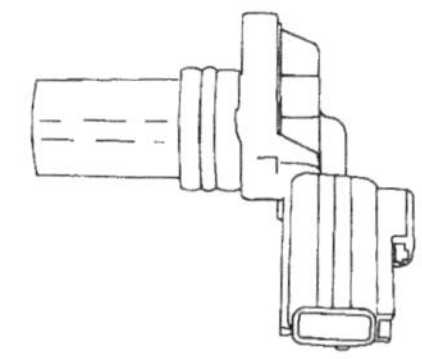

Engine Coolant Temperature (ECT) Sensor

The engine coolant temperature (ECT) sensor is a thermistor device in which resistance changes with temperature. The electrical resistance of a thermistor decreases as the temperature increases, and resistance increases as the temperature decreases. The varying resistance affects the voltage drop across the sensor terminals and provides electrical signals to the PCM corresponding to temperature. If the PCM receives a high engine temperature signal from the ECT, it will adjust fueling rates to protect the engine from damage due to overheating.

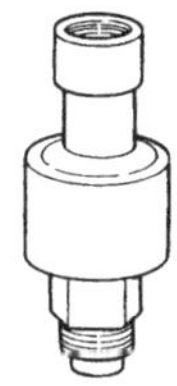

Engine Oil Temperature Sensor

The engine oil temperature (EOT) sensor is a thermistor whose resistance decreases as engine oil temperature increases. The EOT signal is used by the PCM to calculate fuel quantity, injection timing and glow plug operation. At low ambient air temperatures and oil temperature below 70°C (158°F), low idle is increased to maintain stable idle quality. Fuel quantity and timing is controlled throughout the total operating range to provide adequate torque and power.

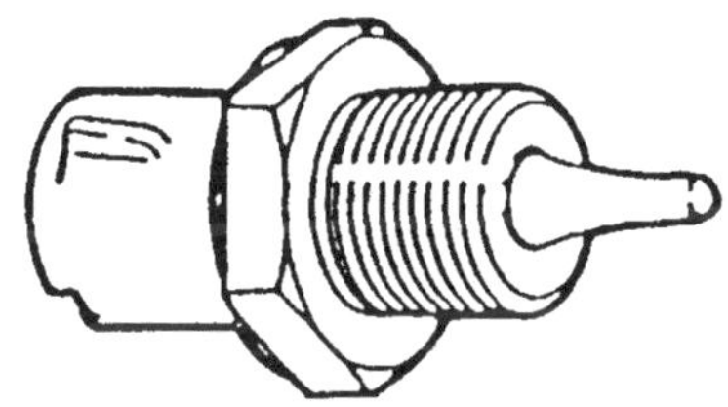

An EOT signal detected out of range, high or low, by the PCM will cause the PCM to substitute a temperature based on ECT to be used for operating purposes.

Exhaust Pressure Sensor

The exhaust pressure (EP) sensor is a variable capacitor sensor that is supplied a 5-volt reference signal by the PCM and returns a linear analog voltage signal that indicates pressure. The EP sensor measures the pressure in the LH exhaust manifold. The sensor feedback signal is used for variable geometry turbo (VGT) and exhaust gas recirculation (EGR) valve control. An open or short in the EP sensor wiring will result in an out-of-range low voltage at the PCM.

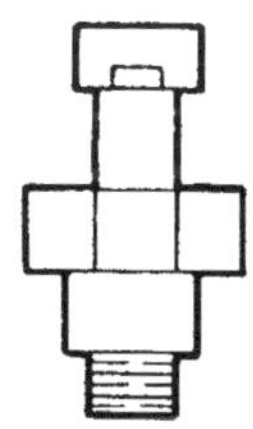

Fan Speed Sensor (FSS)

The fan speed sensor is a Hall effect sensor integral to the visctronic drive fan (VDF). The PCM will monitor the sensor input and control the VDF speed based upon engine coolant temperature (ECT), transmission fluid temperature (TFT) and intake air temperature (IAT) requirements. When an increase in fan speed for vehicle cooling is requested, the PCM will monitor the FSS signal and output the required pulse width modulated (PWM) signal to a fluid port valve within the VDF.

Diesel PCM Inputs (Continued)

Injection Control Pressure Sensor

The injection control pressure (ICP) sensor is a variable capacitor sensor that is supplied a 5-volt reference signal by the PCM and returns a linear analog voltage signal that indicates pressure. The sensor measures the oil pressure in the high pressure pump cover (F-SuperDuty/Excursion — Early Build) or oil rail (E-Series or F-SuperDuty/Excursion — Late Build) and the PCM uses this information to determine injection control pressure. The ICP sensor along with the injection control pressure regulator form a closed loop fuel pressure control system.

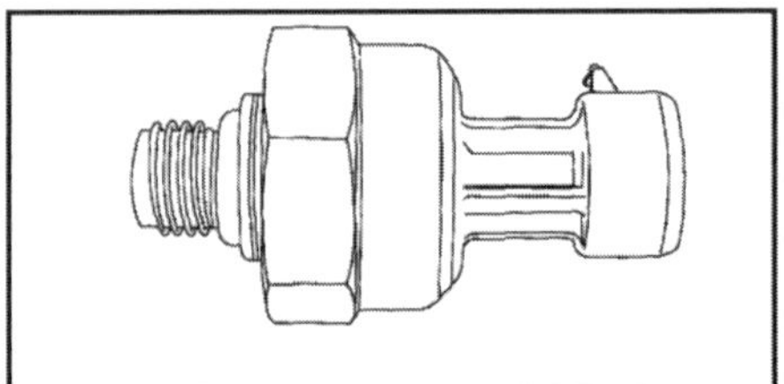

Injection Control Pressure (ICP) Sensor

If the PCM detects an inoperative ICP sensor, the PCM will control injection control pressure from a PCM-estimated injection control pressure.

Intake Air Temperature (IAT) Sensor

The intake air temperature (IAT) sensor is a thermistor device. The electrical resistance of a thermistor decreases as the temperature increases, and resistance increases as the temperature decreases. The varying resistance affects the voltage drop across the sensor providing a signal corresponding to temperature.

The 6.0L diesel utilizes two IAT sensors. The IAT sensor is integrated into the mass air flow (MAF) sensor. The PCM uses the IAT signal to control timing and fuel rate during cold starts and provide an input to the cold idle kicker.

Intake Air Temperature 2 (IAT2) Sensor

The IAT2 or manifold air temperature (MAT) sensor is a thermistor device. The electrical resistance of a thermistor decreases as the temperature increases, and resistance increases as the temperature decreases. The varying resistance affects the voltage drop across the sensor providing a signal corresponding to temperature.

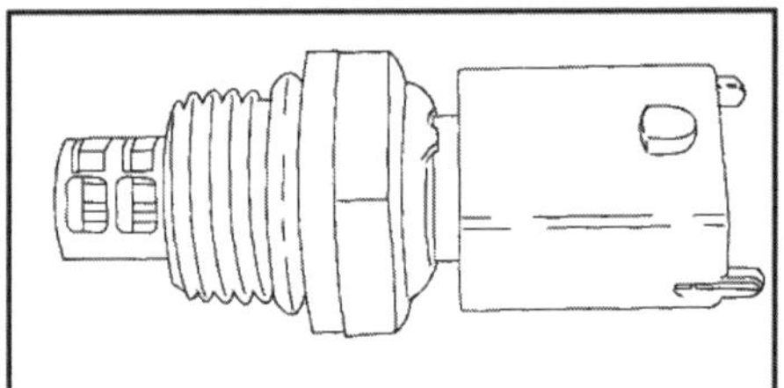

Intake Air Temperature 2 (IAT2) Sensor

The IAT2 sensor is located in the intake manifold. The sensor provides a manifold air temperature signal to the PCM.

Mass Air Flow Sensor

The mass air flow (MAF) sensor provides an analog voltage signal to the PCM proportional to the intake air mass. The MAF sensor utilizes a hot wire sensing element to measure the amount of air entering the engine. The hot wire is maintained at 200° C (392° F) above ambient temperature. Air passing over the hot wire cools the wire. The current required to maintain the temperature of the hot wire is proportional to the air mass flow. The MAF signal is used to calculate exhaust gas recirculation (EGR) and for fault detection.

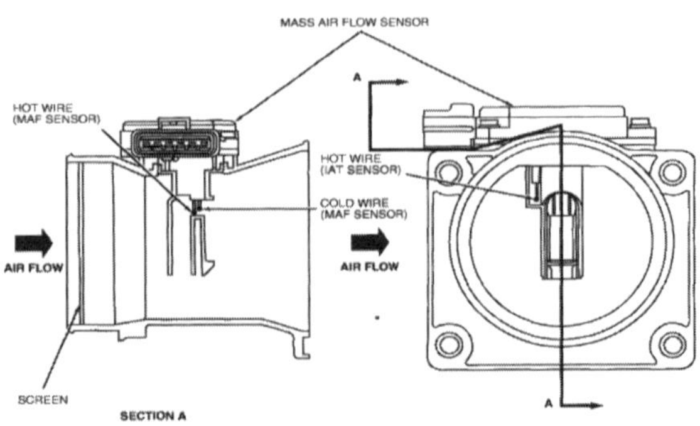

Typical Mass Air Flow (MAF) Sensor

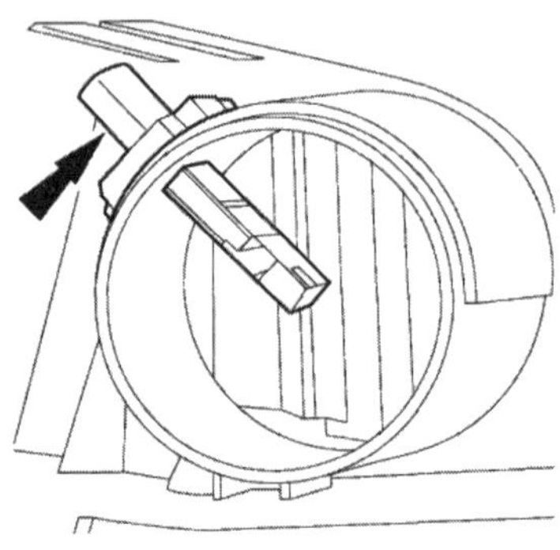

Mass Air Flow (MAF) Sensor (E-Series)

Diesel PCM Inputs (Continued)

Output Shaft Speed Sensor

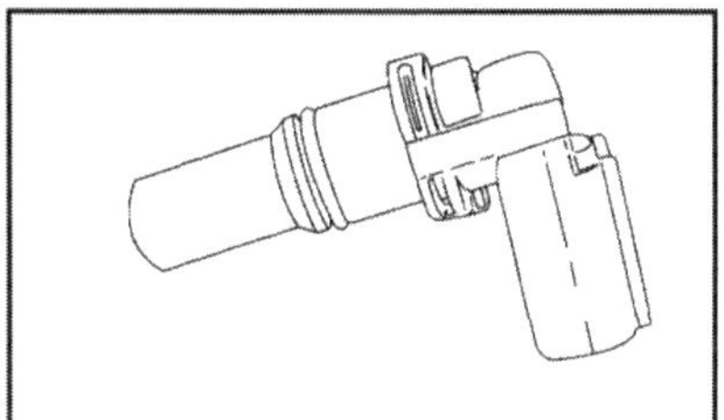

Output Shaft Speed (OSS) Sensor

The OSS sensor detects the speed of the transmission output shaft. The sensor uses variations in a magnetic field caused by the OSS sensor gear to generate an output signal corresponding to shaft speed. The OSS sensor gear uses a repetitive pattern of 3 different angular displacements between the teeth. The magnetic sensor element may be a Hall effect sensor or magnetic pick-up. Due to the open collector design and the RC timing involved in the pull-up, the rising edge of the OSS sensor gear is not a reliable edge for critical timing, only the falling edge. The pull-up resistor is located in the PCM to provide a current for the sensor. This applies specifically to the automatic transmission.

Parking Brake Switch

The parking brake switch signal indicates when the parking brake is applied. On all vehicles, the parking brake switch is located under the instrument panel. The parking brake switch signal will deactivate the speed control if the brake is applied during speed control operation and provides a brake input for the cold idle kicker.

Power Monitor

The diesel engine power monitor (DEPM) strategy resides in the transmission control module (TCM) located inside of the PCM. The function of the DEPM is to monitor engine RPM when there is no power demand from the APP sensor.

Under normal engine idle operation, the DEPM value must always be higher than engine RPM. In the event that RPM does increase to its calibrated idle speed value following completion of deceleration fuel shutoff, due to the injectors not turning off, the DEPM will disable crank and cam output signals, sent by the PCM to the FICMM. The FICMM input line informs the DEPM when the injectors are turned on and when the injectors are turned off. When the FICMM line is either shorted or open, the monitor strategy assumes that the fuel injectors are always turned on and sets a DTC.

Speed Control Command Switches

The speed control command switches are momentary contact switches which are located on the steering wheel. They consist of one ON-OFF switch and one SET/ACCEL-COAST-RESUME switch. These switches, when pressed, select one of several resistance values which is sent to the PCM to select speed control functions.

Transmission Control Switch

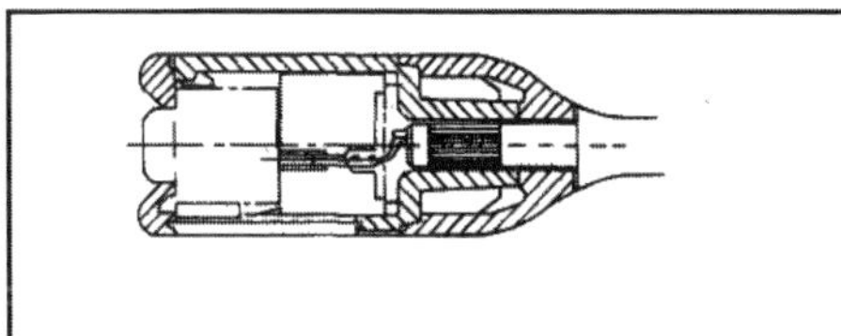

Column-Shift Transmission Control Switch

On automatic transmission equipped vehicles, the transmission control switch provides an input signal to the PCM whenever the switch is pressed. For F-SuperDuty/Excursion, the tow/haul indicator will illuminate when the transmission control switch is cycled to engage and disengage the tow/haul strategy. For E-Series, the overdrive cancel indicator will illuminate when the transmission control switch is cycled to engage or disengage the overdrive function of the transmission.

4x4 Low Switch

The 4x4 Low Switch sends a ground signal to the instrument cluster when in 4x4 low. This input is used to adjust the shift schedule.

Using the standard corporate protocol (SCP) network, the cluster provides a 4x4 status signal to the PCM.

Diesel PCM Outputs

Electronic Air Filter Restriction Gauge

F-SuperDuty and Excursion vehicles are equipped with an electronic air filter restriction gauge located in the air cleaner cover on the clean side of the air filter. The sensor is hardwired to the instrument cluster. When air flow in the air inlet system reaches the maximum allowable restriction limit, a switch will close and the filter restriction indicator will illuminate.

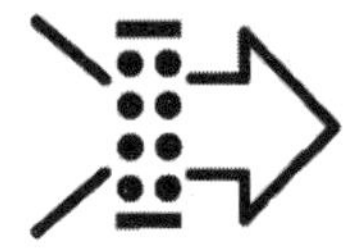

Air Filter Restriction Indicator (F-SuperDuty/Excursion)

Passive Anti-Theft System

Excursion vehicles are equipped with a passive anti-theft system (PATS) which prevents engine operation without the correctly encoded ignition key. For additional information on the operation of this system, refer to Workshop Manual Section 419.

Electronic Variable Geometry Turbocharger (VGT) Control Valve

The electronic variable geometry turbocharger (VGT) control valve is a four-way proportional hydraulic flow control valve with closed center position. The valve controls linear actuator position of a closed loop hydraulic servo by charging and venting flow on both sides of a piston. Linear displacement feedback from the actuator varies a feedback spring force to move the valve spool to the center closed position when the actuator reaches the desired position. Therefore, actuator position is dependent only on control valve current — it is independent of hydraulic fluid temperature and viscosity.

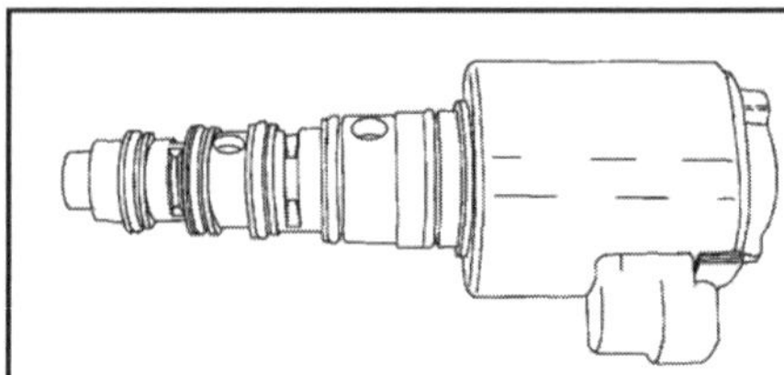

Electronic Variable Geometry Turbocharger (VGT) Control Valve

Exhaust Gas Recirculation Throttle Position Control and Throttle Position Sensor

EGR Throttle Position Actuator

The EGR throttle position (EGRTP) actuator modifies the intake airflow from the charge air cooler into the intake manifold. The EGRTP actuator regulates the rotary motion of the throttle plate located within the throttle body. The control of intake airflow provides increased EGR system efficiency with the throttle plate position determined by a signal from the EGRTP sensor.

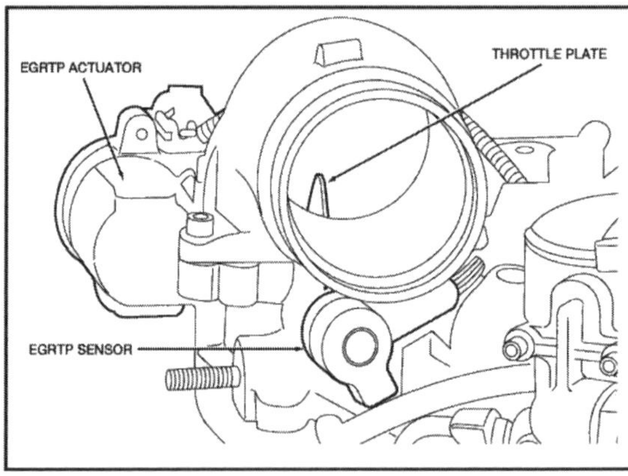

EGR Throttle Position Sensor

The EGR throttle position (EGRTP) sensor is a potentiometer that provides a feedback signal to the PCM. The input signal is an analog voltage proportional to the rotary position (angle) of a throttle plate located within the throttle body.

Exhaust Gas Recirculation Valve Control and Valve Position Sensor

The exhaust gas recirculation valve controls the amount of exhaust gases recirculated back to the intake. It is a proportional solenoid with a built-in position sensor. The valve position sensor is needed to give the control circuit feedback to achieve desired travel position. The exhaust gas recirculation valve position sensor is a variable resistance pintle position sensor that makes position measurements of the exhaust gas recirculation valve control actuator. Input signals from the manifold absolute pressure (MAP), exhaust pressure (EP) and barometric pressure (BARO) sensors are used by the PCM to calculate and control the EGR system flow.

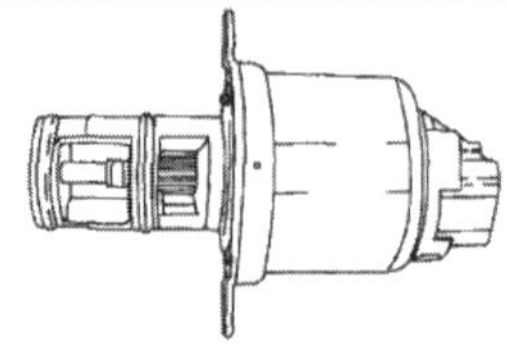

Exhaust Gas Recirculation (EGR) Valve Control and Valve Position Sensor

Diesel PCM Outputs (Continued)

Fuel Injection Control Module

The fuel injection control module (FICM) receives information from the PCM, including volume of fuel desired, RPM, engine oil temperature, injection control pressure and others. The FICM then uses those signals to calculate fuel injection and duration. After calculating injector fuel delivery time, the FICM sends 48 volts at 20-amp pulse to the correct injector so that the correct amount of fuel is delivered to the cylinder at the correct time.

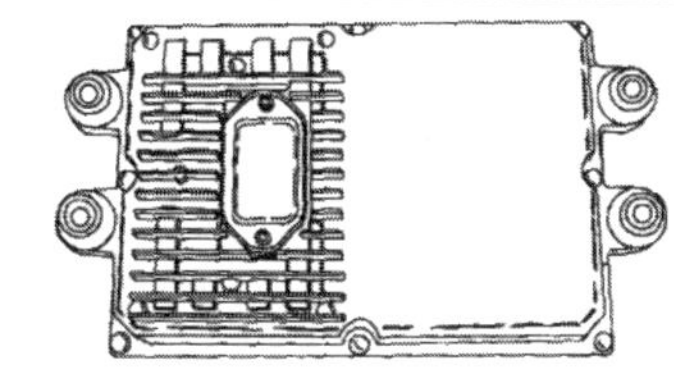

Fuel Injection Control Module (FICM)

Glow Plug Control Module

The glow plug system is composed of solid state glow plug control module (GPCM), glow plugs and the associated wiring harness. The glow plug on-time is controlled by the PCM and is a function of oil temperature, barometric pressure and battery voltage. The PCM enables the GPCM which drives the individual glow plugs. Glow plug on-time normally varies between 1 to 120 seconds. In addition to PCM control, the GPCM internally limits the glow plug operation to 180 seconds regardless of PCM commanded on-time. The power to the glow plugs is provided through the GPCM solid state drivers directly from the vehicle battery. The GPCM monitors and detects individual glow plug functionality, and the control and communication links to the PCM. The failures detected by the GPCM are passed to the PCM using a serial communication signal on the diagnostic line.

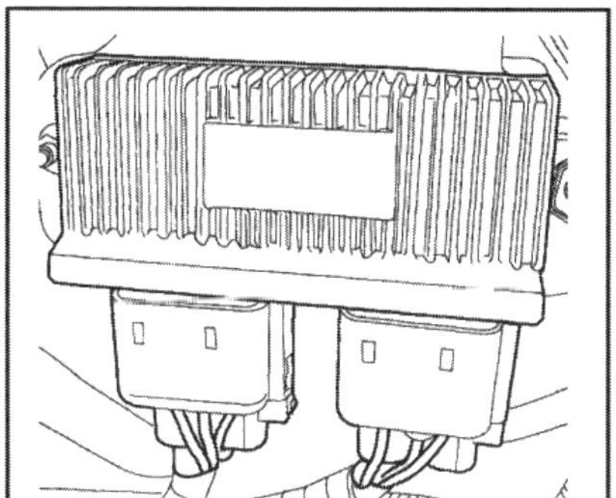

Glow Plug Control Module (GPCM)

Note: Wait-to-start lamp on-time is controlled by the PCM and is independent from GPCM on-time.

Glow Plug Indicator Lamp

The glow plug indicator lamp (GPIL) is located in the instrument cluster and is used to inform the operator when the engine is ready to be started. The indicator is controlled by the instrument cluster based on an electronic command signal from the PCM through the standard corporate protocol (SCP) communications network. On-time normally varies between 1 and 10 seconds and is independent of glow plug relay on-time. As a prove out, the indicator is commanded on at every key cycle even though the glow plug system may not be required.

Glow Plug Indicator

Injection Pressure Regulator

The injection pressure regulator (IPR) controls injection oil pressure. An electrical signal to a solenoid creates a magnetic field which applies a variable force on a valve servo to control pressure. The quantity of fuel delivered to the combustion chamber is proportional to injection control pressure.

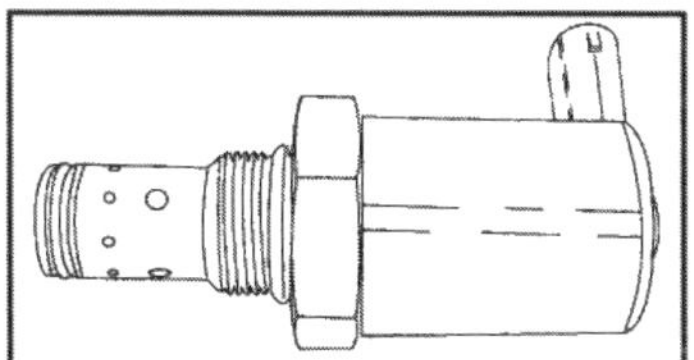

Injection Pressure Regulator

An open circuit will result in minimum oil pressure and a no-start situation. A short to ground in a circuit results in maximum oil pressure, and is limited by a mechanical pop-off valve to 27,580 kPa (4000 psi).

Speed Control Indicator (F-SuperDuty/Excursion)

The speed control indicator is controlled by the instrument cluster. The PCM sends a message through the SCP network to the instrument cluster to illuminate the indicator when the speed control is engaged.

Speed Control Indicator (F-SuperDuty/Excursion)

Diesel PCM Outputs (Continued)

Tachometer

The tachometer is controlled by the instrument cluster. The PCM sends engine RPM data through the standard corporate protocol (SCP) communication network to the instrument cluster.

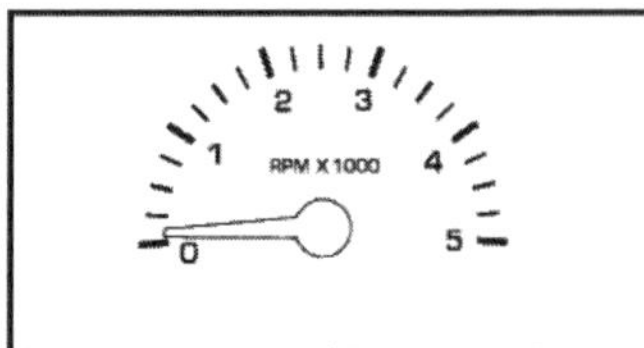

Tachometer (F-SuperDuty)

Visctronic Drive Fan (VDF)

The cooling fan electronic clutch is controlled by the PCM. The PCM monitors engine oil temperature (EOT) and engine coolant temperature (ECT) and sends a duty cycle command to the fan clutch to achieve adequate cooling. When a circuit fault is detected the PCM will set a DTC.

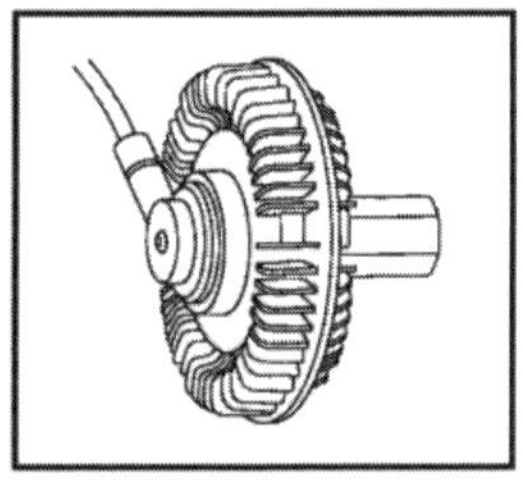

Water in Fuel Indicator

The water in fuel indicator is controlled by the instrument cluster. The cluster receives electronic information from the PCM through the standard corporate protocol (SCP) communication network. If the water in fuel sensor indicates that there is water in the fuel separator/housing (located within the fuel condition module), the PCM sends an SCP message to illuminate the water in fuel indicator.

Water In Fuel Indicator

Diesel Fuel System

Electro-Hydraulic Injector

The electro-hydraulic injector is composed of three major components: The oil control, pressure amplification, and the nozzle assembly. The injector uses two 48-volt at 20-amp coils to control a spool valve that directs oil flow in and out of the injector. The fuel injector has a self-extracting hold-down clamp.

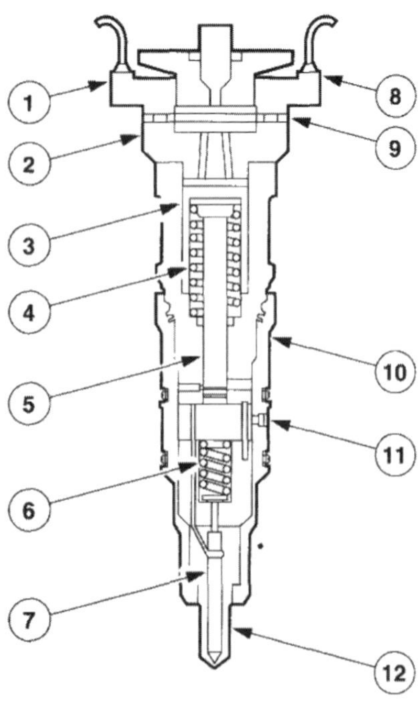

Item	Description
1	Closing coil
2	Control valve body
3	Amplifier piston (7:1 ratio)
4	Piston return spring
5	Plunger
6	Valve opening pressure spring, 260 bar (3,770 psi)
7	Nozzle valve
8	Opening coil
9	Spool valve (control valve)
10	Case nut
11	Fuel strainer (3)
12	Nozzle assembly

Diesel Fuel System (Continued)

Electro-Hydraulic Injector Amplifier Piston
The high pressure oil flows from the oil rails into an electro-hydraulic injector amplifier piston located in the injector. Oil entry and exit to and from the amplifier piston is controlled by a coil-operated spool valve.

Electro-Hydraulic Injector Fuel Plunger
The electro-hydraulic injector fuel plunger is located in the injector and is driven by the amplifier piston. The fuel plunger injects fuel into the combustion chamber at pressures of up to 196,500 kPa (28,500 psi) through the nozzle assembly. Fuel is supplied to the injector at approximately 345 kPa (50 psi) through fuel rails in the cylinder heads.

Engine Timing
The PCM commands the fuel quantity. The FICM controls the duration of the injection event and is shown as parameter ID (PID) "FUELPW" on the scan tool.

The PCM controls the injection pressure and fuel volume by varying the injection oil pressure with the injection pressure regulator (IPR). The command to the IPR is a 12-volt, pulse width modulated (PWM) signal (controlled on the ground side).

The injection oil pressure command is shown as PID IPR which is the percentage ON of the pulse width modulated signal. Injection oil pressure is shown as PID ICP.

The PCM uses camshaft position (CMP) and crankshaft position (CKP) input signals to calculate engine speed and position. The PCM conditions both input signals and supplies the FICM with the CMP and CKP output signals. The FICM uses CMP and CKP output signals to determine the correct sequence for injector firing. The PCM sends information about the fuel demand, engine oil temperature (EOT) and injection control pressure (ICP) through CAN to the fuel injection control module (FICM). The FICM uses this information to calculate the injection cycle.

Fueling Corrections
The PCM adjusts injector output based on oil temperature information received from the engine oil temperature (EOT) sensor and turbo boost information received from the manifold absolute pressure (MAP) sensor and the barometric pressure (BARO) sensor. These corrections are necessary to meet emissions requirements and to optimize power. The engine coolant temperature (ECT) and the EOT sensors are used as the primary input to the PCM to enable adaptive cooling. This provides a means of providing adequate cooling in severe engine temperature conditions. When ECT is greater than 105°C (221°F) or EOT is greater than 123°C (253°F), the fueling rate of the engine is modified to provide cooling protection and prevent engine damage due to overheating. Outputs of these sensors can be monitored using a scan tool reading PIDs.

Fuel Injection Control Module
The coil-operated spool valve requires 48 volts at up to 20 amps to operate, which is more power than the PCM can supply. Therefore, a high power device, the fuel injection control module (FICM), is used to supply power to the solenoid on command from the PCM.

Fuel System
Fuel is drawn from the fuel tank through the primary filter (located in the horizontal fuel conditioning module) by the electric fuel pump. Pressurized fuel, approximately 310-379 kPa (45-55 psi), is supplied to the secondary filter (the fuel filter housing located on the front left side of the engine) by means of the electric pump and regulator valve. The regulator relieves the pressure, sending fuel back to the fuel tank. Only the filtered fuel going through the fuel filter will go to the heads. A check valve is located on both heads to prevent fuel pressure spikes in the fuel rail.

Diesel Fuel System (Continued)

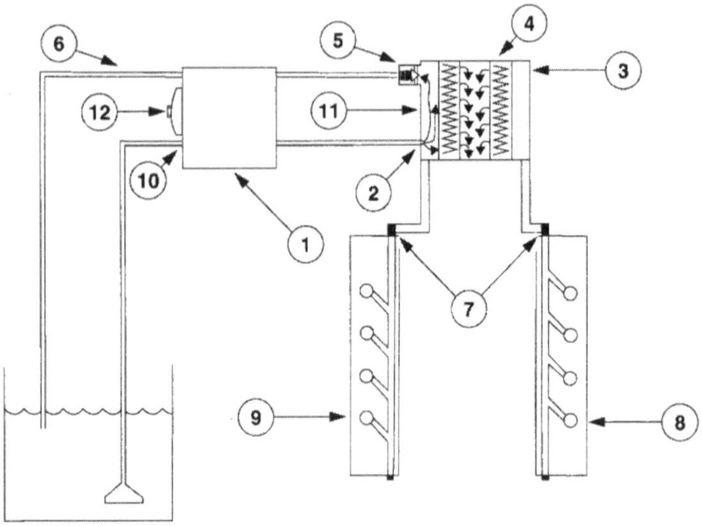

Item	Description
1	Electric fuel pump
2	Fuel inlet
3	Fuel filter housing
4	Fuel filter (secondary)
5	Regulator
6	Fuel return line
7	Check valves
8	Right head
9	Left head
10	Restriction check
11	Fuel pump pressure check
12	Fuel filter (primary)

High Pressure Oil System

The 6.0L Powerstroke diesel injectors are powered by lubricating oil which is pressurized by a swashplate pump (Rexroth pump) in the engine valley. The pump output pressure ranges from 3,102 to 20,685 kPa (450 psi to 4,000 psi). Oil pressure is controlled by the PCM through the injector pressure regulator (IPR) valve. The PCM controls pressure in the oil rail by opening (relieving pressure) and closing (increasing pressure) the IPR valve. The high pressure oil is delivered to oil rails in the cylinder heads. An injection control pressure (ICP) sensor mounted on the high pressure pump cover sends an analog voltage signal (0.5V to 5.0V) to the PCM for feedback control of the oil pressure.

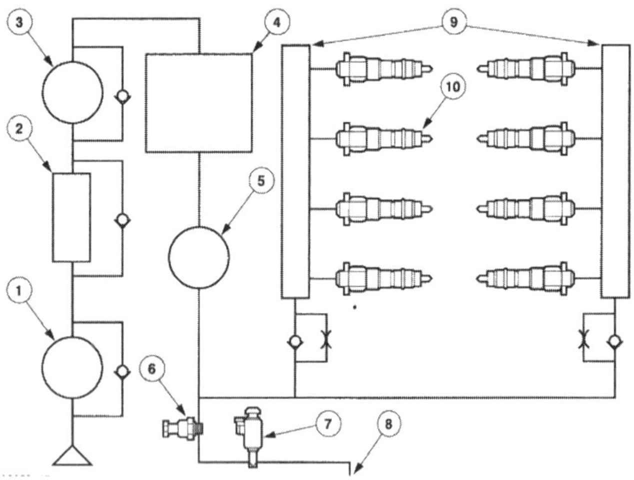

Item	Number	Description
1	6608	Oil pump
2	6A638	Oil cooler
3	6714	Oil filter
4	—	High pressure reservoir (part of 6010)
5	9A543	High pressure oil pump
6	9F838	ICP sensor
7	9C968	IPR valve
8	—	Drain to crankcase (part of 6010)
9	9D280	High pressure oil rails
10	9E527	Fuel injectors

Diesel Intake Air Systems

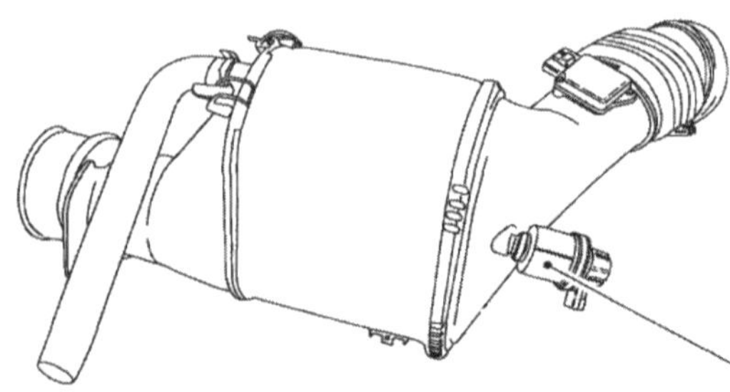

Intake Air System (F-SuperDuty/Excursion)

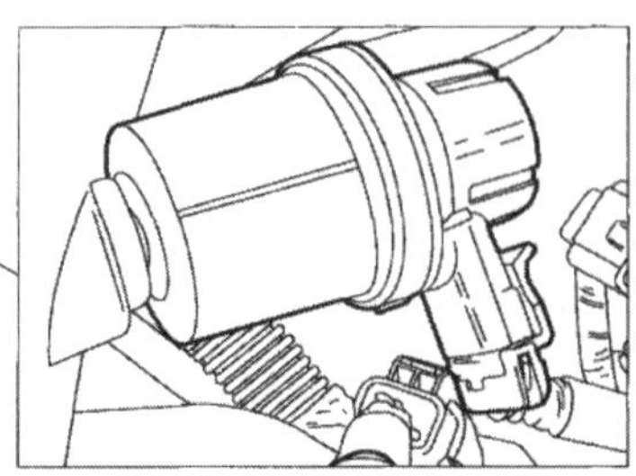

AIR FILTER RESTRICTION GAUGE

Diesel Intake Air Systems (Continued)

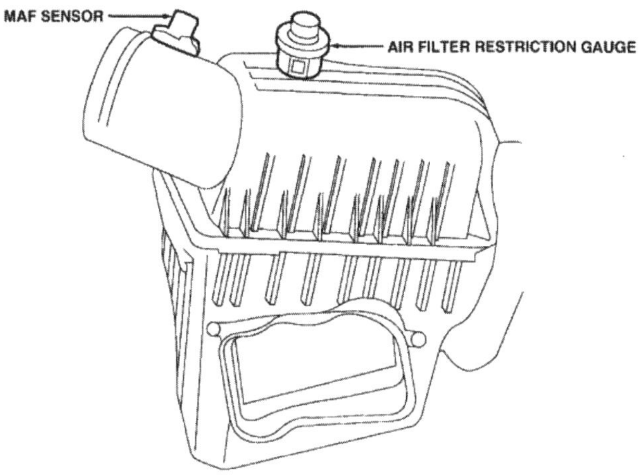

Intake Air System (E-Series)

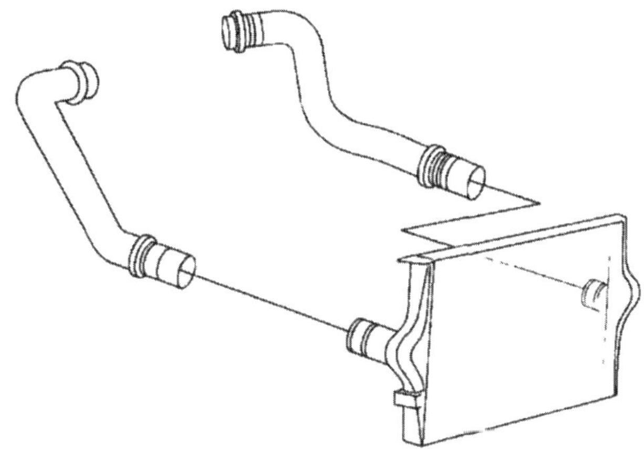

Charge Air Cooler System (F-SuperDuty/Excursion)

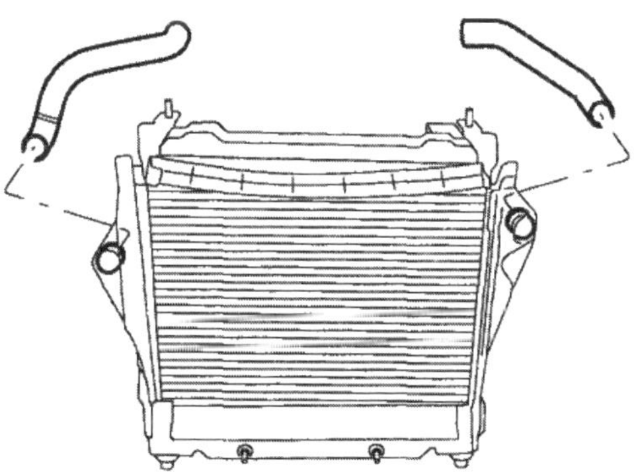

Charge Air Cooler System (E-Series)

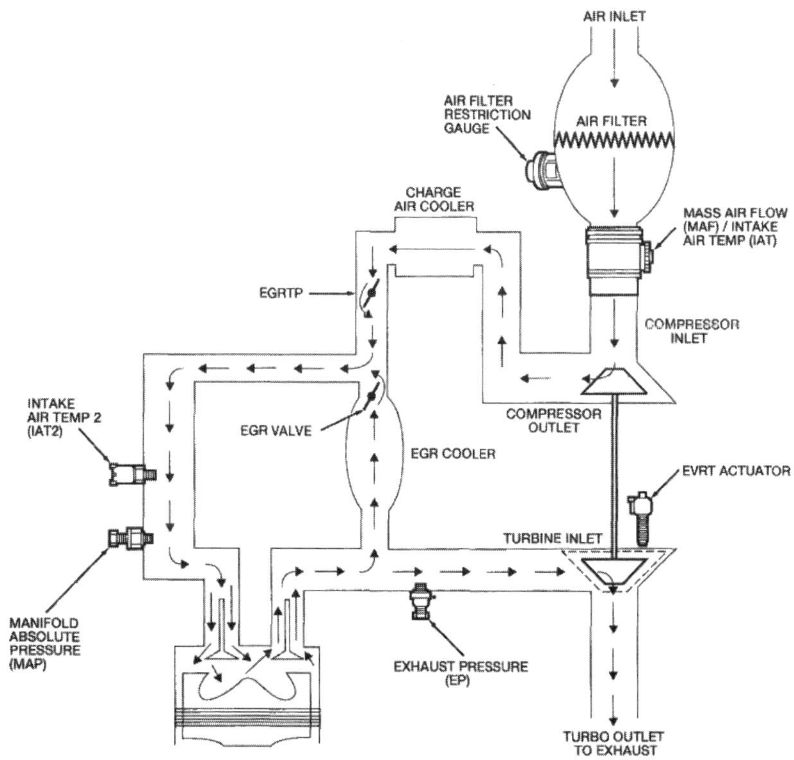

Air Flow Schematic

Diesel Catalyst and Exhaust Systems

Catalytic Converter

All vehicles are equipped with a catalytic converter.

NOTES

DIAGNOSTIC METHODS INDEX

NOTES

Diagnostic Methods

Overview

The Diagnostic Methods section provides detailed instructions on how to access or perform routine diagnostic tasks. This section can be referenced as many times as needed for step-by-step instructions for routine procedures.

Diagnostic Tools

Required Equipment

- 23 Multimeter 105-00050 or equivalent. Input impedance 10 mega-ohm minimum.
- Adapter, Oil Pressure Leak Test 303-756 or equivalent
- Adapter, Compression Test 303-757 or equivalent
- Adapter, High Pressure Pump Test 303-765 or equivalent
- Adapter, High Pressure Rail Test 303-766 or equivalent
- Adapter Set, Oil High Pressure Leakage 303-1071
- ICP/EBC Adapter Cable 418-D003 (D94T-50-A) or equivalent.
- Pressure Test Adapter Kit 014-00761 or equivalent
- Rotunda Smoke Machine, Fuel Evaporative Emission System Tester 218-00001 (522) or equivalent.
- Scan Tool
- Gauge 0-1.1 MPa (0-160 psi) Bar 014-00761 or equivalent

Optional Equipment

- R134A Manifold Gauge Set 176-R932A or equivalent.
- Non-powered test lamp.
- Vacuum/Pressure Tester 164-R0253 or equivalent. Range 101.3 kPa (0.30 in-Hg). Resolution 3.4 kPa (1 in-Hg).

Note: Refer to equipment user manual for details on tool accessories and function.

Scan Tool Hook Up

The Scan Tool must be connected to the Data Link Connector (DLC) for communication with the vehicle.

Data Link Connector

The DLC is located in the passenger compartment. It is attached under the instrument panel and accessible from the driver seat.

The DLC is rectangular in design and capable of accommodating up to 16 terminals. The connector has keying features to allow easy connection. The vehicle connector and the test equipment connector have latching features that ensure the test equipment connector will remain mated when properly connected.

Refer to scan tool manufacturer manual for specific cables and/or adapters required for scan tool hookup.

Communication Error

It is possible to get a test fail error from a scan tool when initiating a test or viewing PIDs. The communication error could be caused by operator error, the vehicle wiring or connectors, or the powertrain control module (PCM) and other control modules connected to the DLC wiring. The PCM will respond to a scan tool whenever the scan tool requests a test. Some are normal responses to valid requests. The others are communication error responses. If the scan tool displays any of the test fail error responses reinitialize the scan tool. If the error is still present, GO to Pinpoint Test AE, after checking scan tool connections, cable/adapters and entry of vehicle information. Verify auxiliary powertrain control (RPM control) is off when trying to perform self-tests.

Quick Test Description

Quick Test is divided into five specialized tests:

1. Retrieve/Clear Continuous DTCs
2. Key On Engine Off (KOEO) On-Demand Self-Test
3. Key On Engine Off (KOEO) Injector Electrical Self-Test (Click Test)
4. Key On Engine Running (KOER) On-Demand Self-Test
5. Key On Engine Running (KOER) Glow Plug Monitoring Self-Test

All are described below.

Quick Test checks the integrity and function of the Powertrain Control system and outputs the test results upon demand. Quick Test also provides a quick end check of the powertrain control system and is usually performed at the start of each diagnostic procedure. It is also performed at the end of most pinpoint tests for verification of repair and to make sure no other faults were incurred while servicing a previous fault.

MIL DTCs

MIL DTCs are generated to alert the driver that there is a concern with the system or the vehicle is in Failure Management Effects Mode (FMEM). MIL DTCs are also used to indicate an emission concern. Non-MIL DTCs indicate a less serious or non-emission related concern with the system.

Quick Test Operation

Quick Test is performed by retrieving KOEO, KOER, and Continuous Memory DTCs (CMDTCs).

Special Notes

Before running Quick Test, always perform the necessary visual checks and safety precautions listed below.

Visual Check

- Inspect the air cleaner and inlet ducting.
- Check system wiring harness for proper connections, bent or broken pins, corrosion, loose wires, proper routing, etc.
- Check the PCM, sensors and actuators for physical damage.
- Check the engine coolant for proper level and mixture.
- Check the transmission fluid level and quality.
- Make all necessary repairs before continuing with Quick Test.

Vehicle Preparation

- Perform ALL safety steps required to start and run vehicle tests. Apply parking brake, place shift lever firmly into PARK position (NEUTRAL on manual transmission), block drive wheels, etc.
- Turn off ALL electrical loads — radios, lights, A/C, blower, fans, etc.
- Start engine and bring up to normal operating temperature before running Quick Test.

Quick Test Operation

For all concerns, refer to the Symptom Charts. Quick Test is performed by retrieving KOEO, KOER and Continuous Memory DTCs. If a DTC is retrieved, you must then go to the appropriate Pinpoint Test. If unable to complete a self-test, GO to Pinpoint Test AE.

Key On Engine Off (KOEO) On-Demand Self-Test

Key On Engine Off (KOEO) On-Demand Self-Test
Key On Engine Off (KOEO) On-Demand Self-Test is a functional test of the PCM performed on demand with the key on and the engine off. This test will check that all inputs and outputs (circuits, sensors, regulators, relays and solenoids) connected to the PCM are electrically operating without fault.

Note: It is necessary to wait 4 seconds after key on to initiate the KOEO self-test. Failure to do so will result in a "test fail PCM" message to be displayed.

Connect the scan tool. Turn off accessories. If vehicle is equipped with an auxiliary powertrain control (RPM control), it must be turned off to perform self-tests.

1. Perform the necessary vehicle preparation and visual inspection. Refer to Quick Test Operation .
2. Refer to the manufacturer manual for scan tool instructions.

Note: The following steps must be performed for the powertrain and transmission control modules.

1. Select KOEO ON DEMAND SELF-TEST.
2. Turn key on.
3. Wait 4 seconds for PCM and FICM to initialize completely.
4. Follow operating instructions from the menu.
5. Record DTCs and follow appropriate pinpoint test.
6. After test, cycle key to off before running other tests or driving vehicle.
7. If performing repeated self-tests, it may be necessary to unplug the GPCM to keep battery from going dead. Ignore any glow plug codes while the GPCM is unplugged.

Key On Engine Running (KOER) On-Demand Self-Test

Key On Engine Running (KOER) On-Demand Self-Test
Key On Engine Running (KOER) On-Demand Self-Test is a functional test of the PCM performed on demand with the engine running. Temperature is not a factor, but A/C must be turned off. A check is made on the injection control pressure (ICP) variable geometry turbocharger (VGT), and exhaust gas recirculation (EGR) systems. During this test, engine RPM can vary. A fault must be present at the time of testing for the KOER On Demand Self-Test to detect a fault. If a fault is detected, a Diagnostic Trouble Code (DTC) will be the output on the data link at the end of the test when requested by a scan tool. Only a hard fault code (DTC) will be displayed.

Connect the scan tool. Turn off accessories. Turn A/C off. If vehicle is equipped with an auxiliary powertrain control (RPM control), it must be turned off to perform self-tests.

Note: Engine may run rough during this test.

1. Perform the necessary vehicle preparation and visual inspection. Refer to Quick Test Operation .
2. Refer to the manufacturer manual for scan tool instructions.

Note: The following steps must be performed for the powertrain and transmission control modules.

1. Select KOER ON DEMAND SELF-TEST.
2. Follow operating instructions from the menu.
3. Record DTCs and follow appropriate pinpoint test.
4. After test, cycle key to off before running other tests or driving vehicle.

Retrieve/Clear Continuous Memory DTCs (CMDTCs)

Retrieve/Clear Continuous DTCs

Retrieve/CMDTCs is a functional test of the PCM. DTCs can be retrieved or cleared with the key on and the engine off or running. Unlike KOEO and KOER self-tests, which can only be activated on demand, the Continuous monitor is always active in monitoring the system. When a fault is detected, a DTC will be stored in memory to be retrieved at a later date, making it possible to diagnose intermittent faults. A P1000 DTC may be the only DTC displayed, indicating an incomplete OBD drive cycle (more drive time needed). The fuel injection control module (FICM) monitors the injector circuits for a fault continuously. The FICM sends failure information via CAN communications and a PCM DTC is set. CMDTCs indicate the injector electrical system failure.

Connect the scan tool. Turn off accessories. If vehicle is equipped with an auxiliary powertrain control (RPM control), it must be turned off to perform self-tests.

1. Perform the necessary vehicle preparation and visual inspection. Refer to Quick Test Operation .
2. Refer to the manufacturer manual for scan tool instructions.

Note: The following steps must be performed for the powertrain and transmission control modules.

1. Select RETRIEVE/CLEAR CONTINUOUS MEMORY DTCs
2. Turn key on.
3. Follow operating instructions from the menu.
4. Record DTCs and follow appropriate pinpoint test for continuous memory DTC diagnostics.
5. After test, cycle key to off before running other tests or driving vehicle.
6. Continuous memory DTCs must be cleared after repair is made.

KOEO Injector Electrical Self-Test (Click Test)

Key On Engine Off (KOEO) Injector Electrical Self-Test (Click Test)

Key On Engine OFF (KOEO) injector electrical self-test (click test) is a functional test of the FICM performed on demand with the key on and the engine off. This test determines if the injector circuits and solenoids are electrically operating without fault. All injectors will click (audible feedback of the injector solenoids energizing the injector valves) together. Then each injector 1 through 8 is cycled approximately one second in sequence. This process is repeated three times during the entire test. If a fault is detected by the FICM, a CAN message is sent to the PCM. Corresponding fault DTCs are displayed at the end of the self-test.

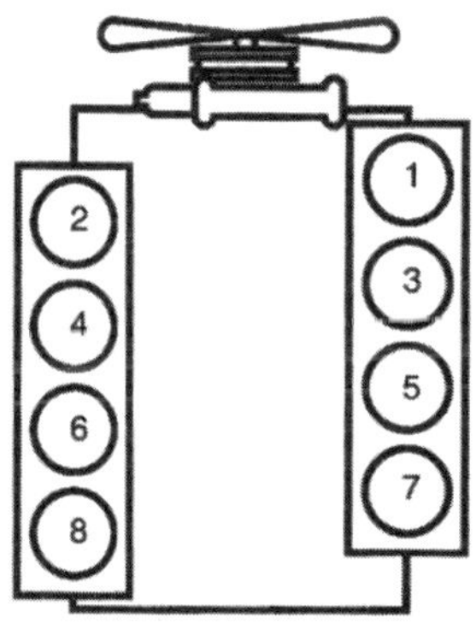

FIRING ORDER
1-2-7-3-4-5-6-8

Note: If no DTCs are present and the KOEO Injector Electrical Self-Test (click test) aborts while trying to perform, Diagnostic Subroutines , Enhanced Injector Diagnostics.

Connect the scan tool. Turn off accessories. If vehicle is equipped with an auxiliary powertrain control (RPM control), it must be turned off to perform self-tests.

1. Perform the necessary vehicle preparation and visual inspection. Refer to Quick Test Operation .
2. Refer to the manufacturer manual for scan tool instructions.
3. Select KOEO INJECTOR ELECTRICAL SELF-TEST (CLICK TEST).
4. Follow operating instructions from the menu.
5. Record DTCs and follow appropriate pinpoint test.
6. After test, cycle key to off before running other tests or driving vehicle.

KOEO/KOER Output State Control

Output State Control
The Output State Control (OSC) aids in servicing output actuators associated with the PCM. It allows the technician to energize and de-energize most of the system output actuators on command.

Connect the scan tool. Turn off accessories. If vehicle is equipped with an auxiliary powertrain control (RPM control), it must be turned off to perform self-tests. The outputs will be limited during KOEO output state control.

1. Perform the necessary vehicle preparation and visual inspection. Refer to Quick Test Operation .
2. Refer to the manufacturer manual for scan tool instructions.
3. Select KOEO/KOER output state control
4. Follow special instructions on screen.
5. Use scan tool to access output PIDs.
6. Cycle output state on EGR, ICP, injectors, vistronic fan, VGT, RPM and fuel pump.
7. Select PIDs.
8. Activate OSC.
9. After test, cycle key to off before running other tests or driving vehicle.
10. Clear the DTCs.

KOER Glow Plug Monitoring Self-Test

Key On Engine Running (KOER) Glow Plug Monitoring Self-Test
Key On Engine Running (KOER) Glow Plug Monitoring Self-Test is a functional test of the glow plug system performed on demand with the engine running and the A/C off. Battery voltage must be maintained at 10-14 volts during this test. If necessary, press the accelerator pedal to increase voltage to the specified level. The PCM will activate the Glow Plug Control Module (GPCM) and monitor the glow plug circuits. A fault must be present at the time of testing for the test to detect a fault. DTCs will be sent to the PCM on the diagnostic line and then output to the scan tool.

Connect the scan tool. Turn off accessories. Turn A/C off. If vehicle is equipped with an auxiliary powertrain control (RPM control), it must be turned off to perform self-test.

Note: When running this self-test, battery voltage must not drop below 10 volts or go above 14 volts. Make sure that batteries are OK and the charging system is working properly. If necessary, press the accelerator pedal to maintain specified voltage.

1. Perform the necessary vehicle preparation and visual inspection. Refer to Quick Test Operation .
2. Using a digital multimeter connected to the battery, monitor voltage. It may be necessary to raise RPM to maintain voltage.
3. Start the engine.
4. Parking brake set, automatic transmission in PARK or manual transmission in NEUTRAL.
5. Refer to the manufacturer manual for scan tool instructions.
6. Select KOER GLOW PLUG MONITORING SELF-TEST.
7. Follow instructions on screen.
8. Maintain system voltage until test is complete.
9. Record DTCs and follow appropriate pinpoint test.

Parameter Identification (PID)

The Parameter Identification (PID) mode allows access to certain data values, analog and digital inputs and outputs, calculated values, and system status information. Throughout the manual, there will be references to PID values. PID Data Monitor and Record can be accessed from the scan tool through the Diagnostic Data Link menu.

Selecting Parameter Identification (PID)
Connect the scan tool to the DLC under the dash. Turn off accessories. If the vehicle is equipped with an auxiliary powertrain control (RPM control), it must be turned off to perform self-tests.

1. Perform the necessary vehicle preparation and visual inspection. Refer to Quick Test Operation .
2. Refer to the manufacturer manual for scan tool instructions.
3. Follow operating instructions from the menu.
4. Select the PIDs, press START to begin monitoring.

Accumulating PCM Data

Accumulating PCM data can be done in a number of ways. Gather as much data as possible when the malfunction is occurring to prevent misdiagnosis. Data should be accumulated in different operating conditions and based on the customer description of the intermittent fault. Acquisition of PCM PID data using a scan tool is one of the easiest ways to gather information.

For information on the functions of your scan tool, refer to the manufacturer manual for scan tool instructions.

Below is a list of the functions available:

- Selecting and viewing PIDs
- Storing PIDs
- Recording measurements along with PIDs
- Playback of stored PIDs
- Peripheral inputs

Comparing PCM Data
After the PCM values have been acquired, it is necessary to determine the fault area. Typically, it will require the comparison of the actual values from the vehicle to the typical values from Reference Value Charts.

Analyzing PCM Data

Various Data Procedures
Once the fault area is identified, the circuit must be checked to determine if the wiring or component is at fault. Use any of the following methods to diagnose a suspected PCM wire circuit or device. Some methods are particular to a certain type of PCM device.

- Change Condition to Cause Response by Input
- Change Input and Verify Output Response
- Click Testing/Output State Control
- Coil Resistance (Solenoids/Relays)
- Harness Opens
- Harness Shorts

Change Condition to Cause Response by Input
The purpose is to verify sensor receives and responds to changes.

1. Select, view and record the appropriate sensor PID(s).
2. Create condition or cause condition to change.
3. If reading changes appropriately, then it should be operating OK.

   Examples:
   - View EOT PID while engine warms up.
   - It should change from a higher voltage (2.6V) for a cold engine, to a lower voltage as the engine warms up (0.6V).
   - Move accelerator pedal, observe APP PID change.
   - Press brake pedal, watch BOO PID change states.

Change Input and Verify Output Response
The purpose is to verify how the PCM and actuator circuit responds to sensor input.

1. Select, view the appropriate sensor PID(s).
2. Create condition to cause input condition to change.
3. Observe change (response) in actuator PID or actuator signal circuit measured by a measuring device.

   Example:
   - Increase accelerator pedal position under load, observe RPM DSD PID and circuit
   - change.

Coil Resistance (Solenoids/Relays)

- The purpose is to measure the correct resistance value of device.
- Key off.
- DLC disconnected, from any scan tools.
- Disconnect the component from vehicle harness.
- Using an ohmmeter and referencing the Static Resistance Value Chart in this section, measure across the component terminals in question.

Harness Opens
The purpose is to check harness for open circuits

1. Key off.
2. DLC disconnected from any scan tools.
3. Disconnect the component from the harness.
4. Disconnect the appropriate PCM harness.
5. Using an ohmmeter, isolate the circuit in question from the harness pin to the component connector signal pin.
6. Reading should be less than 5 ohms.

Analyzing PCM Data (Continued)

Harness Shorts

The purpose is to check harness for short circuits (to ground or power).

1. Key off only.
2. DLC disconnected from any scan tools.
3. Disconnect the component from vehicle harness.
4. Using an ohmmeter, measure the resistance between the signal circuit and signal return circuit, power ground circuit, VREF circuit or vehicle power.
5. If reading is less than 5 ohms, then the two circuits are shorted.

On-Board System Readiness Test

Description

All OBD scan tools should display the On-Board System Readiness (OSR) Test. The OSR will display the supported monitors on the vehicle, the status of all monitors (complete or not complete) and the MIL status. If a monitor is not complete, the scan tool will not identify the monitor that has not completed. Readiness for the 6.0L is reported based on the completeness of the comprehensive component monitors (CCM) and the misfire monitor and the EGR monitor. These monitors must be complete before readiness can be cleared. It is not possible to determine which individual monitors are complete/incomplete.

Accessing On-Board System Readiness Test

Refer to the manufacturer manual for scan tool instructions.

OBD PID Data Monitor

The PID monitor for OBD offers real time evaluation of several emissions-related parameters. Most of these are related to the HO2S for which the diesel has no equivalent. The only parameters which apply to 6.0L diesel applications are DTC CNT, IAT, LOAD, MAP, MIL, RPM and VSS.

Accessing the PID Data Monitor

- Refer to the manufacturer manual for scan tool instructions.

OBD Pending DTCs

Pending DTCs are DTCs that have only set during one drive cycle and may not have set the MIL yet. These DTCs may be identified using the Retrieve Pending DTCs Feature. Additionally, they will be found in the vehicle-specific mode RETRIEVE/CLEAR CONTINUOUS MEMORY DTCs from the instant that the DTC is set.

This function will only report pending failures that have occurred during the present drive cycle, but not indicate single failures that happened on any previous drive cycle.

Accessing the PID Data Monitor

- Refer to the manufacturer manual for scan tool instructions.

Misfire Monitoring Supported

This menu pick can be used to determine if the misfire monitoring system is supported on your particular application. All emission-equipped 6.0L diesel vehicles under 14,000 lbs utilize misfire detection. When selecting this function, a message will be displayed on the screen: "test not supported by this module" or "test supported by this module."

Comprehensive Component Monitoring Supported

This menu pick can be used to determine if the comprehensive component monitoring system is supported on your particular application. All emission-equipped 6.0L diesel vehicles under 14,000 lbs utilize comprehensive component monitoring. When selecting this function, a message will be displayed on the screen: "test not supported by this module" or "test supported by this module."

On-Board System Readiness Test (Continued)

Misfire Monitoring Status
This menu pick can be used to determine the status of the misfire monitoring system on your particular application. All emission-equipped 6.0L diesel vehicles under 14,000 lbs utilize misfire detection. When selecting this function, a message will be displayed on the screen: "test complete, or not applicable" or "test not complete."

Comprehensive Component Monitoring Status
This menu pick can be used to determine the status of the comprehensive component monitoring system on your particular application. All emission-equipped 6.0L diesel vehicles under 14,000 lbs utilize comprehensive component monitoring. When selecting this function, a message will be displayed on the screen: "test complete, or not applicable" or "test not complete."

PTO Status
This menu pick can be used to determine the status of the PTO system on your particular application. Some transmission OBD monitors are disabled during PTO operation. Vehicle must be out of PTO mode to clear P1000. When selecting this function, a message will be displayed on the screen: "PTO active" or "PTO non active."

Freeze Frame Data

Description
Freeze Frame Data allows access to emission related values from specific generic PIDs. These values are stored the instant an emission related DTC is stored in Continuous Memory. This provides a snapshot of the conditions that were present when the DTC was stored. Once one set of freeze frame data is stored, this data will remain in memory even if another emission related DTC is stored. When a DTC associated with the freeze frame is erased or a PCM memory reset is performed, new freeze frame data can be stored again. In the event of multiple emission related DTCs in memory, always note the DTC for the freeze frame data.

FREEZE FRAME DATA TABLE			
PID Number	**Acronym**	**Description**	**Measurement Units**
2	DTCFZ	DTC Which Stored Freeze Frame	#
4	LOAD	Calculated Load Value	Percent
5	ECT	Engine Coolant Temperature	Celsius/Degrees F
0B	MAP	Manifold Absolute Pressure	kPa
0C	RPM	Engine RPM	RPM
0D	VSS	Vehicle Speed	MPH
0F	IAT	Intake Air Temperature	Degrees F
10	MAF	Air Flow Rate	Grams per second
1E	PTO	Power Take Off Status	Mode
1F	RUNTM	Time Since Engine Start	Seconds
2C	EGR_PCT	Commanded EGR	Percent
2D	EGR_ERR	EGR Error	Percent
2F	FLI	Fuel Level Input	Percent
30	WARM_UPS	# Of Warm-ups Since Codes Cleared	#
31	CLR_DIST	Distance Since Codes Cleared	Kilometers
33	BARO	Barometric Pressure	in-Hg
42	VPWR	Control Module Voltage	Volts
49	APP_D	Accelerator Pedal Position D	Percent
4A	APP_E	Accelerator Pedal Position E	Percent

Freeze Frame Data (Continued)

Accessing Freeze Frame PID Data

- Refer to the manufacturer manual for scan tool instructions.

Resetting Vehicle Specific DTCs

- Refer to the manufacturer manual for scan tool instructions.

OBD PID Data Monitor

The PID monitor for OBD offers real time evaluation of several emissions-related parameters. Most of these are related to the HO2S for which the diesel has no equivalent.

Powertrain Control Module (PCM) Reset

Description

The Powertrain Control Module (PCM) Reset allows the scan tool to command the PCM to clear all emission-related diagnostic information. When resetting the PCM, a DTC P1000 will be stored in the PCM until all the OBD system monitors or components have been tested to satisfy a drive cycle, without any other faults occurring. For more information about a drive cycle, refer to Drive Cycles. Clearing DTCs from the vehicle-specific RETRIEVE/CLEAR CONTINUOUS MEMORY DTCs will clear all systems, including the OBD systems. Clearing DTCs from the OBD generic menu will only clear OBD features for the 6.0L diesel.

The following events occur when a PCM reset is performed:

- Clears the number of Diagnostic Trouble Codes (DTC).
- Clears the DTCs.
- Clears the freeze frame data.
- Resets status of the OBD system monitors.
- Sets DTC P1000 as a vehicle-specific DTC. P1000 will not appear as an OBD DTC.

Resetting OBD DTCs

- Refer to the manufacturer manual for scan tool instructions.

Resetting KAM

Disconnect the negative cable from the battery for a minimum of 5 minutes.

After KAM has been reset, DTCs P1000 and P0603 will be stored in the PCM.

Flash Electrically Eraseable Programmable Read Only Memory — Flash EEPROM

Description

E-Series

Flash Electrically Eraseable Programmable Read Only Memory (EEPROM) is contained in an Integrated Circuit (IC) internal to the PCM. The EEPROM contains the vehicle strategy including calibration information specific to the vehicle and is capable of being reprogrammed or reflashed repeatedly.

As part of the calibration there is an area referred to as the Vehicle Identification (VID) block. The VID block must be programmed when replacing the PCM as described under Programming the VID Block for a Replacement PCM. Failure to perform this procedure may generate fault codes: P1635, P1639, VID Block not programmed or is corrupt. The VID block in an existing PCM can also be tailored to accommodate various hardware/parameter changes made to the vehicle since production. Failure to perform this procedure properly may generate fault code: P1635, Tire/Axle Ratio out of Acceptable Range is one of the main causes for code: P1639. This is described under Making Changes to the VID Block and also under Making Changes to the PCM Calibration. The VID block contains many items used by the strategy for a variety of functions.

Flash Electrically Eraseable Programmable Read Only Memory — Flash EEPROM

Some of these items include the VIN number, octane adjust, fuel octane, fuel type, vehicle speed limit, tire size, axle ratio, the presence of speed control and four wheel drive electronic shift on the fly versus manual shift on the fly. Only items applicable to vehicle hardware and supported by the VID block will be displayed on the scan tool.

When changing items in the VID block, the strategy will place range limits on certain items such as tire and axle ratio. The VID block is also limited to the number of times to be reconfigured. When this limit is reached, the scan tool will display a message indicating the need to flash the PCM again to reset the VID block.

Each of the procedures described below use the Worldwide Diagnostic System (WDS). Reprogramming can be performed by a local Ford dealer for any non-Ford facility. There are other Enhanced Scan Tools that may have reprogramming capabilities available. Refer to the manufacturers users manual for details.

Programming the VID Block for a Replacement PCM

A new PCM will contain the latest strategy and calibration level for a particular vehicle. However, the VID block will be blank and will need programming. There are two procedures available. The first is an automatic data transfer from the old PCM to the new PCM and the second is manual data entry into the new PCM.

Automatic data transfer will be performed if the old PCM is capable of communicating. This is done by the use of a scan tool to retrieve data from the old PCM before removing it from the vehicle. The stored data can now be downloaded to the new PCM after it has been replaced.

Manual data entry must be performed if the old module is damaged and/or incapable of communicating. Remove and replace the old PCM. Using a compatible scan tool select and execute Module/Parameter reprogramming referring to the manufacturers users manual. Important, make certain that all parameters are included. Failure to properly program Tire Size in revolutions per mile, (rev/mile - 63,360 divided by the tire circumference in inches) Axle Ratio, 4x4/4x2, and/or Manual/Electronic shift on the fly (MSOF/ESOF) may result in codes: P1635, P1639. You may be instructed to contact the "AS BUILT" data center for the information needed to manually update the VID block with the scan tool. Contact the center ONLY if the old PCM cannot be used or the data is corrupt. For Ford L-M technicians, contact your National Hotline or the Professional Technician Society (PTS) web sight for "As Built" data. Non-Ford technicians, use the Fed World website at "fedworld.gov". Select Auto Service Information and search for "Calibrations" or "Vehicle Calibrations" then specify vehicle manufacturer, model name and model year as required.

Making Changes to the VID Block

A PCM which is programmed may require changes to be made to certain VID information to accommodate vehicle hardware. Refer to PCM/Module Reprogramming on the Scan Tool.

Making Changes to the PCM Calibration

At certain times, the entire EEPROM will need to be completely reprogrammed. This is due to changes made to the strategy or calibration after production or the need to reset the VID block because it has reached its limit. Refer to PCM/Module Reprogramming on the Scan Tool.

Drive Cycles

OBD Drive Cycle
Description
The primary intention of the OBD Drive Cycle is to clear the DTC P1000 and to satisfy the specifications for SAE specification J1979. Each OBD monitor must run during the drive cycle.

If the drive cycle is completed and P1000 is not cleared, repeat the entire drive cycle. If a particular step is interrupted, simply repeat the drive mode. If the drive cycle is interrupted with a key-off, only drive modes that were incomplete must be run.

CAUTION: Strict observance of posted speed limits and attention to driving conditions are mandatory when proceeding through the drive cycle.

Rough road conditions may prevent certain steady state conditions and steady accelerations from validating the transmission- and load-related monitors.

Vehicles equipped with Power Take-Off (PTO) must have that system disengaged before the OBD drive cycle is initiated.

Drive Cycle Procedure:

1. Key on. Do not crank until the WAIT TO START light cycles, or at least 10 seconds, whichever is greater.
2. Start the engine. Idle in PARK or NEUTRAL for 40 seconds.
3. The following outlines the appropriate conditions for running certain OBD monitors that require the engine to be under load:
    a. Accelerate steadily to third gear (M/T use fourth gear) and hold at 1500 RPM for 3 seconds. Accelerate steadily from 35 mph to 65 mph over approximately 15 seconds (M/T 11 seconds minimum).
    b. Repeat Step 3a three times while maintaining the conditions in Step 3a.
    c. Before proceeding, turn all accessories off and disengage tow/haul.
4. Before continuing, EOT must exceed 60°C (140°F).
5. Idle the vehicle for 20 seconds in PARK or NEUTRAL.
6. Key off.
7. Start the engine. Idle in PARK or NEUTRAL for 40 seconds.
8. Rerun Quick Test.

Note: If P1000 is present after running the drive cycle:

- Rerun Step 3a, being certain to maintain a minimum MFDES of 35 mg/stroke above 2,000 RPM for 11 seconds. Also, maintain a minimum MFDES of 30 mg/stroke above 2,800 RPM for at least 6 seconds.
- Rerun Step 8. MFDES must remain below 12 mg/stroke for 11 consecutive seconds.

OBD PID Data Monitor
The PID monitor for OBD offers real time evaluation of several emission-related parameters. The only parameters that apply to 6.0L diesel applications are DTC CNT, MAP, MIL, RPM and VSS.

Accessing the PID Data Monitor

- Refer to the manufacturer manual for scan tool instructions.

OBD Pending DTCs
Pending DTCs are DTCs that have only set during one drive cycle and may not have set the MIL yet. These DTCs can be identified using the Retrieve Pending DTCs feature. Additionally, they will be found in the vehicle-specific mode RETRIEVE/CLEAR CONTINUOUS MEMORY DTCs from the instant the DTC is set.

This function will only report pending failures that have occurred during the present or previous drive cycle, but will not indicate single failures that happened on any two previous drive cycles.

Accessing the PID Data Monitor

- Refer to the manufacturer manual for scan tool instructions.

Intermittent Diagnostic Techniques

Intermittent diagnostic techniques help find and isolate the root cause of intermittent faults associated with the PCM. The material is organized to help find the fault and perform the repair. There are examples that illustrate the diagnostic techniques. The process of finding and isolating an intermittent, starts with recreating a fault symptom, accumulating PCM data and comparing that data to typical values and analyzing the results.

Before proceeding, be sure that:

- Customary mechanical system tests and inspections do not reveal a problem. (Remember, mechanical component problems can make a PCM system react abnormally.)
- Review Technical Service Bulletins (TSBs) and OASIS messages, if available.
- Quick Test and associated Pinpoint diagnosis have been completed without finding a fault, and the symptom is still occurring.

Recreating the Fault

Recreating the fault is the first step in isolating the cause of the intermittent symptom. If Freeze Frame Data is available, it may help in recreating the conditions at the time of a Malfunction Indicator Lamp Diagnostic Trouble Code (MIL-type DTC). Listed below are some of the conditions for recreating the fault.

Conditions To Recreate Fault	
Engine Type Conditions	**Non-Engine Type Conditions**
Engine Temperature	Ambient Temperature
Engine RPM Engine Load Engine idle/accel/decel	Moisture Conditions Road Conditions (smooth-bumpy)

Accumulating PCM Data

PCM data can be accumulated in a number of ways. Gather as much data as possible when the malfunction is occurring to prevent misdiagnosis. Data should be accumulated during different operating conditions and based on the customer description of the intermittent fault. Reference the known good data values located in Reference Values. This will require recording data in four conditions for comparison: 1) KOEO, 2) HOT IDLE, 3) 30 mph and 4) 55 mph. Acquisition of PCM PID data using a scan tool is one of the easiest ways to gather information.

For information on the functions of your scan tool, refer to the manufacturer manual for scan tool instructions.

Below is a list of the functions available:

- Selecting and viewing PIDs
- Storing PIDs
- Recording measurements along with PIDs
- Playback of stored PIDs
- Peripheral inputs

Comparing PCM Data

After the PCM values have been acquired, it is necessary to determine the fault area. Typically, it will require the comparison of the actual values from the vehicle to the typical values from Reference Value Charts.

Intermittent Diagnostic Techniques (Continued)

Analyzing PCM Data

Once the fault area is identified, the circuit must be checked to determine if the wiring or component is at fault. When making circuit and component measurements, make sure all accessories and dome and hood lights are off. Use any of the following methods to diagnose a suspected PCM wire circuit or device. Some methods are particular to a certain type of PCM device.

- Change Condition to Cause Response by Input
- Change Input and Verify Output Response
- Click Testing/Output State Control (Solenoids/Relays)
- Coil Resistance (Solenoids/Relays)
- Harness Opens
- Harness Shorts

Change Conditions to Cause Response by Input

The purpose is to verify the sensor receives and responds to changes.

1. Select, view and record the appropriate sensor PID(s).
2. Create condition or cause condition to change.
3. If reading changes appropriately, then it should be operating OK.
   Examples:
   a. EOT should change from a higher voltage (2.6V) for a cold engine, to a lower voltage as the engine warms up (0.6V).
   b. Move accelerator pedal, observe APP PID change.
   c. Press brake pedal, watch BOO PID change states.

Change Input and Verify Output Response

The purpose is to verify how the PCM and actuator circuit responds to sensor input.

1. Select, view and record the appropriate sensor PID(s).
2. Create condition to cause input condition to change.
3. Observe change (response) in actuator PID or actuator signal circuit measured by a measuring device.

Coil Resistance (Solenoids/Relays)

The purpose is to measure the correct resistance value of a device.

1. Key off.
2. Data Link Connector (DLC) disconnected from any diagnostic tools.
3. Disconnect component from vehicle harness.
4. Using an ohmmeter, measure across the component terminals.

Harness Opens

The purpose is to check harness for open circuits.

1. Key off.
2. DLC disconnected from any scan tools.
3. Disconnect the component and PCM from vehicle harness.
4. Disconnect the appropriate PCM harness.
5. Using an ohmmeter, isolate the circuit in question from the harness pin to the component connector pin.
6. Reading should be less than 5 ohms.

Intermittent Diagnostic Techniques (Continued)

Harness Shorts

The purpose is to check the harness for short circuits (to ground or power).

1. Key off only.
2. DLC disconnected from any scan tools.
3. Disconnect the component and PCM from vehicle harness.
4. Using a digital multimeter, measure the resistance between the signal circuit and signal return circuit, power ground circuit, VREF circuit or vehicle power.
5. If reading is less than 5 ohms, then the two circuits may be shorted.

PCM Graphs and Charts

Glow Plug Lamp Cycles On/Glow Plugs Recycle/Engine Stumble/Stalls/No Accelerator Pedal Authority Until Return to Idle Position

A powertrain control module (PCM) reset occurs when the PCM momentarily "reboots" or is turned off and on while the engine is operating. If the condition occurs a single time, the engine will momentarily stumble and the PCM will go through a normal key on cycle, including turning the glow plug lamp and glow plugs on and will also attempt to validate the accelerator pedal position. If the pedal is not at the idle position when this fault occurs, pedal authority will not be allowed by the PCM until the accelerator pedal is released and the engine returns to idle.

Probable Causes:

- Momentary loss of power to the PCM or FICM: Defective power relays, shorted or open harness, intermittent connectors, poor grounds.
- Momentary short to ground of VREF: Shorted harness or connector, damaged sensor (sensors that use VREF include - exhaust pressure (EP) sensor, injection control pressure (ICP) sensor, manifold absolute pressure (MAP) sensor, barometric pressure (BARO) sensor, accelerator pedal position (APP) sensor, exhaust gas recirculation (EGR) sensor.

Procedures:

- Check all power and ground connections for the PCM and FICM.
- Inspect the CMP sensor and the CKP sensor harness connectors and the harness for a signal short to ground condition.
- Remove and inspect the CMP camshaft position sensor and CKP crankshaft position sensor for possible damage.
- If the PCM reset condition is repeatable, disconnect the following sensors one at a time and operate the engine to determine if the reset will reoccur: EP sensor, ICP sensor, MAP sensor, APP sensor, EGR sensor. Inspect each harness, including the connector upon removal.
- Remove the valve covers and inspect the under valve cover wiring for possible rub through against the valve train.

Basic Circuit Checks

Description

Basic circuit checks help to minimize pinpoint test steps by providing a procedure to diagnose harness faults associated with the Electronic Engine Control (EEC) System. The following techniques provide helpful reminders for diagnosing open circuits (continuity), shorts to ground and shorts to power.

Note

- The suspect circuit must be isolated before testing.
- When disconnecting any harness connector, always inspect for damaged or pushed-out pins, corrosion and loose wires. Repair as necessary.
- The digital multimeter must be set to the correct scale.
- The techniques do no apply in all situations; therefore, it is necessary to perform each pinpoint test step accurately and completely.
- General resistance and voltage values are specified below. Always use the pinpoint test values if they differ.
- Always turn the key to the OFF position unless directed otherwise by the pinpoint test.
- Each of the following procedures will require the powertrain control module (PCM) and component to be disconnected to isolate the harness.

Basic Circuit Checks (Continued)

Open Circuit (Continuity)
Disconnect the appropriate PCM harness. Measure the harness resistance between the suspect circuit at the harness connector and the appropriate PCM harness pin. The resistance must be less than 5 ohms.

Shorts to Ground
Measure the harness resistance between the suspect circuit at the harness connector and a reliable ground (B-, chassis or PWR GND at the PCM harness). The resistance must be greater than 10,000 ohms.

Shorts to Power
Key ON to power up circuit. Measure voltage between the suspect circuit at the harness connector and a reliable ground. The voltage must be less than 1.0 volt.

NOTES

SYMPTOM CHARTS
INDEX

System	Symptom(s): No Crank/Slow Crank (2003-07 6.0L Diesel)	Test
Starting Concern		AA

Test/Step **Result**

AA1 PRELIMINARY CHECKS

Perform The Following Preliminary Checks:
- -Automatic Transmission In Park Or Neutral
- -Anti-Theft System
- -Clutch Fully Depressed
- -Battery Connections
- -Starter Relay Connections
- -Battery Load Test
- -Fuse Links
- -Ignition Switch
- -PCM Connectors Not Fully Seated

Are All Checks OK ?

YES → AA2

NO → Repair As Necessary → STOP

AA2 CHECK STARTER RELAY

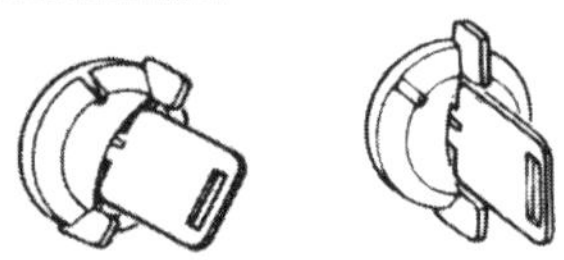

Cycle Ignition Key

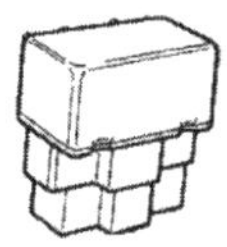

Is A Clicking Sound Heard From The Starter Relay When The Ignition Key Is Turned To Start?

YES → AA4

NO → AA3

AA3 CHECK ELECTRICAL ACCESSORIES

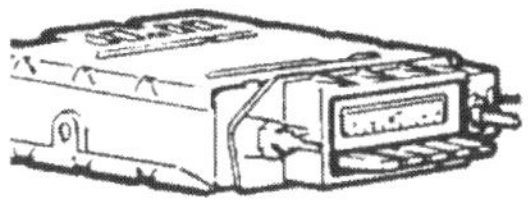

Check Headlamps

Do Any Other Electical Accessories Operate (Headlights, Radio, Ect.)?

YES → AA4

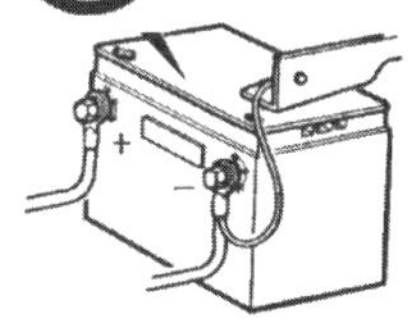

NO → Diagnose And Repair Battery And/Or Charging System. → STOP

AA4 CHECK STARTING SYSTEM SECONDARY CIRCUITS

Measure Voltage Drops In Starter Circuits.

Refer To The Appropriate Service Information.

Is Fault Indicated?

YES → REPAIR As Required → STOP

NO → AA5

AA5 CHECK STARTING SYSTEM COMPONENTS

Check Starter, Starter Relay, Clutch Pedal Position Sensor And Manual Lever Position Sensor.

Is Fault Indicated?

YES → REPAIR As Required → STOP

NO → AA6

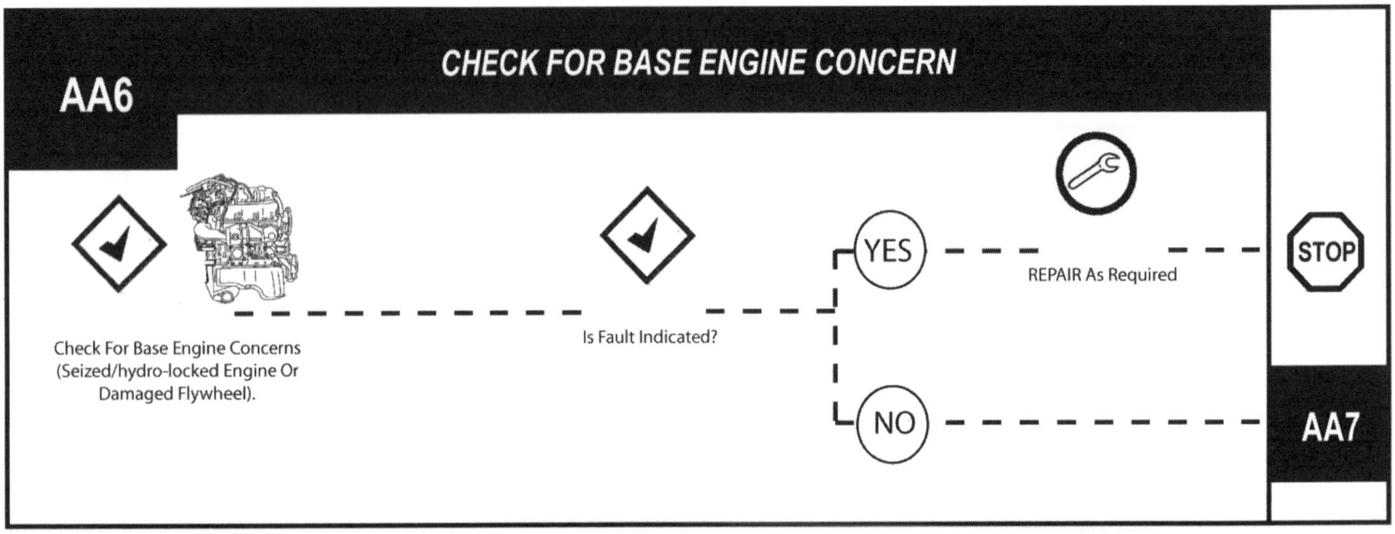
AA6
CHECK FOR BASE ENGINE CONCERN
Check For Base Engine Concerns (Seized/hydro-locked Engine Or Damaged Flywheel).
Is Fault Indicated?
YES
REPAIR As Required
STOP
NO
AA7

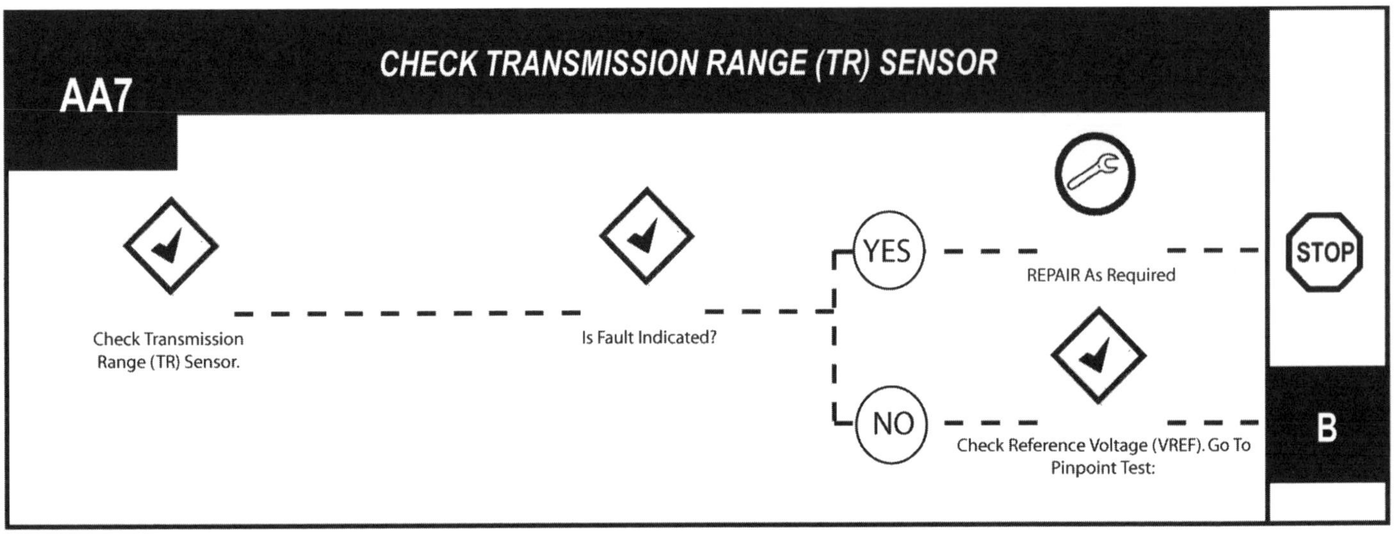
AA7
CHECK TRANSMISSION RANGE (TR) SENSOR
Check Transmission Range (TR) Sensor.
Is Fault Indicated?
YES
REPAIR As Required
STOP
NO
Check Reference Voltage (VREF). Go To Pinpoint Test:
B

System	Symptom(s)	Test
Stalls or Quits	Symptom(s): Stalls/Quits During Idle Acceleration, Cruise, Deceleration (2003-07 6.0L Diesel)	AB

Test/Step AB1 — PRELIMINARY CHECKS

Perform The Following Preliminary Checks:
- -Engine Oil Level/Quality
- -Electrical Connections
- -Automatic Transmission - Torque Converter Operation, Always Applied
- -ATF Level
- -Engine Overheating.

Are All Checks Ok?

- YES → Result: AB2
- NO → REPAIR As Required. VERIFY Symptom No Longer Exists. → Result: STOP

AB2 — CHECK FOR COLD WEATHER STALL

Does The Engine Restart After A Stall?

- YES → Oil Return To Pan May Be Too Slow. CONFIRM Proper Grade Of Oil For Fridged Conditions. CHANGE Oil, If Necessary. → Result: STOP
- NO → Result: AB3

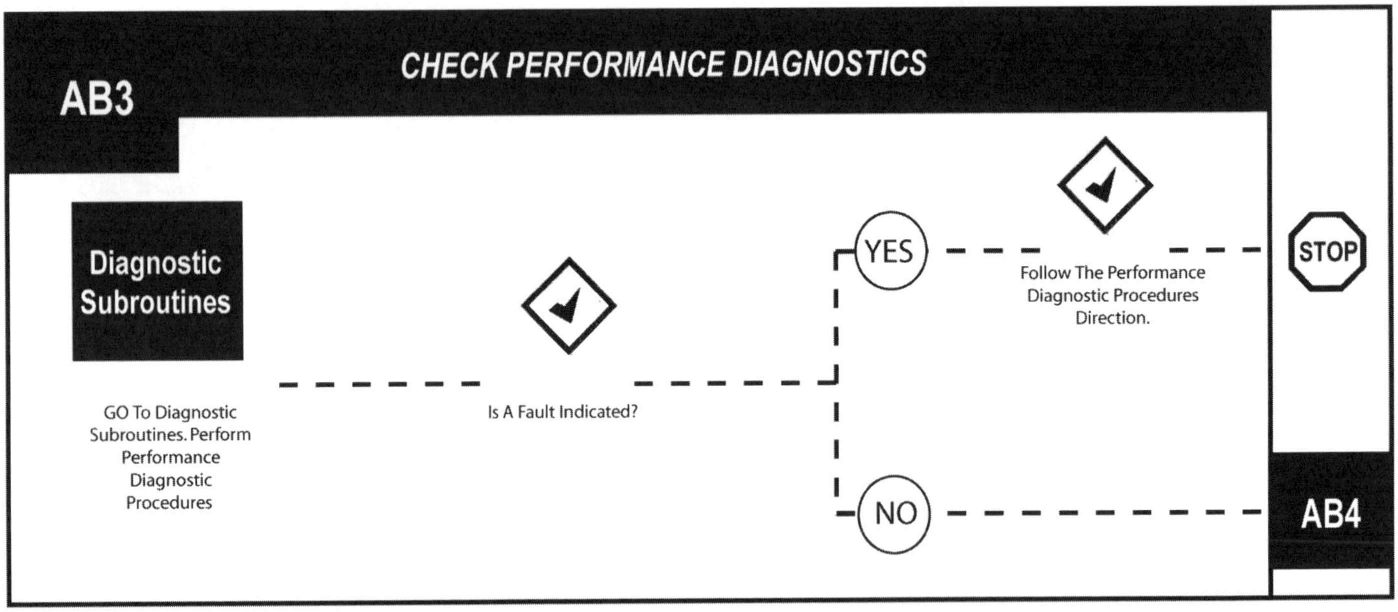
AB3
CHECK PERFORMANCE DIAGNOSTICS
Diagnostic Subroutines
GO To Diagnostic Subroutines. Perform Performance Diagnostic Procedures
Is A Fault Indicated?
YES
Follow The Performance Diagnostic Procedures Direction.
STOP
NO
AB4

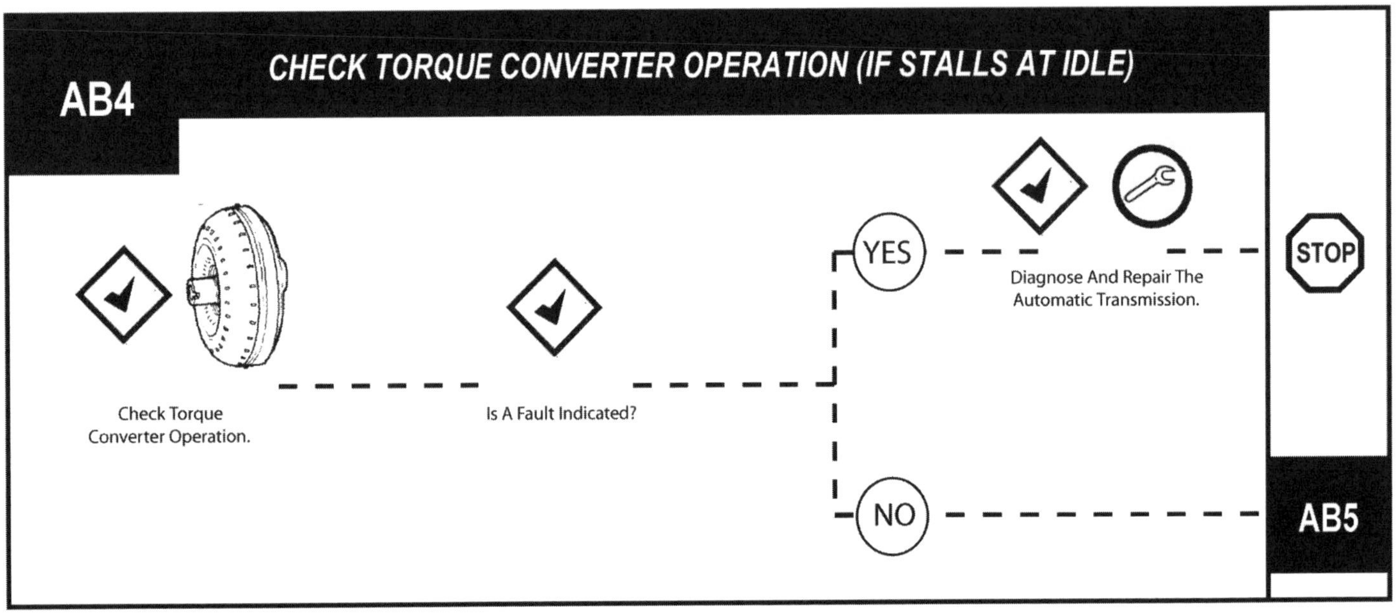
AB4
CHECK TORQUE CONVERTER OPERATION (IF STALLS AT IDLE)
Check Torque Converter Operation.
Is A Fault Indicated?
YES
Diagnose And Repair The Automatic Transmission.
STOP
NO
AB5

AB5

CHECK AIR INTAKE SYSTEM

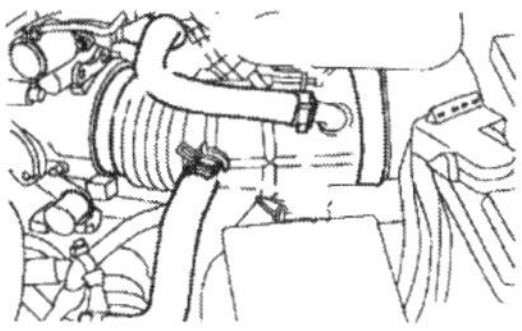

Visually Inspect Air Inlet For Obstructions And/Or Restrictions

Is A Fault Indicated?

YES — REPAIR As Necessary. RESET Air Filter Restriction Gauge — STOP

NO — AB6

AB6

CHECK REFERENCE VOLTAGE (VREF)

B

Go To Pinpoint Test B. Check Reference Voltage.

Is A Fault Indicated?

YES — REPAIR As Necessary. Confirm Problem No Longer Exists. — STOP

NO — Go To Diagnostic Subroutines And Perform The Check RPM Signal While Cranking Test. — Diagnostic Subroutines

System		Test
Starting Concern	Symptom(s): Stall After Start, No Start/Normal Crank Hard Start/Long Crank, Erratic Start/Erratic Crank (2003-07 6.0L Diesel)	AC

Test/Step **Result**

AC1 — PREMLIMINARY CHECKS

Perform The Following Preliminary Checks:
- Check Engine Oil Level And Veiscosity
- Confirm Proper Dipstick Part Number
- Check For Sufficient Clean Fuel
- Check For An Intake Restriction

Are All Checks OK?

YES → AC2

NO → REPAIR As Necessary. VERIFY Symptom No Longer Exists. → STOP

AC2 — CHECK FUEL PUMP PRESSURE

Diagnostic Subroutines

Go To Diagnostic Subroutines, Performance Diagnostic Procedures. Perform Fuel Pump Pressure Test.

Is Fuel Pressure Within Specification?

YES → AC3

NO → REPAIR As Necessary. VERIFY Symptom No Longer Exists. → STOP

AC3 — CHECK PARAMETER IDENTIFICATIONS (PIDS)

Diagnostic Subroutines

Go To Diagnostic Subroutines, Hard Start/No Start Diagnostics, Perform Step 10.

Is A Fault Indicated?

- YES → Appropriate Pinpoint Test
- NO → AC4

AC4 — CHECK ENGINE OIL PRESSURE

Check Engine Oil Pressure.

Is Engine Oil Pressure Within Specification?

- YES → AC5
- NO → REPAIR As Necessary. VERIFY Symptom No Longer Exists. → STOP

AC5 — PERFORM KOEO INJECTOR ELECTRICAL SELF TEST (CLICK TEST)

Is Fault Indicated?

- YES → Go To Appropriate Pinpoint Test
- NO → AC6

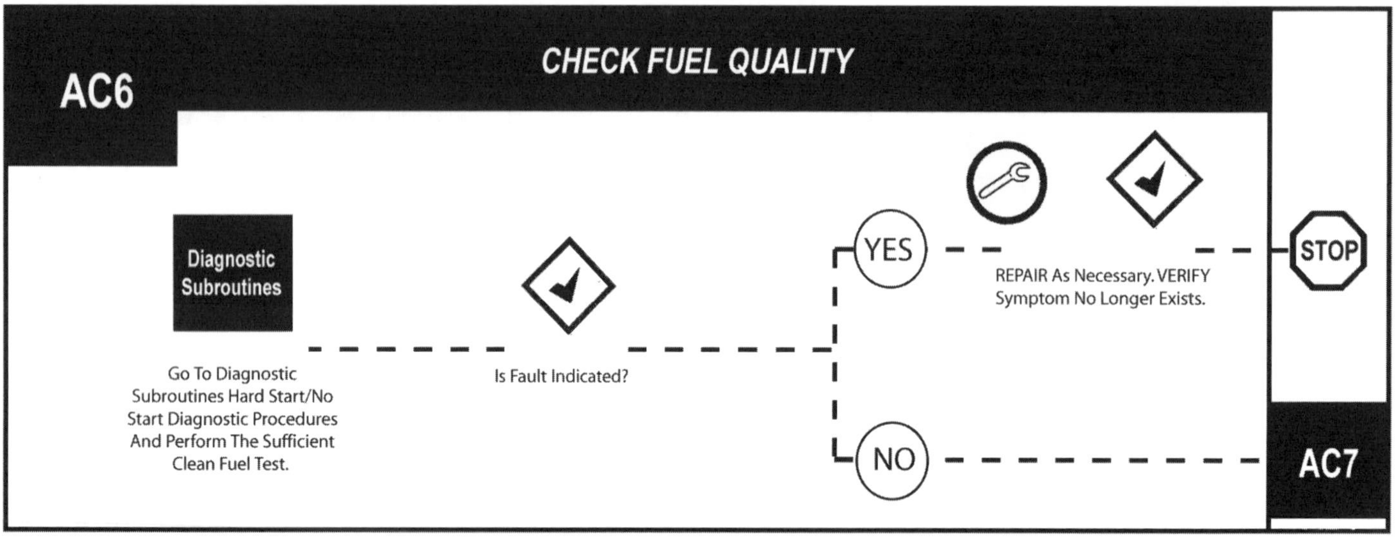
AC6
CHECK FUEL QUALITY
Diagnostic Subroutines
Go To Diagnostic Subroutines Hard Start/No Start Diagnostic Procedures And Perform The Sufficient Clean Fuel Test.
Is Fault Indicated?
YES
REPAIR As Necessary. VERIFY Symptom No Longer Exists.
STOP
NO
AC7

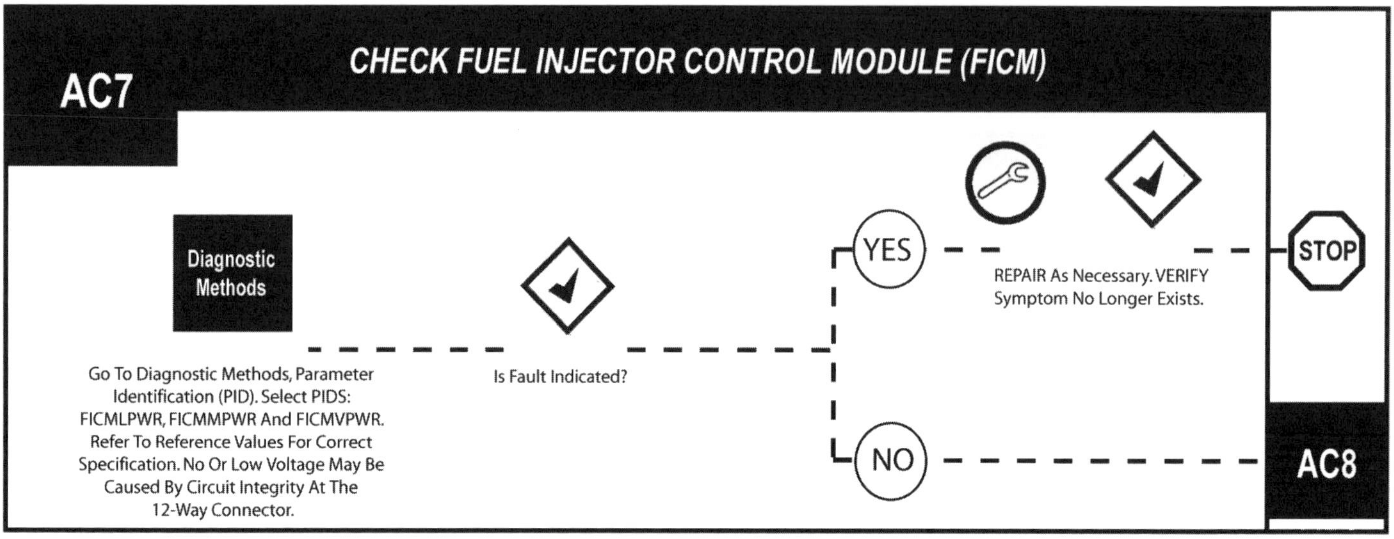
AC7
CHECK FUEL INJECTOR CONTROL MODULE (FICM)
Diagnostic Methods
Go To Diagnostic Methods, Parameter Identification (PID). Select PIDS: FICMLPWR, FICMMPWR And FICMVPWR. Refer To Reference Values For Correct Specification. No Or Low Voltage May Be Caused By Circuit Integrity At The 12-Way Connector.
Is Fault Indicated?
YES
REPAIR As Necessary. VERIFY Symptom No Longer Exists.
STOP
NO
AC8

AC8 — CHECK GLOW PLUG CONTROL MODULE CIRCUIT

CAUTION: Confirm Resistance To Ground Is Above 10,000 Ohms Before Attaching To Starter Relay.

Measure Voltage Between	
Glow Plug Circuit	Chassis
38/38A	Ground

Voltage

Is Battery Voltage Present?

YES → AF

Go To Pinpoint Test AF And Check The Following:
- Glow Plug Control Module
- Glow Plugs
- Glow Plug Connectors

OK → AC9

Not OK → Repair As Necessary. Verify Symptom No Longer Exists. → STOP

NO → Check Fusible Link And Circuitry And Repair As Necessary. Verify Symptom No Longer Exists. → STOP

AC9 — CHECK FOR ICP BIAS SENSOR

Q

Go To Pinpoint Test Q And Check For ICP Sensor Bias

Is Fault Indicated?

YES → REPAIR As Necessary. VERIFY Symptom No Longer Exists. → STOP

NO → Go To Diagnostic Subroutines Performance Diagnostic Procedures. CARRY OUT the EGR Position Test. → Diagnostic Subroutines

System	Symptom(s): Engine Will Only Idle (2003-07 6.0L Diesel)	Test
Engine Will Only Idle		AD

Test/Step

AD1 PRELIMINARY CHECKS

Note: Heavy external load upon cold start may cause a momentary lack of power.

Perform The Following Preliminary Checks:
- APP Sensor Electrical Connection Failure
- Interference From Floor Mats
- Accelerator Pedal Damage
- Transmission Control Module (TCM) DTC's

Are All Checks OK?

- YES → Result: AD2
- NO → REPAIR As Necessary. VERIFY Symptom No Longer Exists. → Result: STOP

AD2 CHECK TRANSMISSION CONTROL MODULE (TCM)

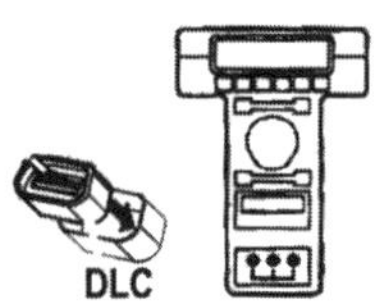

Connect The Scan Tool And Retrieve DTCs.

Are DTCs P0148 Or P2552 Present?

- YES → Result: Go To The Appropriate Pinpoint Test
- NO → Key Off. Inspect Ground Under Vehicle For Objects Blocking Wheels. → Result: STOP

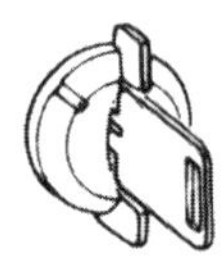

System		Test
Unique Idle Concerns	Symptom(s): Slow Return To Idle, Fast Idle (2003-07 6.0L Diesel)	AE

Test/Step

AE1 — PRELIMINARY CHECKS

Result

Perform The Following Preliminary Checks:
- PCM System Check
- Accelerator Pedal For Binding, Broken Return Spring, Stuck Condition (Floor Mats)
- External Fual Source
- Engine Not Reaching Normal Operating Temperature
- Power Take Off (PTO)
- Auxiliary Idle Control
- Charge Protection Devices Disengaged

Are All Checks Ok?

- YES
  - 2003-04 → **AE2**
  - 2005-07 → **AE3**
- NO → REPAIR As Necessary, VERIFY A Symptom No Longer Exists. → **STOP**

AE2 — PERFORM QUICK TEST OPERATION (2003-04)

Diagnostic Methods

Go To Diagnostic Methods. Perform Quick Test Operation.

Is A Fault Indicated?

- YES → Go To: **Appropriate Pinpoint Test**
- NO → **AE3**

AE3 — CHECK ACCELERATOR PEDAL

- Check For Floor Mat Interference.
- Check Bushing For Damage.
- Check Return Spring.

Is A Fault Indicated?

YES — Repair As Required. — STOP

NO — AE4

AE4 — CHECK FOR CORRECT OIL LEVEL

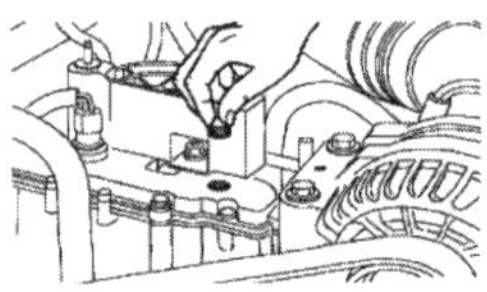

Check For Overfilled Crankcase.

Is Crankcase Overfilled?

YES — DRAIN Crankcase To Correct Level. DETERMINE Cause Of Overfill. — STOP

NO — AE5

AE5 — CHECK ENGINE OIL

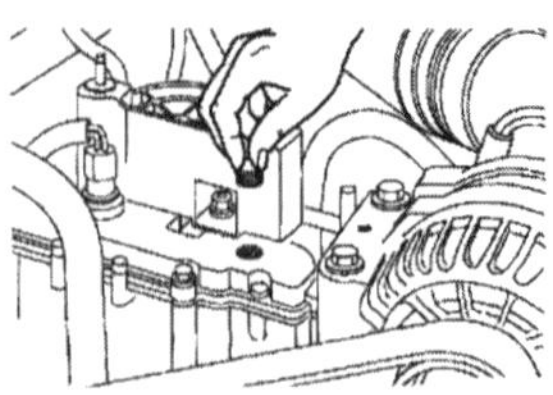

Check For Coolant Or Fuel In Engine Oil.

Is There Coolant Or Fuel In Engine Oil?

YES — REPAIR As Required. — STOP

NO — AE6

AE6 CHECK FOR FUEL CONTAMNIATION

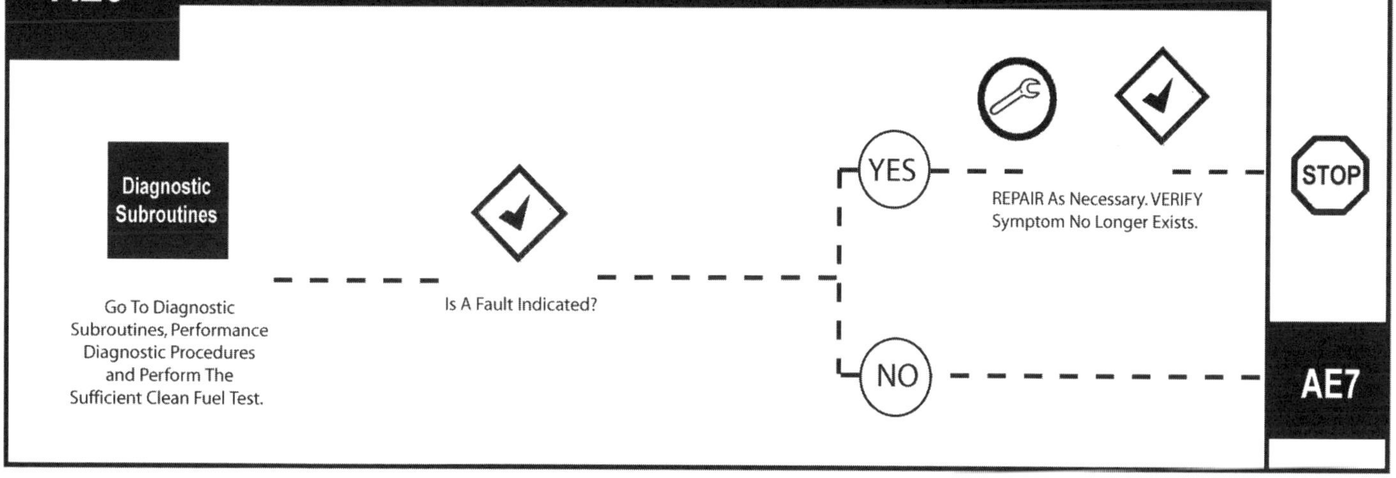

AE7 CHECK PARAMETER IDENTIFICATIONS (PIDS)

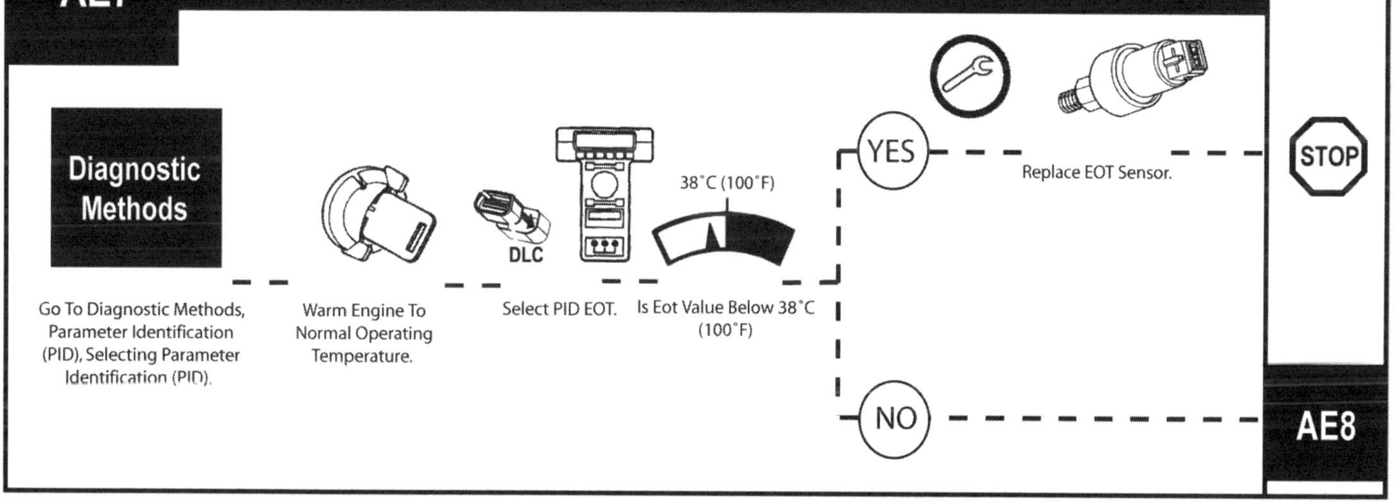

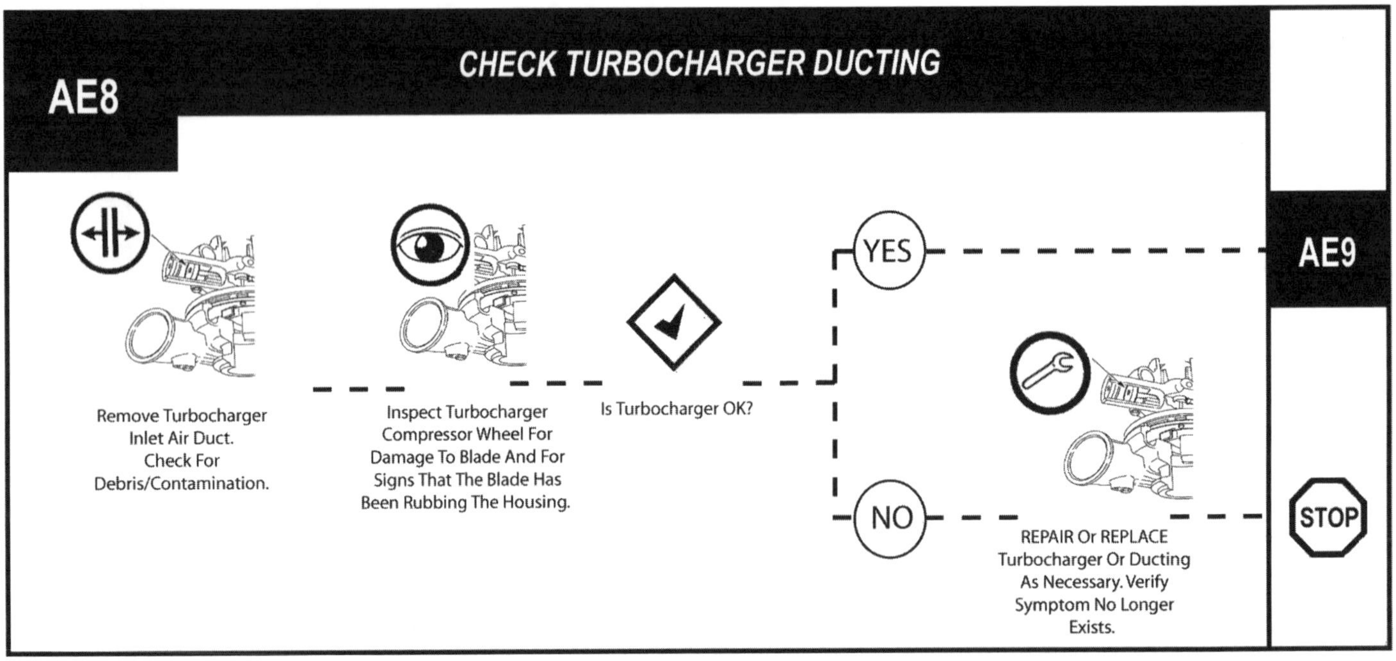
AE8
CHECK TURBOCHARGER DUCTING
Remove Turbocharger Inlet Air Duct. Check For Debris/Contamination.
Inspect Turbocharger Compressor Wheel For Damage To Blade And For Signs That The Blade Has Been Rubbing The Housing.
Is Turbocharger OK?
YES
AE9
NO
REPAIR Or REPLACE Turbocharger Or Ducting As Necessary. Verify Symptom No Longer Exists.
STOP

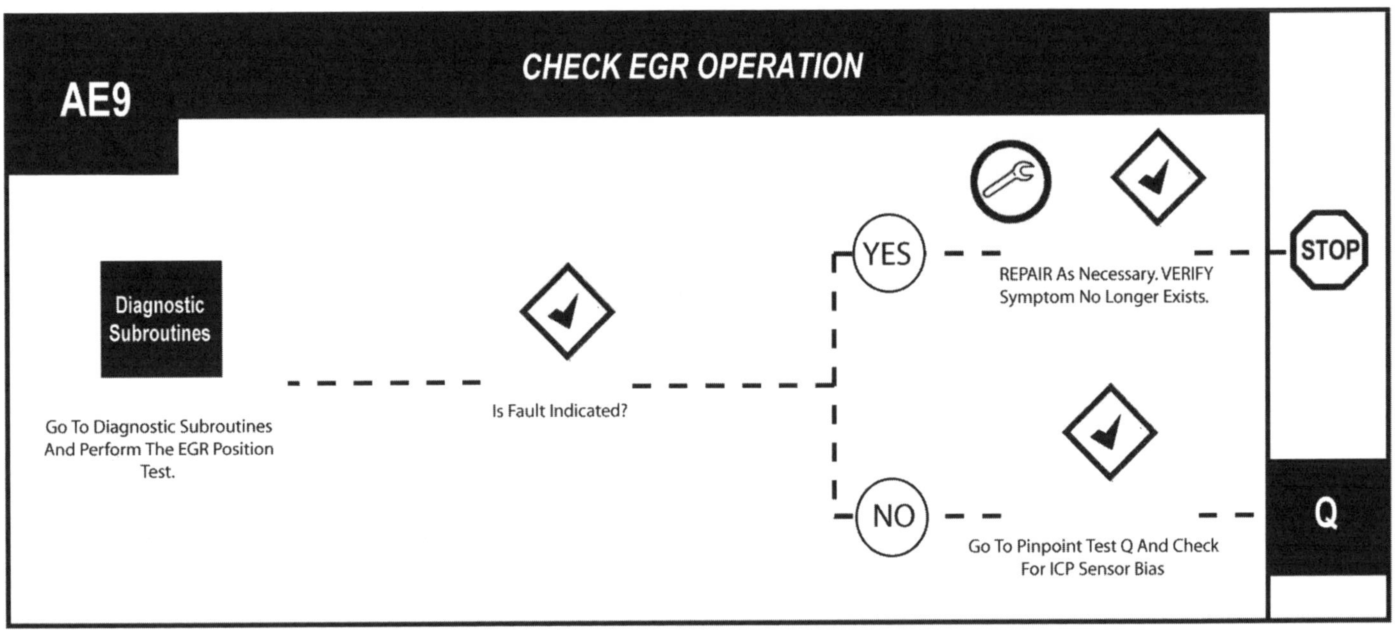
AE9
CHECK EGR OPERATION
Diagnostic Subroutines
Go To Diagnostic Subroutines And Perform The EGR Position Test.
Is Fault Indicated?
YES
REPAIR As Necessary. VERIFY Symptom No Longer Exists.
STOP
NO
Go To Pinpoint Test Q And Check For ICP Sensor Bias
Q

System		Test
Unique Idle Concerns	Symptom(s): Runs Rough: Rolling Idle, Stalls When Engaging Clutch, Rough Idle (2003-07 6.0L Diesel)	AF

Test/Step

AF1 PRELIMINARY CHECKS

Result

Perform The Following Preliminary Checks:
- Check Engine Oil Level
- Confirm Oil Change Within 12,070 Km (7500 Miles)
- Check Air Intake System: Air Filter Restriction Gauge, Inlet For Obstructiont/Restrictions. Reset Air Filter Restriction Gauge
- Confirm Acceptable SAE Oil Viscosity And API Rating Of Oil. Refer To Operater's Guide.
- Check Fuel Quality. Refer To Diagnostic Subroutines

Are All Checks Ok?

YES — 2003-04 → **AF2**

YES — 2005-07 → **AF3**

NO → REPAIR As Necessary. VERIFY Symptom No Longer Exists → **STOP**

AF2 PERFORM KOEO ON-DEMAND SELF TEST (2003-04)

Diagnostic Subroutines

Go To Diagnostic Subroutines, Performance Diagnostic Procedures. Perform KOEO On-Demand Self Test.

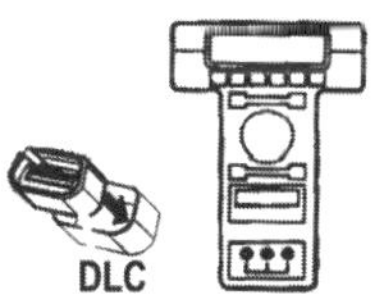

Retrieve And Clear Continuous Codes.

Is A Fault Indicated?

YES → Go To Appropriate Pinpoint Test.

NO → **AF3**

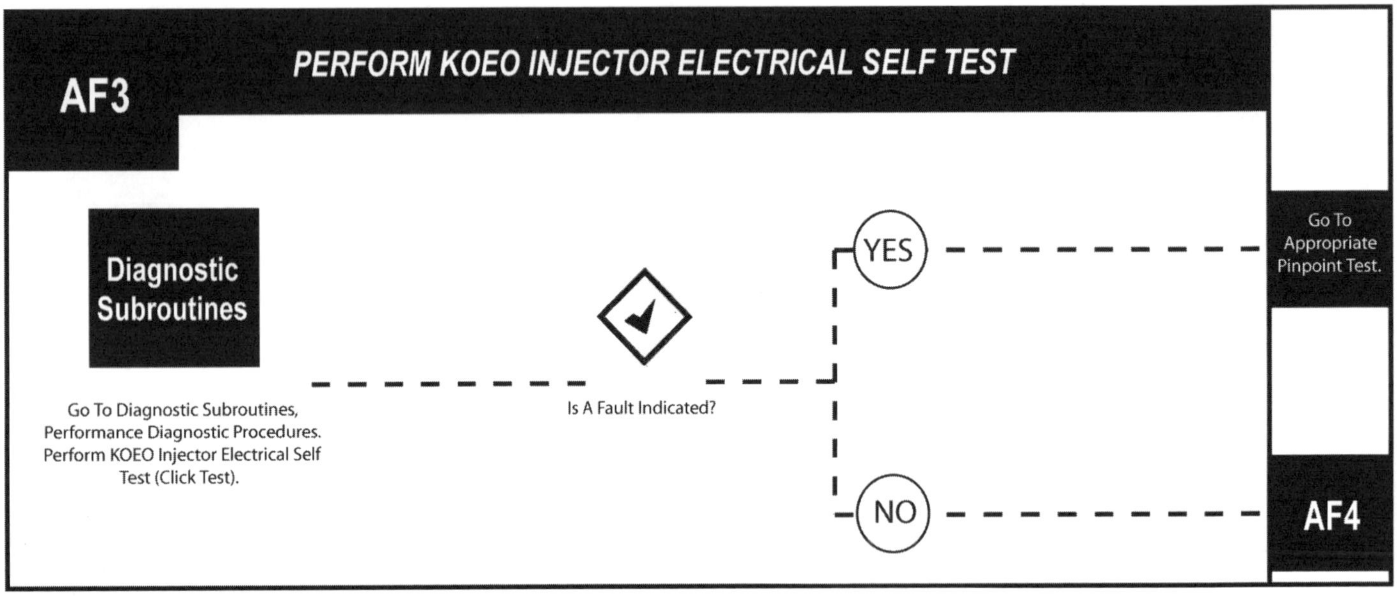
AF3
PERFORM KOEO INJECTOR ELECTRICAL SELF TEST
Diagnostic Subroutines
Go To Diagnostic Subroutines, Performance Diagnostic Procedures. Perform KOEO Injector Electrical Self Test (Click Test).
Is A Fault Indicated?
YES
Go To Appropriate Pinpoint Test.
NO
AF4

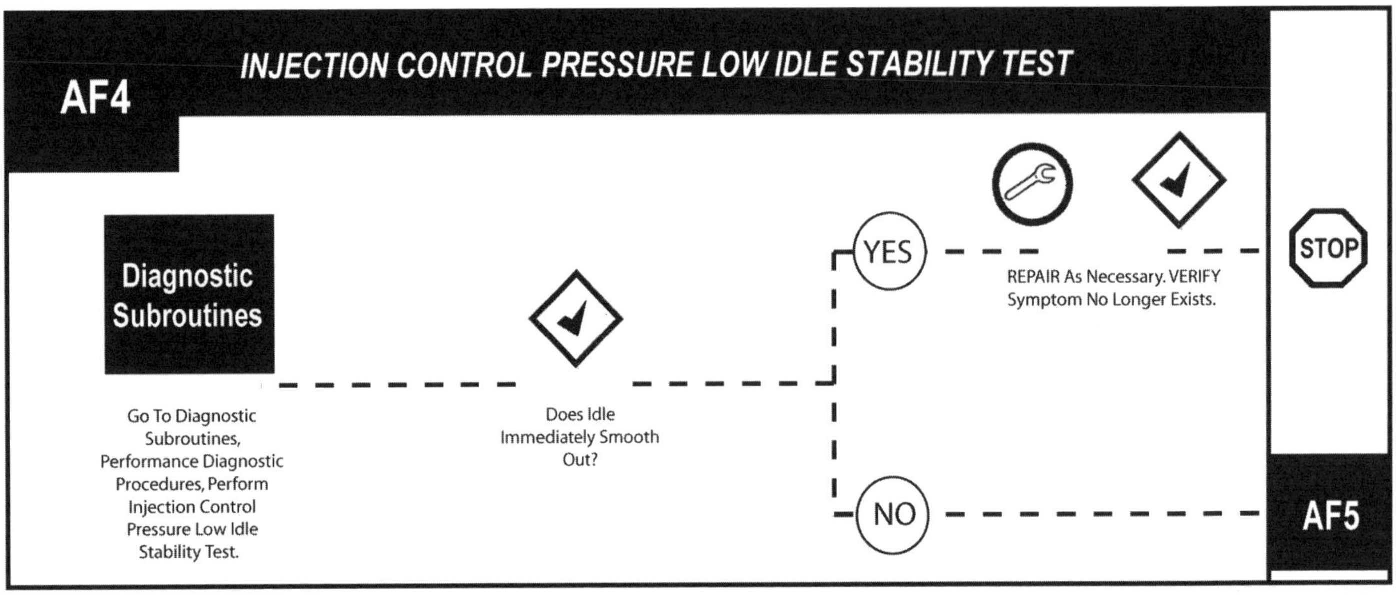
AF4
INJECTION CONTROL PRESSURE LOW IDLE STABILITY TEST
Diagnostic Subroutines
Go To Diagnostic Subroutines, Performance Diagnostic Procedures, Perform Injection Control Pressure Low Idle Stability Test.
Does Idle Immediately Smooth Out?
YES
REPAIR As Necessary. VERIFY Symptom No Longer Exists.
STOP
NO
AF5

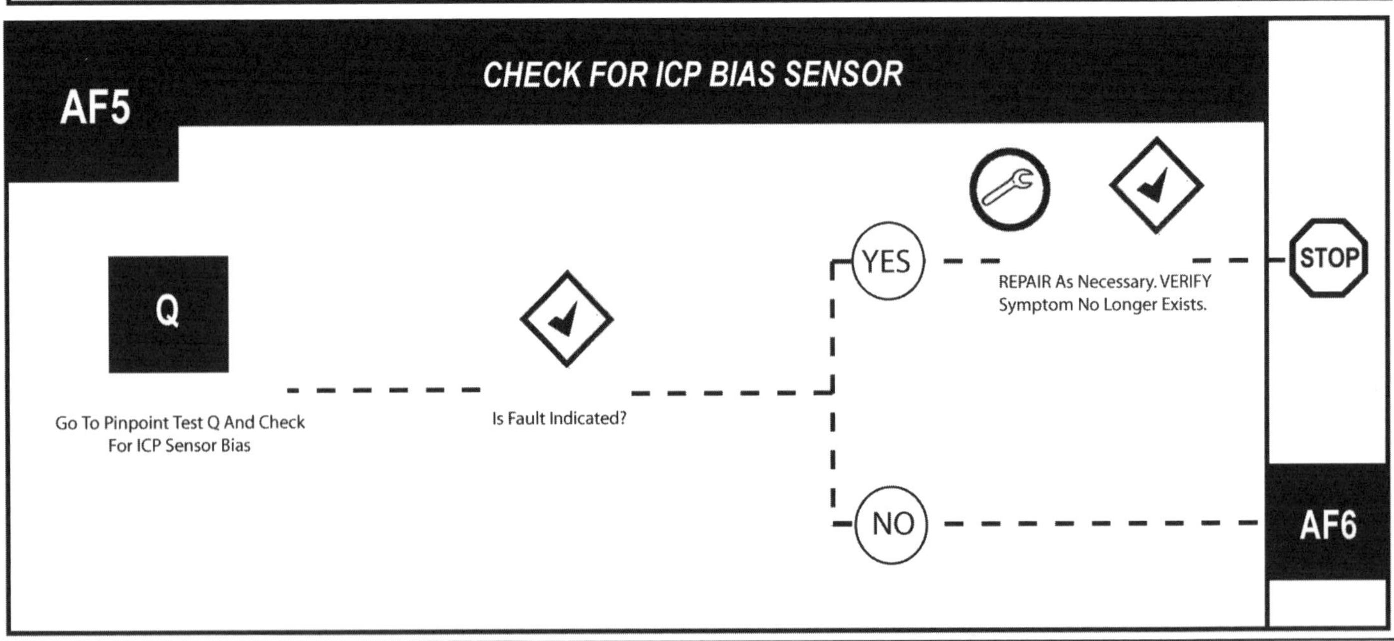
AF5
CHECK FOR ICP BIAS SENSOR
Q
Go To Pinpoint Test Q And Check For ICP Sensor Bias
Is Fault Indicated?
YES
REPAIR As Necessary. VERIFY Symptom No Longer Exists.
STOP
NO
AF6

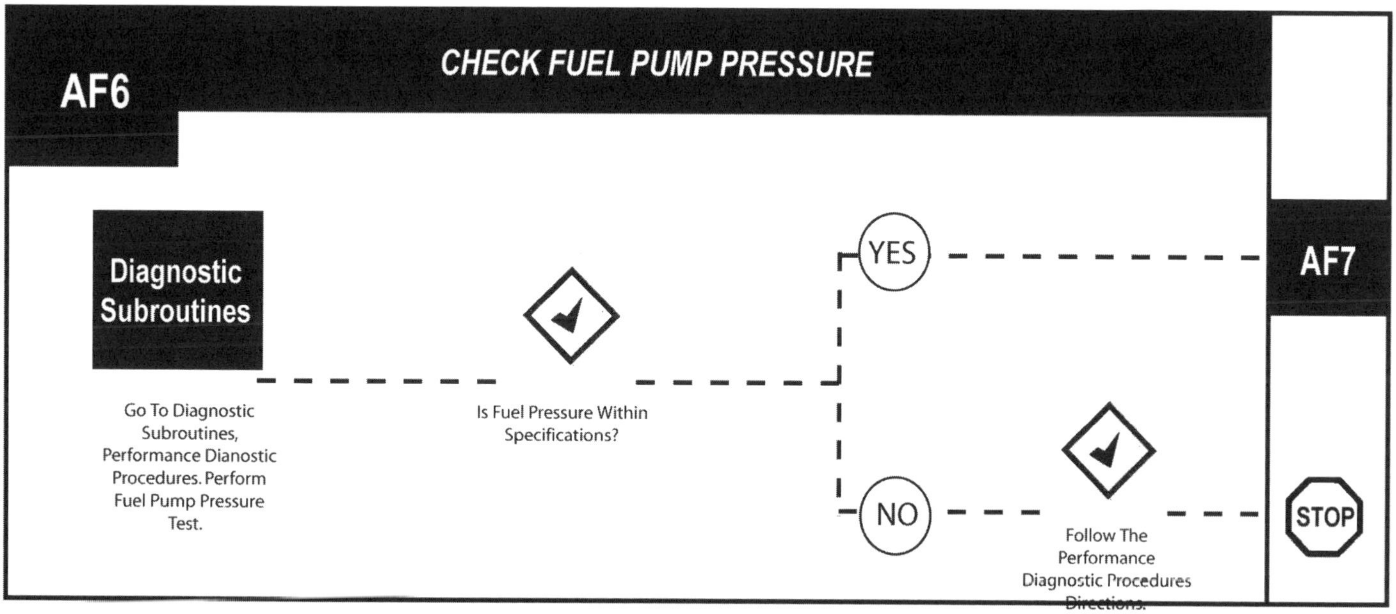
AF6
CHECK FUEL PUMP PRESSURE
Diagnostic Subroutines
Go To Diagnostic Subroutines, Performance Dianostic Procedures. Perform Fuel Pump Pressure Test.
Is Fuel Pressure Within Specifications?
YES
AF7
NO
Follow The Performance Diagnostic Procedures Directions.
STOP

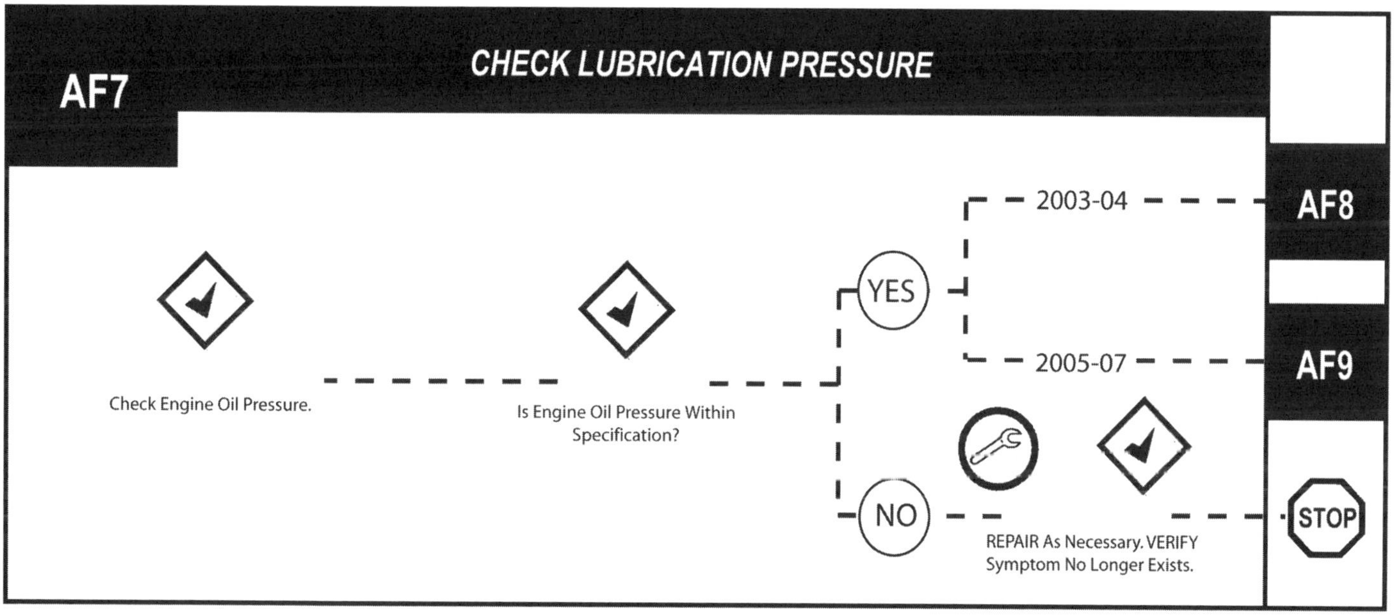
AF7
CHECK LUBRICATION PRESSURE
Check Engine Oil Pressure.
Is Engine Oil Pressure Within Specification?
YES
2003-04
AF8
2005-07
AF9
NO
REPAIR As Necessary. VERIFY Symptom No Longer Exists.
STOP

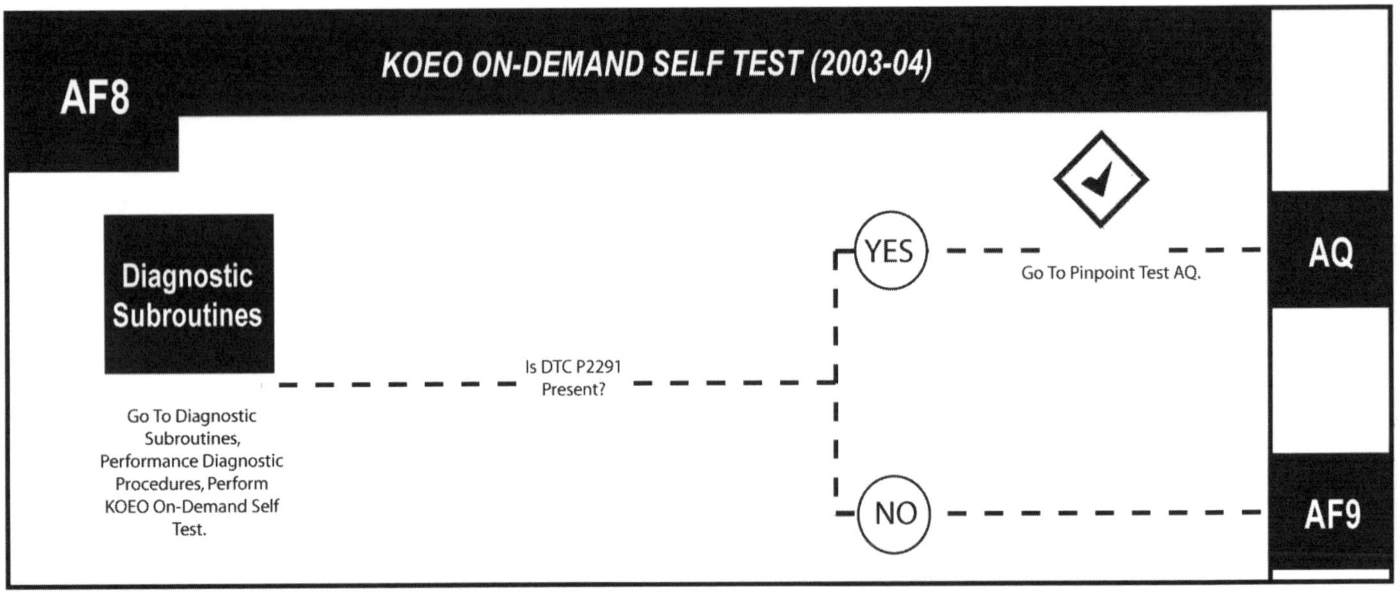
AF8
KOEO ON-DEMAND SELF TEST (2003-04)
Diagnostic Subroutines
Go To Diagnostic Subroutines, Performance Diagnostic Procedures, Perform KOEO On-Demand Self Test.
Is DTC P2291 Present?
YES
Go To Pinpoint Test AQ.
AQ
NO
AF9

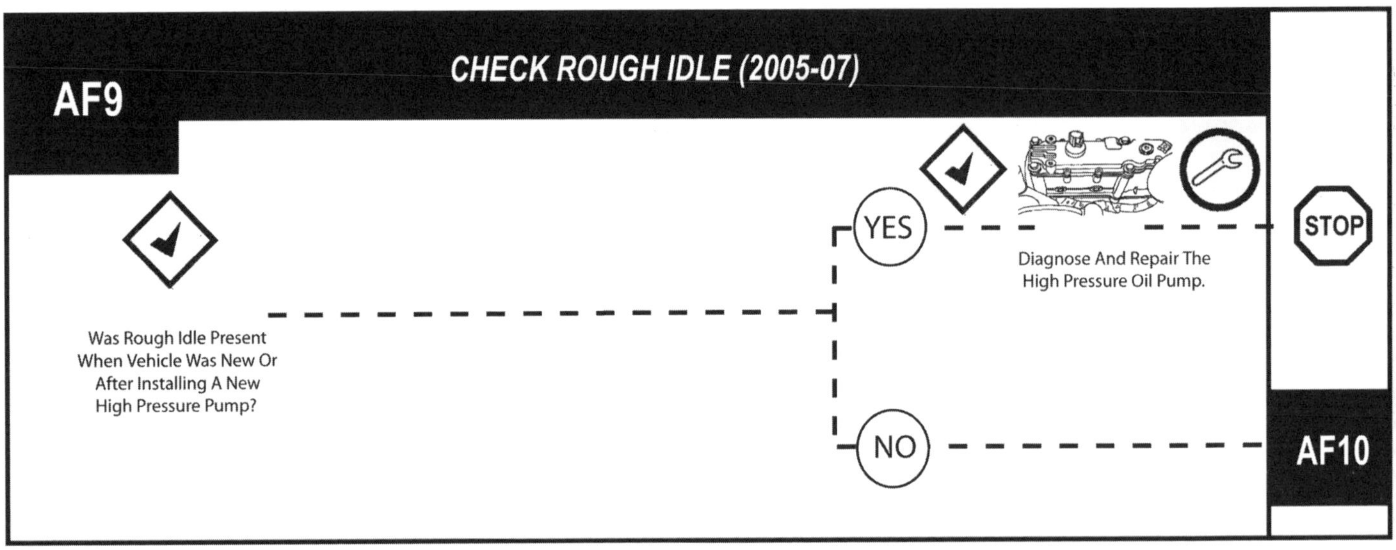
AF9
CHECK ROUGH IDLE (2005-07)
Was Rough Idle Present When Vehicle Was New Or After Installing A New High Pressure Pump?
YES
Diagnose And Repair The High Pressure Oil Pump.
STOP
NO
AF10

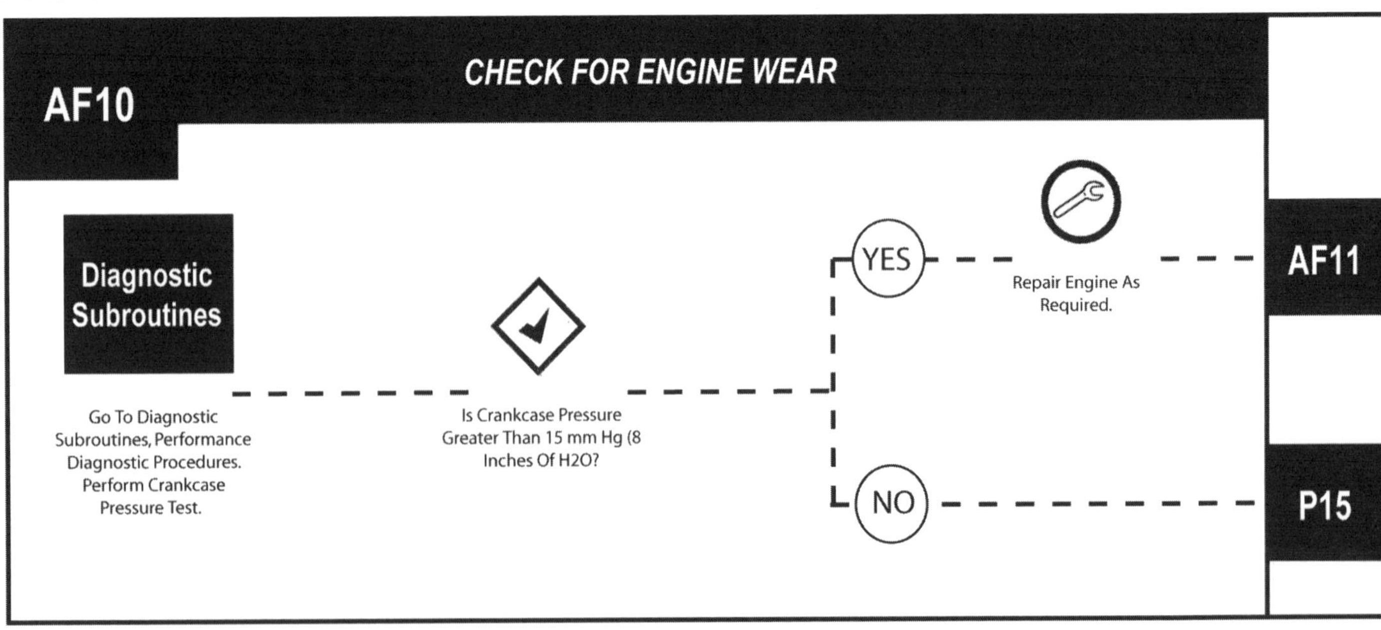
AF10
CHECK FOR ENGINE WEAR
Diagnostic Subroutines
Go To Diagnostic Subroutines, Performance Diagnostic Procedures. Perform Crankcase Pressure Test.
Is Crankcase Pressure Greater Than 15 mm Hg (8 Inches Of H2O?
YES
Repair Engine As Required.
AF11
NO
P15

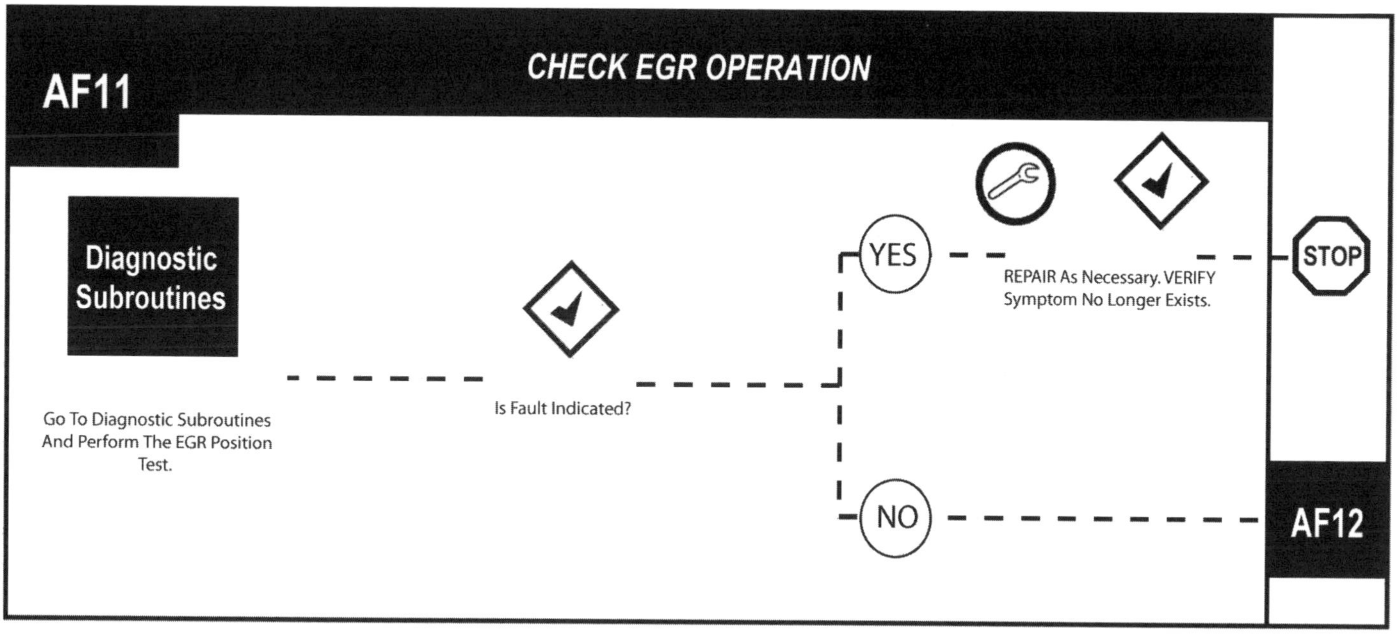
AF11
CHECK EGR OPERATION
Diagnostic Subroutines
Go To Diagnostic Subroutines And Perform The EGR Position Test.
Is Fault Indicated?
YES
REPAIR As Necessary. VERIFY Symptom No Longer Exists.
STOP
NO
AF12

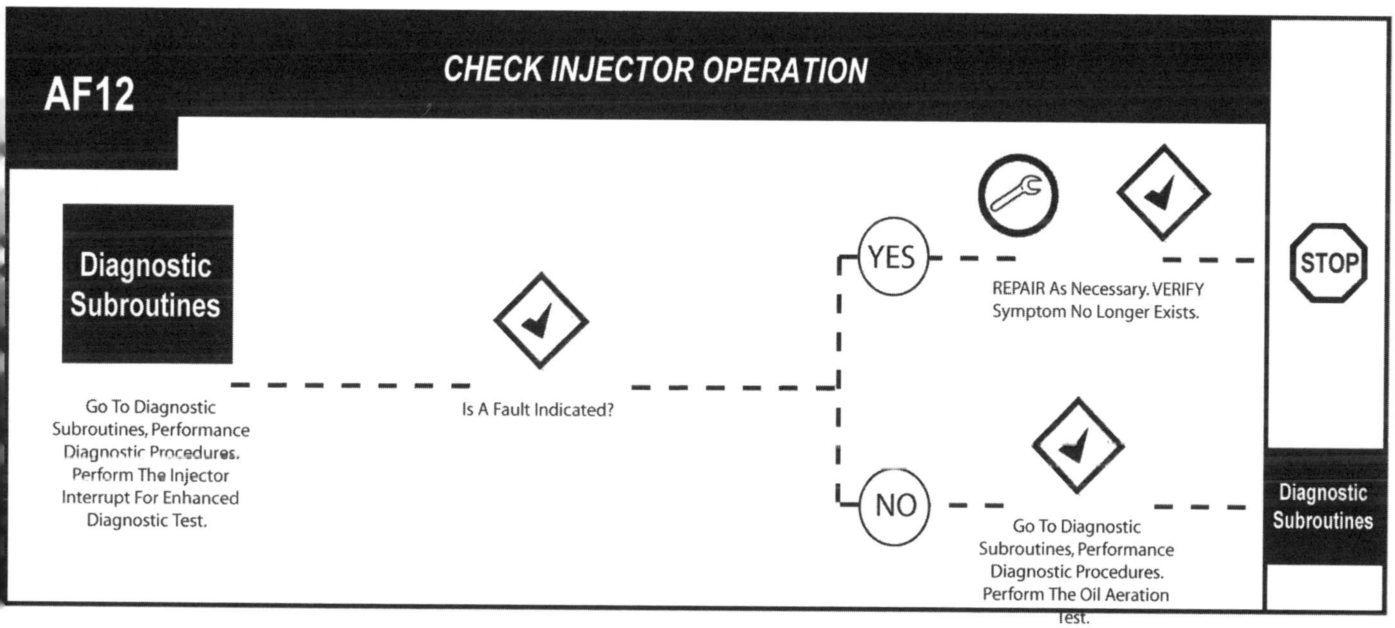
AF12
CHECK INJECTOR OPERATION
Diagnostic Subroutines
Go To Diagnostic Subroutines, Performance Diagnostic Procedures. Perform The Injector Interrupt For Enhanced Diagnostic Test.
Is A Fault Indicated?
YES
REPAIR As Necessary. VERIFY Symptom No Longer Exists.
STOP
NO
Go To Diagnostic Subroutines, Performance Diagnostic Procedures. Perform The Oil Aeration Test.
Diagnostic Subroutines

System	Symptom(s): See Table Below (2003-07 6.0L Diesel)	Test
See Table Below		AG

SYSTEM	SYMPTOM(S)
Runs Rough	Acceleration, Cruise
Misses	Idle, Acceleration, Cruise
Buck/Jerk	Acceleration, Cruise, Deceleration
Hesitation/Stumble	Acceleration
Surge	Acceleration, Cruise
Lack/Loss Of Power	Acceleration, Cruise
Fuel Mileage	Poor Fuel Economy

Test/Step

AG1

PRELIMINARY CHECKS

Result

Perform The Following Preliminary Checks:

- Confirm Breaks Are Not Dragging
- Confirm Transmission And Axle Fluid Levels
- Confirm Transmission And Axle Tube Are Not "Cooked".
- Check For Oil In Coolant.
- Check Engine Oil Level
- Confirm Oil Change Within 12,070 km (7500 Miles)
- Check Air Intake System: Check For Intake Restriction. Air Filter Restriction Gauge/Air Filter
- Check Map Sensor Hose For Holes, Blockage Or Disconnection. Confirm Correct MAP Sensor
- Check Intake System For Leaks
- Confirm Acceptable SAE Oil Viscosity And API Rating Of Oil, Refer To Operator's Guide.
- Check Fuel Quality. Refer To Diagnostic Subroutines.
- Check For Low Boost Pressure
- Check For Over Boost Condition
- Check For Air In Fuel
- Check For Sufficient Clean Fuel
- Check For An Intake Restriction
- Check EP Senosr And Tube For Holes, Blockage, Or Disconnection.
- Compare Loaded Weight Of Vehicle With Performance Expectations
- Power Monitor Active
- Transmission Concers

Are All Checks OK?

YES → 2003-04 → **AG2**

YES → 2005-07 → **AG3**

NO → REPAIR As Necessary. VERIFY Symptom No Longer Exists → **STOP**

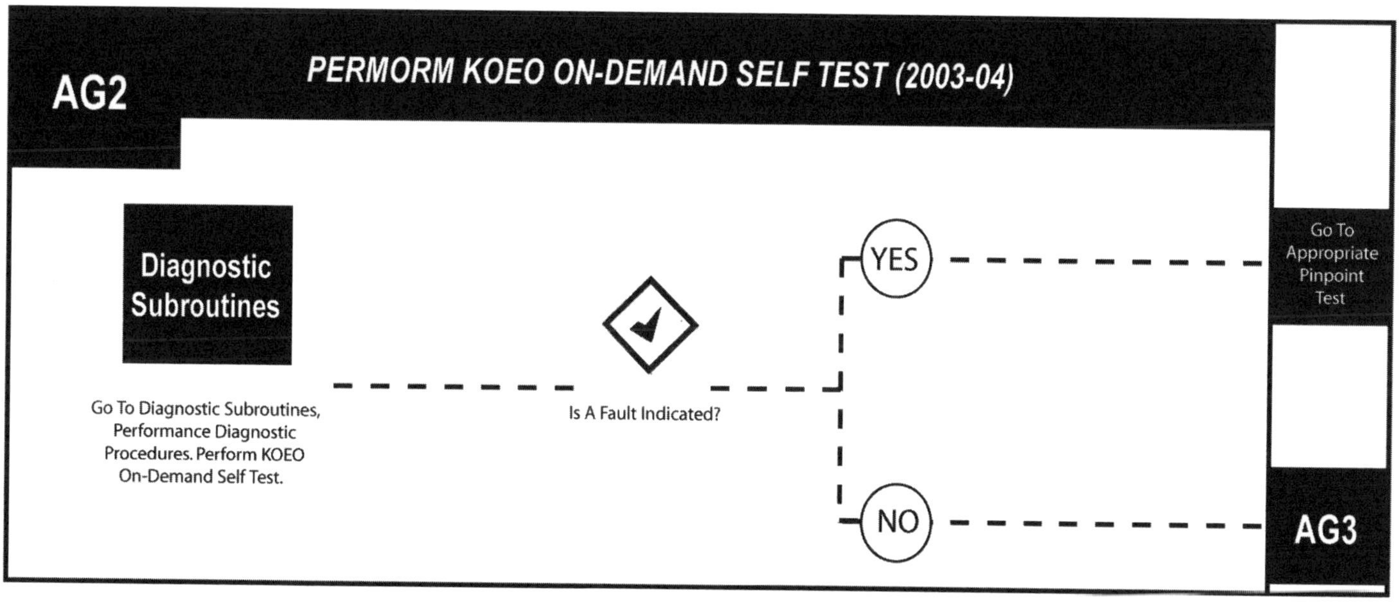
AG2
PERMORM KOEO ON-DEMAND SELF TEST (2003-04)
Diagnostic Subroutines
Go To Diagnostic Subroutines, Performance Diagnostic Procedures. Perform KOEO On-Demand Self Test.
Is A Fault Indicated?
YES
Go To Appropriate Pinpoint Test
NO
AG3

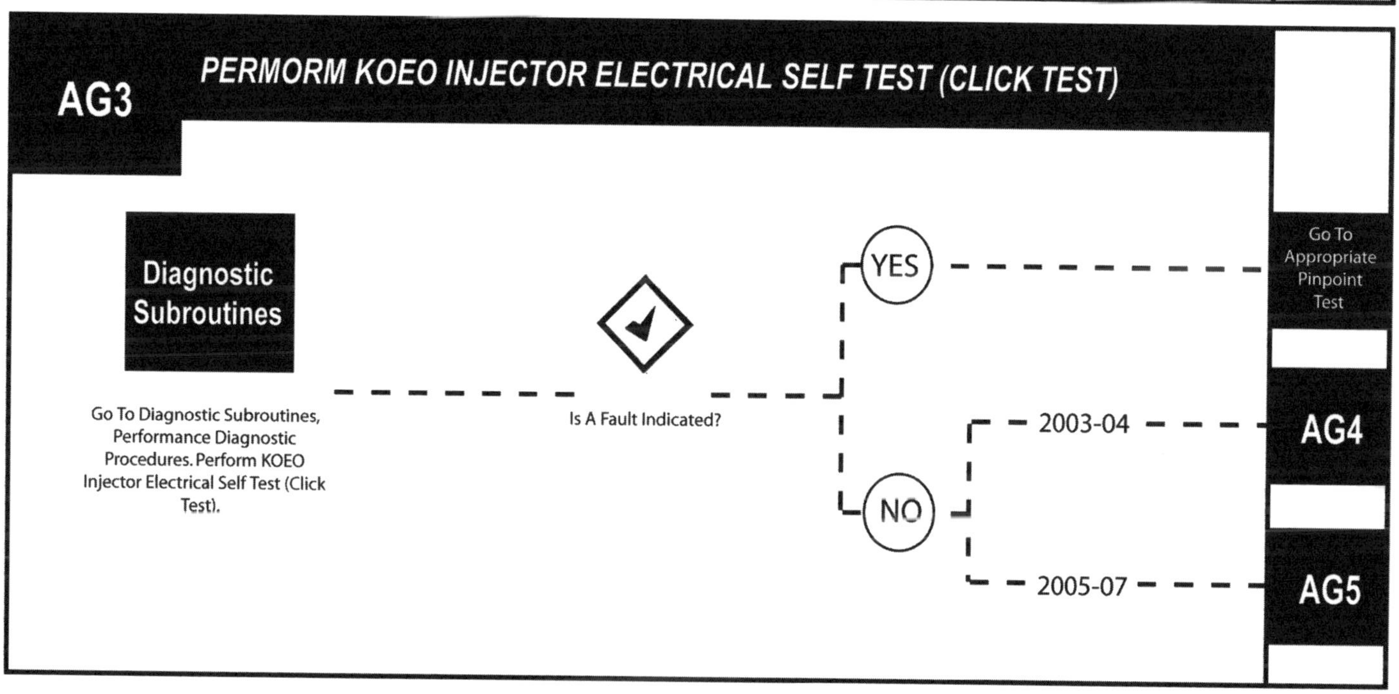
AG3
PERMORM KOEO INJECTOR ELECTRICAL SELF TEST (CLICK TEST)
Diagnostic Subroutines
Go To Diagnostic Subroutines, Performance Diagnostic Procedures. Perform KOEO Injector Electrical Self Test (Click Test).
Is A Fault Indicated?
YES
Go To Appropriate Pinpoint Test
NO
2003-04
AG4
2005-07
AG5

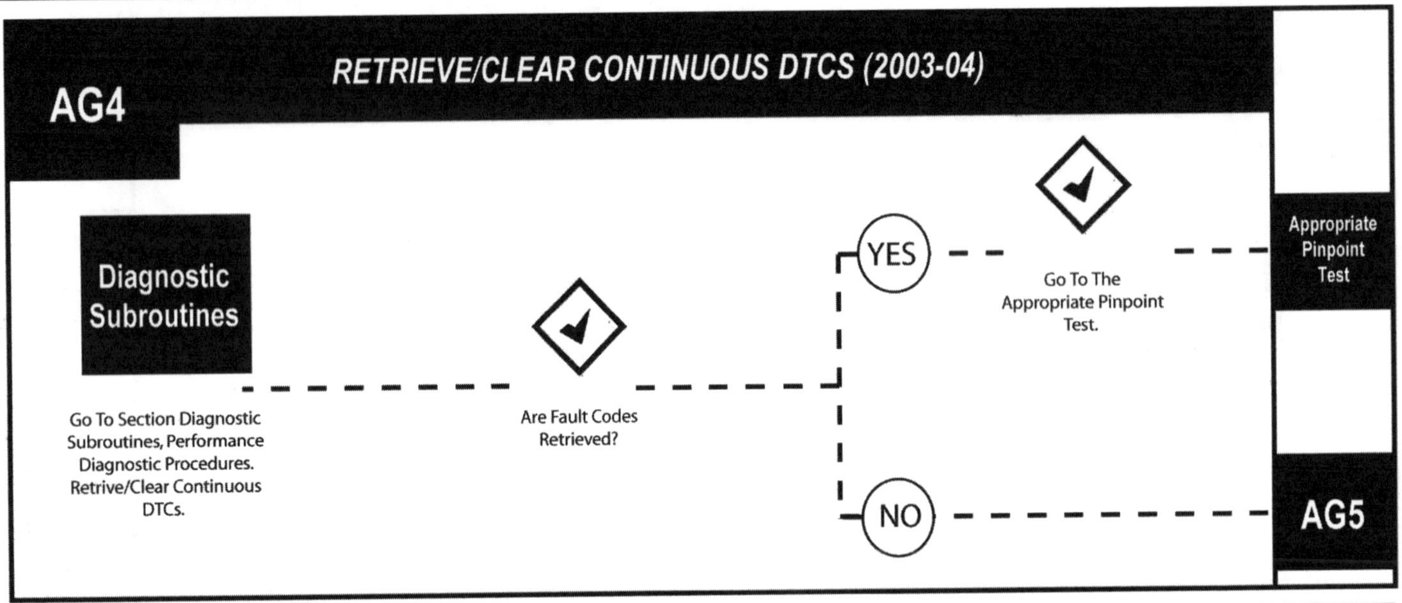
AG4
RETRIEVE/CLEAR CONTINUOUS DTCS (2003-04)
Diagnostic Subroutines
Go To Section Diagnostic Subroutines, Performance Diagnostic Procedures. Retrive/Clear Continuous DTCs.
Are Fault Codes Retrieved?
YES
Go To The Appropriate Pinpoint Test.
Appropriate Pinpoint Test
NO
AG5

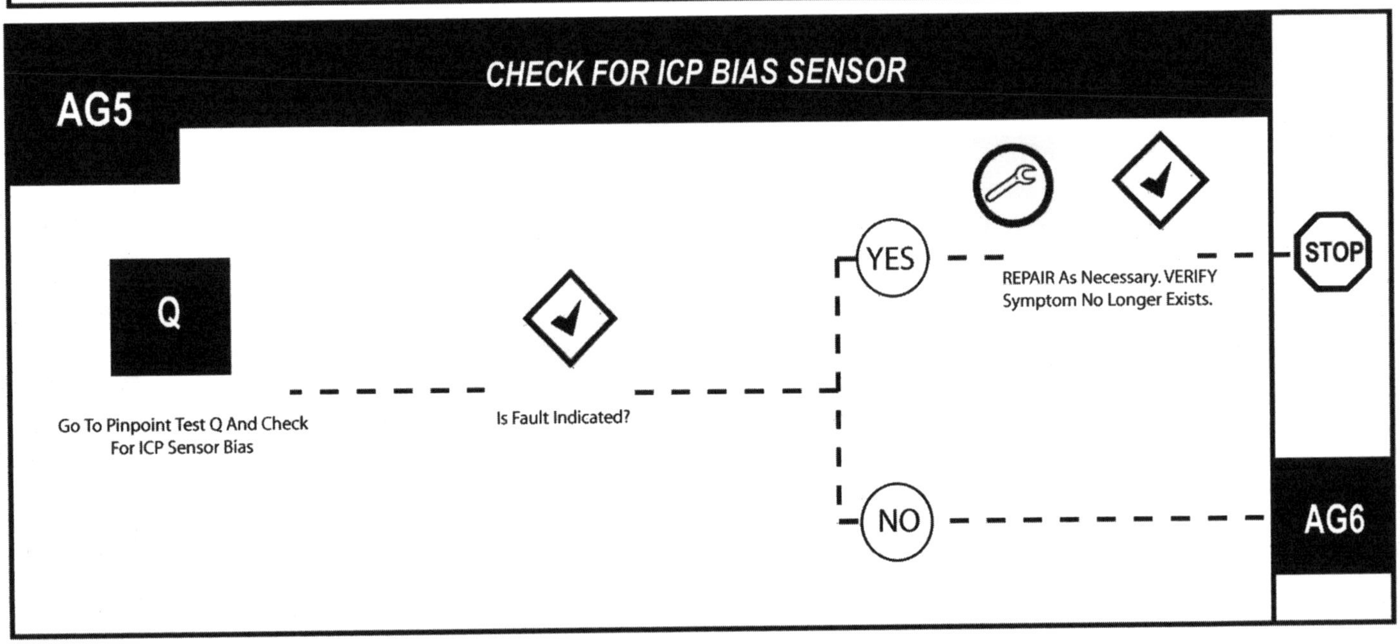
AG5
CHECK FOR ICP BIAS SENSOR
Q
Go To Pinpoint Test Q And Check For ICP Sensor Bias
Is Fault Indicated?
YES
REPAIR As Necessary. VERIFY Symptom No Longer Exists.
STOP
NO
AG6

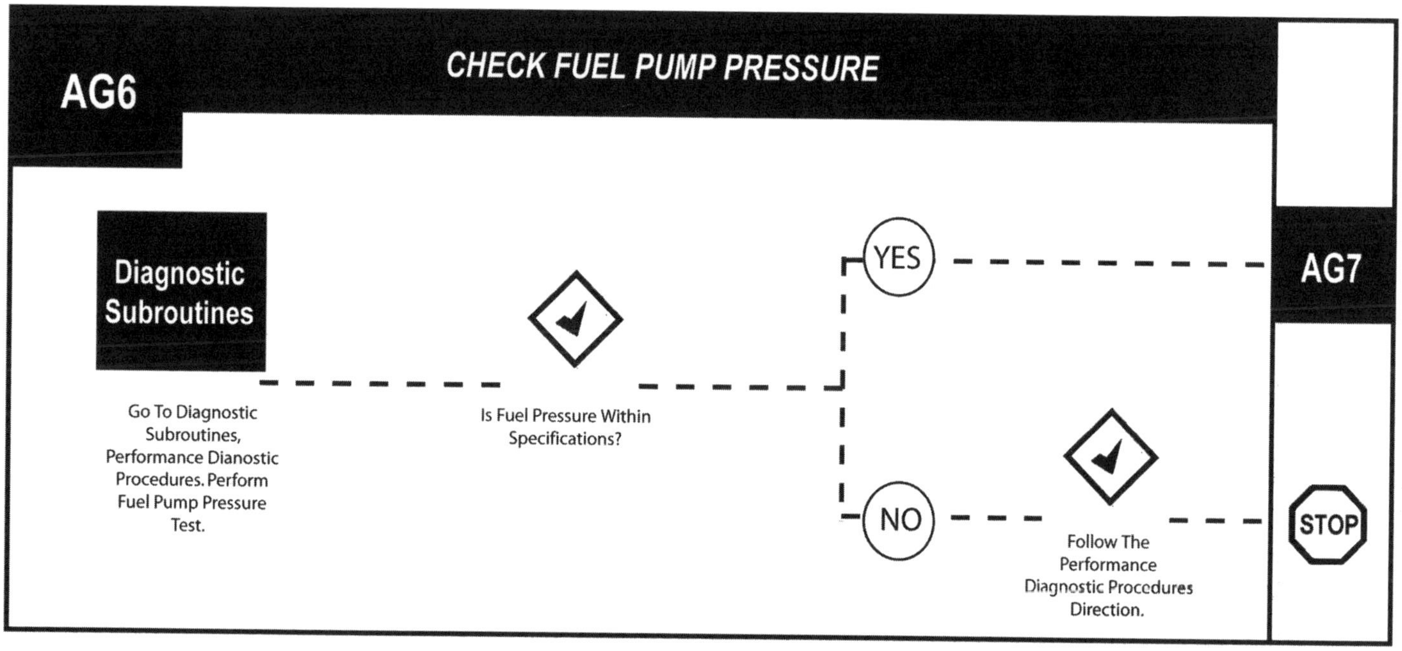
AG6
CHECK FUEL PUMP PRESSURE
Diagnostic Subroutines
Go To Diagnostic Subroutines, Performance Dianostic Procedures. Perform Fuel Pump Pressure Test.
Is Fuel Pressure Within Specifications?
YES
AG7
NO
Follow The Performance Diagnostic Procedures Direction.
STOP

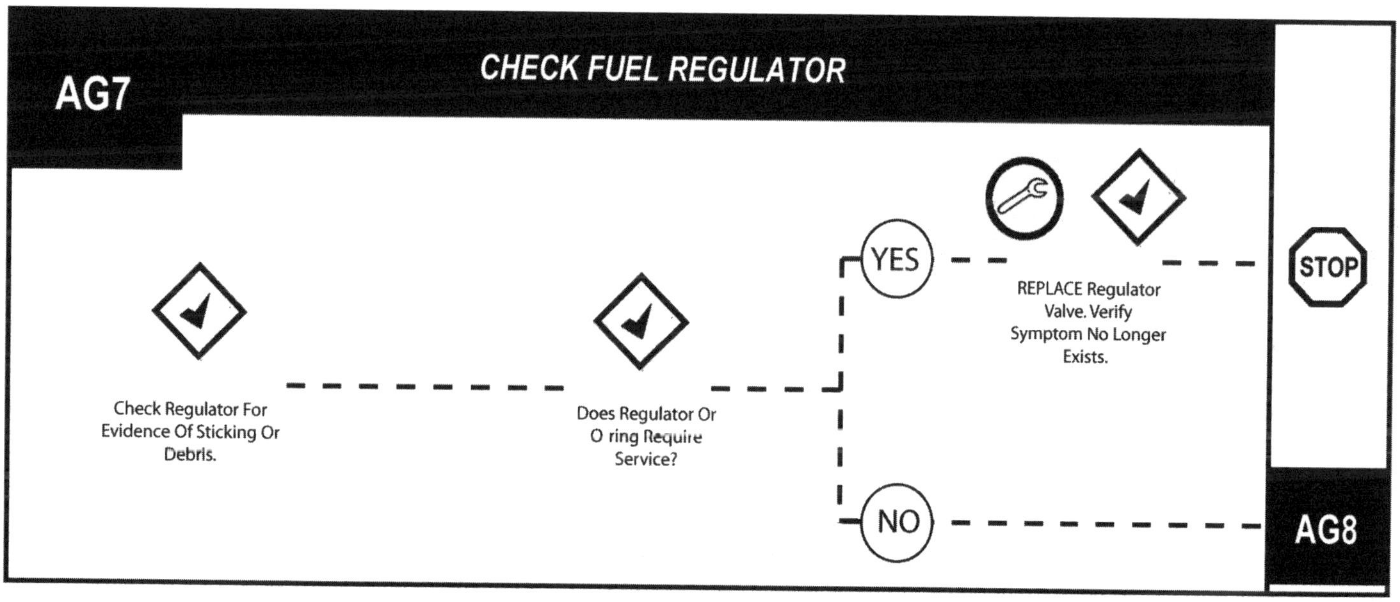
AG7
CHECK FUEL REGULATOR
Check Regulator For Evidence Of Sticking Or Debris.
Does Regulator Or O ring Require Service?
YES
REPLACE Regulator Valve. Verify Symptom No Longer Exists.
STOP
NO
AG8

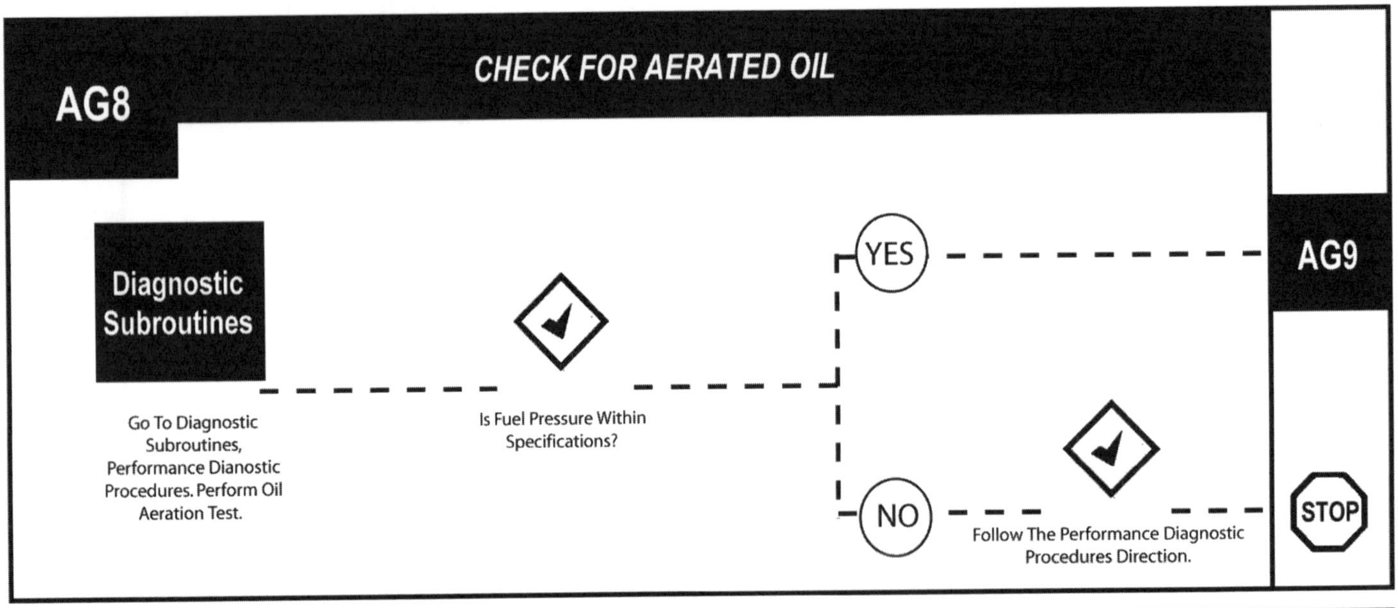
AG8
CHECK FOR AERATED OIL
Diagnostic Subroutines
Go To Diagnostic Subroutines, Performance Dianostic Procedures. Perform Oil Aeration Test.
Is Fuel Pressure Within Specifications?
YES
AG9
NO
Follow The Performance Diagnostic Procedures Direction.
STOP

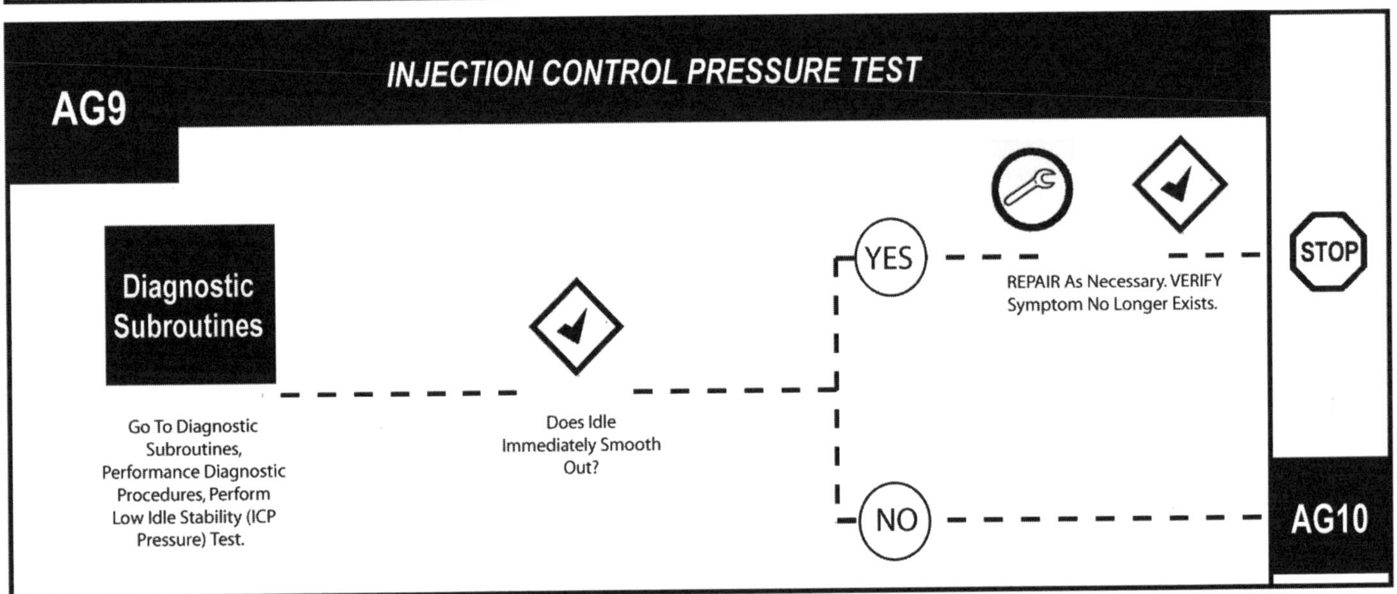
AG9
INJECTION CONTROL PRESSURE TEST
Diagnostic Subroutines
Go To Diagnostic Subroutines, Performance Diagnostic Procedures, Perform Low Idle Stability (ICP Pressure) Test.
Does Idle Immediately Smooth Out?
YES
REPAIR As Necessary. VERIFY Symptom No Longer Exists.
STOP
NO
AG10

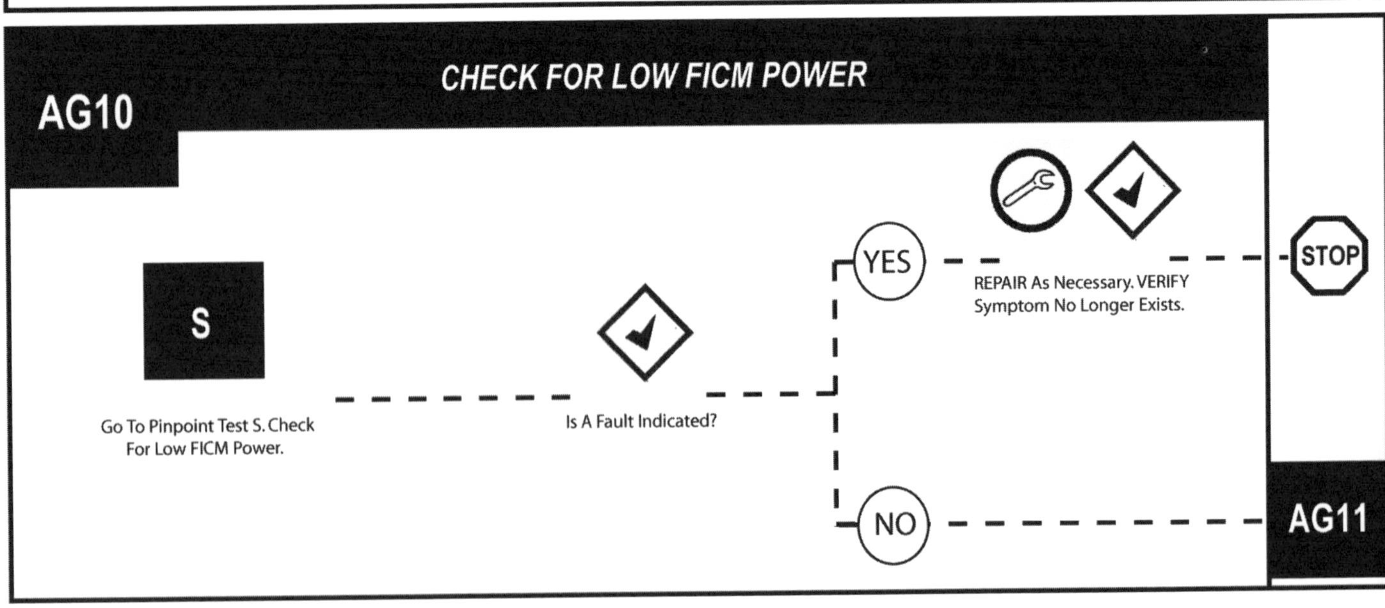
AG10
CHECK FOR LOW FICM POWER
S
Go To Pinpoint Test S. Check For Low FICM Power.
Is A Fault Indicated?
YES
REPAIR As Necessary. VERIFY Symptom No Longer Exists.
STOP
NO
AG11

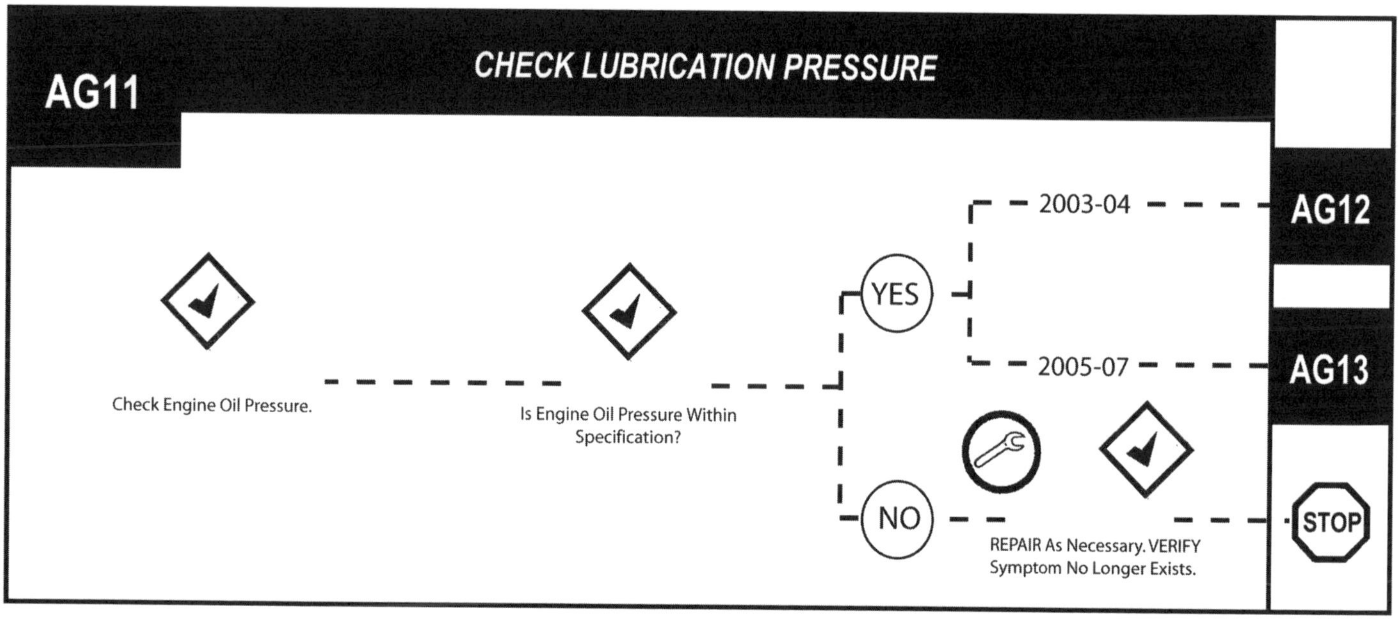
AG11
CHECK LUBRICATION PRESSURE
Check Engine Oil Pressure.
Is Engine Oil Pressure Within Specification?
YES
2003-04
AG12
2005-07
AG13
NO
REPAIR As Necessary. VERIFY Symptom No Longer Exists.
STOP

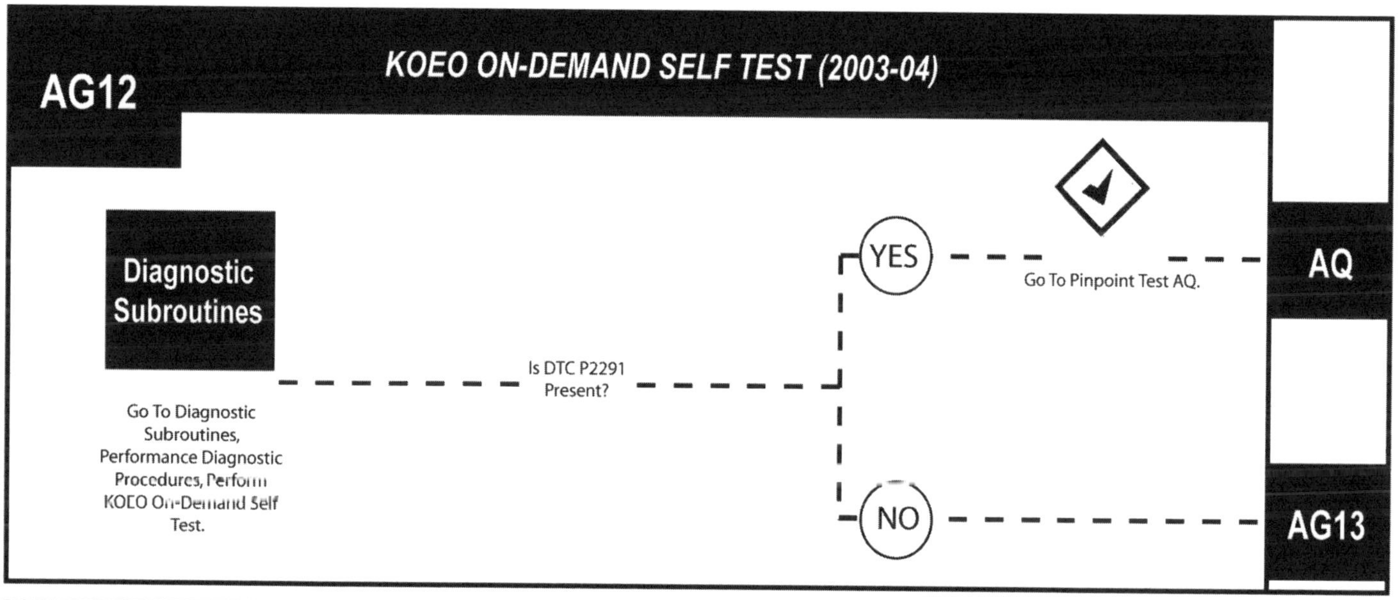
AG12
KOEO ON-DEMAND SELF TEST (2003-04)
Diagnostic Subroutines
Go To Diagnostic Subroutines, Performance Diagnostic Procedures, Perform KOEO On-Demand Self Test.
Is DTC P2291 Present?
YES
Go To Pinpoint Test AQ.
AQ
NO
AG13

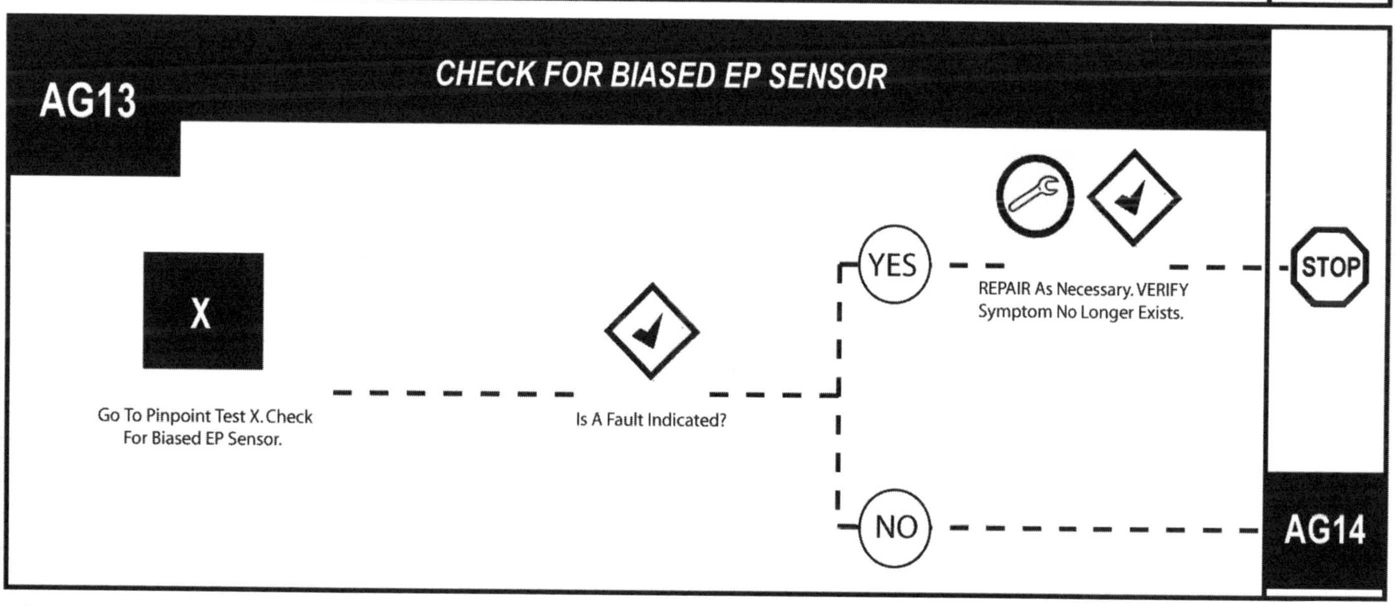
AG13
CHECK FOR BIASED EP SENSOR
X
Go To Pinpoint Test X. Check For Biased EP Sensor.
Is A Fault Indicated?
YES
REPAIR As Necessary. VERIFY Symptom No Longer Exists.
STOP
NO
AG14

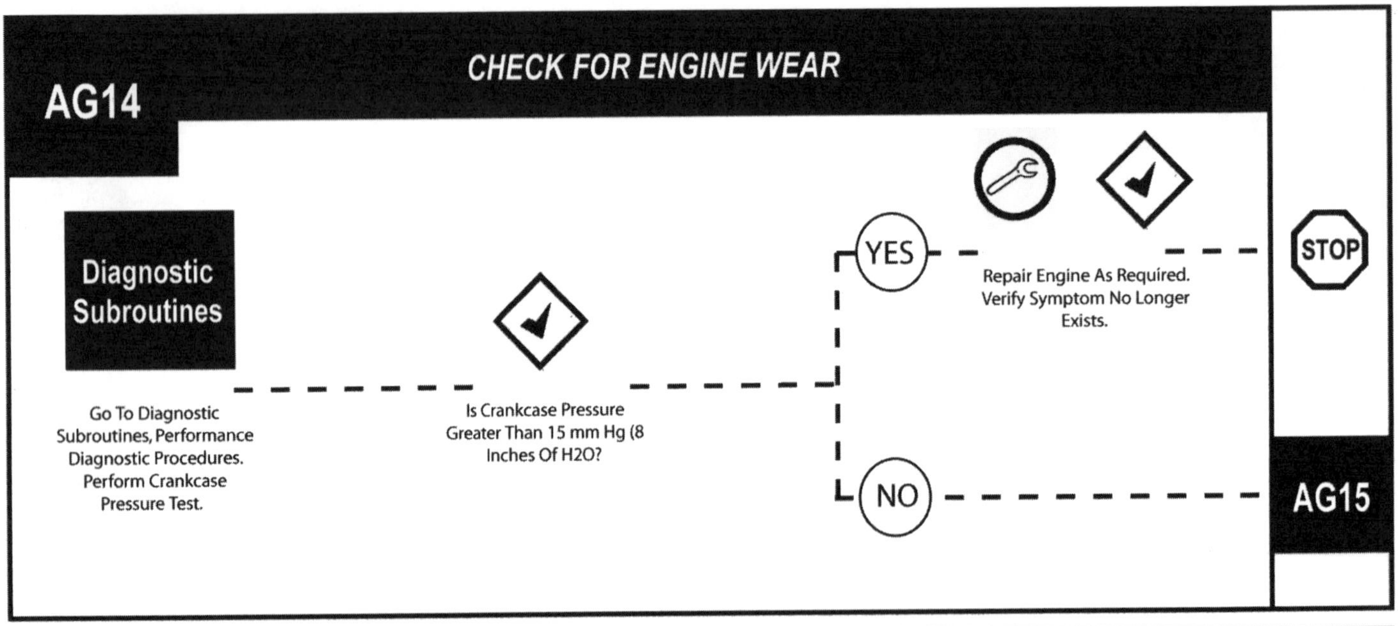
AG14
CHECK FOR ENGINE WEAR
Diagnostic Subroutines
Go To Diagnostic Subroutines, Performance Diagnostic Procedures. Perform Crankcase Pressure Test.
Is Crankcase Pressure Greater Than 15 mm Hg (8 Inches Of H2O?
YES
Repair Engine As Required. Verify Symptom No Longer Exists.
STOP
NO
AG15

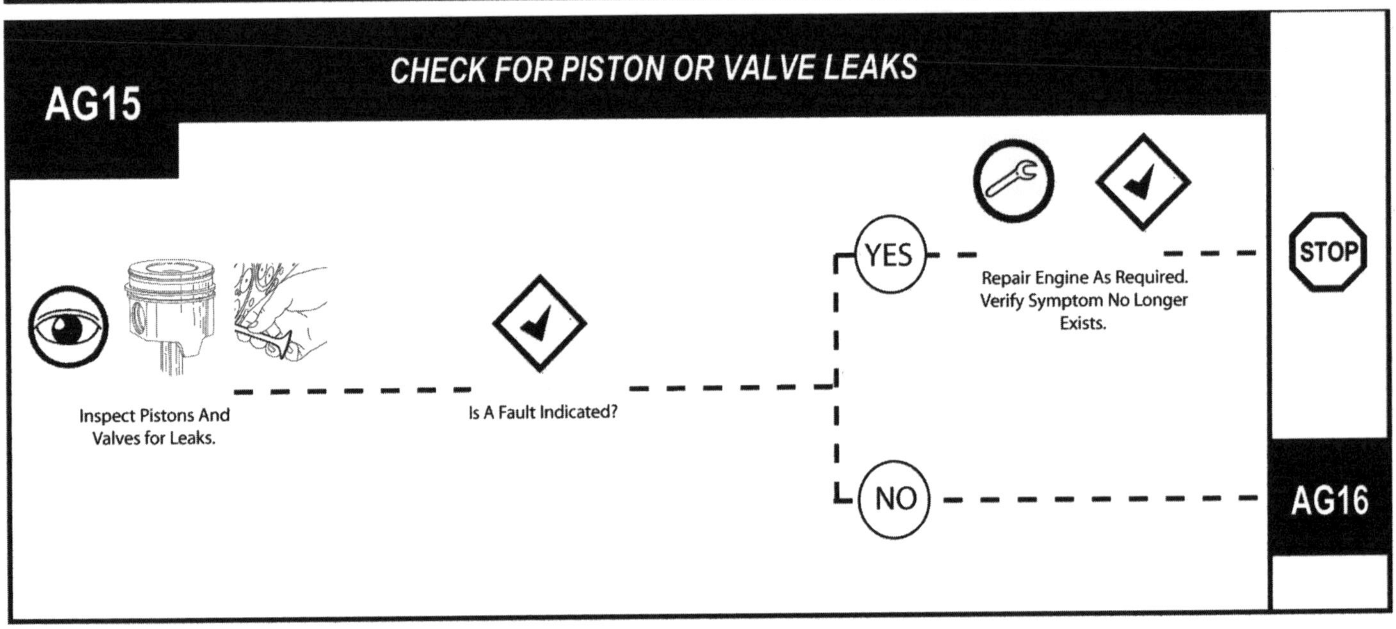
AG15
CHECK FOR PISTON OR VALVE LEAKS
Inspect Pistons And Valves for Leaks.
Is A Fault Indicated?
YES
Repair Engine As Required. Verify Symptom No Longer Exists.
STOP
NO
AG16

AG16 CHECK FOR EXHAUST RESTRICTION

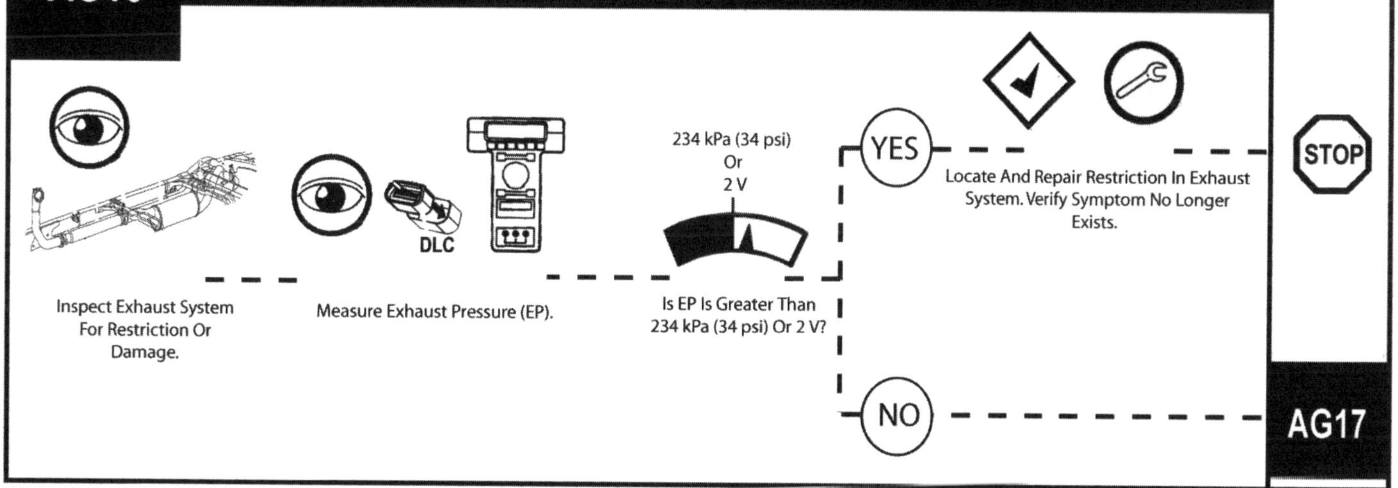

AG17 ATMOSPHERIC PRESSURE RATIONALITY CHECK

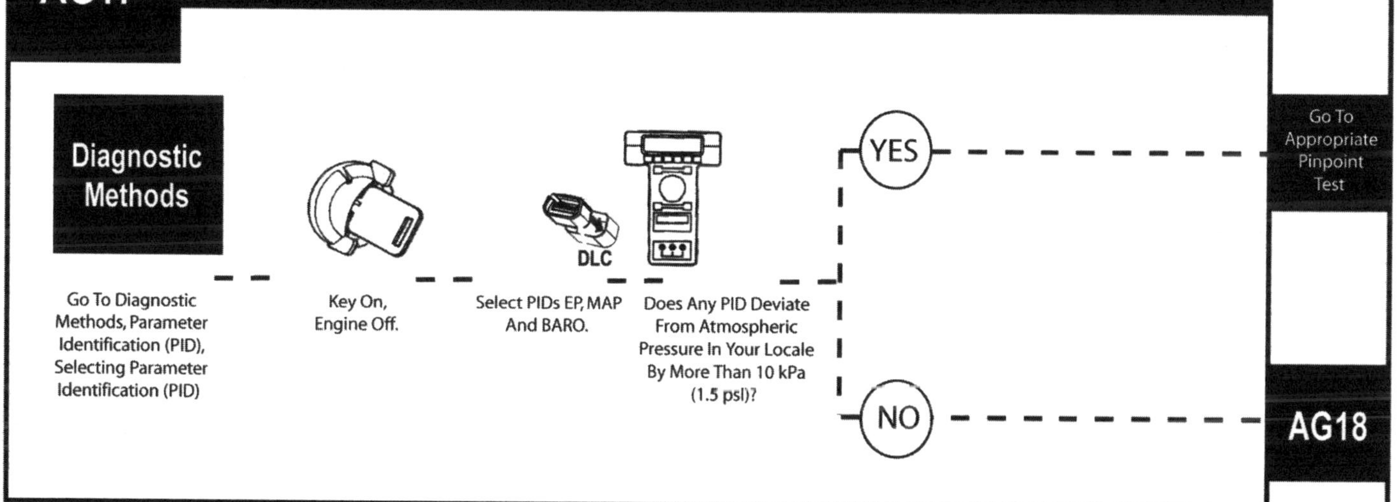

AG18 CHECK FOR BIASED EOT SENSOR

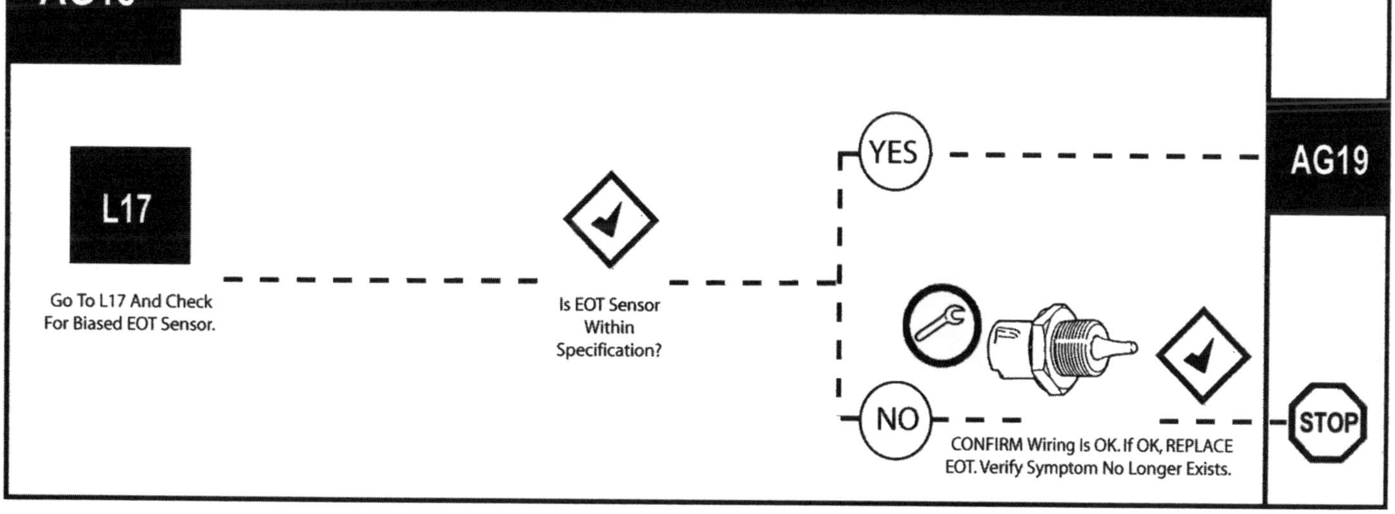

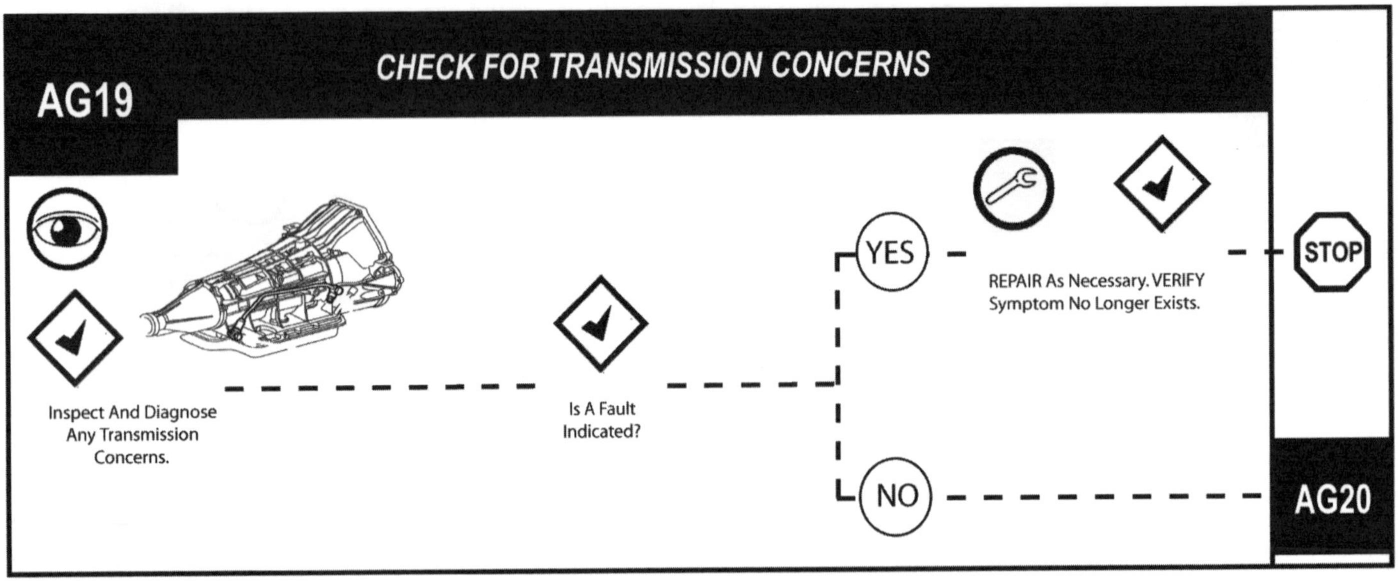
AG19
CHECK FOR TRANSMISSION CONCERNS
Inspect And Diagnose Any Transmission Concerns.
Is A Fault Indicated?
YES
REPAIR As Necessary. VERIFY Symptom No Longer Exists.
STOP
NO
AG20

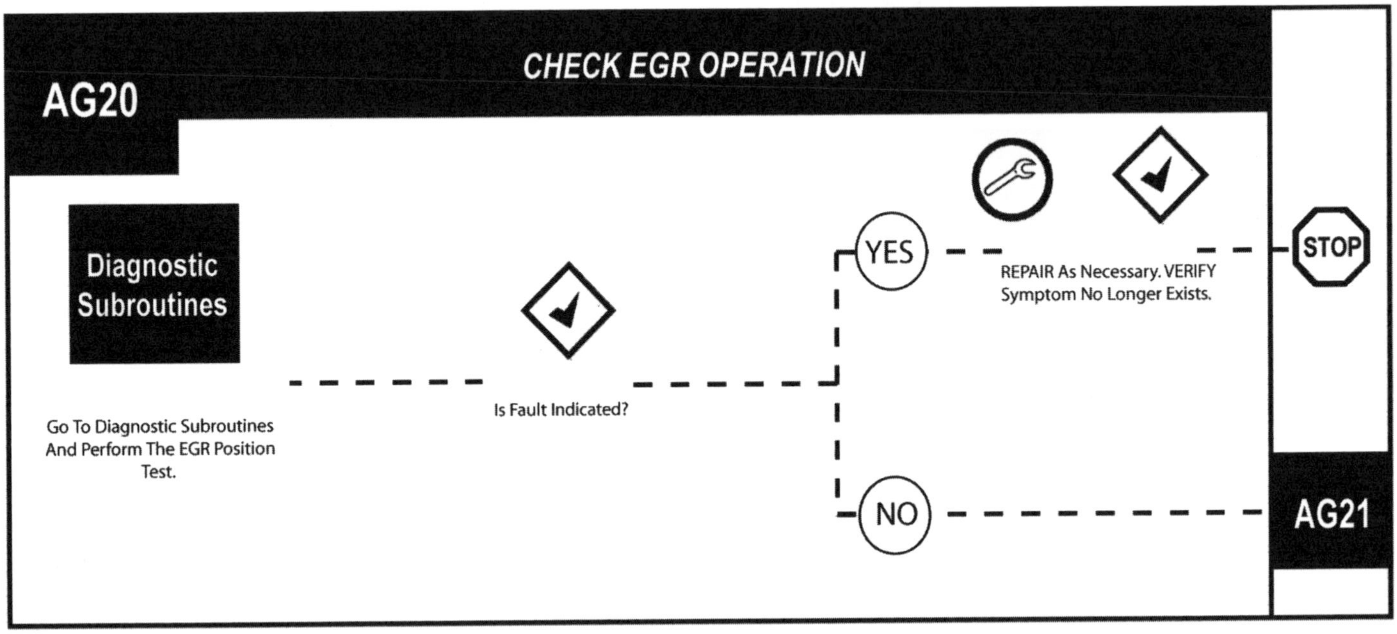
AG20
CHECK EGR OPERATION
Diagnostic Subroutines
Go To Diagnostic Subroutines And Perform The EGR Position Test.
Is Fault Indicated?
YES
REPAIR As Necessary. VERIFY Symptom No Longer Exists.
STOP
NO
AG21

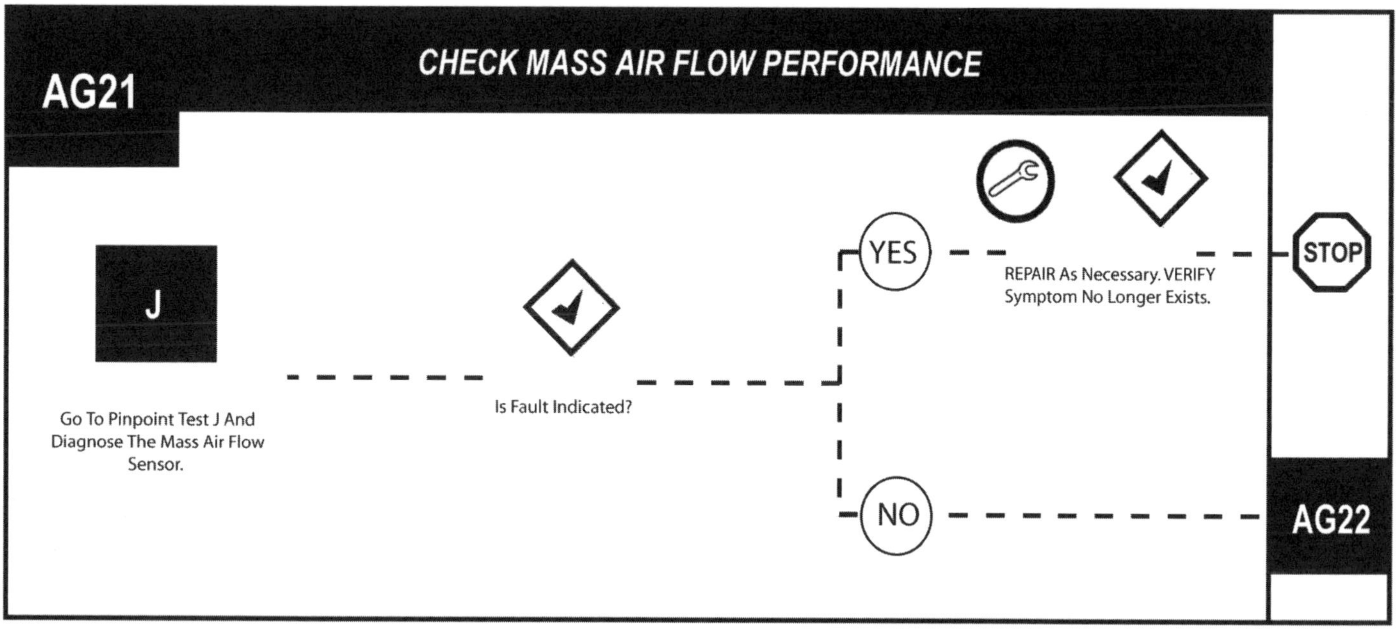
AG21
CHECK MASS AIR FLOW PERFORMANCE
J
Go To Pinpoint Test J And Diagnose The Mass Air Flow Sensor.
Is Fault Indicated?
YES
REPAIR As Necessary. VERIFY Symptom No Longer Exists.
STOP
NO
AG22

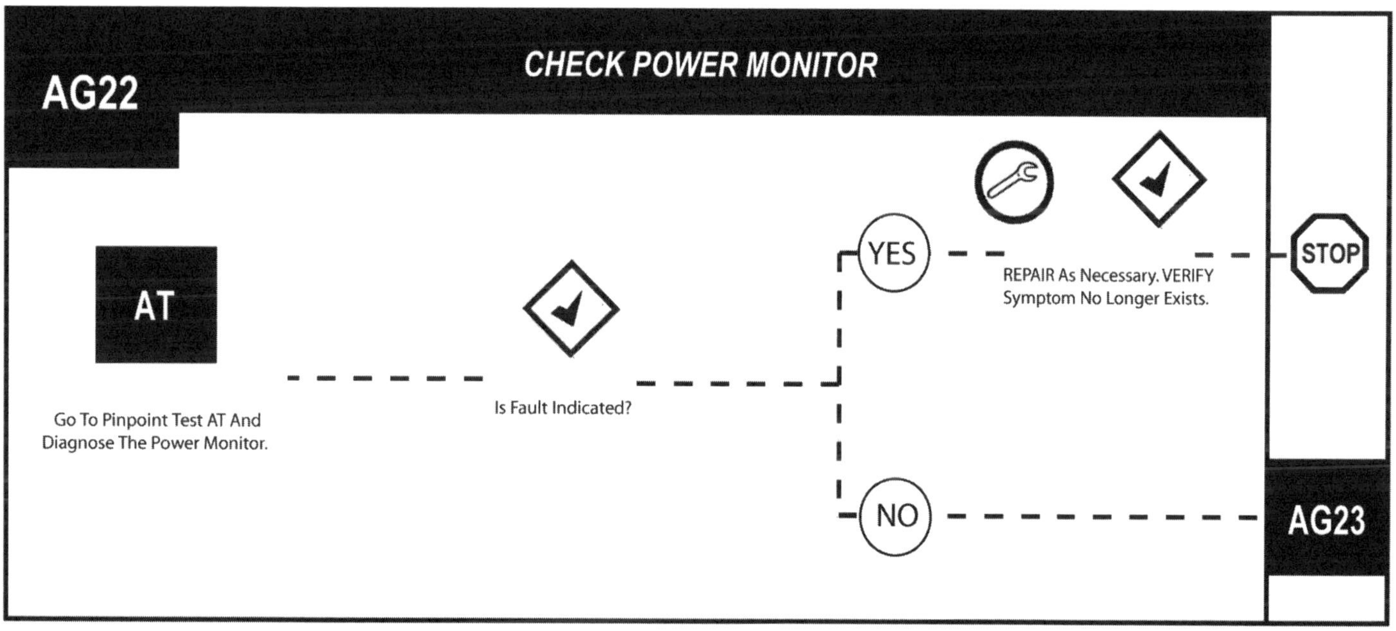
AG22
CHECK POWER MONITOR
AT
Go To Pinpoint Test AT And Diagnose The Power Monitor.
Is Fault Indicated?
YES
REPAIR As Necessary. VERIFY Symptom No Longer Exists.
STOP
NO
AG23

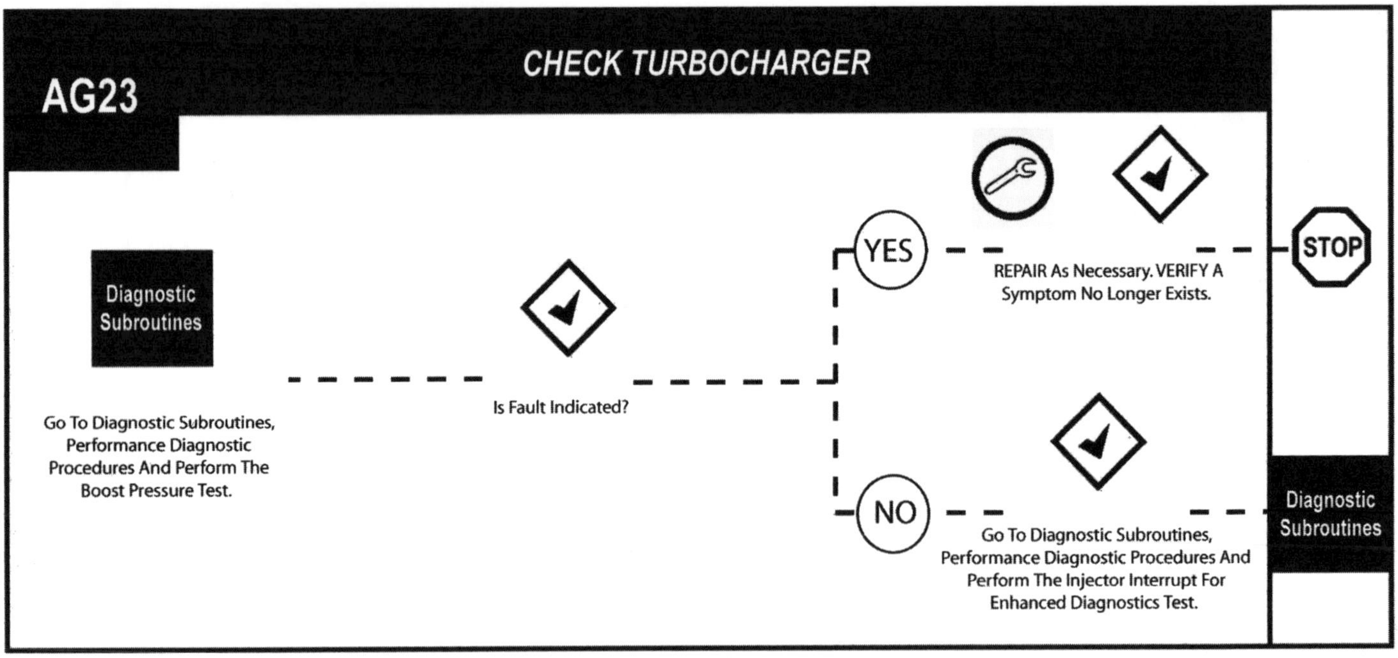
AG23
CHECK TURBOCHARGER
Diagnostic Subroutines
Go To Diagnostic Subroutines, Performance Diagnostic Procedures And Perform The Boost Pressure Test.
Is Fault Indicated?
YES
REPAIR As Necessary. VERIFY A Symptom No Longer Exists.
STOP
NO
Go To Diagnostic Subroutines, Performance Diagnostic Procedures And Perform The Injector Interrupt For Enhanced Diagnostics Test.
Diagnostic Subroutines

NOTES

System		Test
Check Engine Light Concern	Symptom(s): Check Engine Light Concern (2003-07 6.0L Diesel)	AH

Test/Step

AH1 PERFORM KOEO ON-DEMAND SELF TEST AND RETRIEVE/CLEAR CONTINUOUS DTC'S.

Diagnostic Methods

Go To Diagnostic Methods. Perform KOEO On Demand Self Test, And Retrieve/Clear Continuous DTCs

Is Fault Indicated?

	Result
YES	Go To Appropriate Pinpoint Test
NO	AH2

AH2 RUN BULB CHECK

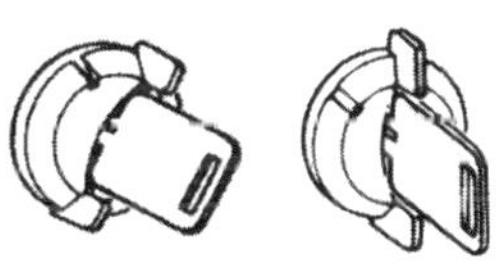

Cycle Ignition Switch.

Does The CHECK ENGINE Light Flash?

		Result
YES	CHECK ENGINE Light Is OK.	STOP
NO	Diagnose And Repair Instrument Cluster.	STOP

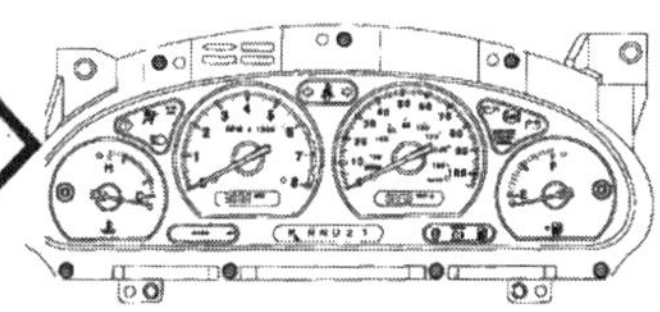

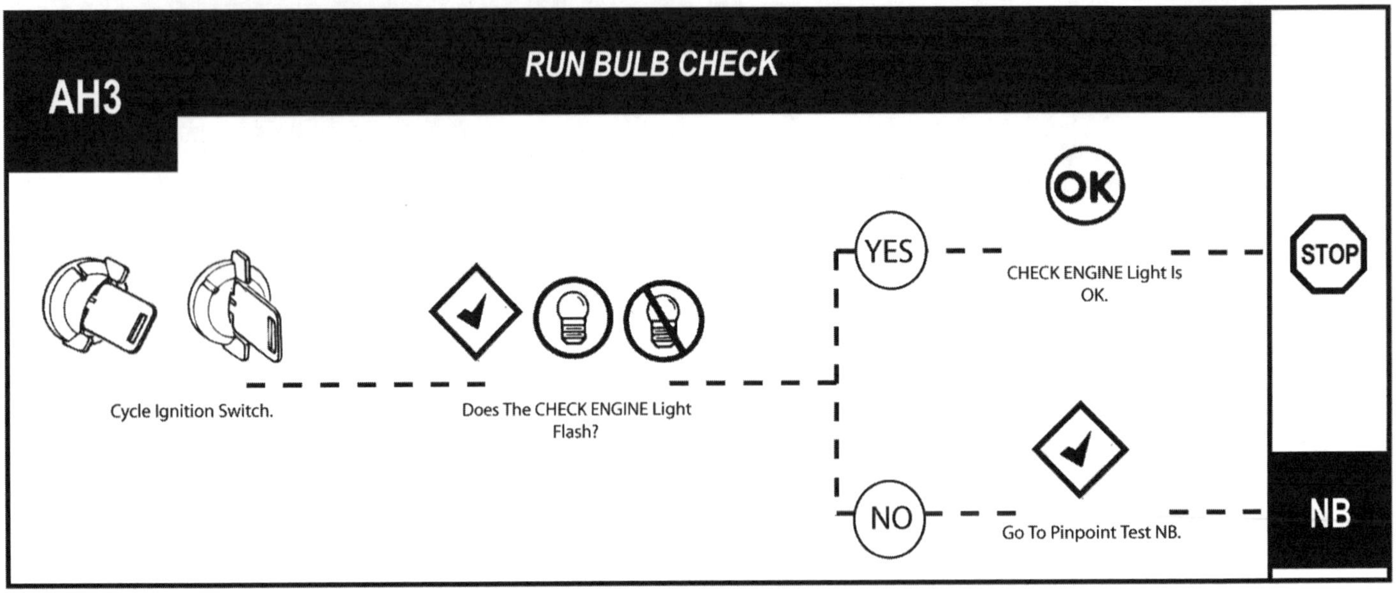
AH3
RUN BULB CHECK
Cycle Ignition Switch.
Does The CHECK ENGINE Light Flash?
YES
OK
CHECK ENGINE Light Is OK.
STOP
NO
Go To Pinpoint Test NB.
NB

NOTES

System	Symptom(s): Visible Smoke (No Oil Consumption) (2003-07 6.0L Diesel)	Test
Exhaust System Concerns		AJ

Test/Step

AJ1 PRELIMINARY CHECKS

Diagnostic Subroutines

Perform Preliminary Checks. Refer To Diagnostic Performance Procedures Steps 1-4 In Diagnostic Subroutines.

Are All Checks Ok?

- YES → AJ2
- NO → REPAIR As Required. VERIFY Symptom No Longer Exists → STOP

Result

AJ2 ATMOSPHERIC PRESSURE RATIONALITY CHECK

Diagnostic Methods

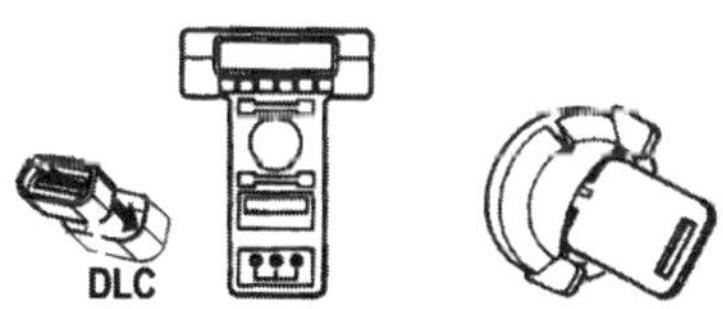

Go To Diagnostic Methods, Parameter Identification (PID), Selecting Parameter Identification (PID)

Select PIDs EP, MAP And BARO (EBP Is Not Used On F-Series 650/750).

Key On, Engine Off.

Does Any PID Deviate From Atmospheric Pressure In Your Locale By More Than 10 kPa (1.5 psi)?

- YES → Go To Appropriate Pinpoint Test
- NO → AJ3

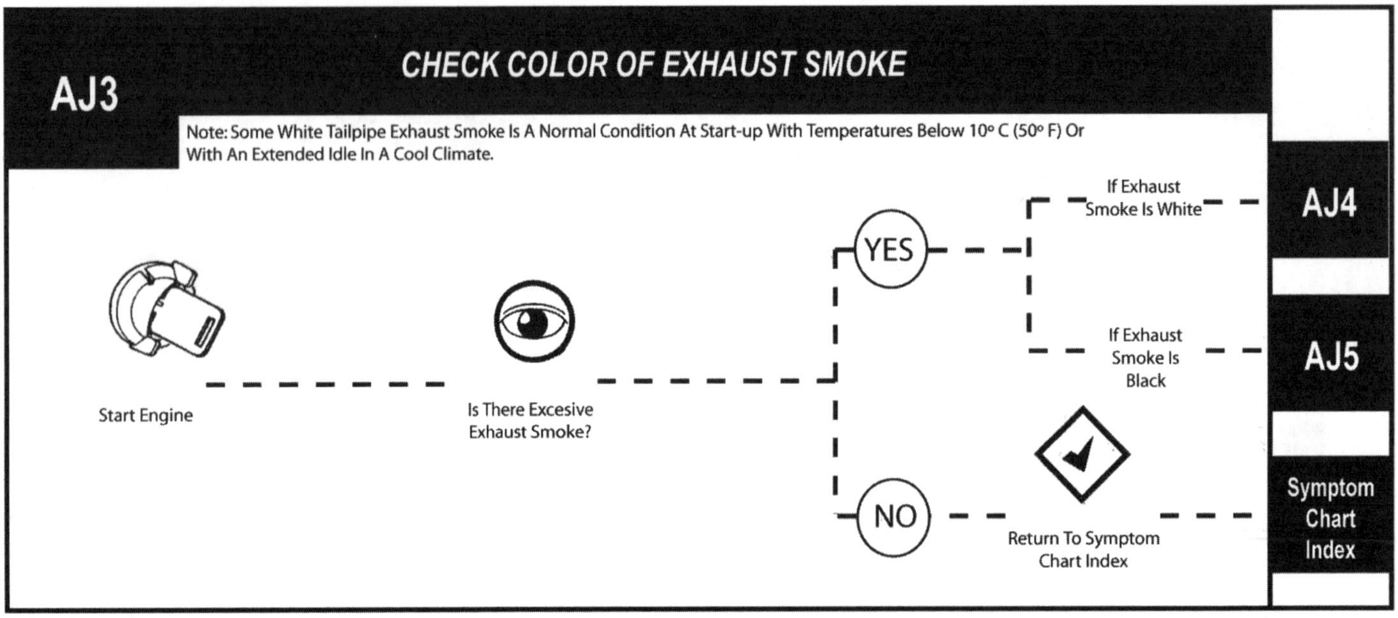
AJ3
CHECK COLOR OF EXHAUST SMOKE
Note: Some White Tailpipe Exhaust Smoke Is A Normal Condition At Start-up With Temperatures Below 10° C (50° F) Or With An Extended Idle In A Cool Climate.
Start Engine
Is There Excesive Exhaust Smoke?
YES
If Exhaust Smoke Is White
AJ4
If Exhaust Smoke Is Black
AJ5
NO
Return To Symptom Chart Index
Symptom Chart Index

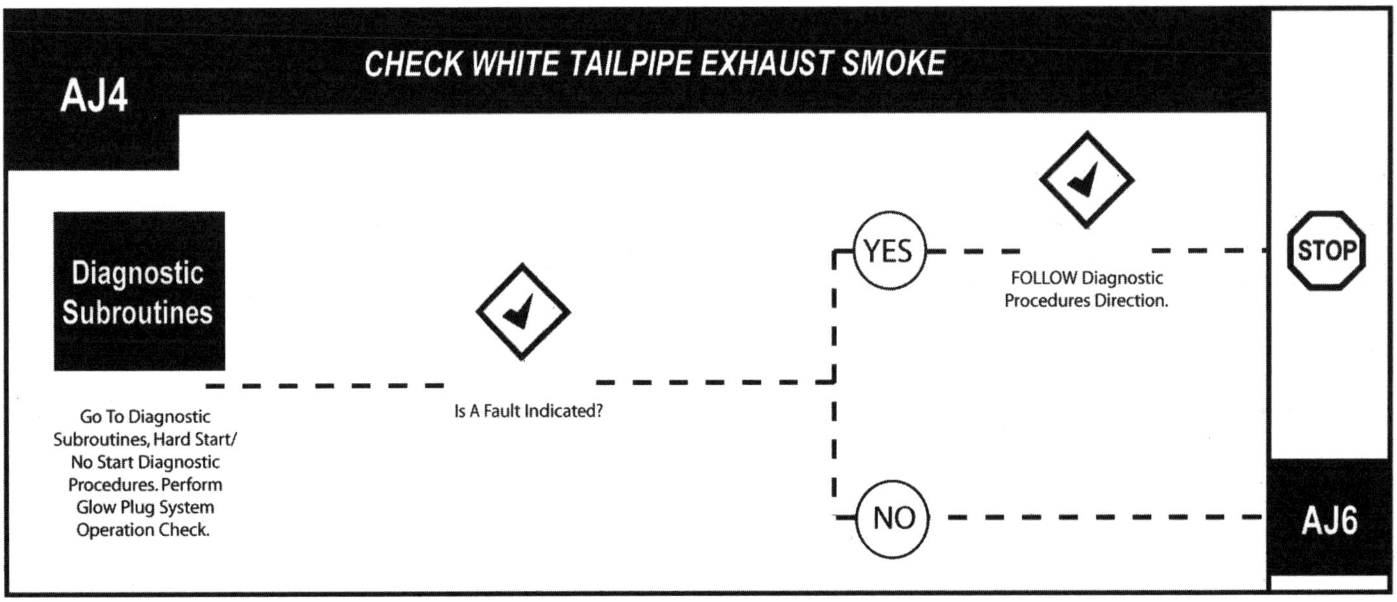
AJ4
CHECK WHITE TAILPIPE EXHAUST SMOKE
Diagnostic Subroutines
Go To Diagnostic Subroutines, Hard Start/ No Start Diagnostic Procedures. Perform Glow Plug System Operation Check.
Is A Fault Indicated?
YES
FOLLOW Diagnostic Procedures Direction.
STOP
NO
AJ6

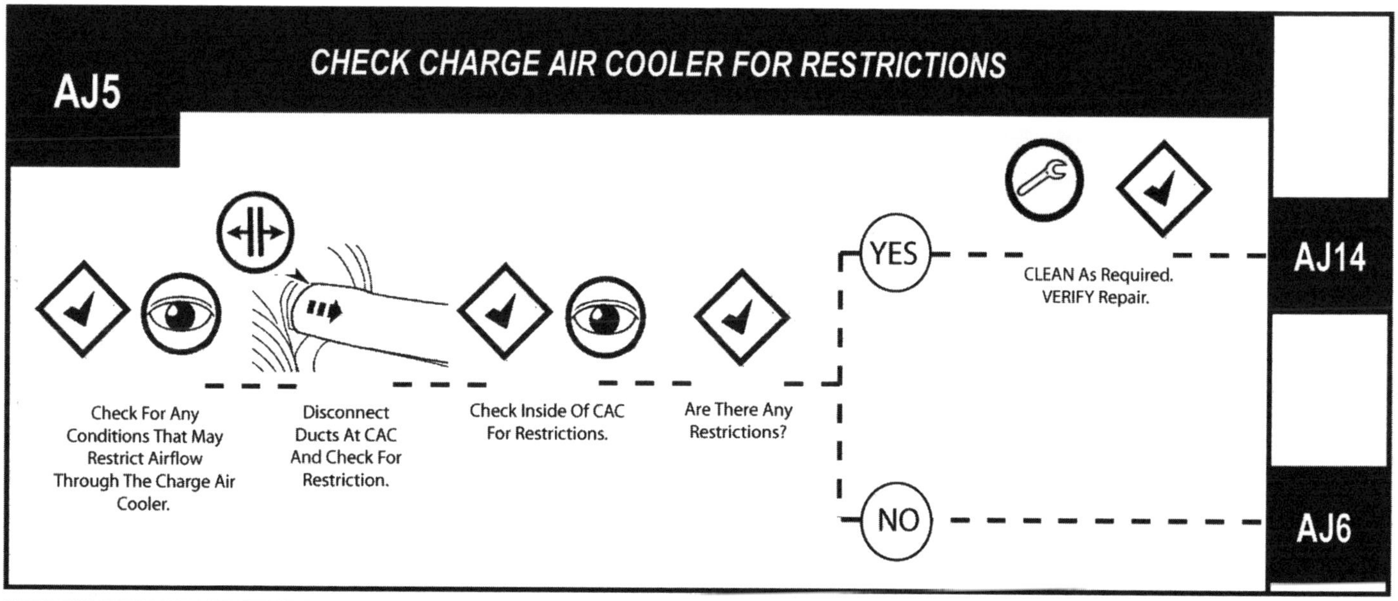
AJ5
CHECK CHARGE AIR COOLER FOR RESTRICTIONS
Check For Any Conditions That May Restrict Airflow Through The Charge Air Cooler.
Disconnect Ducts At CAC And Check For Restriction.
Check Inside Of CAC For Restrictions.
Are There Any Restrictions?
YES
CLEAN As Required. VERIFY Repair.
AJ14
NO
AJ6

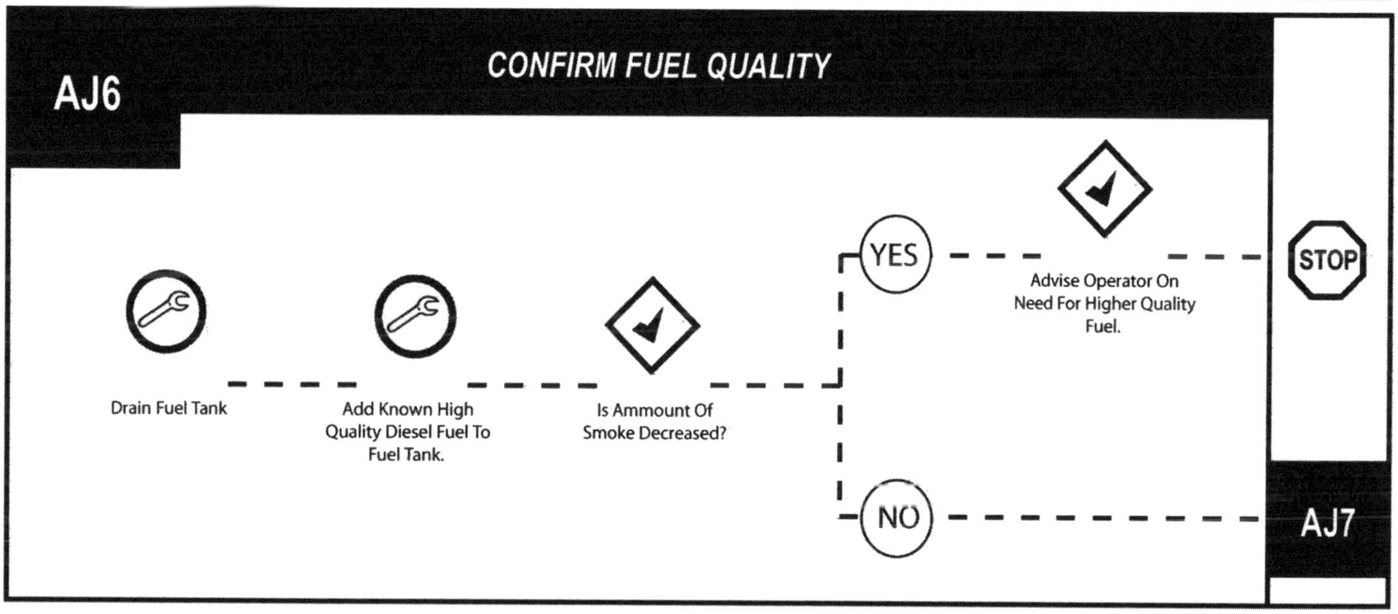
AJ6
CONFIRM FUEL QUALITY
Drain Fuel Tank
Add Known High Quality Diesel Fuel To Fuel Tank.
Is Ammount Of Smoke Decreased?
YES
Advise Operator On Need For Higher Quality Fuel.
STOP
NO
AJ7

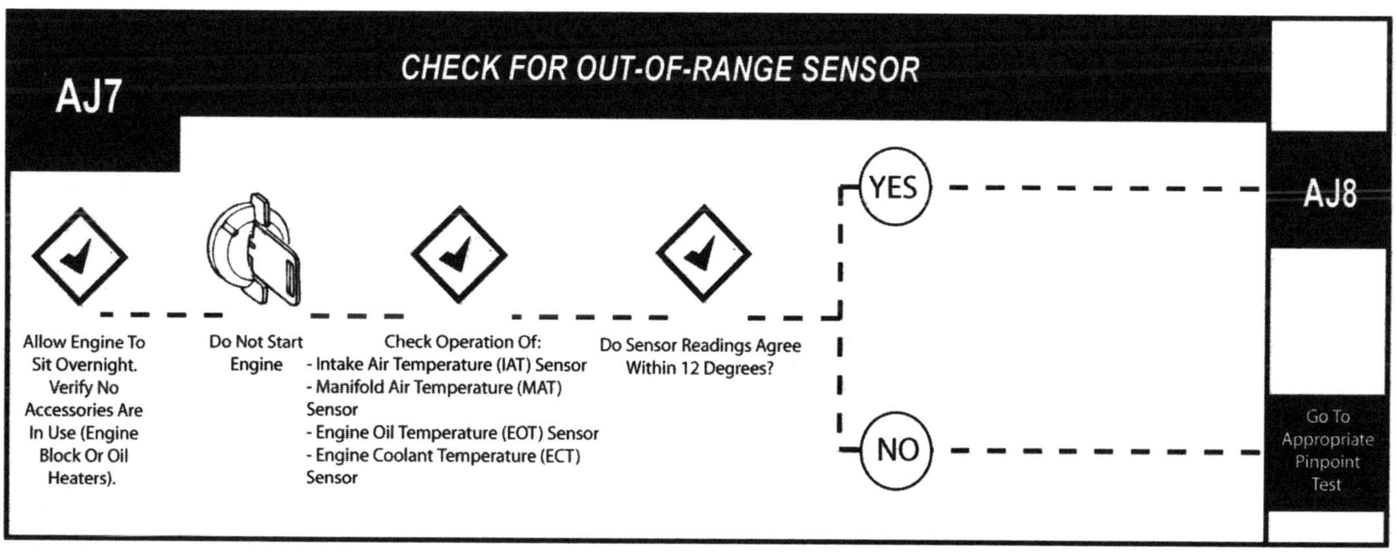
AJ7
CHECK FOR OUT-OF-RANGE SENSOR
Allow Engine To Sit Overnight. Verify No Accessories Are In Use (Engine Block Or Oil Heaters).
Do Not Start Engine
Check Operation Of:
- Intake Air Temperature (IAT) Sensor
- Manifold Air Temperature (MAT) Sensor
- Engine Oil Temperature (EOT) Sensor
- Engine Coolant Temperature (ECT) Sensor
Do Sensor Readings Agree Within 12 Degrees?
YES
AJ8
NO
Go To Appropriate Pinpoint Test

AJ8 — CHECK FOR COOLANT LEAK

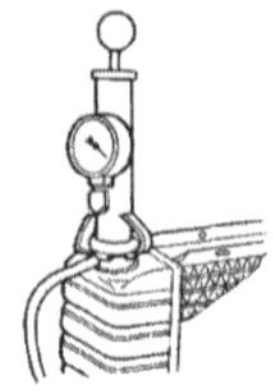

Check Cooling System For Leaks.

Does Coolant Leak Internally?

- YES — REPAIR As Necessary. — STOP
- NO — AJ9

AJ9 — CHECK THERMOSTAT OPERATION

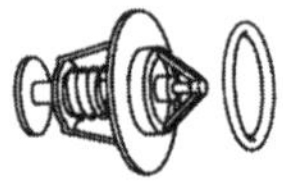

Diagnose Thermostat Operation.

Is Thermostat Operating Properly?

- YES — AJ10
- NO — REPAIR As Necessary — STOP

AJ10 — ISOLATE ENGINE CYLINDERS GENERATING SMOKE

Note: When The Vehicle Is Assembled, Gaskets Are Not Installed Beween The Cylinder Head And The Exhaust Manifold. If The Exhaust Manifold Is Seperated From The Cylinder Head, Gaskets Must Be Installed.

Remove The Exhaust Manifold Attaching Bolts.

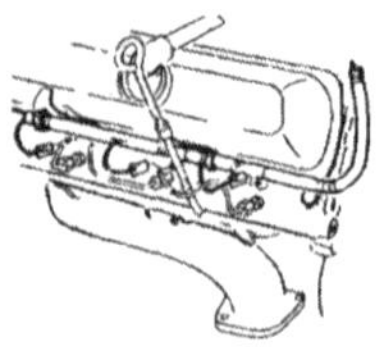

Pry The Manifolds Away From The Cylander Head.

Start The Engine

Are All Cylinders Generating Smoke?

- YES — AJ11
- NO — AJ12

AJ11 — CHECK FOR ENGINE WEAR

Diagnostic Subroutines

Go To Diagnostic Subroutines, Performance Diagnostic Procedures. Perform Crankcase Pressure Test

Is Crankcase Pressure Higher Than 4 Inches Of H_2O?

- YES: REPAIR Engine For Rings or Valves As Required. — STOP
- NO: AJ12

AJ12 — TEST GLOW PLUG

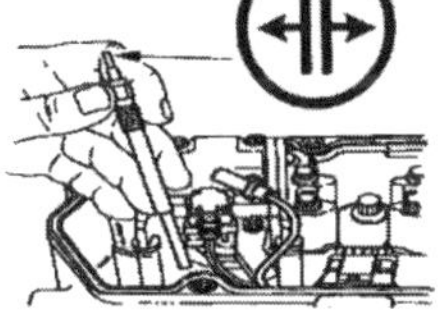

Remove Glowplug From Smoking Cylinder.

Ground Body Of Glow Plug And Install Jumper Wire From B+ Source To Glow Plug Terminal.

Does Glowplug Tip Become Hot?

- YES: AJ13
- NO: REPLACE Glow Plug. VERIFY Repair. — STOP

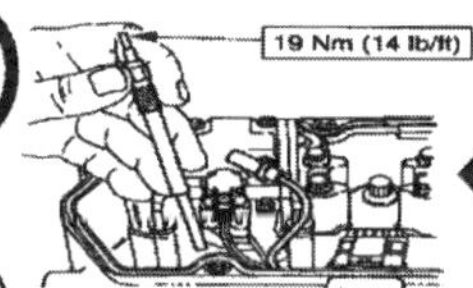

AJ13 — TEST THE FUEL INJECTORS

Diagnostic Subroutines

TEST The Fuel Injectors. REFER The Injector Performance Test. Switch Injectors Between A Smoking And Non-Smoking Cylinder(s).

Is A Fuel Injector Concern Present? Does Smoke Move With Injector?

- YES: INSTALL New Injector(s). VERIFY Repair. — STOP
- NO: REPAIR Base Engine. — STOP

AJ14

CHECK FOR BLACK SMOKE

Diagnostic Subroutines

Go To Diagnostic Subroutines, Performance Diagnostic Procedures And Perform The Fuel Aeration Test.

Is A Fault Indicated?

YES — REPAIR As Necessary. VERIFY Symptom No Longer Exists. — STOP

NO — AJ15

AJ15

CHECK EGR OPERATION

Diagnostic Subroutines

Go To Diagnostic Subroutines And Perform The EGR Position Test.

Is Fault Indicated?

YES — REPAIR As Necessary. VERIFY A Symptom No Longer Exists. — STOP

NO — Go To Pinpoint Test J And Diagnose The Mass Air Flow Sensor. — J

System		Test
Speed Contro Concerns	Symptom(s): Speed Control (2003-07 6.0L Diesel)	AK

Test/Step

AK1 — *CHECK ABS LIGHT*

Result

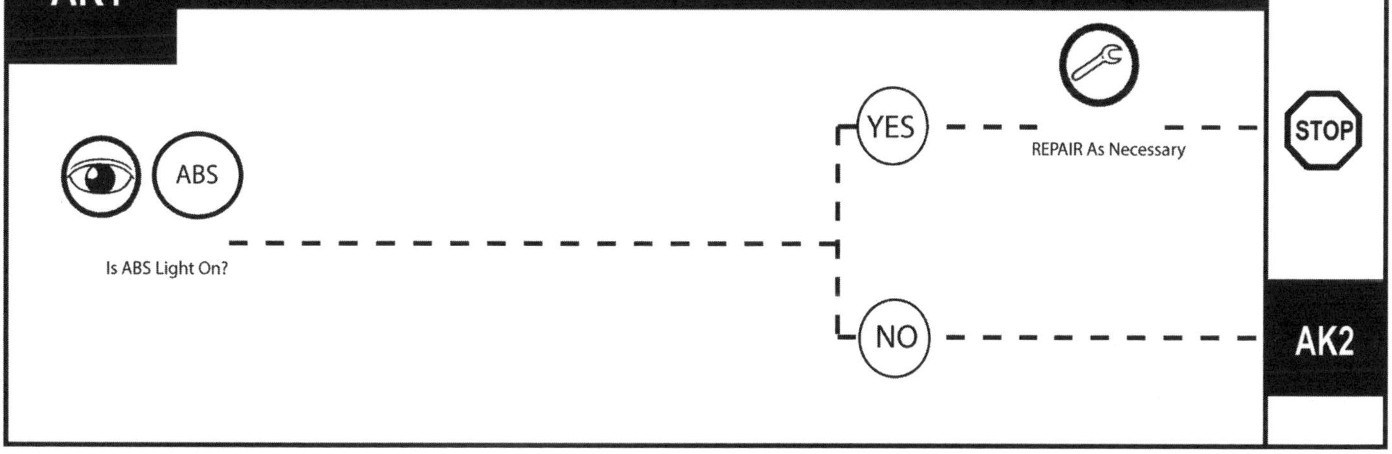

AK2 — *CHECK BRAKE WARNING LIGHT*

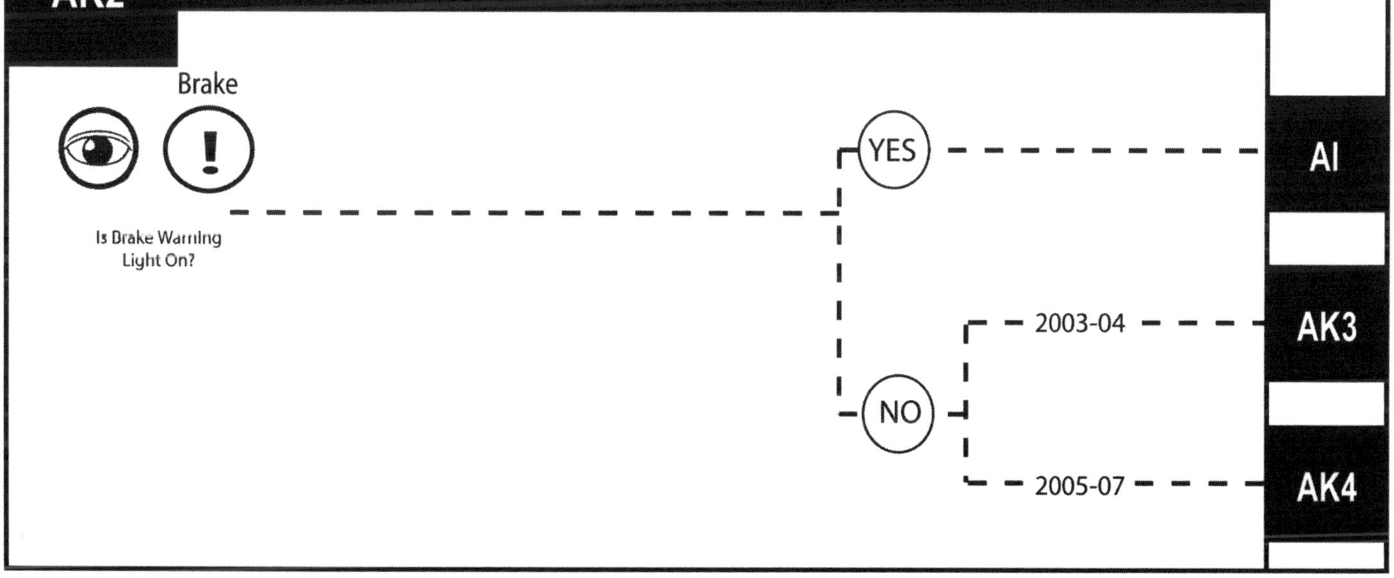

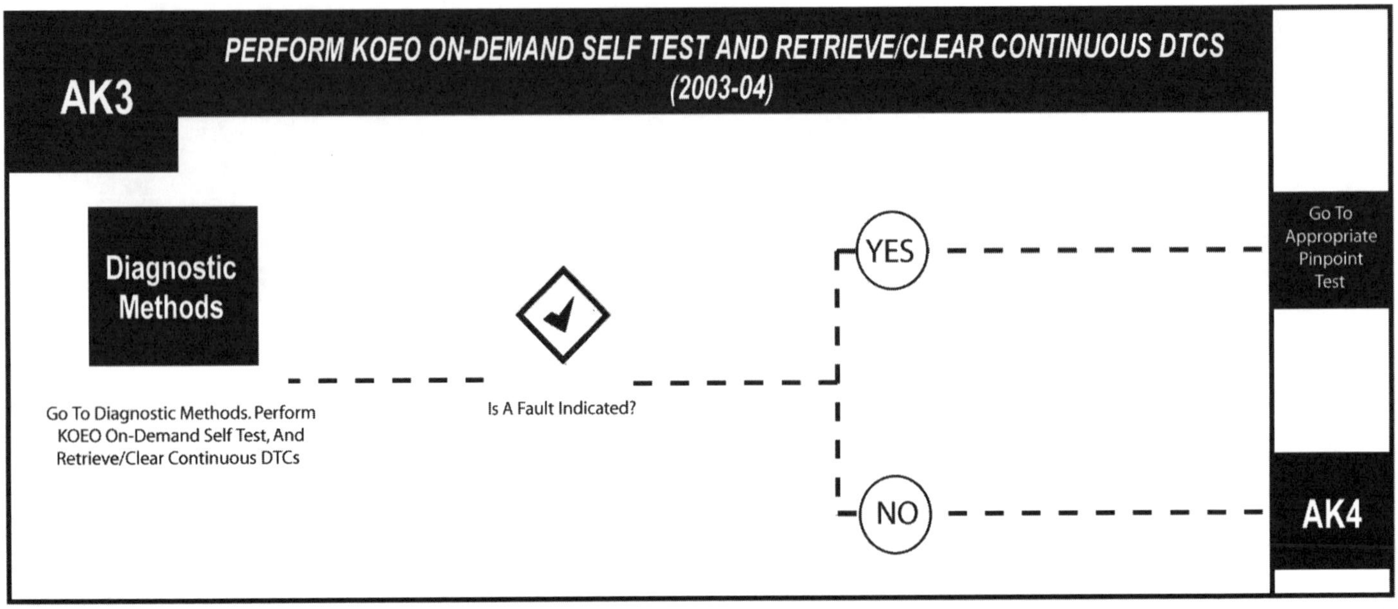
AK3
PERFORM KOEO ON-DEMAND SELF TEST AND RETRIEVE/CLEAR CONTINUOUS DTCS
(2003-04)
Diagnostic Methods
Go To Diagnostic Methods. Perform KOEO On-Demand Self Test, And Retrieve/Clear Continuous DTCs
Is A Fault Indicated?
YES
Go To Appropriate Pinpoint Test
NO
AK4

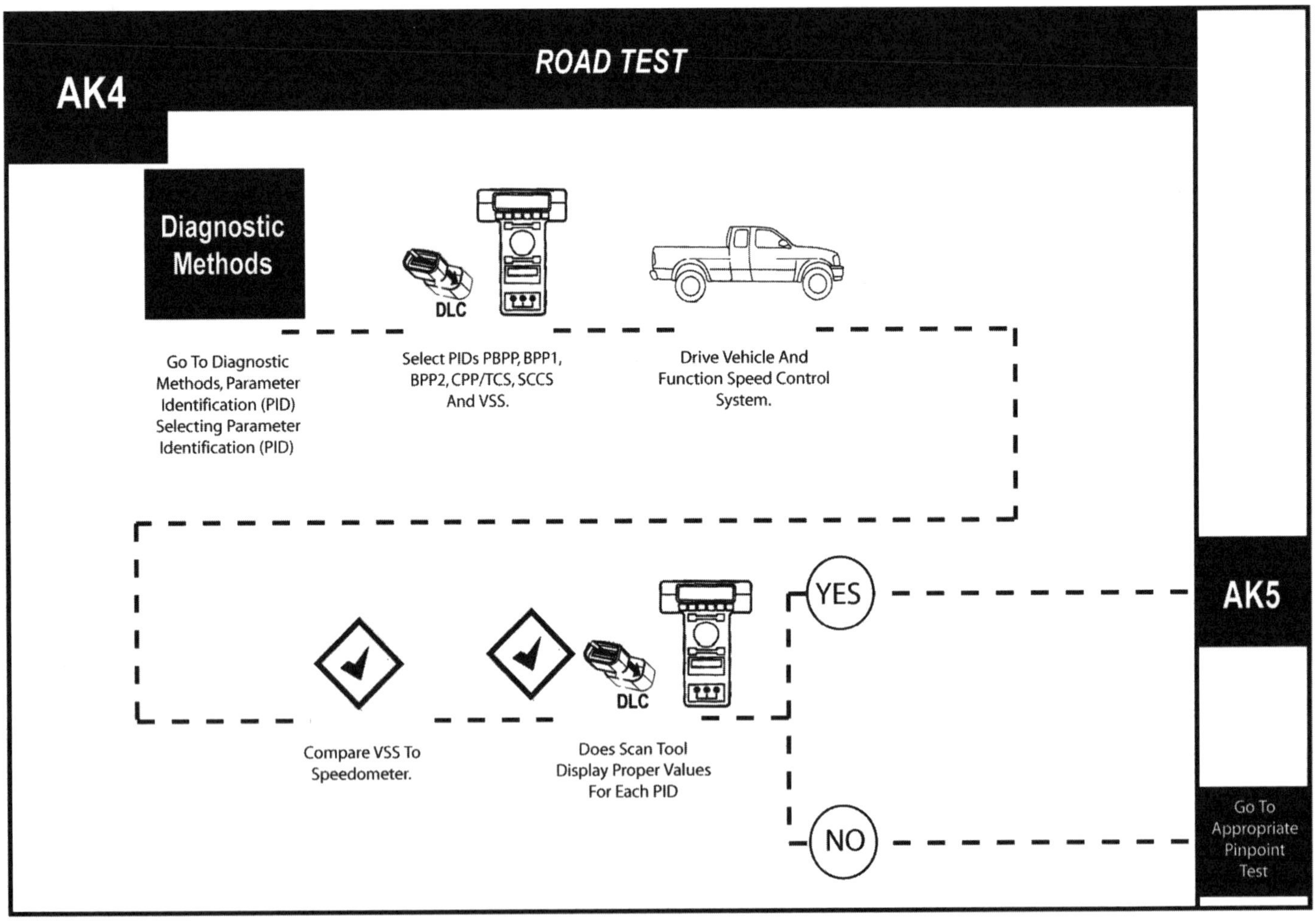
AK4
ROAD TEST
Diagnostic Methods
Go To Diagnostic Methods, Parameter Identification (PID) Selecting Parameter Identification (PID)
DLC
Select PIDs PBPP, BPP1, BPP2, CPP/TCS, SCCS And VSS.
Drive Vehicle And Function Speed Control System.
Compare VSS To Speedometer.
DLC
Does Scan Tool Display Proper Values For Each PID
YES
AK5
NO
Go To Appropriate Pinpoint Test

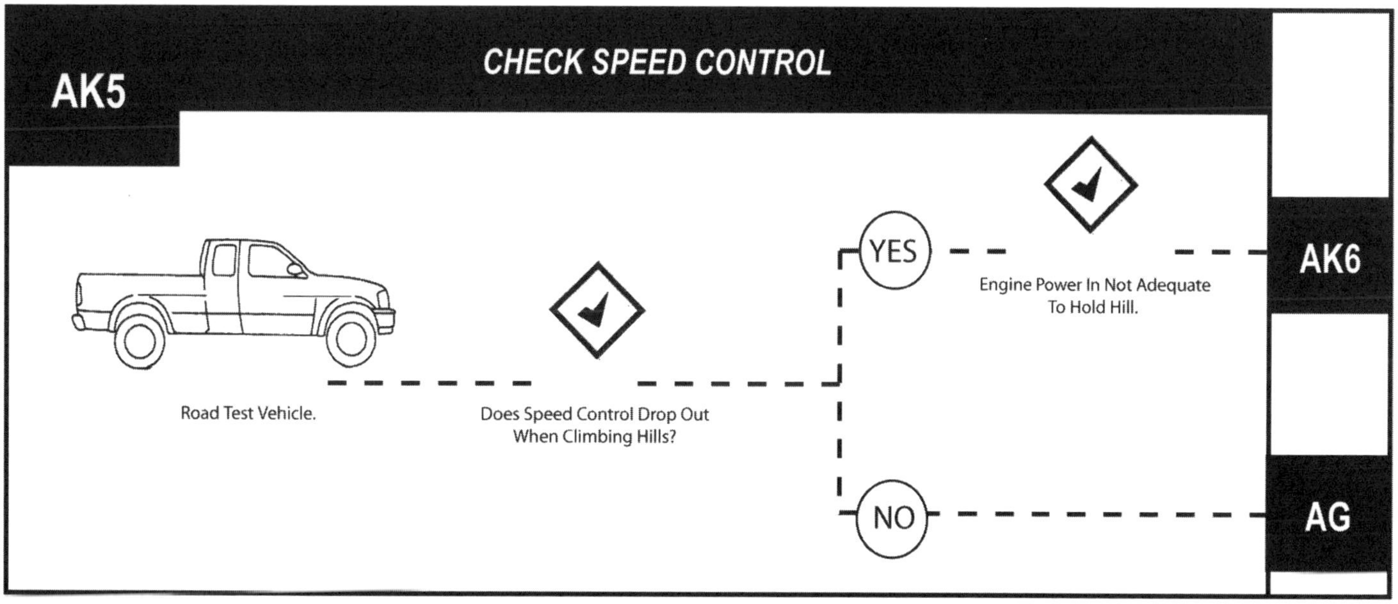
AK5
CHECK SPEED CONTROL
Road Test Vehicle.
Does Speed Control Drop Out When Climbing Hills?
YES
Engine Power In Not Adequate To Hold Hill.
AK6
NO
AG

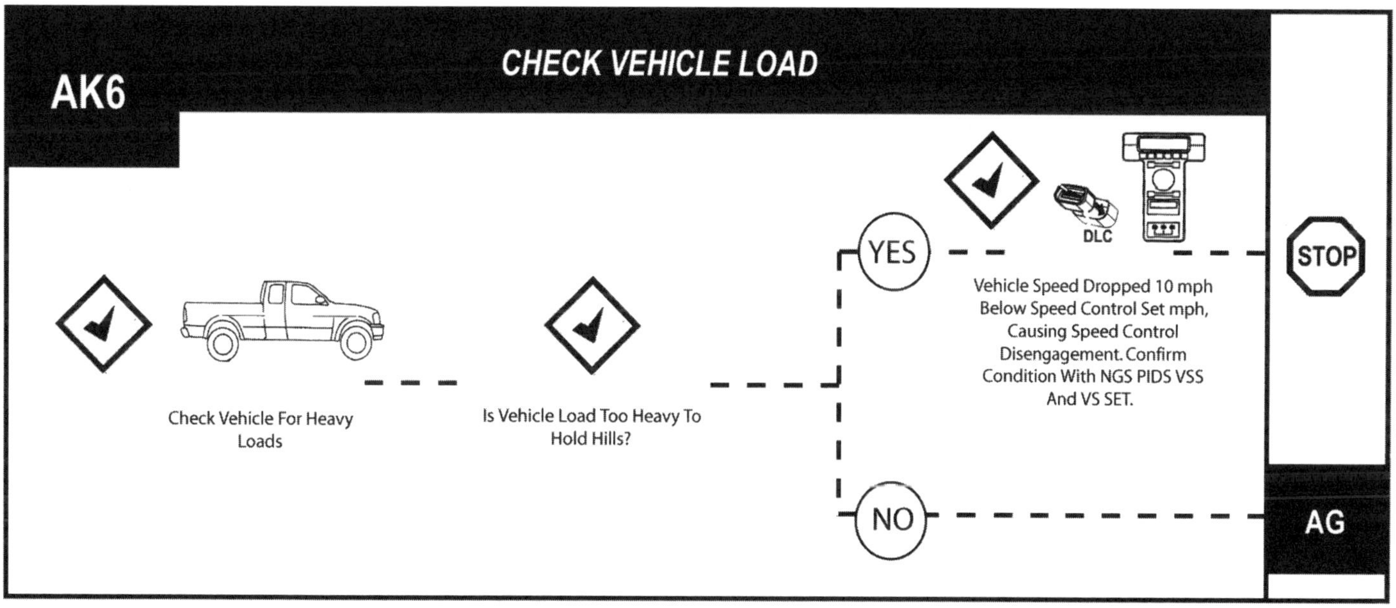
AK6
CHECK VEHICLE LOAD
Check Vehicle For Heavy Loads
Is Vehicle Load Too Heavy To Hold Hills?
YES
DLC
Vehicle Speed Dropped 10 mph Below Speed Control Set mph, Causing Speed Control Disengagement. Confirm Condition With NGS PIDS VSS And VS SET.
STOP
NO
AG

System	Symptom(s)	Test
Warning Indicator Concerns	Symptom(s): Transmission Control Idicator Lamp (TCIL) (2003-07 6.0L Diesel)	AL

Test/Step AL1 — CHECK TCIL

Is Transmission Control Indicator Lamp (TCIL) Flashing?

	Result
YES	AL3
NO	AL2

AL2 — TOGGLE TCIL

Toggle TCIL By Pushing TCS Switch Several Times.

Does TCIL Toggle On And Off?

YES — OK — TCIL Is OK.	STOP
NO	AL3

AL3 — OBSERVE CHECK ENGINE LIGHT

Is Check Engine Light On?

YES	AH
NO	AL4

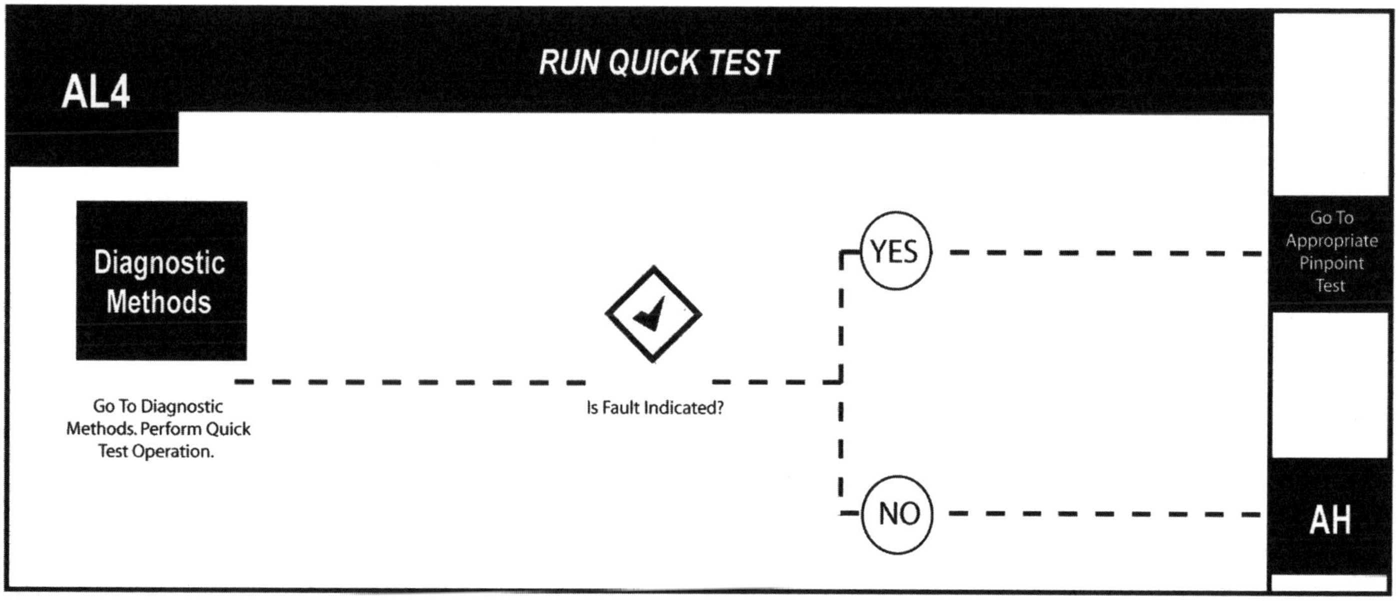
AL4
RUN QUICK TEST
Diagnostic Methods
Go To Diagnostic Methods. Perform Quick Test Operation.
Is Fault Indicated?
YES
Go To Appropriate Pinpoint Test
NO
AH

NOTES

System	Symptom(s): A/T Upshift, Downshift, Engagement (2003-07 6.0L Diesel)	Test
A/T Shift Concerns		AM

Test/Step

AM1 — CHECK AUTOMATIC TRANSMISSION FLUID

Result

Check Automatic Transmission Fluid Level And Quality.

Are Fluid Level And Quality OK?

YES — AM2

NO — REPAIR As Necessary. VERIFY A Symptom No Longer Exists. — STOP

AM2 — CHECK PARAMETER IDENTIFICATIONS (PIDS)

Diagnostic Subroutines

Go To Diagnostic Methods, Parameter Identification (PID), Selecting Parameter Identification (PID).

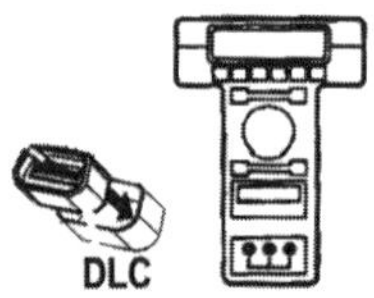

Display PIDS TR, SCCS M, 4x4L And VSS

Does The Scan Tool Display Proper Values For Each PID?

YES — AM3

NO — Go To Appropriate Pinpoint Test

AM3 CHECK PERFORMANCE DIAGNOSTIC

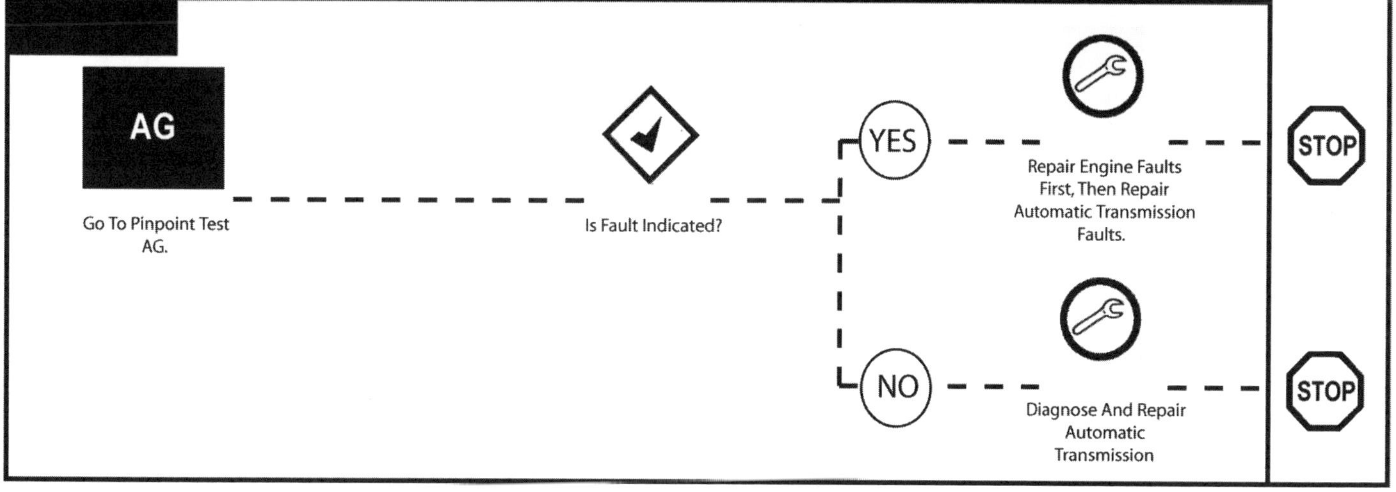

NOTES

System		Test
Oil System Concerns	Symptom(s): High Oil Consumption, Leaks (2003-07 6.0L Diesel)	AN

Test/Step		Result
AN1	**PRELIMINARY CHECKS**	

Perform The Following Preliminary Checks:
- External Leaks (Rocker Cover Gasket, Crankshaft Seals, ect.)
- Proper Oil Viscosity And API Rating. Refer To Operator's Guide.
- Check Air Intake System: Air Filter Restriction Gauge
- Loads Do Not Excede Maximum GVW

Are All Checks OK?

- YES → AN2
- NO → REPAIR As Required. VERIFY Problem No Longer Exists. → STOP

Test/Step		Result
AN2	**VERIFY HIGH OIL CONSUMPTION**	

Confirm Oil Consumption Through Maintenance Records Or Other Documentation.

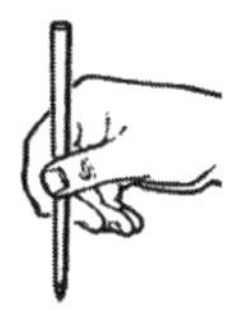

If Records Are Not Available, Have Operator Log Oil Consumption For 1450 km (900 Miles).

Is Oil Consumption Rate Higher Than 0.95 Liter (1 Quart) Per 900 Miles?

- YES → AN3
- NO → OK — No Problem Exhists, RETURN Vehicle To Costumer. → STOP

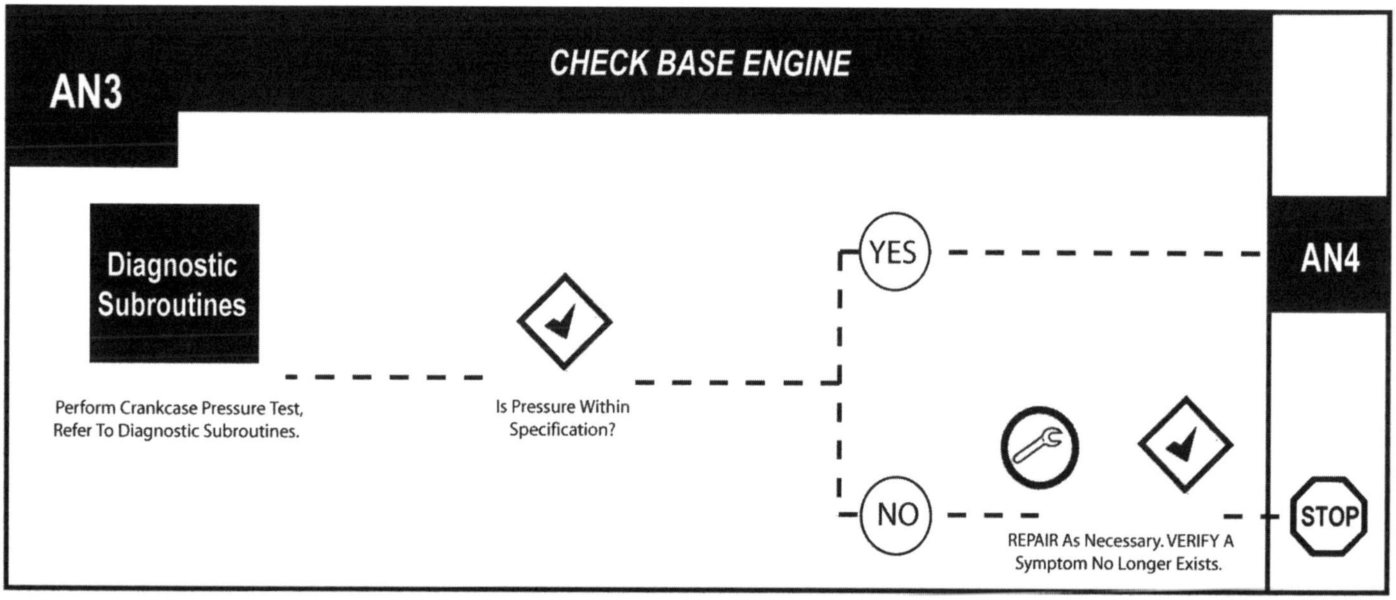
AN3
CHECK BASE ENGINE
Diagnostic Subroutines
Perform Crankcase Pressure Test, Refer To Diagnostic Subroutines.
Is Pressure Within Specification?
YES
AN4
NO
REPAIR As Necessary. VERIFY A Symptom No Longer Exists.
STOP

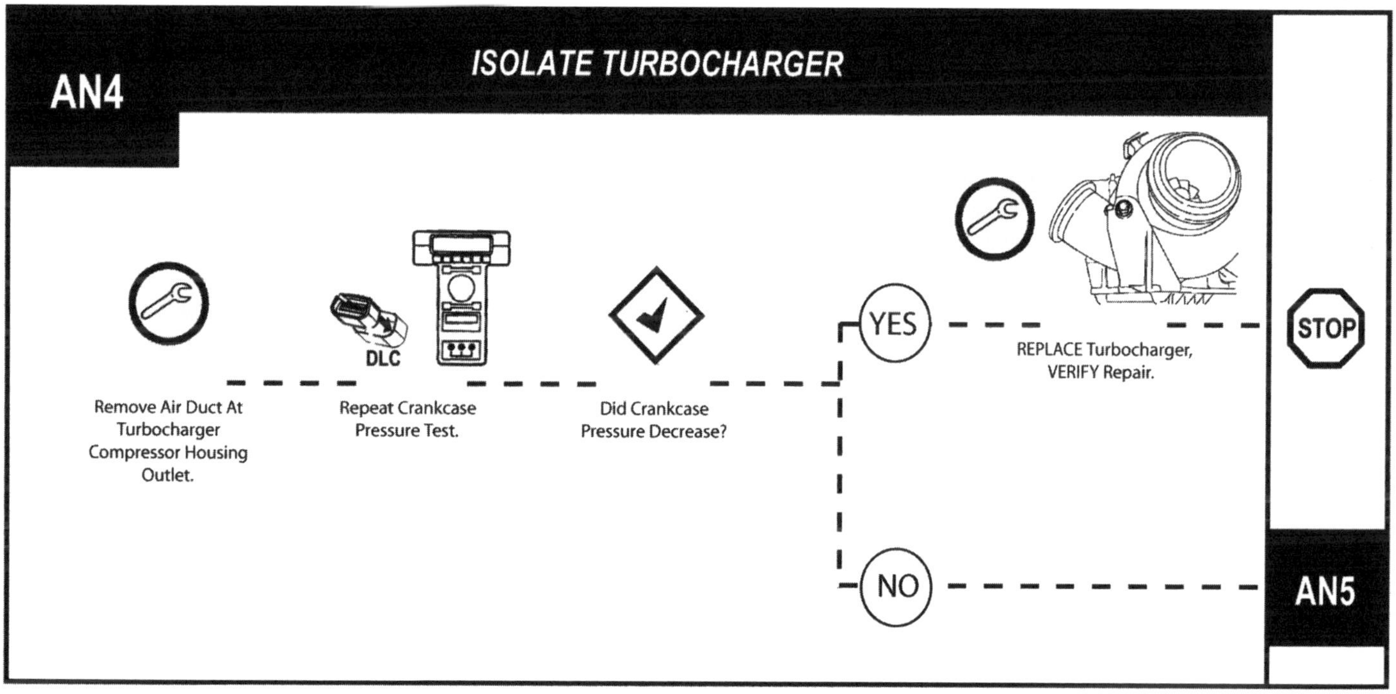
AN4
ISOLATE TURBOCHARGER
Remove Air Duct At Turbocharger Compressor Housing Outlet.
DLC
Repeat Crankcase Pressure Test.
Did Crankcase Pressure Decrease?
YES
REPLACE Turbocharger, VERIFY Repair.
STOP
NO
AN5

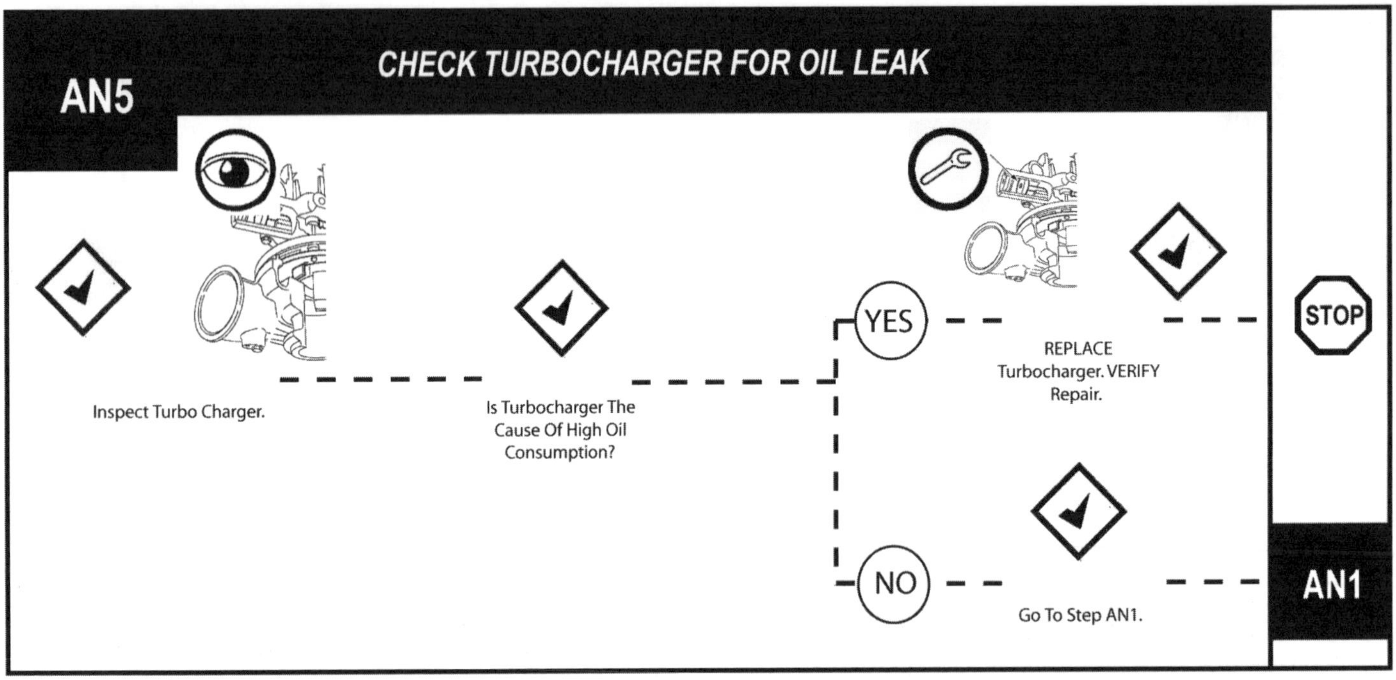
AN5
CHECK TURBOCHARGER FOR OIL LEAK
Inspect Turbo Charger.
Is Turbocharger The Cause Of High Oil Consumption?
YES
REPLACE Turbocharger. VERIFY Repair.
STOP
NO
Go To Step AN1.
AN1

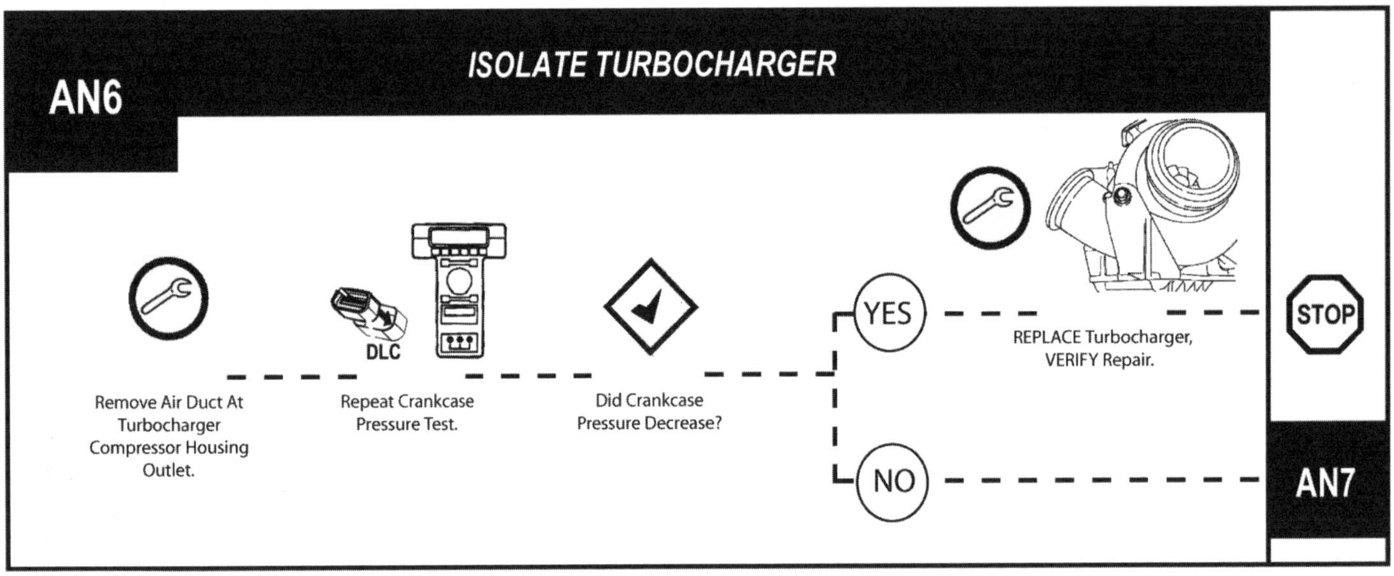
AN6
ISOLATE TURBOCHARGER
Remove Air Duct At Turbocharger Compressor Housing Outlet.
DLC
Repeat Crankcase Pressure Test.
Did Crankcase Pressure Decrease?
YES
REPLACE Turbocharger, VERIFY Repair.
STOP
NO
AN7

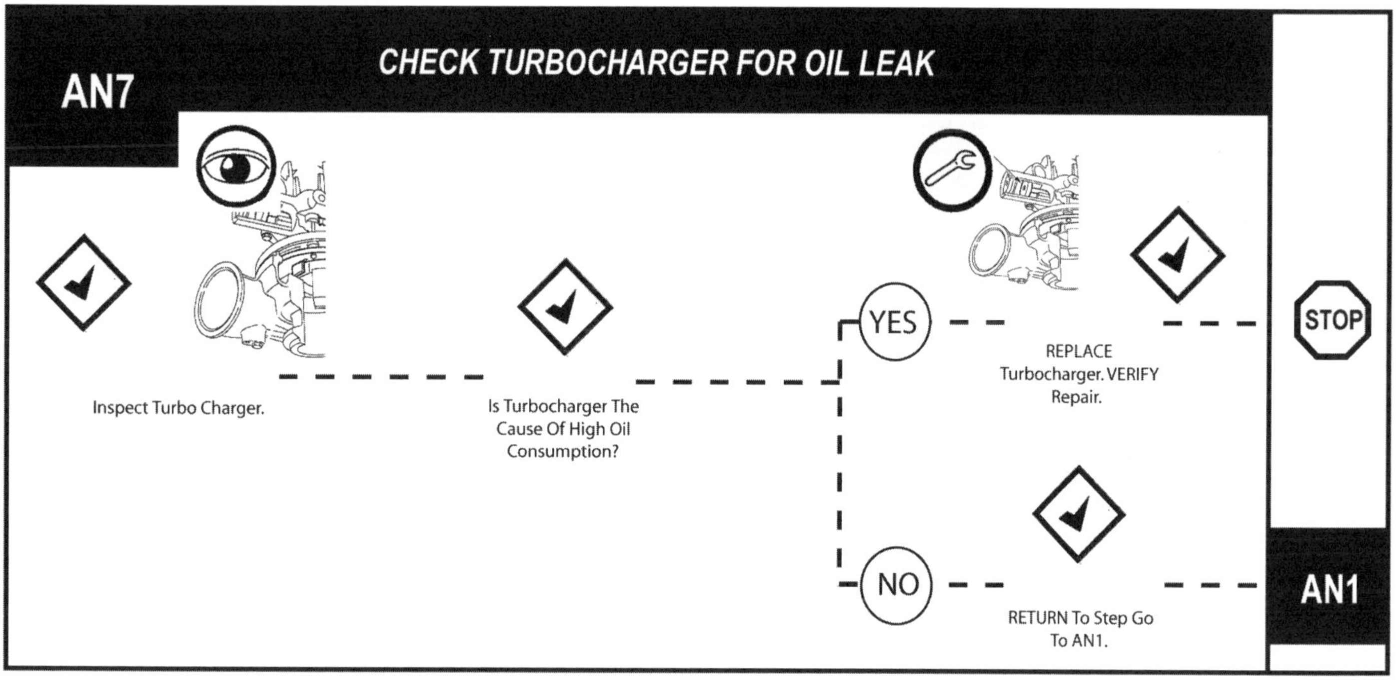
AN7
CHECK TURBOCHARGER FOR OIL LEAK
Inspect Turbo Charger.
Is Turbocharger The Cause Of High Oil Consumption?
YES
REPLACE Turbocharger. VERIFY Repair.
STOP
NO
RETURN To Step Go To AN1.
AN1

NOTES

System	Symptom(s): Cooling Fan Does Not Operate (2003-07 6.0L Diesel)	Test
Cooling System Concerns		AO

Test/Step

AO1

CHECK FAN CLUTCH COMPONENTS

Result

AH

Go To Pinpoint Test AH And Diagnose Fan Clutch Components.

Is A Fault Indicated?

YES — REPAIR As Necessary. VERIFY Symptom No Longer Exists. — STOP

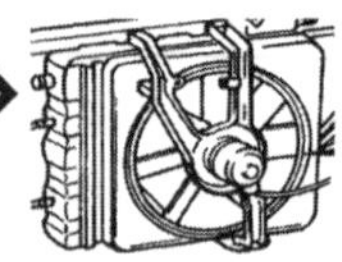

NO — Diagnose And Repair Cooling System. VERIFY Symptom No Longer Exists. — STOP

NOTES

System	Symptom(s): Cooling Always On (2003-07 6.0L Diesel)	Test
Cooling System Concerns		AP

Test/Step

AP1

CHECK FAN CLUTCH COMPONENTS

Note: This test is inteded to only diagnose a cooling fan that always runs with a "cool" Engine and the A/C and defroster off.

AH

Go To Pinpoint Test AH And Diagnose Fan Clutch Components.

Is A Fault Indicated?

YES — REPAIR As Necessary. VERIFY Symptom No Longer Exists. — **Result: STOP**

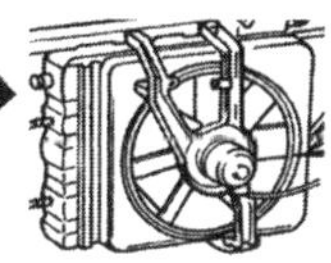

NO — Diagnose And Repair Cooling System. VERIFY Symptom No Longer Exists. — **Result: STOP**

NOTES

NOTES

DIAGNOSTIC SUBROUTINES
INDEX

NOTES

1. Visual Engine/Chassis Inspection

Purpose:
The purpose of this test is to check the general condition of the engine and look for obvious causes of a hard start or no start condition.

Fuel, Oil, Coolant, Electrical, Hoses ,Leaks

Method	Check
Visual	

Recommended Procedure:
Inspect the fuel system including the fuel tank and the fuel lines for kinks, bends and/or leakage. Inspect for coolant leaks at the radiator and the heater hoses and check the coolant level. Inspect the manifold absolute pressure (MAP) sensor and the intercooler for pinched hoses and leaks. Inspect wiring for correct routing and make sure no rubbing or chafing has occurred. Inspect the engine harness, fuel injector control module (FICM), powertrain control module (PCM) and sensor connectors to make sure they are completely seated and in good condition.

Possible Causes:
- Loose or leaking fuel supply lines could cause fuel system to lose prime
- Kinked or blocked fuel supply lines will create fuel restriction
- Fuel or oil leaks could contribute to no start conditions
- Coolant leaks could indicate engine problems
- Electronic connectors may be damaged or not properly installed causing a no start condition. The camshaft position (CMP) sensor, crankshaft position (CKP) sensor, injection pressure regulator (IPR) and the PCM connectors are the most critical electronic sensors/actuators to inspect in no start situations
- Pinched or open MAP sensor hose
- Pinched or open intercooler hose

Tools Required:
Inspection light

2. Check Engine Oil Level

Purpose:
The purpose of this test is to verify the oil quality and determine if there is sufficient oil to operate the injectors.

Check Engine Oil Level
- Check for contaminants (fuel, coolant).
- Correct grade/viscosity.
- Miles/hours on oil, correct level.

Method	Check
Visual	

2. Check Engine Oil Level (Continued)

Recommended Procedure:
Check for correct oil level using the dipstick with the vehicle on level ground. If there is no oil or very little oil in the crankcase, the injectors will not operate.

If the oil level on the dipstick is overfull, it is possible the engine was incorrectly serviced or fuel/coolant is diluting the oil and filling the crankcase.

Inspect the oil for color. A milky white oil indicates possible coolant contamination and will have an ethylene glycol odor.

WARNING: SMOKING OR OPEN FLAME OF ANY TYPE MUST NOT BE PRESENT WHEN WORKING NEAR FUEL OR FUEL VAPOR. FAILURE TO FOLLOW THESE INSTRUCTIONS MAY RESULT IN PERSONAL INJURY.

Oil contaminated with diesel fuel will have a diesel fuel odor and will increase the engine oil level. If the engine oil level is above "Max" due to diesel fuel dilution, the oil will appear thin and watery.

If the oil level is overfilled, drain the oil and the fuel filter housing. Isolate either cylinder head by removing the corresponding fuel line from the fuel filter housing. Remove the fuel pressure test port plug from the secondary filter housing. Install the plug in the outlet port. Install Fuel Pressure Adapter (303-765) and Gauge 0-1.1 MPa (0-160 psi) Bar 014-00761 or equivalent at the test port to confirm constant fuel pressure.

Using the scan tool, access OSC and command the fuel pump ON. Watch for fuel to drain out of the oil pan drain hole. Depending on the severity of the leak, it may take some time before a leak is noticeable. Remove the valve cover at the suspect cylinder head and inspect the injector area for leaks.

Check the service records for correct oil type and viscosity for the vehicle operating temperature. Single weight or 15W-40 oil is not recommended for cold ambient temperatures. 10W-30 oil is recommended for cold ambient temperatures. Oil that has had extended drain intervals will have increased viscosity (become thicker) and will make engine cranking more difficult and starting less reliable at temperatures below freezing. Refer to the lube oil chart in the Workshop Manual or Owner Literature for the correct oil selection for temperature conditions.

Possible Causes:
- Loss of lube oil pressure
- Oil level low — oil leak, oil consumption, incorrect servicing
- Oil level high — incorrect servicing, fuel dilution from injector O-rings
- Oil contamination with coolant — oil cooler, head gasket, porosity

Tools Required:
- Adapter 303-765 or equivalent
- Gauge 0-1.1 MPa (0-160 psi) Bar (part of Pressure Adapter Kit 014-00761 or equivalent)
- Scan tool

3. Intake/Exhaust Restriction

Purpose:
This purpose of this test is to determine if an air intake or exhaust restriction is contributing to a no start or hard start condition. If the engine does start with a high air intake or exhaust restriction, a considerable amount of black/blue smoke is produced.

3. Intake/Exhaust Restriction (Continued)

Intake/Exhaust Restriction

- Inspect the air filter and inlet ducts.
- Inspect the exhaust system.
- Check for illumination of the air filter restriction indicator (F-Super Duty/Excursion)

Method	Check
Visual	

Recommended Procedure:
Inspect the air cleaner inlet and ducting to verify it is not blocked or collapsed. Inspect the air cleaner housing and filter for proper installation. Inspect the air filter restriction gauge to ensure the intake restriction is below the red marks. Inspect the exhaust system for damaged or blocked pipes.

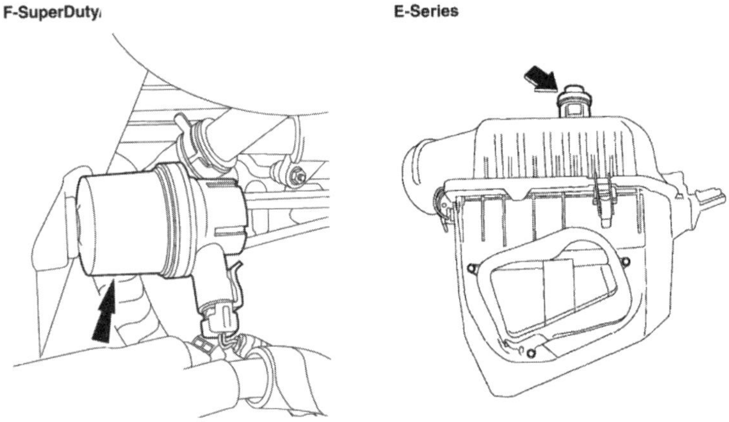

Possible Causes:
Note: Reset the air filter restriction gauge after repairing a restriction concern.

- Snow, plastic bags or other foreign material may restrict airflow at the air inlet
- Misrouted air cleaner ducting
- On engines recently repaired, rags or cap plugs may have been inadvertently left in an air inlet pipe
- Tailpipe or muffler may have collapsed or been damaged

Tools Required:
None

4. Sufficient Clean Fuel

Purpose:
The purpose of this test is to verify the fuel quality.

Sufficient Clean Fuel

- Check for illumination of the WATER IN FUEL indicator.
- After verifying that there is fuel in the tank, drain a sample from the fuel control module.
- A cetane rating between 40 and 50 is recommended for optimum performance.

Method	Check
Visual	

4. Sufficient Clean Fuel (Continued)

Recommended Procedure:
Open the drain valve on the fuel control module and fill a clear container until it is half full. Close the drain valve.

Observe WATER IN FUEL indicator. If the indicator is illuminated, the fuel is probably contaminated with water.

Flow out of the drain should be a steady stream. Insufficient flow could indicate fuel supply or fuel system problems.

Inspect the fuel in the container. It should be clear, not cloudy. It also should be free of water and contaminants. Dyed red or blue fuel indicates off-highway fuel.

Some sediment and water may be present in the fuel sample if the fuel filter has not been serviced for a prolonged period of time and/or if the sediment and water have not been drained recently. If that is the case, a second sample may be required to determine fuel quality.

F-SuperDuty/

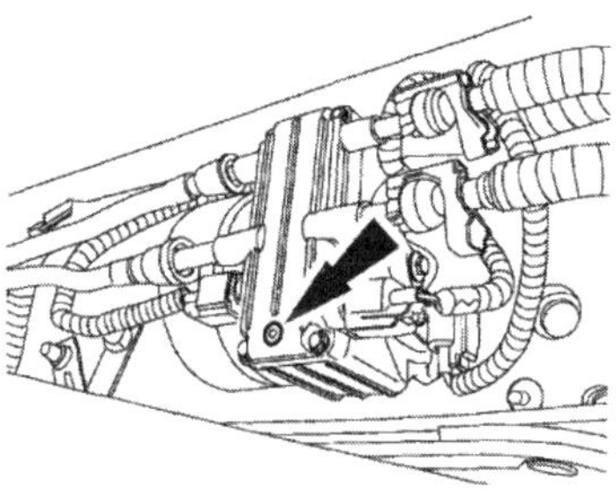

E-Series

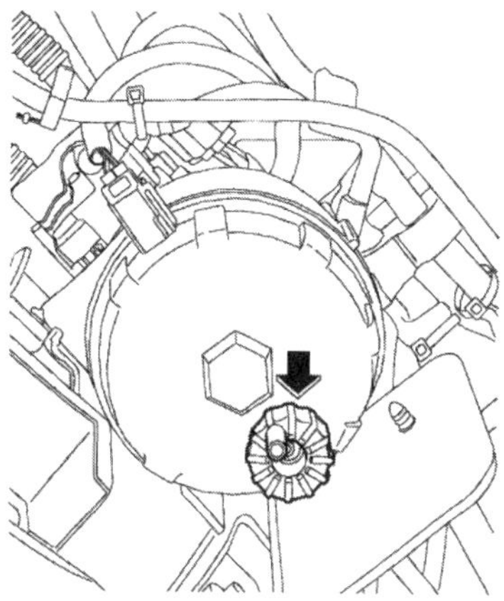

Possible Causes:
- No fuel in the tank
- Fuel supply line could be broken or crimped
- Fuel could be jelled (most likely in cold weather with No. 2 fuel)
- Pickup tube screen in tank could be clogged
- Restricted fuel filters

Cloudy fuel indicates that the fuel may not be a suitable grade for cold temperatures.

Excessive water or contaminants may indicate that the tank and fuel system may need to be flushed and cleaned.

Tools Required:
Clear container — approximately 0.95L (1-quart)

5. Electric Fuel Pump Pressure

Purpose:
The purpose of this test is to verify there is sufficient fuel pressure for starting.

Electric Fuel Pump Pressure
- Verify there is fuel in the tank.
- Check the fuel pump power and ground circuits.
- Measure the fuel pressure at the engine fuel filter housing test port with a 0-101 MPa (0-160 psi) gauge.
- Fuel pump will run for 20 seconds at initial key on and then pressure will decrease.

5. Electric Fuel Pump Pressure (Continued)

Instrument	Specification	Measurement
0-1.1 MPa (0-160 psi) gauge	310 kPa (45 psi minimum)	

Recommended Procedure:
Verify there is fuel in the tank and battery voltage at the fuel pump. Using a digital multimeter, measure the voltage between the fuel pump power and ground circuits. Battery voltage will be present for approximately 20 seconds after the ignition key is turned on. If no voltage is present, GO to Pinpoint Test **M**

F-SuperDuty

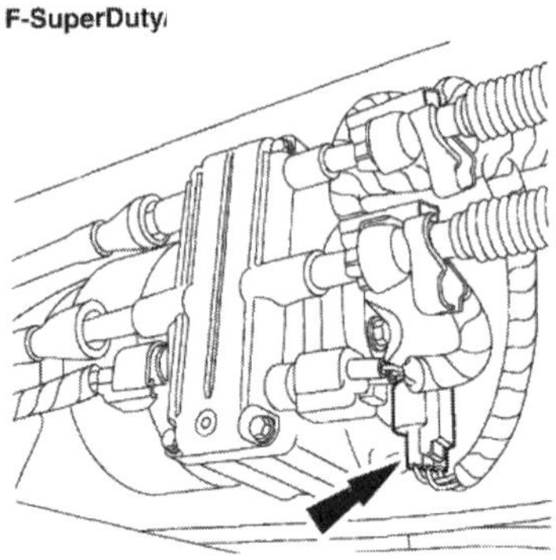

E-Series

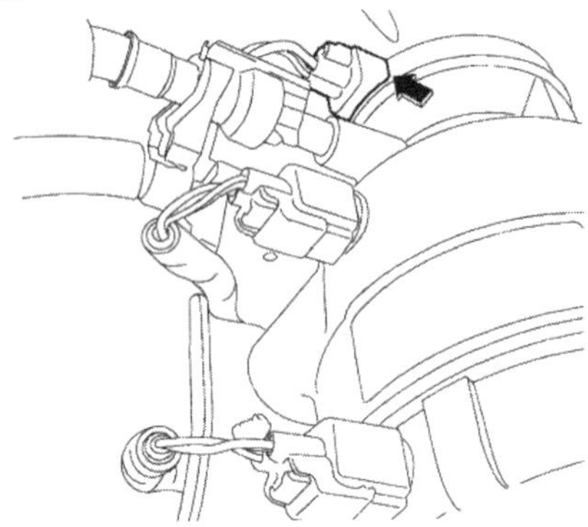

Remove the plug from the front of the fuel filter housing. Install Adapter 303-765 and Gauge 0-1.1 MPa (0-160 psi) Bar 014-00761 or equivalent. Measure fuel pressure in crank or run. If fuel pressure is below the specification of 310 kPa (45 psi), go to the next step to verify no restriction.

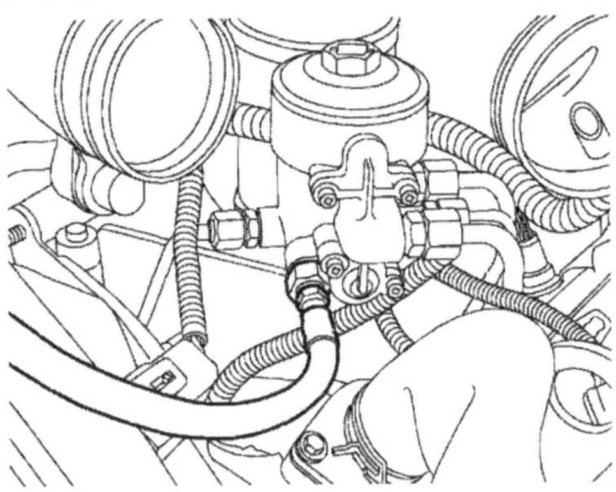

Possible Causes
- Fuel pump relay
- Inertia switch

Tools Required:
- Adapter 303-765 or equivalent
- Gauge 0-1.1 MPa (0-160 psi) Bar (part of Pressure Adapter Kit 014-00761 or equivalent)

6. Electric Fuel Pump Inlet Restriction

Purpose:
The purpose of this test is to isolate the cause of low fuel pressure.

Electric Fuel Pump Inlet Restriction
Measure restriction at fuel pump inlet.

6. Electric Fuel Pump Inlet Restriction (Continued)

Instrument	Specification	Measurement
(0-30 in Hg) Vacuum	6 in Hg maximum	

Recommend Procedure:
Remove the fuel line to the inlet side of the fuel pump. Install Fuel Pump Adapter 310-111 or equivalent between the fuel inlet line and the electric fuel pump. Connect the test adapter to gauge 0-762 mm Hg (0-30 in Hg) vacuum. Measure restriction at wide open throttle (WOT) (maximum engine speed out of gear with the brakes set and the wheels blocked). If restriction is greater than 152 mm Hg (6 in Hg), there is a restriction between the fuel pump and the fuel tank. If restriction is less than 152 mm Hg (6 in Hg), inspect both fuel filters. If filters are OK, inspect the fuel regulator valve. If regulator and filters are OK, install a new fuel pump.

Possible Causes:

- Fuel line restriction
- Fuel pressure regulator valve
- Fuel filters
- Fuel pump

Tools Required:

- Fuel Pump Adapter 310-111 or equivalent
- Vacuum Gauge (part of Pressure Adapter Kit 014-00761 or equivalent)

7. Perform Key On Engine Off (KOEO) On-Demand Self-Test

Purpose:
The purpose of this test is to determine if the PCM has detected any fault conditions that would cause a hard start or no start condition.

Perform KOEO On-Demand Test
Use the scan tool. Diagnostic trouble codes (DTCs) set during this test are current faults.

Diagnostic Trouble Codes (DTCs)	

Recommended Procedure:
Note: To ensure that the DTC is a hard fault, first clear continuous DTCs (be sure to record all DTCs and Freeze Frame information before clearing). Rerun KOEO On-Demand Self Test. If the DTC is set again, a hard fault has occurred.

Connect the scan tool. Turn off all accessories. If the vehicle is equipped with an auxiliary powertrain control module (APCM), it must be turned off to perform the self-test.

1. Perform the necessary vehicle preparation and a visual inspection. Refer to Quick Test Operation .
2. Refer to the scan tool operating manual for instructions.
3. Key on, engine off.
4. Wait 4 seconds for the transmission control module (TCM), PCM and the FICM to initialize.
5. Follow the operating instructions from the diagnostic menu.
6. Perform a KOEO On Demand Self-Test.
7. Record the DTCs and Freeze Frame information and refer to the appropriate pinpoint test.

7. Perform Key On Engine Off (KOEO) On-Demand Self-Test (Continued)

Possible Causes:
The most likely PCM detectable faults that will cause a no start or hard start condition are:
- CMP sensor inactive fault
- CKP sensor inactive fault
- IPR output circuit check fault
- FICM ENABLE circuit fault

Tools Required:
- Scan tool

8. Retrieve Continuous DTCs

Purpose:
The purpose of this test is to determine if the PCM has detected any historical or intermittent fault conditions that would cause a hard start/no start symptom. The condition that caused a continuous DTC may no longer exist.

Retrieve Continuous DTCs
DTCs retrieved during this test are historical faults.

Diagnostic Trouble Codes (DTCs)	

Recommended Procedure:
Connect the scan tool. Turn off all accessories. If the vehicle is equipped with an APCM, it must be turned off to perform self tests.

1. Perform the necessary vehicle preparation and a visual inspection. Refer to Quick Test Operation .
2. Refer to the scan tool operating manual for instructions.
3. Key on, engine off.
4. Follow operating instructions from the menu.
5. Record the continuous DTCs and Freeze Frame information and perform the appropriate pinpoint test for continuous DTC diagnostics.
6. Continuous DTCs must be cleared after a repair.

Tools Required:
Scan tool

9. KOEO Injector Electrical Self-Test (Click Test)

Purpose:
Note: If unable to perform KOEO Injector Electrical Self-Test (Click Test), disconnect the FICM connector and check injector for shorts or opens.

The purpose of this test is to determine if the injector solenoids and valves are functioning by clicking all injectors together and then each injector in numerical sequence (1 through 8).

KOEO Injector Electrical Self-Test (Click Test)
- Use the scan tool. Injector DTCs will be displayed after the self-test is completed.
- Note: Sequence repeats three times.

All injectors will momentarily click, then each injector will click in sequence 1 through 8.

9. KOEO Injector Electrical Self-Test (Click Test) (Continued)

Diagnostic Trouble Codes (DTCs)	

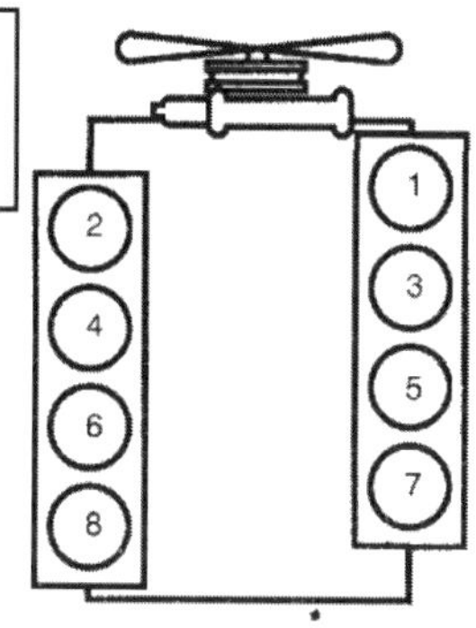

FIRING ORDER
1-2-7-3-4-5-6-8

Recommended Procedure:
This test determines if the injector circuits and solenoids are electrically operating without fault. All injectors will first click together for approximately 2 seconds, then each injector will click for approximately 1 second in numerical order (1 through 8). If a fault is detected, a DTC will be output on the data link at the end of the test when requested by a scan tool. Only a hard fault DTC will be displayed.

Connect the scan tool. Turn off accessories. If vehicle is equipped with an APCM, it must be turned off to perform self tests.

1. Perform the necessary vehicle preparation and a visual inspection. Refer to Quick Test Operation .
2. Refer to the scan tool operating manual for instructions.
3. Key on, engine off.
4. Follow the operating instructions from the diagnostic menu.
5. Perform the KOEO Injector Electrical Self-Test (Click Test).
6. Record the DTCs and refer to the appropriate pinpoint test.

Possible Causes:
- Open or shorted injector circuit
- Injector connector
- Injector solenoid
- FICM power or ground circuit
- FICM

Tools Required:
- Scan tool

10a. Check VPWR During Cranking

Purpose:
The purpose of this test is to verify PCM power during cranking.

Scan Tool — Data List Monitoring
- Scan tool may reset below 9.5 volts.
- Select the parameters indicated from the scan tool parameter list and monitor while cranking the engine.

Note: The 8 volt specification represents the minimum battery voltage required for engine starting. Greater than average crank times will be encountered if the battery voltage is less than 9.5 volts. If excessive crank time is a concern, verify battery voltage is greater than 9.5 volts.

Note: You may need to use an outside power source for the scan tool.

Parameter	Specification	Measurement
B+	8 volts minimum	

Note: If a low voltage condition is present, check the battery, charging system or power and ground circuits to the PCM.

10a. Check VPWR During Cranking (Continued)

Recommended Procedure:

1. Connect the scan tool.
2. Key on, engine off.
3. Access and monitor the B+ PID while cranking the engine.

Possible Causes:

- Battery cables
- Low battery voltage
- Charging system problem
- Power circuit and ground faults to the PCM
- PCM relay
- GO to Pinpoint Test A to diagnose a voltage concern.

Note: Battery voltage below 9.5 volts may cause the scan tool to reset. If the scan tool resets during a self-test or while PID monitoring, it may be necessary to install a battery charger to maintain correct system voltage.

Tools Required:
Scan tool

10b and 10c. Check FICM Power During Cranking

Purpose:
The purpose of this test is to verify FICM power during cranking.

Scan Tool — Data List Monitoring

- Scan tool may reset below 9.5 volts.
- Select the parameters indicated from the scan tool parameter list and monitor while cranking the engine.

Note: You may need to use an outside power source for the scan tool.

Parameter	Specification	Measurement	NOTES
FICMLPWR	8 volts minimum		No or low voltage indicated could be caused by a 12-way connector issue or logic power fuse. GO to Pinpoint Test S
FICMVPWR	8 volts minimum		No or low voltage indicated could be caused by a 12-way connector issues

Recommended Procedure:

1. Connect the scan tool.
2. Key on, engine off.
3. Access and monitor the FICMLPWR and FICMVPWR PIDS while cranking the engine.

Possible Causes:

- Low or no battery voltage to the FICM
- High resistance or an open FICM power circuit
- FICM power relay

GO to Pinpoint Test AS for FICMVPWR diagnosis.

Tools Required:
Scan tool

10d. Check The RPM Signal While Cranking

Purpose:
The purpose of this test is to verify the CMP and CKP sensors and circuits are functioning.

Scan Tool — Data List Monitoring
- Scan tool may reset below 9.5 volts.
- Select the parameters indicated from the scan tool parameter list and monitor while cranking engine.

Note: You may need to use an outside power source for the scan tool.

Parameter	Specification	Measurement
RPM	100 RPM minimum	

Note:Low RPM may be caused by starting or charging system concerns. No RPM indicated while cranking may be a CMP or CKP fault.

Recommended Procedure:
1. Connect the scan tool.
2. Key on, engine off.
3. Access and monitor the RPM PID while cranking the engine.
4. No FICM synchronization while cranking the engine with the ICP, RPM and VPWR signals correct usually indicates a loss of the CMP or CKP synchronization signal. Refer to the appro priate pinpoint test for diagnosis. Refer to Section 6, Control System Diagnostic Sheet Refernce for normal operating values.

Possible Causes:
- Weak battery or starter
- Circuitry
- CKP sensor
- CMP sensor

GO to Pinpoint Test **D** for CKP sensor diagnosis.

GO to Pinpoint Test **V** for CMP sensor diagnosis.

Tools Required:
- Scan tool

10e. and 10f. Monitor The ICP While Cranking

Purpose:
The purpose of this test is to determine if the injection control system can supply enough injection control pressure to sustain starting.

Scan Tool — Data List Monitoring
- Scan tool may reset below 9.5 volts.
- Select the parameters indicated from the scan tool parameter list and monitor while cranking engine.

Parameter	Specification	Measurement	Notes
ICP	3.5 MPa (500 psi) minimum		A minimum of 3.5 MPa (500 psi) is required before the injectors are enabled. No or low oil in the system, system leakage, injector O-rings, faulty IPR or high pressure pump could cause low pressure. **IPR duty cycle defaults to 14% [2.1 MPa (300 psi)] without a CKP signal.**
ICP VOLTS	0.80 volts minimum		Voltage reading below specification indicates low ICP during crank.

10e. and 10f. Monitor The ICP While Cranking (Continued)

Recommended Procedure:

1. Connect the scan tool.
2. Key on, engine off.
3. Access and monitor the ICP and IPR PIDS while cranking the engine.

Note: A CKP signal is required before the IPR is commanded above 14%.

If the ICP does not meet the minimum specification of 3.5 mPa (500 psi), the injectors will not be enabled by the PCM because of insufficient rail pressure.

If the IPR duty cycle is greater than 14%, ICP pressure should increase above 3.5 MPa (500 psi) provided that the IPR valve is not stuck open, the high pressure pump is building pressure and there is not an injection control pressure leak between the high pressure pump and the injectors.

Possible Causes:

- Low base engine oil pressure
- Injection control pressure system leak
- IPR failure
- Faulty high pressure pump
- Injector O-ring leaking

Purpose:
Isolate the cause of low injection control pressure.

Recommended Procedure: (F-Super Duty/Excursion — Early Build)
Air Pressure Check (F-Super Duty/Excursion — Early Build)
Note: An air leak on the high pressure oil pump shaft lip seal is normal while performing this procedure. This is not an indication of a high pressure oil leak and the high pressure pump should not be replaced for this condition.

Verify base engine oil pressure. Check the operation of the IPR valve. Remove one of the valve covers from the engine. Apply regulated shop-air pressure to the high pressure oil rail using the adapter from the Oil High Pressure Leakage Test Adapter Set 303-766. With air pressure applied, an air leak will result from the IPR valve's normally open state. With KOEO, use the scan tool to increase the IPR valve duty cycle. The IPR valve should close and block the air leak. If no change is heard, the IPR valve is not functioning as commanded. GO to Pinpoint Test R for IPR circuit DTCs. If no DTCs are present, install a new IPR valve and retest. With the IPR commanded closed, check the high pressure oil system for leaks. A leak may exist in the following areas:

High Pressure Oil System (F-SuperDuty/Excursion — Early Build)

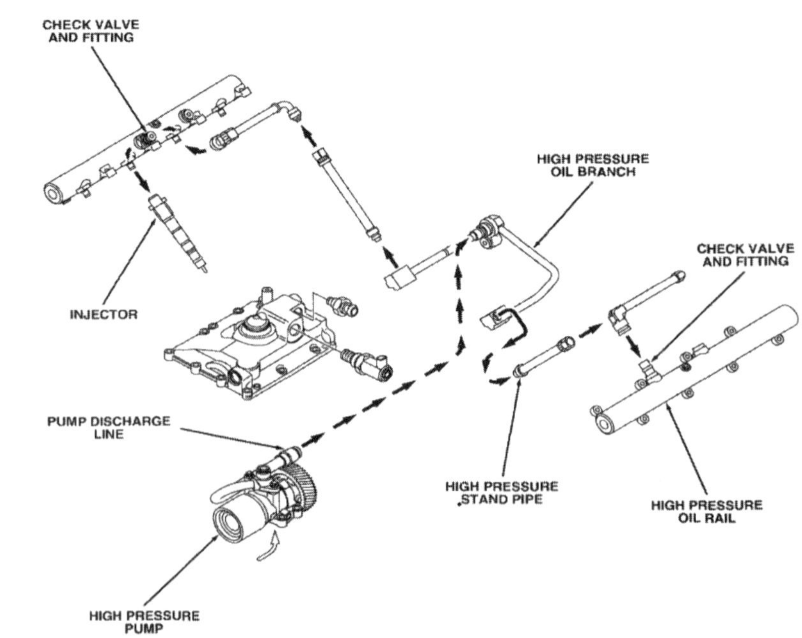

10e. and 10f. Monitor The ICP While Cranking (Continued)

- Cracked/broken check valve and fitting in high pressure oil rail, under either valve cover.
- Disconnected high pressure hoses under the valve covers.
- Standpipe from high pressure hose to tappet gallery (O-ring at the bottom may be cut/torn or missing).
- O-ring in-between discharge tube and high pressure pump.
- O-ring around discharge tube that fits inside high pressure pump cover.
- High pressure inlet O-ring.
- Branch tube from high pressure pump to stand pipe in tappet gallery.

Right Cylinder Head Check (F-Super Duty/Excursion — Early Build)
Remove the RH valve cover. Remove the high pressure hose from the right oil manifold using Quick Disconnect Tool 303-755, or equivalent, and cover the fitting on the oil manifold. Install the plug from the Oil High Pressure Leakage Test Adapter Set 303-756, or equivalent, into the high pressure hose to block it off. Crank the engine and monitor the ICP PID. If the ICP/EBC Adapter Cable D94T-50-A is connected to the ICP sensor, connect a digital multimeter between signal return and ICP signal wires on the adapter cable. Crank the engine and monitor the signal. The digital multimeter should read 1 to 4 volts.

F-SuperDuty/

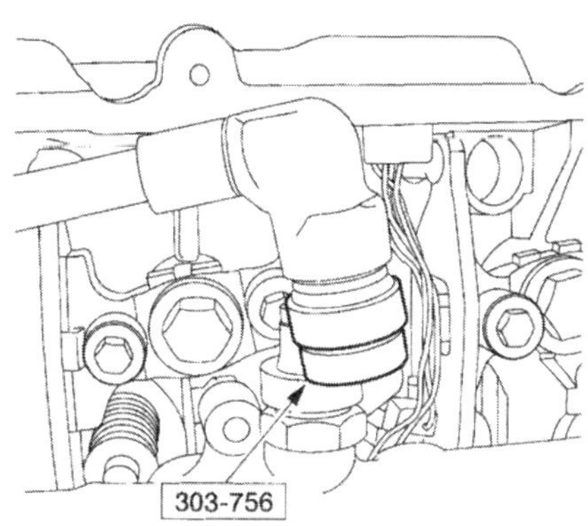

CAUTION: The engine may start.
If the engine starts or if injection control pressure is now within specification, the injection control pressure leak has been isolated to the right cylinder head.

CAUTION: Oil is under high pressure.

Left Cylinder Head Check (F-Super Duty/Excursion — Early Build)
Remove the LH valve cover. Remove the cap and plug, then reinstall the high pressure hose to the right oil manifold. Remove the high pressure hose from the left oil manifold using Quick Disconnect Tool 303-755, or equivalent, and cover the fitting on the oil manifold. Install 303-756, or equivalent into the high pressure hose. Crank the engine and monitor the ICP PID. If the ICP/EBC Adapter Cable D94T-50-A is connected to the ICP sensor, connect a digital multimeter between the signal return and ICP signal wires of the adapter cable. Crank the engine and monitor the signal. The digital multimeter should read 1 to 4 volts.

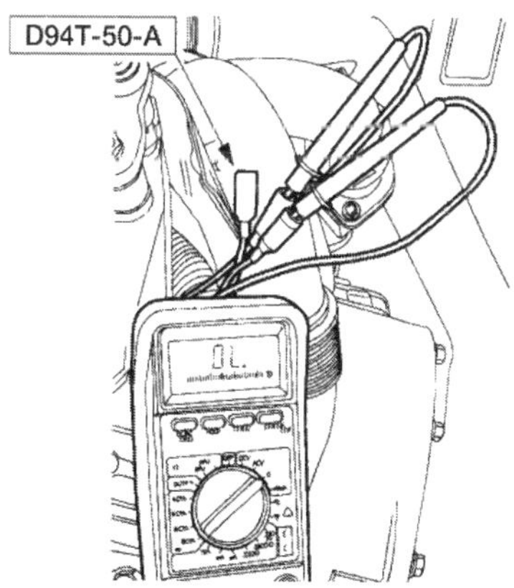

CAUTION: The engine may start.

If the engine starts or if injection control pressure is now within specification, the injection control pressure leak has been isolated to the left cylinder head.

IPR and High Pressure Pump Test (F-Super Duty/Excursion — Early Build)
If injection control pressure is still low after ruling out both cylinder heads as the source of injection control pressure leakage, perform the following steps to isolate the cause. Remove the high pressure hose from the right oil manifold and cover the fitting. Install the plug from the Oil High Pressure Leakage Test Adapter Set 303-756, or equivalent, into the high pressure hose to block it off. With the high pressure pump effectively deadheaded, crank the engine and monitor the ICP PID. If a low pressure condition still exists, the problem is most likely with the high pressure pump or the high pressure pump drive gear.

10e. and 10f. Monitor The ICP While Cranking (Continued)

Recommended Procedure (E-Series or F-Super Duty/Excursion — Late Build)
Air Pressure Check (E-Series or F-Super Duty/Excursion — Late Build)
Note: An air leak on the high pressure oil pump shaft lip seal is normal while performing this procedure. This is not an indication of a high pressure oil leak and the high pressure oil pump should not be replaced for this condition.

Verify base engine oil pressure. Remove the left valve cover from the engine. Apply regulated shop-air pressure to the high pressure oil rail using the adapter from the Test Adapter Set 303-766, or equivalent. With air pressure applied, an air leak will result from the IPR valve's normally open state. With KOEO, use the scan tool to increase the IPR valve duty cycle. The IPR valve should close and block the air leak. If no change is heard, the IPR valve is not functioning as commanded. GO to Pinpoint Test R for IPR circuit DTCs. If no DTCs are present, install a new IPR valve and retest.
With the IPR commanded closed, check the high pressure oil system for leaks. A stethoscope may be used to help identify a leak. A leak may exist in the following areas:

High Pressure Oil System (E-Series or F-SuperDuty/Excursion — Late Build)

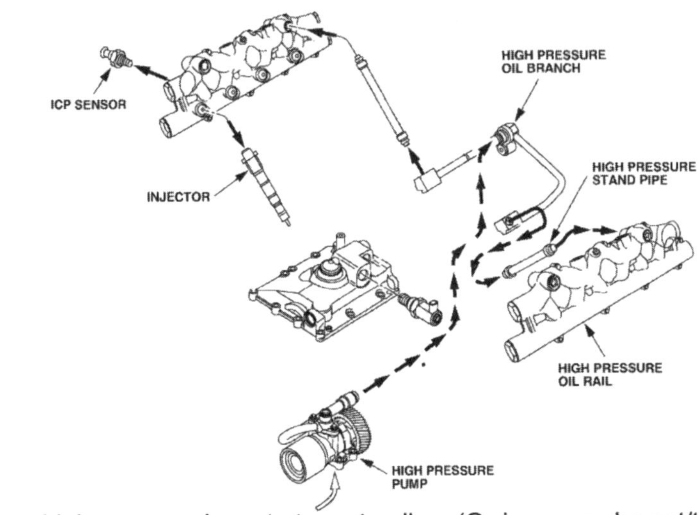

- Standpipe from high pressure hose to tappet gallery (O-rings may be cut/torn or missing).
- O-ring between discharge tube and high pressure pump.
- O-ring around discharge tube that fits inside high pressure pump cover.
- High pressure inlet O-ring.
- Branch tube and O-ring from high pressure pump to stand pipe in tappet gallery.
- O-ring on the top of the injector where the high pressure rail seats into the injector.

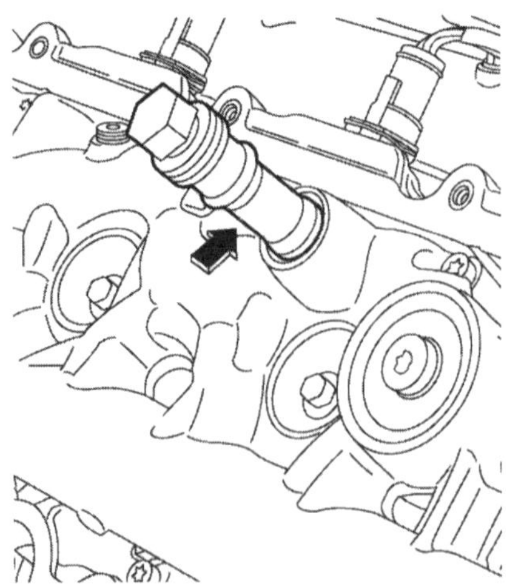

Left Cylinder Head Check (E-Series or F-Super Duty/Excursion — Late Build)
CAUTION: The engine may start.
CAUTION: Oil is under high pressure.

1. Remove the LH valve cover.
2. Remove the existing pipe plug.
3. Install the Oil High Pressure Leakage Test Adapter 303-1071/2, or equivalent, (solid block-off tool).
4. Access and monitor the ICP PID while cranking the engine.
5. If the engine starts or if injection control pressure is now within specification, the injection control pressure leak has been isolated to the left cylinder head. If the pressure is not within specification, remove the test adapter (303-1071/2), reinstall the original pipe plug and perform the right cylinder head check.

10e. and 10f. Monitor The ICP While Cranking (Continued)

Right Cylinder Head Check (E-Series or F-Super Duty/Excursion — Late Build)

CAUTION: The engine may start.
CAUTION: Oil is under high pressure.

1. Remove the RH valve cover.
2. Remove the existing pipe plug.
3. Remove the ICP sensor from the high pressure rail.
4. Install the ICP sensor into the Oil High Pressure Leakage Test Adapter 303-1071/1, or equivalent, (open-ended block-off tool).
5. Install the Oil High Pressure Leakage Test Adapter 303-1071/1, or equivalent.
6. Connect the ICP/EBC Adapter Cable 418-D003, or equivalent, to the wiring harness and the ICP sensor.
7. Access and monitor the ICP PID while cranking the engine.

If the engine starts or if injection control pressure is now within specifications, the injection control pressure leak has been isolated to the right cylinder head. If the pressure is not within specifications, perform the IPR and high pressure pump test.

IPR And High Pressure Pump Test (E-Series or F-Super Duty/Excursion — Late Build)
If injection control pressure is still low after ruling out both cylinder heads as the source of injection control pressure leakage, perform the following to isolate the cause.

1. Reinstall the left cylinder head Oil High Pressure Leakage Test Adapter 303-1071/2, or equivalent, (solid block-off tool).
2. With the high pressure pump effectively deadheaded, crank the engine and monitor the ICP PID.
3. If a low pressure condition still exists, the problem is most likely with the high pressure pump or the high pressure drive gear. If the injection control pressure is now within specifications, inspect all components carrying high pressure oil on both engine banks.
4. The leak could also be in the discharge tube, branch tube, or stand pipes. Perform the air pressure check if this procedure has not already been completed. If the concern is related to a leaking standpipe, branch tube or discharge tube, this test should identify the leak.

Tools Required:
- ICP/EBC Adapter Cable 418-D003 (D94T-50-A) or equivalent
- Oil High Pressure Leakage Test Adapter Set 303-756 or equivalent
- Oil High Pressure Leakage Test Adapter Set 303-1071 or equivalent
- Multimeter 105-00050 or equivalent
- Quick Disconnect Tool 303-755 or equivalent

10g. Check Fuel Pulse Width While Cranking

Purpose:
The purpose of this test is to verify the fuel delivery signal is correct.

Scan Tool — Data List Monitoring
- Scan tool may reset below 9.5 volts.
- Select the parameters indicated from the scan tool parameter list and monitor while cranking engine

Parameter	Specification	Measurement
FUEL PW	5 m S to 2 mS	

Pulse width defaults to 0 with no CMP or CKP signal.

10g. Check Fuel Pulse Width While Cranking (Continued)

Recommended Procedure:

1. Connect the scan tool.
2. Key on, engine off.
3. Access and monitor the FUEL PW PID while cranking the engine.
4. No fuel command signal when the ICP, RPM and VPWR signals are correct usually indicates a
5. loss of the CMP signal. GO to Pinpoint Test **V**.

A 5 m s to 2 ms fuel pulse width (FUEL PW) will be sent to the FICM if the system voltage is greater than 8 volts during cranking, engine cranking speed is above 100 RPM, injection control pressure is above 3.5 MPa (500 psi) and is in sync. Note that low fuel pressure or no glow plug operation could still be the cause of the No Start or Hard Start condition. A 0 ms fuel pulse width (a no fueling pulse) will be sent by the PCM when a sync pulse has not been received from the CMP sensor and if insufficient injection control pressure is present.

Possible Causes:

- PCM
- FICM
- CKP signal
- CMP signal

Tools Required:

- Scan tool

10h. Check FICM Synchronization While Cranking

Purpose:
The purpose of this test is to verify the PCM and FICM synchronization.

Scan Tool — Data List Monitoring

- Scan tool may reset below 9.5 volts.
- Select the parameters indicated from the scan tool parameter list and monitor while cranking the engine.

Parameter	Spec.	Measurement
FICMSYNC	YES/NO	

No synchronization could be caused by a CMP or CKP fault.

Recommended Procedure:

1. Connect the scan tool.
2. Key on, engine off.
3. Access and monitor the FICMSYNC PID while cranking the engine.
4. No FICM synchronization while cranking the engine with the ICP, RPM and VPWR signals correct usually indicates a loss of the CMP or CKP synchronization signal. Refer to the appro priate pinpoint test for diagnosis. Refer to Control System Diagnostic Sheet Reference for normal operating values.

Possible Causes:

- PCM
- FICM
- CKP synchronization signal
- CMP synchronization signal

Tools Required:

- Scan tool

11. Glow Plug System Operation

Purpose:
The purpose of this test is to verify the glow plug system operation.

Glow Plug System Operation

Glow Plug Control Module (GPCM) Operation

1. Glow plug ON time is dependent on oil temperature and altitude. The GPCM will command the glow plugs on for 1 to 120 seconds. The GPCM does not operate if the oil temperature is above 55°C (131°F).
2. Connect the scan tool. Access and retrieve the KOEO and continuous DTCs. If GPCM DTCs are present, GO to Pinpoint Test AF .
3. Verify B+ voltage is supplied to the GPCM.
4. Access and monitor the GPCTM and EOT PIDS to verify sufficient glow plug "ON" time.
5. Turn the key to the ON position and measure the glow plug voltage ("ON" time).

Note: The Wait To Start indicator "ON" time (1-10 seconds) is independent from glow plug "ON" time.

On Time	Specification	Measurement
1 to 120 seconds	B+	

Glow Plug Resistance

1. Disconnect the glow plug bus bar connector.
2. Measure the resistance between the glow plug bus bar connector, component side and battery ground.
3. Disconnect the GPCM.
4. Measure the resistance between the GPCM connector, harness side and the glow plug bus bar connector, harness side.

Glow Plug Number	Glow Plug to Ground (0.1 to 2 ohms)	Glow Plug Connector to GPCM Connector (less than 5 ohms)
1		
3		
5		
7		
2		
4		
6		
8		

Recommended Procedure:
Note: Check for poor connections or loose fitting pins.
Note: Incorrect measurements will result if all glow plug connectors are not disconnected.

The glow plug monitor self-test is a KOER functional test of the glow plug system. The self-test is performed on-demand with the engine running and the A/C off. The PCM will activate the GPCM which monitors the glow plugs. The pedal may be used to increase the engine speed to increase voltage if needed. For fault detection, the self-test requires a fault present at the time of testing.

11. Glow Plug System Operation (Continued)

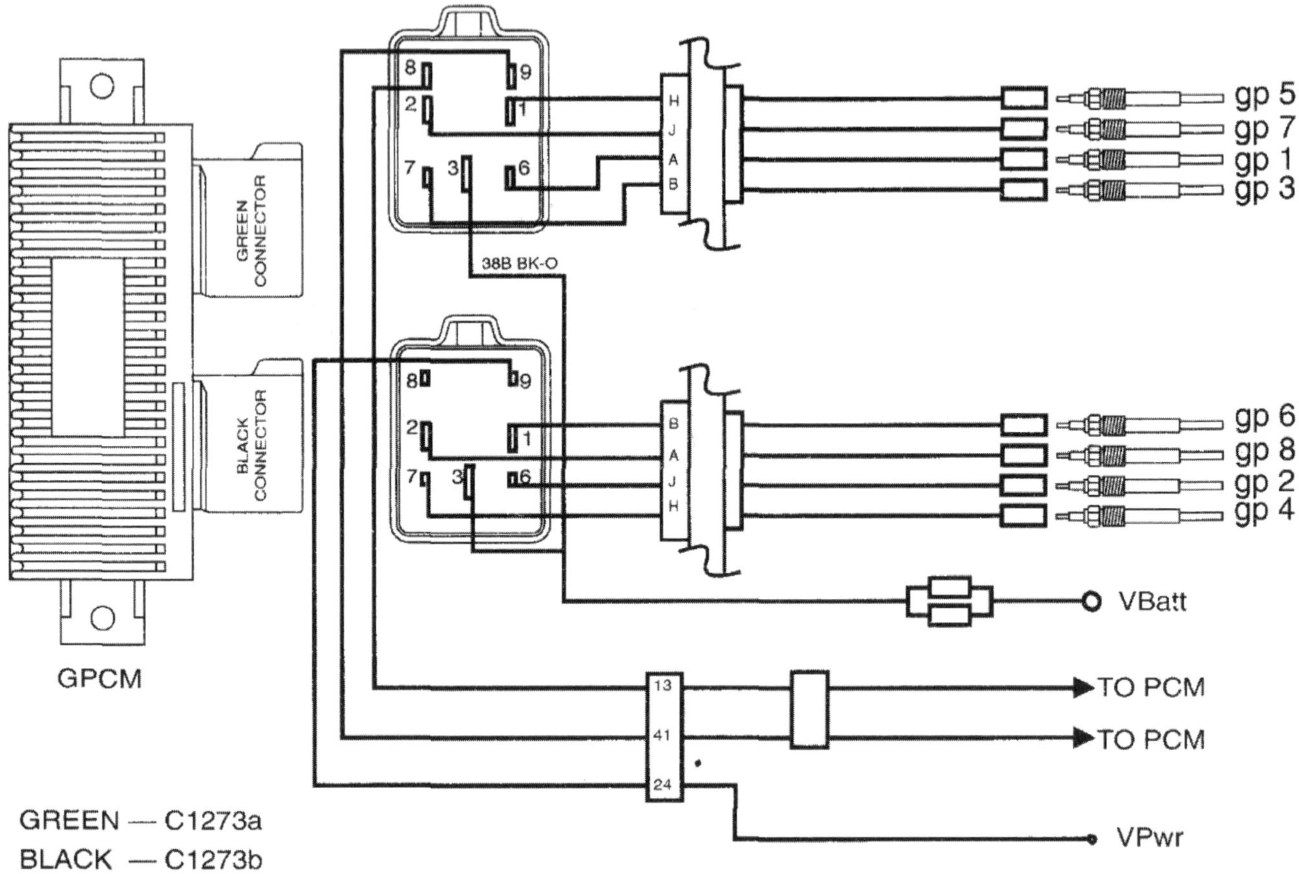

Possible Causes:

- GPCM power circuits
- Glow plugs
- Glow plug bus bar
- GPCM
- Circuitry

Tools Required:

- Multimeter 105-00050 or equivalent
- Scan tool

1. Visual Engine/Chassis Inspection

Purpose:
The purpose of this test is to check the general condition of the engine and chassis.

Visual Engine/Chassis Inspection
- Verify that there are no fluid or pressure leaks.
- Inspect all wire connections for damage.
- Inspect the MAP hose, intercooler hoses and intake for leaks.

Fuel, Oil, Coolant, Electrical, Hoses ,Leaks

Method	Check
Visual	

Recommended Procedure:
1. Inspect for a hole in the MAP sensor hose or a pinched hose.
2. Inspect the fuel system, including the fuel tank, fuel pump, fuel filter housing and fuel lines, for kinks, bends or leakage.
3. Inspect for coolant leaks at the radiator and coolant hoses. Check the coolant level.
4. Inspect wiring for correct routing, and verify no rubbing or chafing has occurred.
5. Inspect all sensors, and verify the connectors are properly secured.
6. Inspect the intercooler hoses for leaks.

Possible Causes:
- Loose or leaking fuel lines
- Kinked or blocked fuel lines
- Fuel or oil leaks
- Coolant leaks could indicate engine problems
- Pinched or open MAP sensor hose
- Pinched or open intercooler hose
- Improper connections

Tools Required:
- Inspection light

2. Sufficient Clean Fuel

Purpose:
The purpose of this test is to verify the fuel quality.

Sufficient Clean Fuel
- Check for illumination of the WATER IN FUEL indicator.
- Drain a fuel sample from the fuel control module.
- A cetane rating between 40 and 50 is recommended for optimum performance.

Method	Check
Visual	

2. Sufficient Clean Fuel (Continued)

Recommended Procedure:
Open the drain valve on the fuel control module and fill a clear container until it is half full. Close the drain valve.

Observe the WATER IN FUEL indicator. If the indicator is illuminated, the fuel is probably contaminated with water.

Flow out of the drain should be a steady stream. Insufficient flow could indicate fuel supply or fuel system problems.

Inspect the fuel in the container. It should be clear, not cloudy. It also should be free of water and contaminants. Dyed red or blue fuel indicates off-highway fuel.

Some sediment and water may be present in the fuel sample if the fuel filter has not been serviced for a prolonged period of time and/or if the sediment and water have not been drained recently. If that is the case, a second sample may be required to determine fuel quality.

F-SuperDuty/Excursion

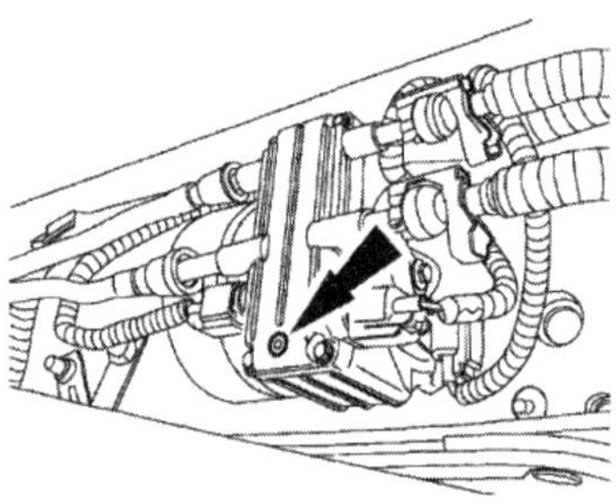

E-Series

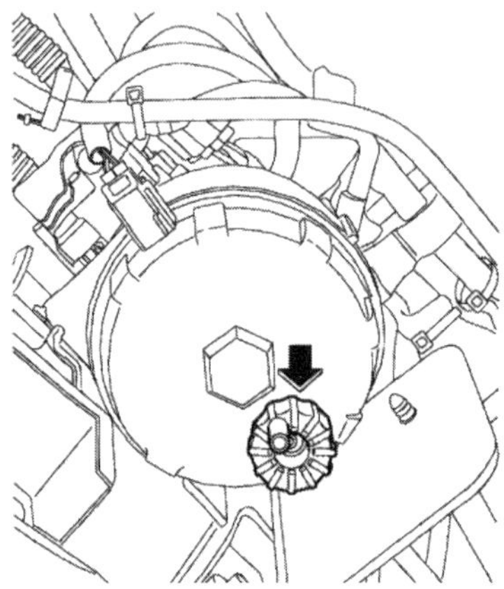

Possible Causes:

- No fuel in tank
- Fuel supply line could be broken or crimped
- Fuel could be jelled (most likely in cold weather with No. 2 fuel)
- Pickup tube screen in tank could be clogged

Cloudy fuel indicates that the fuel may not be a suitable grade for cold temperatures.

Excessive water or contaminants may indicate that the tank and fuel system may need to be flushed and cleaned.

Tools Required:

- Clear container — approximately 0.95L (1-quart)

3. Check Engine Oil Level

Purpose:
The purpose of this test is to verify oil quality and determine if there is sufficient oil to operate the injectors.

Check Engine Oil Level

- Check for contaminants (fuel, coolant).
- Correct grade/viscosity.
- Miles/hours on oil, correct level.

3. Check Engine Oil Level (Continued)

Method	Check
Visual	

Recommended Procedure:
Check for correct oil level using the dipstick with the vehicle on level ground. If there is no oil or very little oil in the crankcase, the injectors will not operate.

If the oil level is overfull, it is possible the engine was incorrectly serviced or fuel is diluting the oil and filling the crankcase.

Inspect oil for color. A milky white oil indicates possible coolant contamination which will have an ethylene glycol odor.

WARNING: SMOKING OR OPEN FLAME OF ANY TYPE MUST NOT BE PRESENT WHEN WORKING NEAR FUEL OR FUEL VAPOR. FAILURE TO FOLLOW THESE INSTRUCTIONS MAY RESULT IN PERSONAL INJURY.

Oil contaminated with diesel fuel will have a diesel fuel odor and will increase the engine oil level. If the engine oil level is above "Max" due to diesel fuel dilution, the oil will appear thin and watery.

If the oil level is overfilled, drain the oil and the fuel filter housing. Isolate either cylinder head by removing the corresponding fuel line from the fuel filter housing. Remove the fuel pressure test port plug from the secondary filter housing. Install the plug in the outlet port. Install Fuel Pressure Adapter (303-765) and Gauge 0-1.1 MPa (0-160 psi) Bar 014-00761 or equivalent at the test port to confirm constant fuel pressure.

Using the scan tool, access OSC and command the fuel pump ON. Watch for fuel to drain out of the oil pan drain hole. Depending on the severity of the leak, it may take some time before a leak is noticeable. Remove the valve cover at the suspect cylinder head and inspect the injector area for leaks.

Check service records for correct oil type and viscosity for the vehicle operating temperature. Single weight or 15W-40 oil is not recommended for cold ambient temperatures. 10W-30 oil is recommended for cold ambient temperatures. Oil that has had extended drain intervals will have increased viscosity (become thicker) and will make engine cranking more difficult and starting less reliable at temperatures below freezing. Refer to the lube oil chart for the correct oil selection for temperature conditions.

Possible Causes:
- Oil level low — oil leak, oil consumption, incorrect servicing
- Oil level high — incorrect servicing, fuel dilution from injector O-rings
- Oil contamination with coolant — oil cooler, head gasket, porosity

Tools Required:
- Adapter 303-765 or equivalent
- Gauge 0-1.1 MPa (0-160 psi) Bar (part of Pressure Adapter Kit 014-00761 or equivalent)
- Scan tool

4. Perform KOEO On-Demand Self-Test

Purpose:
The purpose of this test is to determine if the PCM has detected any fault conditions that would cause a performance problem.

Perform KOEO On-Demand Test

- Use the scan tool.
- DTCs set during this test are current faults.

Diagnostic Trouble Codes (DTCs)	

Recommended Procedure:
Note: To ensure that the DTC is a hard fault, first clear Continuous DTCs (be sure to record all DTCs and Freeze Frame information before clearing). Rerun KOEO On-Demand Self Test. If the DTC is set again, a hard fault has occurred.

Connect the scan tool. Turn off all accessories. If the vehicle is equipped with an auxiliary powertrain control module (APCM), it must be turned off to perform the self-test.

1. Perform the necessary vehicle preparation and a visual inspection. Refer to Quick Test Operation .
2. Refer to the scan tool operating manual for instructions.
3. Key on, engine off.
4. Wait 4 seconds for the transmission control module (TCM), powertrain control module (PCM) and the fuel injector control module (FICM) to initialize.
5. Follow the operating instructions from the diagnostic menu.
6. Perform a KOEO On Demand Self-Test.
7. Record the DTCs and Freeze Frame information and refer to the appropriate pinpoint test.

Tools Required:

- Scan tool

5. Retrieve Continuous DTCs

Purpose:
The purpose of this test is to determine if the PCM has detected any historical or intermittent fault conditions that would cause a performance problem. The condition that caused a continuous DTC may no longer exist.

Retrieve Continuous DTCs

- Use the scan tool.
- DTCs retrieved during this test are historical faults.

Diagnostic Trouble Codes (DTCs)	

Recommended Procedure:
Connect the scan tool. Turn off all accessories. If the vehicle is equipped with an APCM, it must be turned off to perform self-tests.

5. Retrieve Continuous DTCs (Continued)

1. Perform the necessary vehicle preparation and a visual inspection. Refer to Quick Test Operation .
2. Refer to the scan tool operating manual for instructions.
3. Key on, engine off.
4. Follow operating instructions from the menu.
5. Record the continuous DTCs and Freeze Frame information from the PCM and TCM and refer to the appropriate pinpoint test for continuous DTC diagnostics.
   Continuous DTCs must be cleared after a repair is made.

Tools Required:
Scan tool

6. KOEO Injector Electrical Self-Test (Click Test)

Purpose:
Note: If unable to perform KOEO Injector Electrical Self-Test (Click Test), disconnect the FICM connector and check injectors for shorts or opens.

The purpose of this test is to determine if the injector solenoids and valves are functioning by clicking all injectors together and then each injector in numerical sequence (1 through 8).

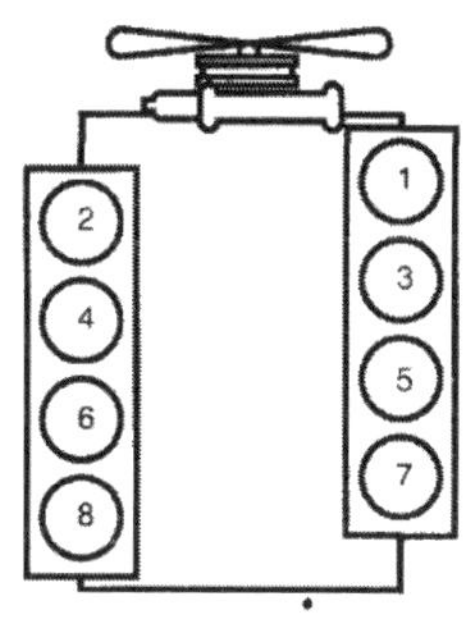

FIRING ORDER
1-2-7-3-4-5-6-8

KOEO Injector Electrical Self-Test (Click Test)

1. Use the scan tool. Injector DTCs will be transmitted after the self-test is completed.
2. All injectors will momentarily click, then each injector will click in sequence 1 through 8.

Note: Sequence repeats three times.

NOTE: DTCs can be historical if not cleared from a previous test.

Diagnostic Trouble Codes (DTCs)	

Recommended Procedure:
This test determines if the injector circuits and solenoids are electrically operating without fault. All injectors will first click together for approximately 2 seconds, then each injector will click for approximately 1 second in numerical order (1 through 8). If a fault is detected, a DTC will be output on the data link at the end of the test when requested by a scan tool. Only a hard fault DTC will be displayed.

Connect the scan tool. Turn off accessories. If vehicle is equipped with an APCM, it must be turned off to perform self-tests.

1. Perform the necessary vehicle preparation and a visual inspection. Refer to Quick Test Operation .
2. Refer to the scan tool operating manual for instructions.
3. Key on, engine off.
4. Follow the operating instructions from the menu.
5. Perform the KOEO Injector Electrical Self-Test (Click Test).
6. Record the DTCs and perform the appropriate pinpoint test.

Possible Causes:

- Open or shorted injector circuit
- Injector connector
- Injector solenoid
- FICM power or ground circuit
- FICM

Tools Required:

- Scan tool

7. Intake Restriction

Purpose:
This purpose of this test is verify an air intake or exhaust restriction is not contributing to a low power condition. If the engine does have a high air intake restriction, a considerable amount of black or blue smoke may be produced.

Intake Restriction
- Check air filter restriction gauge/indicator.
- Measure the vacuum on clean side of air inlet system at WOT with Magnehelic® gauge, or equivalent.

Instrument	Specification	Measurement
Magnehelic/Filter Restriction Gauge	3.7-46.7 mm Hg (2-25 H 2 O)	

Recommended Procedure:
Inspect the air cleaner inlet and ducting to ensure that it is not blocked or collapsed. Inspect the air cleaner housing and filter for proper installation.

Inspect for any indication of water intrusion into air cleaner/filter.

If necessary, use Pressure Adapter Kit 014-00761 or equivalent to install a Magnehelic® gauge on the port on the air cleaner and measure restriction at high idle.

F-SuperDuty/Excursion

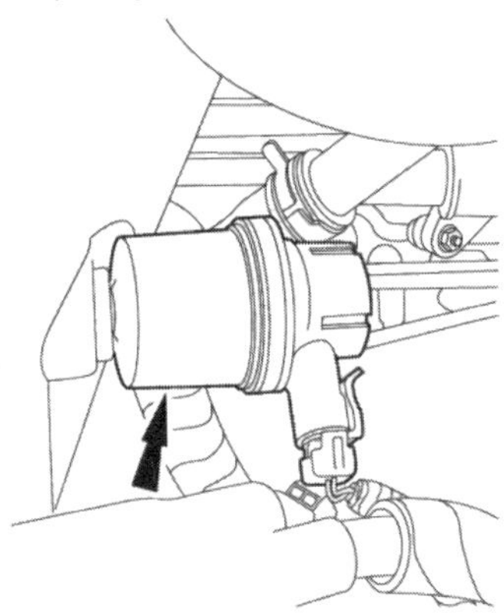

E-Series

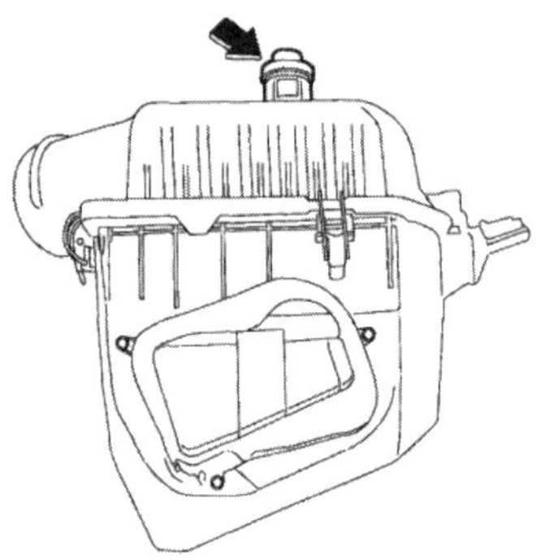

Magnehelic® Gauge

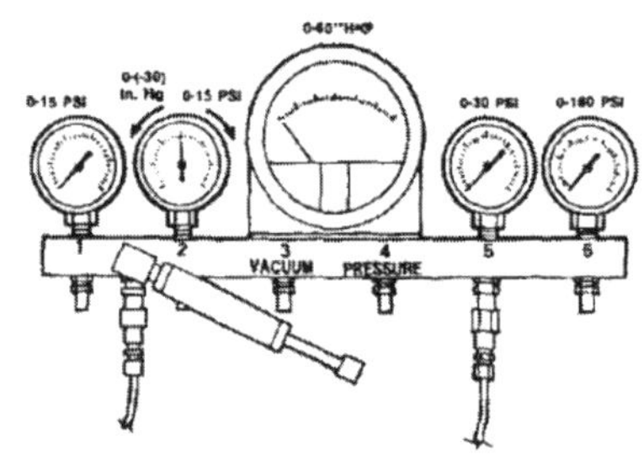

Possible Causes:
- Snow, plastic bags or other foreign material may restrict airflow at the air inlet
- Misrouted air cleaner ducting
- On engines recently repaired, rags or cap plugs may have been inadvertently left in an air inlet pipe

Tools Required:
Magnehelic® gauge (part of Pressure Adapter Kit 014-00761 or equivalent)

8. EGR Position

Purpose:
The purpose of this test is to check the EGR valve closed, open, and travel positions.

EGR Position
1. Key on, engine off.
2. Use the diagnostic tool to command the output state control (OSC) for the EGR valve.
3. Monitor the EGR valve position sensor PID to calculate travel

8. EGR Position (Continued)

Instrument	Specification	Actual
Diagnostic Tool	0% (Closed) 0.4 - 1.2 volts	____ Closed
	100% (Open) 3.3 - 4.7 volts	____ Open
	Travel greater than 2.71 volts	____ Travel

Correct concern causing out of specification values before continuing.

Recommended Procedure:

1. Connect the diagnostic tool.
2. Key on, engine off.
3. Access the EGR valve position sensor PID and record the closed EGR valve position voltage.
4. Access OSC and open the EGR valve 100%.
5. Record the EGR valve open position voltage.
6. Subtract the closed EGR valve position voltage from the open EGR valve position voltage.
7. If the valve travel and travel voltage are below specification, install a new EGR valve and retest the system.

Possible Causes:

- EGR control circuit
- EGR valve
- PCM

Tools Required:
Scan tool

9. Exhaust Restriction

Purpose:
The purpose of this test is to determine if an exhaust system restriction is causing a performance problem.

Exhaust Restriction

1. Visually inspect the exhaust system for damage.
2. Monitor the EP PID with scan tool and the engine temperature greater than 70°C (158°F) at 3,800 RPM with the vehicle in PARK/NEUTRAL.

Parameter	Model(s)	Specification	Measurement
EP	(F-Super Duty/ Excursion — Early Build)	234 kPa (34 psi) MAX @ 3,800 RPM	
	(E-Series or F-Super Duty/ Excursion — Late Build)	244 kPa (35 psi) MAX @ 3,800 RPM	

Recommended Procedure:

1. Connect the scan tool. Access and monitor the EP and RPM PIDS.
2. Run engine at 3,800 RPM.
3. Record the EP PID value.
4. A low RPM or an EP PID greater than 234 kPa (34 psi) (F-Super Duty/Excursion — Early Build) or 244 kPa (35 psi) (E-Series or F-Super Duty/Excursion — Late Build) indicates a restricted exhaust condition.

9. Exhaust Restriction (Continued)

Possible Causes:

- Collapsed tail pipe
- Clogged tail pipe
- Clogged catalytic converter
- Damaged muffler
- Turbocharger

Tools Required:

- Scan tool

10a. Electric Fuel Pump Pressure

Purpose:
The purpose of this test is to verify the fuel system pressure.

Electric Fuel Pump Pressure

- Measure the fuel system pressure at the engine filter housing test port.
- Road test the vehicle with the engine at full load condition.

Instrument	Specification	Measurement
0-1.1 MPa (0-160 psi) gauge	310-379 kPa (45-55 psi) minimum	

If pressure fails low, go to step 10b.
If fuel pressure is above specification, check fuel return lines for restriction.
If no restriction is present, install a new fuel pressure regulator valve.

Recommended Procedure:

1. Verify there is fuel in the tank.
2. Remove the plug from the front of the engine fuel filter housing. Install Adapter 303-765 and Gauge 0-1.1 MPa (0-160 psi) Bar 014-00761 or equivalent.
3. Road test the vehicle at a full load condition.
4. If pressure fails low, go to step 10b.

If fuel pressure is above specification, check the fuel return lines for a restriction.
If no restriction is present, install a new fuel pressure regulator vave.

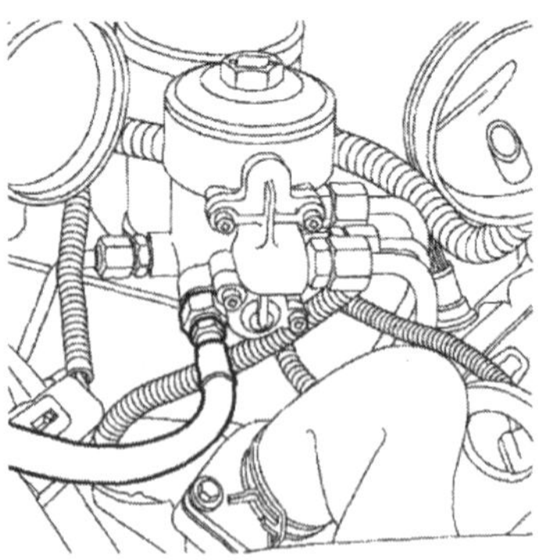

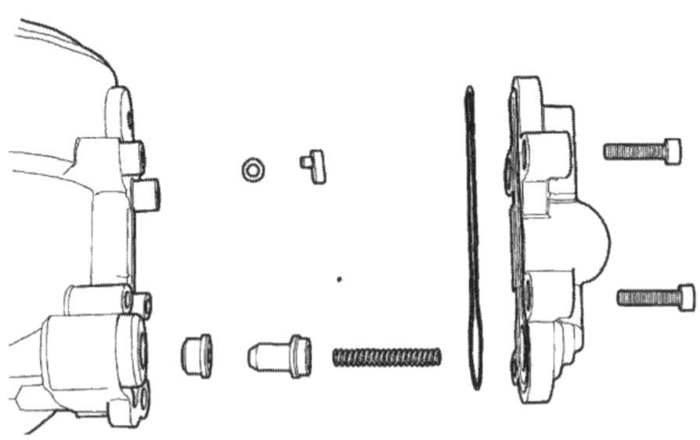

Possible Causes:

- Fuel return line restriction
- Fuel pressure regulator valve

Tools Required:

- Adapter 303-765 or equivalent
- Gauge 0-1.1 MPa (0-160 psi) Bar (part of Pressure Adapter Kit 014-00761 or equivalent)

10b. Electric Fuel Pump Inlet Restriction

Purpose:
The purpose of this test is to isolate the cause of low fuel pressure.

Electric Fuel Pump Inlet Restriction
Measure restriction at fuel pump inlet.

Instrument	Specification	Measurement
(0-30 in Hg) Vacuum gauge	6 in Hg MAX	

Recommended Procedure:
Remove the fuel line to the inlet side of the fuel pump. Install Fuel Pump Adapter 310-111 or equivalent between the fuel inlet line and the electric fuel pump. Connect test adapter to a 0-762 mm Hg (0-30 in Hg) vacuum gauge. Measure restriction at WOT (maximum engine speed out of gear with the brakes set and the wheels blocked). If restriction is greater than 152 mm Hg (6 in Hg), there is a restriction between the fuel pump and the fuel tank. If restriction is less than 152 mm Hg (6 in Hg), inspect both fuel filters. If the filters are OK, inspect the fuel regulator valve. If the regulator and filters are OK, install a new fuel pump.

Possible Causes:

- Fuel line restriction
- Fuel pressure regulator valve
- Fuel filters
- Fuel pump

Tools Required:

- Fuel Pump Adapter 310-111 or equivalent
- Vacuum Gauge (part of Pressure Adapter Kit 014-00761 or equivalent)

10c: Fuel Aeration Test

Purpose
To determine if the fuel is aerated and causing a poor idle condition.

Fuel Aeration Test

- Install a clear hose on the fuel return line at the fuel control module.
- Run at the engine at WOT for 2 minutes.
- Return fuel should be free of bubbles.

Method	Check
Visual	

1. Remove the fuel return line at the fuel control module.
2. Attach a 1/4-inch diameter hose (clear) to the fuel control module and place the other end of the hose into Rotunda Tool 034-00005 or equivalent.
3. Turn the ignition on. The fuel pump will cycle on, then off. Diesel fuel may not flow out through the attached hose after only one key cycle. Repeat the key cycle until fuel flows from the attached hose.
4. Start the engine.
5. Run at WOT for 2 minutes. Return fuel should be free of bubbles.
6. If the fuel has air in it, the fuel returning to the tank will appear white, foamy or non-transparent. If no or very little air is present in the fuel, the fuel will appear clear/transparent.
7. If air is present in the return fuel, inspect the fuel system for leaks.

10c: Fuel Aeration Test (Continued)

Possible Causes:
Note: Air can enter the fuel system at any point where there is suction in the fuel system.

- Fuel pickup in the fuel tank
- Fuel supply line entering the fuel module from the fuel tank
- Fuel return line entering the fuel module from the engine

Tools Required:

- Rotunda Tool 034-00005 or equivalent

11. Perform KOER On-Demand Self Test

Purpose:
The purpose of this test is to determine if the PCM has detected any fault conditions that would cause a performance problem while the engine is running. The KOER self-test will perform step tests on the injection control pressure system and the exhaust back pressure system.

Step tests are PCM controlled tests where the PCM commands a specific exhaust back pressure or injection control pressure and then measures the result. If a predetermined threshold is not reached, a DTC will be generated. This test can be performed at any engine temperature.

Perform KOER On-Demand Self Test
Note: Before performing the KOER On-Demand Self-Test make sure the high pressure oil passages are free of air. A recent repair to the system may allow air to enter the oil passages. To remove the air, drive the vehicle (with deceleration and acceleration cycles) at highway speeds for 32 km (20 miles).

The self-test will verify the ICP, EGR and VGT systems performance.

Diagnostic Trouble Codes (DTCs)	

Recommended Procedure:
Note: Engine will run rough during this test.
Connect the scan tool. Turn off all accessories. If the vehicle is equipped with an APCM, it must be turned off to perform the self-test.

1. Perform the necessary vehicle preparation and visual inspection. Refer to Quick Test Operation .
2. Refer to the scan tool operating manual for instructions.
3. Key on, engine running.
4. Follow the operating instructions from the diagnostic menu.
5. Perform a KOER On Demand Self-Test.
6. Record the DTCs and refer to the appropriate pinpoint test.

Tools Required:

- Scan tool

12a. Low Idle Stability (ICP Pressure)

Purpose:
The purpose of this test is to determine if idle stability or low power is caused by a stuck or dirty IPR or faulty ICP signal.

Low Idle Stability (ICP Pressure)

- Check at low idle and EOT above 70°C (158° F)
- Monitor the ICP and RPM PIDS with the scan tool.

12a. Low Idle Stability (ICP Pressure) (Continued)

Parameter	Models	Specification @ 670 RPM	Measurement
	Take reading before disconnecting ICP sensor.		
ICP	F-Super Duty/ Excursion — Early Build	4.5 - 5.5 MPa ± 0.3 MPa (650-800 psi ± 45 psi)	
	E-Series or F-Super Duty/ Excursion — Late Build	4.0 - 5.0 MPa ± 0.3 MPa (580-725 psi ±45 psi)	

If engine RPM is unstable, disconnect the ICP sensor.

ICP will default to 5.0 MPa (725 psi) when disconnected.
If RPM smooths out, the ICP sensor or circuit is suspect.
If RPM is still unstable, reconnect the ICP sensor and go to step 12b.

Recommended Procedure:
Note: The engine must be at operating temperature.

1. Connect the scan tool.
2. Key on, engine running.
3. Access and monitor the ICP and RPM PIDS.
4. Operate the engine at low idle speed.
5. If engine does not stabilize, disconnect the ICP sensor. If low idle speed stabilizes with the ICP sensor disconnected, the problem is most likely in the ICP sensor circuit. GO to Pinpoint Test **Q** If RPM does not stabilize, reconnect the sensor and GO to 12b.

Possible Causes:

- Debris stuck in the IPR
- ICP sensor
- Circuitry

Tools Required:

- Scan tool

12b. Injection Pressure Regulator Test

Purpose:
The purpose of this test is to determine if idle stability or low power is caused by a stuck or dirty IPR.

Low Idle Stability (ICP Pressure)

- Check at low idle and EOT above 70° C (158° F)
- Monitor the IPR PID with the scan tool.

Parameter	Specification @ 670 RPM	Measurement
IPR	30% MAX	

Recommended Procedure:
Note: The engine must be at operating temperature.

1. Connect the scan tool.
2. Key on, engine running.
3. Access and monitor the IPR and RPM PIDS.
4. Operate the engine at low idle speed.

If the IPR duty cycle is less than the MAX specification, go to the next step. If the IPR duty cycle is greater than the MAX specification, go to Test 10e of the Hard Start/No Start Diagnostics and check for an jection control system leak.

12b. Injection Pressure Regulator Test (Continued)

Possible Causes:

- Debris stuck in the IPR
- Injection control system leak

Tools Required:

- Scan tool

13. Boost Pressure Test

Purpose:

The purpose of this test is to determine if the engine can develop sufficient boost to obtain specific power.

Boost Pressure Test

1. Carefully inspect the intercooler tubes/connections, turbocharger connections, exhaust system and MAP hose for signs of damage or leaks.
2. Carry out the boost pressure test at 1,200 RPM.
3. Monitor the EGR duty cycle, MGP, RPM, and VGT duty cycle PIDS with the diagnostic tool.
4. Control the engine speed to 1,200 RPM and EGR duty cycle to 0% with the diagnostic tool.
5. Control the VGT duty cycle to 0% and 85% with the diagnostic tool.

Parameter	Specification	Measurement
MGP	3 kPa (0.44 psi) maximum at 0% VGT duty cycle	
MGP	6 kPa (0.87 psi) minimum at 85% VGT duty cycle	

Recommended Procedure:

Note: If the test fails, GO to Pinpoint Test KA.

1. Key on, engine running.
2. Access and monitor the MGP, EGR_DC, RPM, and VGTDC PIDS.
3. Control the RPM_DSD PID to 1,200 RPM and EGRDC PID to 0%.
4. Decrease the VGTDC PID to 0% and record the MGP PID value.
5. Increase the VGTDC PID to 85% and record the MGP PID value.
6. Decrease the VGTDC PID to 0% and record the MGP PID value.
7. If the MGP PID values are within specifications, go to the next step. If the MGP PID values are not within specifications, GO to Pinpoint Test KA.

Possible Causes:

- MAP hose pinched or open
- Leaking intake, hoses or fittings
- Leaking exhaust system
- Turbocharger
- Base engine
- Intercooler hoses leaking
- Intercooler
- EGR sticking

Tools Required

- Diagnostic tool

14. Crankcase Pressure Test

Purpose:
The purpose of this test is to measure the crankcase pressure. Crankcase pressure is a measure of how well the cylinders are sealing.

Crankcase Pressure Test - F-Super Duty/Excursion

- Measure the pressure at the oil fill tube with 6.0L Crankcase Pressure Tester p/n 303-758, with engine at 70° C (158° F) minimum.
- Block the breather tube on the left valve cover.
- Measure with no load at 3,000 RPM.

Instrument	Specification.	Measurement
0-112 mm Hg (0 to 60 H_2O) Magnehelic® gauge	Less than 15 mm Hg (8 H_2O)	

If greater than 15 mm Hg (8 H 2 O), diagnose and repair base engine concern.

Recommended Procedure - F-Super Duty/Excursion:
Note: Do not plug hole on Crankcase Pressure Test Adapter 303-758.

F-SuperDuty/Excursion

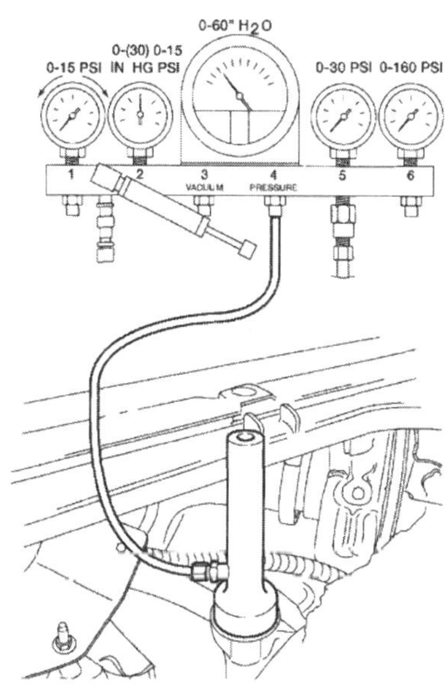

Make sure the engine is at operating temperature. A cold engine will give higher readings. Remove the ducting to the turbocharger inlet pipe and remove the inlet pipe and elbow that connects to the breather box. Block the outlet at the valve cover with the cap provided in Pressure Test Adapter Kit 014-00761 or equivalent. Install a protective screen over the turbocharger inlet.

Screw the Crankcase Pressure Test Adapter 303-758 in the oil fill cap hole. Plumb to the Magnehelic® gauge in the gauge block. Make sure the Magnehelic® gauge has been zeroed.

Start the engine and operate at 3,000 RPM. Hold for 30 seconds minimum and take a stabilized reading. Do not block the hole at the top of the restrictor tool.

Crankcase Pressure Test - E-Series
Measure the pressure at the oil fill tube with 6.0L Crankcase Pressure Tester p/n 303-758, with engine at 70°C (158°F) minimum.
Block the breather tube on the left valve cover.
Measure with no load at 3,000 RPM.

E-Series

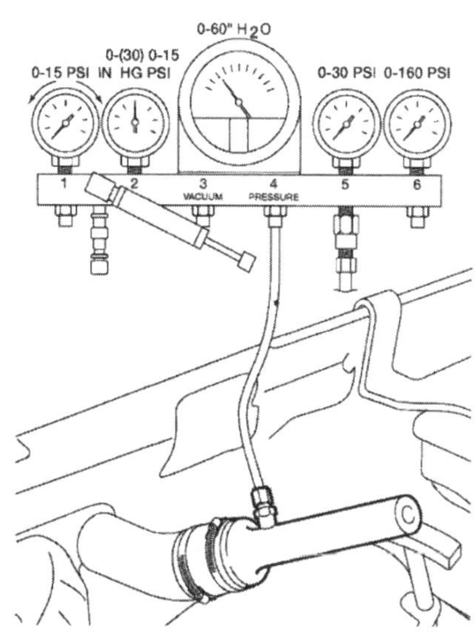

Recommended Procedure - E-Series:
Note: Do not plug hole on Crankcase Pressure Test Adapter 303-758.

Make sure the engine is at operating temperature. A cold engine will give higher readings. Remove the air cleaner assembly. Remove the ducting to the turbocharger inlet pipe. Remove the inlet pipe and elbow that connects to the breather box. For additional information, refer to Workshop Manual Section 303. Block the outlet at the valve cover with the cap provided in Pressure Test Adapter Kit 014-00761 or equivalent. Install a protective screen over the turbocharger inlet.

14. Crankcase Pressure Test (Continued)

CAUTION: Excessive screw clamp torque may damage the oil fill tube.

Install the Crankcase Pressure Test Adapter 303-758 to the oil fill tube using an appropriate diameter hose and secure with screw clamps. Start by connecting one end of the hose to the knurled surface of Crankcase Pressure Test Adapter 303-758 and the opposite end of the hose to the oil fill tube and secure with screw clamps.

Plumb to the Magnehelic® gauge in the gauge block. Make sure the Magnehelic® gauge has been zeroed.

Start the engine and operate at 3,000 RPM. Hold for 30 seconds minimum and take a stabilized reading. Do not block the hole at the top of the restrictor tool.

Possible Causes:

- Broken or worn compression rings
- Polished cylinder bores
- Leaking or bent valves
- Inspect the air induction system. If the air induction system allows dirt to enter the cylinders, it will quickly "dust" the engine causing high crankcase pressure.

Tools Required:

- Crankcase Pressure Test Adapter 303-758 or equivalent
- Magnehelic® gauge (part of Pressure Test Kit 014-00761) or equivalent
- Protective screen
- Hose
- Screw clamps

15. Oil Aeration Test

Purpose:
The purpose of this test is to determine if the engine lube oil is aerated and causing poor idle quality.

Oil Aeration Test

1. Run engine at 3000 RPM for 1 minute.
2. Take an oil sample from the engine oil temperature (EOT) sensor port at idle.
3. Inspect the sample for presence of air bubbles.

Excessive oil aeration can be caused by depleted oil additives, pick-up tube leak, front cover seal leak, or upper pan seal leak.

Method	Check
Visual	

Recommended Procedure:
Note: The engine must be at operating temperature.

1. Remove the EOT sensor.
2. Install the High Pressure Pump Test Adapter 303-765 or equivalent and Gauge 0-1.1 MPa (0-160 psi) Bar 014-00761 or equivalent.
3. Increase the engine speed to 3,000 RPM and maintain for 1 minute.
4. Return the engine speed to idle.
5. Obtain an oil sample with the engine at idle.

15. Oil Aeration Test (Continued)

Possible Causes:

- Extended oil drain intervals — the anti-foam additives in the oil may be depleted from severe use or extended intervals.
- Air may be present due to recent engine repair on injection control pressure system. It may be necessary to operate the vehicle aggressively for 24-32 kilometers (15-20 miles) to remove excess air.
- Incorrect type or grade of oil.

Tools Required:

- Clear container — approximately 0.95L (1-quart)
- Gauge 0-1.1 MPa (0-160 psi) Bar 014-00761 or equivalent
- High Pressure Pump Test Adapter 303-765 or equivalent

Injector Interrupt for Enhanced Diagnostics

Diagnose Rough Idle Only
This test will help diagnose weak cylinders/injectors by allowing the diagnostic tool operator to interrupt injectors individually. This is carried out by using the diagnostic tool to interrupt the injector output to the engine cylinder that corresponds to the number on the injector parameter identification (PID).

A performance diagnostic sheet should be completed before using this tool. High injection control pressure (ICP) after the completion of the high idle test (test 12 on the performance diagnostics sheet to indicate oil aeration), low fuel pressure, the dual-mass flywheel, and high crankcase pressure are just a few of the possibilities that can cause a rough idle. After completing the performance diagnostic sheet, if there is no direction given for corrective action and rough idle is present, use the enhanced diagnostic ability of the Injector Interrupt.

Start the engine and monitor the engine oil temperature (EOT) PID. The EOT is to be maintained within -15°C (5°F) during the test. The test must be carried out while the rough idle is occurring. Record the mass fuel desired (MFDES) reading before interrupting any injectors; this will be the baseline for the rest of the test. The MFDES reading will not steady and will fluctuate rapidly. Take a five second snapshot of the PID information and record the high and low MFDES readings on the Injector Interrupt Tool Enhanced diagnostics sheet, using only MFDES numbers that have appeared at lease two times during the recording. After recording a baseline, interrupt each injector individually and repeat the above steps to obtain the high and low readings while each injector is interrupted. These numbers should be recorded on the Injector Interrupt Tool Enhanced diagnostic sheet.

The enhanced diagnostic sheet is intended to be photocopied and then completed for each vehicle that is tested. A completed performance diagnostic sheet and enhanced diagnostic sheet must accompany any returned warranty parts and cores.

The following example illustrates the use of a diagnostic tool to monitor the MFDES, as described in the paragraph above.

Injector Off	MFDES	Average MFDES	Deviation
Baseline = 0	10.1-10.1 mg	10.1	—
1	10.7-10.9 mg	10.8	0.7
2	10.6-10.9 mg	10.9	0.7
3	10.5-11.0 mg	10.8	0.7
4	9.8-10.5 mg (Low Cylinder Contributor)	10.2	0.1
5	10.5-10.9 mg	10.7	0.6
6	10.2-10.6 mg (Low Cylinder Contributor)	10.4	0.3
7	10.4-11.0 mg	10.7	0.6
8	10.6-10.8 mg	10.7	0.6

EOT = 80°C (176°F)
mg = milligrams

Complete the enhanced diagnostic graph to illustrate the difference between cylinders and help identify weak cylinders/injectors. To start this process, derive the average for each of the MFDES readings by adding the high and low numbers together and dividing by two. For example, cylinder number 5 has a reading of 10.5-10.9. Therefore, the equation would be: 10.5 + 10.9 = 21.4 divided by 2 = 10.7 (if needed, round the number up to the nearest tenth). Record this number on the enhanced diagnostic sheet for the appropriate cylinder. Now, subtract his number from the average MFDES recorded for the baseline. For example, cylinder No. 5 has an average MFDES of 10.7 and if the baseline MFDES was 10.1, the equation would be: 10.7-10.1 = 0.6. Therefore, 0.6 would be the deviation from the baseline. Carry out these calculations for all cylinder readings.

Injector Interrupt for Enhanced Diagnostics (Continued)

In order to get the cutoff number, average all of the deviations. For example, as in the sample, we would add the following numbers: 0.7 + 0.7 + 0.7 + 0.1 + 0.6 + 0.3 + 0.6 + 0.6 = 4.3. Divide the sum by eight and round to the nearest tenth: 4.3 ÷ 8 = 0.5375 ~ 0.5. Therefore, 0.5 would be the cutoff line for weak cylinders/injectors. By using this test, cylinders No. 4 and 6 have been identified as being weak.

Carry out a relative compression test using a diagnostic tool or check compression with a mechanical gauge on the weak cylinders and one or two of the strong contributing cylinders. Compare the weak cylinder readings against the strong contributing cylinders to verify that the power cylinder is mechanically sound. If the compression readings are within 20% of each other, the power cylinders can be considered good.

After injector replacement, test drive the vehicle for at least 20 miles and confirm that the rough idle concern has been corrected, do not retest.

If a rough idle persists, the test should be carried out again. After all of the tests are carried out, all DTCs should be cleared before returning the vehicle to the customer.

Injector Interrupt Enhanced Diagnostics Sheet

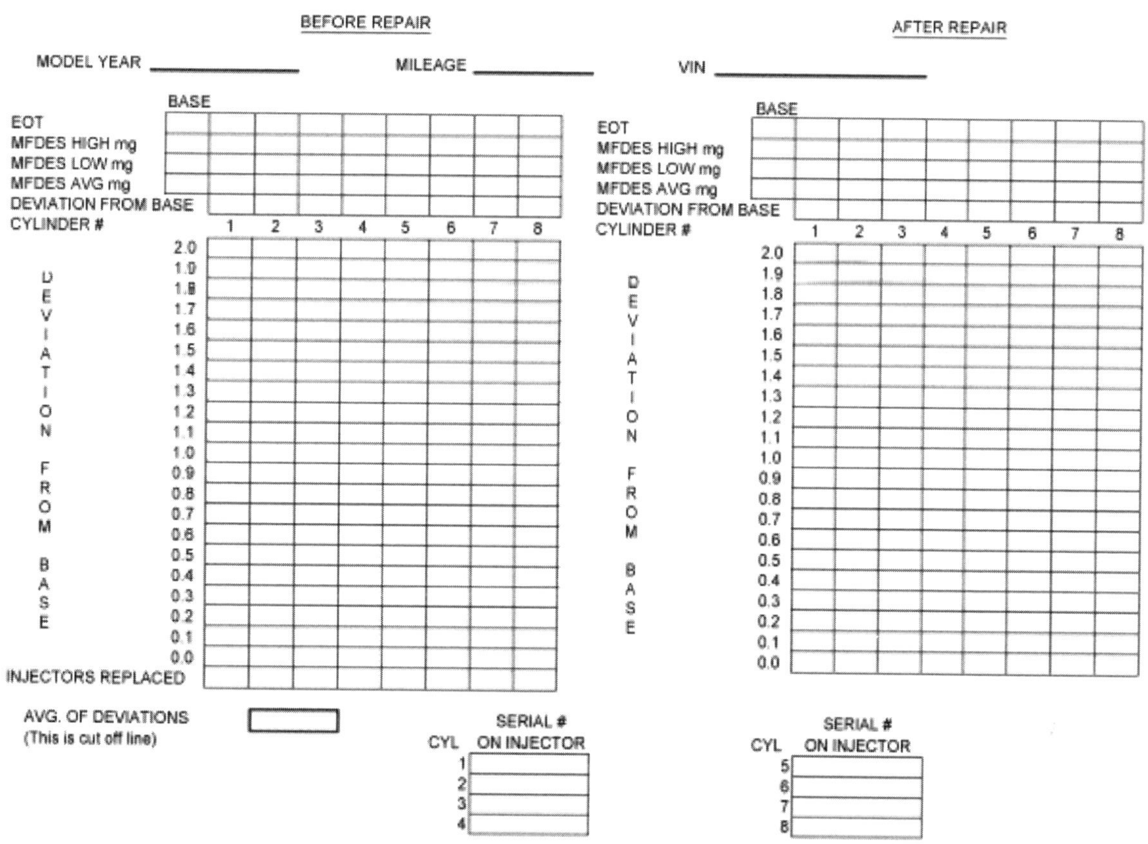

NOTES

DTC	YEAR(S)	Description	KOEO	KOER	Continuous	PAGE
B1213	2006-07	Less than two keys programmed to the system	DIAGNOSE And REPAIR The Anti-Theft System			NA
B1342	2006-07	ECU damaged (EEPROM in PCM not working, replace PCM)	DIAGNOSE And REPAIR The Anti-Theft System			NA
B1600	2006-07	Ignition Key Transponder Signal Is Not Received	DIAGNOSE And REPAIR The Anti-Theft System			NA
B1601	2006-07	PATS Received Incorrect Key-Code From Ignition Key Transponder	DIAGNOSE And REPAIR The Anti-Theft System			NA
B1602	2006-07	PATS Received Invalid Format Of Key-Code From Ignition Key Transponder	DIAGNOSE And REPAIR The Anti-Theft System			NA
B1681	2006-07	PATS Transceiver Module Signal Is Not Received	DIAGNOSE And REPAIR The Anti-Theft System			NA
B2103	2006-07	Antenna Not Connected	DIAGNOSE And REPAIR The Anti-Theft System			NA
B2431	2006-07	Transponder Programming Failed	DIAGNOSE And REPAIR The Anti-Theft System			NA
P0046	2006-07	Turbo/Super Charger Boost Control Solenoid Circuit Range/Performance	GO To Test AK			5-67
P0069	2006-07	MAP - Barometric Pressure Correlation	GO To Test AU			5-125
P0096	2006-07	Intake Air Temperature Sensor 2 Circuit Range/Performance	GO To Test G			5-183
P0097	2006-07	Intake Air Temperature Sensor 2 Circuit Low Input	GO To Test G			5-183
P0098	2006-07	Intake Air Temperature Sensor 2 Circuit High Input	GO To Test G			5-183
P0101	2006-07	Mass or Volume Air Flow Circuit Range/Performance	GO To Test J			5-205
P0103	2006-07	Mass or Volume Air Flow Circuit High Input	GO To Test J			5-205
P0107	2006-07	Manifold Absolute Pressure/BARO Sensor Low Input	GO To Test H			5-191
P0108	2006-07	Manifold Absolute Pressure/BARO Sensor High Input	GO To Test H			5-191
P0112	2006-07	Intake Air Temperature Circuit Low Input	GO To Test F			5-177
P0113	2006-07	Intake Air Temperature Circuit High Input	GO To Test F			5-177
P0117	2006-07	Engine Coolant Temperature Circuit Low Input	GO To Test K			5-215
P0118	2006-07	Engine Coolant Temperature High Input	GO To Test K			5-215
P0148	2006-07	Fueling Error	GO To Test AT			5-119
P0196	2006-07	Engine Oil Temperature Sensor Circuit Range/Performance	GO To Test L			5-231
P0197	2006-07	Engine Oil Temperature Sensor Circuit Low Input	GO To Test L			5-231
P0198	2006-07	Engine Oil Temperature Sensor Circuit High Input	GO To Test L			5-231
P0219	2006-07	Engine Overspeed Condition	REPEAT The Self-Test			NA
P0230	2006-07	Fuel Pump Primary Circuit	GO To Test M			5-245
P0231	2006-07	Fuel Pump Secondary Circuit Low	GO To Test M			5-245
P0232	2006-07	Fuel Pump Secondary Circuit High	GO To Test M			5-245
P0236	2006-07	Turbo/Super Charger Boost Sensor A Circuit Range/Performance	GO To Test E			5-167

DTC	YEAR(S)	Description	KOEO	KOER	Continuous	PAGE
P0237	2006-07	Turbo/Super Charger Boost Sensor A Circuit Low	GO To Test E			5-167
P0238	2006-07	Turbo/Super Charger Boost Sensor A Circuit High	GO To Test E			5-167
P0261	2006-07	Cylinder #1 Injector Circuit Low	GO To Test P			5-267
P0262	2006-07	Cylinder #1 Injector Circuit High	GO To Test P			5-267
P0263	2006-07	Cylinder #1 Contribution/Balance	GO To Test P			5-267
P0264	2006-07	Cylinder #2 Injector Circuit Low	GO To Test P			5-267
P0265	2006-07	Cylinder #2 Injector Circuit High	GO To Test P			5-267
P0266	2006-07	Cylinder #2 Contribution/Balance	GO To Test P			5-267
P0267	2006-07	Cylinder #3 Injector Circuit Low	GO To Test P			5-267
P0268	2006-07	Cylinder #3 Injector Circuit High	GO To Test P			5-267
P0269	2006-07	Cylinder #3 Contribution/Balance	GO To Test P			5-267
P0270	2006-07	Cylinder #4 Injector Circuit Low	GO To Test P			5-267
P0271	2006-07	Cylinder #4 Injector Circuit High	GO To Test P			5-267
P0272	2006-07	Cylinder #4 Contribution/Balance	GO To Test P			5-267
P0273	2006-07	Cylinder #5 Injector Circuit Low	GO To Test P			5-267
P0274	2006-07	Cylinder #5 Injector Circuit High	GO To Test P			5-267
P0275	2006-07	Cylinder #5 Contribution/Balance	GO To Test P			5-267
P0276	2006-07	Cylinder #6 Injector Circuit Low	GO To Test P			5-267
P0277	2006-07	Cylinder #6 Injector Circuit High	GO To Test P			5-267
P0278	2006-07	Cylinder #6 Contribution/Balance	GO To Test P			5-267
P0279	2006-07	Cylinder #7 Injector Circuit Low	GO To Test P			5-267
P0280	2006-07	Cylinder #7 Injector Circuit High	GO To Test P			5-267
P0281	2006-07	Cylinder #7 Contribution/Balance	GO To Test P			5-267
P0282	2006-07	Cylinder #8 Injector Circuit Low	GO To Test P			5-267
P0283	2006-07	Cylinder #8 Injector Circuit High	GO To Test P			5-267
P0284	2006-07	Cylinder #8 Contribution/Balance	GO To Test P			5-267
P0297	2006-07	Vehicle Overspeed Condition	REPEAT The Self-Test			NA
P0298	2006-07	Engine Oil Overtemperature Condition	GO To Test L			5-231
P0299	2006-07	Turbo/Supercharger Under Boost	GO To Test KA			5-221
P0300	2006-07	Random Misfire Detected	GO To Test AR			5-101
P0301	2006-07	Cylinder #1 Misfire Detected	GO To Test P			5-267
P0302	2006-07	Cylinder #2 Misfire Detected	GO To Test P			5-267

DTC	YEAR(S)	Description	KOEO	KOER	Continuous	PAGE
P0303	2006-07	Cylinder #3 Misfire Detected	GO To Test P			5-267
P0304	2006-07	Cylinder #4 Misfire Detected	GO To Test P			5-267
P0305	2006-07	Cylinder #5 Misfire Detected	GO To Test P			5-267
P0306	2006-07	Cylinder #6 Misfire Detected	GO To Test P			5-267
P0307	2006-07	Cylinder #7 Misfire Detected	GO To Test P			5-267
P0308	2006-07	Cylinder #8 Misfire Detected	GO To Test P			5-267
P0335	2006-07	Crankshaft Position Sensor A Circuit	GO To Test D			5-155
P0336	2006-07	Crankshaft Position Sensor A Circuit Range/Performance	GO To Test D			5-155
P0340	2006-07	Camshaft Position Sensor A Circuit (Bank 1 or single sensor)	GO To Test V			5-307
P0341	2006-07	Camshaft Position Sensor A Circuit Range/Performance (Bank 1 or single sensor)	GO To Test V			5-307
P0381	2006-07	Glow Plug/Heater Indicator Circuit	DIAGNOSE And REPAIR The Instrument Cluster			NA
P0401	2006	Exhaust Gas Recirculation Flow Insufficient Detected	GO To Test W			5-315
P0401	2007	Exhaust Gas Recirculation Flow Insufficient Detected	GO To Test W			5-325
P0402	2006	Exhaust Gas Recirculation Flow Excessive Detected	GO To Test W			5-315
P0402	2007	Exhaust Gas Recirculation Flow Excessive Detected	GO To Test W			5-325
P0403	2006	Exhaust Gas Recirculation Control Circuit	GO To Test W			5-315
P0403	2007	Exhaust Gas Recirculation Control Circuit	GO To Test W			5-325
P0404	2006	Exhaust Gas Recirculation Control Circuit Range/Performance	GO To Test W			5-315
P0405	2006	Exhaust Gas Recirculation Sensor A Circuit Low	GO To Test W			5-316
P0405	2007	Exhaust Gas Recirculation Sensor A Circuit Low	GO To Test W			5-325
P0406	2006	Exhaust Gas Recirculation Sensor A Circuit High	GO To Test W			5-315
P0406	2007	Exhaust Gas Recirculation Sensor A Circuit High	GO To Test W			5-325
P0460	2006-07	Fuel Level Sensor A Circuit	GO To Test QH			5-289
P0462	2006-07	Fuel Level Sensor A Circuit Low Input	GO To Test QH			5-289
P0463	2006-07	Fuel Level Sensor A Circuit High Input	GO To Test QH			5-289
P0470	2006-07	Exhaust Pressure Sensor	GO To Test X			5-335
P0471	2006-07	Exhaust Pressure Sensor Range/Performance	GO To Test X			5-335
P0472	2006-07	Exhaust Pressure Sensor Low Input	GO To Test X			5-335
P0473	2006-07	Exhaust Pressure Sensor High Input	GO To Test X			5-335
P0478	2006-07	Exhaust Pressure Control Valve High Input	GO To Test KA			5-221
P0480	2006-07	Fan 1 Control Circuit	GO To Test AH			5-49
P0500	2006-07	Vehicle Speed Sensor A	For F-Super Duty, GO To Test DF. For E-Series, DIAGNOSE And REPAIR The Transmission			5-163

DTC	YEAR(S)	Description	KOEO	KOER	Continuous	PAGE
P0528	2006-07	Fan Speed Sensor Circuit No Signal	GO To Test AH			5-49
P0562	2006	System Voltage Low	GO To Test A			5-1
P0562	2007	System Voltage Low	GO To Test A			5-7
P0563	2006	System Voltage High	GO To Test A			5-1
P0563	2007	System Voltage High	GO To Test A			5-7
P0565	2006-07	Cruise Control ON Signal	DIAGNOSE And REPAIR The Speed Control System			NA
P0566	2006-07	Cruise Control OFF Signal	DIAGNOSE And REPAIR The Speed Control System			NA
P0567	2006-07	Cruise Control RESUME Signal	DIAGNOSE And REPAIR The Speed Control System			NA
P0568	2006-07	Cruise Control SET Signal	DIAGNOSE And REPAIR The Speed Control System			NA
P0569	2006-07	Cruise Control COAST Signal	DIAGNOSE And REPAIR The Speed Control System			NA
P0605	2006-07	Powertrain Control Module Read Only Memory (ROM) Error	GO To Test AA			5-13
P0606	2006-07	ECM/PCM Processor	GO To Test Z			5-347
P0611	2006-07	Fuel Injector Control Module Performance	GO To Test S			5-301
P0620	2006-07	Generator 1 Control Circuit	DIAGNOSE And REPAIR The Charging System			NA
P0623	2006-07	Generator Lamp Control Circuit	DIAGNOSE And REPAIR The Charging System			NA
P0645	2006-07	A/C Clutch Relay Control Circuit	GO To Test AV			5-129
P0649	2006-07	Cruise Control Lamp Control Circuit	DIAGNOSE And REPAIR The Instrument Cluster			NA
P0657	2006-07	Actuator Supply Voltage A Circuit Open	DIAGNOSE And REPAIR The Transmission			NA
P0670	2006-07	Glow Plug Module Control Circuit	GO To Test AF			5-25
P0671	2006-07	Cylinder 1 Glow Plug Circuit	GO To Test AF			5-25
P0672	2006-07	Cylinder 2 Glow Plug Circuit	GO To Test AF			5-25
P0673	2006-07	Cylinder 3 Glow Plug Circuit	GO To Test AF			5-25
P0674	2006-07	Cylinder 4 Glow Plug Circuit	GO To Test AF			5-25
P0675	2006-07	Cylinder 5 Glow Plug Circuit	GO To Test AF			5-25
P0676	2006-07	Cylinder 6 Glow Plug Circuit	GO To Test AF			5-25
P0677	2006-07	Cylinder 7 Glow Plug Circuit	GO To Test AF			5-25
P0678	2006-07	Cylinder 8 Glow Plug Circuit	GO To Test AF			5-25
P0683	2006-07	Glow Plug Control Module to PCM Communication Circuit	GO To Test AF			5-25
P0700	2006-07	Transmission Control System (MIL Request)	DIAGNOSE And REPAIR The Transmission			NA
P0703	2006-07	Brake Switch B Input Circuit	GO To Test AY			5-131
P0830	2006-07	Clutch Pedal Switch A Circuit	GO To Test C			5-149
P0833	2006-07	Clutch Pedal Switch B Circuit	GO To Test C			5-149

DTC	YEAR(S)	Description	KOEO KOER Continuous	PAGE
P1000	2006-07	OBD Systems Readiness Test Not Complete	GO To Test AC	5-15
P1001	2006-07	KOER Not Able To Complete, KOER Aborted	REPEAT The Self-Test	NA
P1102	2006-07	Mass Air Flow Sensor In Range But Lower Than Expected	GO To Test J	5-205
P1148	2006-07	Generator 2 Control Circuit	DIAGNOSE And REPAIR The Charging System	NA
P1149	2006-07	Generator 2 Control Circuit High	DIAGNOSE And REPAIR The Charging System	NA
P115A	2006-07	Low Fuel Level - Forced Limited Power	GO To Test QH	5-289
P1184	2006-07	Engine Oil Temperature Sensor Out Of Self-Test Range	GO To Test L	5-231
P1260	2006-07	Theft Detected, Vehicle Immobilized	DIAGNOSE And REPAIR The Anti-Theft System	NA
P1284	2006-07	Aborted KOER — Injector Control Pressure Regulator	Carry out the KOEO self-test. Diagnose all other DTCs first. Clear the DTCs. Repeat the KOER self test. If	NA
P1335	2006	EGR Position Sensor Minimum Stop Performance	GO To Test W	5-315
P1335	2007	EGR Position Sensor Minimum Stop Performance	GO To Test W	5-325
P1378	2006-07	FICM Supply Voltage Circuit Low	GO To Test S	5-301
P1379	2006-07	FICM Supply Voltage Circuit High	GO To Test S	5-301
P1397	2006-07	System Voltage Out Of Self-Test Range	DTC will set if system voltage is greater than 18 volts or less than 10 volts during the KOER Glow Plug self-	NA
P1408	2006-07	Exhaust Gas Recirculation Flow Out Of Self-Test Range	Refer to the Performance Diagnostic Procedures for further diagnostics.	NA
P1464	2006-07	A/C Demand Out Of Self-Test Range	GO To Test AM	5-77
P1501	2006-07	Vehicle Speed Sensor Out Of Self-Test Range	For F-Super Duty, GO To Test DF. For E-Series, DIAGNOSE And REPAIR The Transmission	5-163
P1502	2006-07	Invalid Test — Auxiliary Power Control Functioning	GO To Test AN	5-79
P1531	2006-07	Invalid Test — Accelerator Pedal Movement	REPEAT The Self-Test	NA
P1536	2006-07	Parking Brake Switch Circuit	GO To Test I	5-199
P1610	2006-07	Interactive Reprogramming Code — Replace PCM module	REFLASH The PCM/TCM/FICM.	NA
P1611	2006-07	Interactive Reprogramming Code — Diagnose Further	REFLASH The PCM/TCM/FICM.	NA
P1615	2006-07	Interactive Reprogramming Code — Flash Erase Error	REFLASH The PCM/TCM/FICM.	NA
P1616	2006-07	Interactive Reprogramming Code — Flash Erase Error, Low Voltage	REFLASH The PCM/TCM/FICM.	NA
P1617	2006-07	Interactive Reprogramming Code — Block Programming Error	REFLASH The PCM/TCM/FICM.	NA
P1618	2006-07	Interactive Reprogramming Code — Block Programming Error, Low Voltage	REFLASH The PCM/TCM/FICM.	NA
P1633	2006-07	Keep Alive Power Voltage Too Low	GO To Test Z	5-347
P1635	2006-07	Tire/Axle Out Of Acceptable Range	Refer to "Flash Electrically Erasable Programmable Read Only Memory (EEPROM)"	NA
P1639	2006-07	Vehicle ID Block Corrupted, Not Programmed	Refer to "Flash Electrically Erasable Programmable Read Only Memory (EEPROM)"	NA
P1703	2006-07	Brake Switch Out Of Self-Test Range	GO To Test AI	5-61
P1705	2006-07	Transmission Range Circuit Not Indicating Park/Neutral During Self-Test	GO To Test AL	5-75
P1725	2006-07	Insufficient Engine Speed Increase During Self-Test	If the Glow Plug Monitor Self-Test aborts and this DTC is present, repeat the Glow Plug Monitor Self-Test.	NA

DTC	YEAR(S)	Description	KOEO KOER Continuous	PAGE
P1726	2006-07	Insufficient Engine Speed Decrease During Self-Test	If the Glow Plug Monitor Self-Test aborts and this DTC is present, repeat the Glow Plug Monitor Self-Test.	NA
P2067	2006-07	Fuel Level Sensor B Circuit Low Input	GO To Test QH	5-289
P2068	2006-07	Fuel Level Sensor B Circuit High Input	GO To Test QH	5-289
P2104	2006-07	Throttle Actuator Control System Forced Idle	GO To Test AG	5-39
P2122	2006-07	Throttle/Pedal Position Sensor/Switch D Circuit Low Input	GO To Test AG	5-39
P2123	2006-07	Throttle/Pedal Position Sensor/Switch D Circuit High Input	GO To Test AG	5-39
P2127	2006-07	Throttle/Pedal Position Sensor/Switch E Circuit Low Input	GO To Test AG	5-39
P2128	2006-07	Throttle/Pedal Position Sensor/Switch E Circuit High Input	GO To Test AG	5-39
P2132	2006-07	Throttle/Pedal Position Sensor/Switch F Circuit Low Input	GO To Test AG	5-39
P2133	2006-07	Throttle/Pedal Position Sensor/Switch F Circuit High Input	GO To Test AG	5-39
P2138	2006-07	Throttle/Pedal Position Sensor/Switch D/E Voltage Correlation	GO To Test AG	5-39
P2139	2006-07	Throttle/Pedal Position Sensor/Switch D/F Voltage Correlation	GO To Test AG	5-39
P2140	2006-07	Throttle/Pedal Position Sensor/Switch E/F Voltage Correlation	GO To Test AG	5-39
P2199	2006-07	Intake Air Temperature 1/2 Correlation	GO To Test AW	5-127
P2262	2006-07	Turbo/Super Charger Boost Pressure Not Detected — Mechanical	GO To Test KA	5-221
P2263	2006-07	Turbo/Super Charger System Performance	GO To Test KA	5-221
P2269	2006-07	Water in Fuel Condition	GO To Test O	5-259
P2284	2006-07	Injector Control Pressure Sensor Circuit Range/Performance	GO To Test AO	5-83
P2285	2006-07	Injector Control Pressure Sensor Circuit Low	GO To Test Q	5-281
P2286	2006-07	Injector Control Pressure Sensor Circuit High	GO To Test Q	5-281
P2287	2006-07	Injector Control Pressure Sensor Circuit Intermittent	GO To Test Q	5-281
P2288	2006-07	Injector Control Pressure Too High	GO To Test AO	5-83
P2289	2006-07	Injector Control Pressure Too High — Engine Off	GO To Test AP	5-91
P2290	2006-07	Injector Control Pressure Too Low	GO To Test AO	5-83
P2291	2006-07	Injector Control Pressure Too Low — Engine Cranking	GO To Test AQ	5-95
P2457	2006-07	EGR Cooler Range/Performance	GO To Test AZ	5-135
P2552	2006-07	FICMM Circuit — Throttle/Fuel Inhibit Circuit	GO To Test AT	5-119
P2614	2006-07	Camshaft Position Output Circuit	GO To Test V	5-307
P2617	2006-07	Crankshaft Position Output Circuit	GO To Test D	5-155
P2623	2006-07	Injector Control Pressure Regulator Circuit	GO To Test R	5-295
U0100	2006-07	Lost Communication With ECM/PCM A	Clear the DTCs. Repeat the self-test. If the DTC is present again, install a new PCM.	NA
U0101	2006-07	Lost Communication With TCM	Clear the DTCs. Repeat the self-test. If the DTC is present again, install a new PCM.	NA

DTC	YEAR(S)	Description	KOEO	KOER	Continuous	PAGE
U0105	**2006**	Lost Communication with FICM	**GO To Test AS**			**5-103**
U0105	**2007**	Lost Communication with FICM	**GO To Test AS**			**5-111**
U0155	**2006-07**	Lost Communication with Instrument Cluster	**DIAGNOSE And REPAIR The Module Communications Network**			**NA**
U0306	**2006-07**	Software Incompatibility with Fuel Injector Control Module	**REFLASH The PCM, TCM, FICM**			**NA**

NOTES

NOTES

Trouble Code (s)	Problem: Vehicle Battery	Test
P0562, P0563		A 2006

REFERENCE LIST

DIAGNOSTIC TROUBLE CODE INDEX			
CODES	KOEO	KOER	CM
P0562	A1	A1	A1
P0563	A1	A1	A1

P0562 = System Voltage Low
P0563 = System Voltage High

Circuit Function:
With the ignition switch in the START or RUN position, voltage is applied to the coil of the powertrain control module (PCM) power relay. The applied voltage energizes the relay and closes the internal relay contacts. With the relay contacts closed, vehicle power (VPWR) is supplied to the PCM.

Fault Detection:
Vehicle power is monitored by the PCM. When the voltage is above or below a pre-calibrated value, an internal counter increments until a DTC is set.

CAUTION!:
The powertrain control module (PCM) harness connectors must be properly seated and the connector latch properly attached to eliminate possible driveability concerns or a no start condition. Installing the PCM connectors on an angle may cause an improper connection, misdiagnosis and damaged components. Install the connector until the lever pivots and seats itself. Apply light pressure to get the connector into position on the PCM and then fully seat the connector.

Note: Visually inspect the harness connectors for corrosion, damage, proper mating and correct pin tension.

Note: When the PCM is disconnected additional DTCs will be set. Clear all DTCs after restoring the vehicle.

PINPOINT TEST SCHEMATICS AND CONNECTORS

Powertrain Control Module Power
(PCM Power) Relay Connector

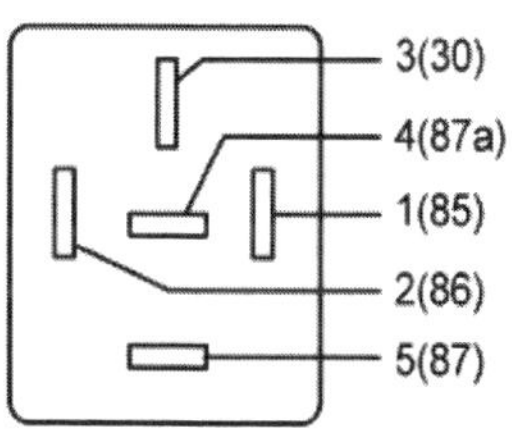

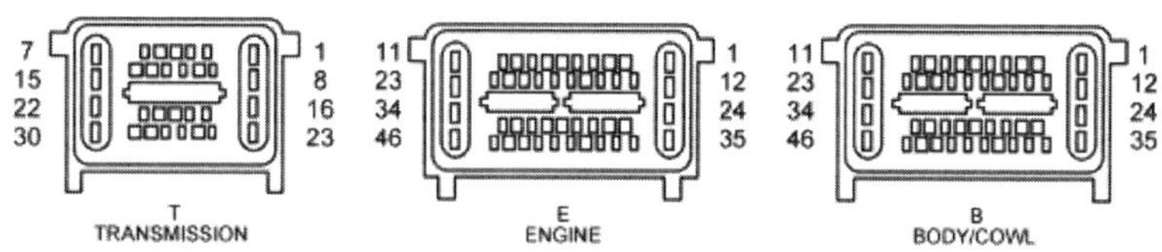

Vehicle	Connector	Pin	Circuit
E-Series	A	87	VPWR
		86	GND
		85	IGN START/RUN
		30	B+
F-Super Duty	A	87	VPWR
		85	GND
		86	IGN START/RUN
		30	B+

Pin	Circuit
B10, B11, B23	PWRGND (Power Ground)
B34, B46	VPWR (Vehicle Power)

A1 — DIAGNOSTIC TROUBLE CODES (DTCS) P0562 OR P0563

Note:DTC P0563 may set during a 24-volt jump start. DTC P0562 may set during crank with a low battery condition.

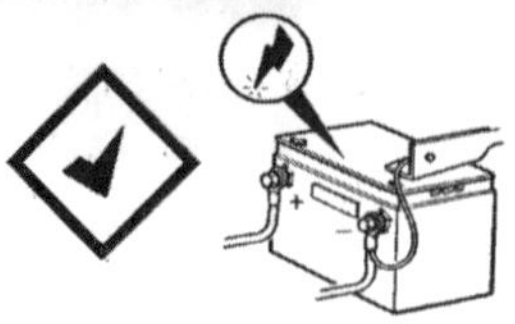

Check The Battery And Charging System Voltages.

Are The Battery And Charging System Voltages Within Specifications?

YES → A2

NO →

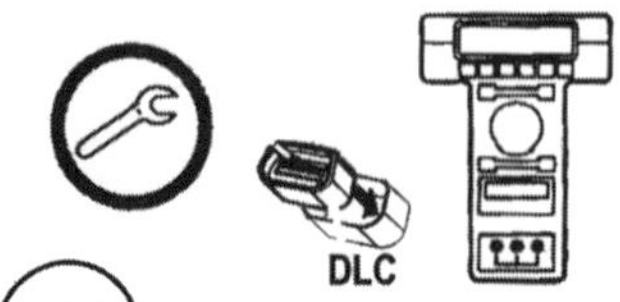

REPAIR As Necessary. CLEAR The DTC's. Repeat The Self-Test.

A2 — CHECK THE PCM RELAY VOLTAGE CIRCUITS

Note: For E-Series, the PCM diode is internal to the battery junction box (BJB).

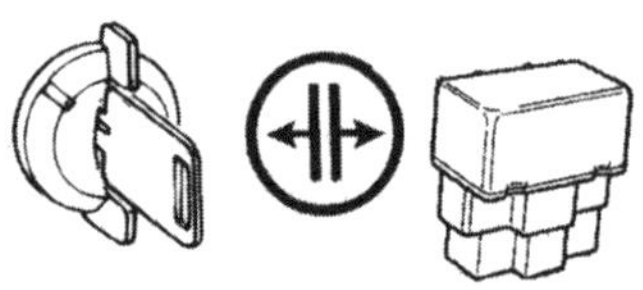

Key Off.

Disconnect The PCM Relay.

Key On, Engine Off.

Measure the voltage between:	
(+) PCM Power Relay Connector, Harness Side	(-)
B+	Ground
IGN START/RUN	Ground

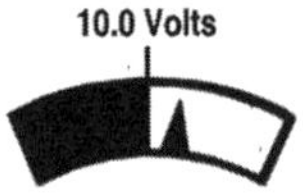

Are The Voltages Greater Than 10.0 Volts?

YES → A3

NO →

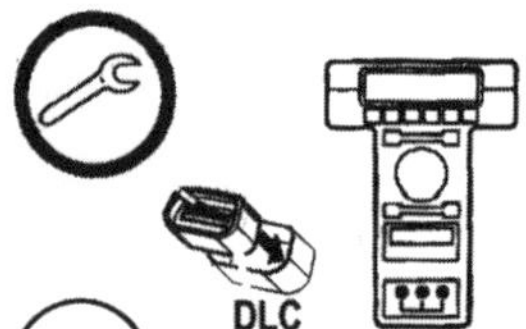

REPAIR The Circuit In Question. CLEAR The DTC's. Repeat The Self-Test.

A3 CHECK THE PCM RELAY COIL GROUND CIRCUIT

Note: For F-Super Duty, the PCM diode is internal to the central junction box (CJB).

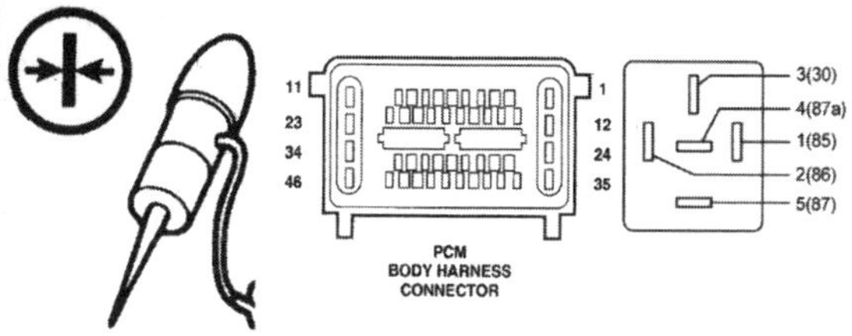

Connect a non-powered test lamp between:	
(+) PCM Power Relay Connector, Harness Side	(-) PCM Power Relay Connector, Harness Side
B+	GND
SEE CONNECTOR VIEWS ON INTRO PAGE	

Is The Test Lamp Illuminated?

YES → A4

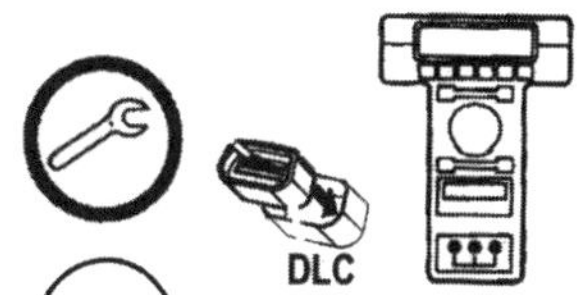

NO → STOP

REPAIR The Open Circuit. CLEAR The DTCs. REPEAT The Self-Test.

A4 CHECK THE PCM POWER RELAY

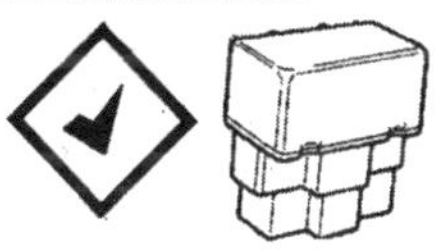

Perform The PCM Relay Component Test.

Does The PCM Relay Pass The Component Test?

YES → A5

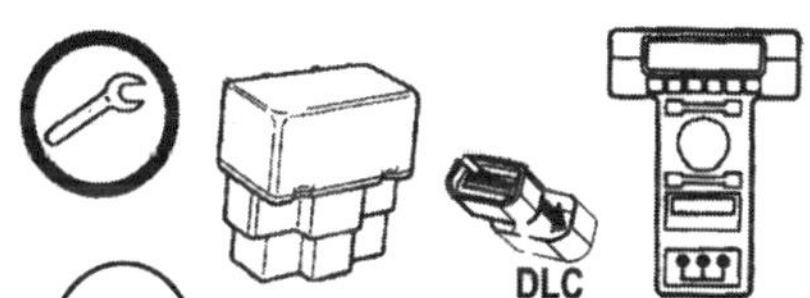

NO → STOP

INSTALL A New PCM Relay. CLEAR The DTCs. REPEAT The Self-Test.

A5 CHECK THE PCM VOLTAGE CIRCUITS FOR AN OPEN

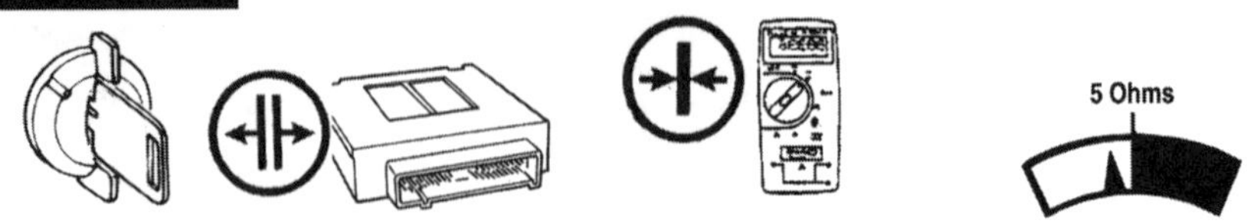

Key Off.

Disconnect The PCM.

Measure the resistance between:	
(+) PCM Power Relay Connector, Harness Side	(-) PCM Connector, Harness Side
VPWR	VPWR - Pin B34, B46
SEE CONNECTOR VIEWS ON INTRO PAGE	

Are The Resistances Less Than 5 Ohms?

YES → A6

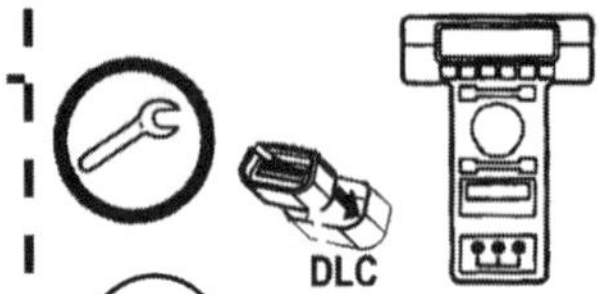

NO → STOP

REPAIR The Circuit In Question. CLEAR The DTCs. REPEAT The Self-Test.

A6 CHECK THE PCM GROUND CIRCUITS FOR AN OPEN

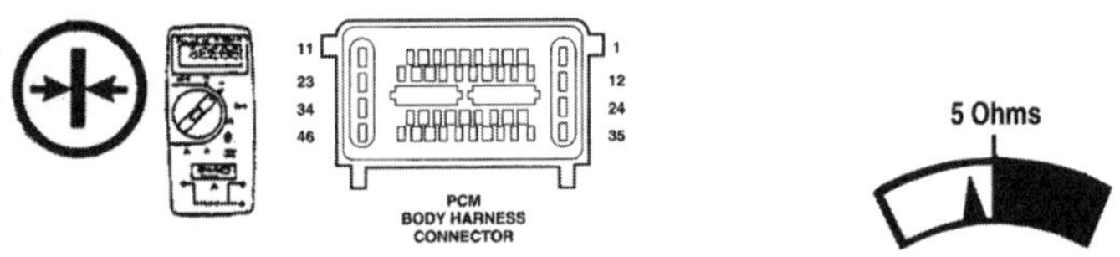

Measure the resistance between:	
(+) PCM Connector, Harness Side	(-)
PWRGND - Pin B10, B11, B23	Ground
SEE CONNECTOR VIEWS ON INTRO PAGE	

Are The Resistances Less Than 5 Ohms?

YES → A7

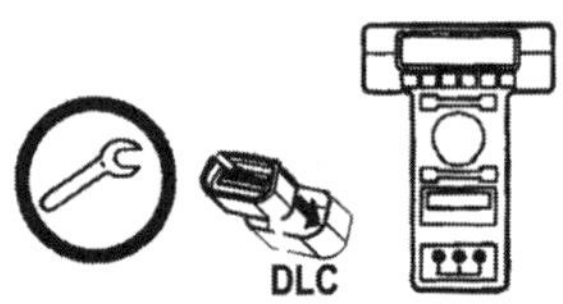

NO → STOP

REPAIR The Circuit In Question. CLEAR The DTCs. REPEAT The Self-Test.

A7 CHECK FOR CORRECT PCM OPERATION

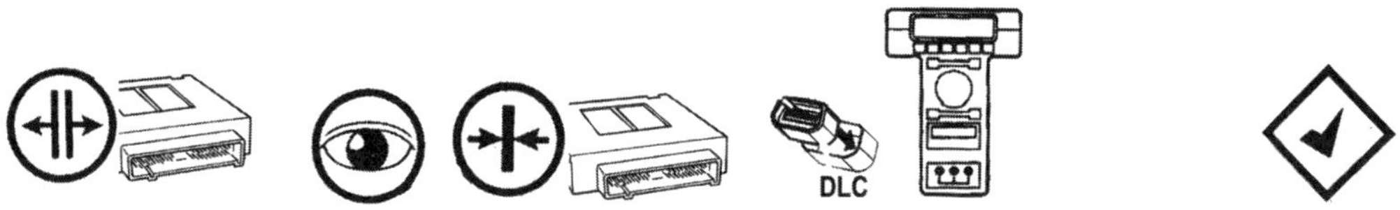

Disconnect All PCM Connectors.

Visually Inspect For Pushed Out Pins, Corrosion.

Connect All PCM Connectors And Make Sure They Seat Correctly.

Perform The PCM Self-Test And Verify The Concern Is Still Present.

Is The Concern Still Present?

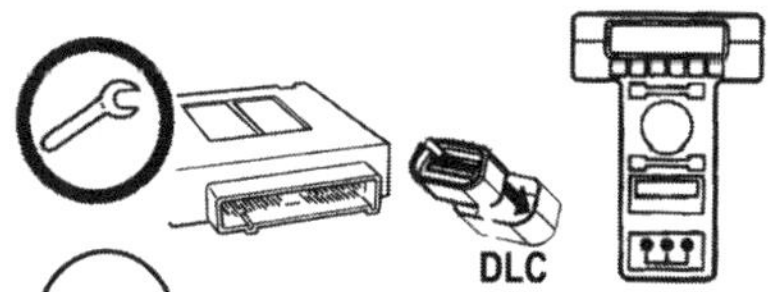

YES: INSTALL A New PCM. CLEAR The DTC's And REPEAT The Self-Test. STOP

OK

NO: The System Is Operating Correctly At This Time. The Concern May Have Been Caused By A Loose Or Corroded Connector. STOP

NOTES

Trouble Code (s)	Problem: Powertrain Control Module (PCM) Power Relay	Test
P0562, P0563		A 2007

REFERENCE LIST

This pinpoint test is intended to diagnose the following:

- Harness circuits: B+, IGN START/RUN, VPWR, GND and PWRGND
- PCM power relay
- PCM

DIAGNOSTIC TROUBLE CODE INDEX			
CODES	KOEO	KOER	CM
P0562	A1	A1	A1
P0563	A1	A1	A1

P0562 = System Voltage Low
P0563 = System Voltage High

PINPOINT TEST SCHEMATICS AND CONNECTORS

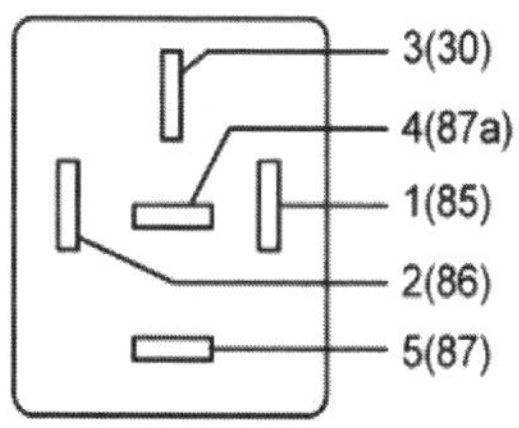

Powertrain Control Module Power (PCM Power) Relay Connector

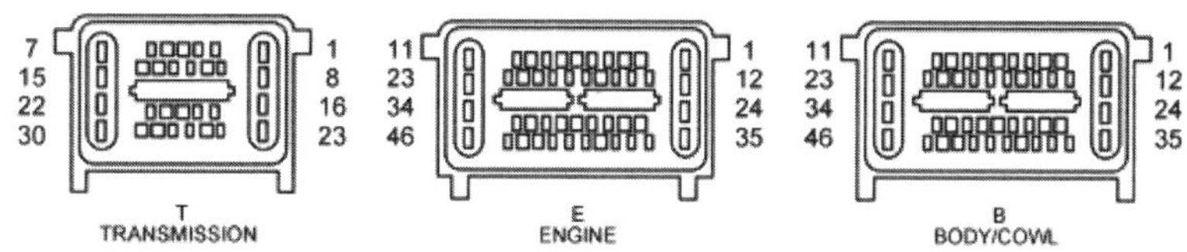

Vehicle	Connector	Pin	Circuit
E-Series	A	87	VPWR
		86	GND
		85	IGN START/RUN
		30	B+
F-Super Duty	A	87	VPWR
		85	GND
		86	IGN START/RUN
		30	B+

Pin	Circuit
B10, B11, B23	PWRGND (Power Ground)
B34, B46	VPWR (Vehicle Power)

A1 DIAGNOSTIC TROUBLE CODES (DTCS) P0562 OR P0563

Note: DTC P0563 may set during a 24-volt jump start. DTC P0562 may set during crank with a low battery condition.

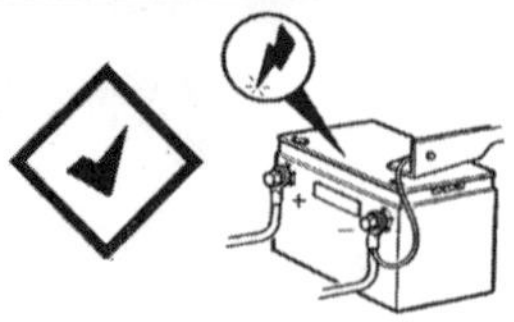

Check The Battery And Charging System Voltages.

Are The Battery And Charging System Voltages Within Specifications?

YES → A2

NO → STOP

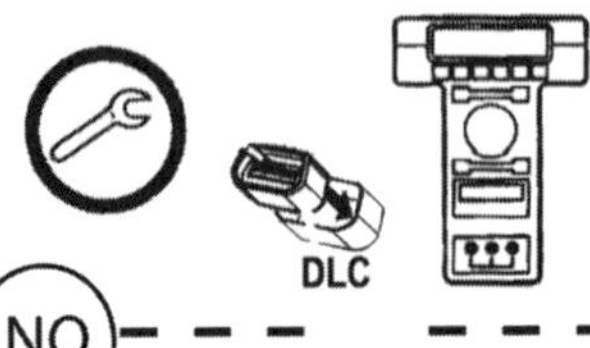

REPAIR As Necessary. CLEAR The DTC's. Repeat The Self-Test.

A2 CHECK THE B+ AND IGN START/RUN CIRCUITS FOR VOLTAGE

Note: For E-Series, the PCM diode is internal to the battery junction box (BJB).

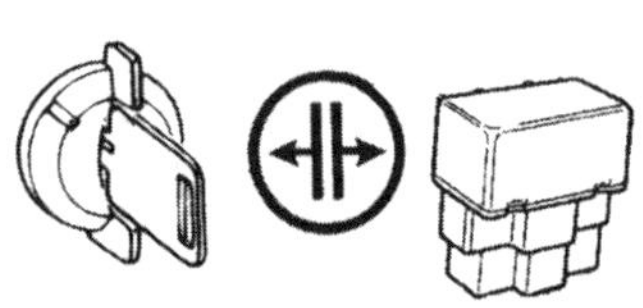

Key Off.

Disconnect The PCM Relay.

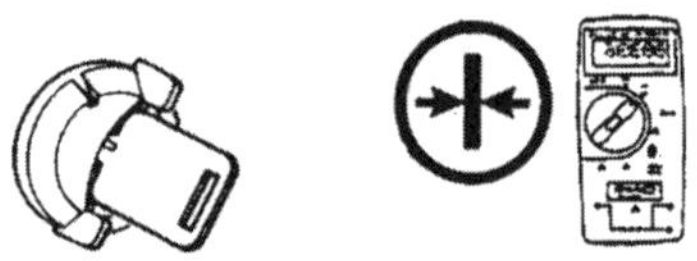

Key On, Engine Off.

Measure the voltage between:	
(+) PCM Power Relay Connector, Harness Side	(-)
B+	Ground
IGN START/RUN	Ground

Are The Voltages Greater Than 10.0 Volts?

YES → A3

NO → STOP

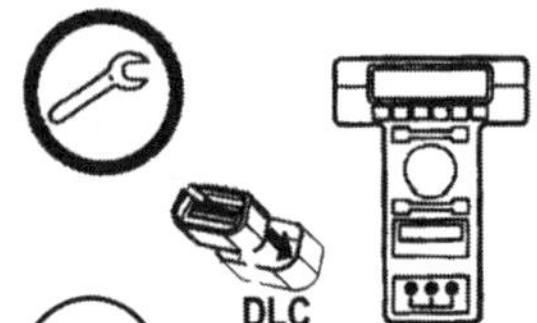

REPAIR The Circuit In Question. CLEAR The DTC's. Repeat The Self-Test.

A3 CHECK THE PCM RELAY COIL GROUND CIRCUIT

Note: For F-Super Duty, the PCM diode is internal to the central junction box (CJB).

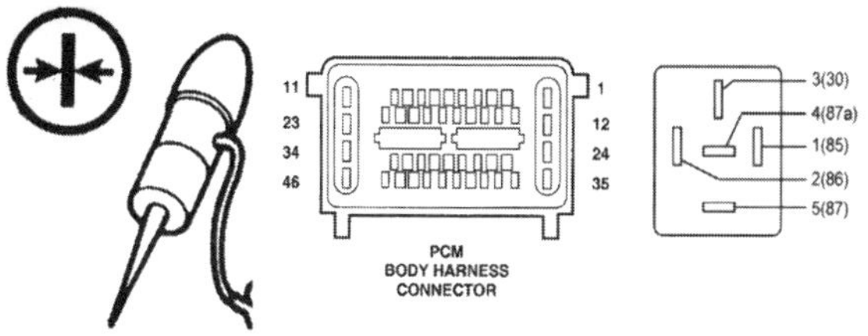

Connect a non-powered test lamp between:	
(+) PCM Power Relay Connector, Harness Side	(-) PCM Power Relay Connector, Harness Side
B+	GND
SEE CONNECTOR VIEWS ON INTRO PAGE	

Is The Test Lamp Illuminated?

YES → A4

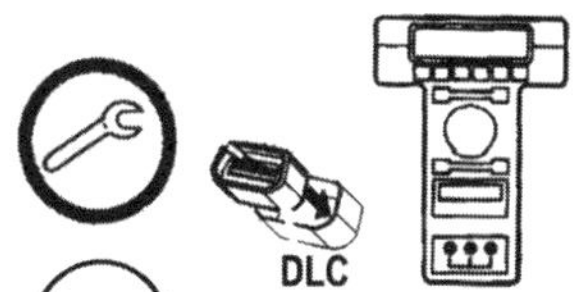

NO → STOP

REPAIR The Open Circuit. CLEAR The DTCs. REPEAT The Self-Test.

A4 CHECK THE PCM POWER RELAY

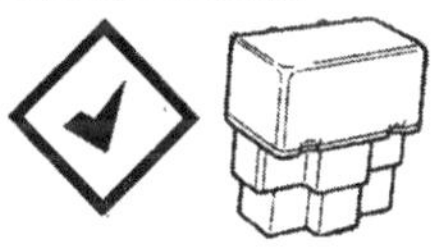

Perform The PCM Relay Component Test.

Does The PCM Relay Pass The Component Test?

YES → A5

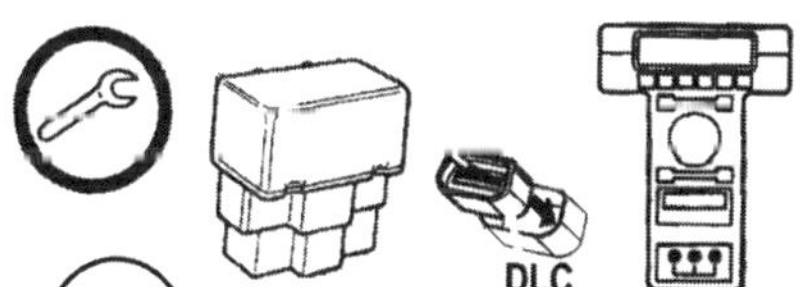

NO → STOP

INSTALL A New PCM Relay. CLEAR The DTCs. REPEAT The Self-Test.

A5 CHECK THE VPWR CIRCUITS FOR AN OPEN

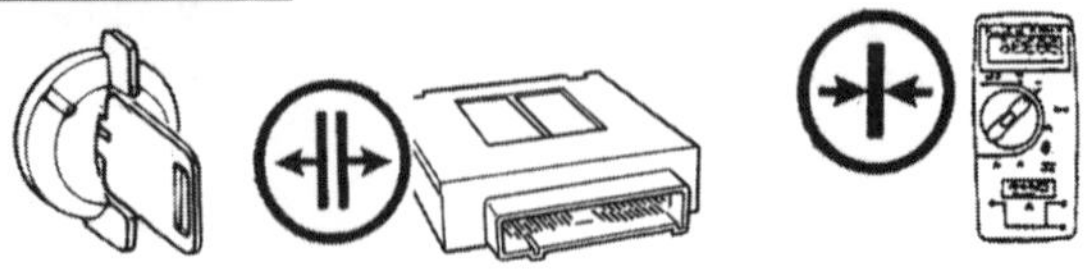

Key Off.

Disconnect The PCM.

Measure the resistance between:	
(+) PCM Power Relay Connector, Harness Side	(-) PCM Connector, Harness Side
VPWR	VPWR - Pin B34, B46
SEE CONNECTOR VIEWS ON INTRO PAGE	

Are The Resistances Less Than 5 Ohms?

YES — A6

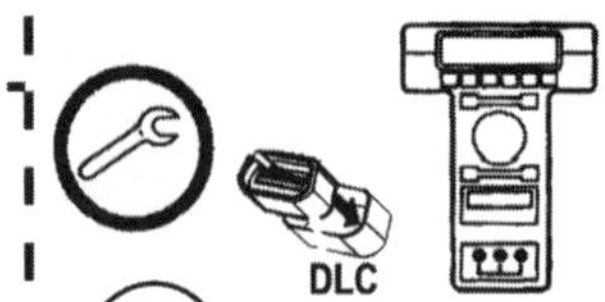

NO — STOP

REPAIR The Circuit In Question. CLEAR The DTCs. REPEAT The Self-Test.

A6 CHECK THE PCM GROUND CIRCUITS FOR AN OPEN

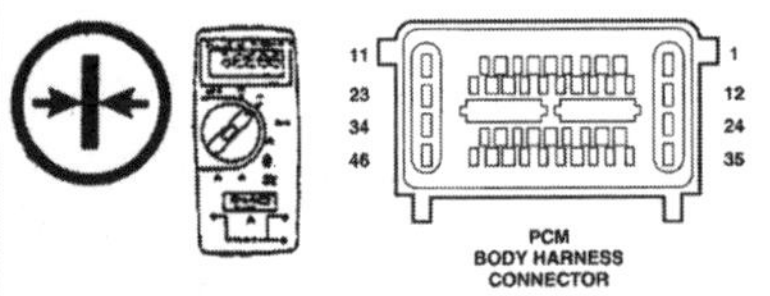

Measure the resistance between:	
(+) PCM Connector, Harness Side	(-)
PWRGND - Pin B10, B11, B23	Ground
SEE CONNECTOR VIEWS ON INTRO PAGE	

Are The Resistances Less Than 5 Ohms?

YES — A7

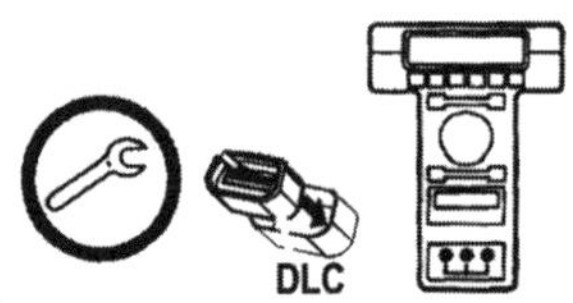

NO — STOP

REPAIR The Circuit In Question. CLEAR The DTCs. REPEAT The Self-Test.

A7 CHECK FOR CORRECT PCM OPERATION

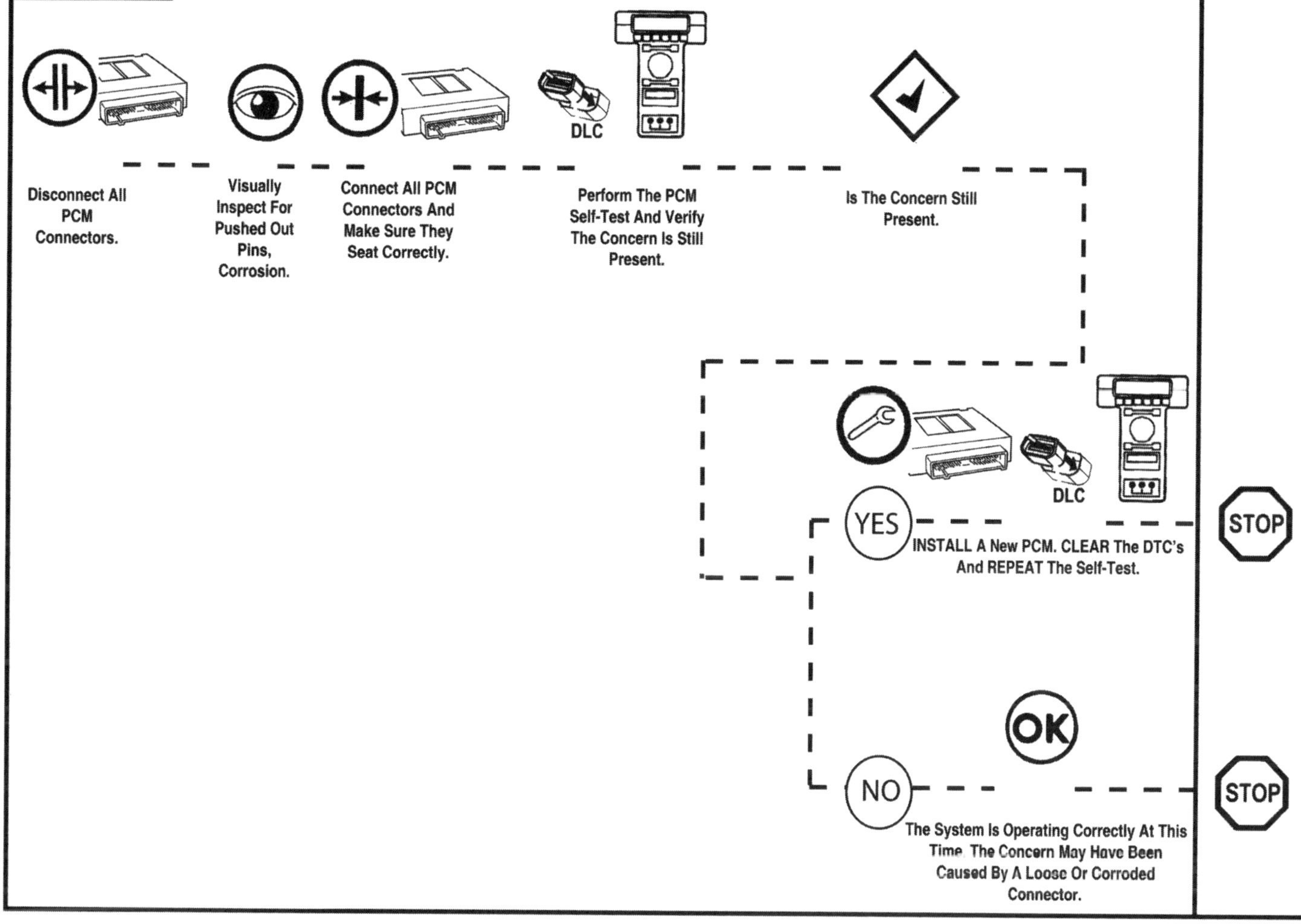

NOTES

Trouble Code (s)	Problem: Powertrain Control Module (PCM) Read Only Memory (ROM) Error	Test
P0605		AA 2006-07

REFERENCE LIST

This pinpoint test is intended to diagnose the following:
- Incorrect PCM calibration
- PCM

DIAGNOSTIC TROUBLE CODE INDEX			
Codes	KOEO	KOER	CM
P0605	AA1	AA1	AA1

CAUTION!: The powertrain control module (PCM) harness connectors must be properly seated and the connector latch properly attached to eliminate possible driveability concerns or a no start condition. Installing PCM connectors on an angle may cause an improper connection, misdiagnosis and damaged components. Install the connector until the lever pivots and seats itself. Apply light pressure to get the connector into position on the PCM and then fully seat the connector.

Note: Visually inspect the harness connectors for corrosion, damage, proper mating and correct pin tension.

Note: When the PCM is disconnected additional DTCs will be set. Clear all DTCs after restoring the vehicle.

AA1 DIAGNOSTIC TROUBLE CODE (DTC) P0605

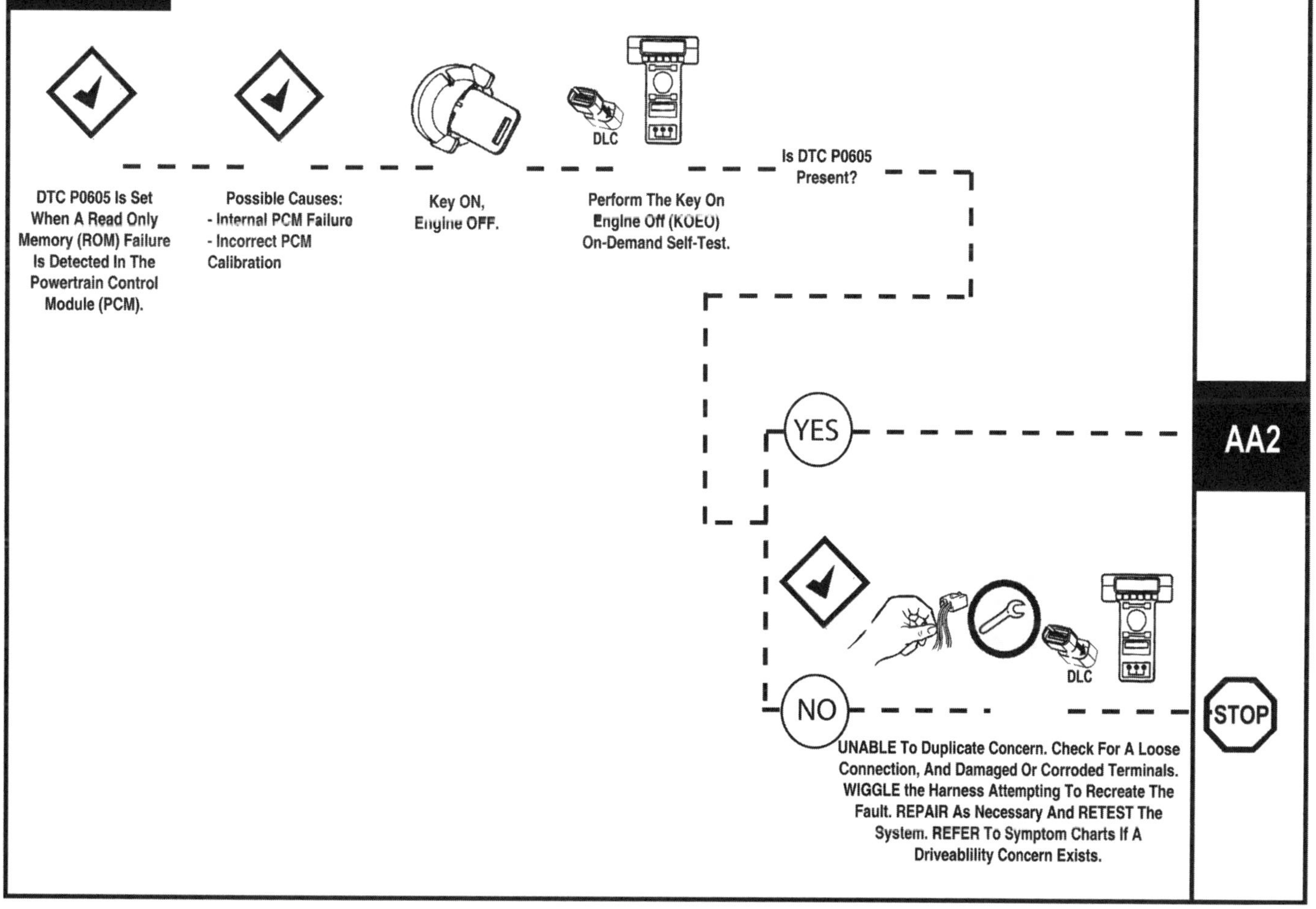

AA2 FLASH THE PCM

DLC

Flash The PCM. Refer To Chapter 2 Diagnostic Methods Flash Electrically Eraseable Programmable Read Only Memory - Flash EEPROM.

Clear The DTC's.

DLC

Perform The Key On Engine Off (KOEO) On-Demand Self-Test.

Is DTC P0605 Present?

YES — AA3

NO — OK — Test Is Complete. — STOP

AA3 CHECK FOR CORRECT PCM OPERATION

Disconnect All PCM Connectors.

Visually Inspect For Pushed Out Pins, Corrosion.

Connect All PCM Connectors And Make Sure They Seat Correctly.

DLC

Perform The PCM Self-Test And Verify The Concern Is Still Present.

Is The Concern Still Present.

DLC

YES — INSTALL A New PCM. CLEAR The DTC's And REPEAT The Self-Test. — STOP

NO — OK — The System Is Operating Correctly At This Time. The Concern May Have Been Caused By A Loose Or Corroded Connector. — STOP

Trouble Code (s)	Problem: OBD II Readiness Code Test Not Complete	Test
P1000		AC 2006-07

REFERENCE LIST

This pinpoint test is intended to diagnose the following:

- Drive conditions not satisfied
- Keep alive memory (KAM) cleared

DIAGNOSTIC TROUBLE CODE INDEX			
Codes	KOEO	KOER	CM
P1000	AC1	AC1	AC1

P1000 = OBD Systems Readiness Test Not Complete

AC1 — DIAGNOSTIC TROUBLE CODE (DTC) P1000

Note: If the voltage to the powertrain control module (PCM) is disconnected, P1000 is set until the OBD monitors have run and cleared.

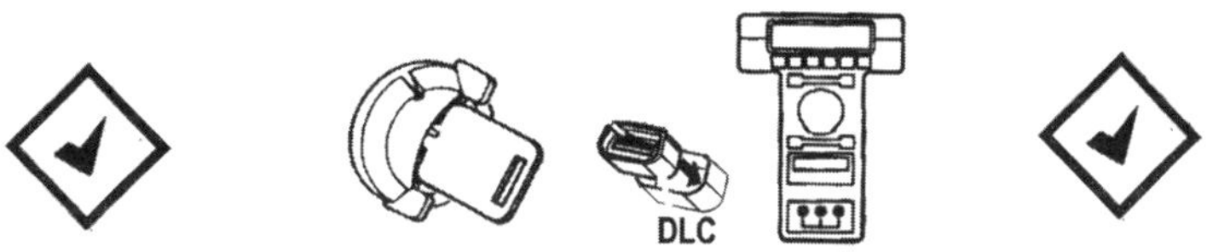

Diagnostic Code P1000 Indicates That Operating Conditions Have Not Been Satisfied For All Of The OBD II Monitors To Run.
Possible Causes:
- Drive Conditions Not Satisfied
- KAM Cleared
- Damaged PCM

Key On, Engine Off

Carry Out The Self-Test.

Are DTCs Other Than P1000 Set?

YES — DISREGARD the current diagnostic trouble code (DTC) at this time. DIAGNOSE the next DTC. → DTC INDEX

NO → AC2

AC2 — CARRY OUT THE DIESEL DRIVE CYCLE TEST

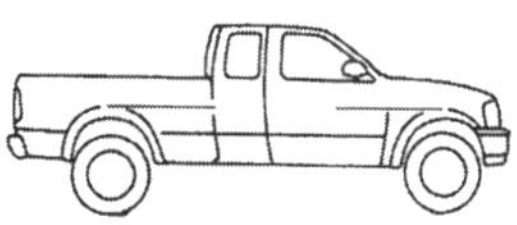

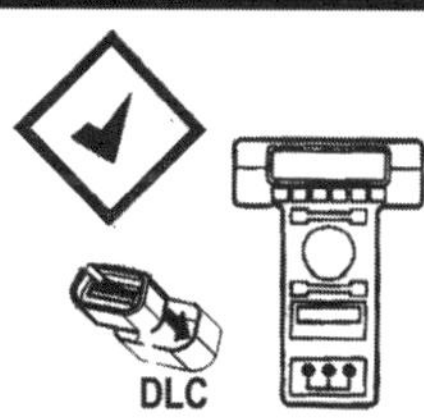

Perform Diesel Drive Cycle. Refer To DRIVE CYCLE

Is P1000 Cleared?

YES — No Issue; DTC P1000 Cleared. → STOP

NO → AC3

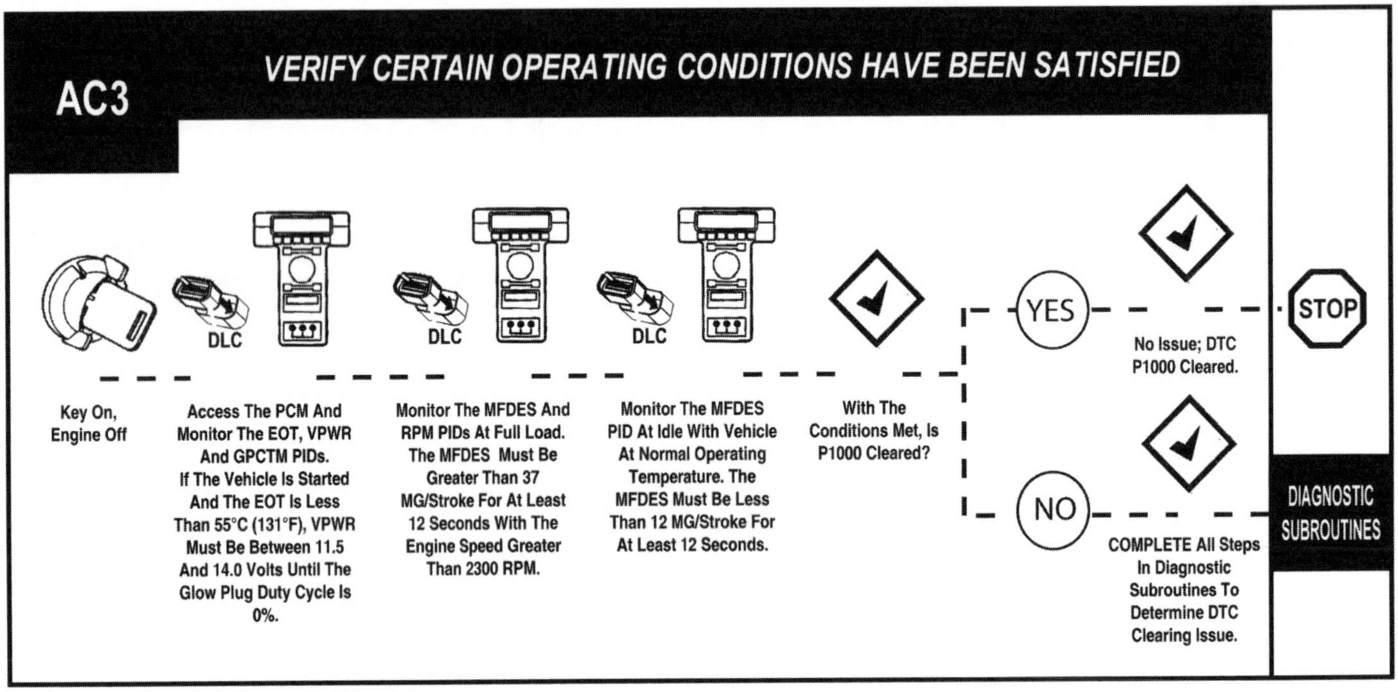
AC3
VERIFY CERTAIN OPERATING CONDITIONS HAVE BEEN SATISFIED
DLC
DLC
DLC
Key On, Engine Off
Access The PCM And Monitor The EOT, VPWR And GPCTM PIDs. If The Vehicle Is Started And The EOT Is Less Than 55°C (131°F), VPWR Must Be Between 11.5 And 14.0 Volts Until The Glow Plug Duty Cycle Is 0%.
Monitor The MFDES And RPM PIDs At Full Load. The MFDES Must Be Greater Than 37 MG/Stroke For At Least 12 Seconds With The Engine Speed Greater Than 2300 RPM.
Monitor The MFDES PID At Idle With Vehicle At Normal Operating Temperature. The MFDES Must Be Less Than 12 MG/Stroke For At Least 12 Seconds.
With The Conditions Met, Is P1000 Cleared?
YES
No Issue; DTC P1000 Cleared.
STOP
NO
COMPLETE All Steps In Diagnostic Subroutines To Determine DTC Clearing Issue.
DIAGNOSTIC SUBROUTINES

Trouble Code (s)	Problem: Powertrain Control Module (PCM) Reset	Test
NA		AD 2006-07

REFERENCE LIST

Description:

Note: The PCM reset occurs when the PCM momentarily reboots or is turned off and on while the engine is operating. If the condition occurs a single time, the engine momentarily stumbles and the PCM goes through a normal key on cycle, including turning the glow plug lamp and glow plugs on, and attempts to validate the accelerator pedal position. If the pedal is not at the idle position when this concern occurs, pedal authority is not allowed by the PCM until the accelerator pedal is released and the engine returns to idle.

This pinpoint test is intended to diagnose the following:

- Loose, damaged or corroded connections
- Harness circuits: VPWR and PWRGND

CAUTION!:

The PCM harness connectors must be properly seated and the connector latch properly attached to eliminate possible driveability concerns or a no start condition. Installing the PCM connectors on an angle may cause an improper connection, misdiagnosis and damaged components. Install the connector until the lever pivots and seats itself. Apply light pressure to get the connector into position on the PCM and then fully seat the connector.

Note: Visually inspect the harness connectors for corrosion, damage, proper mating and correct pin tension.

Note: When the PCM is disconnected additional DTCs will be set. Clear all DTCs after restoring the vehicle.

PINPOINT TEST SCHEMATICS AND CONNECTORS

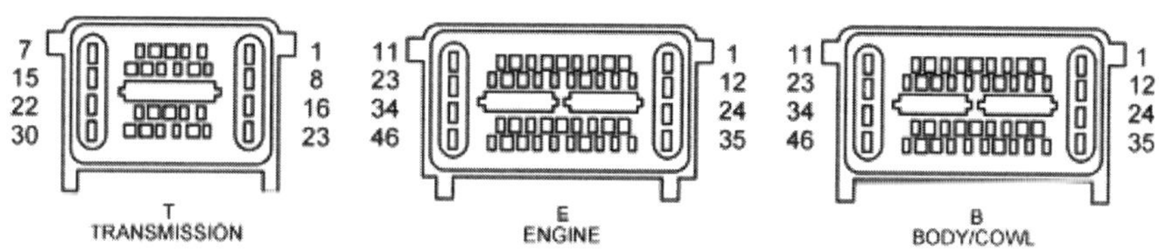

Pin	Circuit
B10, B11, B23	PWRGND (Power Ground)
B34, B46	VPWR (Vehicle Power)

AD1 CHECK FOR DTCS

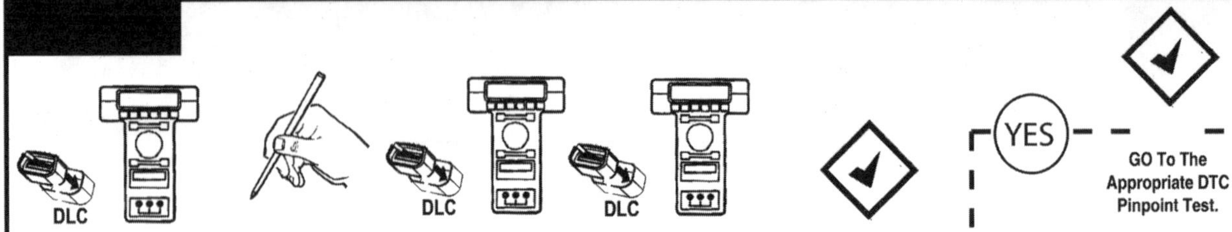

1. Connect The Scan Tool.
2. Access And Record Any DTCs. Record Freeze Frame Data.
3. Clear DTCs.
4. Perform The KOEO And KOER On-Demand Self Tests.
5. Are Any DTCs Present?

YES — GO To The Appropriate DTC Pinpoint Test. STOP

NO — AD2

AD2 CHECK THE VPWR CIRCUITS FOR VOLTAGE

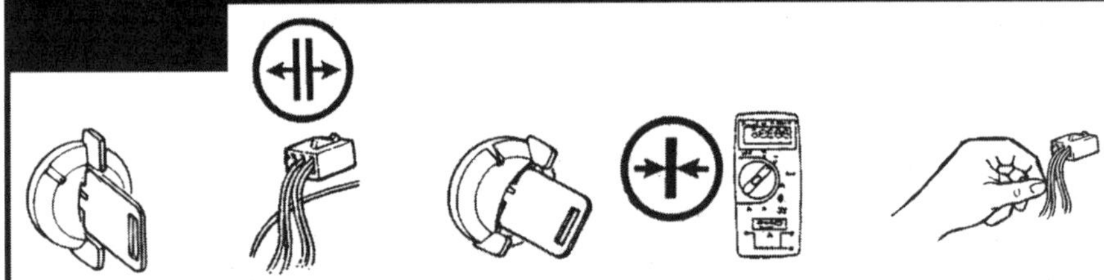

1. Key Off.
2. Disconnect The PCM Body Harness Connector.
3. Key On. Engine Off.

Measure the voltage between:	
(+) PCM Connector, Harness Side	(-)
VPWR - Pin B34, B46	Ground

4. Check For Loose Connections. Check For Damaged Pins Or Corrosion. Wiggle The Harness While Taking Measurements.

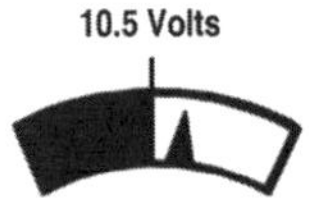

5. Are The Voltages Greater Than 10.5 Volts?

YES — AD3

NO —

REPAIR Open In The Circuit. CLEAR The DTC's. REPEAT The Self-Test. STOP

AD3

CHECK PCM POWER RETURN CIRCUIT

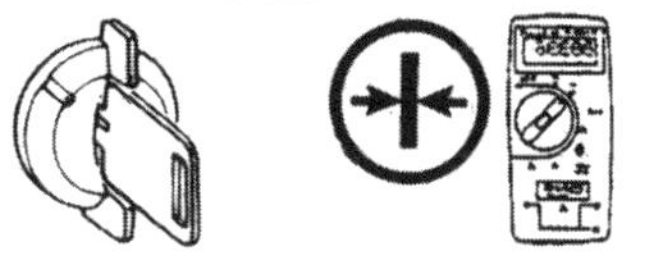

Key Off.

Measure the resistance between:

(+) PCM Connector, Harness Side	(-)
PWRGND - Pin B10, B11, B23	Ground

Are The Resistances Less Than 5 Ohms?

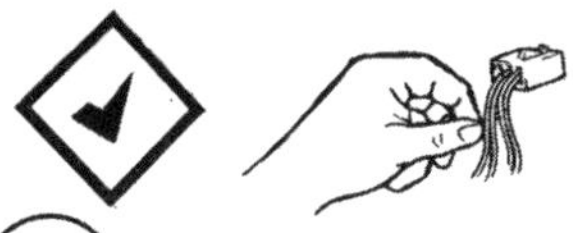

YES — Unable To Duplicate Condition. CHECK For Loose Connections, Damaged Or Corroded Pins. WIGGLE Harness Attempting To Recreate Fault. REFER To Symptom Charts. **STOP**

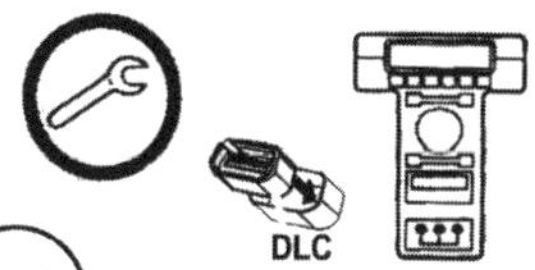

NO — REPAIR The Open Circuit. CLEAR The DTC's. REPEAT The Self-Test. **STOP**

NOTES

Trouble Code (s)	Problem: Unable To Activate Self-Test Network Communication Error	Test
NA		AE 2006-07

REFERENCE LIST

This pinpoint test is intended to diagnose the following:

- Unable to communicate with the PCM
- Unable to activate the PCM self-test
- Harness circuits: VPWR, VREF and PWRGND

CAUTION!:
The powertrain control module (PCM) harness connectors must be properly seated and the connector latch properly attached to eliminate possible driveability concerns or a no start condition. Installing the PCM connectors on an angle may cause an improper connection, misdiagnosis and damaged components. Install the connector until the lever pivots and seats itself. Apply light pressure to get the connector into position on the PCM and then fully seat the connector.

Note: Visually inspect the harness connectors for corrosion, damage, proper mating and correct pin tension.

PINPOINT TEST SCHEMATICS AND CONNECTORS

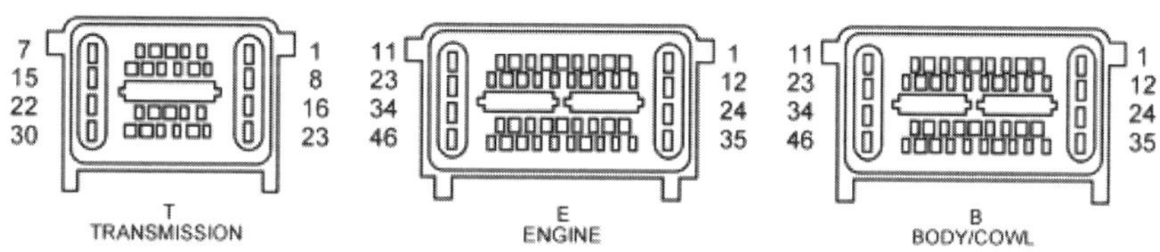

Pin	Circuit
B29, B45, E36	VREF (Reference Voltage)
B10, B11, B23	PWRGND (Power Ground)
B34, B46	VPWR (Vehicle Power)

AE1 — CARRY OUT A VEHICLE INSPECTION AND VERIFY THE SELF-TEST PROCEDURE

Note: The scan tool resets if battery voltage is less than 9.5 volts. If carrying out repeated self-tests, disable the glow plug control module (GPCM) and disregard the glow plug codes while the GPCM is disabled. If equipped with a power take-off (PTO) system, it must be disabled while carrying out the PCM self-tests.

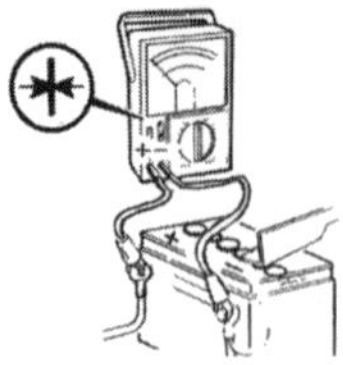

Check The Battery Voltage.

DIAGNOSTIC METHODS

Verify That The Correct Procedure Was Used To Activate The Self Test For The Diagnostic Tool. REFER To Diagnostic Methods.

Was The Correct Self-Test Procedure Used?

- YES → AE2
- NO → DIAGNOSTIC METHODS — REFER To Diagnostic Methods For Correct Operating Procedures. → STOP

AE2 — CARRY OUT THE DATA LINK DIAGNOSTIC TEST

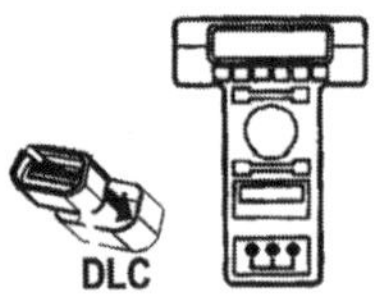

Connect The Diagnostic Tool.

Key On, Engine Off.

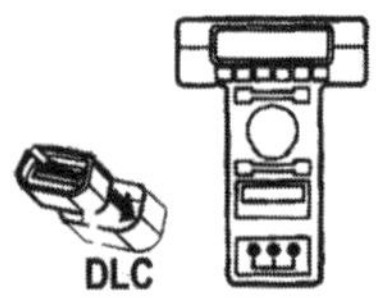

Carry Out The Data Link Diagnostics Test.

Is System Passed Obtained?

- YES → QUICK TEST
- NO →
  - If All Electronic Control Units No Response/not Equipped: → AE3
  - If No Response From The Diagnostic Tool, Diagnose And Repair The Module Communications Network → STOP

AE3 CHECK THE PCM VPWR CIRCUITS FOR VOLTAGE

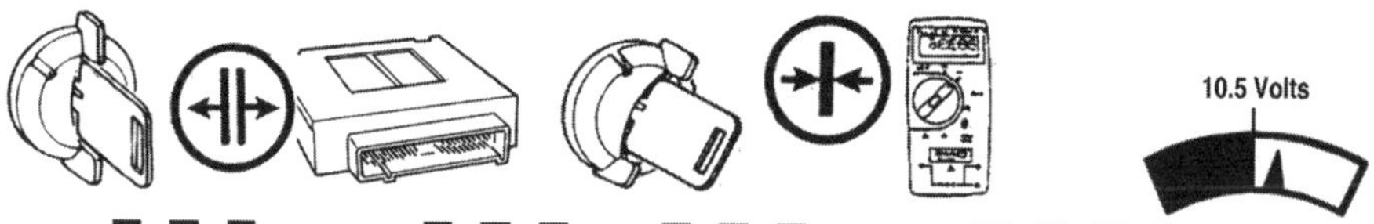

Key Off.

Disconnect The PCM Body Harness Connector.

Key On, Engine Off

Measure the voltage between:	
(+) PCM Connector, Harness Side	(-)
VPWR - Pin B34, B46	Ground

Are The Voltages Greater Than 10.5 Volts?

YES — AE4

NO — Go To Pinpoint Test: A

AE4 CHECK THE PCM GROUND CIRCUITS FOR AN OPEN

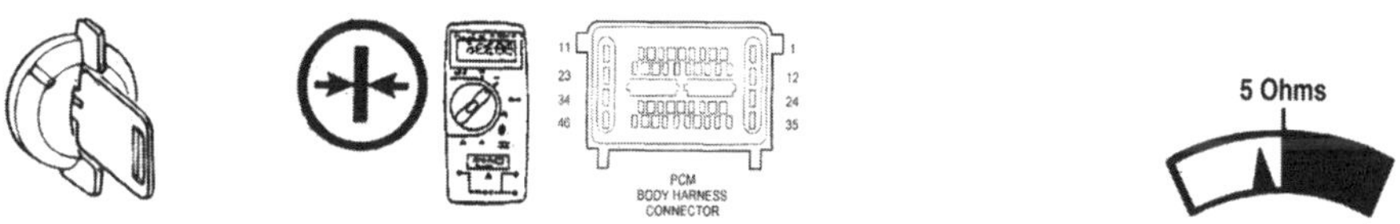

Key Off.

Measure the resistance between:	
(+) PCM Connector, Harness Side	(-)
PWRGND - Pin B10, B11, B23	Ground

Are The Resistances Less Than 5 Ohms?

YES — AE5

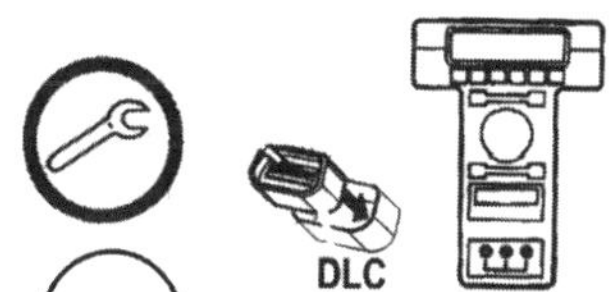

NO — REPAIR The Circuit In Question. CLEAR The DTCs. REPEAT The Self-Test. STOP

AE5 CHECK THE VREF CIRCUITS FOR A SHORT TO GROUND

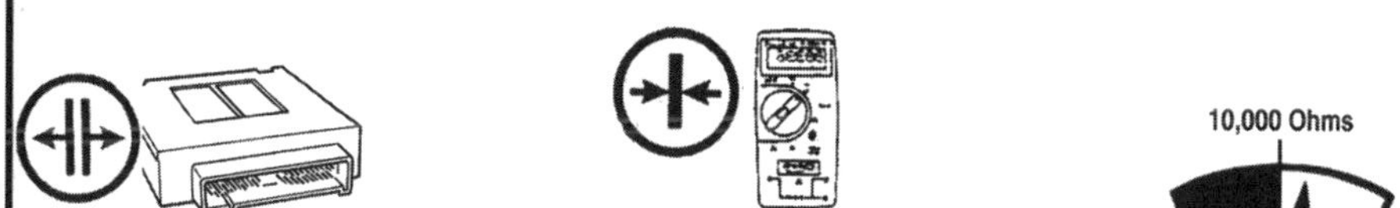

Disconnect The PCM Engine Harness Connector.

Measure the resistance between:	
(+) PCM Connector, Harness Side	(-)
VREF - Pin B29, B45	Ground

Measure the resistance between:	
(+) PCM Connector, Harness Side	(-)
VREF - Pin E36	Ground

Are The Resistances Greater Than 10,000 Ohms?

YES — DIAGNOSE And REPAIR The Module Communications Network. STOP

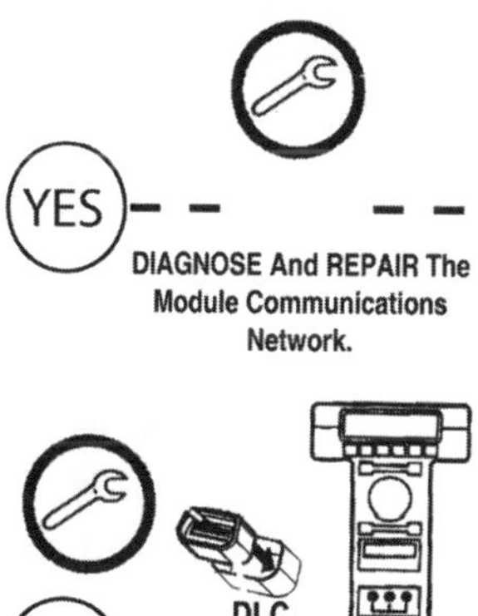

NO — REPAIR The Circuit In Question. CLEAR The DTCs. REPEAT The Self-Test. STOP

NOTES

Trouble Code (s)	Problem: Glow Plug System	Test
P0670, P0671, P0672, P0673, P0674, P0675, P0676, P0677, P0678, P0683		AF 2006-07

REFERENCE LIST

This pinpoint test is intended to diagnose the following:

- Glow plugs
- Glow plug bus bar
- GPCM
- PCM

P0670 = Glow Plug Module Control Circuit
P0671 = Cylinder 1 Glow Plug Circuit
P0672 = Cylinder 2 Glow Plug Circuit
P0673 = Cylinder 3 Glow Plug Circuit
P0674 = Cylinder 4 Glow Plug Circuit
P0675 = Cylinder 5 Glow Plug Circuit
P0676 = Cylinder 6 Glow Plug Circuit
P0677 = Cylinder 7 Glow Plug Circuit
P0678 = Cylinder 8 Glow Plug Circuit
P0683 = Glow Plug Control Module To PCM Communication Circuit

DIAGNOSTIC TROUBLE CODE INDEX			
CODES	KOEO	KOER	CM
P0670	AF1	AF1	AF1
P0671	AF1	AF1	AF1
P0672	AF1	AF1	AF1
P0673	AF1	AF1	AF1
P0674	AF1	AF1	AF1
P0675	AF1	AF1	AF1
P0676	AF1	AF1	AF1
P0677	AF1	AF1	AF1
P0678	AF1	AF1	AF1
P0683	AF1	AF1	AF1

Description:

The glow plug system is composed of a solid state glow plug control module (GPCM) and the glow plugs. The GPCM provides direct battery voltage to the glow plugs. The GPCM monitors the individual glow plug operation. Faults detected by the GPCM are transmitted to the PCM using a serial communication signal on a diagnostic line. The powertrain control module (PCM) enables and controls the GPCM. The glow plug ON time is a function of oil temperature, barometric pressure, and battery voltage. The ON time normally varies between 1 to 120 seconds and is limited by the GPCM to 180 seconds regardless of PCM commands.

Note: The wait-to-start indicator ON time is independent from GPCM ON time.

Glow Plug Monitor Self-Test:

The glow plug monitor self-test is a functional key on, engine running (KOER) self-test of the glow plug system. The PCM activates the GPCM, which monitors the glow plugs. A fault must be present at the time of testing to detect a fault. If the GPCM detects a fault, a message is sent to the PCM through the diagnostic line and then transmitted to the diagnostic tool.

Note: The accelerator pedal may be used to increase the engine RPM and maintain battery voltage during the self-test. Make sure that the vehicle batteries are fully charged prior to carrying out the glow plug monitor self-test.

CAUTION!:

GPCM connectors are not interchangeable. The system generates DTCs if the connectors are mismatched. Make sure the green harness connector is plugged into the green module connector and the black harness connector into the black module connector. There is a black stripe on the GPCM housing for visual verification. The stripe should line up with the black connector.

When disconnecting the green and black connectors at the GPCM it is suggested that the module be unbolted from the mounting bracket first and then the connectors disconnected. This allows for easier disconnect and prevents damage to the connector terminals.

CAUTION!:

The powertrain control module (PCM) harness connectors must be properly seated and the connector latch properly attached to eliminate possible driveability concerns or a no start condition. Installing the PCM connectors on an angle may cause an improper connection, misdiagnosis and damaged components. Install the connector until the lever pivots and seats itself. Apply light pressure to get the connector into position on the PCM and then fully seat the connector.

Note: Visually inspect the harness connectors for corrosion, damage, proper mating and correct pin tension.

Note: When the PCM is disconnected additional DTCs will be set. Clear all DTCs after restoring the vehicle.

Trouble Code (s)	Problem: Glow Plug System (Continued)	Test
P0670, P0671, P0672, P0673, P0674, P0675, P0676, P0677, P0678, P0683		AF 2006-07

PINPOINT TEST SCHEMATICS AND CONNECTORS

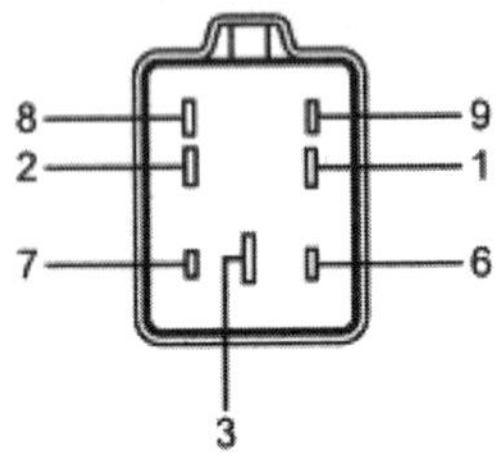

Glow Plug Control Module-Green (GPCM-Green) Connector

Pin	Circuit
2	Glow Plug Number 7
1	Glow Plug Number 5
7	Glow Plug Number 3
6	Glow Plug Number 1
9	Diagnostic Communication (Diagnostic Communication)
8	GPE (Glow Plug Enable)

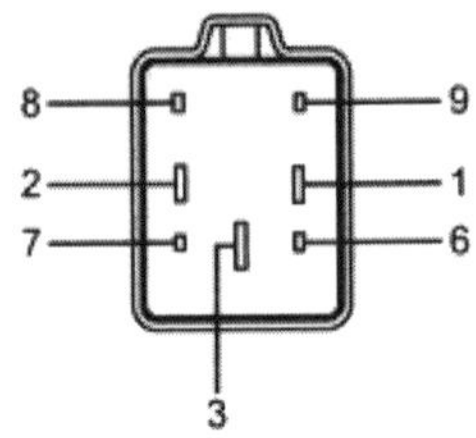

Glow Plug Control Module-Black (GPCM-Black) Connector

Pin	Circuit
2	Glow Plug Number 8
1	Glow Plug Number 6
7	Glow Plug Number 4
6	Glow Plug Number 2
9	VPWR (Vehicle Power)

Glow Plug Bus Bar Connector

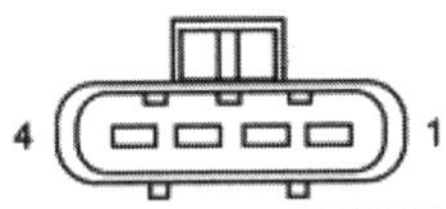

BANK 1	
Pin	Circuit
4	Glow Plug Number 7
3	Glow Plug Number 5
2	Glow Plug Number 3
1	Glow Plug Number 1

BANK 2	
Pin	Circuit
4	Glow Plug Number 8
3	Glow Plug Number 6
2	Glow Plug Number 4
1	Glow Plug Number 2

Powertrain Control Module (PCM) Connector

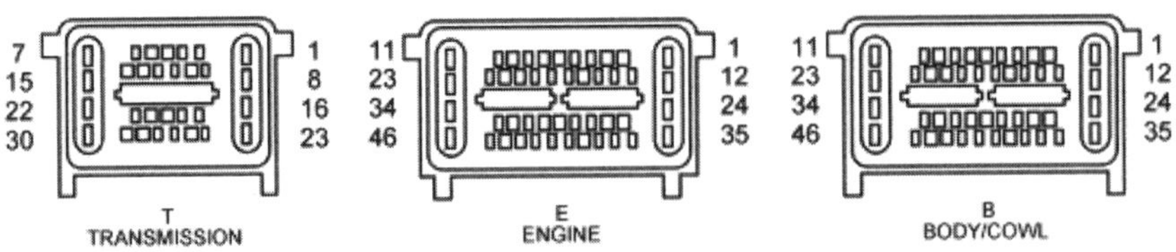

Pin	Circuit
E17	GPD (GPCM / PCM Communication)
E3	GPE (Glow Plug Enable)

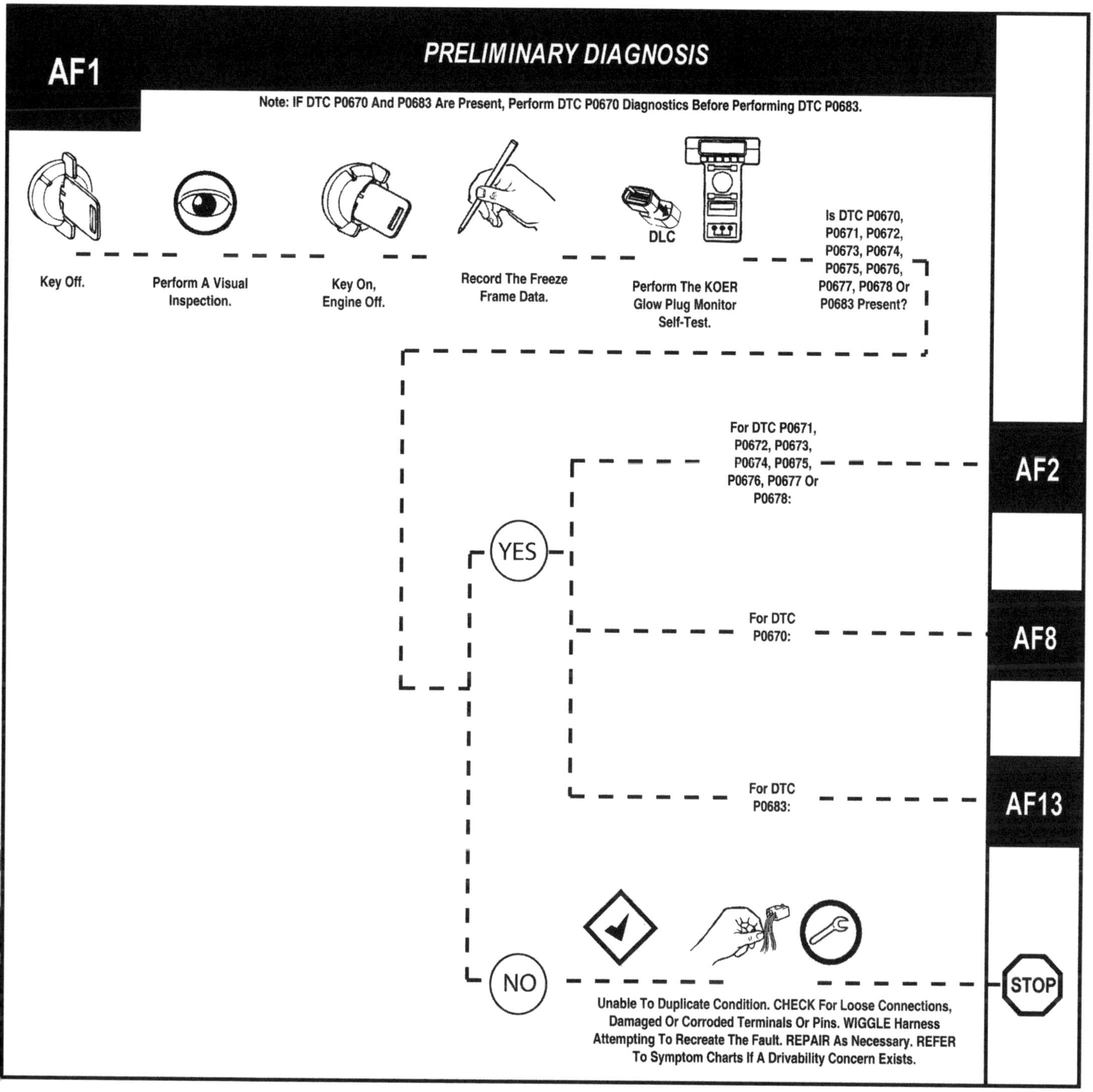
AF1
PRELIMINARY DIAGNOSIS
Note: IF DTC P0670 And P0683 Are Present, Perform DTC P0670 Diagnostics Before Performing DTC P0683.
Key Off.
Perform A Visual Inspection.
Key On, Engine Off.
Record The Freeze Frame Data.
DLC
Perform The KOER Glow Plug Monitor Self-Test.
Is DTC P0670, P0671, P0672, P0673, P0674, P0675, P0676, P0677, P0678 Or P0683 Present?
YES
For DTC P0671, P0672, P0673, P0674, P0675, P0676, P0677 Or P0678:
AF2
For DTC P0670:
AF8
For DTC P0683:
AF13
NO
Unable To Duplicate Condition. CHECK For Loose Connections, Damaged Or Corroded Terminals Or Pins. WIGGLE Harness Attempting To Recreate The Fault. REPAIR As Necessary. REFER To Symptom Charts If A Drivability Concern Exists.
STOP

AF2

DIAGNOSTIC TROUBLE CODES P0671, P0672, P0673, P0674, P0675, P0676, P0677 OR P0678

Note: Refer To The Wiring Diagram At The Beginning Of This Pinpoint Test.

Note: Glow Plugs Are Numbered Front To Rear With Even Numbers On Left Side Of Engine.

Note: Use The Wiring Diagram To Match The Connector Pins With The DTCs.

Possible Causes:
- Glow Plug Circuit Open
- Glow Plug Circuit Shorted
- GPCM
- Glow Plug

Key Off.

Disconnect The Suspect Glow Plug Bus Bar.

Check For Loose, Damaged Or Backed Out Terminals.

Measure the resistance between:	
(+) Glow Plug Bus Bar Connector, Component Side	(-)
Suspect Glow Plug	Ground

0.5 And 2 Ohms

Is The Resistance Between 0.5 And 2 Ohms?

YES → AF4

NO → AF3

AF3

CHECK THE GLOW PLUG CIRCUIT BETWEEN THE BUS BAR CONNECTOR AND THE GLOW PLUG

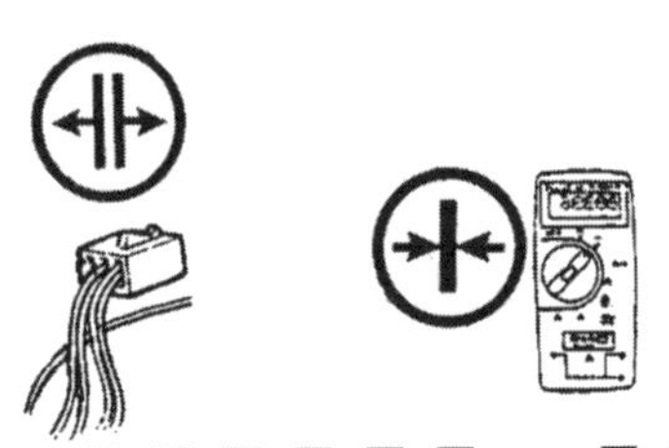

Disconnect The Bus Bar Harness From The Glow Plugs.

Measure the resistance between:	
(+) Glow Plug Bus Bar Connector, Harness Side	(-) Glow Plug Connector, Harness Side
Suspect Glow Plug	Suspect Glow Plug

5 Ohms

Is The Resistance Less Than 5 Ohms?

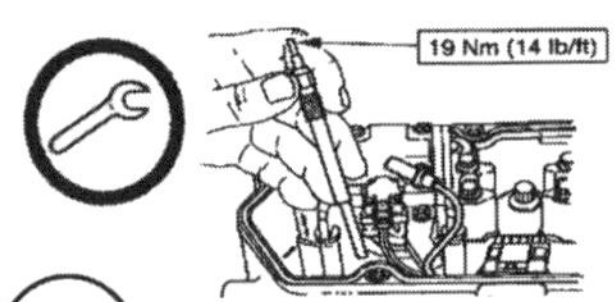

YES → STOP
INSTALL A New Glow Plug. CLEAR The DTCs. REPEAT The Self-Test.

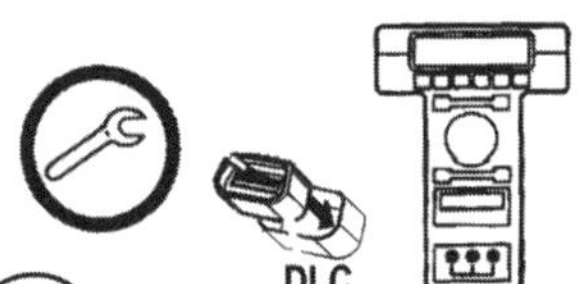

NO → STOP
INSTALL A New Bus Bar. CLEAR The DTCs. REPEAT The Self-Test.

AF4 — CHECK THE GLOW PLUG VOLTAGE

Disconnect The Suspect GPCM Connector.

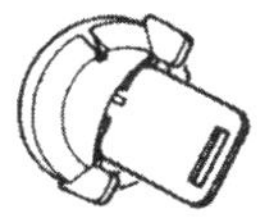

Key On, Engine Off.

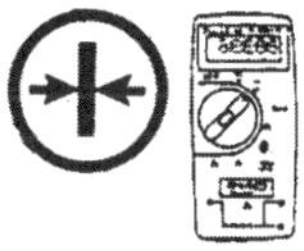

Measure the voltage between:	
(+) Suspect GPCM Connector, Harness Side	(-)
VBAT	Ground

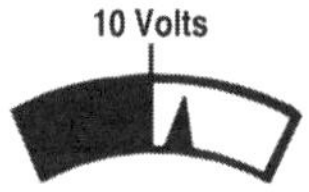

Is The Voltage Greater Than 10 Volts?

YES → AF5

NO →

REPAIR The Circuit. CLEAR The DTCs. REPEAT The Self-Test. STOP

AF5 — CHECK THE GLOW PLUG CIRCUIT FOR AN OPEN

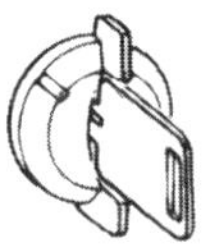

Key Off.

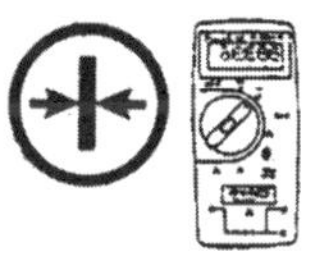

Measure the resistance between:	
(+) Suspect GPCM Connector, Harness Side	(-) Glow Plug Bus Bar Connector, Harness Side
Suspect GPCM	Suspect Glow Plug

Is The Resistance Less Than 5 Ohms?

YES → AF6

NO →

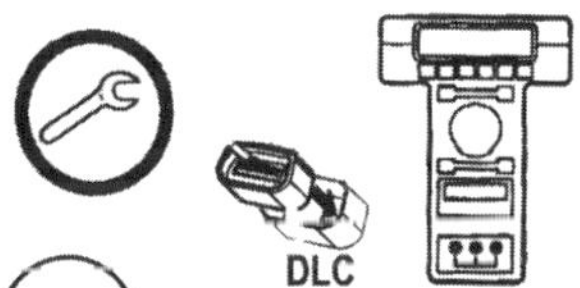

REPAIR The Circuit. CLEAR The DTCs. REPEAT The Self-Test. STOP

AF6 — CHECK THE GLOW PLUG CIRCUIT FOR A SHORT TO GROUND

10,000 Ohms

Measure the resistance between:	
(+) Suspect GPCM Connector, Harness Side	(-)
Suspect GPCM Pin	Ground

Is The Resistance Greater Than 10,000 Ohms?

YES — AF7

NO — DLC — REPAIR The Circuit. CLEAR The DTCs. REPEAT The Self-Test. — STOP

AF7 — CHECK THE GLOW PLUG CIRCUIT FOR A SHORT TO VOLTAGE

Key On, Engine Off.

Measure the voltage between:	
(+) Suspect GPCM Connector, Harness Side	(-)
Suspect GPCM Pin	Ground

Voltage

Is Any Voltage Indicated?

YES — DLC — REPAIR The Circuit. CLEAR The DTCs. REPEAT The Self-Test. — STOP

NO — INSTALL A New GPCM. CLEAR The DTCs. REPEAT The Self-Test. — STOP

AF8 — DIAGNOSTIC TROUBLE CODE P0670

Note: Refer To The PCM And Component Connector Pin Numbers At The Beginning Of This Pinpoint Test.

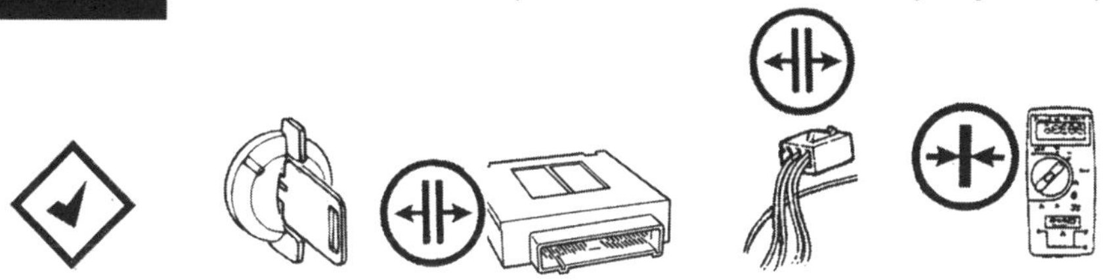

Possible Causes:
- Control Line Open
- Control Line Shorted To Ground Or Power
- PCM
- GPCM

Key Off.

Disconnect The PCM.

Disconnect The GPCM (Green Connector).

Measure the resistance between:	
(+) GPCM-Green Connector, Harness Side	(-) PCM Connector, Harness Side
GPE - Pin 8	GPE - Pin E3

Is The Resistance Less Than 5 Ohms?

YES → AF9

NO →

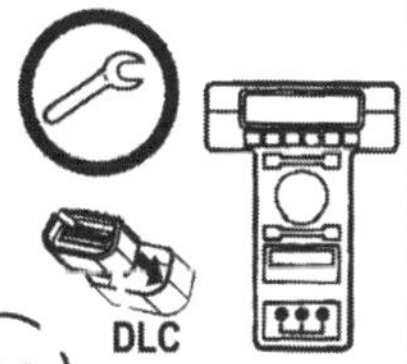

REPAIR The Circuit. CLEAR The DTCs. REPEAT The Self-Test. → STOP

AF9 — CHECK THE CONTROL CIRCUIT FOR A SHORT TO GROUND

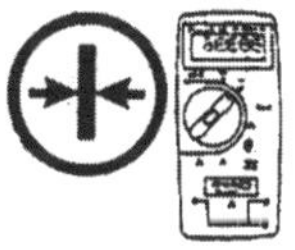

Measure the resistance between:	
(+) GPCM-Green Connector, Harness Side	(-)
GPE - Pin 8	Ground

Is The Resistance Greater Than 10,000 Ohms?

YES → AF10

NO →

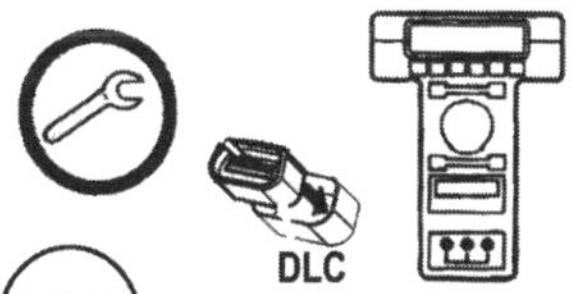

REPAIR The Circuit. CLEAR The DTCs. REPEAT The Self-Test. → STOP

AF10

CHECK THE CONTROL CIRCUIT FOR A SHORT TO VOLTAGE

Key On, Engine Off.

Measure the voltage between:	
(+) GPCM-Green Connector, Harness Side	(-)
GPE - Pin 8	Ground

Voltage

Is Any Voltage Indicated?

YES — REPAIR The Circuit. CLEAR The DTCs. REPEAT The Self-Test. DLC — STOP

NO — AF11

AF11

CHECK THE GPCM FOR VOLTAGE

Key Off.

Disconnect The GPCM (Black Connector).

Key On, Engine Off.

Measure the voltage between:	
(+) GPCM-Black Connector, Harness Side	(-)
VPWR - Pin 9	Ground

10 Volts

Is The Voltage Greater Than 10 Volts?

YES — AF12

NO — REPAIR The Circuit In Question. CLEAR The DTCs. REPEAT The Self-Test. DLC — STOP

AF12 — CHECK THE GPCM CONTROL CIRCUIT

Note: With the key ON, engine OFF the PCM provides a path to ground for the glow plug enable (GPE) circuit.

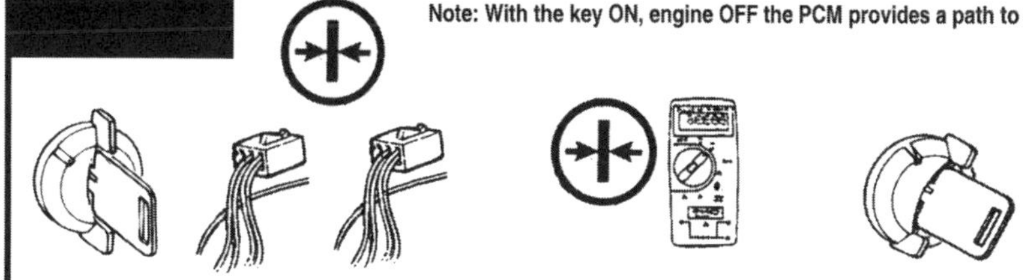

Key Off.

GPCM-Black Connector Connected.

PCM Connector Connected.

Measure the voltage between:	
(+) Vehicle Battery	(-) GPCM-Green Connector, Harness Side
Positive terminal	GPE - Pin 8

Key On, Engine Off.

10 Volts

Is The Voltage Greater Than 10 Volts?

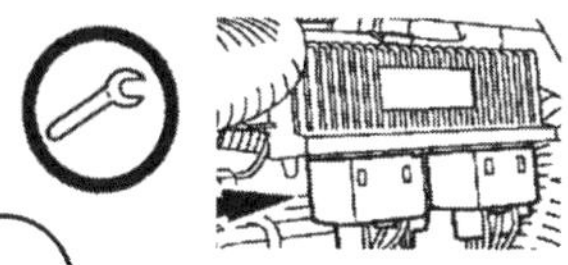

YES — INSTALL A New GPCM. CLEAR The DTCs. REPEAT The Self-Test. → STOP

NO → AF18

AF13 — DIAGNOSTIC TROUBLE CODE P0683

Note: Refer To The PCM And Component Connector Pin Numbers At The Beginning Of This Pinpoint Test.

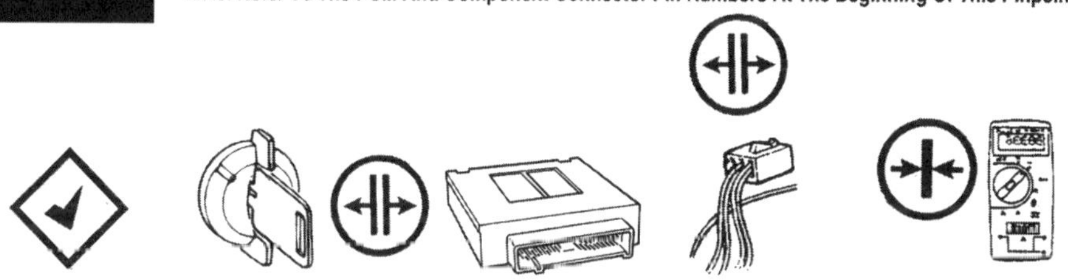

Possible Causes:
- Diagnostic Line Open
- Diagnostic Line Shorted To Ground
- Diagnostic Line Shorted To Power
- PCM
- GPCM

Key Off.

Disconnect The PCM.

Disconnect The GPCM (Green Connector).

Measure the resistance between:	
(+) GPCM-Green Connector, Harness Side	(-) PCM Connector, Harness Side
Diagnostic Communication - Pin 9	GPD - Pin E17

5 Ohms

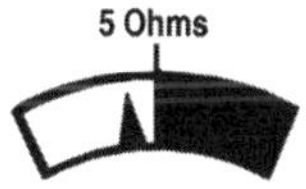

Is The Resistance Less Than 5 Ohms?

YES → AF14

NO — REPAIR The Circuit. CLEAR The DTCs. REPEAT The Self-Test. → STOP

AF14

CHECK THE DIAGNOSTIC CIRCUIT FOR A SHORT TO GROUND

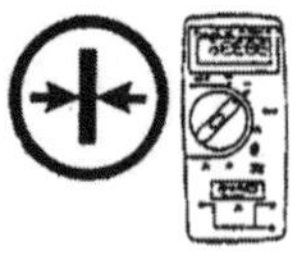

Measure the resistance between:	
(+) GPCM-Green Connector, Harness Side	(-)
Diagnostic Communication - Pin 9	Ground

Is The Resistance Greater Than 10,000 Ohms?

YES → AF15

NO → REPAIR The Circuit. CLEAR The DTC's. Repeat The Self-Test. → STOP

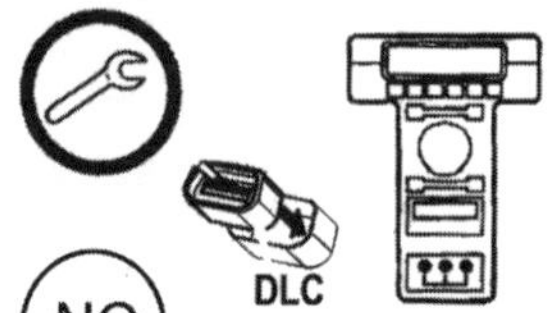

AF15

CHECK THE DIAGNOSTIC CIRCUIT FOR A SHORT TO VOLTAGE

Key On, Engine Off.

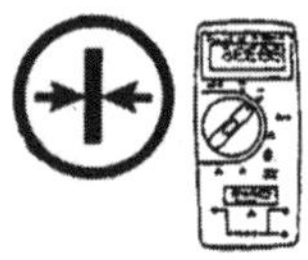

Measure the voltage between:	
(+) GPCM-Green Connector, Harness Side	(-)
Diagnostic Communication - Pin 9	Ground

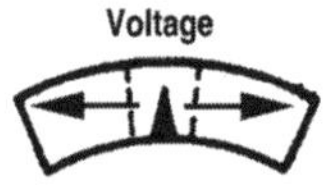

Is Any Voltage Indicated?

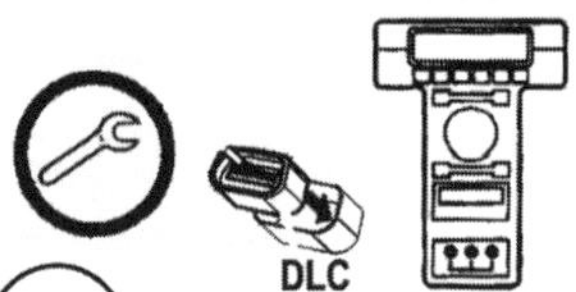

YES → REPAIR The Circuit. CLEAR The DTC's. Repeat The Self-Test. → STOP

NO → AF16

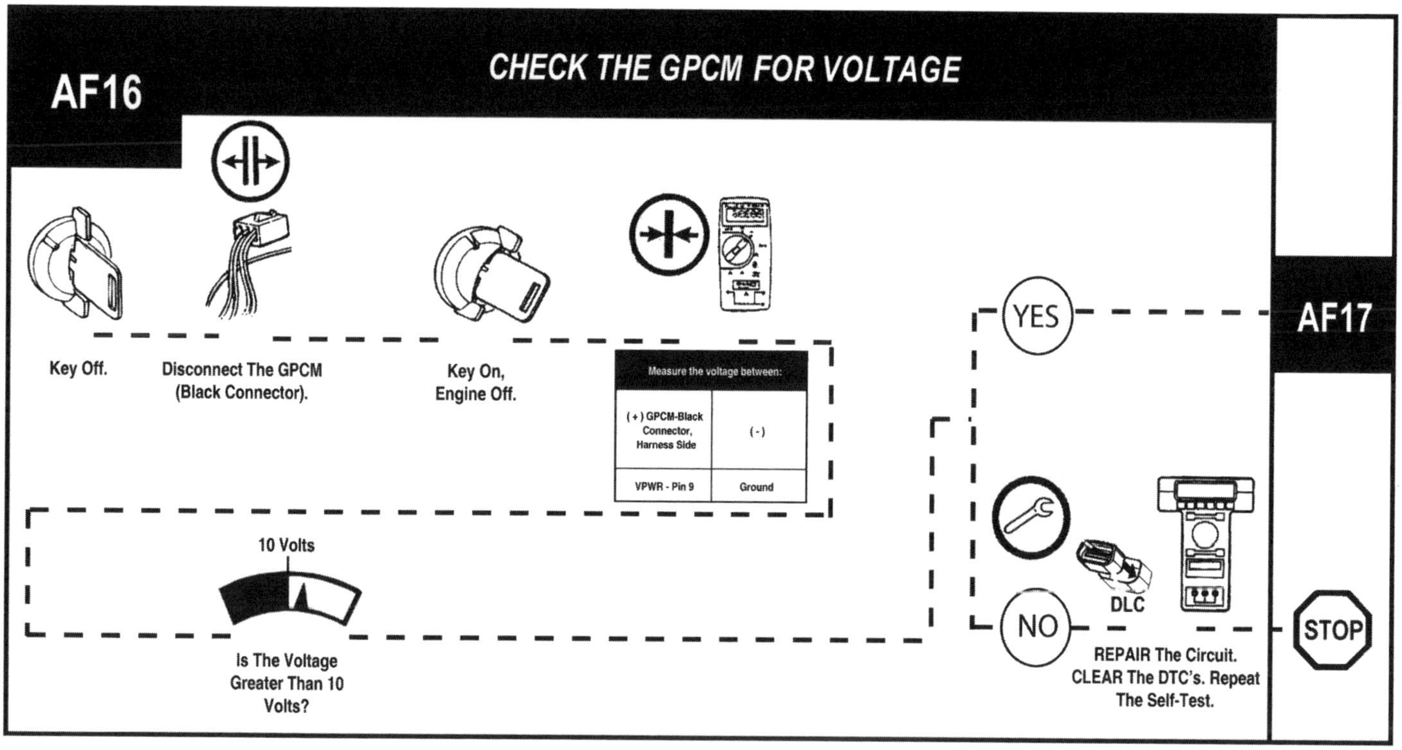
AF16
CHECK THE GPCM FOR VOLTAGE
Key Off.
Disconnect The GPCM (Black Connector).
Key On, Engine Off.
Measure the voltage between:
(+) GPCM-Black Connector, Harness Side
(-)
VPWR - Pin 9
Ground
10 Volts
Is The Voltage Greater Than 10 Volts?
YES
AF17
NO
DLC
REPAIR The Circuit. CLEAR The DTC's. Repeat The Self-Test.
STOP

AF17

CHECK THE GPCM OPERATION

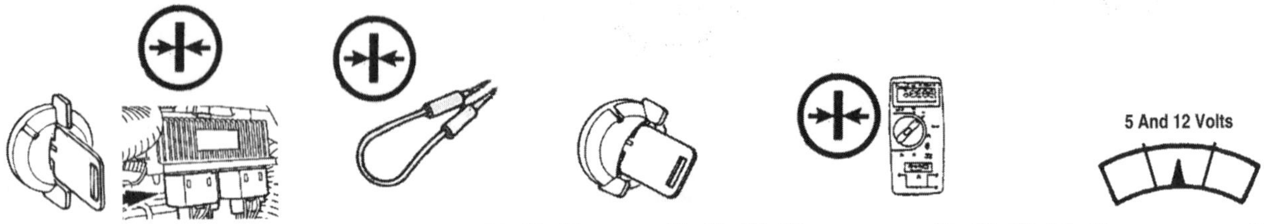

Key Off.

Reconnect The Black And Green GPCM Connectors.

Connect a 5 amp fused jumper wire between the following:	
Point A PCM Connector, Harness Side	Point B
GPE - Pin E3	Ground

Key On, Engine Off.

Measure the voltage between:	
(+) PCM Connector, Harness Side	(-)
GPD - Pin E17	Ground

Is The Voltage Between 5 And 12 Volts?

YES → AF18

NO →

INSTALL A New GPCM. CLEAR The DTCs. REPEAT The Self-Test.

STOP

AF18 CHECK FOR CORRECT PCM OPERATION

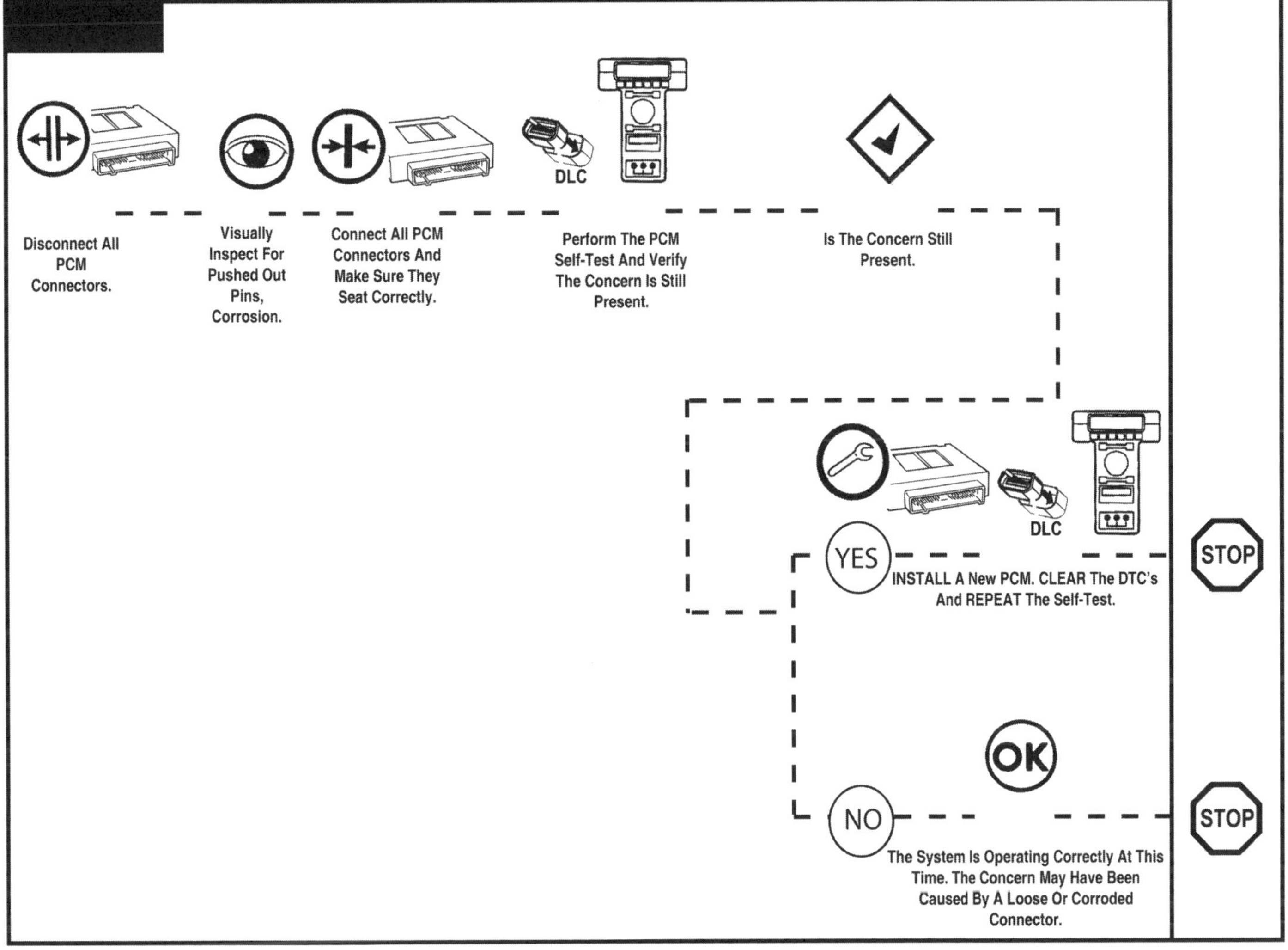

NOTES

Trouble Code (s)		Test
P2104, P2122, P2123, P2127, P2128, P2132, P2133, P2138, P2139, P2140	**Problem: Accelerator Pedal Position (APP) Sensor**	**AG** 2006-07

REFERENCE LIST

This pinpoint test is intended to diagnose the following:

- APP sensor (9F836)
- Harness circuits: APP1, APP2, APP3, SIGRTN, and VREF
- Powertrain control module (PCM) (12A650)

P2104 = Throttle Actuator Control System Forced Idle
P2122 = APP Sensor D Low Input
P2123 = APP Sensor D High Input
P2127 = APP Sensor E Low Input
P2128 = APP Sensor E High Input
P2132 = APP Sensor F Low Input
P2133 = APP Sensor F High Input
P2138 = APP Sensor D/E Voltage Correlation
P2139 = APP Sensor D/F Voltage Correlation
P2140 = APP Sensor E/F Voltage Correlation

DIAGNOSTIC TROUBLE CODE INDEX			
Codes	**KOEO**	**KOER**	**CM**
P2104	AG1	AG1	AG1
P2122	AG1	AG1	AG1
P2123	AG1	AG1	AG1
P2127	AG1	AG1	AG1
P2128	AG1	AG1	AG1
P2132	AG1	AG1	AG1
P2133	AG1	AG1	AG1
P2138	AG1	AG1	AG1
P2139	AG1	AG1	AG1
P2140	AG1	AG1	AG1

Note:
Base idle should be no lower than 0.5 Volt.

Signal Functions:
The APP sensor provides 3 independent input signals to the powertrain control module (PCM) indicating the driver's demand for power and is based upon the rotation angle of the pedal. The PCM determines the APP position by processing and uses the input signals to calculate the desired fuel quantity, injection timing, and the correct injection control pressure.

Fault Detection/Management:
Any detected malfunction of the APP sensor illuminates the electronic throttle control (wrench) indicator. A PCM detected fault on one of the 3 sensor signals will permit normal operation. A fault with 2 sensor signals only allows the engine to operate at idle.

CAUTION!:
The powertrain control module (PCM) harness connectors must be properly seated and the connector latch properly attached to eliminate possible driveability concerns or a no start condition. Installing the PCM connectors on an angle may cause an improper connection, misdiagnosis and damaged components. Install the connector until the lever pivots and seats itself. Apply light pressure to get the connector into position on the PCM and then fully seat the connector.

Note: Visually inspect the harness connectors for corrosion, damage, proper mating and correct pin tension.

PINPOINT TEST SCHEMATICS AND CONNECTORS

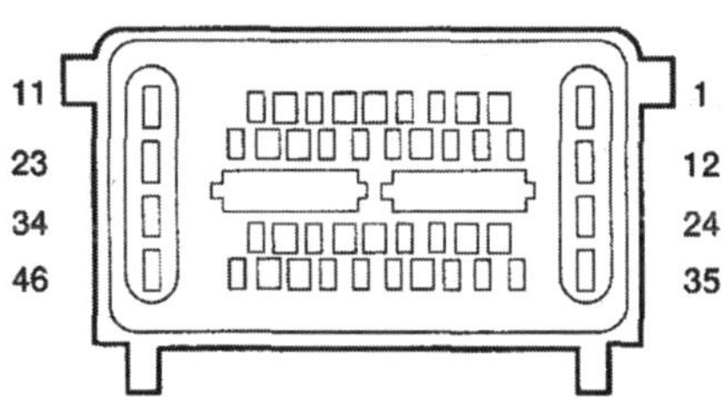

PCM
BODY HARNESS
CONNECTOR

Pin	Circuit
B20, B33	SIGRTN (Signal Return)
B25	APPS3 (Accelerator Pedal Position Sensor 3)
B37	APPS2 (Accelerator Pedal Position Sensor 2)
B26	APPS1 (Accelerator Pedal Position Sensor 1)

Accelerator Pedal Position (APP) Sensor Connector

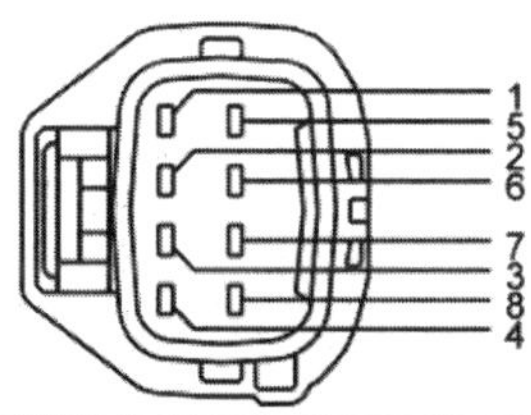

Pin	Circuit
8	APPS3 (Accelerator Pedal Position Sensor 3)
5	APPS2 (Accelerator Pedal Position Sensor 2)
2	APPS1 (Accelerator Pedal Position Sensor 1)
1, 3	SIGRTN (Signal Return)
6, 7	VREF (Reference Voltage)

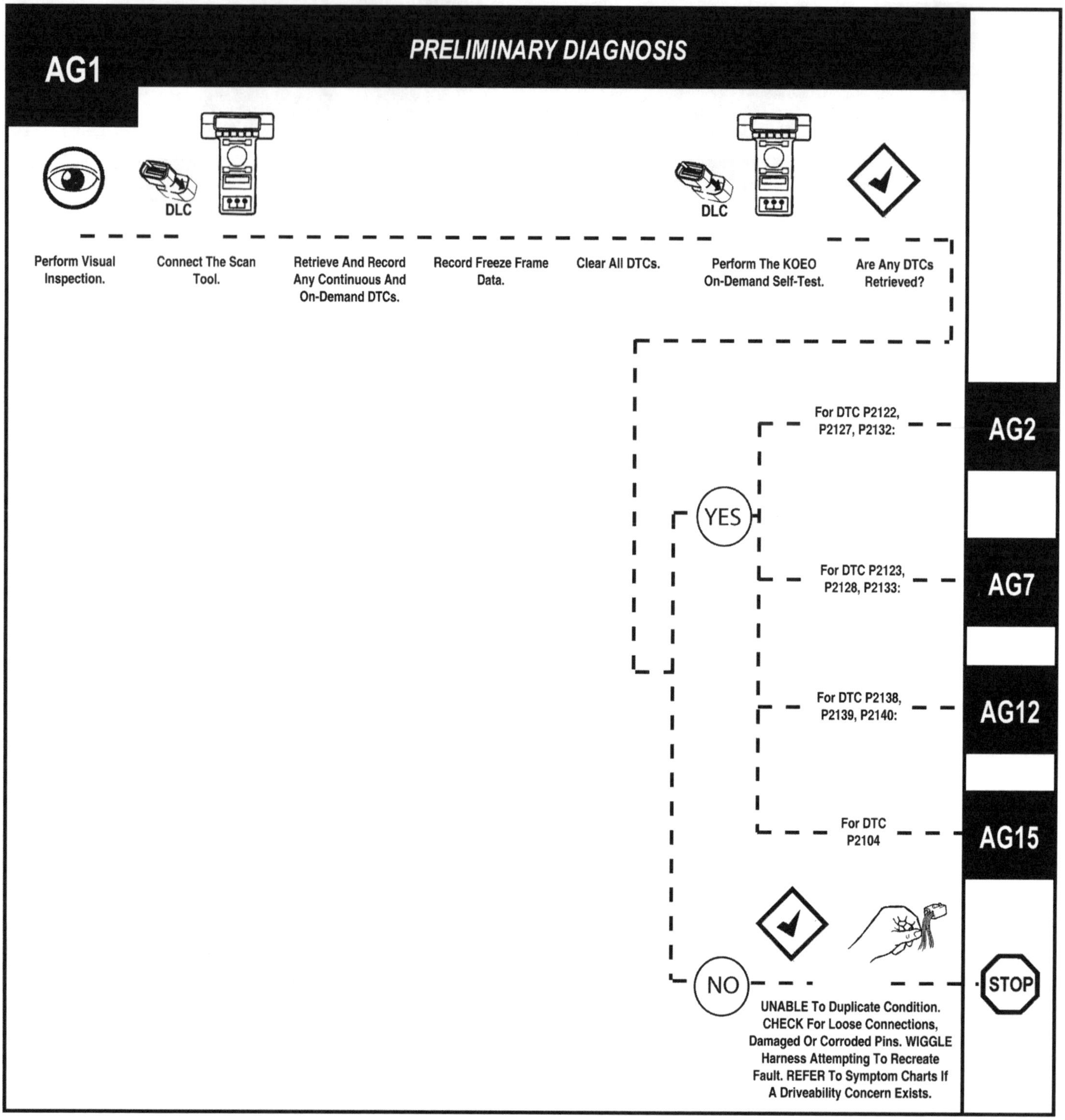
AG1
PRELIMINARY DIAGNOSIS
DLC
DLC
Perform Visual Inspection.
Connect The Scan Tool.
Retrieve And Record Any Continuous And On-Demand DTCs.
Record Freeze Frame Data.
Clear All DTCs.
Perform The KOEO On-Demand Self-Test.
Are Any DTCs Retrieved?
YES
For DTC P2122, P2127, P2132:
AG2
For DTC P2123, P2128, P2133:
AG7
For DTC P2138, P2139, P2140:
AG12
For DTC P2104
AG15
NO
UNABLE To Duplicate Condition. CHECK For Loose Connections, Damaged Or Corroded Pins. WIGGLE Harness Attempting To Recreate Fault. REFER To Symptom Charts If A Driveability Concern Exists.
STOP

AG2 DIAGNOSTIC TROUBLE CODE (DTC) P2122/P2127/P2132

Note: Refer To The PCM And Component Connector Pin Numbers At The Beginning Of This Pinpoint Test.

Possible Causes:
- APP Not Seated Correctly
- APP Sensor
- Open In Signal Circuit
- Signal Circuit Shorted To Ground
- PCM

Key On. Engine Off.

Access PIDs APP1, APP2, APP3.

Apply And Release The Accelerator Pedal.

Observe PID While Pressing And Releasing The Accelerator Pedal.

Do All Signal Values Change When The Accelerator Is Pressed And Released?

YES — Unable To Duplicate Condition. CHECK For Loose Connections, Damaged Or Corroded Pins. WIGGLE Harness Attempting To Recreate Fault. REPAIR As Necessary. REFER To Symptom Charts If A Driveablility Concern Exists. STOP

NO — AG3

AG3 CHECK VREF SENSOR CIRCUITS

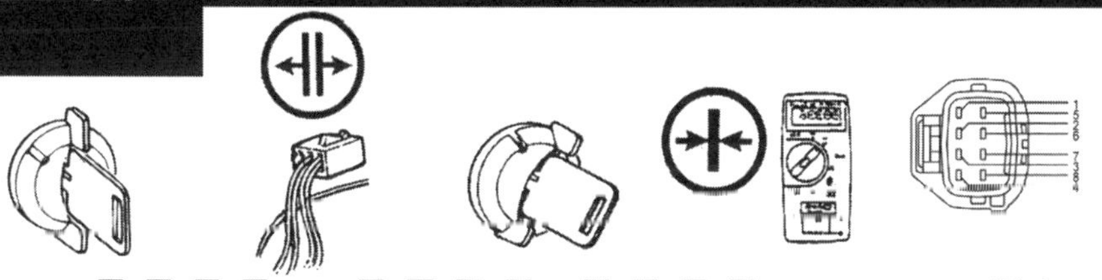

Key Off.

Disconnect The APP Sensor Electrical Connector.

Key On. Engine Off.

Measure the voltage between:	
(+) APP Sensor Connector, Harness Side	(-) APP Sensor Connector, Harness Side
VREF - Pin 6, 7	SIGRTN - Pin 1, 3

4.5-5.5 Volts

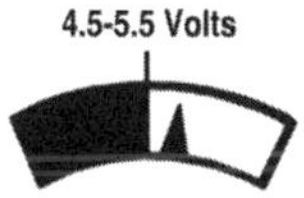

Is The Voltage Greater Than 4.5-5.5 Volts?

YES — AG4

NO —

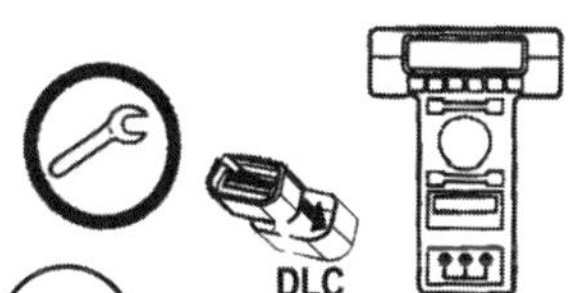

REPAIR The Open Circuit. CLEAR The DTCs. REPEAT The Self-Test. STOP

AG4 CHECK APP SENSOR OPERATION

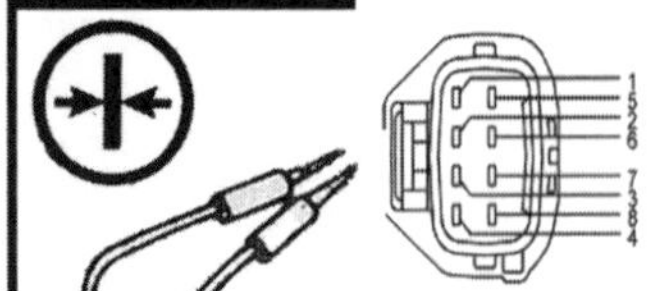

Connect a 5 amp fused jumper wire between the following:	
Point A APP Sensor Connector, Harness Side	Point B APP Sensor Connector, Harness Side
VREF - Pin 6	APPS1 - Pin 2
VREF - Pin 6	APPS2 - Pin 5
VREF - Pin 6	APPS3 - Pin 8

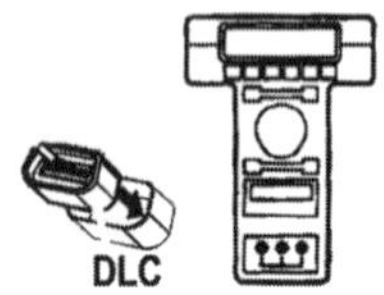

Access APP1, APP2 And APP3 PID Values.

Is The Voltage Of The Faulted Signal Circuit 4.5-5.5 Volts?

YES: 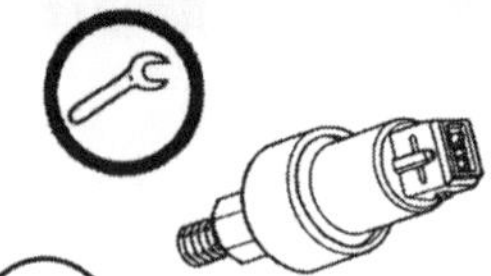

INSTALL A New APP Sensor Assembly. CLEAR DTCs. REPEAT The Self-Test.

NO: AG5

AG5 CHECK APP SENSOR SIGNAL FOR AN OPEN

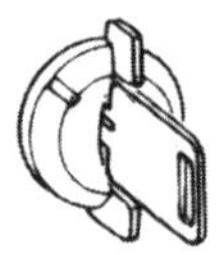

Key Off.

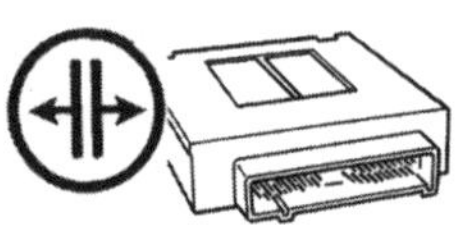

Disconnect The PCM Body Harness Connector.

Measure the resistance between:	
(+) APP Sensor Connector, Harness Side	(-) PCM Connector, Harness Side
APPS1 - Pin 2	APPS1 - Pin B26
APPS2 - Pin 5	APPS2 - Pin B37
APPS3 - Pin 8	APPS3 - Pin B25

Is The Resistance Less Than 5 Ohms?

YES: AG6

NO: 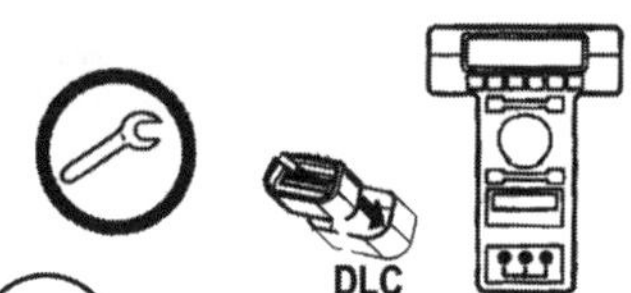

REPAIR Open In Faulted APP Signal Circuit. CLEAR DTCs. REPEAT The Self-Test.

AG6 — CHECK THE SUSPECT APP SENSOR SIGNAL FOR A SHORT TO GROUND

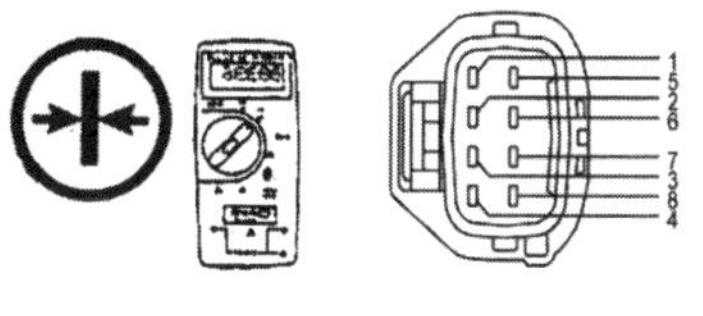

10,000 Ohms

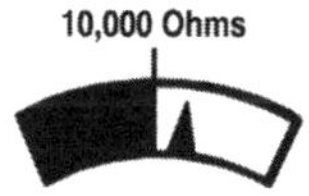

Measure the resistance between:	
(+) APP Sensor Connector, Harness Side	(-)
APPS1 - Pin 2	Ground
APPS2 - Pin 5	Ground
APPS3 - Pin 8	Ground

Is The Resistance Greater Than 10,000 Ohms?

YES → AG18

NO → STOP

REPAIR The Short Circuit To GND. CLEAR The DTCs. REPEAT The Self-Test.

AG7 — DIAGNOSTIC TROUBLE CODE (DTCS) P2123/P2128/P2133

Note: Refer To The PCM And Component Connector Pin Numbers At The Beginning Of This Pinpoint Test.

Slowly Press And Release Accelerator Pedal

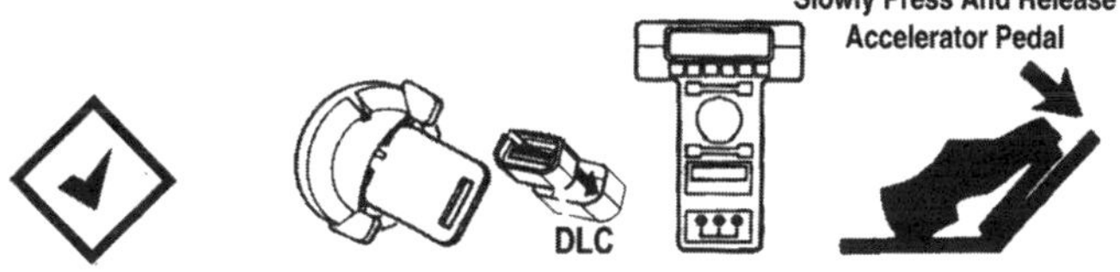

Possible Causes:
Signal Shorted To Voltage
- Signal Return Open
- APP Sensor
- PCM

Key On. Engine Off.

Access The APP1, APP2, And APP3 PIDs

Apply And Release The Accelerator Pedal.

Do All The Signal Values Change When The Accelerator Is Applied And Released?

YES → STOP

Unable To Duplicate Condition. CHECK For Loose Connections, Damaged Or Corroded Pins. WIGGLE Harness Attempting To Recreate Fault. REFER To Symptom Charts If A Driveablility Concern Exists.

NO → AG8

AG8 — SIMULATE THE OPPOSITE SIGNAL

Note: The APP sensor is disconnected to simulate the opposite DTC.

1. Key Off.
2. Disconnect The APP Sensor Electrical Connector.
3. Key On. Engine Off.
4. Access The APP1, APP2, AND APP3 PIDs.
5. Do All PID Signal Values Read 0 Volts?

- YES → AG11
- NO → AG9

AG9 — CHECK APP SIGNAL CIRCUITS FOR A SHORT TO VOLTAGE

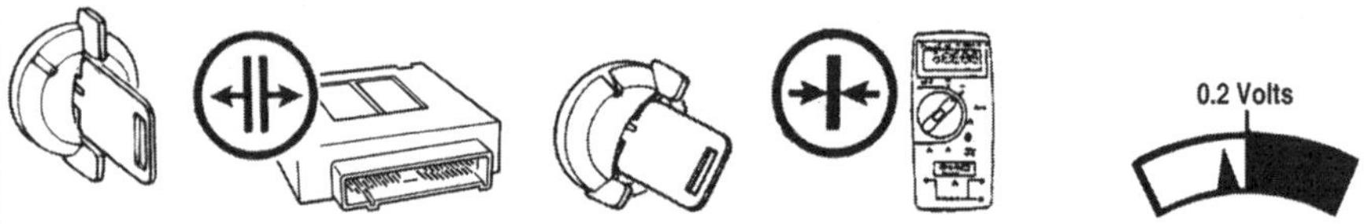

1. Key Off.
2. Disconnect The PCM Electrical Connector.
3. Key On. Engine Off.
4. Measure the voltage between:

(+) APP Sensor Connector, Harness Side	(-)
APPS1 - Pin 2	Ground
APPS2 - Pin 5	Ground
APPS3 - Pin 8	Ground

5. Is The Voltage Less Than 0.2 Volts?

- NO → AG10
- NO → STOP

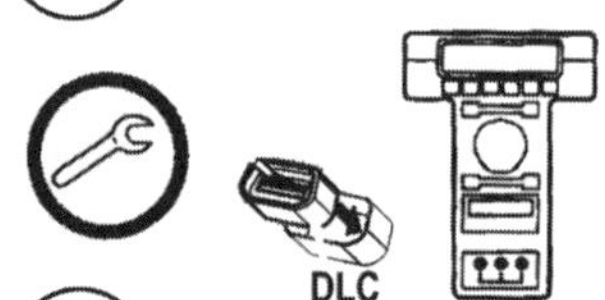

REPAIR Short To Voltage In Faulted APP Signal Circuit. CLEAR DTCs. REPEAT The Self-Test.

AG10 — CHECK THE APP SIGNAL CIRCUITS FOR A SHORT TO GROUND

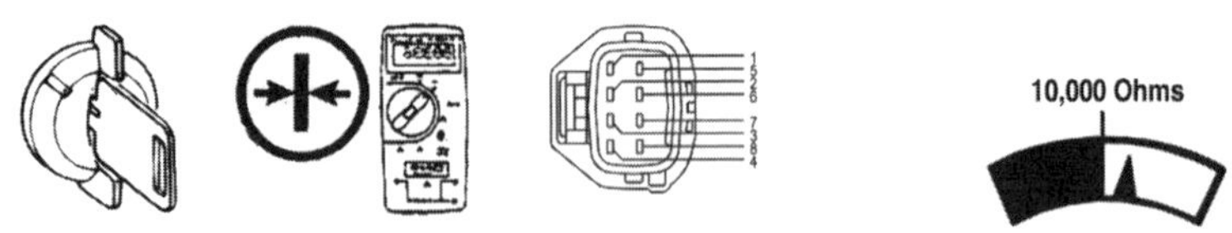

1. Key Off.
2. Measure the resistance between:

(+) APP Sensor Connector, Harness Side	(-)
APPS1 - Pin 2	Ground
APPS2 - Pin 5	Ground
APPS3 - Pin 8	Ground

3. Is The Resistance Greater Than 10,000 Ohms?

- YES → AG18
- NO → STOP

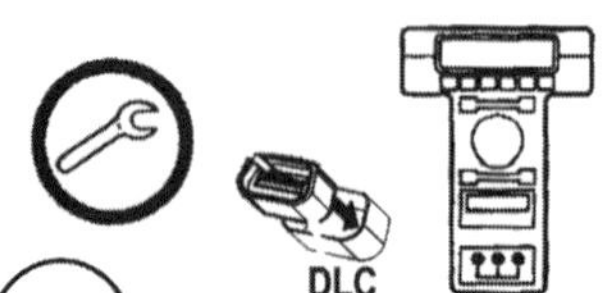

REPAIR Short To Ground In The Faulted Signal Circuit. CLEAR DTCs. REPEAT The Self-Test.

AG11 CHECK APP SIGNAL RETURN CIRCUIT FOR AN OPEN

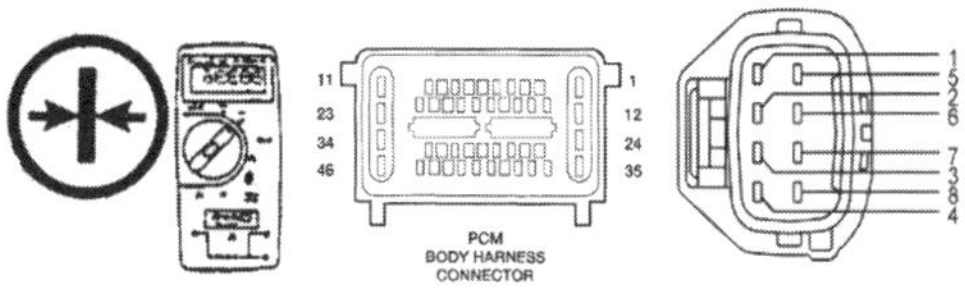

Measure the resistance between:	
(+) PCM Connector, Harness Side	(-) APP Sensor Connector, Harness Side
SIGRTN - Pin B20, B33	SIGRTN - Pin 1, 3

Is The Resistance Less Than 5 Ohms?

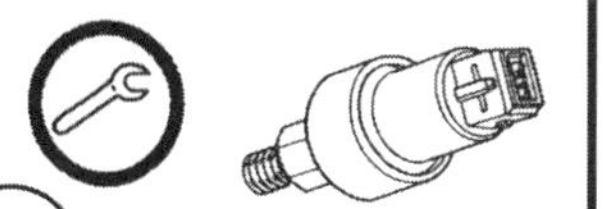

YES — INSTALL A New APP Sensor Assembly. CLEAR The DTCs. REPEAT The Self-Test. — STOP

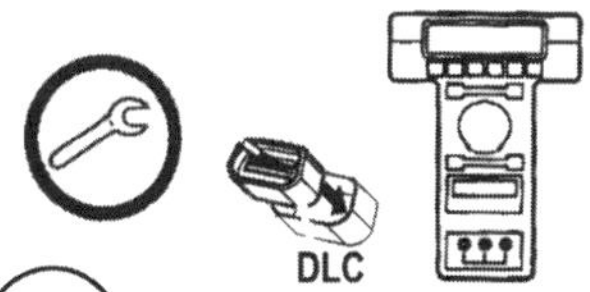

NO — REPAIR Open In Signal Return Circuit. CLEAR DTCs. REPEAT The Self-Test. — STOP

AG12 DIAGNOSTIC TROUBLE CODES (DTCS) P2138/P2139/P2140

Note: Refer To The PCM And Component Connector Pin Numbers At The Beginning Of This Pinpoint Test.

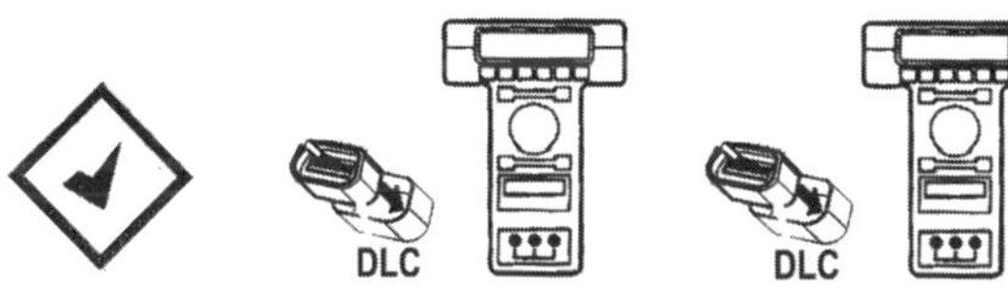

Possible Causes:
- APP Sensor
- PCM
- High Circuit Resistance

Clear All DTCs.

Carry Out The KOEO And KOER Self-Test.

Are DTCs P2138, P2139 Or P2140 Present?

YES — AG13

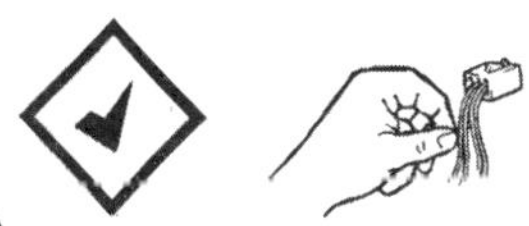

NO — UNABLE To Duplicate The Condition. CHECK For A Loose Connection, And Damaged Or Corroded Terminals. WIGGLE The Harness Attempting To Recreate The Fault. REPAIR As Necessary. REFER To Symptom Charts If A Driveability Concern Exists. — STOP

AG13 CHECK ACCELERATOR PEDAL ASSEMBLY FOR BINDING

Press The Accelerator Pedal Slowly To The Floor And Release.

Does The Accelerator Pedal Move To The Floor And Return Without Binding?

YES → AG14

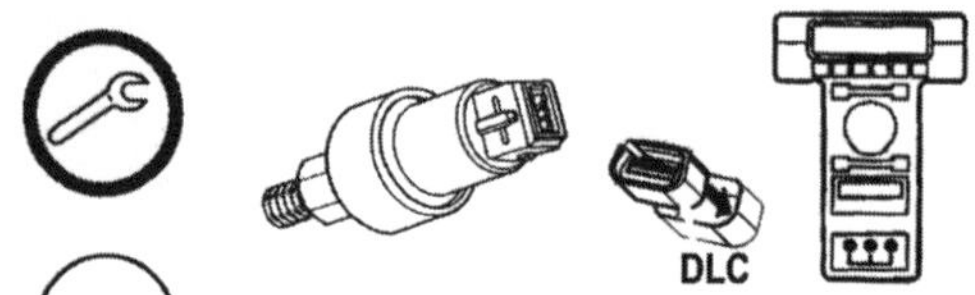

NO → STOP

INSTALL A New APP Sensor Assembly. CLEAR The DTCs. REPEAT The Self-Test.

AG14 CHECK APP SENSOR SIGNAL VOLTAGE RANGES

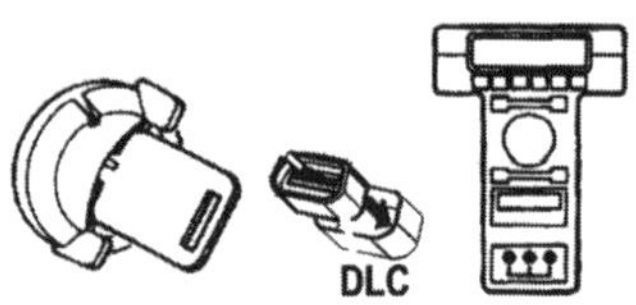

Key On. Engine Off.

Access And Monitor APP1. APP2, And APP3 PIDs.

Press The Accelerator Pedal Slowly To The Floor And Release.

2006			
Pedal Position	APP1	APP2	APP3
Closed Throttle	3.43 - 4.69	1.13 - 1.88	0.64 - 1.28
Wide Open Throttle (E-Series)	0.73 - 1.28	3.03 - 4.55	2.59 - 3.89
Wide Open Throttle (F-Super Duty)	0.39 - 0.97	3.45 - 4.71	2.95 - 4.10
2007			
Pedal Position	APP1	APP2	APP3
Closed Throttle	3.43 - 4.69	1.13 - 1.88	0.64 - 1.28
Wide Open Throttle	0.73 - 1.28	3.03 - 4.55	2.59 - 3.89

Are The PID Voltages Within The Range Specified In The Table?

YES → AG18

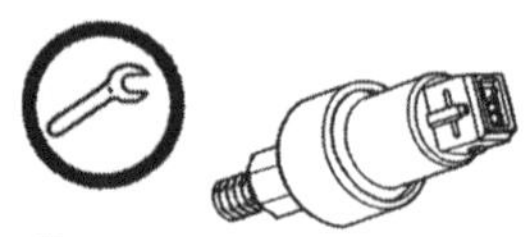

NO → STOP

INSTALL A New APP Sensor Assembly. CLEAR DTCs. REPEAT The Self-Test.

AG15

DIAGNOSTIC TROUBLE CODE P2104

Note: Refer To The PCM And Component Connector Pin Numbers At The Beginning Of This Pinpoint Test

2006		
Pedal Position	APP2	APP3
Closed Throttle	1.13 - 1.88	0.64 - 1.28
Wide Open Throttle (E-Series)	3.03 - 4.55	2.59 - 3.89
Wide Open Throttle (F-Super Duty)	3.45 - 4.71	2.95 - 4.10
2007		
Pedal Position	APP2	APP3
Closed Throttle	1.13 - 1.88	0.64 - 1.28
Wide Open Throttle	3.45 - 4.71	2.95 - 4.10

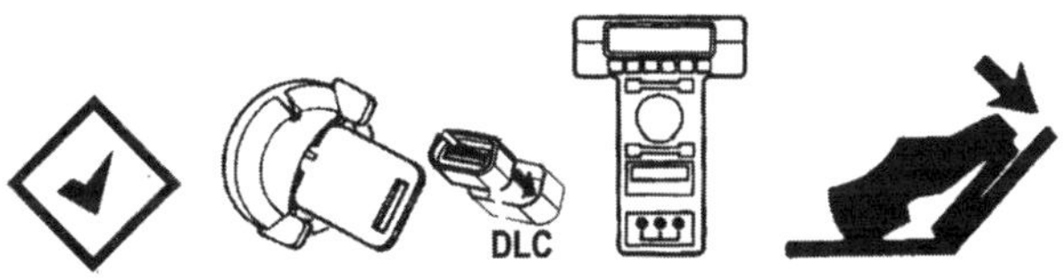

Possible Causes:
- APP Sensor
- APP Track 2 And 3 Circuits Shorted
- PCM

Key On. Engine Off.

Access The APP2, And APP3 PIDs

Slowly Apply And Release The Accelerator Pedal While Observing The PID Voltage Values.

Are The PID Voltage Values Within The Range Specified In The Table

YES → STOP

Unable To Duplicate Condition. CHECK For Loose Connections, Damaged Or Corroded Pins. WIGGLE Harness Attempting To Recreate Fault. REFER To Symptom Charts If A Driveablility Concern Exists.

NO → AG16

AG16

CHECK THE APP SENSOR

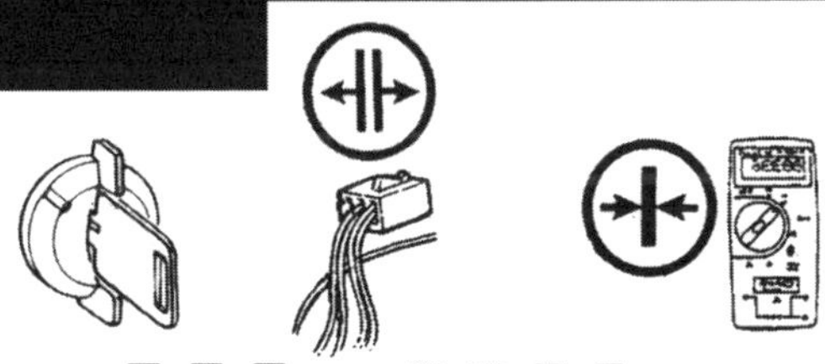

Key Off.

Disconnect The APP Sensor.

Measure the resistance between:	
(+) APP Sensor Connector, Component Side	(-) APP Sensor Connector, Component Side
APPS1 - Pin 2	APPS3 - Pin 8

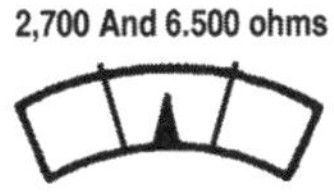

Is The Resistance Between 2,700 And 6.500 ohms?

YES → AG17

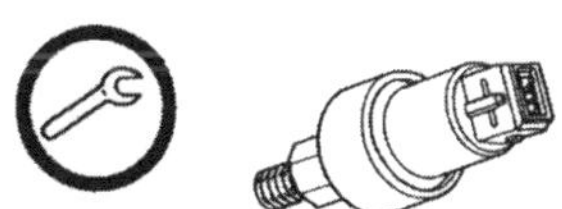

NO → STOP

INSTALL A New APP Sensor Assembly. CLEAR The DTCs. REPEAT The Self-Test.

AG17

CHECK THE APP SENSOR SIGNAL CIRCUITS

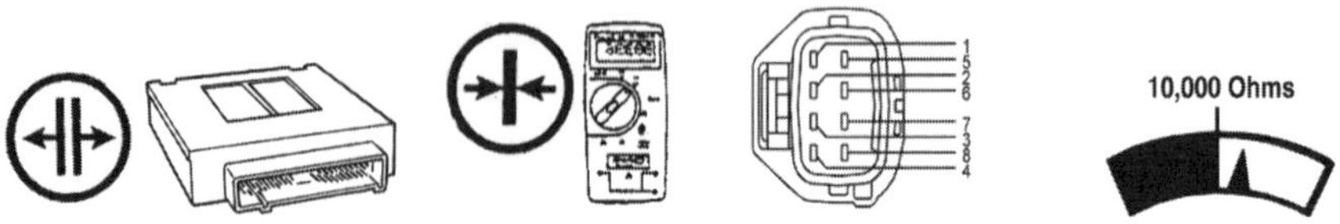

Disconnect The PCM Harness Electrical Connector.

Measure the resistance between:	
(+) APP Sensor Connector, Harness Side	(-) APP Sensor Connector, Harness Side
APPS1 - Pin 2	APPS3 - Pin 8

Is The Resistance Greater Than 10,000 Ohms?

YES → AG18

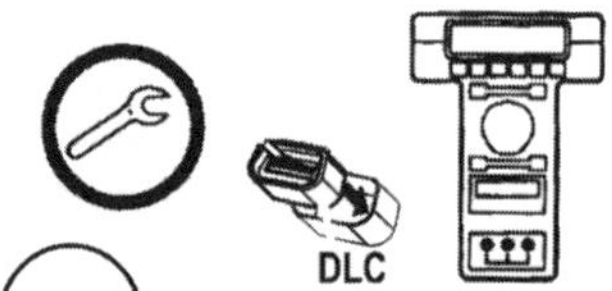

NO → STOP

REPAIR Short The Short Circuit. CLEAR The DTCs. REPEAT The Self-Test.

AG18

CHECK FOR CORRECT PCM OPERATION

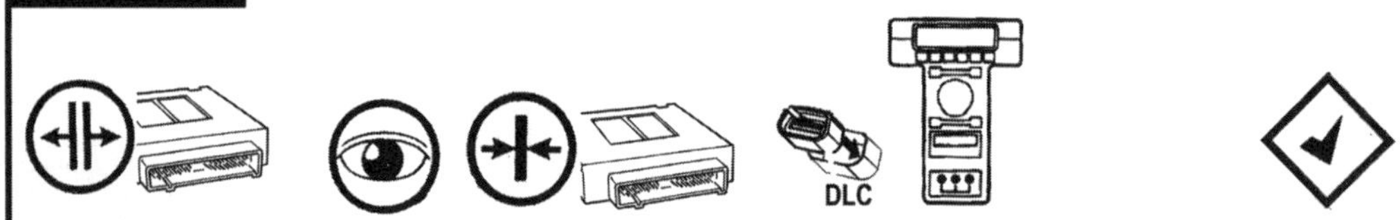

Disconnect All PCM Connectors.

Visually Inspect For Pushed Out Pins, Corrosion.

Connect All PCM Connectors And Make Sure They Seat Correctly.

Perform The PCM Self-Test And Verify The Concern Is Still Present.

Is The Concern Still Present.

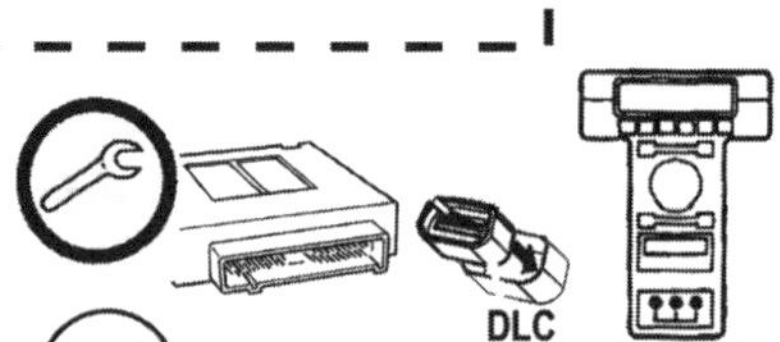

YES → STOP

INSTALL A New PCM. CLEAR The DTC's And REPEAT The Self-Test.

NO → STOP

The System Is Operating Correctly At This Time. The Concern May Have Been Caused By A Loose Or Corroded Connector.

Trouble Code (s)	Problem: Visctronic Drive Fan (VDF)	Test
P0480 P0528		AH 2006-07

REFERENCE LIST

This pinpoint test is intended to diagnose the following:
- Visctronic drive fan
- Fan speed sensor
- Harness circuits
- Powertrain control module (PCM)

DTC	KOEO	KEOR	CM
P0480	AH1	AH1	AH1
P0528	AH1	AH1	AH1

P0480 = Fan 1 Control Circuit
P0528 = Fan Speed Sensor Circuit No Signal

WARNING: TO AVOID THE POSSIBILITY OF PERSONAL INJURY OR DAMAGE TO THE VEHICLE, DO NOT OPERATE THE ENGINE UNTIL THE FAN BLADE HAS BEEN FIRST EXAMINED FOR POSSIBLE CRACKS AND SEPARATION.

Signal Functions:
The visctronic drive fan (VDF) is an electrically actuated viscous clutch that consists of three main elements:
- A working chamber
- A reservoir chamber
- An actuator (electromechanical valve and speed sensor)

The actuator valve controls the fluid flow from the reservoir into the working chamber. Once viscous fluid is in the working chamber, "shearing" of the fluid will result in fan rotation. The valve is activated via a pulse width modulated (PWM) output signal from the PCM. By opening and closing the fluid port valve, the PCM can control fan speed. Fan speed is monitored through a Hall-Effect sensor and is monitored by the PCM during closed loop operation.

The PCM will optimize fan speed based on engine coolant temperature (ECT), engine oil temperature (EOT), transmission fluid temperature (TFT), intake air temperature (IAT), or air conditioning requirements. When an increased demand for fan speed is requested for vehicle cooling, the PCM will monitor the fan speed through the Hall-Effect sensor. If a fan speed increase is required, the PCM will output the PWM signal to the fluid port, providing the required fan speed increase .

Detection/Management:
During KOER Self-Test, the PCM commands 100% duty cycle. A DTC is set if the PCM detects the voltage on the valve control circuit is not within the expected range or if the fan speed is less than a calibrated value.

CAUTION!:
The powertrain control module (PCM) harness connectors must be properly seated and the connector latch properly attached to eliminate possible driveability concerns or a no start condition. Installing the PCM connectors on an angle may cause an improper connection, misdiagnosis and damaged components. Install the connector until the lever pivots and seats itself. Apply light pressure to get the connector into position on the PCM and then fully seat the connector.

Note: Visually inspect the harness connectors for corrosion, damage, proper mating and correct pin tension.

Note: When the PCM is disconnected additional DTCs will be set. Clear all DTCs after restoring the vehicle.

PINPOINT TEST SCHEMATICS AND CONNECTORS

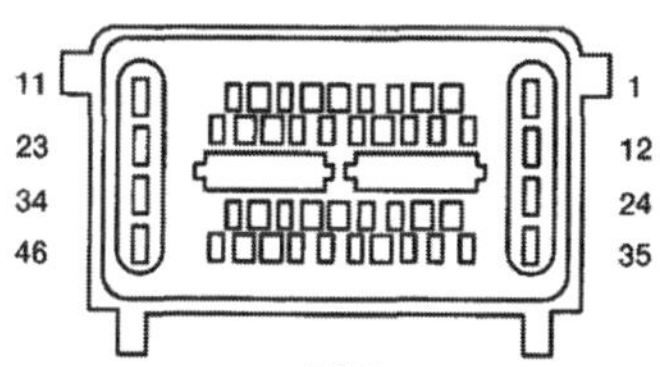

PCM
ENGINE HARNESS
CONNECTOR

Pin	Circuit
E6	FSS (Fan Speed Sensor)
E22	FSS SIGRTN
E46	VBPWR (Vehicle Buffered Power)
E14	VDF (Visctronic Drive Fan)

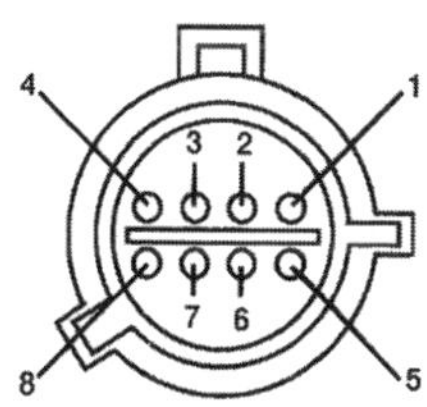

DRIVE FAN
CONNECTOR

Pin	Circuit
1	FSS (Fan Speed Sensor)
2	FSS SIGRTN
6	VBPWR (Vehicle Buffered Power)
4	VDF (Visctronic Drive Fan)
5	VPWR (Vehicle Power)

AH1 — PRELIMINARY DIAGNOSIS

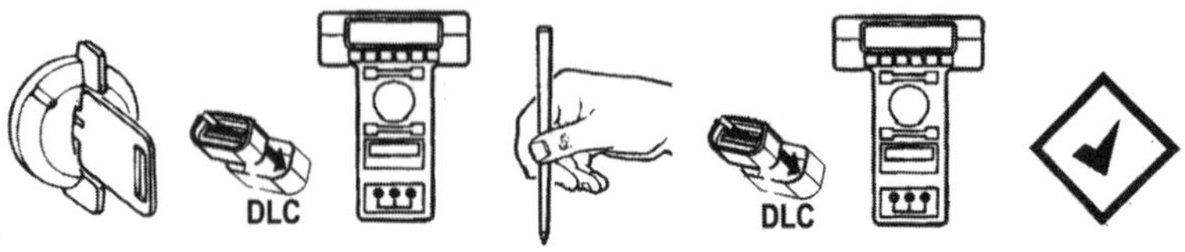

Key Off. — Connect The Scan Tool. — Record The Freeze Frame Data. — Carry Out The PCM Self-Test. — Are Any DTCs Present?

- YES:
  - For DTC P0480: → AH4
  - For DTC P0528: → AH3
  - For All Other DTCs, REFER To Diagnostic Trouble Code (DTC) Charts. → DTC INDEX
- NO: → AH2

AH2 — CHECK THE COOLING FAN FOR MECHANICAL BINDING

Note: The VDF Is A Viscous Coupling. The Viscous Drag Should Be Smooth During Fan Rotation. The Amount Of Resistance Is Dependant Upon The Final VDF Operational State Before Engine Shutdown.

Key Off. — Manually Rotate The Cooling Fan. — Does The Fan Clutch Rotation Feel Rough Or Binding?

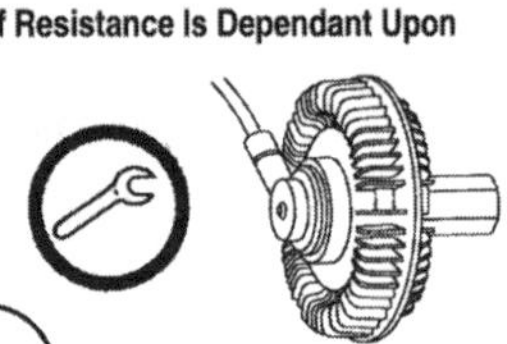

- YES: INSTALL A New VDF. CLEAR The DTCs. REPEAT The Selft-Test. → STOP
- NO: → AH3

AH3

CHECK THE VDF OPERATION

Note: Check And Verify The Most Current PCM Calibration Level.

Note: Inherent To Viscous Clutches Is A Delay In Transitioning From A Low To A High Speed Or From A High To A Low Speed. This Delay Can Vary Between 10 Seconds And 3 Minutes And Is Dependant Upon Engine Speed And Clutch Temperature. With The Engine At Operating Temperature And At 2500 RPM, The Delay Should Be Less Than 30 Seconds.

Key On, Engine Running. → Access Output Test Mode. → Access The FANSS And RPM PIDs. → Increase The Engine Speed To 2500 RPM. → Monitor The FANSS PID And The VDF Duty Cycle. → Command The VDF To LOW. → Allow The Fan Speed To Stabilize Below 600 RPM.

Command The VDF Duty Cycle To HIGH.

Does The Fan Speed Change And Does The FANSS PID Indicate An RPM Change?

YES

If The FANSS PID Stabilizes At Less Than 2,800 RPM With The Engine Speed At 2,500 RPM. INSTALL A New VDF. CLEAR The DTCs. REPEAT The Self-Test. For Cooling System Concerns, Diagnose And Repair The Cooling System — **STOP**

If The FANSS PID Stabilizes At Greater Than 2,800 RPM With The Engine Speed At 2,500 RPM, The System Is Operating Correctly. CLEAR the DTCs. REPEAT the Self-Test. For Cooling System Concerns, Diagnose And Repair The Cooling System — **STOP**

NO

If The Fan Speed Did Not Change: → **AH4**

If The Fan Speed Did Change And The FANSS PID Did Not Indicate An RPM Change: → **AH10**

AH4 CHECK THE VDF SOLENOID RESISTANCE

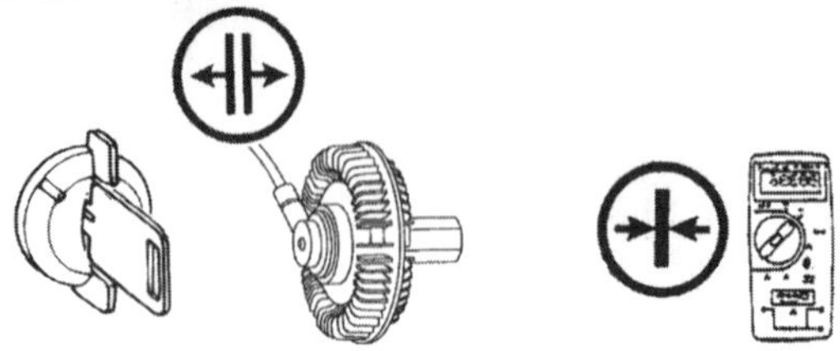

Key Off.

Disconnect The VDF.

Measure the resistance between:	
(+) VDF Connector, Component Side	(-) VDF Connector, Component Side
VPWR - Pin 5	VDF - Pin 4

6.0 And 10.0 Ohms

Is The Resistance Between 6.0 And 10.0 Ohms?

YES — AH5

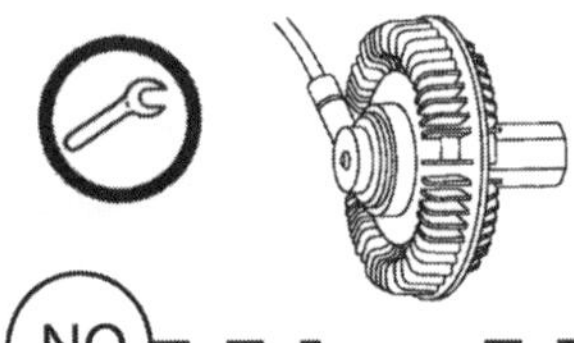

NO — INSTALL A New VDF. CLEAR The DTCs. REPEAT The Self-Test. STOP

AH5 CHECK THE VDF SOLENOID FOR A SHORT TO GROUND

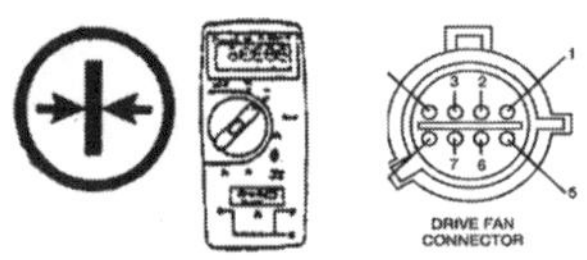

Measure the resistance between:	
(+) VDF Connector, Component Side	(-)
VDF - Pin 4	Ground

10,000 Ohms

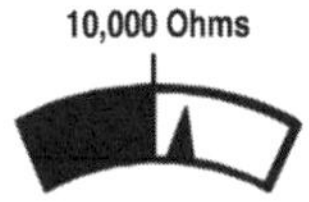

Is The Resistance Greater Than 10,000 Ohms?

YES — AH6

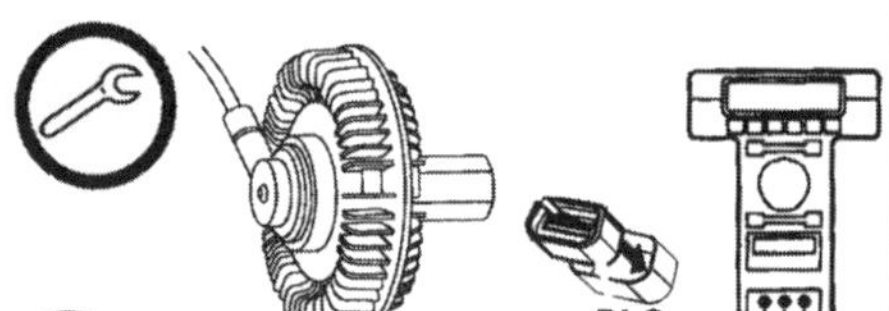

NO — INSTALL A New VDF. CLEAR The DTCs. REPEAT The Self-Test. STOP

AH6 CHECK THE VDF SOLENOID VPWR CIRCUIT FOR VOLTAGE

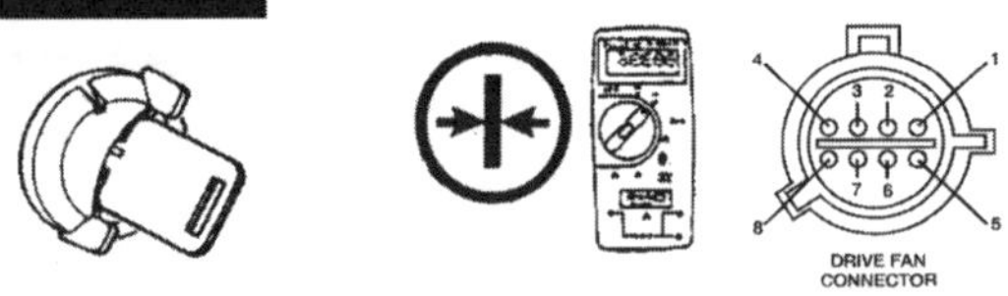

Key On, Engine Off.

Measure the voltage between:	
(+) VDF Connector, Harness Side	(-)
VPWR - Pin 5	Ground

10.5 Volts

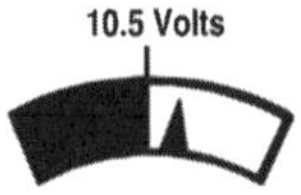

Is The Voltage Greater Than 10.5 Volts?

YES — AH7

NO — REPAIR The Circuit. CLEAR The DTCs. REPEAT The Self-Test. STOP

AH7 CHECK VDF SOLENOID CONTROL CIRCUIT FOR AN OPEN

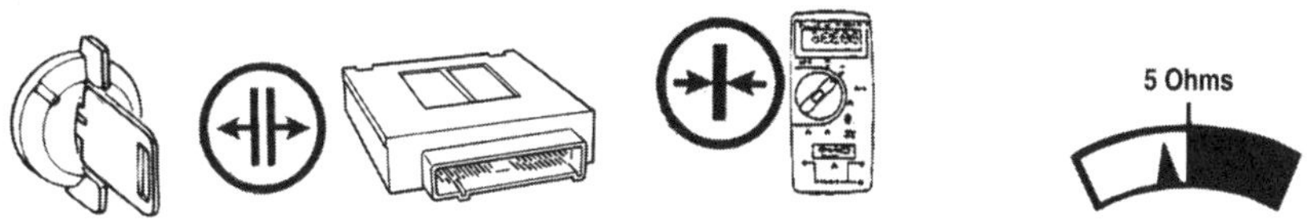

Key Off.

Disconnect The PCM.

Measure the resistance between:	
(+) VDF Connector, Harness Side	(-) PCM Connector, Harness Side
VDF - Pin 4	VDF - Pin E14

Is The Resistance Less Than 5 Ohms?

YES → AH8

NO → STOP

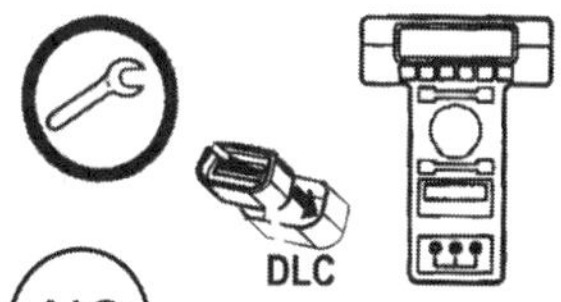

REPAIR The Circuit. CLEAR The DTCs. REPEAT The Self-Test.

AH8 CHECK THE VDF SOLENOID CONTROL CIRCUIT FOR A SHORT TO GROUND

Measure the resistance between:	
(+) VDF Connector, Harness Side	(-)
VDF - Pin 4	Ground

Is The Resistance Greater Than 10,000 Ohms?

YES → AH9

NO → STOP

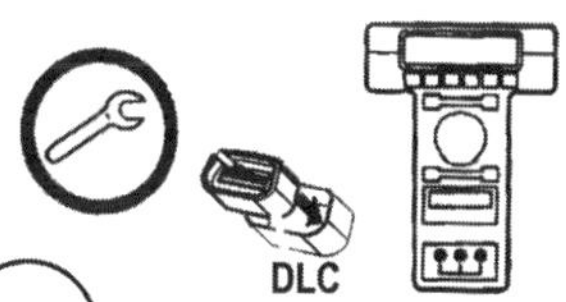

REPAIR The Circuit. CLEAR The DTCs. REPEAT The Self-Test.

AH9 CHECK THE VDF CONTROL CIRCUIT FOR SHORT TO POWER

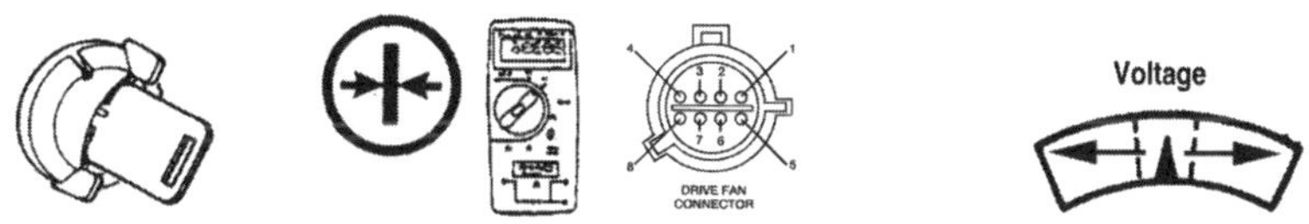

Key On, Engine Off.

Measure the voltage between:	
(+) VDF Connector, Harness Side	(-)
VDF - Pin 4	Ground

Is Any Voltage Indicated?

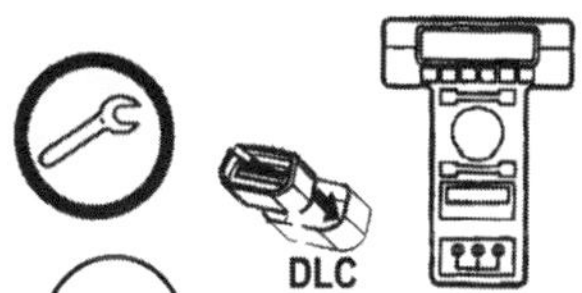

YES → STOP

REPAIR The Short Circuit To PWR. CLEAR The DTCs. REPEAT The Self-Test.

NO → AH19

AH10 — CHECK THE FAN SPEED SENSOR FOR VOLTAGE

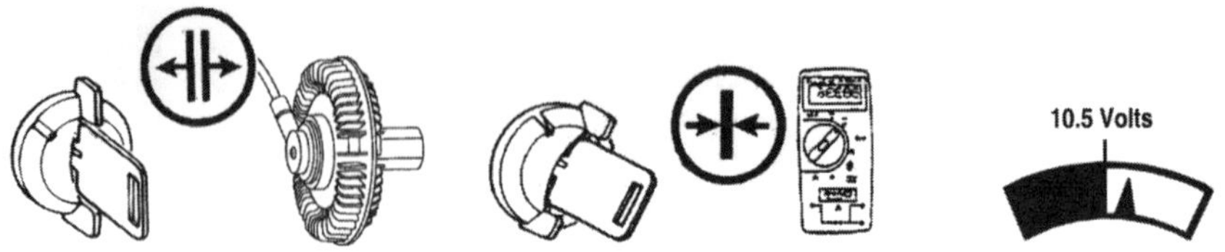

Key Off.

Disconnect The VDF.

Key On, Engine Off.

Measure the voltage between:	
(+) VDF Connector, Harness Side	(-) VDF Connector, Harness Side
VBPWR - Pin 6	FSS SIGRTN - Pin 2

Is The Voltage Greater Than 10.5 Volts?

YES — AH15

NO — AH11

AH11 — CHECK THE VBPWR CIRCUIT FOR VOLTAGE

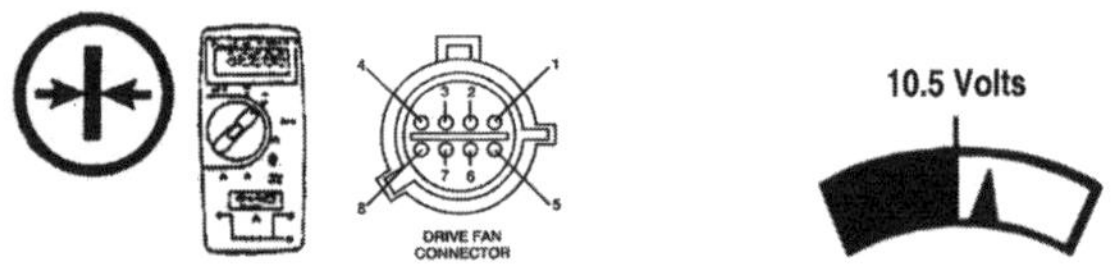

Measure the voltage between:	
(+) VDF Connector, Harness Side	(-)
VBPWR - Pin 6	Ground

Is The Voltage Greater Than 10.5 Volts?

YES — AH14

NO — AH12

AH12 — CHECK THE VBPWR CIRCUIT FOR AN OPEN

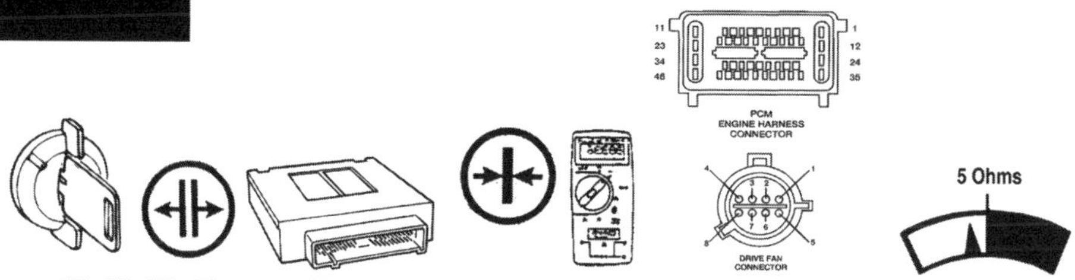

Key Off

Disconnect The PCM.

Measure the resistance between:	
(+) VDF Connector, Harness Side	(-) PCM Connector, Harness Side
VBPWR - Pin 6	VBPWR - Pin E46

Is Resistance Less Than 5 Ohms?

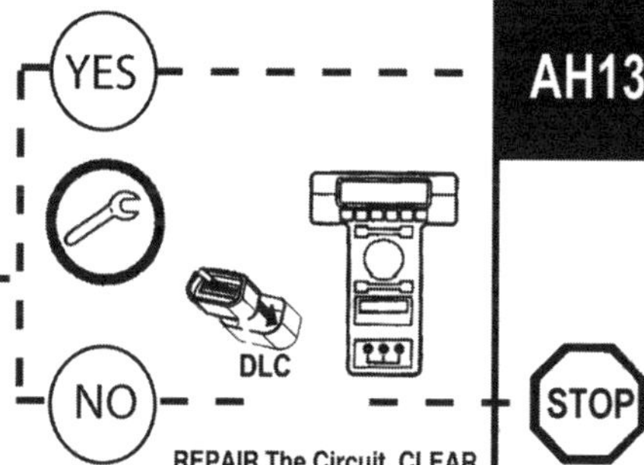

REPAIR The Circuit. CLEAR The DTCs. REPEAT The Self-Test.

AH13 — CHECK THE VBPWR CIRCUIT FOR A SHORT TO GROUND

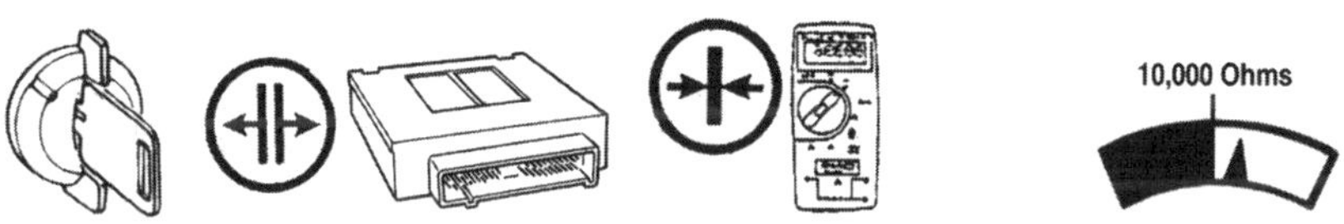

Key Off

Disconnect The PCM.

Measure the resistance between:	
(+) VDF Connector, Harness Side	(-)
VBPWR - Pin 6	Ground

Is The Resistance Greater Than 10,000 Ohms?

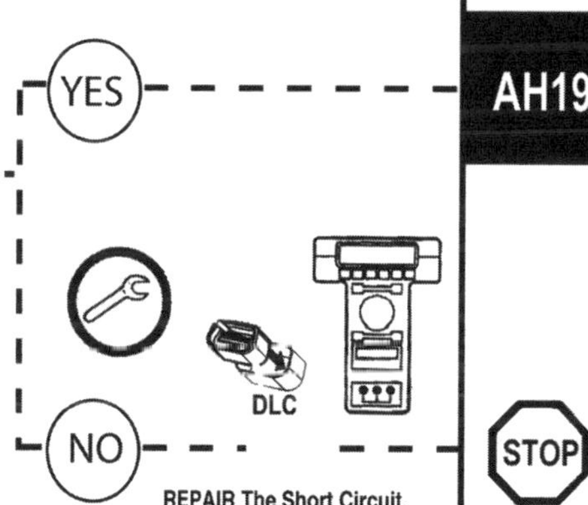

REPAIR The Short Circuit. CLEAR The DTCs. REPEAT The Self-Test.

AH14 — CHECK THE FAN SPEED SENSOR FSS SIGRTN CIRCUIT FOR AN OPEN

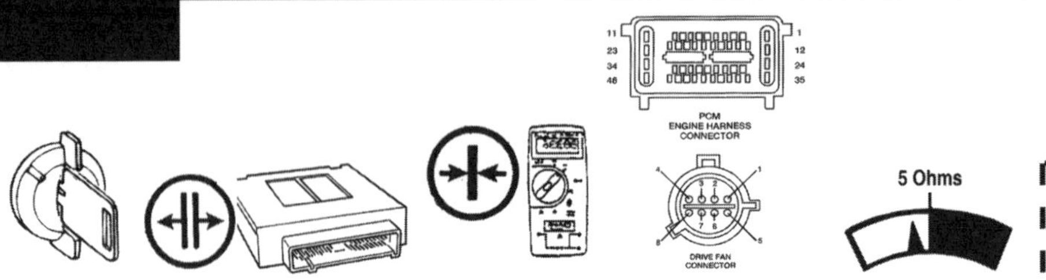

Key Off.

Disconnect The PCM.

Measure the resistance between:	
(+) VDF Connector, Harness Side	(-) PCM Connector, Harness Side
FSS SIGRTN - Pin 2	FSS SIGRTN - Pin E22

Is The Resistance Less Than 5 Ohms?

YES — AH19

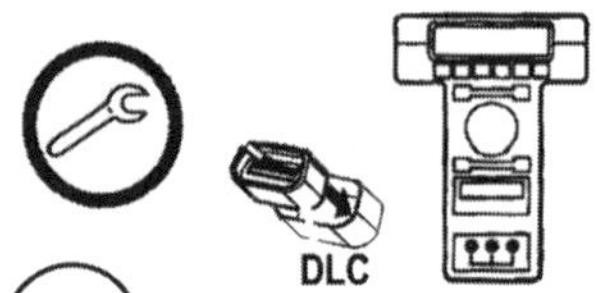

NO — REPAIR The Circuit. CLEAR The DTCs. REPEAT The Self-Test. STOP

AH15 — CHECK THE FAN SPEED SENSOR SIGNAL CIRCUIT FOR AN OPEN

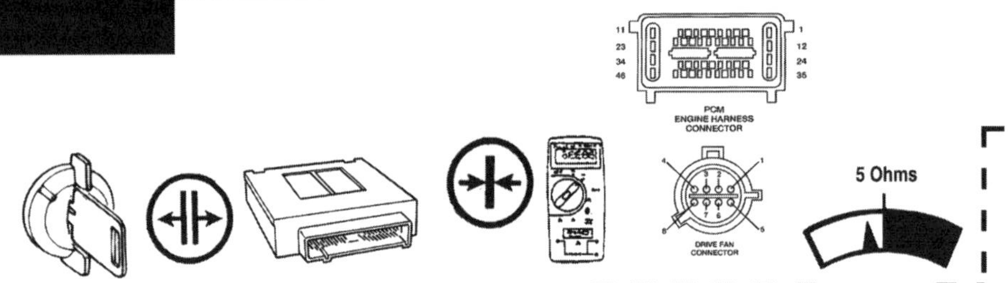

Key Off.

Disconnect The PCM.

Measure the resistance between:	
(+) VDF Connector, Harness Side	(-) PCM Connector, Harness Side
FSS - Pin 1	FSS - Pin E6

Is The Resistance Less Than 5 Ohms?

YES — AH16

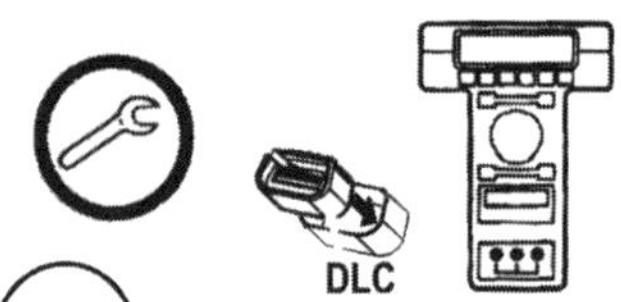

NO — REPAIR The Circuit. CLEAR The DTCs. REPEAT The Self-Test. STOP

AH16 CHECK THE SIGNAL CIRCUIT FOR A SHORT TO GROUND

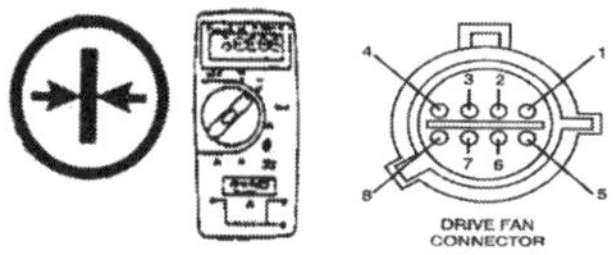

Measure the resistance between:	
(+) VDF Connector, Harness Side	(-)
FSS - Pin 1	Ground

Is The Resistance Greater Than 10,000 Ohms?

YES → AH17

NO → REPAIR The Circuit. CLEAR The DTCs. REPEAT The Self-Test. STOP

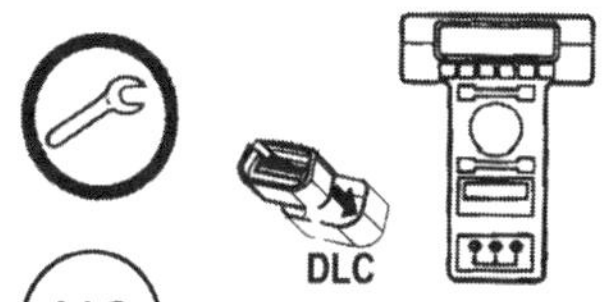

AH17 CHECK THE SIGNAL CIRCUIT FOR A SHORT VOLTAGE

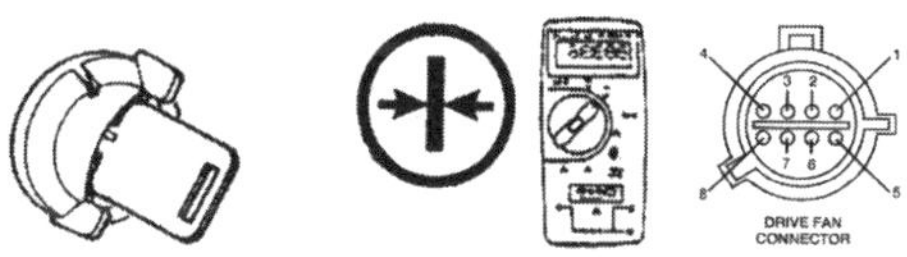

Key On, Engine Off.

Measure the voltage between:	
(+) VDF Connector, Harness Side	(-)
FSS - Pin 1	Ground

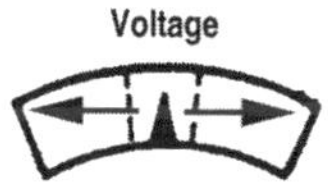

Is Any Voltage Indicated?

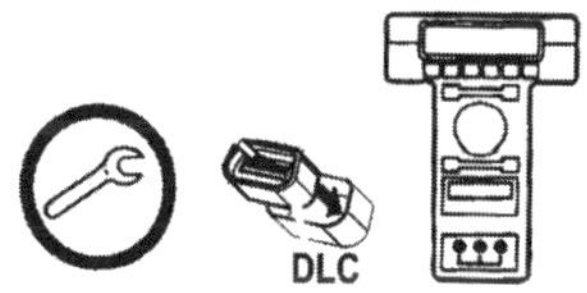

YES → REPAIR The Circuit. CLEAR The DTCs. REPEAT The Self-Test. STOP

NO → AH18

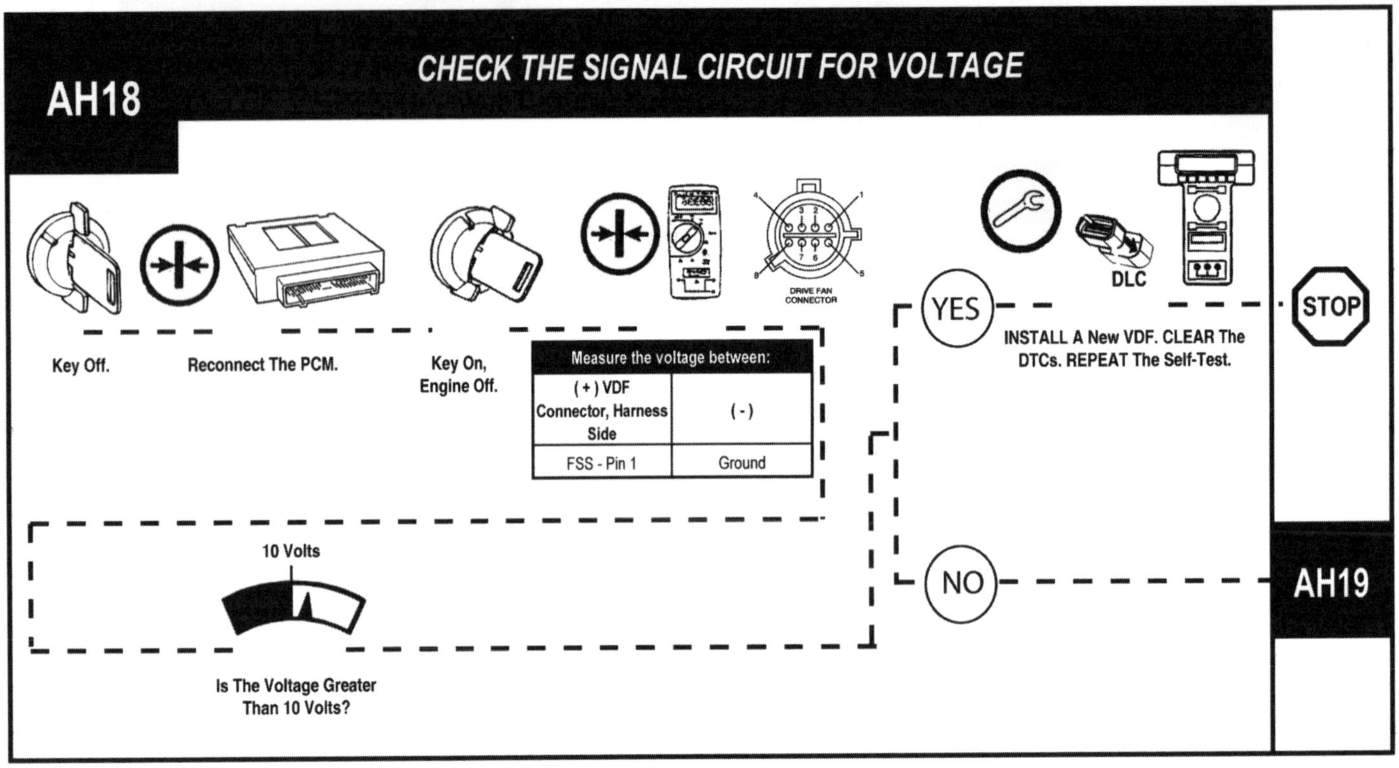

Measure the voltage between:	
(+) VDF Connector, Harness Side	(-)
FSS - Pin 1	Ground

AH19 CHECK FOR CORRECT PCM OPERATION

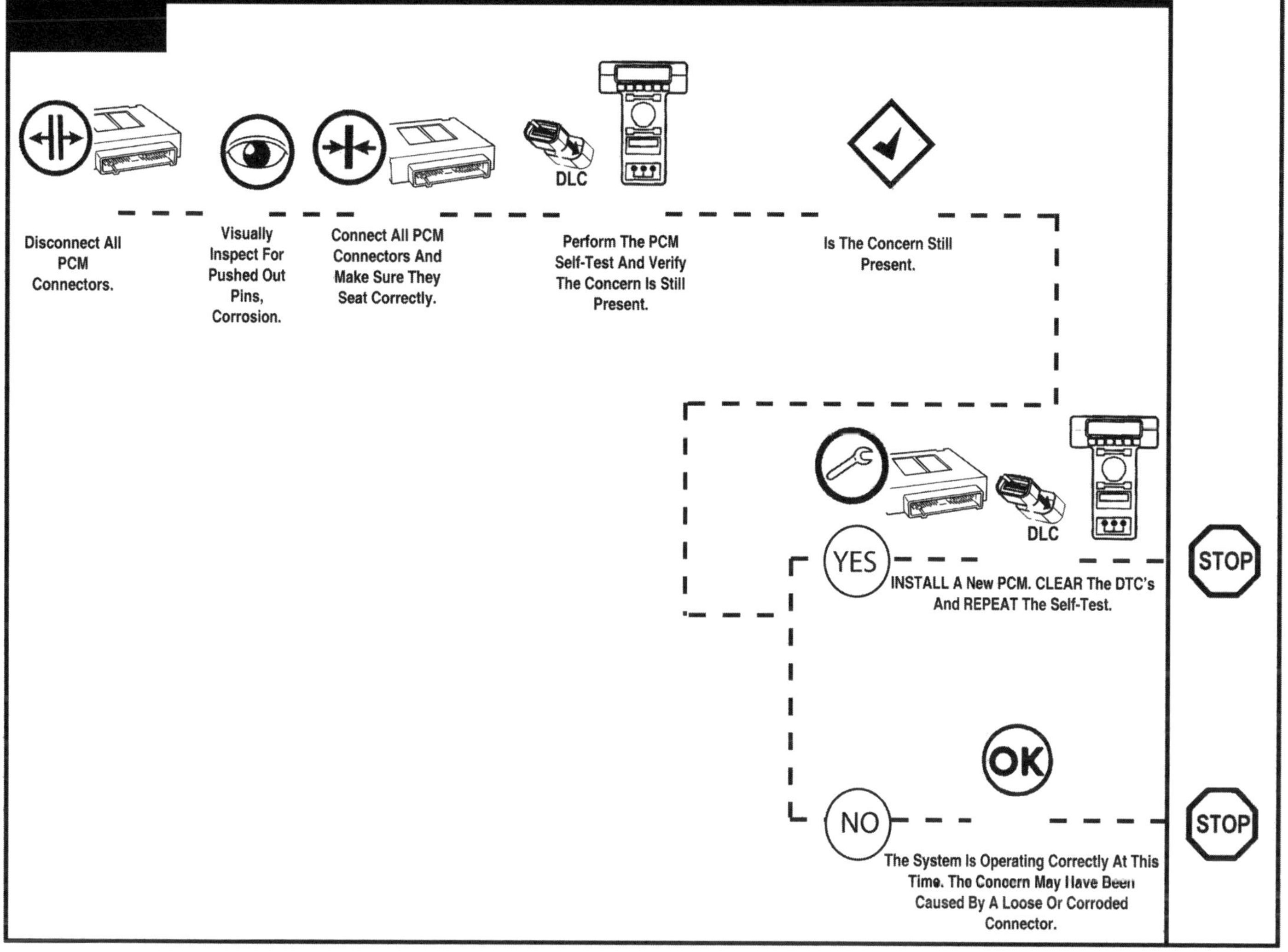

NOTES

Trouble Code (s)		Test
P1703	**Problem: Brake Pedal Position Switch**	AI 2006-07

REFERENCE LIST

This pinpoint test is intended to diagnose the following:

- **Brake pedal position (BPP) switch**
- **Harness circuits BPP and B+**
- **Powertrain control module (PCM) (12A650)**

DIAGNOSTIC TROUBLE CODE INDEX			
Codes	KOEO	KOER	CM
P1703	AI1	AI1	AI1

P1703 = Brake Switch Out Of Self-Test Range

Signal Functions:

The brake pedal position (BPP) switch provides a 12-volt input signal to the powertrain control module (PCM) whenever the brake pedal is applied. The BPP switch input is used to deactivate the speed control system and the auxiliary powertrain control module (APCM).

CAUTION!:

The powertrain control module (PCM) harness connectors must be properly seated and the connector latch properly attached to eliminate possible driveability concerns or a no start condition. Installing the PCM connectors on an angle may cause an improper connection, misdiagnosis and damaged components. Install the connector until the lever pivots and seats itself. Apply light pressure to get the connector into position on the PCM and then fully seat the connector.

Note: Visually inspect the harness connectors for corrosion, damage, proper mating and correct pin tension.

Note: When the PCM is disconnected additional DTCs will be set. Clear all DTCs after restoring the vehicle.

PINPOINT TEST SCHEMATICS AND CONNECTORS

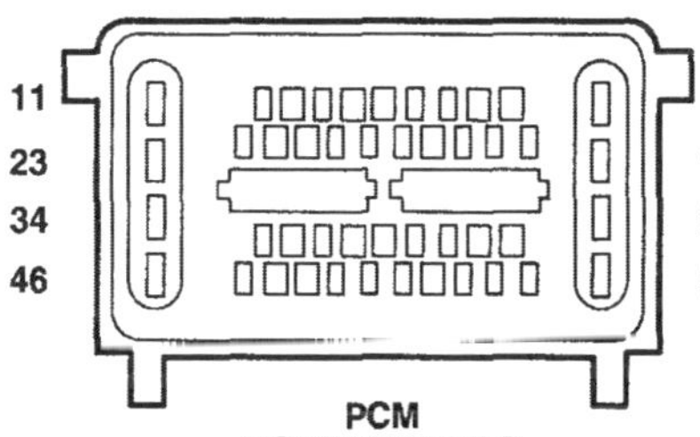

PCM
BODY HARNESS
CONNECTOR

Pin	Circuit
B18	BPP (Brake Pedal Position)

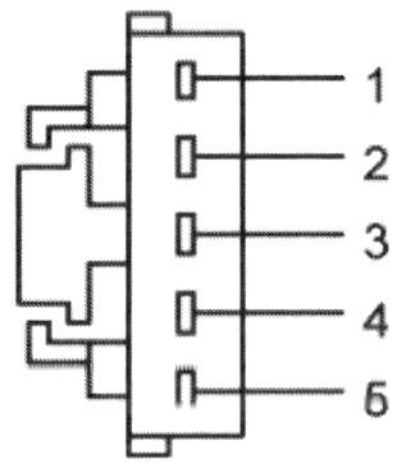

F-SUPER DUTY

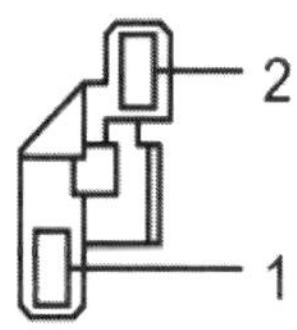

E-SERIES

Vehicle	Pin	Circuit
E-Series	1	B+
	2	BPP
F-Super Duty	3	B+
	4	BPP

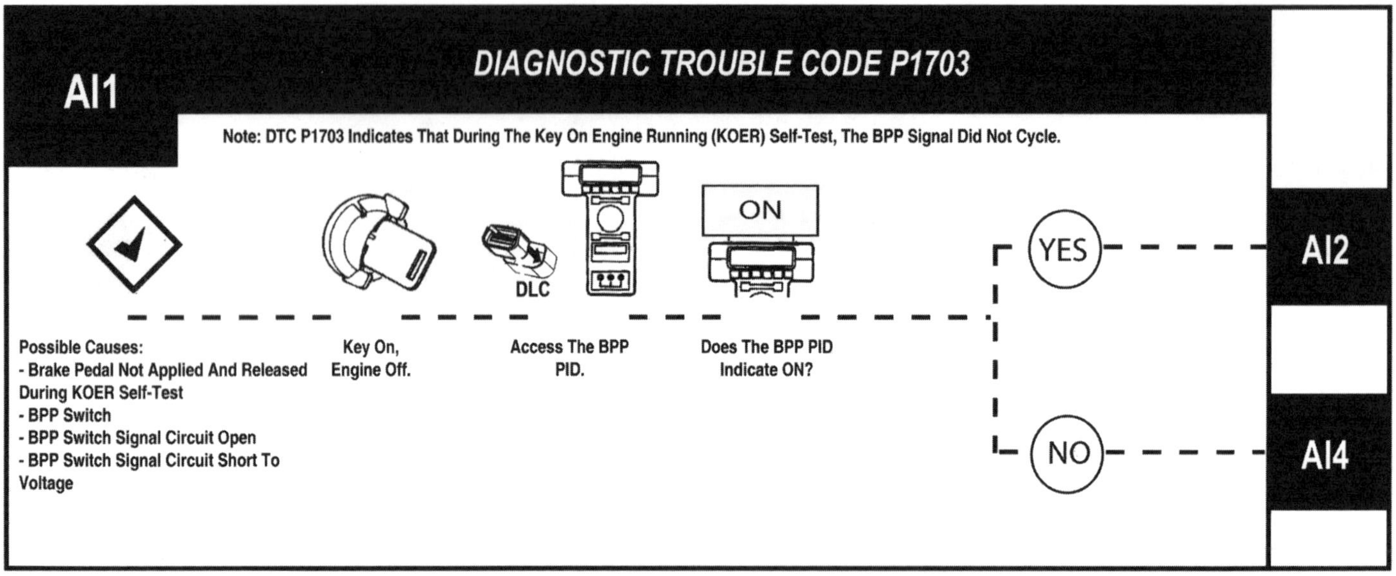
AI1
DIAGNOSTIC TROUBLE CODE P1703
Note: DTC P1703 Indicates That During The Key On Engine Running (KOER) Self-Test, The BPP Signal Did Not Cycle.
DLC
ON
Possible Causes:
- Brake Pedal Not Applied And Released During KOER Self-Test
- BPP Switch
- BPP Switch Signal Circuit Open
- BPP Switch Signal Circuit Short To Voltage
Key On, Engine Off.
Access The BPP PID.
Does The BPP PID Indicate ON?
YES
AI2
NO
AI4

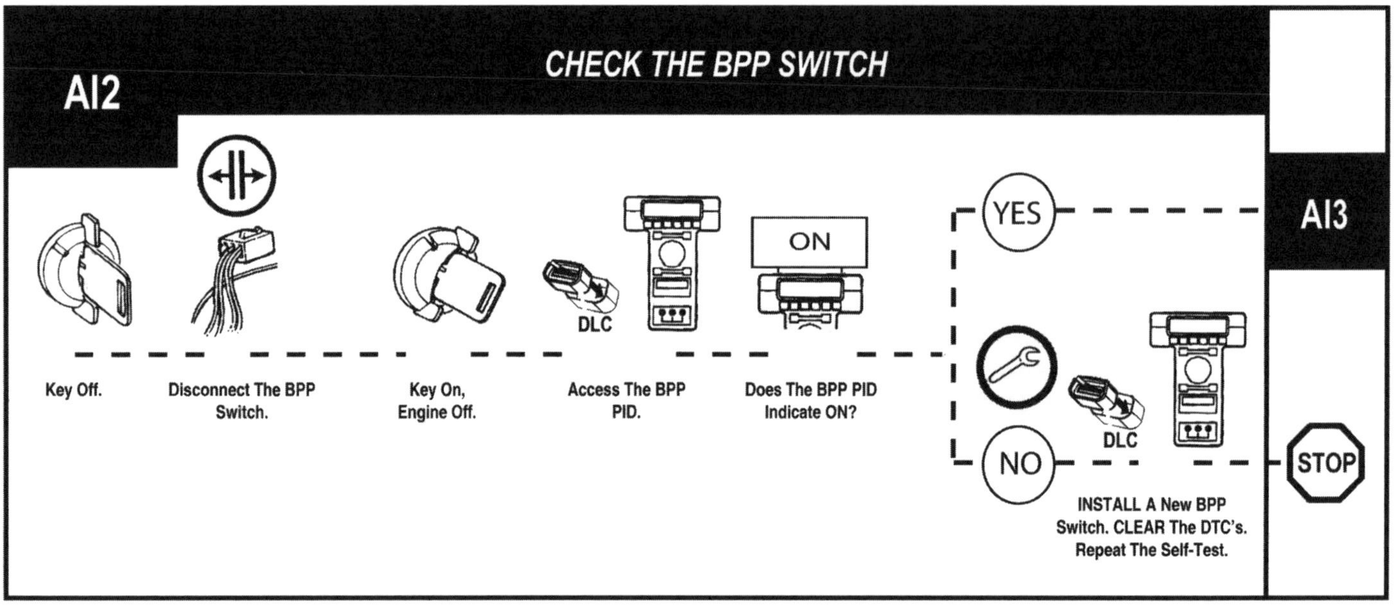
AI2
CHECK THE BPP SWITCH
DLC
ON
Key Off.
Disconnect The BPP Switch.
Key On, Engine Off.
Access The BPP PID.
Does The BPP PID Indicate ON?
YES
AI3
NO
DLC
INSTALL A New BPP Switch. CLEAR The DTC's. Repeat The Self-Test.
STOP

AI3 CHECK THE BPP SWITCH SIGNAL CIRCUIT FOR A SHORT TO VOLTAGE

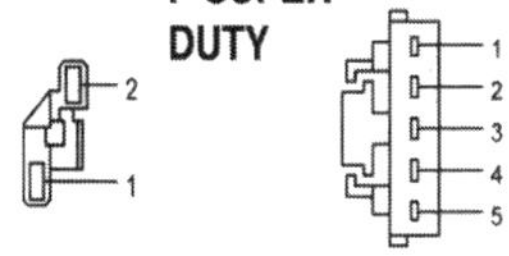

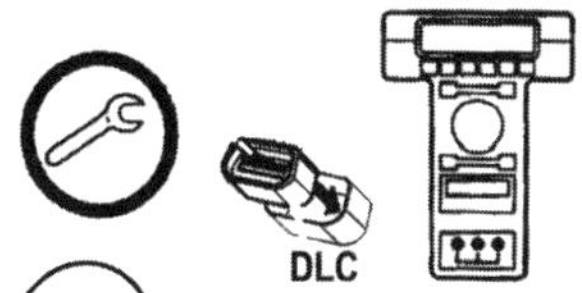

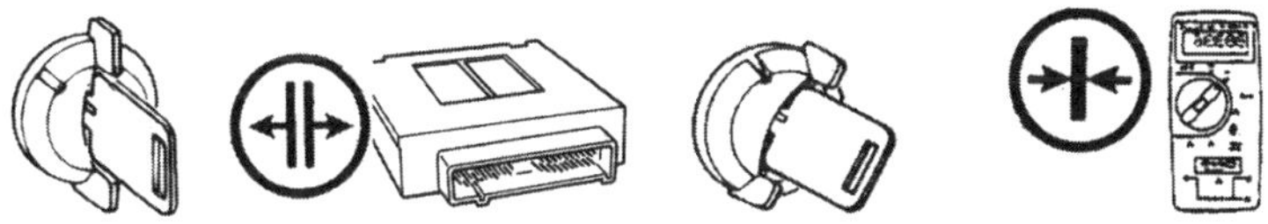

Key Off.

Disconnect The PCM.

Key On, Engine Off.

Measure the voltage between:	
(+) BPP Switch Connector, Harness Side	(-)
BPP	Ground

Voltage

Is Any Voltage Indicated?

YES — REPAIR The Circuit. CLEAR The DTCs. REPEAT The Self-Test. STOP

NO — AI8

AI4 CHECK THE BPP SWITCH OPERATION

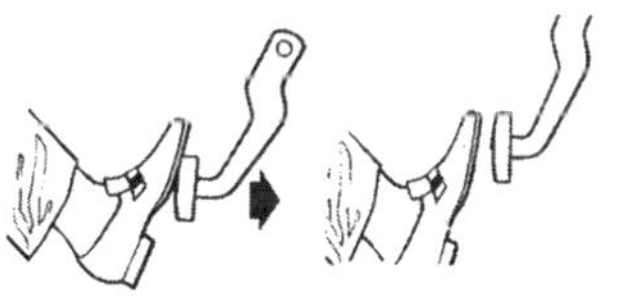

Monitor The BPA PID While Applying And Releasing The Brake Pedal.

BPA PID State Change

Does The BPA PID State Change When The Brake Pedal Is Fully Applied And Released?

YES — UNABLE To Duplicate Concern. Check For A Loose Connection, And Damaged Or Corroded Terminals. WIGGLE the Harness Attempting To Recreate The Fault. REPAIR As Necessary And RETEST The System. REFER To Symptom Charts If A Driveablility Concern Exists. STOP

NO — AI5

AI5 — CHECK THE BPP SWITCH FOR VOLTAGE

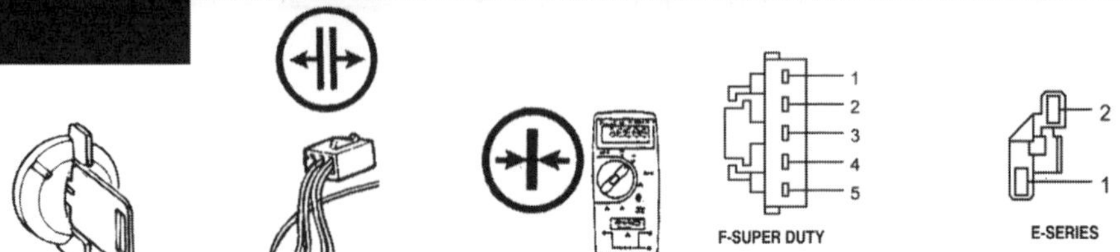

Key Off.

Disconnect The BPP Switch.

Measure the voltage between:	
(+) BPP Switch Connector, Harness Side	(-)
B+	Ground

Vehicle	Pin	Circuit
E-Series	1	B+
	2	BPP
F-Super Duty	3	B+
	4	BPP

Is The Voltage Greater Than 10.5 Volts?

YES — AI6

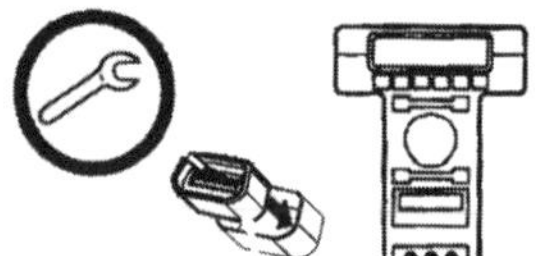

NO — STOP: REPAIR The Circuit. CLEAR The DTCs. REPEAT The Self-Test.

AI6 — CHECK THE BPP SWITCH SIGNAL

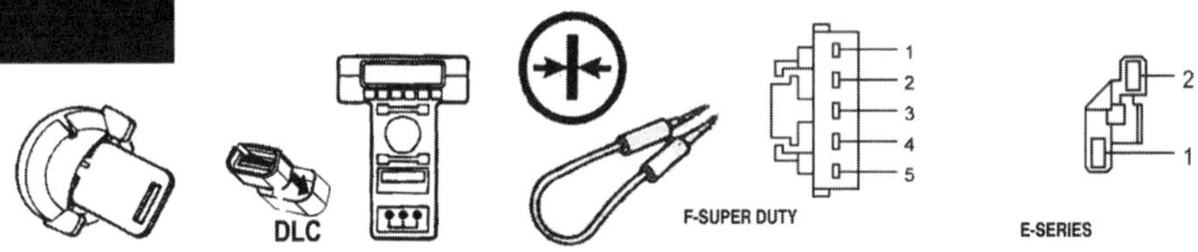

Key On, Engine Off.

Access The BPP PID.

Connect a 5 amp fused jumper wire between the following:	
Point A BPP Switch Connector, Harness Side	Point B BPP Switch Connector, Harness Side
B+	BPP

Vehicle	Pin	Circuit
E-Series	1	B+
	2	BPP
F-Super Duty	3	B+
	4	BPP

Does The BPP PID Indicate ON?

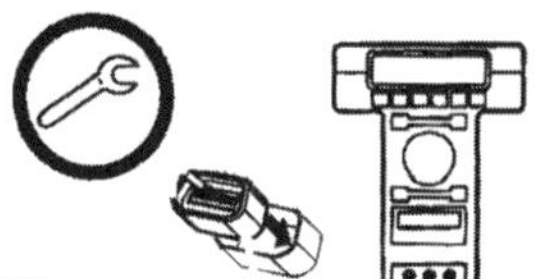

YES — STOP: INSTALL A New BPP Switch. CLEAR The DTCs. REPEAT The Self-Test.

NO — AI7

AI7 CHECK THE BPP SWITCH INPUT CIRCUIT FOR AN OPEN

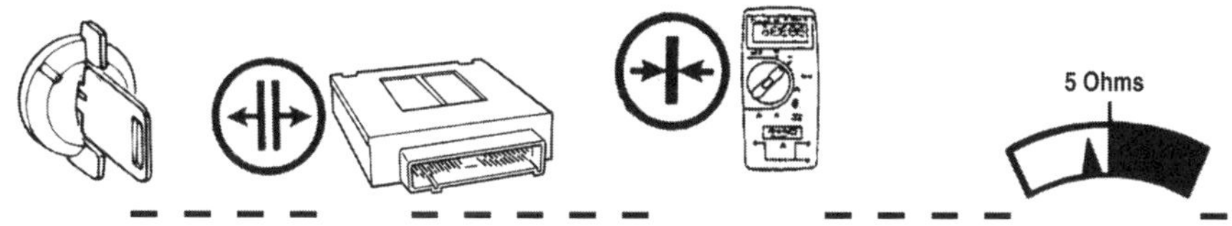

Key Off.

Disconnect The PCM Connector.

Measure the resistance between:	
(+) BPP Switch Connector, Harness Side	(-) PCM Connector, Harness Side
BPP	BPP - Pin B18

Is The Resistance Less Than 5 Ohms?

YES → AI8

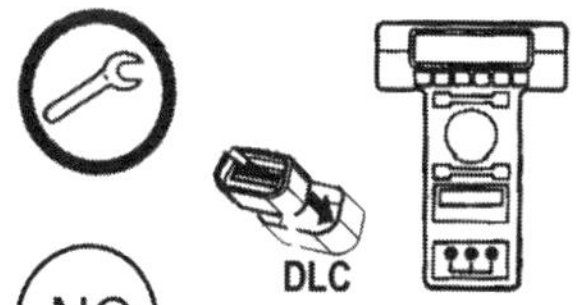

NO → REPAIR The Circuit. CLEAR The DTCs. REPEAT The Self-Test. STOP

AI8 CHECK FOR CORRECT PCM OPERATION

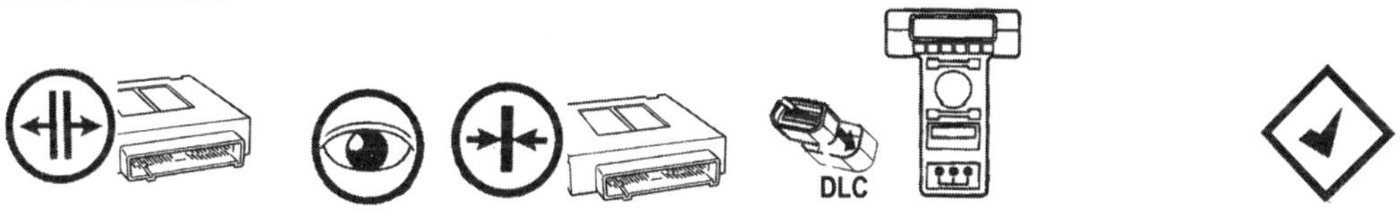

Disconnect All PCM Connectors.

Visually Inspect For Pushed Out Pins, Corrosion.

Connect All PCM Connectors And Make Sure They Seat Correctly.

Perform The PCM Self-Test And Verify The Concern Is Still Present.

Is The Concern Still Present.

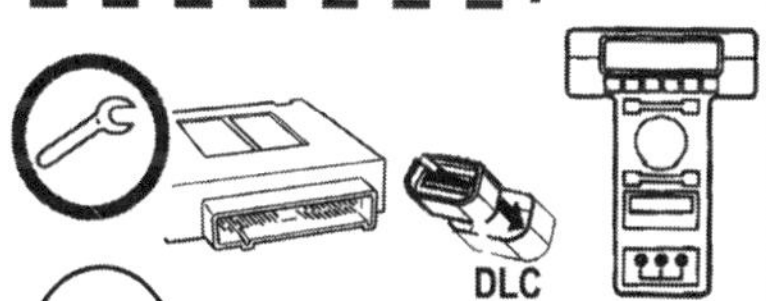

YES → INSTALL A New PCM. CLEAR The DTC's And REPEAT The Self-Test. STOP

OK

NO → The System Is Operating Correctly At This Time. The Concern May Have Been Caused By A Loose Or Corroded Connector. STOP

NOTES

Trouble Code (s)	Problem: Variable Geometry Turbo (VGT) Output	Test
P0046		AK 2006-07

REFERENCE LIST

This pinpoint test is intended to diagnose the following:

- Harness circuits: VGT, and VGTCH
- VGT actuator
- Powertrain control module (PCM) (12A650)

DIAGNOSTIC TROUBLE CODE INDEX			
Codes	KOEO	KOER	CM
P0046	AK1	AK1	AK1

P0046 = Turbo/Supercharger Boost Solenoid Circuit Range/Performance Fault

Circuit Function:
The variable geometry turbo (VGT) actuator is a variable position valve that controls vane position in the turbine housing. The VGT is located on top of the turbocharger. The valve position is controlled by switching voltage source inside of the PCM.

Fault Detection:
The powertrain control module (PCM) will detect an open or short in the VGT control circuit after the KOEO test has run and set a DTC.

CAUTION!:
The powertrain control module (PCM) harness connectors must be properly seated and the connector latch properly attached to eliminate possible driveability concerns or a no start condition. Installing the PCM connectors on an angle may cause an improper connection, misdiagnosis and damaged components. Install the connector until the lever pivots and seats itself. Apply light pressure to get the connector into position on the PCM and then fully seat the connector.

Note: Visually inspect the harness connectors for corrosion, damage, proper mating and correct pin tension.

Note: When the PCM is disconnected additional DTCs will be set. Clear all DTCs after restoring the vehicle.

PINPOINT TEST SCHEMATICS AND CONNECTORS

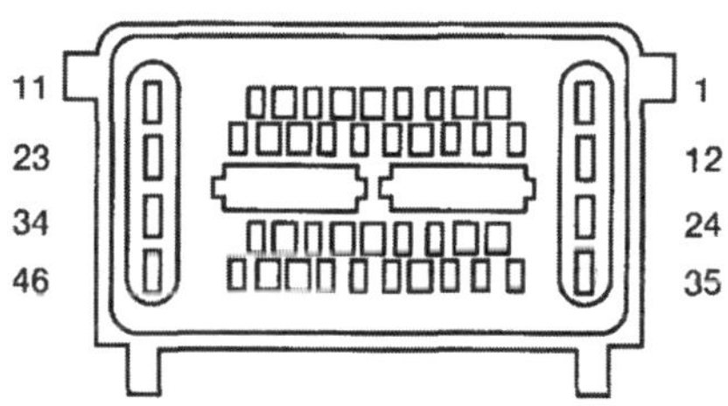

PCM
ENGINE HARNESS
CONNECTOR

Pin	Circuit
E10	VGT (Variable Geometry Turbocharger)

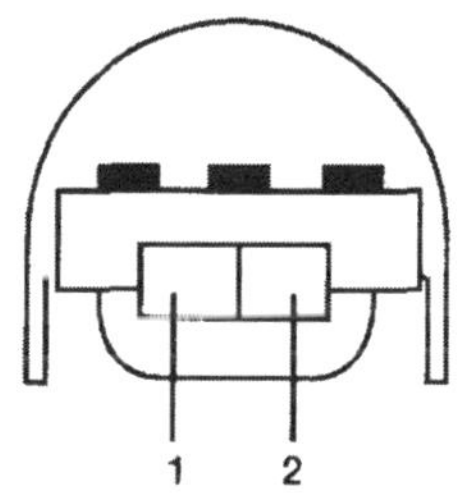

ACTUATOR
CONNECTOR

Pin	Circuit
2	VGTCH (Variable Geometry Turbocharger Actuator Voltage)
1	VGT (Variable Geometry Turbocharger)

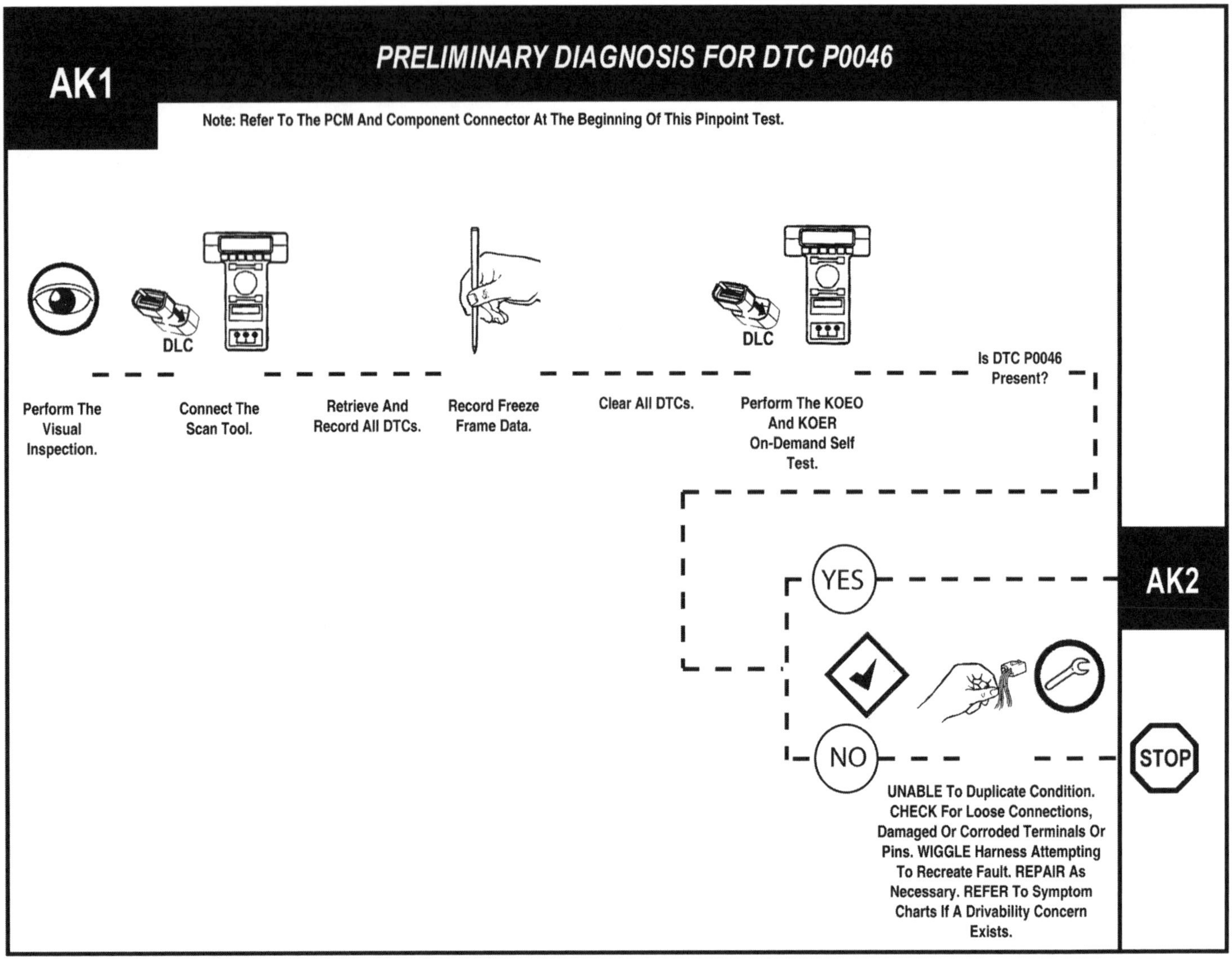
AK1
PRELIMINARY DIAGNOSIS FOR DTC P0046
Note: Refer To The PCM And Component Connector At The Beginning Of This Pinpoint Test.
DLC
DLC
Perform The Visual Inspection.
Connect The Scan Tool.
Retrieve And Record All DTCs.
Record Freeze Frame Data.
Clear All DTCs.
Perform The KOEO And KOER On-Demand Self Test.
Is DTC P0046 Present?
YES
AK2
NO
STOP
UNABLE To Duplicate Condition. CHECK For Loose Connections, Damaged Or Corroded Terminals Or Pins. WIGGLE Harness Attempting To Recreate Fault. REPAIR As Necessary. REFER To Symptom Charts If A Drivability Concern Exists.

AK2

DIAGNOSTIC TROUBLE CODE (DTC) P0046

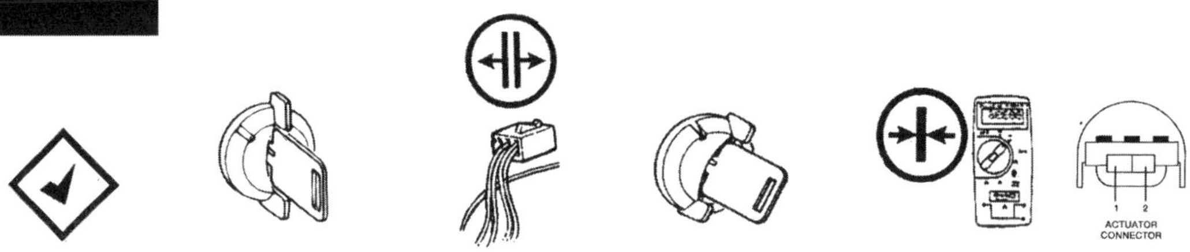

Possible Causes:
- Control Circuit Open
- Control Circuit Shorted To Voltage
- Control Circuit Shorted To Ground
- VGT Actuator
- PCM.

Key Off.

Disconnect The VGT Actuator Electrical Connector.

Key On. Engine Off.

Measure the voltage between:	
(+) VGT Actuator Connector, Harness Side	(-)
VGTCH - Pin 1	Ground

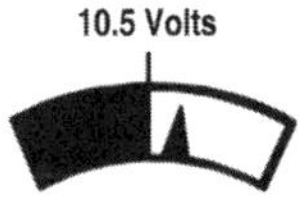

Is The Voltage Greater Than 10.5 Volts?

YES → AK3

NO →

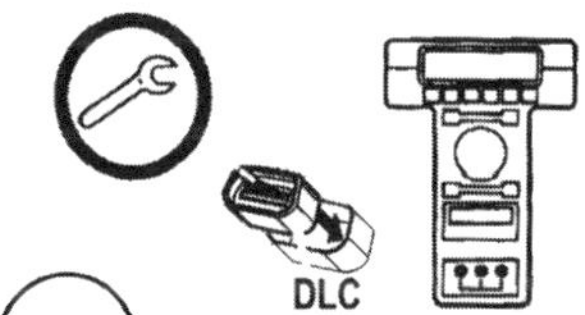

REPAIR Open In Circuit. RESTORE The Vehicle. CLEAR DTCs And RETEST The System.

STOP

AK3

CHECK VGT ACTUATOR

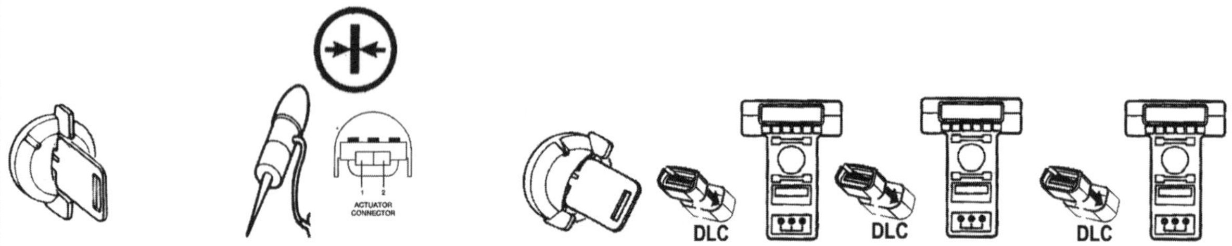

Key Off.

Connect a non-powered test lamp between:

(+) VGT Actuator Connector, Harness Side	(-) VGT Actuator Connector, Harness Side
VGTCH - Pin 1	VGT - Pin 2

Key On. Engine Off.

Access VGT PIDS.

Command VGT To Below 10% Duty Cycle.

Command The VGT To 85% Duty Cycle.

Duty Cycle Driven Low

Duty Cycle Driven High

Does The Test Light Go Out As VGT Duty Cycle Is Driven Low And Lights As The Duty Cycle Is Driven High?

YES → **AK7**

NO → **AK4**

AK4

CHECK VGT CONTROL CIRCUIT FOR A SHORT TO VOLTAGE

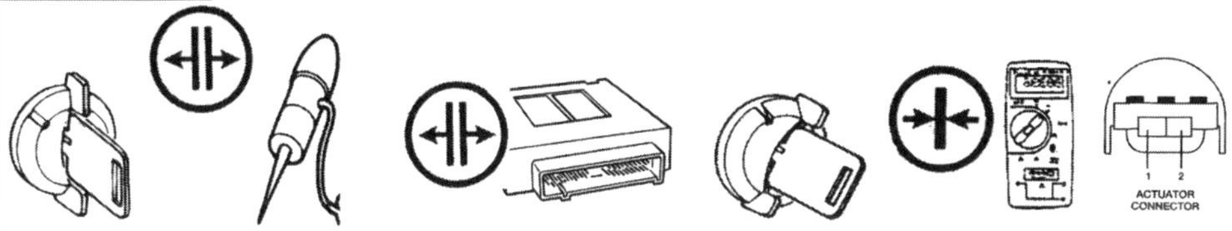

Key Off.

Remove The Test Light.

Disconnect The PCM Electrical Connector.

Key On. Engine Off.

Measure the voltage between:	
(+) VGT Actuator Connector, Harness Side	(-)
VGT - Pin 2	Ground

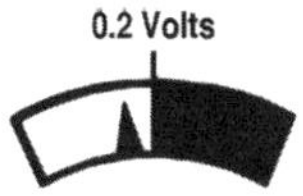

Is The Voltage Less Than 0.2 Volts?

YES → AK5

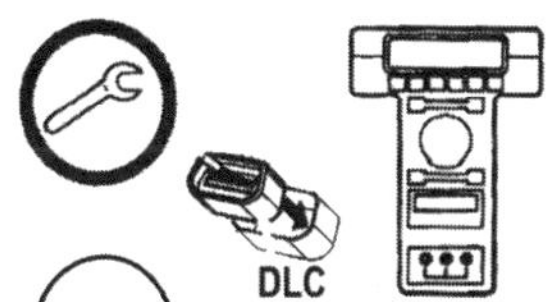

NO → STOP

REPAIR Short In Control Circuit. CLEAR DTCs And REPEAT The Self-Test.

AK5

CHECK VGT CONTROL CIRCUIT FOR A SHORT TO GROUND

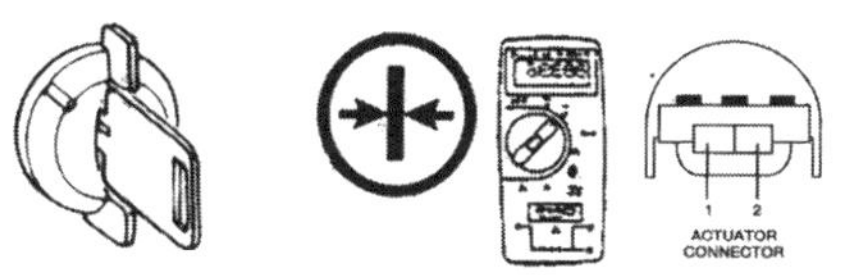

Key Off.

Measure the resistance between:	
(+) VGT Actuator Connector, Harness Side	(-)
VGT - Pin 2	Ground

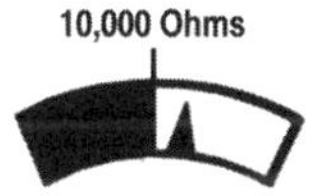

Is The Resistance Greater Than 10,000 Ohms?

YES → AK6

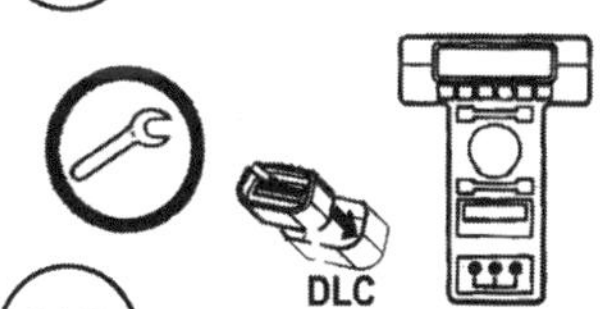

NO → STOP

REPAIR Short In Control Circuit. CLEAR DTCs And REPEAT The Self-Test.

AK6

CHECK VGT CONTROL CIRCUIT FOR AN OPEN

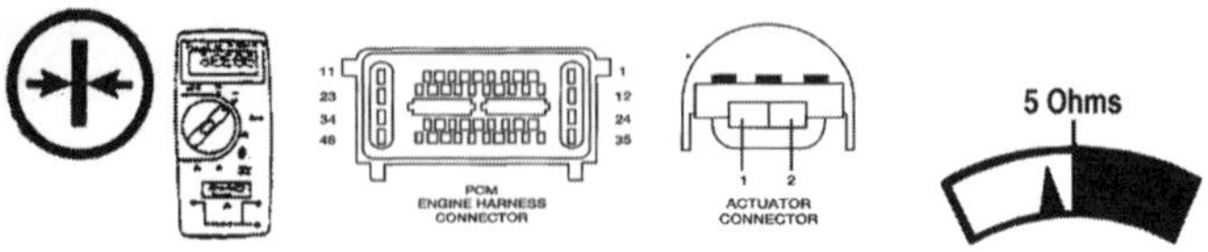

Is The Resistance Less Than 5 Ohms?

Measure the resistance between:	
(+) VGT Actuator Connector, Harness Side	**(-) PCM Connector, Harness Side**
VGT - Pin 2	VGT - Pin E10

YES → **AK8**

NO → **STOP**

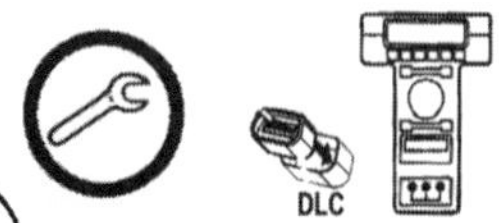

REPAIR Open In Control Circuit. CLEAR DTCs And REPEAT The Self-Test.

AK7 CLEAN THE VGT ACTUATOR VALVE AND RETEST

CAUTION: The VGT actuator valve is sensitive to contamination. All work areas must be clean before starting this procedure. Do not allow contamination to enter the valve ports, cam follower, or turbocharger housing. Handle the VGT actuator by the solenoid body only. Do not attempt to wipe oil off the valve. Do not let the valve come in contact with materials that could contaminate the valve mechanism.

CAUTION: Use only the Carburetor Tune-Up Cleaner PM-2 to spray the actuator valve. Brake cleaner or other solvents may damage the actuator valve.

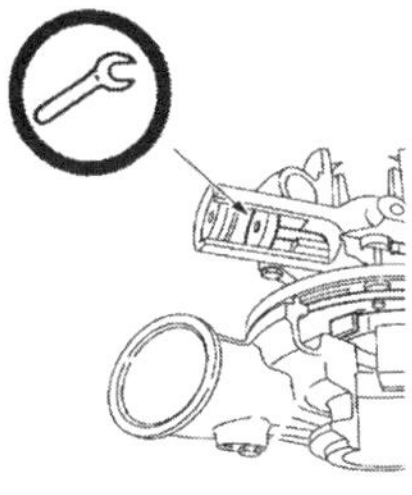

Remove The VGT Actuator From The Turbocharger Assembly.
Note: Do Not Spray The Zinc Coated Larger Diameter Housing, Or Electrical Connector.
Spray The Valve With Carburetor Tune-up Cleaner Pm-2 While Pressing The Cam Follower (Tip Of The Actuator).

Install The VGT Actuator.

VGT Actuator Connector Connected.

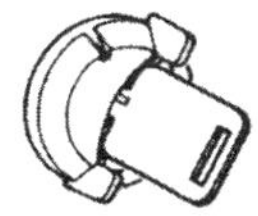

Key On, Engine Running.

Access the PCM and control the VGT PID.

Increase The Commanded Engine Speed To 1,250 RPM Using Output State Control. The Calibration May Limit The Actual RPM Between 1,150 And 1,200 RPM.

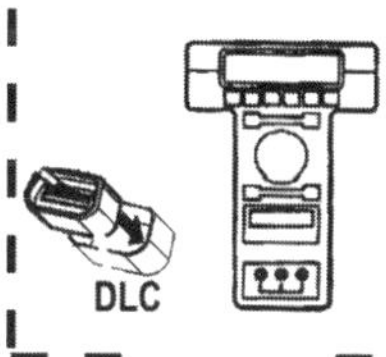

Increase The VGT Duty Cycle To Greater Than 70%.

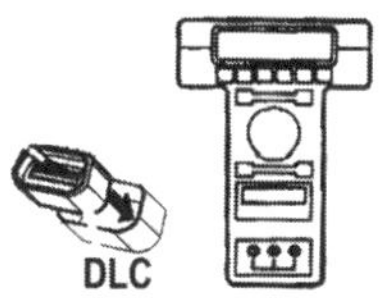

Decrease The VGT Duty Cycle To 0%.

Does The Turbocharger Pitch Change When The VGT Duty Cycle Is Increased And Decreased.

YES

The System Is Operating Correctly At This Time. The Concern May Have Been Caused By A Contaminated VGT Actuator Valve. Change The Oil And Oil Filter Assembly.

STOP

NO

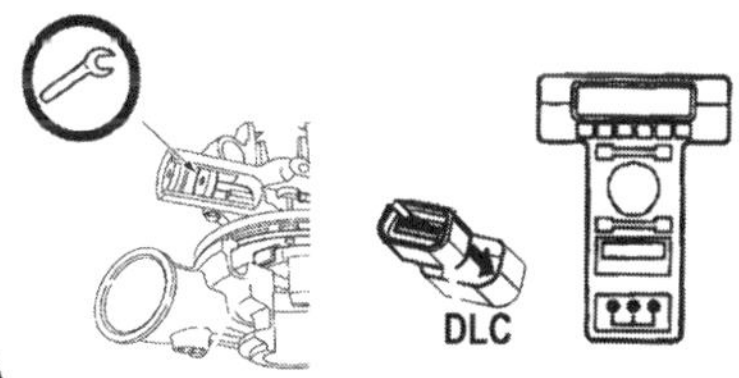

INSTALL A New VGT Actuator. CLEAR The DTCs. REPEAT The Self-Test.

STOP

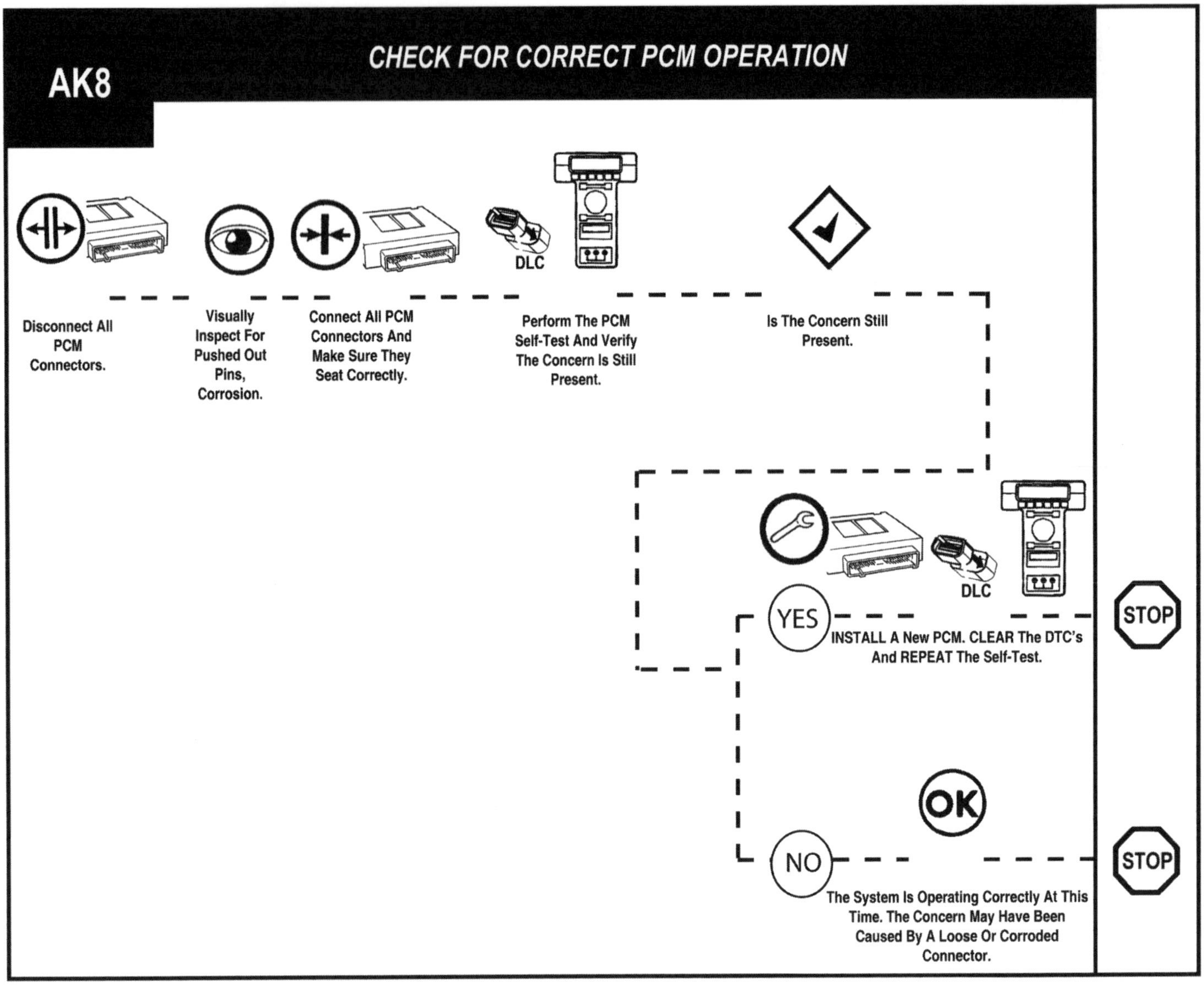
AK8
CHECK FOR CORRECT PCM OPERATION
DLC
Disconnect All PCM Connectors.
Visually Inspect For Pushed Out Pins, Corrosion.
Connect All PCM Connectors And Make Sure They Seat Correctly.
Perform The PCM Self-Test And Verify The Concern Is Still Present.
Is The Concern Still Present.
DLC
YES
INSTALL A New PCM. CLEAR The DTC's And REPEAT The Self-Test.
STOP
OK
NO
The System Is Operating Correctly At This Time. The Concern May Have Been Caused By A Loose Or Corroded Connector.
STOP

Trouble Code (s)	PROBLEM: Transmission Range Circuit	Test
P1705		AL 2006-07

REFERENCE LIST

DIAGNOSTIC TROUBLE CODE INDEX			
Codes	KOEO	KOER	CM
P1705	AL1	AL1	AL1

AL1 PRELIMINARY DIAGNOSIS FOR DTC P1705

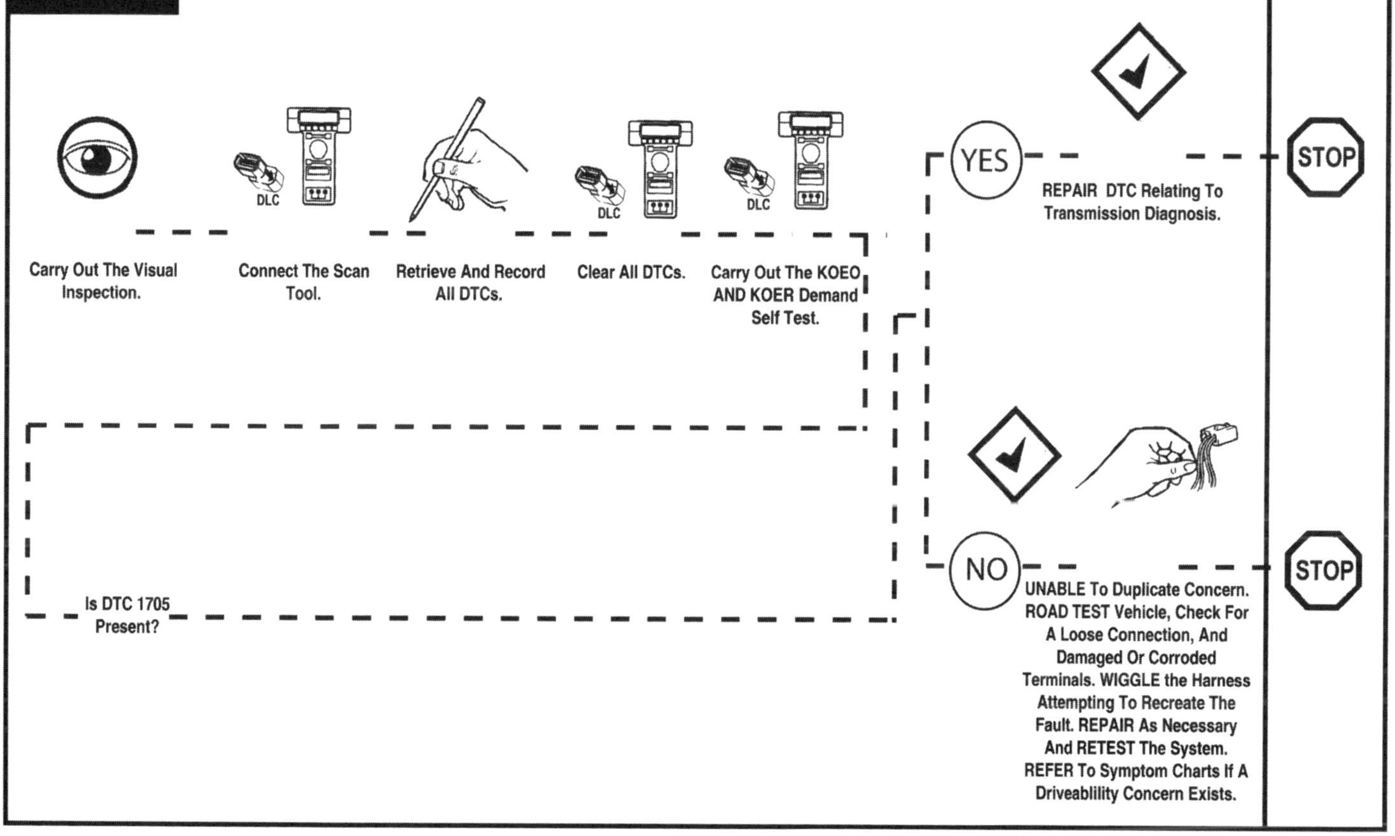

NOTES

Trouble Code (s)		Test
P1464	Problem: A/C Clutch Relay, A/C Circuits	AM 2006-07

REFERENCE LIST

DIAGNOSTIC TROUBLE CODE INDEX			
Codes	KOEO	KOER	CM
P1464	AM1	AM1	AM1

CAUTION!:
The powertrain control module (PCM) harness connectors must be properly seated and the connector latch properly attached to eliminate possible driveability concerns or a no start condition. Installing the PCM connectors on an angle may cause an improper connection, misdiagnosis and damaged components. Install the connector until the lever pivots and seats itself. Apply light pressure to get the connector into position on the PCM and then fully seat the connector.

Note: Visually inspect the harness connectors for corrosion, damage, proper mating and correct pin tension.

AM1

DTC P1464: CHECK ACCS PID

Note: Verify A/C And Defrost Were Off During Self-Test. If A/C Or Defrost Were On, Turn Off And Rerun Self-Test.

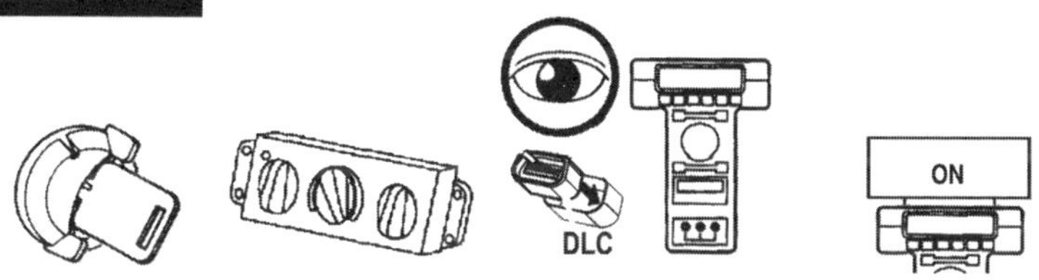

Key On, Engine Off.

A/C And Defrost Off.

Access And Monitor ACCS PID.

Is ACCS PID On?

YES — ACCS Input To PCM Is Requesting A/C. Diagnose And Repair A/C System. STOP

NO — The ACCS PID Indicates That The ACCS Input To The PCM OFF. VERIFY Self Test Results. With A/C And Defrost Off, RERUN Self-Test. STOP

NOTES

Trouble Code (s)	Problem: Power Take-Off (PTO) Control	Test
P1502		AN 2006-07

REFERENCE LIST

This pinpoint test is intended to diagnose the following:

- Harness circuit PTO
- Powertrain control module (PCM) (12A650)

DIAGNOSTIC TROUBLE CODE INDEX			
Codes	KOEO	KOER	CM
P1502	AN1	AN1	AN1

P1502 = Invalid Test - Auxiliary Powertrain Control Functioning

Circuit Function:
The PTO system provides an input signal to the powertrain control module (PCM) indicating there is an additional load being applied to the engine. The PCM disables the on-board diagnostic (OBD) monitors and increases the engine RPM based on the PTO system or auxiliary idle control input.

The following entry conditions are required to enable the PTO or auxiliary idle control:

- Accelerator Pedal Released
- Automatic transmission in PARK or manual transmission in NEUTRAL with the clutch pedal released
- Brake pedal released
- Engine at normal operating temperature
- Parking brake applied

CAUTION!: The powertrain control module (PCM) harness connectors must be properly seated and the connector latch properly attached to eliminate possible driveability concerns or a no start condition. Installing PCM connectors on an angle may cause an improper connection, misdiagnosis and damaged components. Install the connector until the lever pivots and seats itself. Apply light pressure to get the connector into position on the PCM and then fully seat the connector.

Note: Visually inspect the harness connectors for corrosion, damage, proper mating and correct pin tension.

AN1 — *VERIFY CORRECT PTO OPERATION*

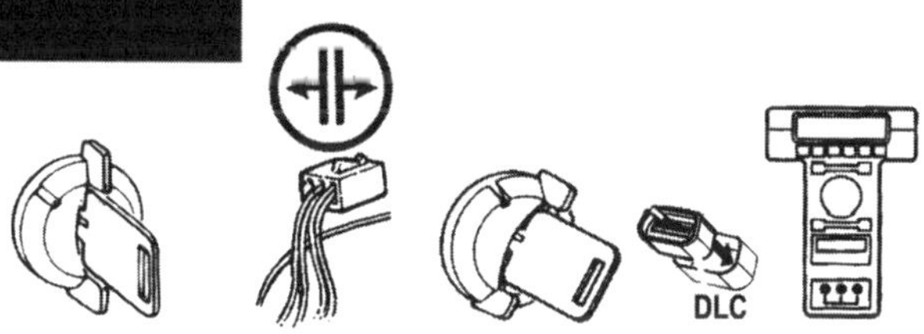

Key Off. — Disconnect The PTO System Or Auxiliary Idle Control. — Key ON, Engine OFF. — Carry Out The PCM Self-Tests.

Is DTC P1502 Present?

- **YES** → **AN2**
- **NO** → **STOP** — REFER to The Owner's Literature Provided With The PTO System Or Auxiliary Idle Control To Verify Correct Operating Procedures, Entry Conditions, And Diagnostic Information.

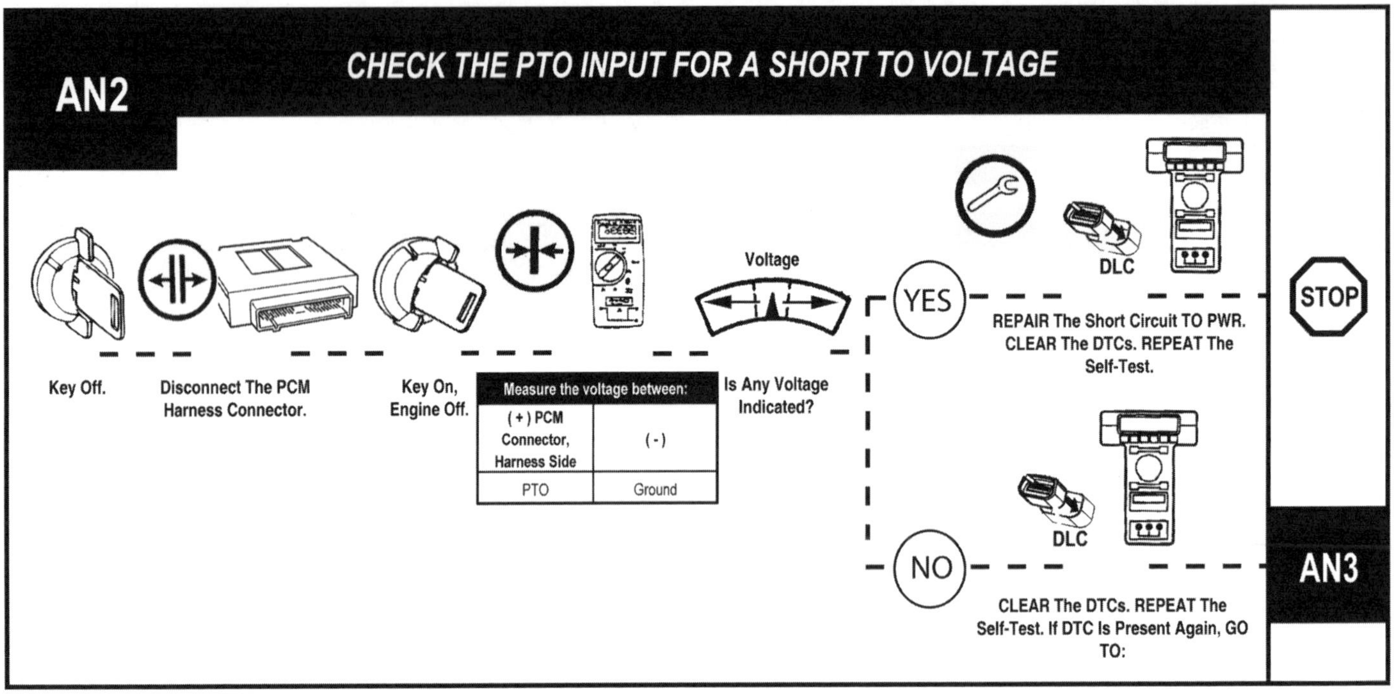

Measure the voltage between:	
(+) PCM Connector, Harness Side	(-)
PTO	Ground

AN3 CHECK FOR CORRECT PCM OPERATION

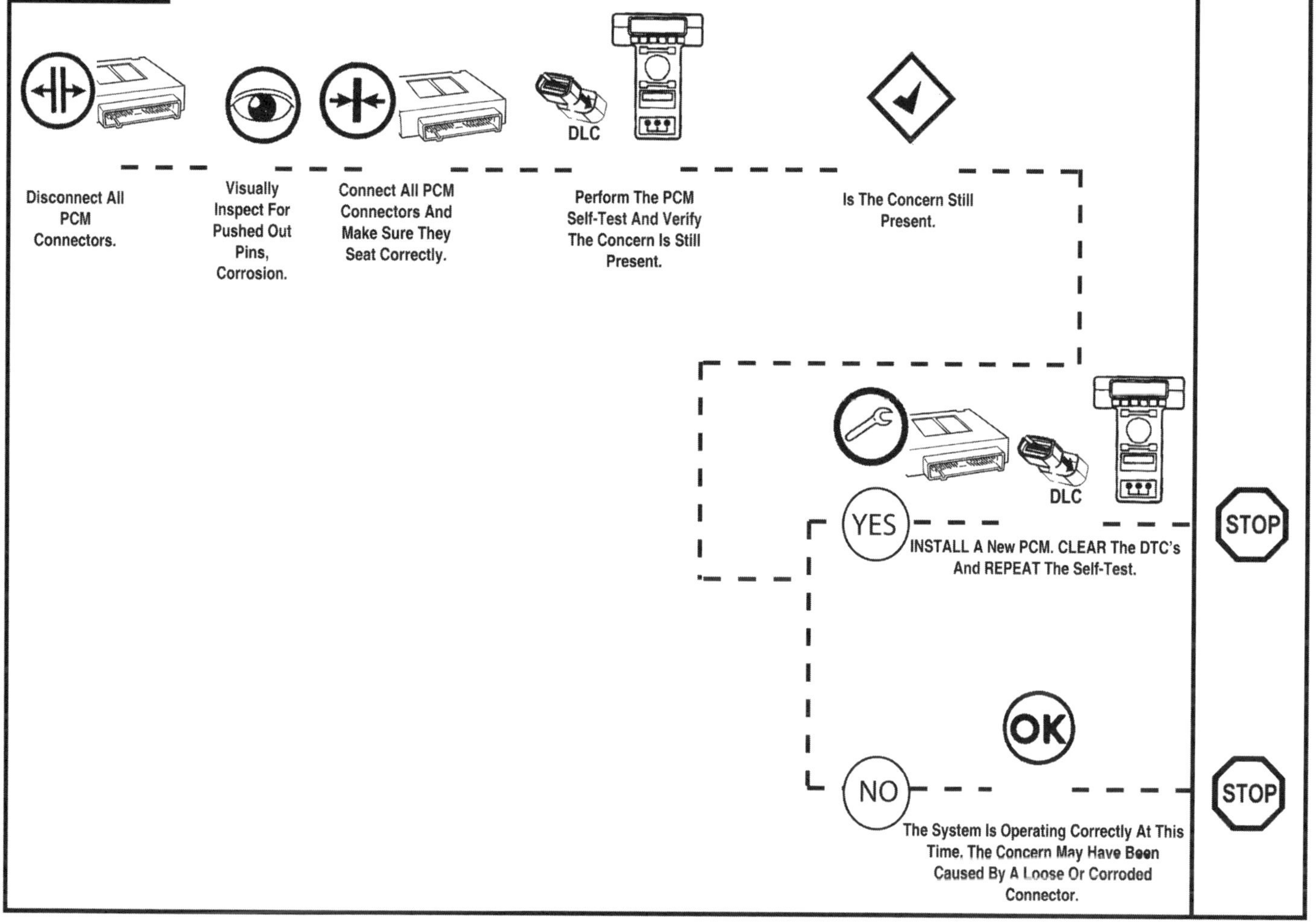

NOTES

Trouble Code (s)	Problem: Injector Control Pressure Too High	Test
P2284, P2288, P2290		AO 2006-07

REFERENCE LIST

This pinpoint test is intended to diagnose the following:

- Fuel system
- High pressure oil system
- ICP sensor
- IPR valve

DIAGNOSTIC TROUBLE CODE INDEX			
Codes	KOEO	KOER	CM
P2284	AO1	AO1	AO1
P2288	AO1	AO1	AO1
P2290	AO1	AO1	AO1

P2284 = Injector Control Pressure Sensor Circuit Range/Perfomance
P2288 = Injector Control Pressure Too High
P2290 = Injector Control Pressure Too Low

Signal Function:
The injector control pressure (ICP) is controlled by injection pressure regulator (IPR). The powertrain control module (PCM) uses many input variables to determine the desired control pressure.
Battery voltage is supplied to the IPR when the ignition switch is the ON position. The valve position is controlled by switching the output signal circuit to ground. The ON/OFF time is modulated from 0% to 65%, depending upon the desired control pressure.

Injection Control Pressure (ICP) Sensor Bias:
The ICP sensor bias varies between sensors. The amount of voltage the sensor deviates from a calculated reference value (sensor bias) may cause an injection control pressure calculation error.

Detection/Management:
An open or short to ground control circuit can be detected by an on demand output circuit check performed during the KOEO test. When the starter is engaged and the engine fails to start, the IPR command increases toward its maximum until the engine starts. The normal range for the IPR at idle is between 16-24%. An above 24% range indicates high effort or higher commanded ICP to achieve idle.
A below 16% indicates low effort to achieve idle.

The PCM is capable of detecting whether the desired injection control pressure is equal to the measured injection control pressure while the engine is running. If the measured injection control pressure does not reasonably compare to the desired injection control pressure, the PCM ignores the measured ICP sensor signal and attempts to control the engine with the desired value. (If the problem is the sensor circuit, this strategy causes little performance deterioration. If the problem is in the control circuit, engine performance will probably still be unsatisfactory).

An injection control pressure step test, in which the PCM commands and then measures calibrated pressures, is performed during the KOER test.
Note: The engine does not operate with an IPR circuit that is not functioning.

CAUTION!:
The powertrain control module (PCM) harness connectors must be properly seated and the connector latch properly attached to eliminate possible driveability concerns or a no start condition. Installing the PCM connectors on an angle may cause an improper connection, misdiagnosis and damaged components. Install the connector until the lever pivots and seats itself. Apply light pressure to get the connector into position on the PCM and then fully seat the connector.

Note: Visually inspect the harness connectors for corrosion, damage, proper mating and correct pin tension.

Note: When the PCM is disconnected additional DTCs will be set. Clear all DTCs after restoring the vehicle.

PINPOINT TEST SCHEMATICS AND CONNECTORS

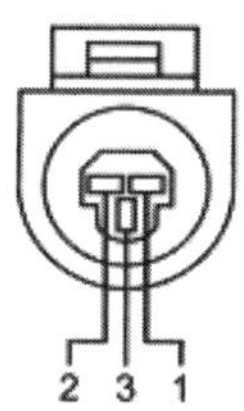

Pin	Circuit
3	ICP (Injection Control Pressure)

AO1 — PRELIMINARY DIAGNOSIS FOR (DTC) P2284 P2288 OR P2290

DLC — Connect The Scan Tool.

Perform KOEO, KOER And Continuous On-Demand Self Tests.

DLC — Retrieve And Record All DTCs. Record The Freeze Frame Data.

Is DTC P2284, P2288 Or P2290 Present?

- **YES**
  - For KOEO And Continuous DTC P2284 Or P2288: → **AO2**
  - For KOER DTC P2284 Or P2288: → **AO8**
  - Fo DTC P2290: → **AO8**
- **NO** → **STOP**
  UNABLE To Duplicate Condition. WIGGLE the Harness Attempting To Recreate The Fault. REPAIR As Necessary And RETEST The System. REFER To Symptom Charts If A Driveablility Concern Exists.
  Symptom Charts

AO2 — DIAGNOSTIC TROUBLE CODE (DTC) P2284 OR P2288

Possible Causes:
- Incorrect Oil Or Viscosity
- Poor Oil Quality
- Gelled Fuel/No Fuel
- Low Fuel Pressure
- Biased ICP Sensor
- IPR Valve
- High-Pressure Oil System Leak
- High-Pressure Oil Pump
- PCM

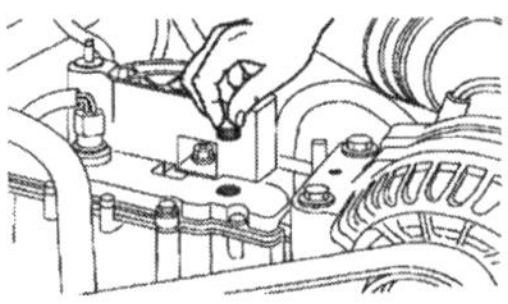

Verify Correct Oil Quality/Viscosity And Correct Fuel Grade Are Being Used For The Temperature Conditions.

Is The Oil And Fuel Quality OK?

- **YES** → **AO3**
- **NO** → **STOP**
  DLC
  REPAIR The Fuel Or Oil Condition. CLEAR The DTC's. REPEAT The Self-Test.

AO3 — CHECK FUEL PRESSURE

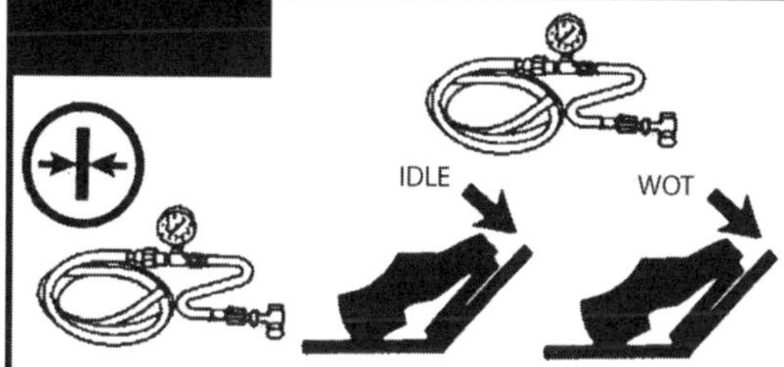

Measure The Fuel Pressure At The Regulator Block.

Measure The Fuel Pressure At Idle And On The Road At Wide Open Throttle (WOT) At Full Load. The Fuel Pressure At Idle Should Be 310 kPa (45 psi) Minimum And 310 - 482 kPa (45-70 psi) At WOT Full Load.

Is The Fuel Pressure Within Specifications?

- YES → AO4

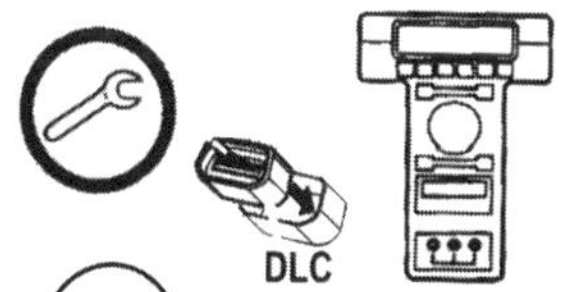

- NO → REPAIR The Fuel System Concern. CLEAR The DTCs. REPEAT The Self Test.

AO4 — PERFORM THE KOER ON-DEMAND SELF-TEST

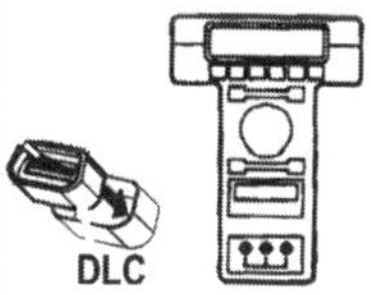

Perform The KOER On-Demand Self-Test.

Is DTC P2284 Or P2288 Present?

- YES → AO5
- NO → AO10

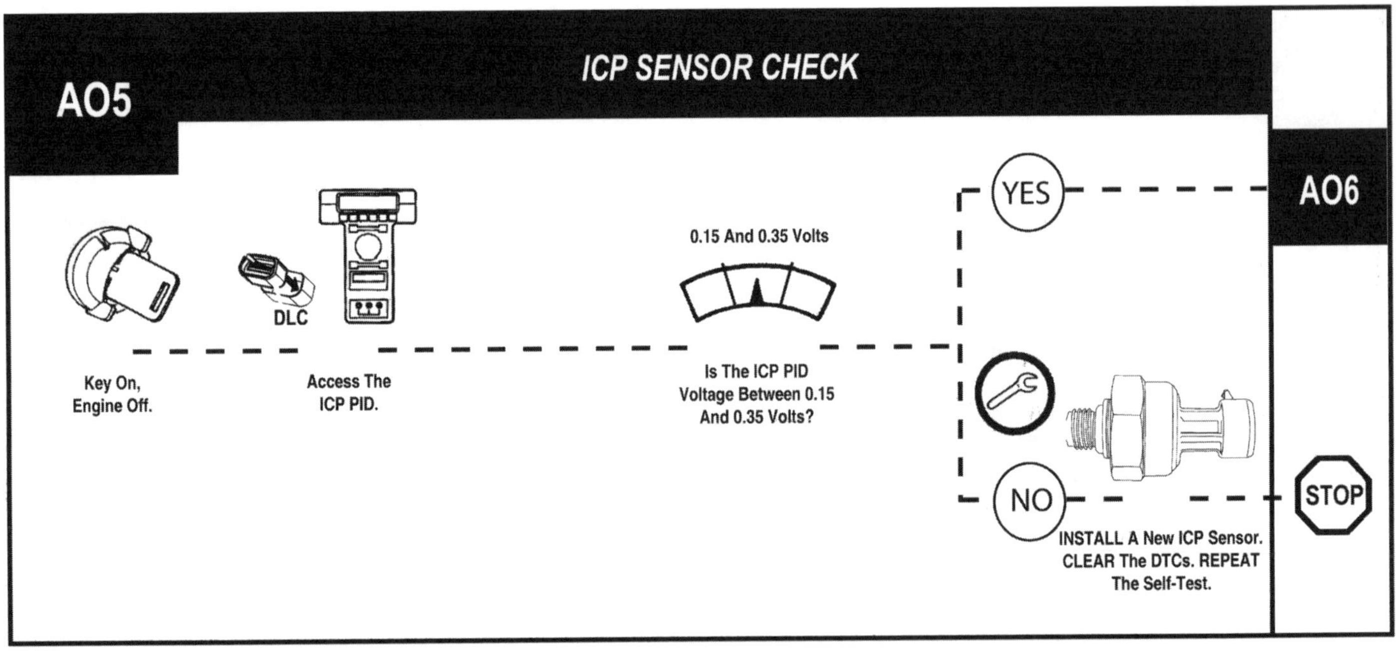
AO5
ICP SENSOR CHECK
DLC
0.15 And 0.35 Volts
Key On, Engine Off.
Access The ICP PID.
Is The ICP PID Voltage Between 0.15 And 0.35 Volts?
YES
AO6
NO
INSTALL A New ICP Sensor. CLEAR The DTCs. REPEAT The Self-Test.
STOP

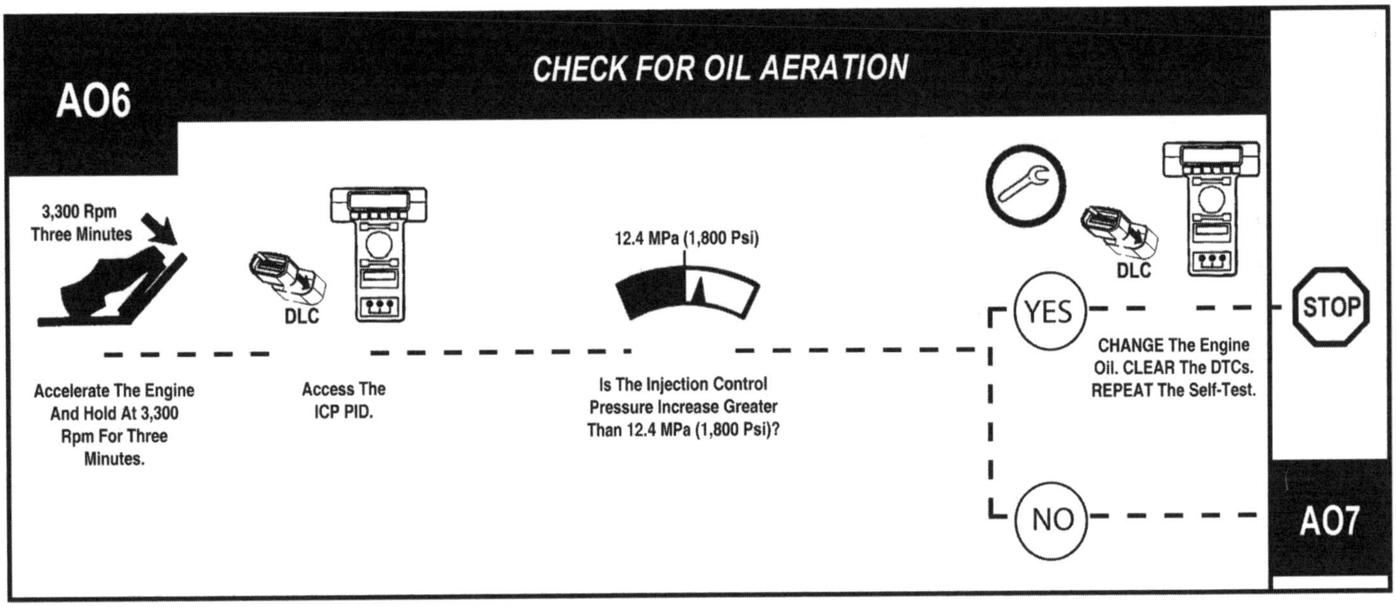
AO6
CHECK FOR OIL AERATION
3,300 Rpm
Three Minutes
DLC
12.4 MPa (1,800 Psi)
Accelerate The Engine And Hold At 3,300 Rpm For Three Minutes.
Access The ICP PID.
Is The Injection Control Pressure Increase Greater Than 12.4 MPa (1,800 Psi)?
DLC
YES
CHANGE The Engine Oil. CLEAR The DTCs. REPEAT The Self-Test.
STOP
NO
AO7

AO7 — CHECK THE IPR DUTY CYCLE UNDER VEHICLE LOAD

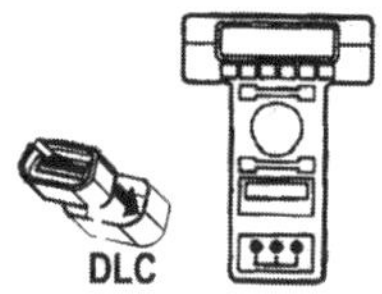

Access The IPR PID.

Operate The Vehicle At WOT And A Full Load Condition.

Does The IPR Duty Cycle Increase To 65%?

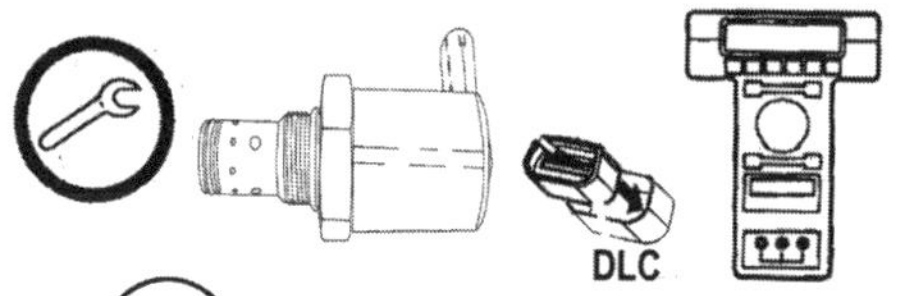

YES → INSTALL A New IPR Valve. CLEAR The DTCs. REPEAT The Self-Test. → **STOP**

NO → UNABLE To Duplicate Concern. Check For A Loose Connection, And Damaged Or Corroded Terminals. WIGGLE the Harness Attempting To Recreate The Fault. REPAIR As Necessary And RETEST The System. REFER To Symptom Charts If A Driveablility Concern Exists. → **STOP**

AO8 — KOER ON-DEMAND DIAGNOSTIC TROUBLE CODE (DTC) P2284, P2288 OR P2290

DTC P2288 Indicates That Injection Control Pressure Was Above Commanded Desired Pressure During Self Test Mode. DTC P2284 Or P2290 Indicate A Difference In Commanded And Actual Injection Control Pressure.

Possible Causes:

- Incorrect Oil Or Viscosity
- Poor Oil Quality
- Gelled Fuel/No Fuel
- Low Fuel Pressure
- Biased ICP Sensor
- High-Pressure Oil System Leak
- High-Pressure Oil Pump
- PCM

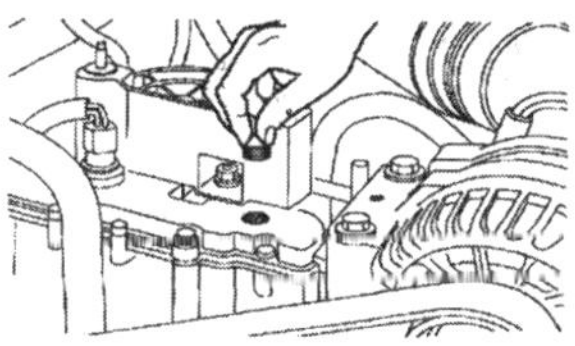

Verify Correct Oil Quality/Viscosity And Correct Fuel Grade Are Being Used For The Temperature Conditions.

Is The Oil And Fuel Quality OK?

YES → **AO9**

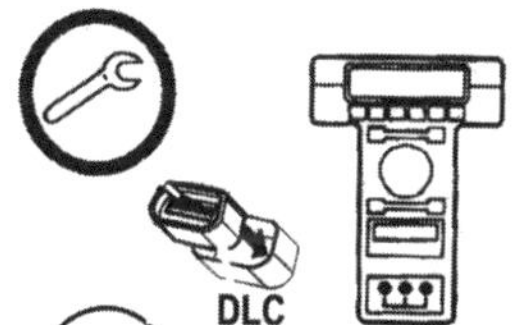

NO → REPAIR The Fuel Or Oil Condition. CLEAR The DTCs. REPEAT The Self-Test. → **STOP**

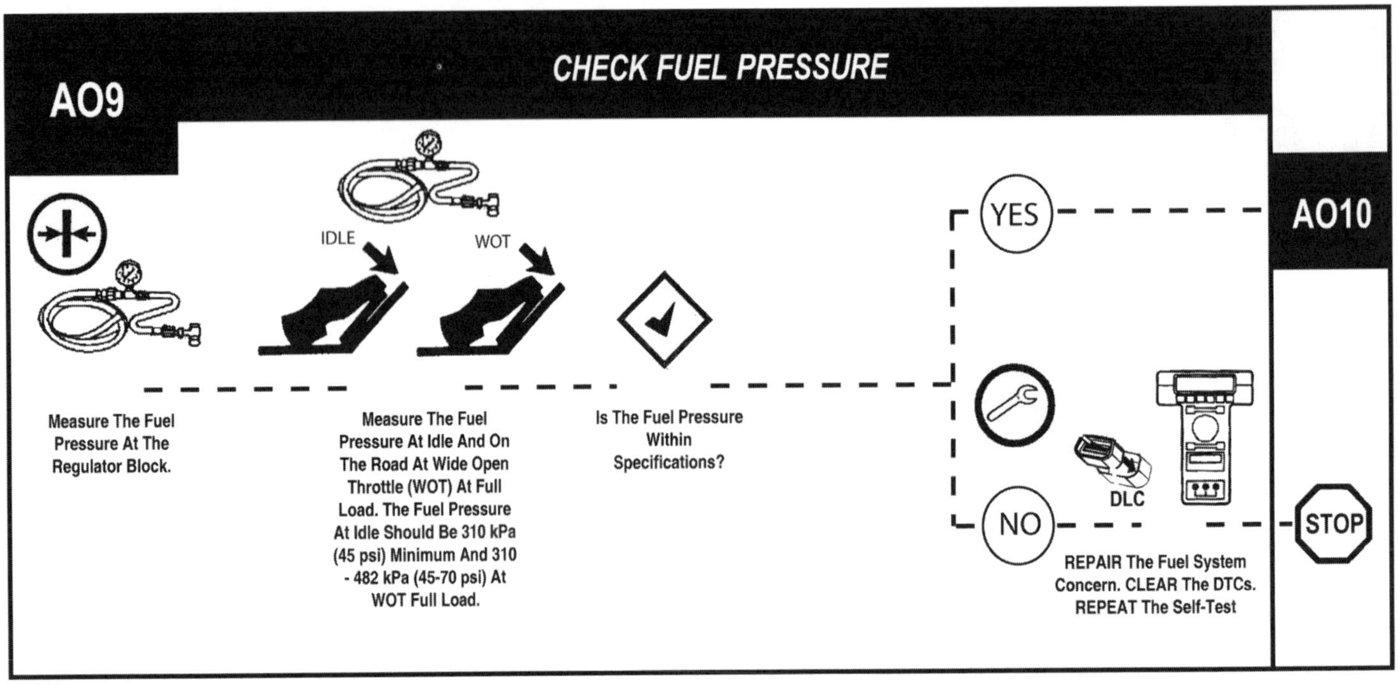
AO9
CHECK FUEL PRESSURE
IDLE
WOT
Measure The Fuel Pressure At The Regulator Block.
Measure The Fuel Pressure At Idle And On The Road At Wide Open Throttle (WOT) At Full Load. The Fuel Pressure At Idle Should Be 310 kPa (45 psi) Minimum And 310 - 482 kPa (45-70 psi) At WOT Full Load.
Is The Fuel Pressure Within Specifications?
YES
AO10
DLC
NO
STOP
REPAIR The Fuel System Concern. CLEAR The DTCs. REPEAT The Self-Test

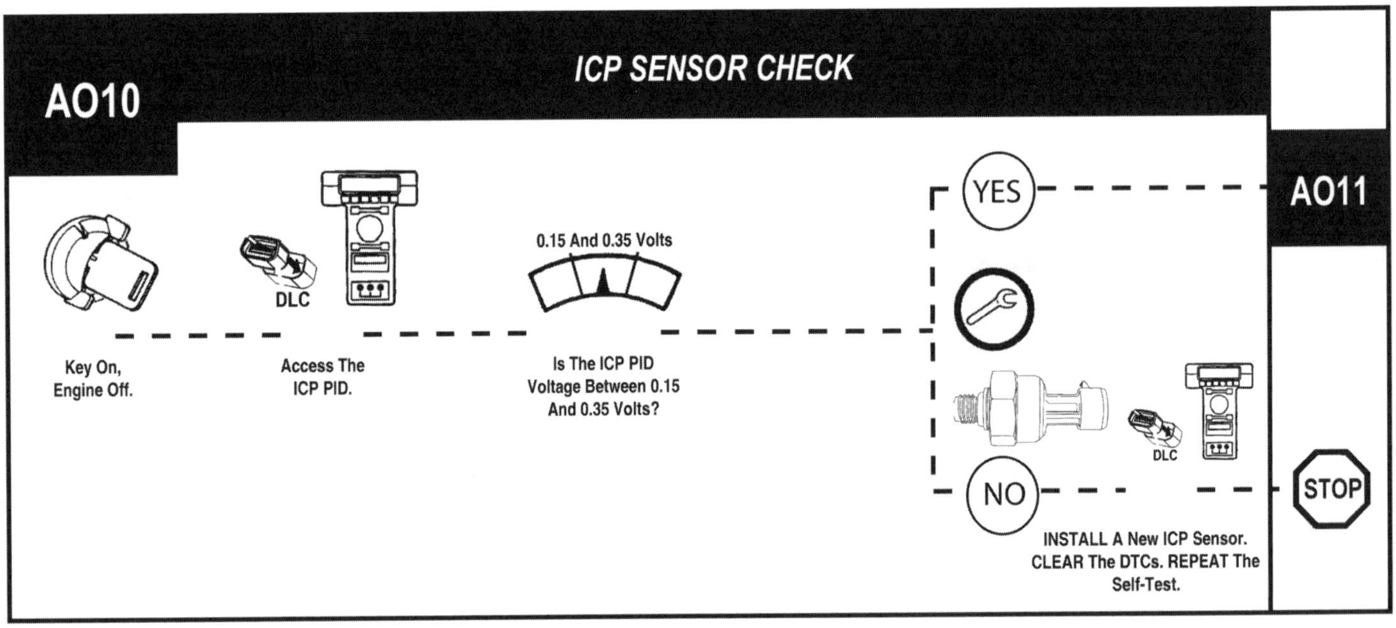
AO10
ICP SENSOR CHECK
DLC
0.15 And 0.35 Volts
Key On, Engine Off.
Access The ICP PID.
Is The ICP PID Voltage Between 0.15 And 0.35 Volts?
YES
AO11
DLC
NO
STOP
INSTALL A New ICP Sensor. CLEAR The DTCs. REPEAT The Self-Test.

AO11 — OIL AERATION CHECK

Wide-Open Throttle Hold For Three Minutes

Accelerate The Engine To Wide-Open Throttle And Hold For Three Minutes.

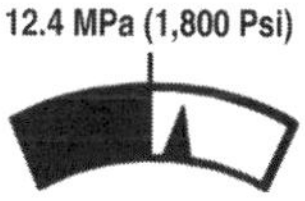

Is The Injection Control Pressure Increase Greater Than 12.4 MPa (1,800 Psi)?

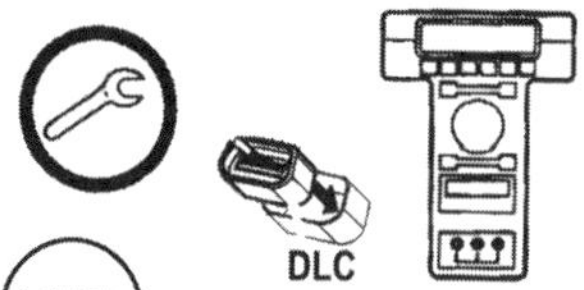

YES — CHANGE The Engine Oil. CLEAR The DTCs. REPEAT The Self-Test. — STOP

NO — AO12

AO12 — CHECK THE ICP SENSOR VOLTAGE WHILE PERFORMING SELF TEST

Note: During KOER On-Demand Self Test, ICP Will Be Commanded High 17 MPa (2465 Psi), Then Low 5 MPa (725 Psi) And Then The EP Will Be Commanded High And Low.

Install The ICP/EBC Adapter Cable D94T-50-A Or Equivalent Between The ICP Sensor And The Engine Harness.

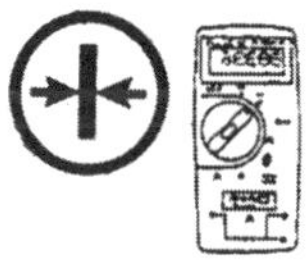

Measure the voltage between:	
(+) ICP Sensor Connector, Harness Side	(-)
ICP - Pin 3	Ground

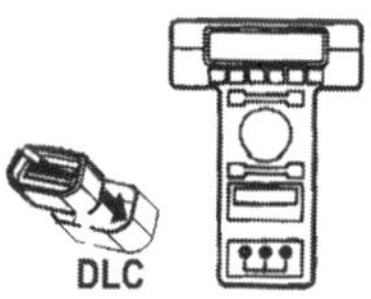

Perform The KOER On-Demand Self-Test.

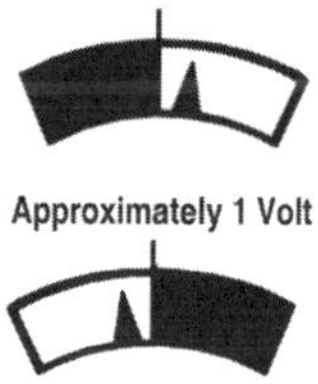

Does The Voltage Reading Increase Above 3 Volts And Then Decrease To Approximately 1 Volt During The ICP Portion Of The KOER Test?

YES — If DTC P2284 Or P2288 Was Displayed, GO TO: AO13

DIAGNOSTIC SUBROUTINES

NO — GO To Diagnostic Routines, Hard Start/No Start Diagnostic Procedures. PERFORM The Monitor The ICP While Cranking Test. — STOP

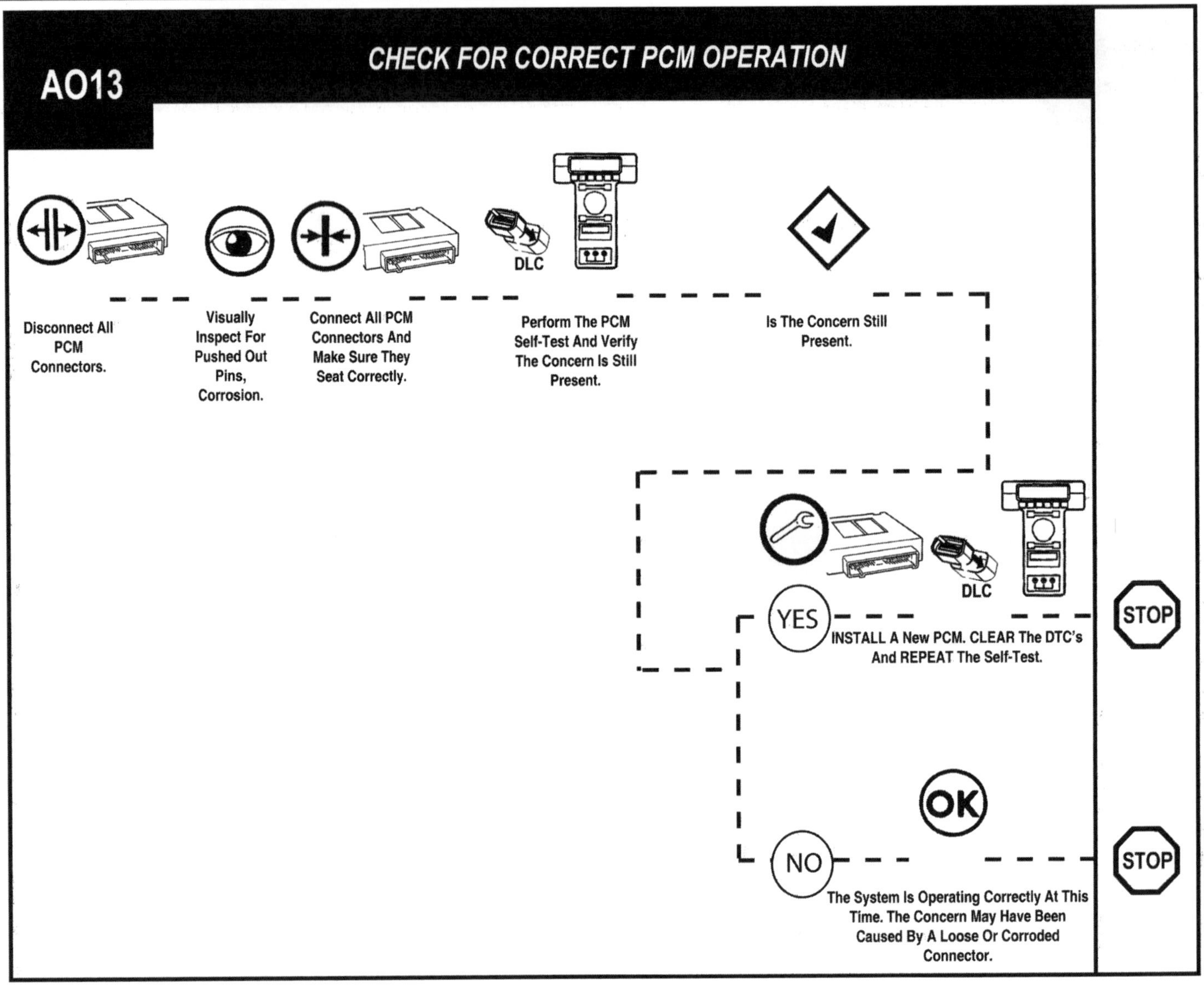
AO13
CHECK FOR CORRECT PCM OPERATION
DLC
Disconnect All PCM Connectors.
Visually Inspect For Pushed Out Pins, Corrosion.
Connect All PCM Connectors And Make Sure They Seat Correctly.
Perform The PCM Self-Test And Verify The Concern Is Still Present.
Is The Concern Still Present.
DLC
YES
INSTALL A New PCM. CLEAR The DTC's And REPEAT The Self-Test.
STOP
OK
NO
The System Is Operating Correctly At This Time. The Concern May Have Been Caused By A Loose Or Corroded Connector.
STOP

Trouble Code (s)	Problem: Injector Control Pressure Too High-Engine Off	Test
P2289		AP 2006-07

REFERENCE LIST

This pinpoint test is intended to diagnose the following:
- Harness circuits: ICP and SIGRTN
- ICP sensor
- Powertrain control module (PCM)

DIAGNOSTIC TROUBLE CODE INDEX			
Codes	KOEO	KOER	CM
P2289	AP1	AP1	AP1

Signal Functions: The injection control pressure (ICP) sensor is a variable capacitance sensor that, when supplied with a 5-volt reference signal from the powertrain control module (PCM), produces a linear analog voltage signal that indicates pressure.

The ICP sensor's primary function is to provide a feedback signal to indicate rail pressure so that the PCM can command the correct injector timing and pulse width and the correct injection control pressure for proper fuel delivery at all speed and load conditions.

Injection Control Pressure (ICP) Sensor Bias: The ICP sensor bias varies between sensors. The amount of voltage the sensor deviates from a calculated reference value (sensor bias) may cause an injection control pressure calculation error.

Detection/Management: If the PCM detects an out of range high or low ICP sensor, the malfunction indicator (MIL) is illuminated and the PCM will function from an estimated injection control pressure (open loop control of injection control pressure).

!CAUTION!:
The powertrain control module (PCM) harness connectors must be properly seated and the connector latch properly attached to eliminate possible driveability concerns or a no start condition. Installing the PCM connectors on an angle may cause an improper connection, misdiagnosis and damaged components. Install the connector until the lever pivots and seats itself. Apply light pressure to get the connector into position on the PCM and then fully seat the connector.

Note: Visually inspect the harness connectors for corrosion, damage, proper mating and correct pin tension.

Note: When the PCM is disconnected additional DTCs will be set. Clear all DTCs after restoring the vehicle.

PINPOINT TEST SCHEMATICS AND CONNECTORS

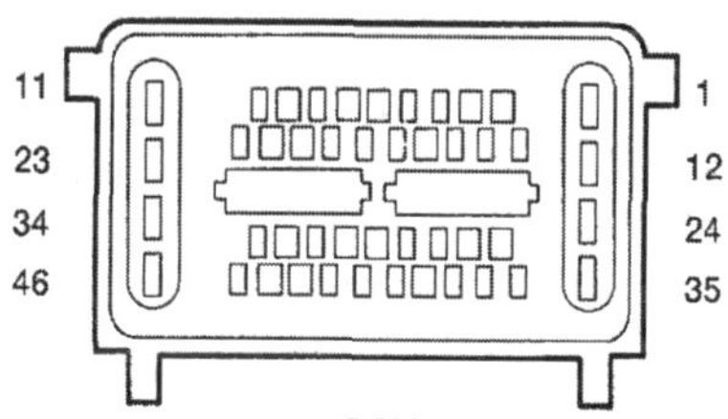

PCM ENGINE HARNESS CONNECTOR

Pin	Circuit
E25	SIGRTN (Signal Return)

SENSOR CONNECTOR

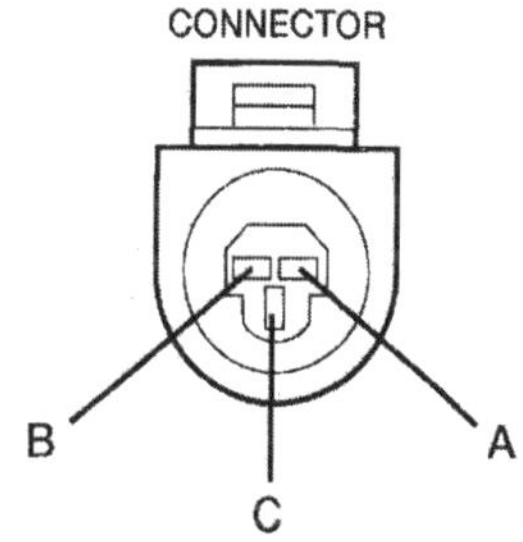

Pin	Circuit
C	ICP (Injection Control Pressure)
A	SIGRTN (Signal Return)

AP1

PRELIMINARY DIAGNOSIS FOR CODE (DTC) P2289

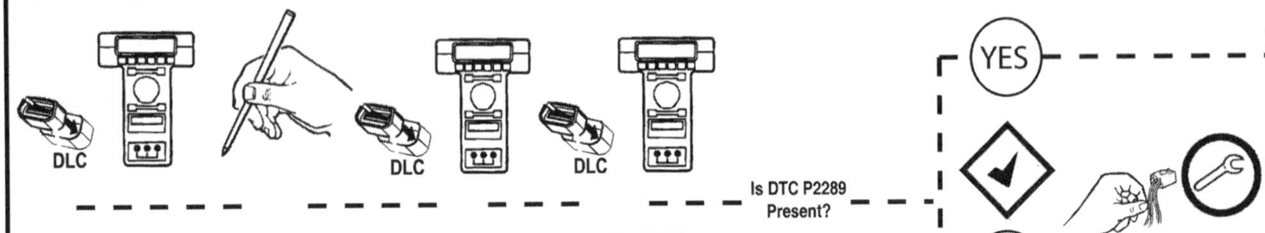

Connect The Scan Tool.

Retrieve And Record All DTCs. Record Freeze Frame Data.

Clear The DTCs.

Perform The KOEO And Continuous On-Demand Self-Tests.

Is DTC P2289 Present?

YES → AP2

NO → STOP: UNABLE To Duplicate Condition. Check For A Loose Connection And Damaged Or Corroded Terminals Wiggle The Harness Attempting To Recreate The Fault. REPAIR As Necessary. REFER To Symptom Charts If A Drivability Concern Exists.

AP2

DIAGNOSTIC TROUBLE CODE (DTC) P2289

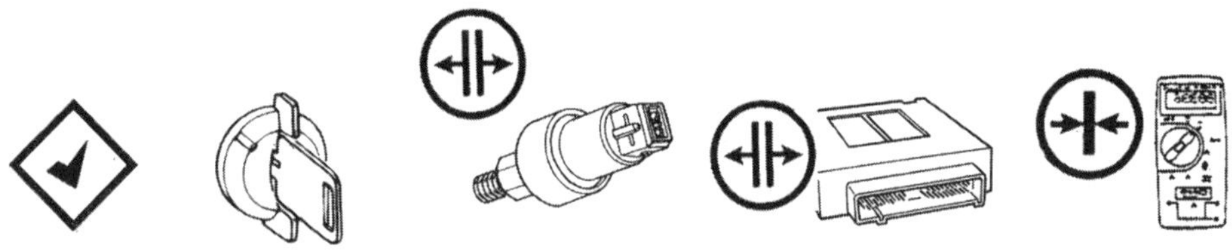

Possible Causes:
- Biased ICP Sensor
- Open SIG RTN Circuit
- Signal Circuit Shorted To Voltage
- PCM

Key Off.

Disconnect The ICP Sensor.

Disconnect The PCM.

Measure the resistance between:	
(+) ICP Sensor Connector, Harness Side	(-) PCM Connector, Harness Side
SIGRTN - Pin A	SIGRTN - Pin E25

Is Resistance Less Than 5 Ohms?

YES → AP3

NO → STOP: REPAIR The Circuit. CLEAR The DTCs. RETEST The System

AP3 — CHECK THE SIGNAL CIRCUIT FOR A SHORT TO VOLTAGE

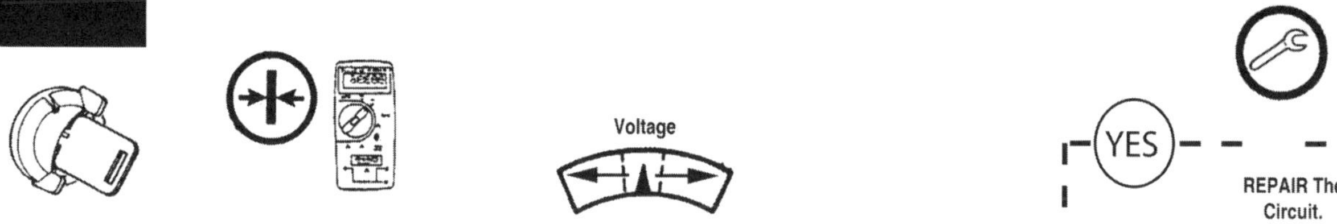

Key On, Engine Off.

Measure the voltage between:	
(+) ICP Sensor Connector, Harness Side	(-)
ICP - Pin C	Ground

Is Any Voltage Indicated?

YES — REPAIR The Circuit. — STOP

NO — AP4

AP4 — ICP SENSOR CHECK

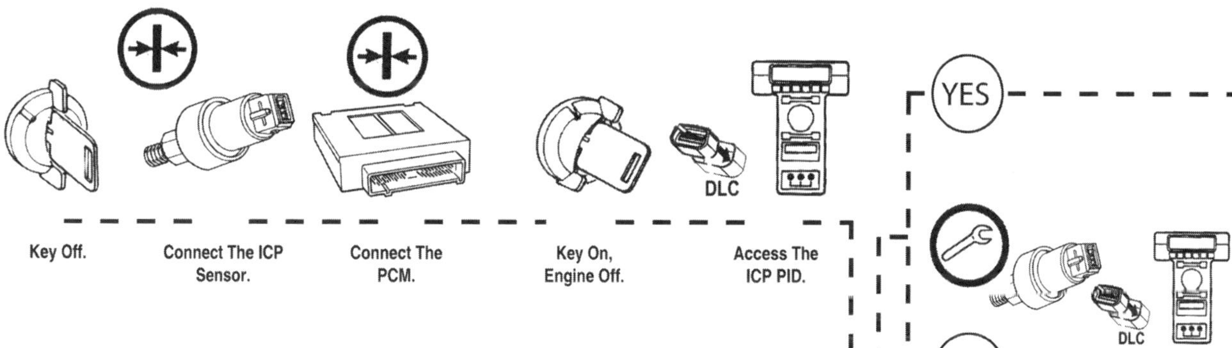

Key Off.

Connect The ICP Sensor.

Connect The PCM.

Key On, Engine Off.

Access The ICP PID.

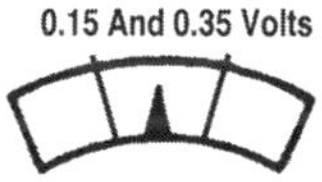

Is The ICP PID Voltage Between: 0.15 And 0.35 Volts?

YES — AP5

NO — INSTALL A New ICP Sensor. CLEAR The DTC's. REPEAT The Self-Test. — STOP

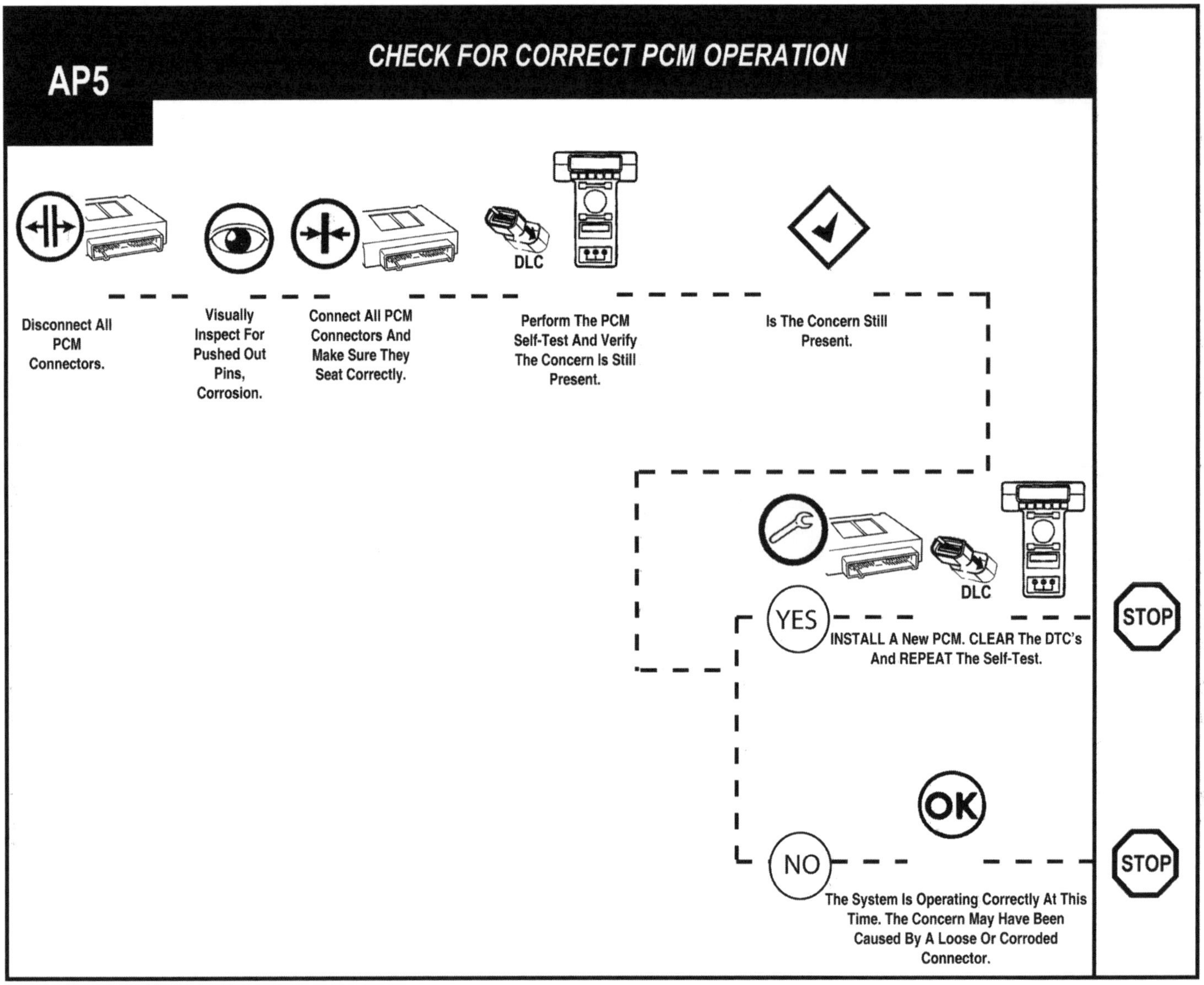

AP5
CHECK FOR CORRECT PCM OPERATION
DLC
Disconnect All PCM Connectors.
Visually Inspect For Pushed Out Pins, Corrosion.
Connect All PCM Connectors And Make Sure They Seat Correctly.
Perform The PCM Self-Test And Verify The Concern Is Still Present.
Is The Concern Still Present.
DLC
YES
INSTALL A New PCM. CLEAR The DTC's And REPEAT The Self-Test.
STOP
OK
NO
The System Is Operating Correctly At This Time. The Concern May Have Been Caused By A Loose Or Corroded Connector.
STOP

Trouble Code (s)	Problem: Injection Control Pressure Too Low - Engine Cranking	Test
P2291		AQ 2006-07

REFERENCE LIST

This pinpoint test is intended to diagnose the following:
- Harness circuits: ICP and IPR
- High-pressure oil pump
- ICP sensor
- IPR valve

DIAGNOSTIC TROUBLE CODE INDEX			
Codes	KOEO	KOER	CM
P2291	AQ1	AQ1	AQ1

P2291 = Injection Control Pressure Too Low - Engine Cranking

Signal Function:
The injection control pressure (ICP) sensor is a variable capacitance sensor. A 5-volt reference voltage provided by the powertrain control module (PCM) is utilized by the sensor to produce a linear analog voltage proportional to hydraulic pressure.

The PCM monitors the ICP sensor signal to control the injection pressure regulator (IPR) duty cycle. The PCM will modulate the IPR based on the ICP signal to adjust the injector timing and injection pressure providing optimum fuel delivery at all speed, load, and temperature conditions.

Detection/Management:
If the PCM detects an out of range high or low ICP sensor, the malfunction indicator (MIL) is illuminated and the PCM will function from an estimated injection control pressure (open loop control of injection control pressure).

CAUTION!:
The powertrain control module (PCM) harness connectors must be properly seated and the connector latch properly attached to eliminate possible driveability concerns or a no start condition. Installing the PCM connectors on an angle may cause an improper connection, misdiagnosis and damaged components. Install the connector until the lever pivots and seats itself. Apply light pressure to get the connector into position on the PCM and then fully seat the connector.

Note: Visually inspect the harness connectors for corrosion, damage, proper mating and correct pin tension.

Note: When the PCM is disconnected additional DTCs will be set. Clear all DTCs after restoring the vehicle.

PINPOINT TEST SCHEMATICS AND CONNECTORS

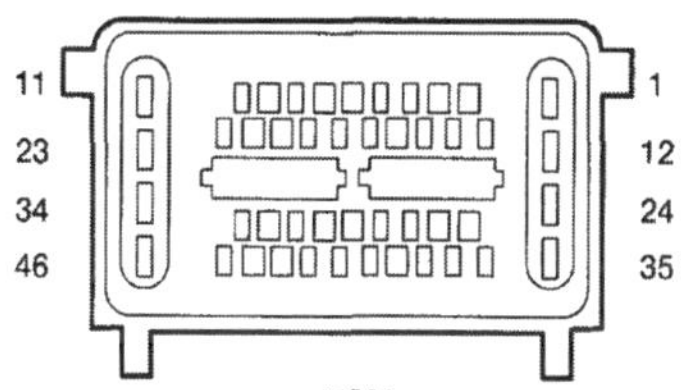

PCM
ENGINE HARNESS
CONNECTOR

Pin	Circuit
E29	ICP (Injection Control Pressure)

SENSOR
CONNECTOR

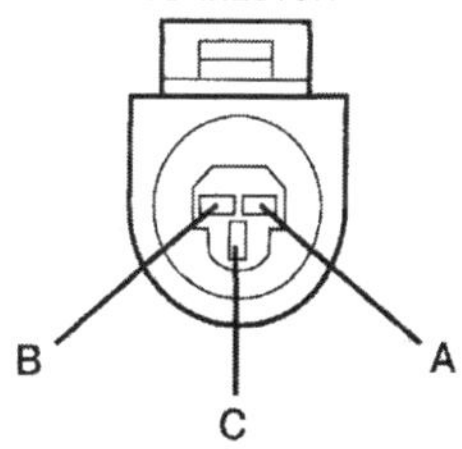

Pin	Circuit
B	VREF (Reference Voltage)
A	SIGRTN (Signal Return)
C	ICP (Injection Control Pressure)

AQ1

DIAGNOSTIC TROUBLE CODE (DTC) P2291

Note: This DTC May Set After A High-Pressure Oil System Repair (I.e., Injector Change). Drive The Vehicle To Clear Air From The System After Repair And Clear The DTC.

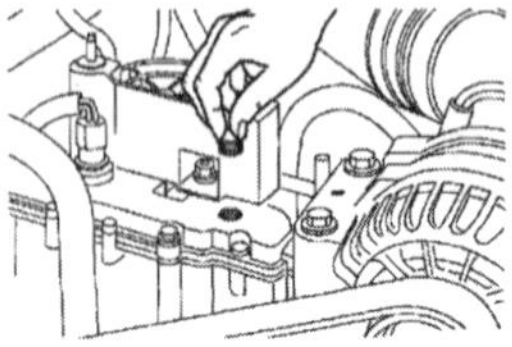

Continuous DTC P2291 Indicates That There Was No ICP Detected During Crank (Long Crank Time).
Possible Causes:
- High-pressure Oil System Repair
- Empty/low Oil Reservoir At Crank
- Low Engine Oil Level Or Incorrect Viscosity
- ICP Circuit Fault
- Low-Pressure Oil Pump
- IPR Valve
- ICP Sensor
- High-Pressure Oil Pump

Check The Engine Oil Level.

Record Freeze Frame Data.

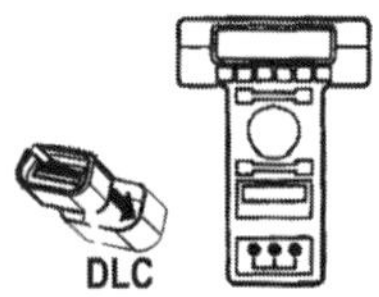

Perform The KOEO On-Demand Self-Test To Verify ICP And IPR Circuit Faults Are Not Present.

Are Any DTCs Present?

YES

For DTC P2291, GO to AQ2 .

For All Others, DISREGARD The Current Diagnostic Trouble Code (DTC) At This Time. DIAGNOSE The Next DTC. GO To Diagnostic Trouble Code (DTC) Descriptions.

NO → AQ2

AQ2
DTC INDEX
AQ2

AQ2 — CHECK THE OIL LEVEL

Note: If The Vehicle Stalls After Start Oil Is Not Being Supplied From The Low Pressure Oil System.

- Verify Correct Oil Level, Quality And Viscosity.
- Check Base Engine Oil Pressure.
- Verify Steps 5 And 10 On The Hard Start/No Start Diagnostic Guide Have Been Performed.
- Clear The DTCs.
- Monitor The ICP PID While Cranking The Engine.
- Is The ICP PID Minimum Voltage 0.85 Volts Within 5 Seconds Of Crank?

YES — The DTC Is Intermittent. → **AQ6**

NO → **AQ3**

AQ3 — CHECK THE ICP SENSOR

- Key Off.
- Disconnect The ICP Sensor.
- Does The Vehicle Start?

YES — INSTALL A New ICP Sensor. CLEAR The DTC's. Repeat The Self-Test. → STOP

NO → **AQ4**

AQ4 HIGH-PRESSURE LEAKAGE TEST

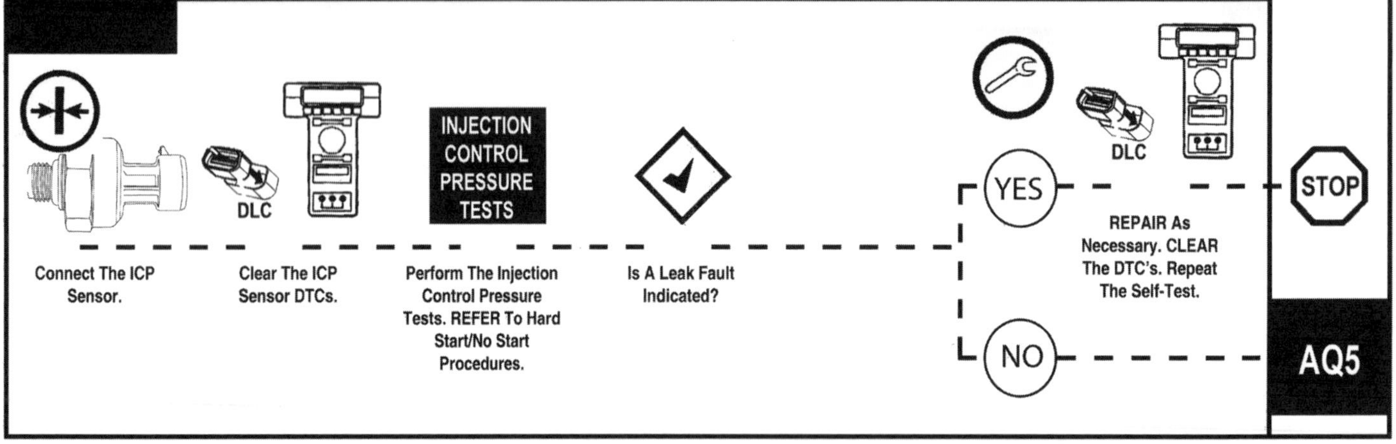

AQ5 IPR VALVE CHECK

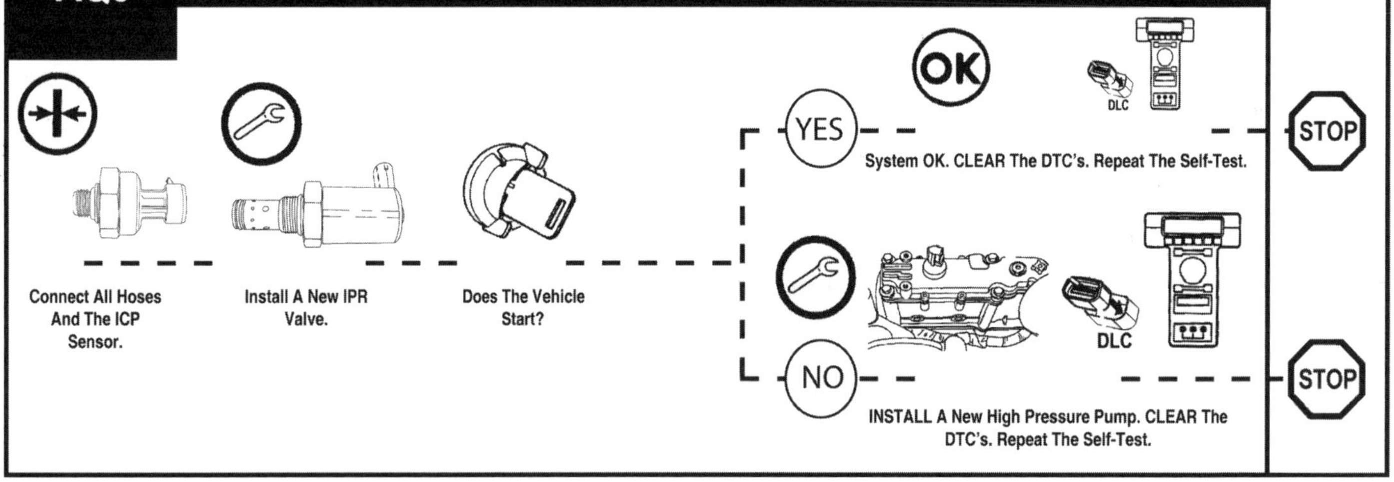

AQ6 CHECK FOR INTERMITTENT CIRCUIT FAULTS

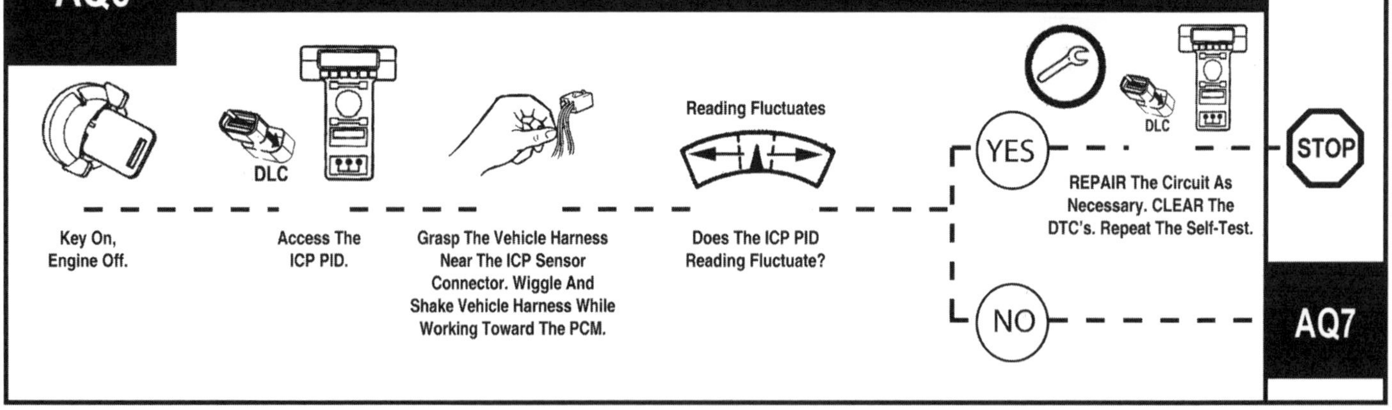

AQ7

INSPECT THE IPR CIRCUIT

Inspect The IPR Circuit From The IPR Valve Through The 12-Way Connector To The PCM.

Are Any Faults Indicated?

YES — DLC — REPAIR The Circuit As Necessary. CLEAR The DTC's. Repeat The Self-Test. — STOP

NO — UNABLE To Duplicate Concern. Check For A Loose Connection, And Damaged Or Corroded Terminals. WIGGLE the Harness Attempting To Recreate The Fault. REPAIR As Necessary And RETEST The System. REFER To Symptom Charts If A Driveablility Concern Exists. — STOP

NOTES

Trouble Code (s)	Problem: Random Misfire Detected	Test
P0300		AR 2006-07

REFERENCE LIST

DIAGNOSTIC TROUBLE CODE INDEX			
Codes	KOEO	KOER	CM
P0300	AR1	AR1	AR1

P0300 = Random Misfire Detected

Signal function:
The misfire detection monitor is an on-board strategy designed to monitor engine misfire and identify the specific cylinder in which the misfire occurred. Misfire is detected using the crankshaft and camshaft position sensors. To determine if a misfire has occurred the powertrain control module (PCM) monitors deceleration of the crankshaft and compares the information received to engine speed and engine load. If a misfire event is determined, the PCM uses the crankshaft position and the camshaft sensor signal information to determine which cylinder misfired.

Detection/Management:
The misfire detection monitor is enabled only when certain engine conditions are satisfied. If the PCM is unable to detect which cylinder is misfiring or multiple cylinders are misfiring the PCM sets a random misfire DTC.

CAUTION!: The PCM harness connectors must be properly seated and the connector latch properly attached to eliminate possible driveability concerns or a no start condition. Installing the PCM connectors on an angle may cause an improper connection, misdiagnosis and damaged components. Install the connector until the lever pivots and seats itself. Apply light pressure to get the connector into position on the PCM and then fully seat the connector.

Note: Visually inspect the harness connectors for corrosion, damage, proper mating and correct pin tension.

Note: When the PCM is disconnected additional DTCs will be set. Clear all DTCs after restoring the vehicle .

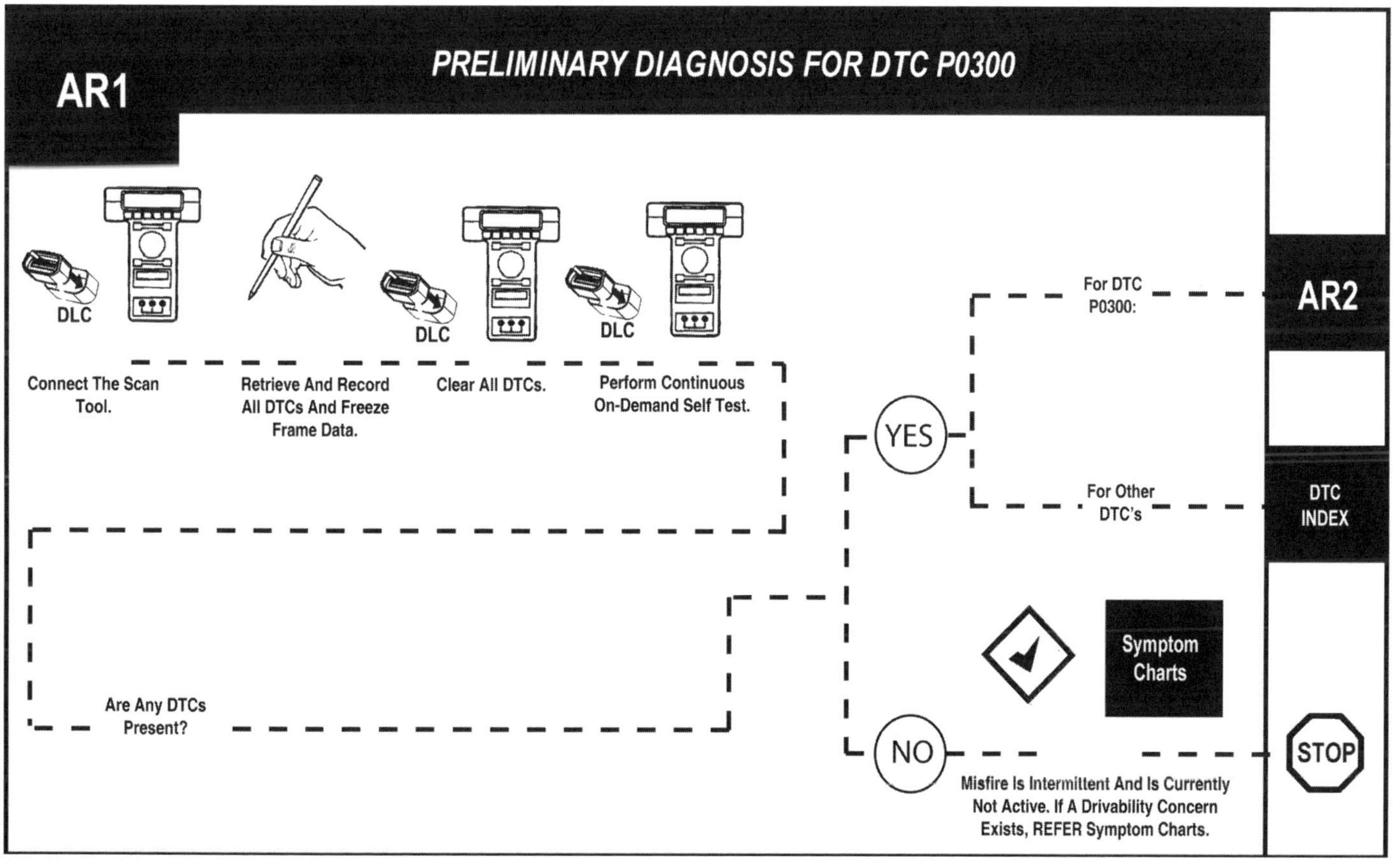

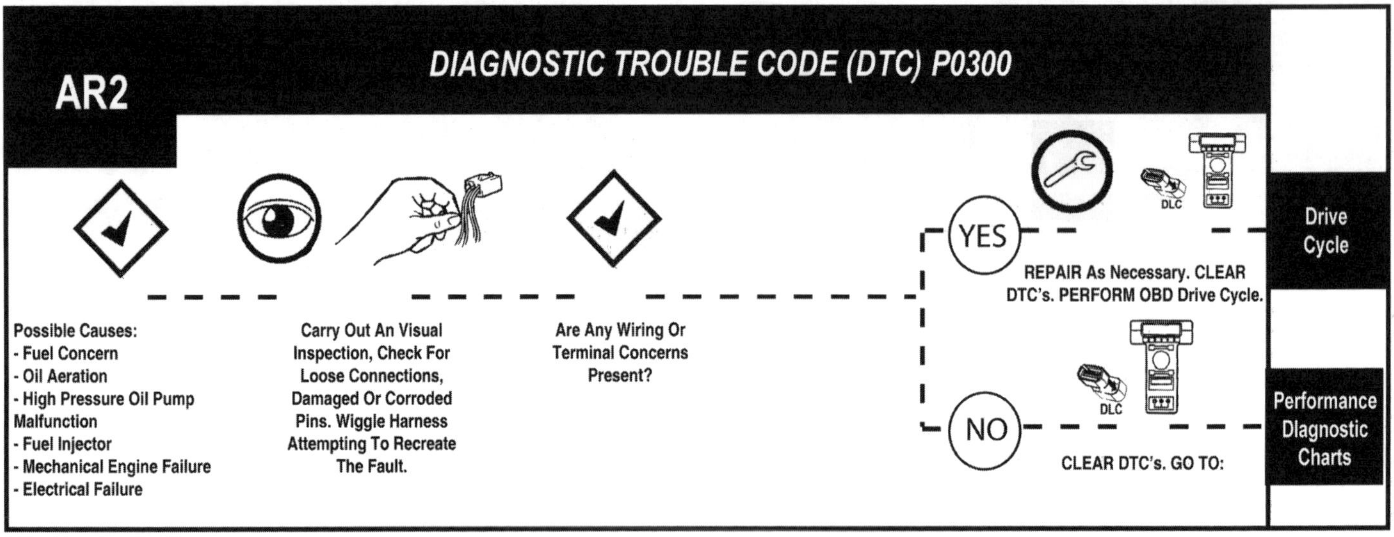

AR2
DIAGNOSTIC TROUBLE CODE (DTC) P0300
Possible Causes:
- Fuel Concern
- Oil Aeration
- High Pressure Oil Pump Malfunction
- Fuel Injector
- Mechanical Engine Failure
- Electrical Failure
Carry Out An Visual Inspection, Check For Loose Connections, Damaged Or Corroded Pins. Wiggle Harness Attempting To Recreate The Fault.
Are Any Wiring Or Terminal Concerns Present?
YES
DLC
REPAIR As Necessary. CLEAR DTC's. PERFORM OBD Drive Cycle.
Drive Cycle
NO
DLC
CLEAR DTC's. GO TO:
Performance Diagnostic Charts

Trouble Code (s)	Problem: Loss Of Communication With The Fuel Injector Control Module (FICM)	Test
U0105		AS 2006

REFERENCE LIST

This pinpoint test is intended to diagnose the following:

- FICM relay
- Harness circuits: B+, CAN2+, CAN2-, IGN START/RUN, FICM Vehicle Power, FICM Control and GND
- FICM
- Powertrain control module (PCM)

DIAGNOSTIC TROUBLE CODE INDEX			
Codes	KOEO	KOER	CM
U0105	AS1	AS1	AS1

U0105 = Loss Of Communication With The FICM

Circuti Function:
The fuel injector control module (FICM) requires a 12-volt power source. The FICM receives power from the vehicle batteries through the FICM relay contacts each time the ignition is turned on. The powertrain control module (PCM) communicates with the FICM utilizing controller area network (CAN) protocol. The CAN protocol is an international standards organization (ISO) standard for serial data communication. The CAN standard includes a physical layer using differential transmission on a twisted palr of wires and a data link layer that defines a few different message types, arbitration rules for bus access and methods for fault detection and fault confinement.

Detection/Management:
The PCM monitors the communication circuits to the FICM for an open or short circuit.

CAUTION!:
The powertrain control module (PCM) harness connectors must be properly seated and the connector latch properly attached to eliminate possible driveability concerns or a no start condition. Installing the PCM connectors on an angle may cause an improper connection, misdiagnosis and damaged components. Install the connector until the lever pivots and seats itself. Apply light pressure to get the connector into position on the PCM and then fully seat the connector.

Note: Visually inspect the harness connectors for corrosion, damage, proper mating and correct pin tension.

Note: When the PCM is disconnected additional DTCs will be set. Clear all DTCs after restoring the vehicle.

PINPOINT TEST SCHEMATICS AND CONNECTORS

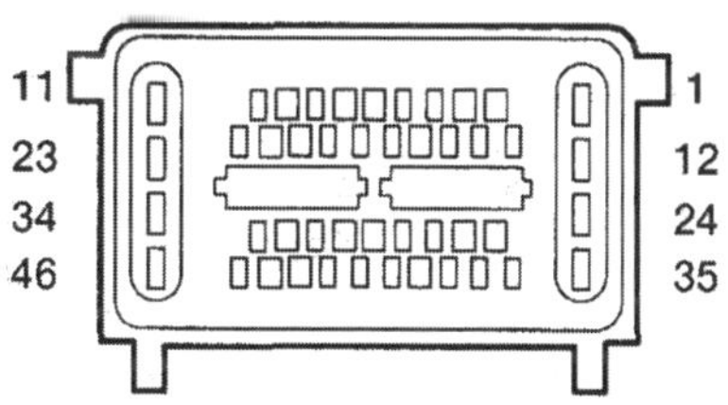

PCM ENGINE HARNESS CONNECTOR

Pin	Circuit
E26	CAN2L (Controller Area Network 2)
E37	CAN2H (Controller Area Network 2)

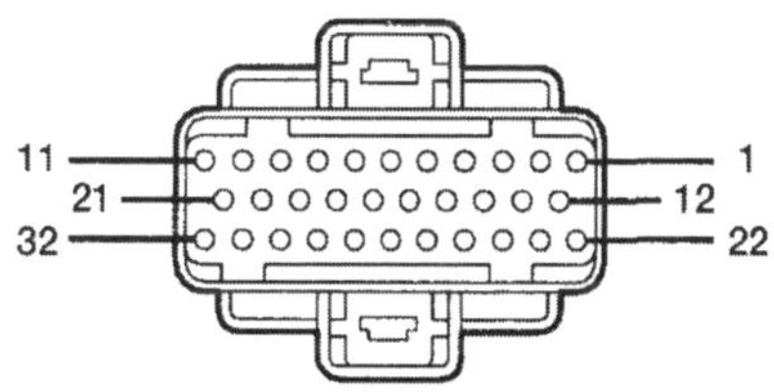

FICM HARNESS CONNECTOR

Pin	Circuit
31	CAN2L (Controller Area Network 2)
30	CAN2H (Controller Area Network 2)
1, 2, 3, 22, 26	GND (Ground)
27	FICM Control
8	FICM Logic Power
7	IGN START/RUN

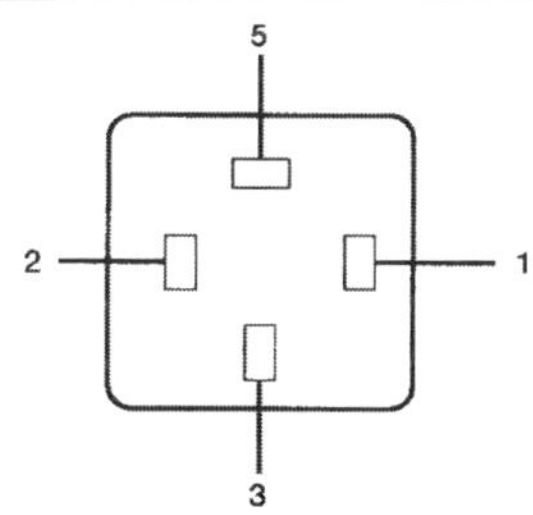

FICM RELAY AUXILIARY RELAY BOX

Pin	Circuit
1	FICM Control
3	FICM Vehicle Power
5	B+ (contacts)
2	B+ (coil)

AS1 — PRELIMINARY DIAGNOSIS FOR DTC U0105

NOTE: Listen For Fuel Injector Buzz At The Beginning Of The On-Demand Self-Test.

Connect The Scan Tool.

Retrieve And Record All DTCs. Record The Freeze Frame Data.

Clear All DTCs.

Carry Out The KOEO And KOER Self-Test.

Is DTC U0105 Present?

YES → AS2

NO → STOP

Unable To Duplicate Condition. CHECK For Loose Connection, Damaged Or Corroded Pins. WIGGLE Harness Attempting To Recreate The Fault. REPAIR As Necessary. REFER To Symptom Charts If A Driveability Concern Exists.

AS2 — DIAGNOSTIC TROUBLE CODE (DTC) U0105

Possible Causes:
- Loose Connection
- CAN Circuit
- FICM Power Circuit
- FICM Ground Circuit
- FICM Relay
- FICM
- PCM

Key Off.

Disconnect The FICM Relay.

Perform The Relay Component Test.

Is The Relay OK?

YES → AS3

NO → STOP

INSTALL A New FICM Relay. CLEAR The DTCs. REPEAT The Self-Test.

AS3 CHECK THE FICM RELAY VOLTAGE CIRCUITS

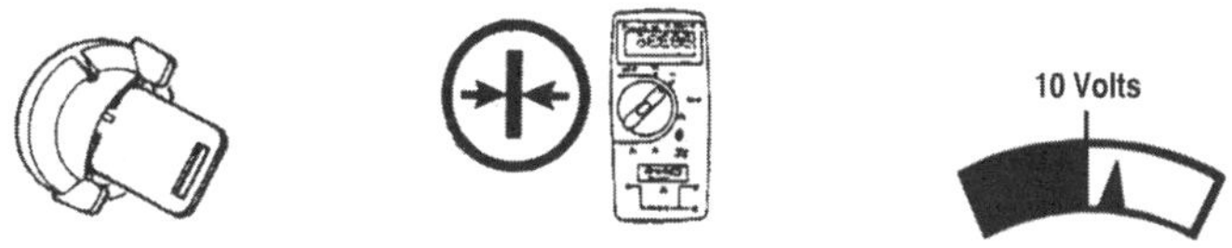

Key On, Engine Off.

Measure the voltage between:	
(+) FICM Relay Connector, Harness Side	(-)
B+ (coil) - Pin 2	Ground
B+ (contacts) - Pin 5	Ground

10 Volts

Are The Voltages Greater Than 10 Volts?

YES → AS4

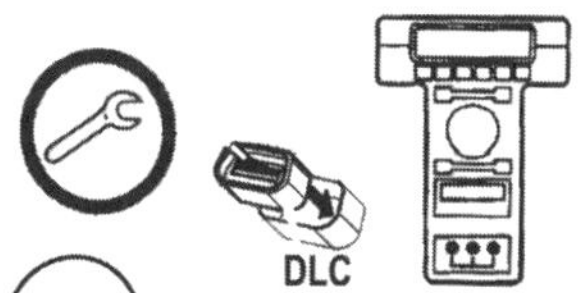

DLC

NO → REPAIR The Circuit In Question. CLEAR The DTCs. REPEAT The Self-Test. STOP

AS4 CHECK THE IGNITION POWER CIRCUIT FOR VOLTAGE

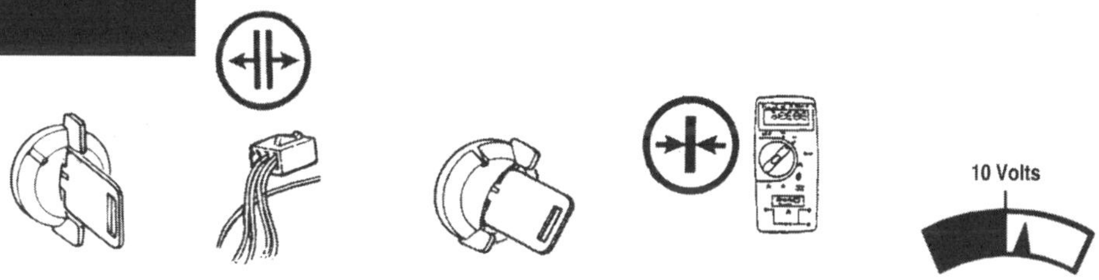

Key Off.

Disconnect The FICM C Connector.

Key On, Engine Off.

Measure the voltage between:	
(+) FICM-C Connector, Harness Side	(-)
IGN START/RUN - Pin 7	Ground

10 Volts

Is The Voltage Greater Than 10 Volts?

YES → AS5

DLC

NO → REPAIR The Circuit. CLEAR The DTCs. REPEAT The Self-Test. STOP

AS5

CHECK THE LOGIC POWER CIRCUIT FOR AN OPEN

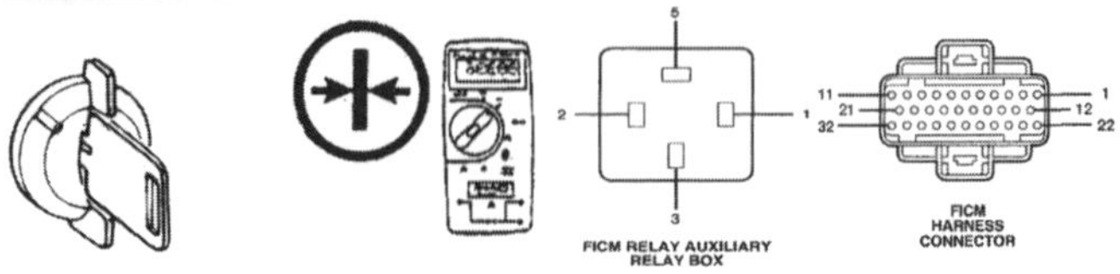

Key Off.

Measure the resistance between:	
(+) FICM Relay Connector, Harness Side	**(-) FICM-C Connector, Harness Side**
FICM Vehicle Power - Pin 3	FICM Vehicle Power - Pin 8

Is The Resistance Less Than 5 Ohms?

YES → **AS6**

NO → REPAIR The Circuit. CLEAR The DTCs. REPEAT The Self-Test. → **STOP**

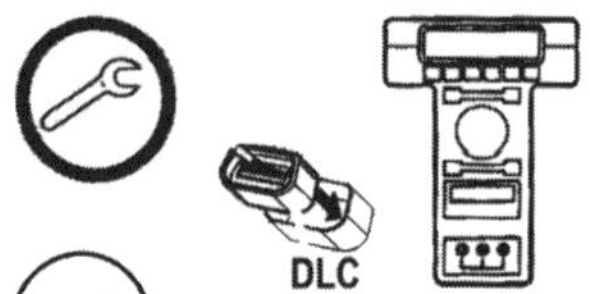

AS6

CHECK THE RELAY CONTROL CIRCUIT FOR AN OPEN

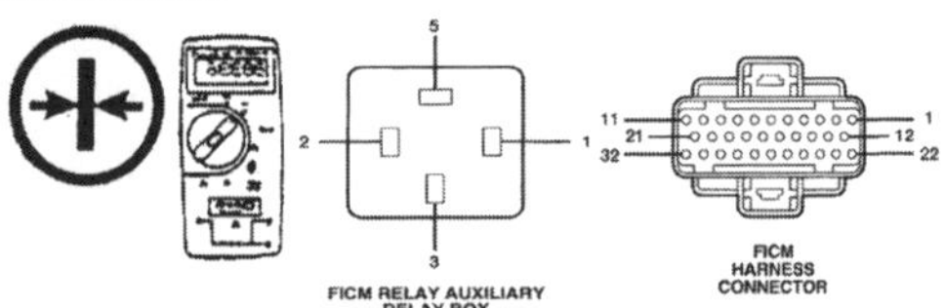

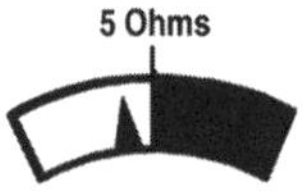

Measure the resistance between:	
(+) FICM Relay Connector, Harness Side	**(-) FICM-C Connector, Harness Side**
FICM Control - Pin 1	FICM Control - Pin 27

Is The Resistance Less Than 5 Ohms?

YES → **AS7**

NO → REPAIR The Circuit. CLEAR The DTCs. REPEAT The Self-Test. → **STOP**

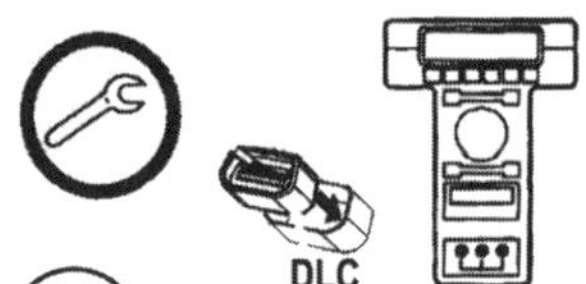

AS7

CHECK THE FICM GROUND CIRCUITS FOR AN OPEN

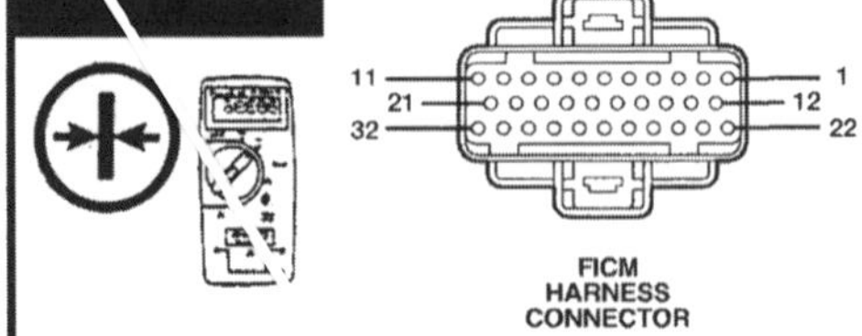

Measure the resistance between:	
(+) FICM-C Connector, Harness Side	**(-)**
GND - Pin 1, 2, 3, 22, 26	Ground

Are The Resistances Less Than 5 Ohms?

YES → **AS8**

NO → REPAIR The Circuit In Question. CLEAR The DTCs. REPEAT The Self-Test. → **STOP**

AS8 — CHECK THE FICM CAN CIRCUITS

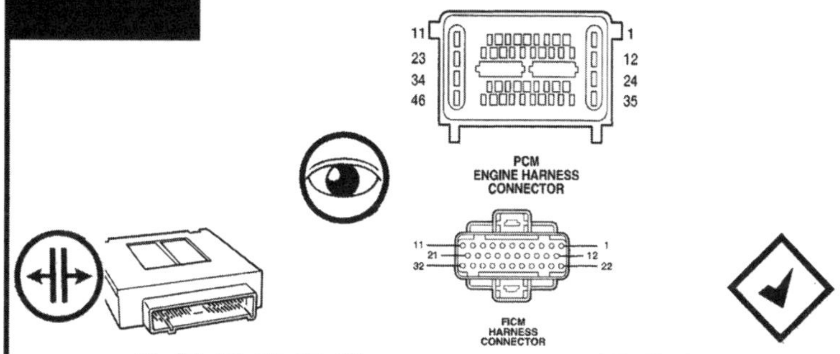

Disconnect The PCM Engine Harness Connector.

Check for Loose Connections or Damage between	
PCM Engine Connector Harness	FICM Relay Connector Harness
Pins 37 & 26	Pins 30 & 31

Are The Wiring And Shield OK?

YES → AS9

NO → STOP

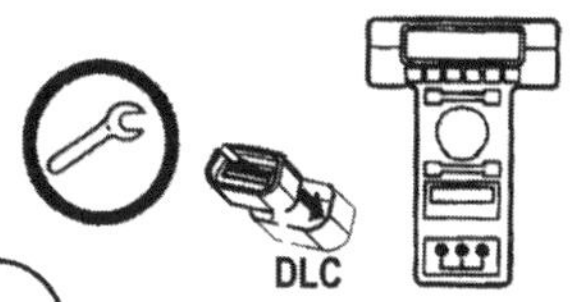

REPAIR As Necessary. RESTORE The Vehicle. CLEAR DTCs And RETEST The System.

AS9 — CHECK FICM CAN CIRCUITS FOR A SHORT TO VOLTAGE

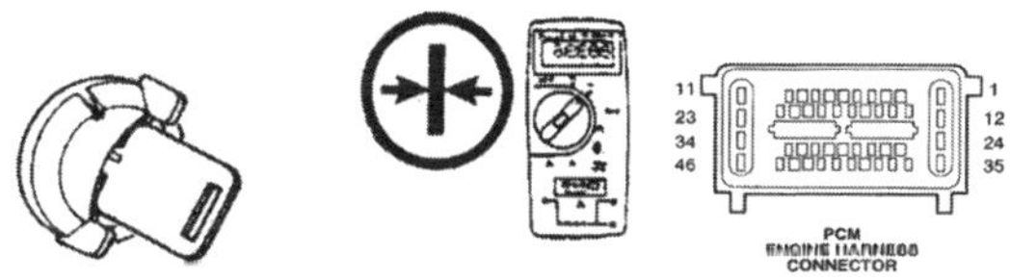

Key On, Engine Off.

Measure the voltage between:	
(+) PCM Connector, Harness Side	(-)
CAN2H - Pin E37	Ground
CAN2L - Pin E26	Ground

Voltage Indicated

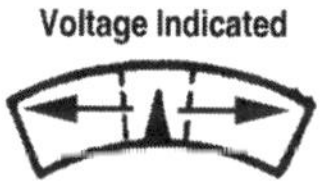

Is Any Voltage Indicated?

YES → STOP

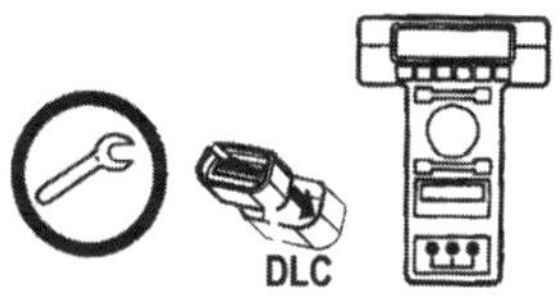

REPAIR Short To Voltage. RESTORE The Vehicle. CLEAR DTCs And RETEST The System.

NO → AS10

AS10 — CHECK FICM CAN CIRCUITS FOR A SHORT TOGETHER

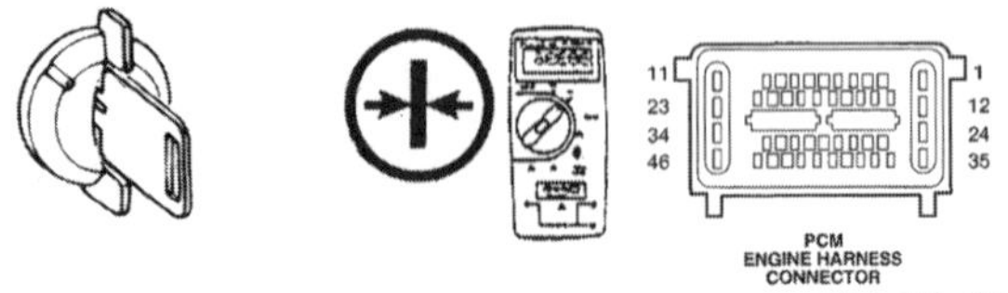

Key Off.

Measure the resistance between:	
(+) PCM Connector, Harness Side	(-) PCM Connector, Harness Side
CAN2H - Pin E37	CAN2L - Pin E26

10,000 Ohms

Is The Resistance Greater Than 10,000 Ohms?

YES → AS11

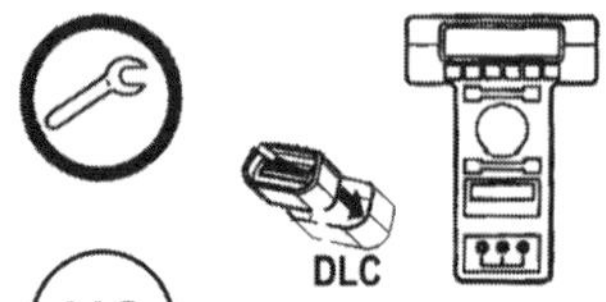

NO → REPAIR Shorted Circuits. RESTORE The Vehicle. CLEAR DTCs And RETEST The System. STOP

AS11 — CHECK FICM CAN CIRCUITS FOR A SHORT TO GROUND

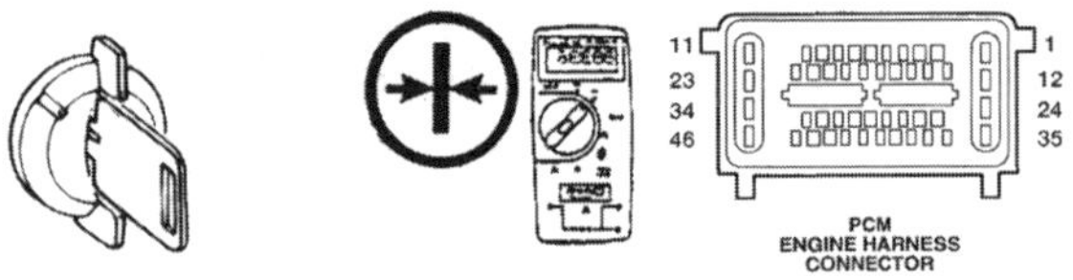

Key Off.

Measure the resistance between:	
(+) PCM Connector, Harness Side	(-)
CAN2H - Pin E37	Ground
CAN2L - Pin E26	Ground

10,000 Ohms

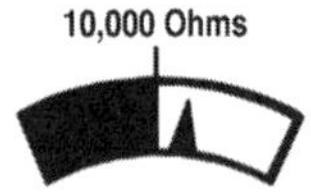

Are The Resistances Greater Than 10,000 Ohms?

YES → AS12

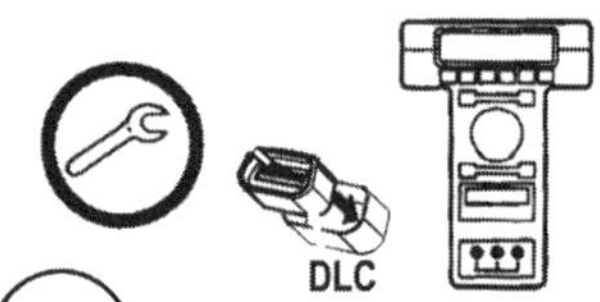

NO → REPAIR Short To Ground. RESTORE The Vehicle. CLEAR DTCs And RETEST The System. STOP

AS12 — CHECK FICM CAN CIRCUITS FOR AN OPEN

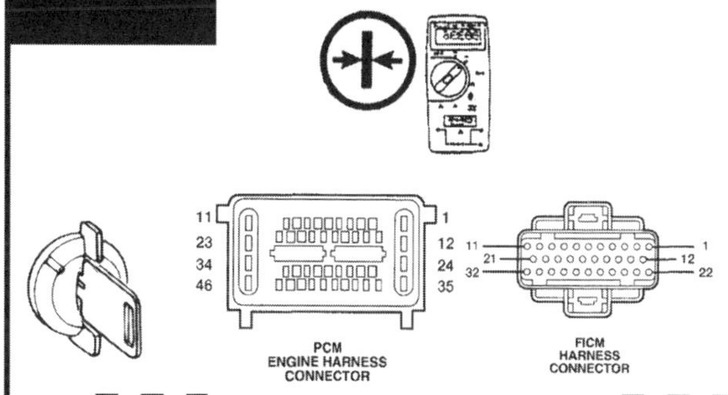

Key Off.

Measure the resistance between:	
(+) PCM Connector, Harness Side	(-) FICM-C Connector, Harness Side
CAN2H - Pin E37	CAN2H - Pin 30
CAN2L - Pin E26	CAN2L - Pin 31

Are The Resistances Less Than 5 Ohms?

YES → AS13

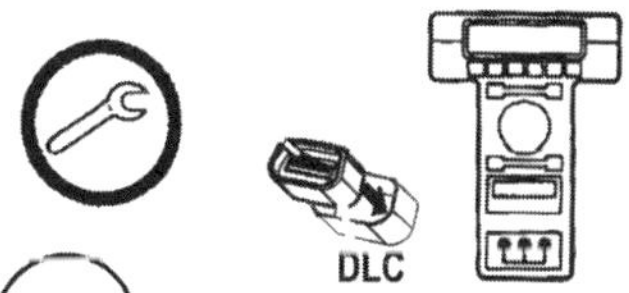

NO → STOP

REPAIR Open In The Faulted CAN Circuit. RESTORE The Vehicle. CLEAR DTCs And RETEST The System.

AS13 — CHECK PCM TO FICM COMMUNICATION

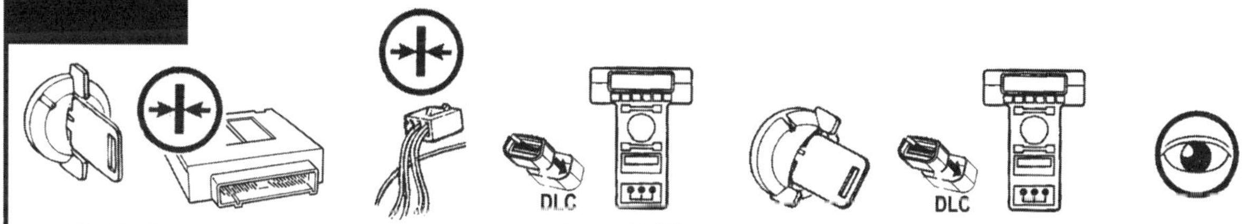

Key Off. → Connect The PCM Engine Harness Connector. → Connect The FICM Harness Connector C. → Connect The Scan Tool. → Key On, Engine Off. → Access The PCM. → Monitor The FICM LPWR, FICM MPWR And FICM VPWR PIDs. Refer To Reference Values.

Do The PIDs Indicate Correctly?

YES → AS14

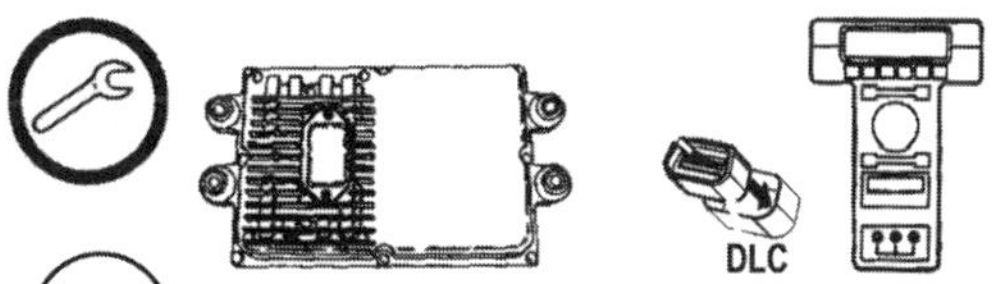

NO → STOP

INSTALL A New FICM. CLEAR The DTCs. REPEAT The Self-Test.

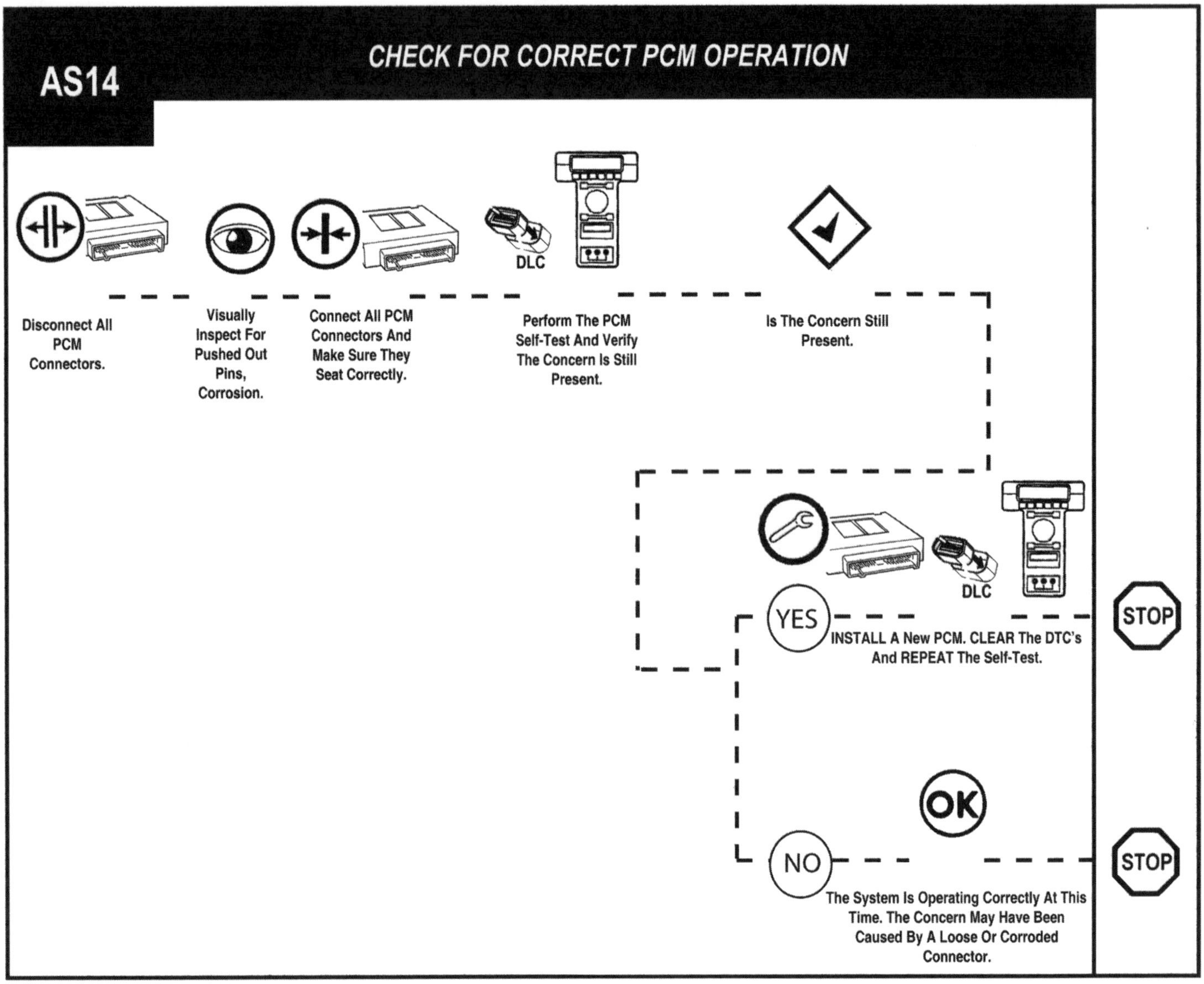
AS14
CHECK FOR CORRECT PCM OPERATION
Disconnect All PCM Connectors.
Visually Inspect For Pushed Out Pins, Corrosion.
Connect All PCM Connectors And Make Sure They Seat Correctly.
DLC
Perform The PCM Self-Test And Verify The Concern Is Still Present.
Is The Concern Still Present.
DLC
YES
INSTALL A New PCM. CLEAR The DTC's And REPEAT The Self-Test.
STOP
OK
NO
The System Is Operating Correctly At This Time. The Concern May Have Been Caused By A Loose Or Corroded Connector.
STOP

Trouble Code (s)	Problem: Loss Of Communication With The Fuel Injector Control Module (FICM)	Test
U0105		AS 2007

REFERENCE LIST

This pinpoint test is intended to diagnose the following:

- FICM relay
- Harness circuits: B+, CAN2+, CAN2-, IGN START/RUN, FICM Vehicle Power, FICM Control and GND
- FICM
- Powertrain control module (PCM)

DIAGNOSTIC TROUBLE CODE INDEX			
Codes	KOEO	KOER	CM
U0105	AS1	AS1	AS1

U0105 = Loss Of Communication With The FICM

Circuti Function:
The fuel injector control module (FICM) requires a 12-volt power source. The FICM receives power from the vehicle batteries through the FICM relay contacts each time the ignition is turned on. The powertrain control module (PCM) communicates with the FICM utilizing controller area network (CAN) protocol. The CAN protocol is an international standards organization (ISO) standard for serial data communication. The CAN standard includes a physical layer using differential transmission on a twisted pair of wires and a data link layer that defines a few different message types, arbitration rules for bus access and methods for fault detection and fault confinement.

Detection/Management:
The PCM monitors the communication circuits to the FICM for an open or short circuit.

CAUTION!:
The powertrain control module (PCM) harness connectors must be properly seated and the connector latch properly attached to eliminate possible driveability concerns or a no start condition. Installing the PCM connectors on an angle may cause an improper connection, misdiagnosis and damaged components. Install the connector until the lever pivots and seats itself. Apply light pressure to get the connector into position on the PCM and then fully seat the connector.

Note: Visually inspect the harness connectors for corrosion, damage, proper mating and correct pin tension.

Note: When the PCM is disconnected additional DTCs will be set. Clear all DTCs after restoring the vehicle.

PINPOINT TEST SCHEMATICS AND CONNECTORS

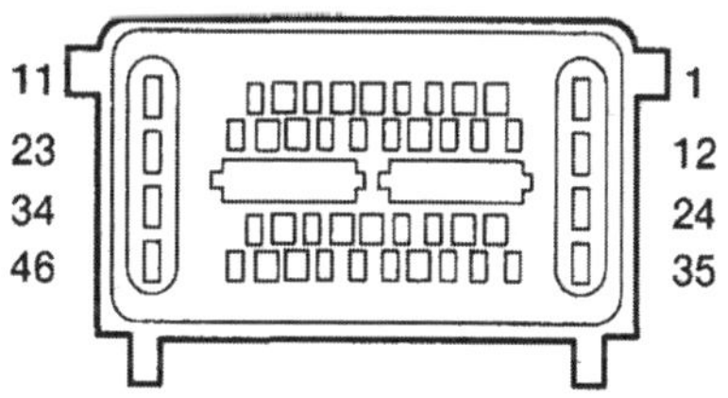

PCM ENGINE HARNESS CONNECTOR

Pin	Circuit
E26	CAN2- (Controller Area Network 2)
E37	CAN2+ (Controller Area Network 2)

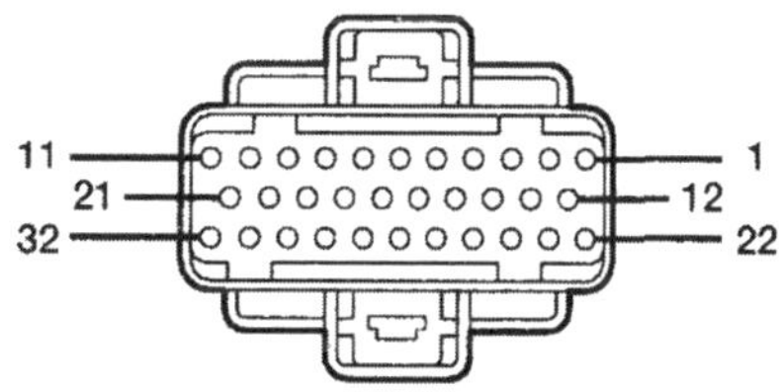

FICM HARNESS CONNECTOR

Pin	Circuit
31	CAN2- (Controller Area Network 2)
30	CAN2+ (Controller Area Network 2)
1, 2, 3, 22, 26	GND (Ground)
27	FICM Control
8	FICM Vehicle Power
7	IGN START/RUN

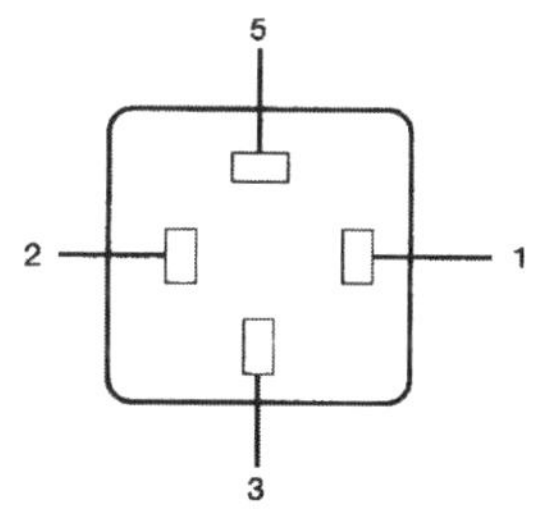

FICM RELAY AUXILIARY RELAY BOX

Pin	Circuit
1	FICM Control
3	FICM Vehicle Power
5	B+ (contacts)
2	B+ (coil)

AS1 — PRELIMINARY DIAGNOSIS FOR DTC U0105

NOTE: Listen For Fuel Injector Buzz At The Beginning Of The On-Demand Self-Test.

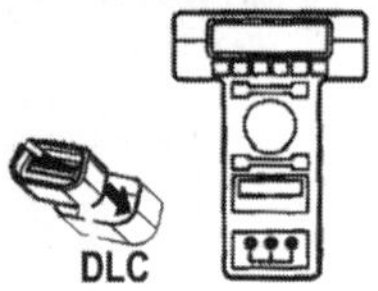

Connect The Scan Tool.

Retrieve And Record All DTCs. Record The Freeze Frame Data.

Clear All DTCs.

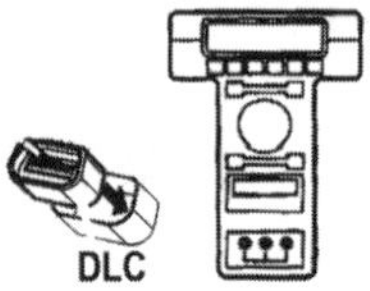

Carry Out The KOEO And KOER Self-Test.

Is DTC U0105 Present?

YES → AS2

NO → STOP

Unable To Duplicate Condition. CHECK For Loose Connection, Damaged Or Corroded Pins. WIGGLE Harness Attempting To Recreate The Fault. REPAIR As Necessary. REFER To Symptom Charts If A Driveability Concern Exists.

AS2 — CHECK THE FICM RELAY

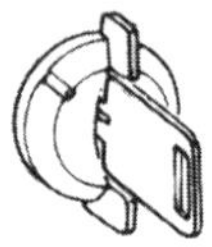

Key Off.

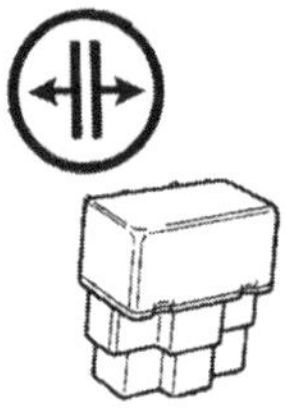

Disconnect The FICM Relay.

Perform The Relay Component Test.

Is The Relay OK?

YES → AS3

NO → STOP

INSTALL A New FICM Relay. CLEAR The DTCs. REPEAT The Self-Test.

AS3 — CHECK THE B+ CIRCUITS FOR VOLTAGE

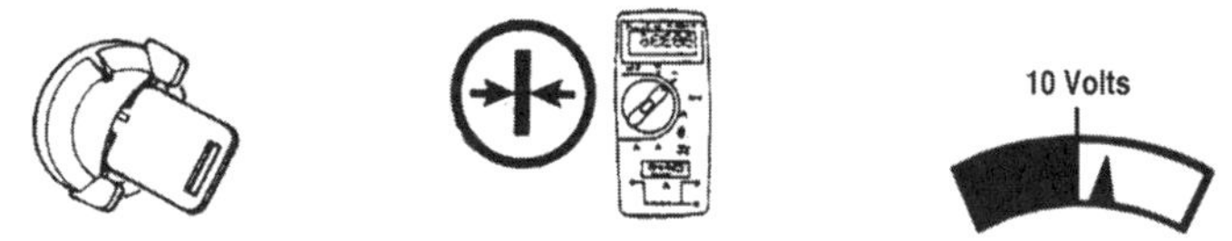

Key On, Engine Off.

Measure the voltage between:	
(+) FICM Relay Connector, Harness Side	**(-)**
B+ (coil) - Pin 2	Ground
B+ (contacts) - Pin 5	Ground

Are The Voltages Greater Than 10 Volts?

YES → AS4

NO → STOP

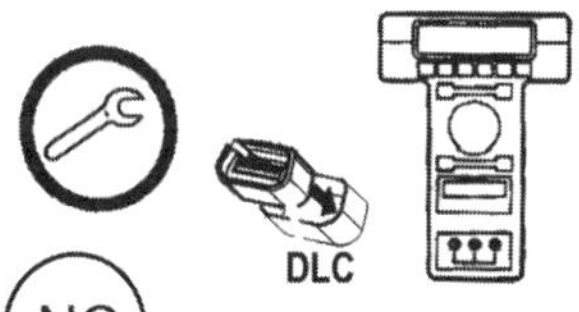

REPAIR The Circuit In Question. CLEAR The DTCs. REPEAT The Self-Test.

AS4 — CHECK THE IGN START/RUN CIRCUIT FOR VOLTAGE

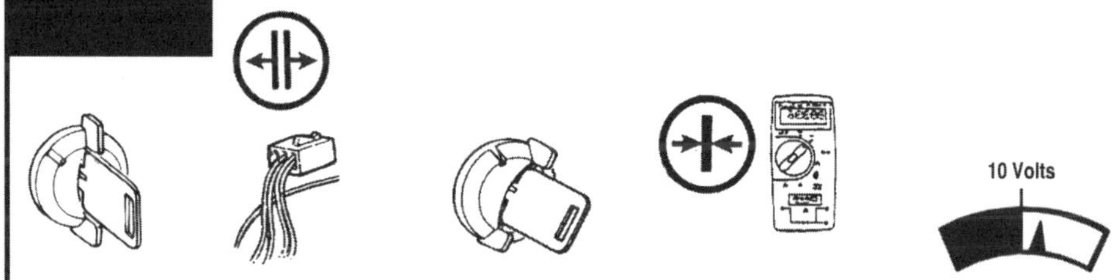

Key Off.

Disconnect The FICM-C Connector.

Key On, Engine Off.

Measure the voltage between:	
(+) FICM-C Connector, Harness Side	**(-)**
IGN START/RUN - Pin 7	Ground

Is The Voltage Greater Than 10 Volts?

YES → AS5

NO → STOP

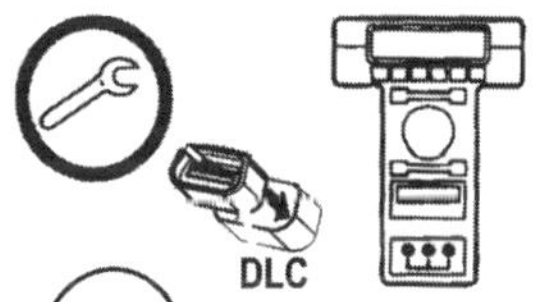

REPAIR The Circuit. CLEAR The DTCs. REPEAT The Self-Test.

AS5

CHECK THE FICM VEHICLE POWER CIRCUIT FOR AN OPEN

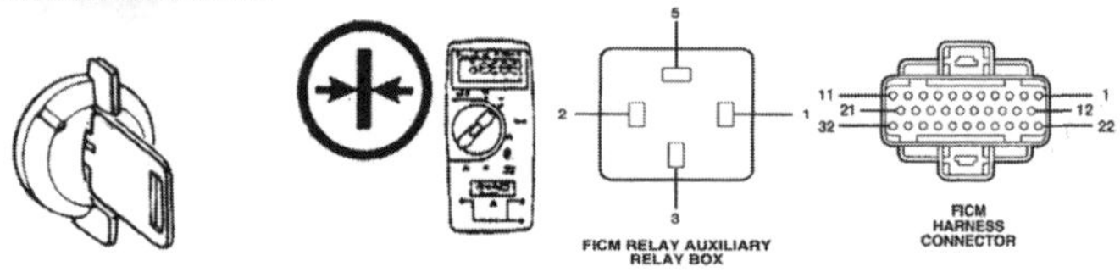

Key Off.

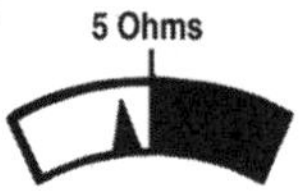

Measure the resistance between:	
(+) FICM Relay Connector, Harness Side	**(-) FICM-C Connector, Harness Side**
FICM Vehicle Power - Pin 3	FICM Vehicle Power - Pin 8

Is The Resistance Less Than 5 Ohms?

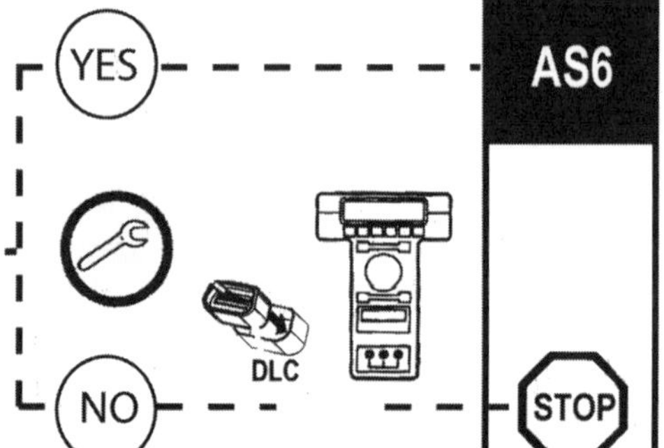

REPAIR The Circuit. CLEAR The DTCs. REPEAT The Self-Test.

AS6

CHECK THE FICM CONTROL CIRCUIT FOR AN OPEN

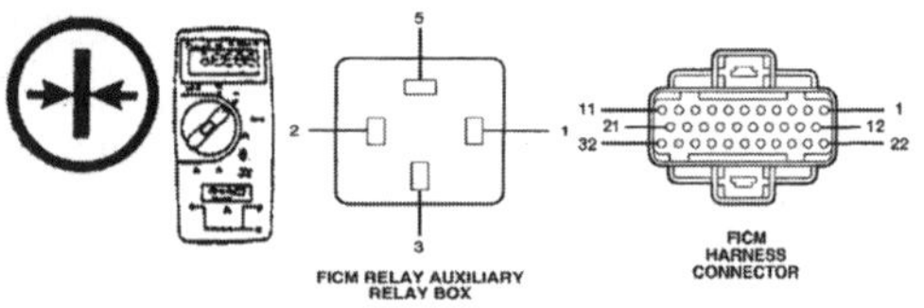

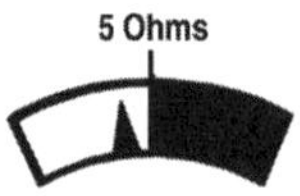

Measure the resistance between:	
(+) FICM Relay Connector, Harness Side	**(-) FICM-C Connector, Harness Side**
FICM Control - Pin 1	FICM Control - Pin 27

Is The Resistance Less Than 5 Ohms?

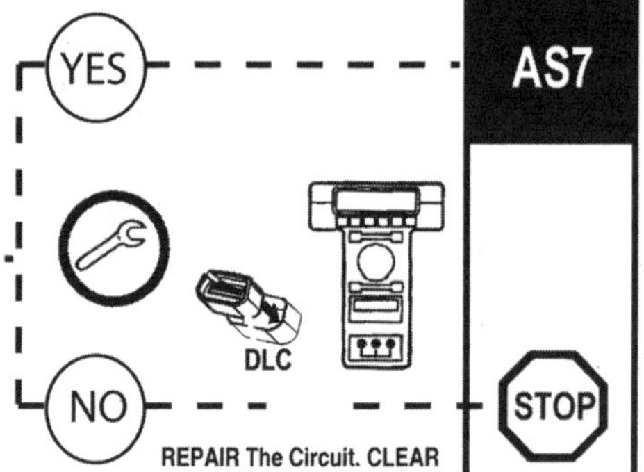

REPAIR The Circuit. CLEAR The DTCs. REPEAT The Self-Test.

AS7

CHECK THE FICM GROUND CIRCUITS FOR AN OPEN

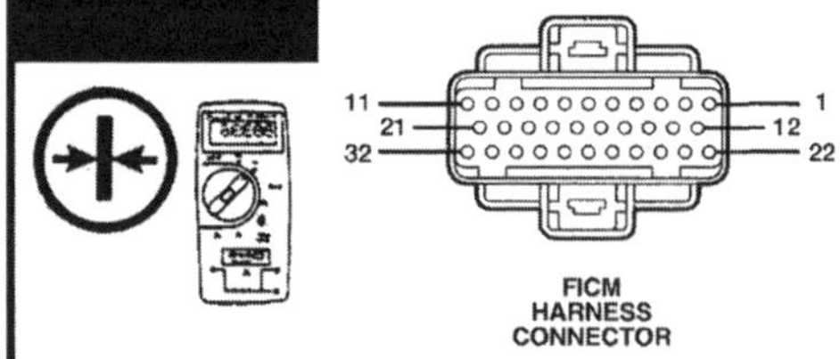

Measure the resistance between:	
(+) FICM-C Connector, Harness Side	**(-)**
GND - Pin 1, 2, 3, 22, 26	Ground

Are The Resistances Less Than 5 Ohms?

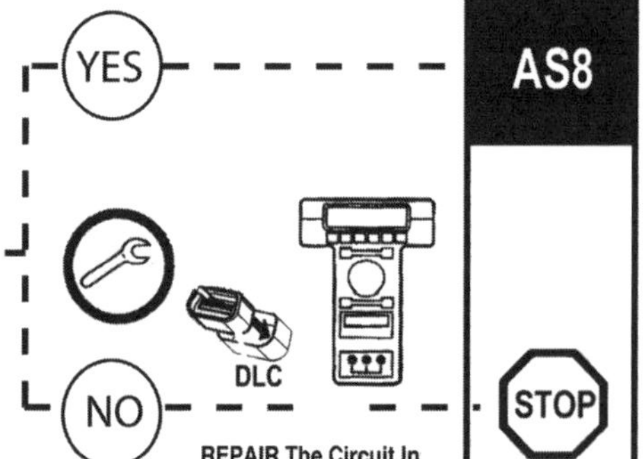

REPAIR The Circuit In Question. CLEAR The DTCs. REPEAT The Self-Test.

AS8 — CHECK THE FICM CAN CIRCUITS

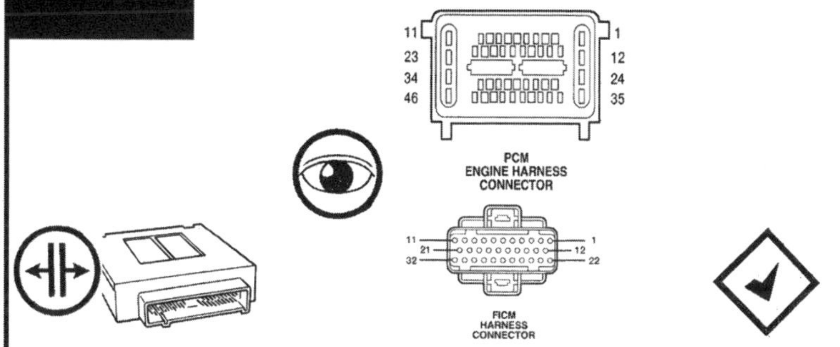

Disconnect The PCM Engine Harness Connector.

Check for Loose Connections or Damage between	
PCM Engine Connector Harness	FICM Relay Connector Harness
Pins 37 & 26	Pins 30 & 31

Are The Wiring And Shield OK?

YES → AS9

NO →

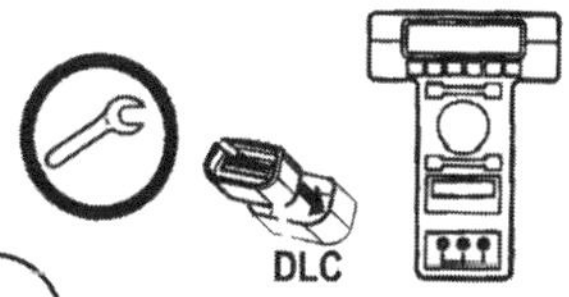

REPAIR As Necessary. RESTORE The Vehicle. CLEAR DTCs And RETEST The System. → STOP

AS9 — CHECK FICM CAN CIRCUITS FOR A SHORT TO VOLTAGE

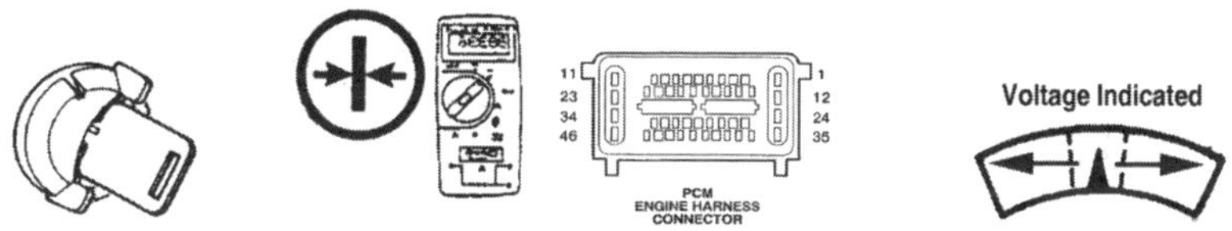

Key On, Engine Off.

Measure the voltage between:	
(+) PCM Connector, Harness Side	(-)
CAN2+ - Pin E37	Ground
CAN2- - Pin E26	Ground

Is Any Voltage Indicated?

YES →

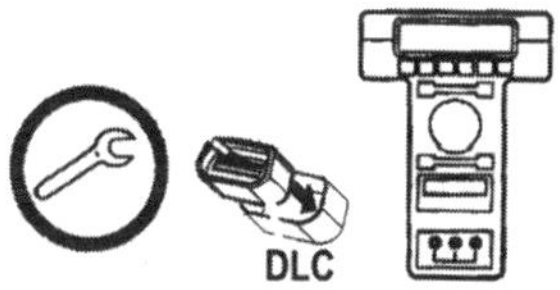

REPAIR Short To Voltage. RESTORE The Vehicle. CLEAR DTCs And RETEST The System. → STOP

NO → AS10

AS10

CHECK FICM CAN CIRCUITS FOR A SHORT TOGETHER

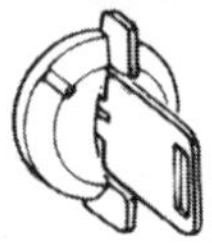

Key Off.

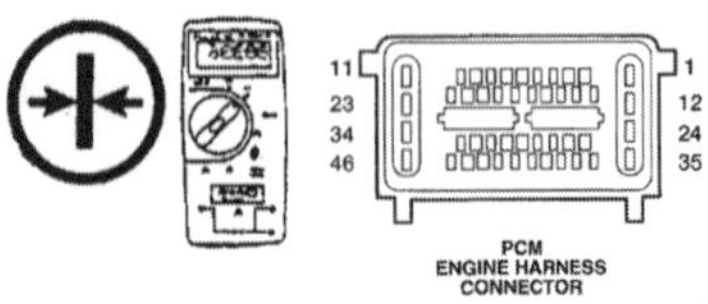

Measure the resistance between:	
(+) PCM Connector, Harness Side	(-) PCM Connector, Harness Side
CAN2+ - Pin E37	CAN2- - Pin E26

10,000 Ohms

Is The Resistance Greater Than 10,000 Ohms?

YES → AS11

NO →

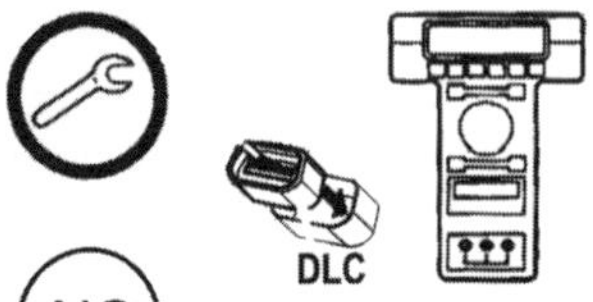

REPAIR Shorted Circuits. RESTORE The Vehicle. CLEAR DTCs And RETEST The System.

STOP

AS11

CHECK FICM CAN CIRCUITS FOR A SHORT TO GROUND

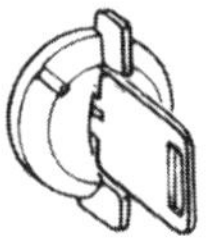

Key Off.

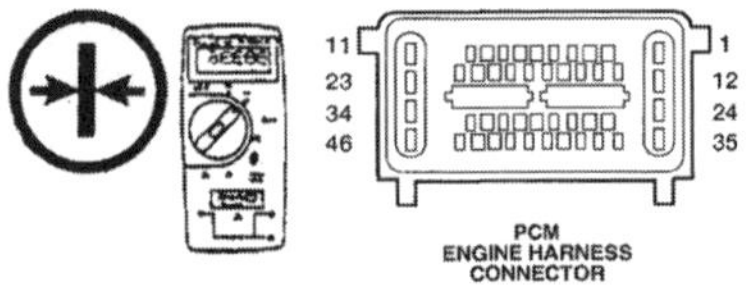

Measure the resistance between:	
(+) PCM Connector, Harness Side	(-)
CAN2+ - Pin E37	Ground
CAN2- - Pin E26	Ground

10,000 Ohms

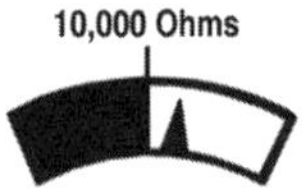

Are The Resistances Greater Than 10,000 Ohms?

YES → AS12

NO →

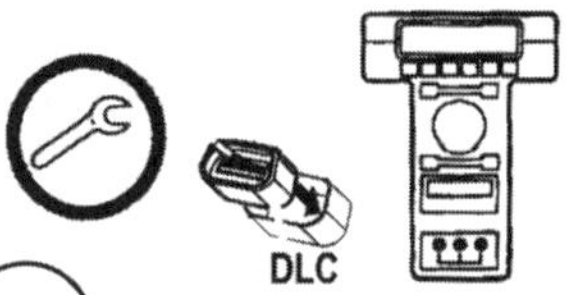

REPAIR Short To Ground. RESTORE The Vehicle. CLEAR DTCs And RETEST The System.

STOP

AS12 CHECK FICM CAN CIRCUITS FOR AN OPEN

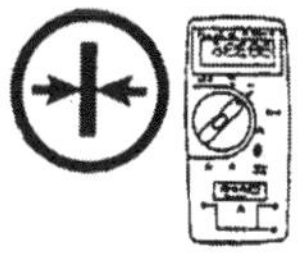

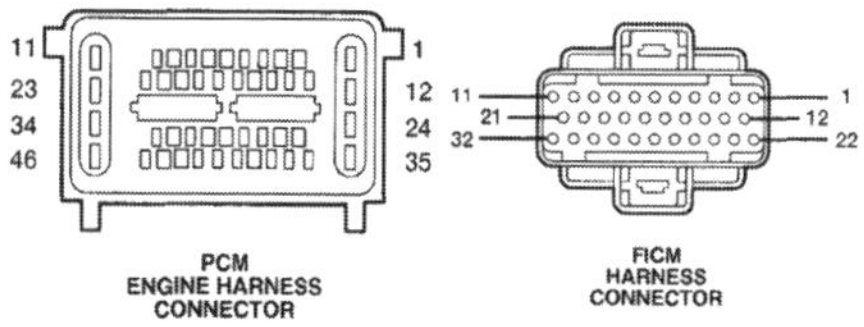

Measure the resistance between:	
(+) PCM Connector, Harness Side	(-) FICM-C Connector, Harness Side
CAN2+ - Pin E37	CAN2+ - Pin 30
CAN2- - Pin E26	CAN2- - Pin 31

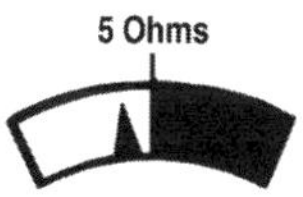

Are The Resistances Less Than 5 Ohms?

YES → AS13

NO →

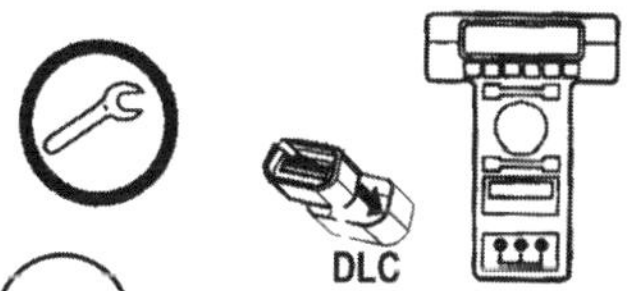

REPAIR Open In The Faulted CAN Circuit. RESTORE The Vehicle. CLEAR DTCs And RETEST The System. STOP

AS13 CHECK PCM TO FICM COMMUNICATION

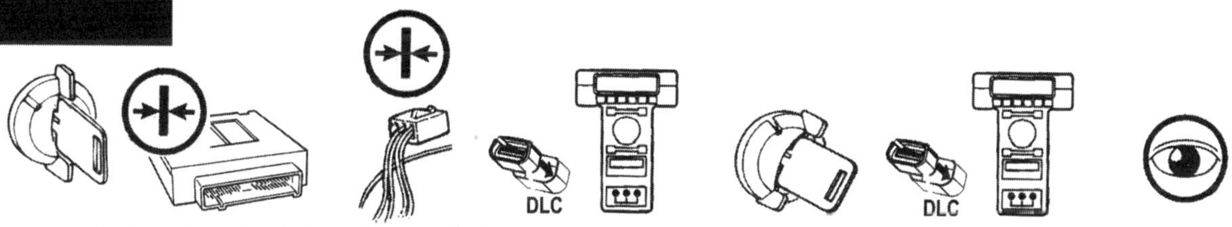

Key Off. → Connect The PCM Engine Harness Connector. → Connect The FICM Harness Connector C. → Connect The Scan Tool. → Key On, Engine Off. → Access The PCM. → Monitor The FICM LPWR, FICM MPWR And FICM VPWR PIDs. Refer To Reference Values.

Do The PIDs Indicate Correctly?

YES → AS14

NO →

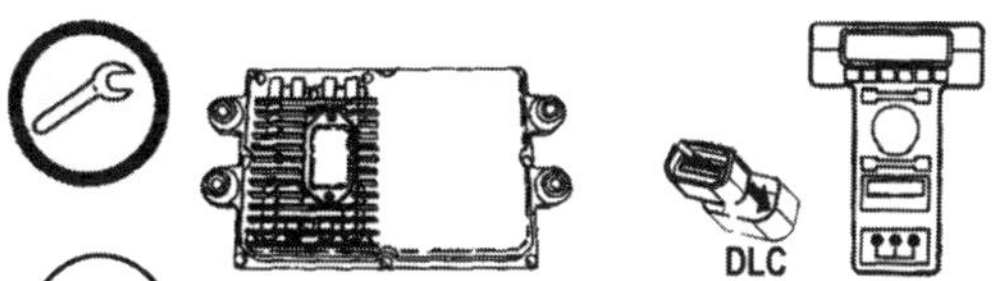

INSTALL A New FICM. CLEAR The DTCs. REPEAT The Self-Test. STOP

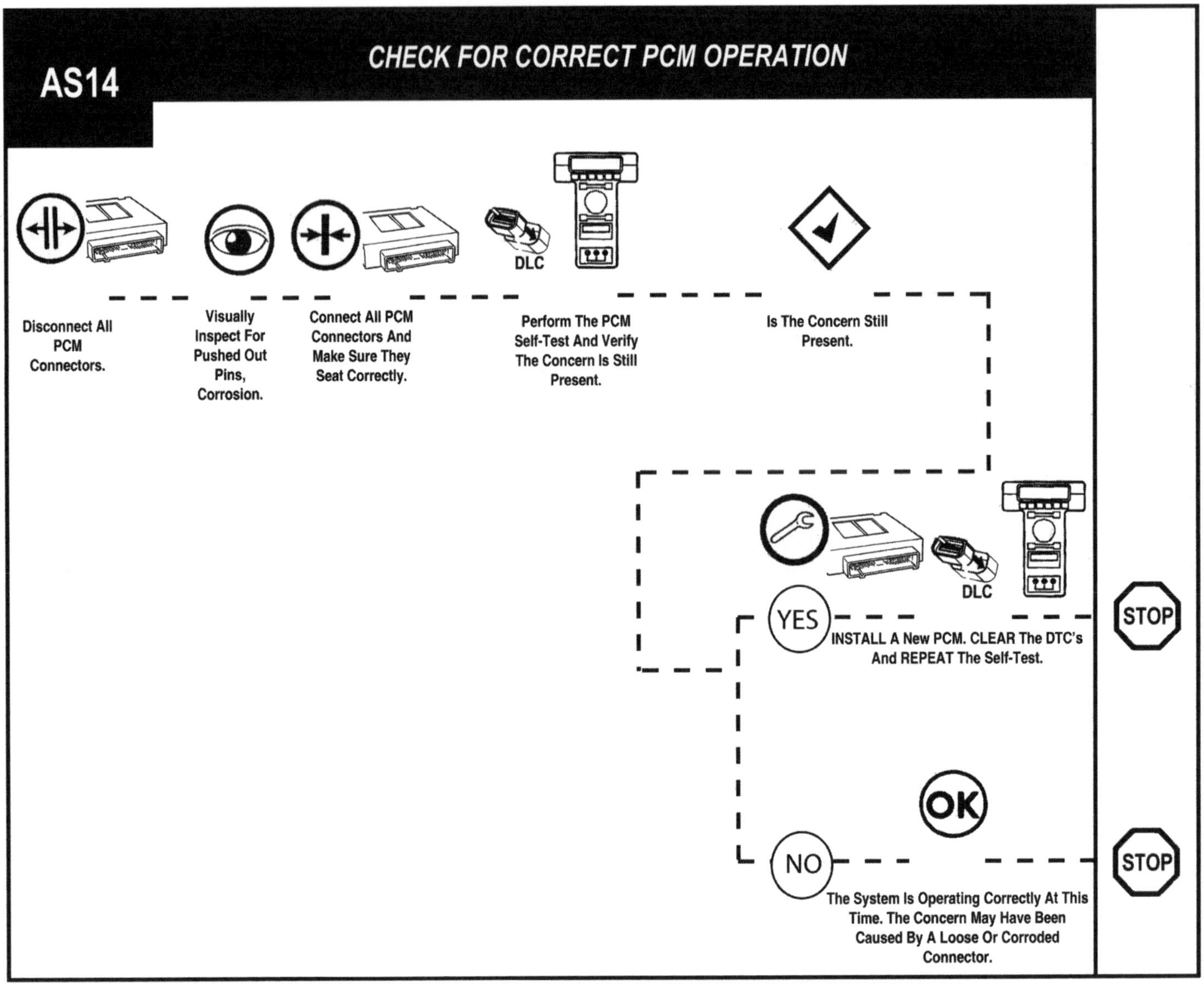
AS14
CHECK FOR CORRECT PCM OPERATION
Disconnect All PCM Connectors.
Visually Inspect For Pushed Out Pins, Corrosion.
Connect All PCM Connectors And Make Sure They Seat Correctly.
DLC
Perform The PCM Self-Test And Verify The Concern Is Still Present.
Is The Concern Still Present.
DLC
YES
INSTALL A New PCM. CLEAR The DTC's And REPEAT The Self-Test.
STOP
OK
NO
The System Is Operating Correctly At This Time. The Concern May Have Been Caused By A Loose Or Corroded Connector.
STOP

Trouble Code (s)		Test
P0148 P2552	**Problem: Fuel Injection Control Module (FICM) Circuit Fuel Delivery Error**	AT 2006-07

REFERENCE LIST

This pinpoint test is intended to diagnose the following:

- FICMM circuitry open
- FICMM circuitry short to ground
- Crankshaft position (CKP) or camshaft position (CMP) signal noise
- Powertrain control module (PCM)

DIAGNOSTIC TROUBLE CODE INDEX			
Codes	KOEO	KOER	CM
P0148	AT1	AT1	AT1
P2552	AT1	AT1	AT1

P0148 = Fuel Error/Engine Speed higher Than Expected
P2552 = FICMM Circuit - Throttle/Fuel Inhibit Circuit

Signal Functions:
The diesel engine power monitor (DEPM) strategy resides in the transmission control module (TCM) located inside the powertrain control module (PCM). The function of the DEPM is to monitol engine RPM when there is no power demand from the accelerator pedal position (APP) sensor.

Under normal engine idle operation, the DEPM RPM value must always be higher than engine RPM. In the event that the engine RPM does exceed the DEPM limit, the DE PM will disable the crank and cam output signal sent by the PCM to the fuel injection control module (FICM).

Detection/Management :
The fuel injection control module monitor (FICMM) input line informs the diesel engine power monitor (DEPM) when the injectors are turned on and when the injectors are turned off. When the FICMM line is either shorted or open, the monitor strategy assumes that the fuel injectors are always turned on and sets DTC P2552. When DTC P2552 is set, the DEPM may, in some instances, set DTC P0148. When DTC P0148 is set, DTCs P2614 (CMPO failure) and P2617 (CKPO failure) may also set. If DTCs P2614 or P2617 are set with P0148, do not attempt to diagnose the CMPO or CKPO circuits. DTCs P1378 or P1379 (FICM voltage) may also cause DTCs P2552, P0148, P2614 and/or P2617 to set.

CAUTION!:
The powertrain control module (PCM) harness connectors must be properly seated and the connector latch properly attached to eliminate possible driveability concerns or a no start condition. Installing the PCM connectors on an angle may cause an improper connection, misdiagnosis and damaged components. Install the connector until the lever pivots and seats itself. Apply light pressure to get the connector into position on the PCM and then fully seat the connector.

Note: Visually inspect the harness connectors for corrosion, damage, proper mating and correct pin tension.

Note: When the PCM is disconnected additional DTCs will be set. Clear all DTCs after restoring the vehicle.

PINPOINT TEST SCHEMATICS AND CONNECTORS

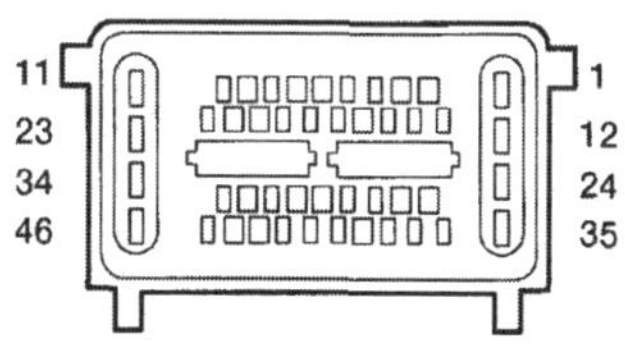

PCM
ENGINE HARNESS
CONNECTOR

Pin	Circuit
E28	FICMM (Fuel Injector Control Module Monitor)

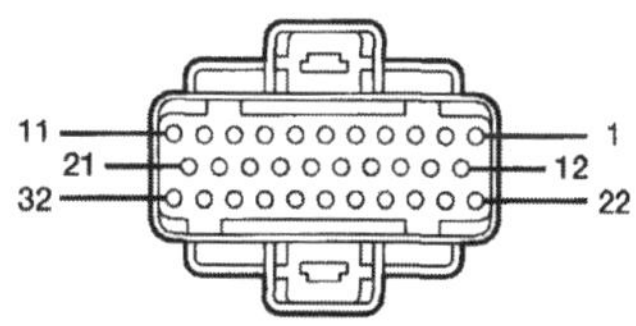

FICM
HARNESS
CONNECTOR

Pin	Circuit
9	FICMM (Fuel Injector Control Module Monitor)

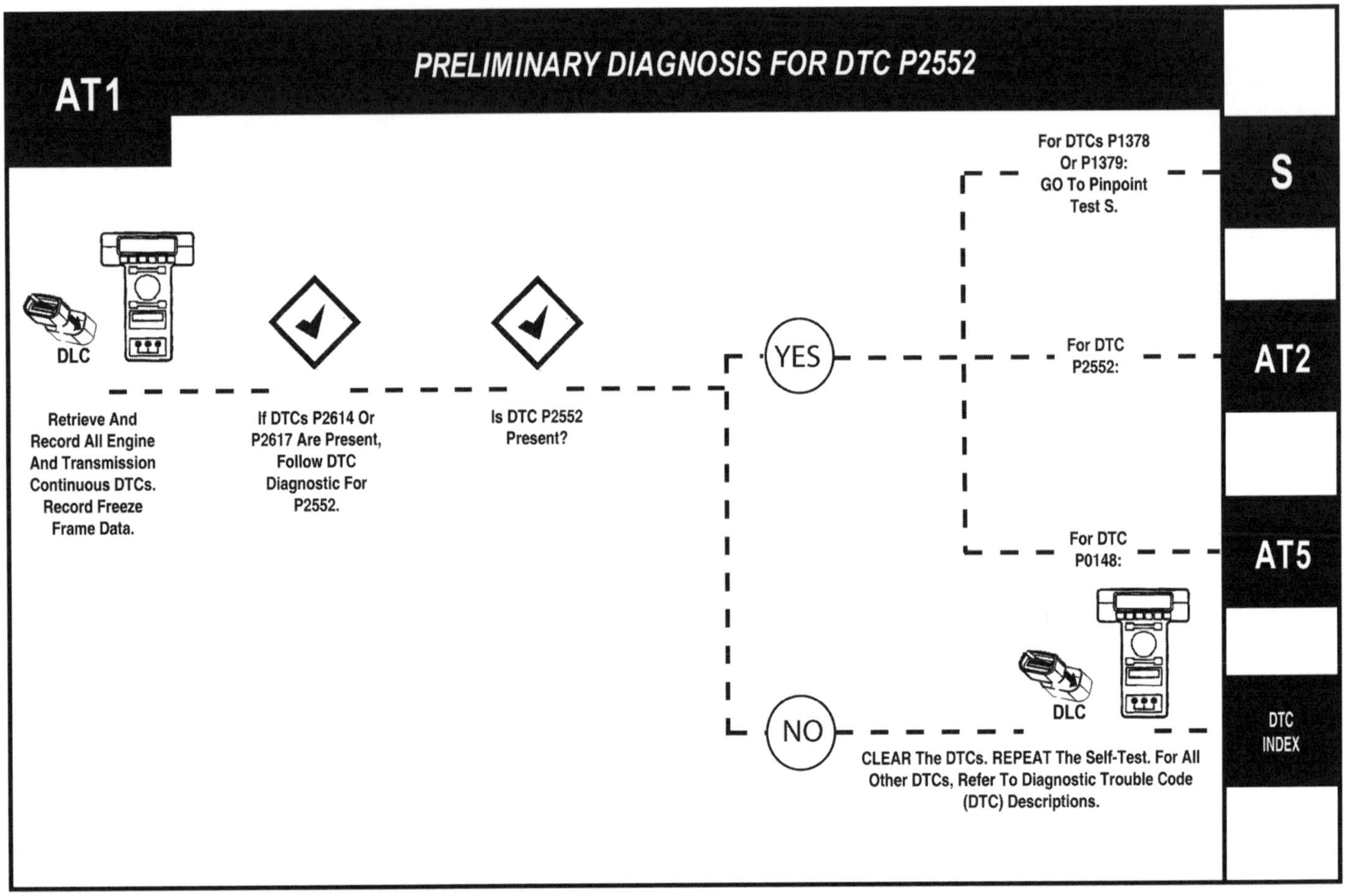
AT1
PRELIMINARY DIAGNOSIS FOR DTC P2552
DLC
Retrieve And Record All Engine And Transmission Continuous DTCs. Record Freeze Frame Data.
If DTCs P2614 Or P2617 Are Present, Follow DTC Diagnostic For P2552.
Is DTC P2552 Present?
YES
For DTCs P1378 Or P1379: GO To Pinpoint Test S.
S
For DTC P2552:
AT2
For DTC P0148:
AT5
NO
DLC
CLEAR The DTCs. REPEAT The Self-Test. For All Other DTCs, Refer To Diagnostic Trouble Code (DTC) Descriptions.
DTC INDEX

AT2 DIAGNOSTIC TROUBLE CODE (DTC) P2552

Note: Refer to the PCM and component connector pin numbers at the beginning of this pinpoint test.

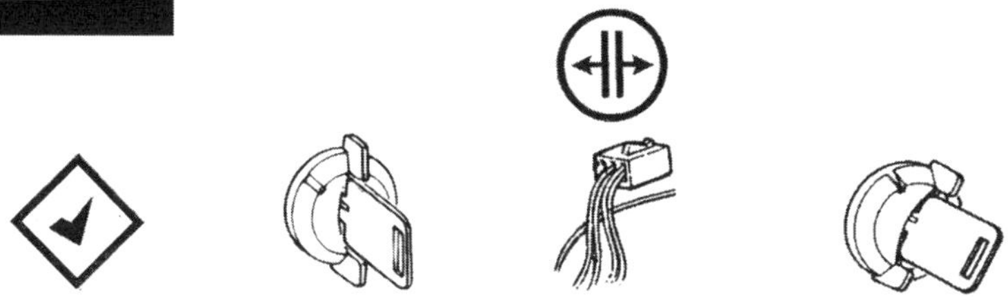

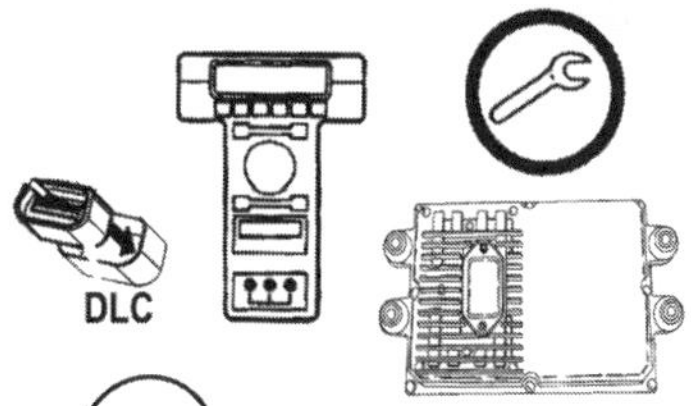

Possible Causes:
- FICMM Line Open
- FICMM Line Shorted To Ground

Key Off.

Disconnect The FICM Electrical Harness Connector C.

Key On, Engine Off.

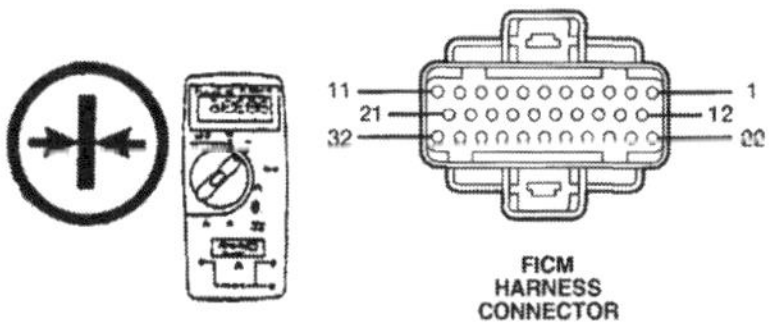

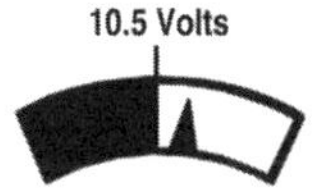

Measure the voltage between:	
(+) FICM-C Connector, Harness Side	(-)
FICMM - Pin 9	Ground

Is The Voltage Greater Than 10.5 Volts?

YES — CLEAR All Engine And Transmission DTCs. REPEAT The Self-Test. If DTC P2552 Is Present Again, INSTALL A New FICM. STOP

NO — AT3

AT3 CHECK FICMM CIRCUIT FOR A SHORT TO GROUND

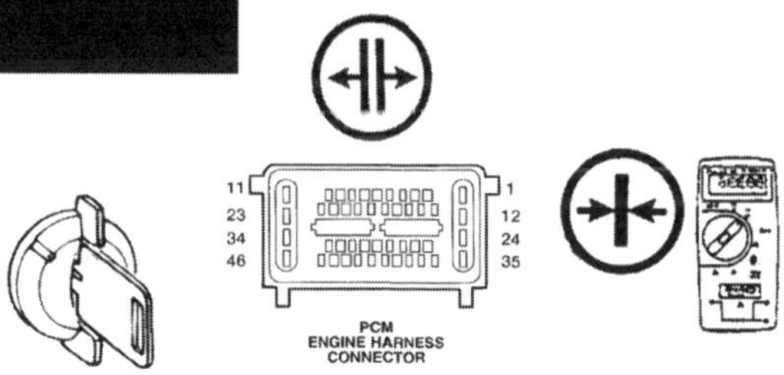

Key Off.

Disconnect The PCM Engine Harness Connector.

Measure the resistance between:	
(+) PCM Connector, Harness Side	(-)
FICMM - Pin E28	Ground

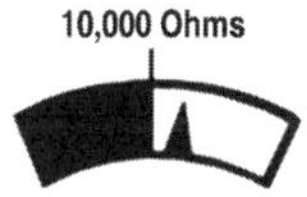

Is The Resistance Greater Than 10,000 Ohms?

YES — AT4

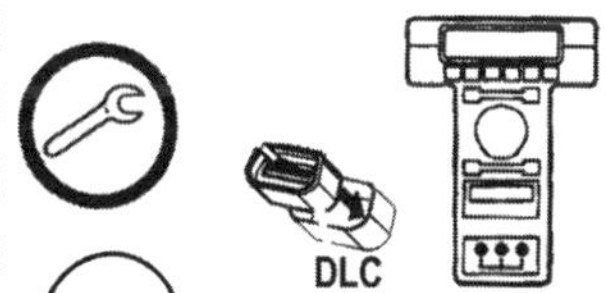

NO — REPAIR The Circuit. CLEAR The DTC's. Repeat The Self-Test. STOP

AT4 — CHECK FICMM CIRCUIT FOR AN OPEN

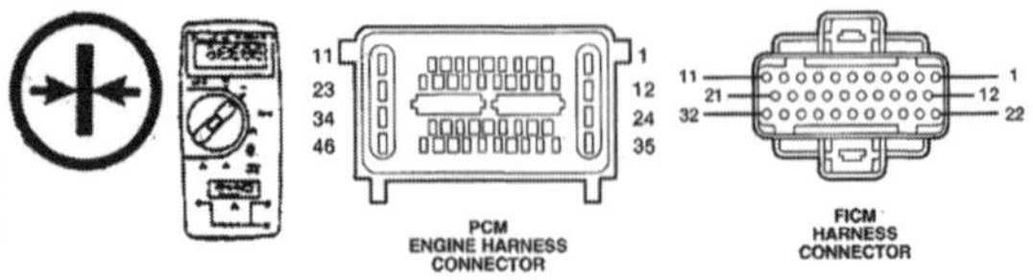

Measure the resistance between:	
(+) PCM Connector, Harness Side	(-) FICM-C Connector, Harness Side
FICMM - Pin E28	FICMM - Pin 9

Is the resistance greater than 10K ohms?

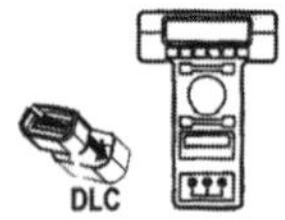

YES → CLEAR All Engine And Transmission DTCs. REPEAT The Self-Test. If DTC P2552 Is Present Again, GO TO AT6

NO → REPAIR The Circuit. CLEAR The DTC's. Repeat The Self-Test. STOP

AT5 — DIAGNOSTIC TROUBLE CODE (DTC) P0148

Possible Causes:
- Fuel Quality
- CKP Or CMP Signal Noise

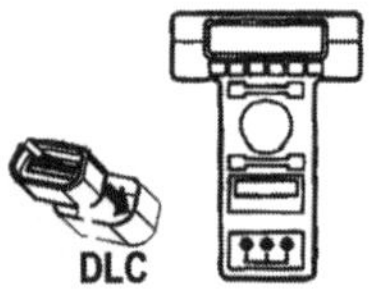

Retrieve And Record All Engine And Transmission DTCs.

Is DTC P0148 Present?

YES → REFER To Performance Diagnostic Procedures. PERFORM A Drive Cycle. REFER To Diagnostic Methods. REFLASH The PCM. If DTC P0148 Is Present Again, GO TO AT6.

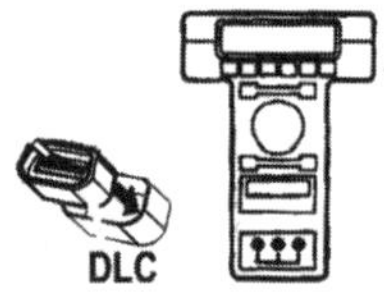

NO → CLEAR The DTCs. REFER To Symptom Charts If A Drivability Concern Exists. STOP

AT6 CHECK FOR CORRECT PCM OPERATION

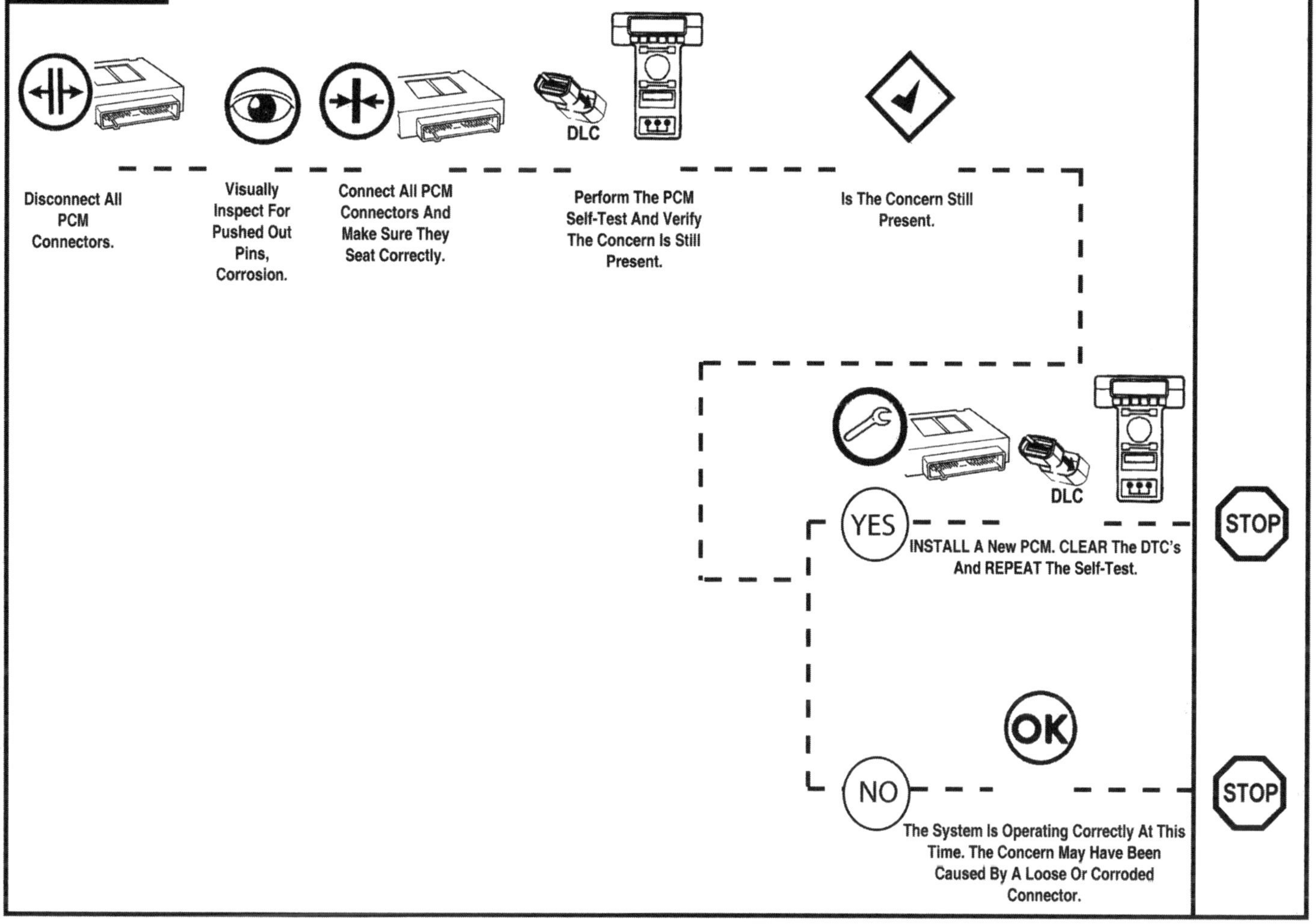

NOTES

Trouble Code (s)		Test
P0069	**Problem: Manifold Absolute Pressure (MAP)/Barometric Pressure (BARO) Correlation**	AU 2006-07

REFERENCE LIST

This pinpoint test is intended to diagnose the following:

- Barometric pressure (BARO) sensor
- Manifold absolute pressure (MAP) sensor
- Harness circuits: MAP, SIGRTN, VREF
- Powertrain control module (PCM) (12A650)

DIAGNOSTIC TROUBLE CODE INDEX			
Codes	KOEO	KOER	CM
P0069	AU1	AU1	AU1

P0069 = BARO/MAP Correlation

Signal Functions:
The powertrain control module (PCM) compares the BARO sensor and the MAP sensor reading at idle.

Detection/Management:
When the difference between the BARO sensor and MAP sensor is greater than a specified value, an increment couter advances until a DTC is set

CAUTION!:
The powertrain control module (PCM) harness connectors must be properly seated and the connector latch properly attached to eliminate possible driveability concerns or a no start condition. Installing the PCM connectors on an angle may cause an improper connection, misdiagnosis and damaged components. Install the connector until the lever pivots and seats itself. Apply light pressure to get the connector into position on the PCM and then fully seat the connector.

Note: Visually inspect the harness connectors for corrosion, damage, proper mating and correct pin tension.

Note: When the PCM is disconnected additional DTCs will be set. Clear all DTCs after restoring the vehicle.

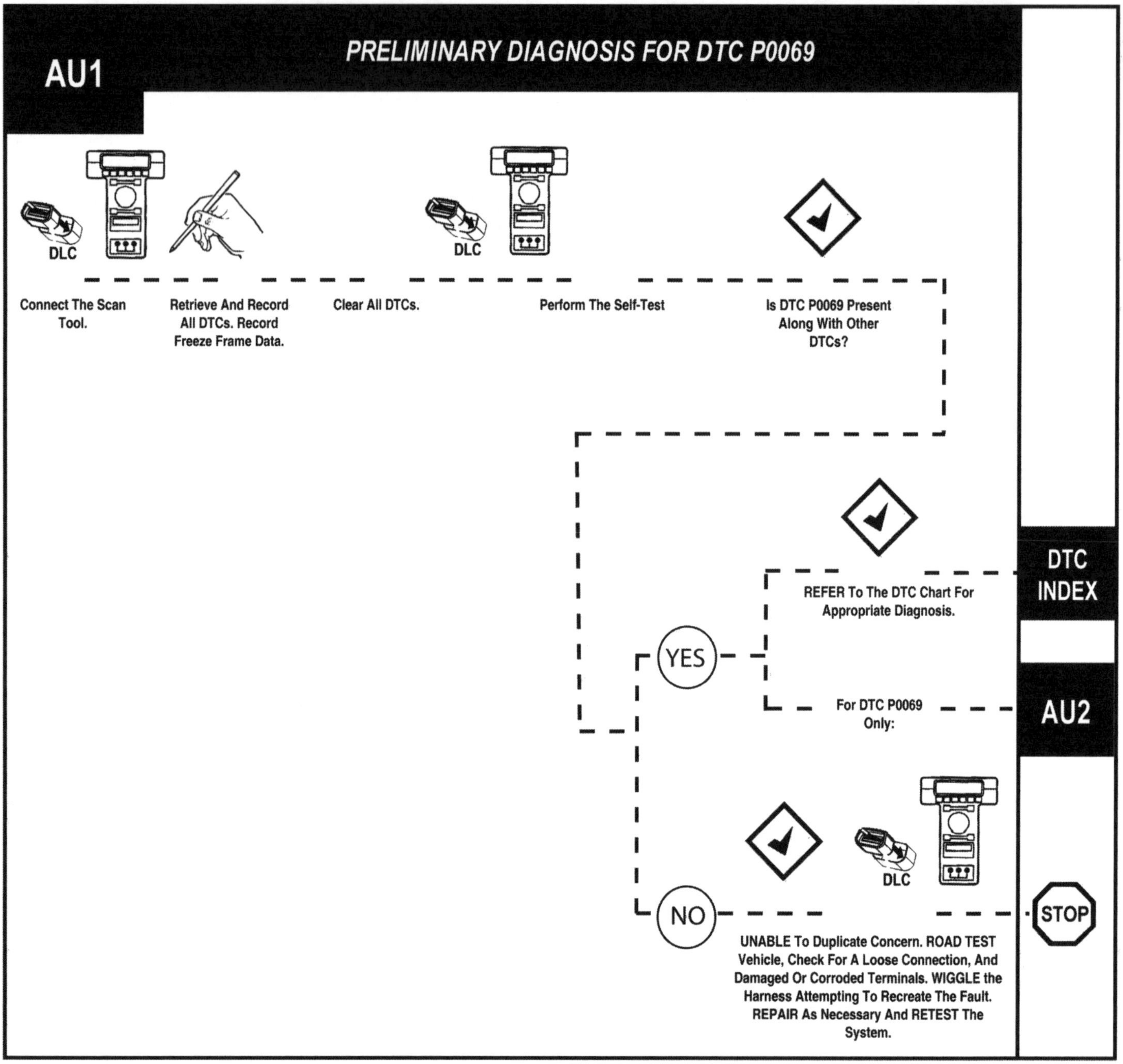
AU1
PRELIMINARY DIAGNOSIS FOR DTC P0069
DLC
DLC
Connect The Scan Tool.
Retrieve And Record All DTCs. Record Freeze Frame Data.
Clear All DTCs.
Perform The Self-Test
Is DTC P0069 Present Along With Other DTCs?
YES
REFER To The DTC Chart For Appropriate Diagnosis.
DTC INDEX
For DTC P0069 Only:
AU2
NO
DLC
UNABLE To Duplicate Concern. ROAD TEST Vehicle, Check For A Loose Connection, And Damaged Or Corroded Terminals. WIGGLE the Harness Attempting To Recreate The Fault. REPAIR As Necessary And RETEST The System.
STOP

AU2

DIAGNOSTIC TROUBLE CODE (DTC) P0069

Note: Most Weather Services Report A Local Barometric Pressure That Has Been Corrected To Factor In Sea Level. However, The BARO PID Reports The Barometric Pressure For The Vehicle's Actual Altitude. Local Weather Conditions (High Or Low Pressure) Will Change The Local Barometric Pressure By Several Inches Of Mercury.

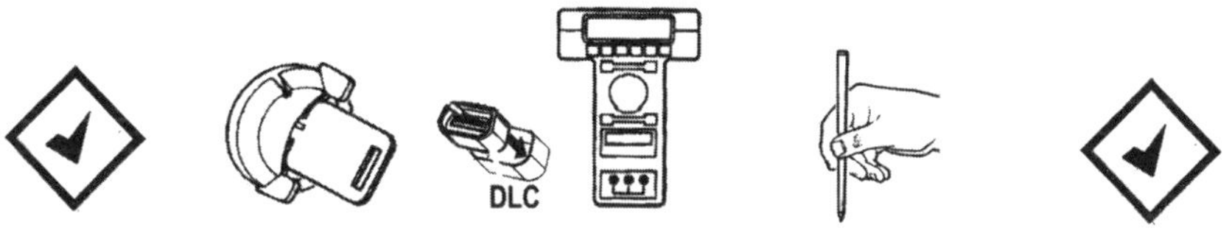

Possible Causes:
- BARO Sensor
- MAP Sensor
- PCM

Key On. Engine Off.

Access BARO/MAP PIDs.

Record BARO/MAP Values.

Check That The BARO/PID Reading Is Approximately The Same As The Barometric Pressure Reading For The Location, Day, And Altitude Where The Vehicle Is Being Serviced.

Does The BARO PID Reading Compare With Daily Barometric Report For Service Area?

YES → AU3

NO →

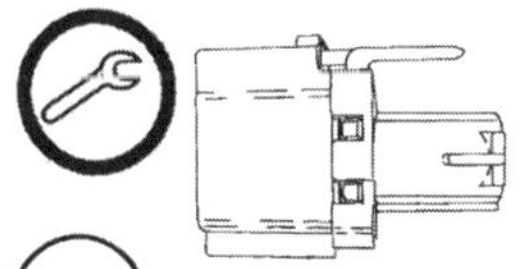

INSTALL A New BARO Sensor. RESTORE The Vehicle. CLEAR DTCs And RETEST The System.

AU3

COMPARE BARO KOER AND KOEO READINGS

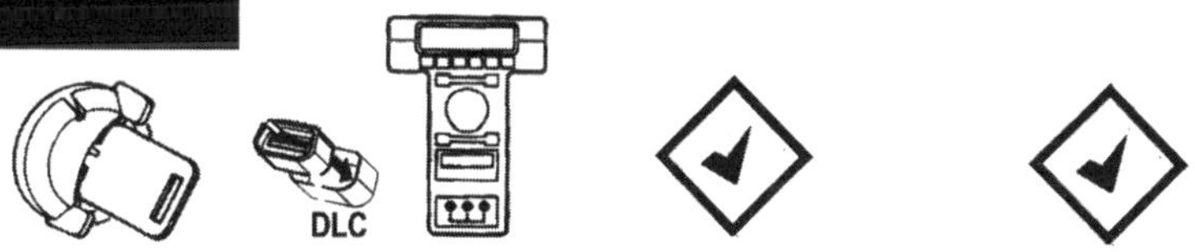

Key ON. Engine Running.

Access BARO PIDs.

Compare Engine BARO KOER Reading With KOEO Reading.

Is The BARO PID Value Equal To The Recorded Value?

YES → AU4

NO →

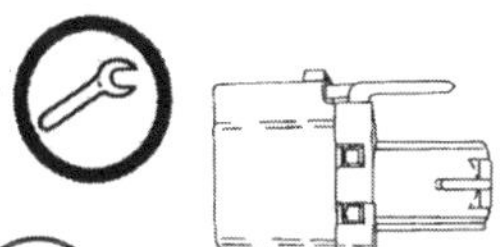

INSTALL A New BARO Sensor. RESTORE The Vehicle. CLEAR DTCs And RETEST The System.

AU4 — COMPARE THE KOER BARO AND MAP PID VALUES

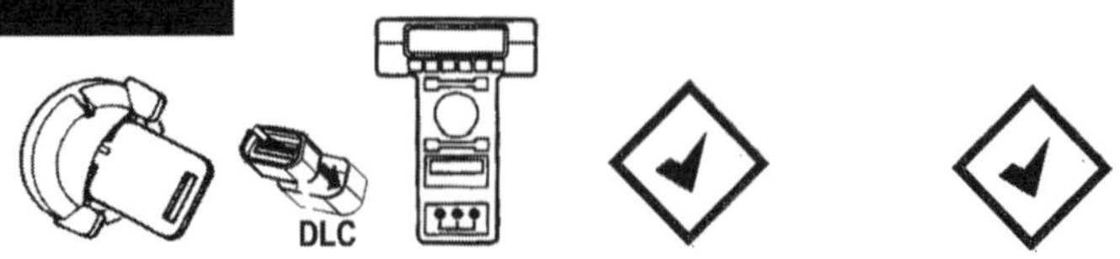

Key ON. Engine Running.

Access BARO/MAP PIDs.

Compare The BARO/MAP Reading With KOEO Values.

Is The MAP Sensor Reading Within 30 KPA (4.3 PSI) Of The BARO Reading?

YES → AU5

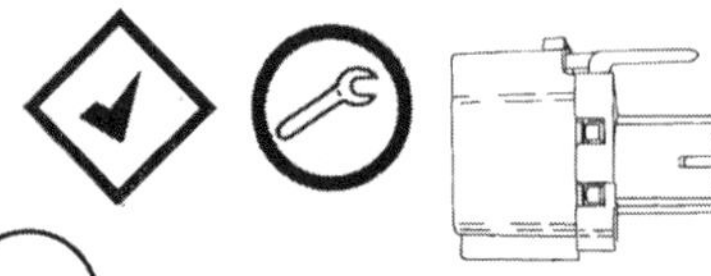

NO → STOP

CHECK MAP Sensor Hose For Damage. REPLACE The Hose If A Concern Is Found. If No Hose Concern Is Detected, INSTALL A New MAP Sensor. RESTORE The Vehicle. CLEAR DTCs And RETEST The System.

AU5 — CHECK FOR CORRECT PCM OPERATION

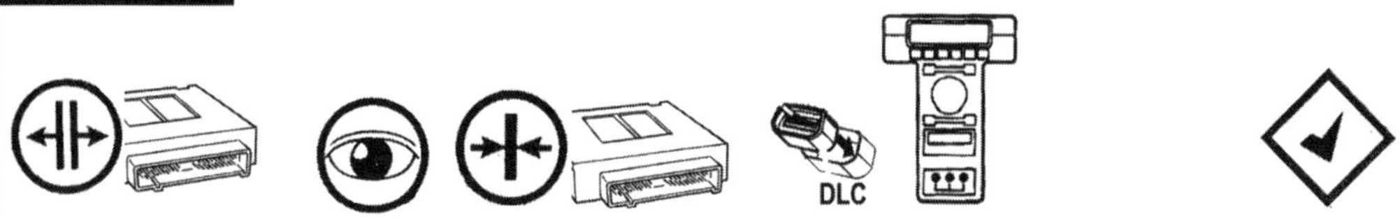

Disconnect All PCM Connectors.

Visually Inspect For Pushed Out Pins, Corrosion.

Connect All PCM Connectors And Make Sure They Seat Correctly.

Perform The PCM Self-Test And Verify The Concern Is Still Present.

Is The Concern Still Present.

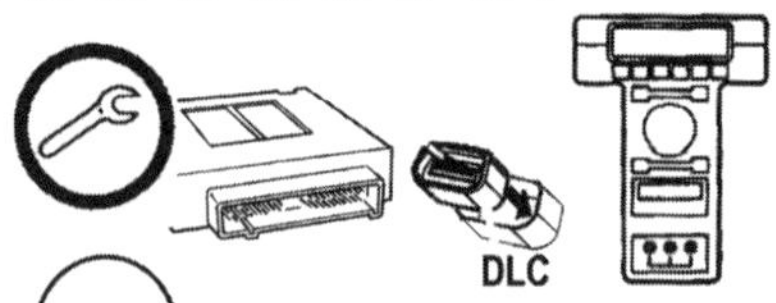

YES → STOP

INSTALL A New PCM. CLEAR The DTC's And REPEAT The Self-Test.

OK

NO → STOP

The System Is Operating Correctly At This Time. The Concern May Have Been Caused By A Loose Or Corroded Connector.

Trouble Code (s)	Problem: Air Conditioning (A/C) Relay Control Circuit	Test
P0645		AV 2006-07

REFERENCE LIST

This pinpoint test is intended to diagnose only the following:

- A/C clutch relay
- A/C clutch relay control circuit
- Powertrain control module (PCM)

DIAGNOSTIC TROUBLE CODE INDEX			
Codes	KOEO	KOER	CM
P0645	AV1	AV1	AV1

P0645 = A/C Clutch Relay Control Circuit

CAUTION!:
The powertrain control module (PCM) harness connectors must be properly seated and the connector latch properly attached to eliminate possible driveability concerns or a no start condition. Installing the PCM connectors on an angle may cause an improper connection, misdiagnosis and damaged components. Install the connector until the lever pivots and seats itself. Apply light pressure to get the connector into position on the PCM and then fully seat the connector.

Note: Visually inspect the harness connectors for corrosion, damage, proper mating and correct pin tension.

PINPOINT TEST SCHEMATICS AND CONNECTORS

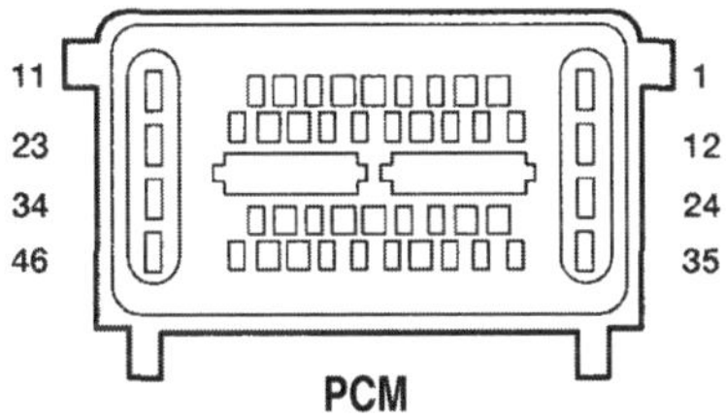

PCM BODY HARNESS CONNECTOR

Pin	Circuit
B2	A/CCR (Air Conditioning Clutch Relay [PCM output])

A/C CLUTCH RELAY

Pin	Circuit
1	A/CCR (Air Conditioning Clutch Relay [PCM output])
5	A/CC PWR (Air Conditioning Clutch Power)
2	VPWR (Vehicle Power)

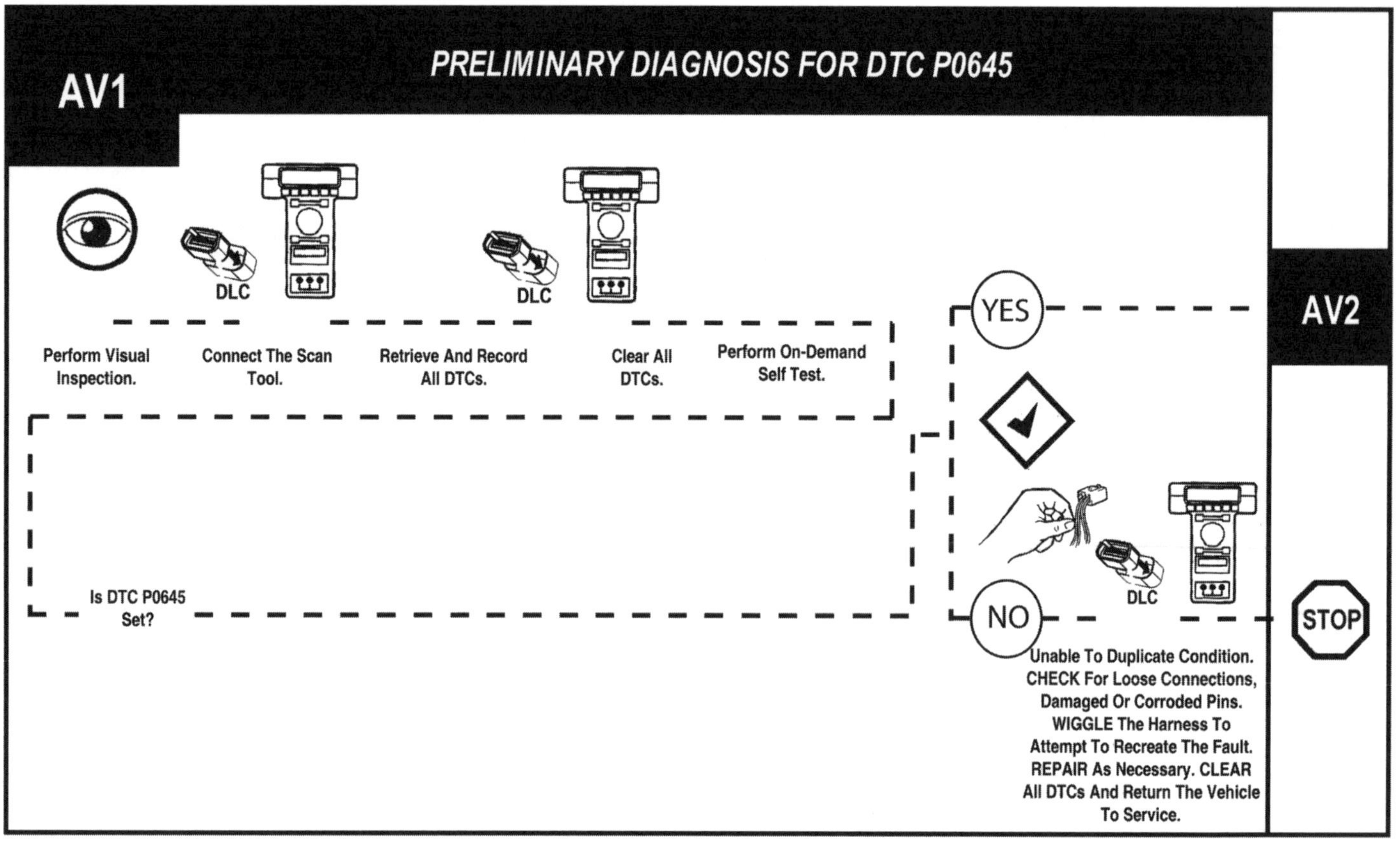
AV1
PRELIMINARY DIAGNOSIS FOR DTC P0645
DLC
DLC
Perform Visual Inspection.
Connect The Scan Tool.
Retrieve And Record All DTCs.
Clear All DTCs.
Perform On-Demand Self Test.
Is DTC P0645 Set?
YES
AV2
NO
DLC
STOP
Unable To Duplicate Condition. CHECK For Loose Connections, Damaged Or Corroded Pins. WIGGLE The Harness To Attempt To Recreate The Fault. REPAIR As Necessary. CLEAR All DTCs And Return The Vehicle To Service.

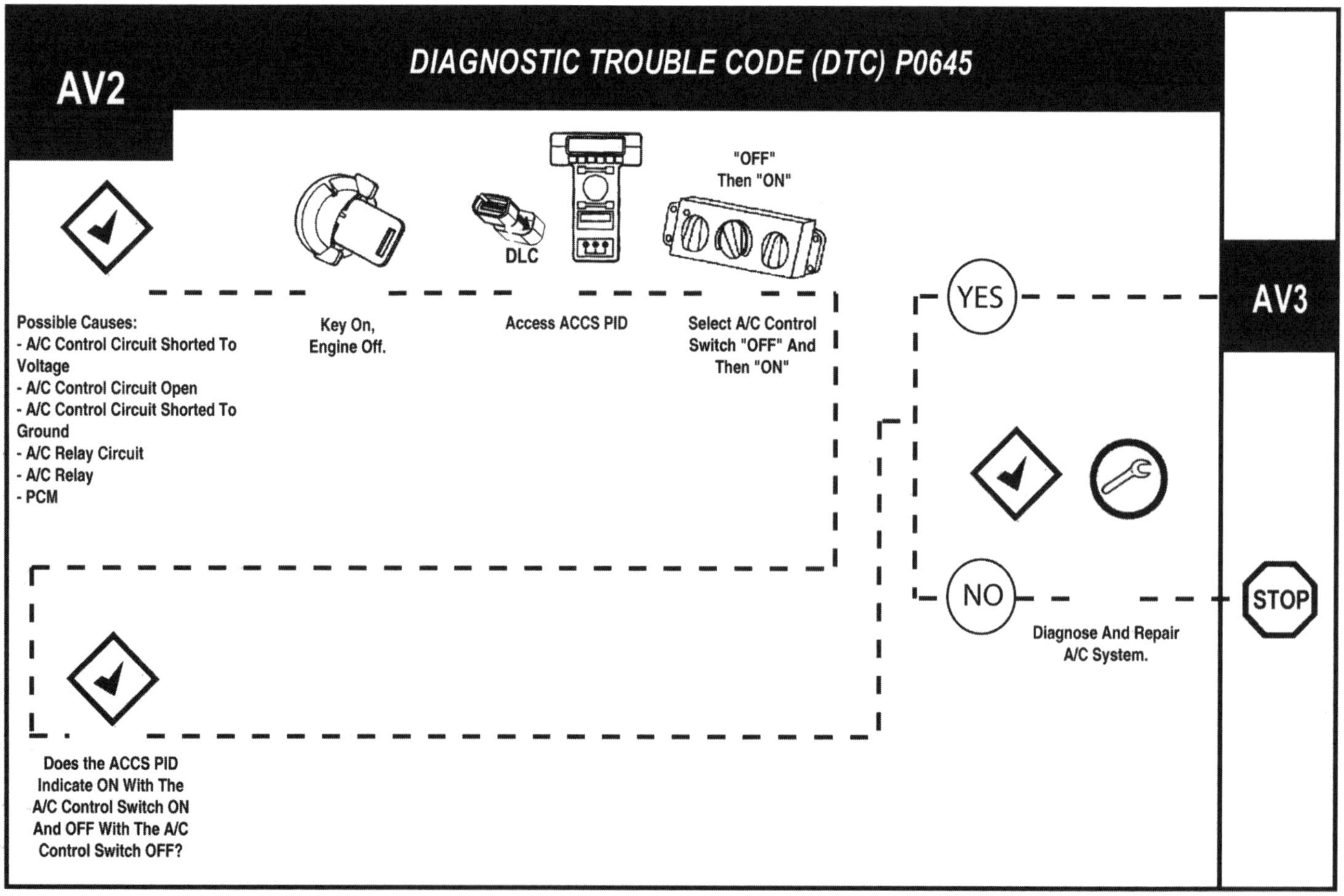
AV2
DIAGNOSTIC TROUBLE CODE (DTC) P0645
DLC
"OFF" Then "ON"
Possible Causes:
- A/C Control Circuit Shorted To Voltage
- A/C Control Circuit Open
- A/C Control Circuit Shorted To Ground
- A/C Relay Circuit
- A/C Relay
- PCM
Key On, Engine Off.
Access ACCS PID
Select A/C Control Switch "OFF" And Then "ON"
Does the ACCS PID Indicate ON With The A/C Control Switch ON And OFF With The A/C Control Switch OFF?
YES
AV3
NO
STOP
Diagnose And Repair A/C System.

AV3 — CHECK THE A/C CLUTCH RELAY COIL POWER CIRCUIT FOR VOLTAGE

Note: Refer To The Wiring Diagram At The Beginning Of This Pinpoint Test.

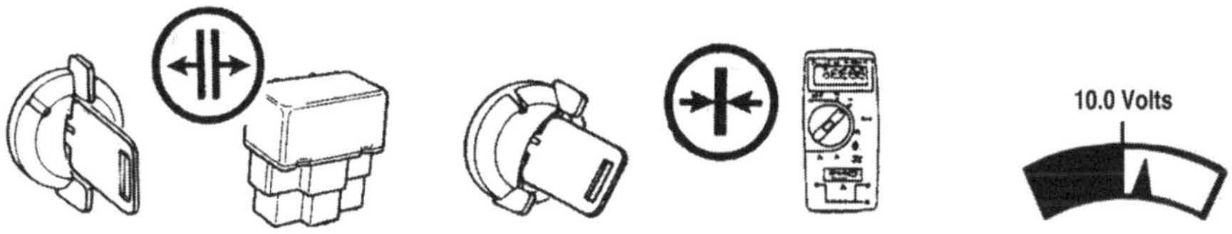

Key Off.

Disconnect The A/C Relay.

Key On. Engine Off.

Measure the voltage between:	
(+) A/CC Relay Connector, Harness Side	(-)
VPWR - Pin 2	Ground

Is Voltage Greater Than 10.0 Volts?

YES → AV4

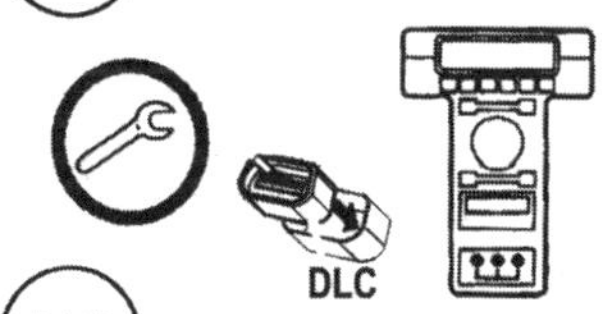

NO → REPAIR The Circuit. CLEAR DTCs. REPEAT The Self-Test. STOP

AV4 — CHECK THE A/C CLUTCH RELAY CONTROL CIRCUIT

Connect a non-powered test lamp between:	
(+) A/CC Relay Connector, Harness Side	(-) A/CC Relay Connector, Harness Side
A/CC PWR - Pin 5	A/CCR - Pin 1

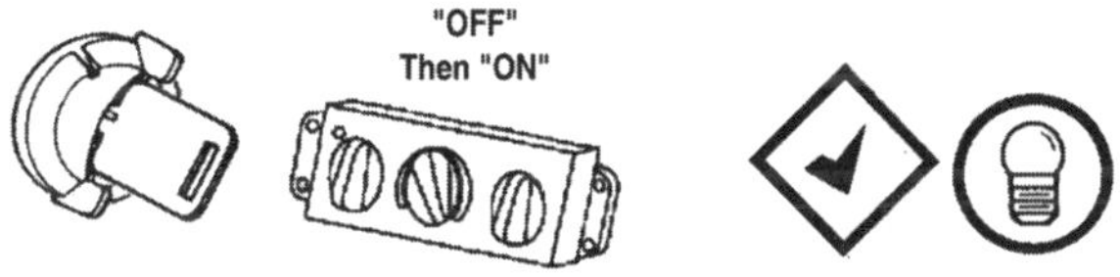

Key ON. Engine Running.

Select A/C On And Then Off.

Does The Test Light Illuminate With A/C On And Then Off When A/C Is Turned Off?

YES → Diagnose And Repair A/CSystem. STOP

NO → AV5

AV5

CHECK A/C CLUTCH RELAY CONTROL CIRCUIT FOR A SHORT TO VOLTAGE

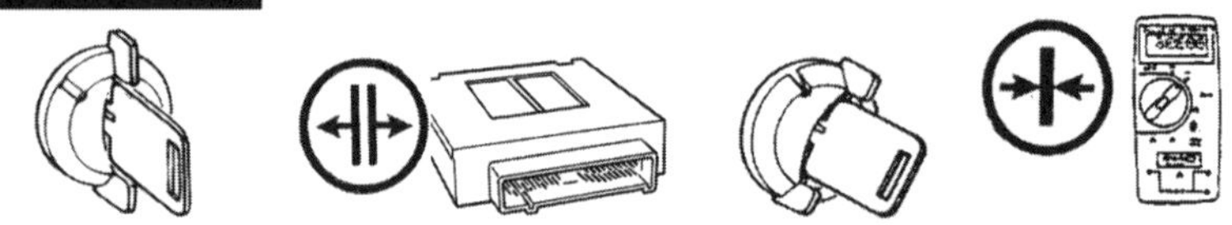

Key Off.

Disconnect The PCM Body Harness Connector.

Key On. Engine Off.

Measure the voltage between:	
(+) PCM Connector, Harness Side	(-)
A/CCR - Pin B2	Ground

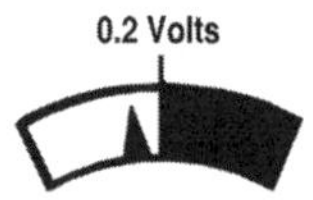

Is The Voltage Less Than 0.2 Volts?

YES → AV6

NO → REPAIR The Short Circuit To PWR. CLEAR The DTCs. REPEAT The Self-test. STOP

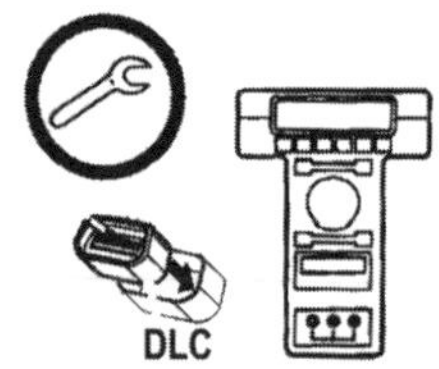

AV6

CHECK A/C RELAY CONTROL CIRCUIT FOR A SHORT TO GROUND

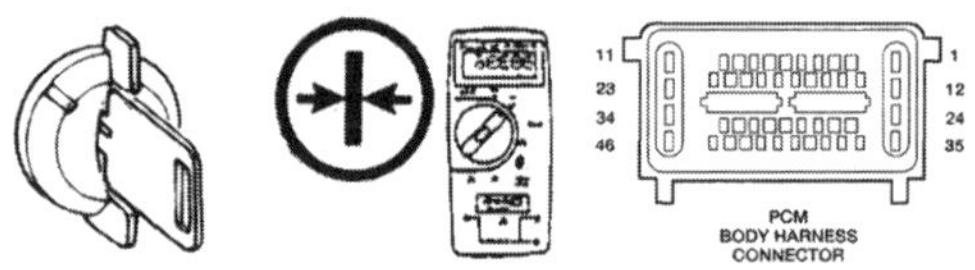

Key OFF.

Measure the resistance between:	
(+) PCM Connector, Harness Side	(-)
A/CCR - Pin B2	Ground

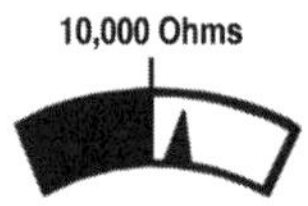

Is The Resistance Greater Than 10,000 Ohms?

YES → AV7

NO → REPAIR Short To Ground. CLEAR DTCs. REPEAT The Self-Test. STOP

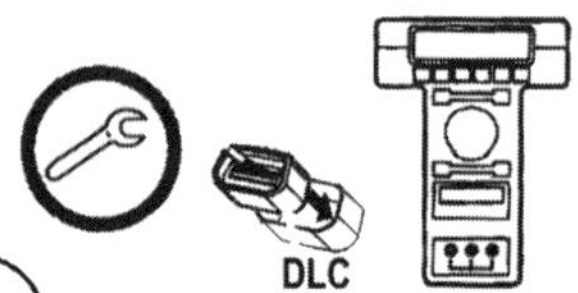

AV7 CHECK A/C RELAY CONTROL CIRCUIT FOR AN OPEN

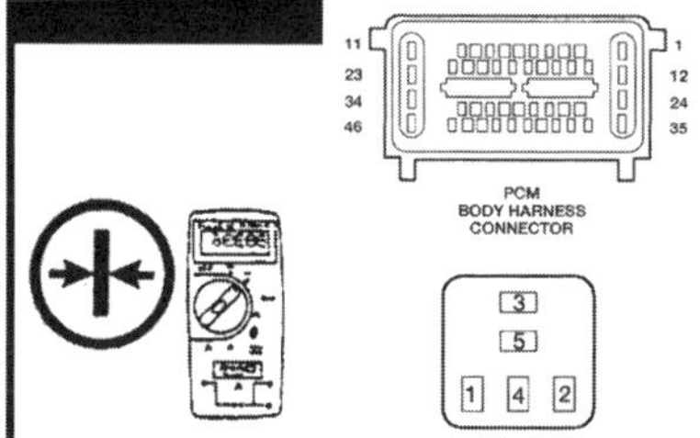

5 Ohms

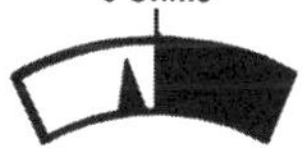

Measure the resistance between:	
(+) PCM Connector, Harness Side	(-) A/CC Relay Connector, Harness Side
A/CCR - Pin B2	A/CCR - Pin 1

Is The Resistance Less Than 5 Ohms?

YES → AV8

NO →

REPAIR The Circuit. CLEAR The DTCs. REPEAT The Self Test. STOP

AV8 CHECK FOR CORRECT PCM OPERATION

Disconnect All PCM Connectors.

Visually Inspect For Pushed Out Pins, Corrosion.

Connect All PCM Connectors And Make Sure They Seat Correctly.

Perform The PCM Self-Test And Verify The Concern Is Still Present.

Is The Concern Still Present.

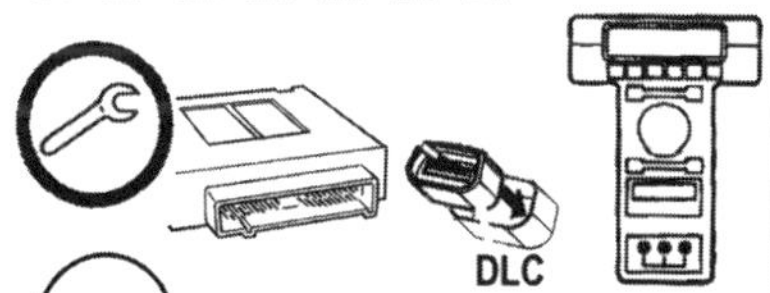

YES → INSTALL A New PCM. CLEAR The DTC's And REPEAT The Self-Test. STOP

NO → The System Is Operating Correctly At This Time. The Concern May Have Been Caused By A Loose Or Corroded Connector. STOP

NOTES

Trouble Code (s)		Test
P2199	**Problem: Intake Air Temperature (IAT)/ Intake Air Temperature 2 (IAT2) Sensor Correlation**	AW 2006-07

REFERENCE LIST

This pinpoint test is intended to diagnose the following:

- Harness circuits: IAT2, and SIGRTN
- IAT2 sensor
- Powertrain control module (PCM) (12A650)

DIAGNOSTIC TROUBLE CODE INDEX			
Codes	KOEO	KOER	CM
P2199	AW1	AW1	AW1

P2199 = Intake Air Temperature 1/2 Correlation

Signal Functions:
The IAT sensor is a thermistor-type sensor with a variable resistance that changes when exposed to different temperatures. When interfaced with the powertrain control module (PCM), it produces a 0-5 volt analog signal that measures temperature.

The 6.0L diesel uses 2 IAT sensors. For vehicles with a mass air flow (MAF) sensor, the IAT sensor is integrated into the MAF sensor. For vehicles without a MAF sensor, the IAT sensor is a stand alone sensor.

The IAT2 sensor or manifold air temperature (MAT) sensor is located in the intake manifold. The IAT2 sensor provides a manifold air temperature signal to the PCM.

Detection/Management:
The PCM compares the IAT2 signal to the IAT signal at initial key on. When the difference between the IAT2 and IAT is greater then the specified value for a given time, a diagnostic trouble code (DTC) is set.

AW1 PRELIMINARY DIAGNOSIS FOR DTC P2199

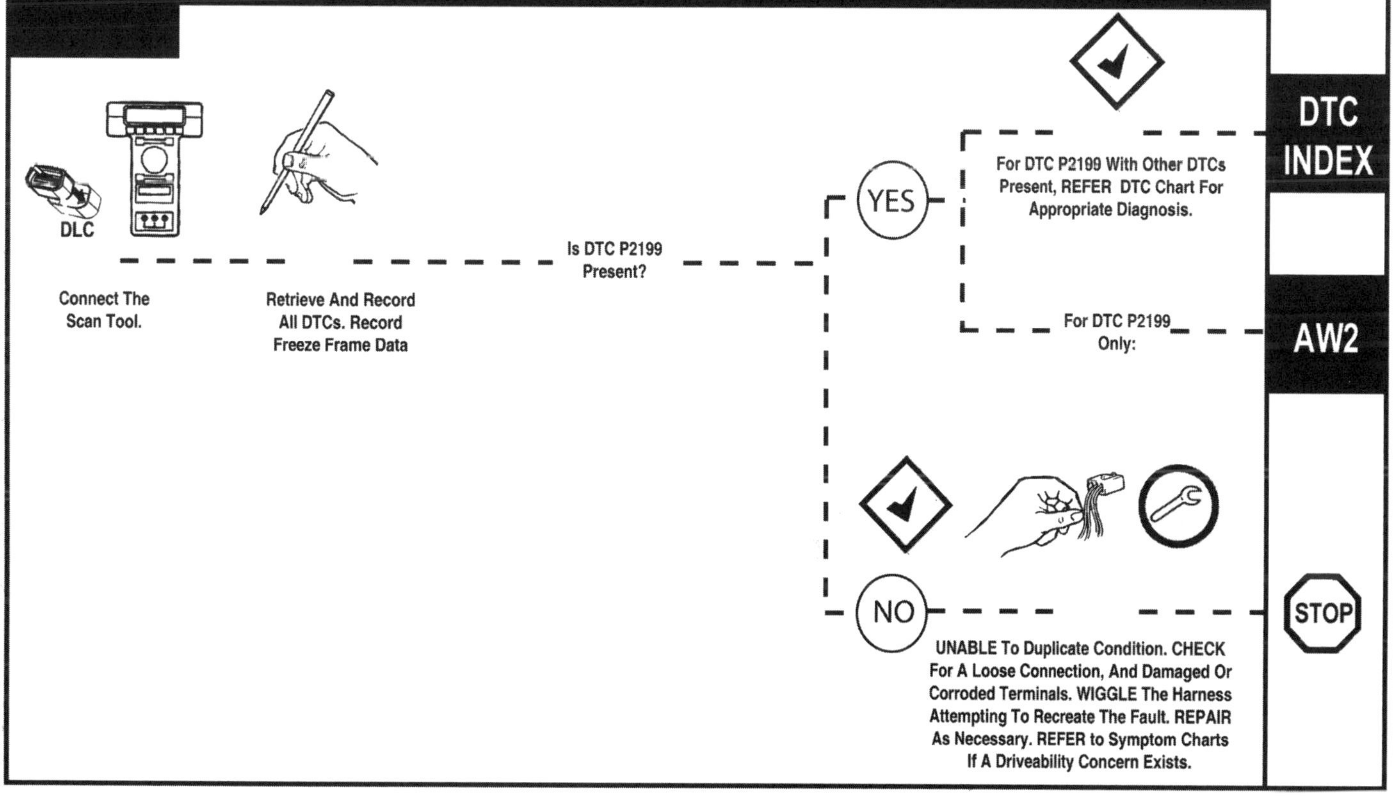

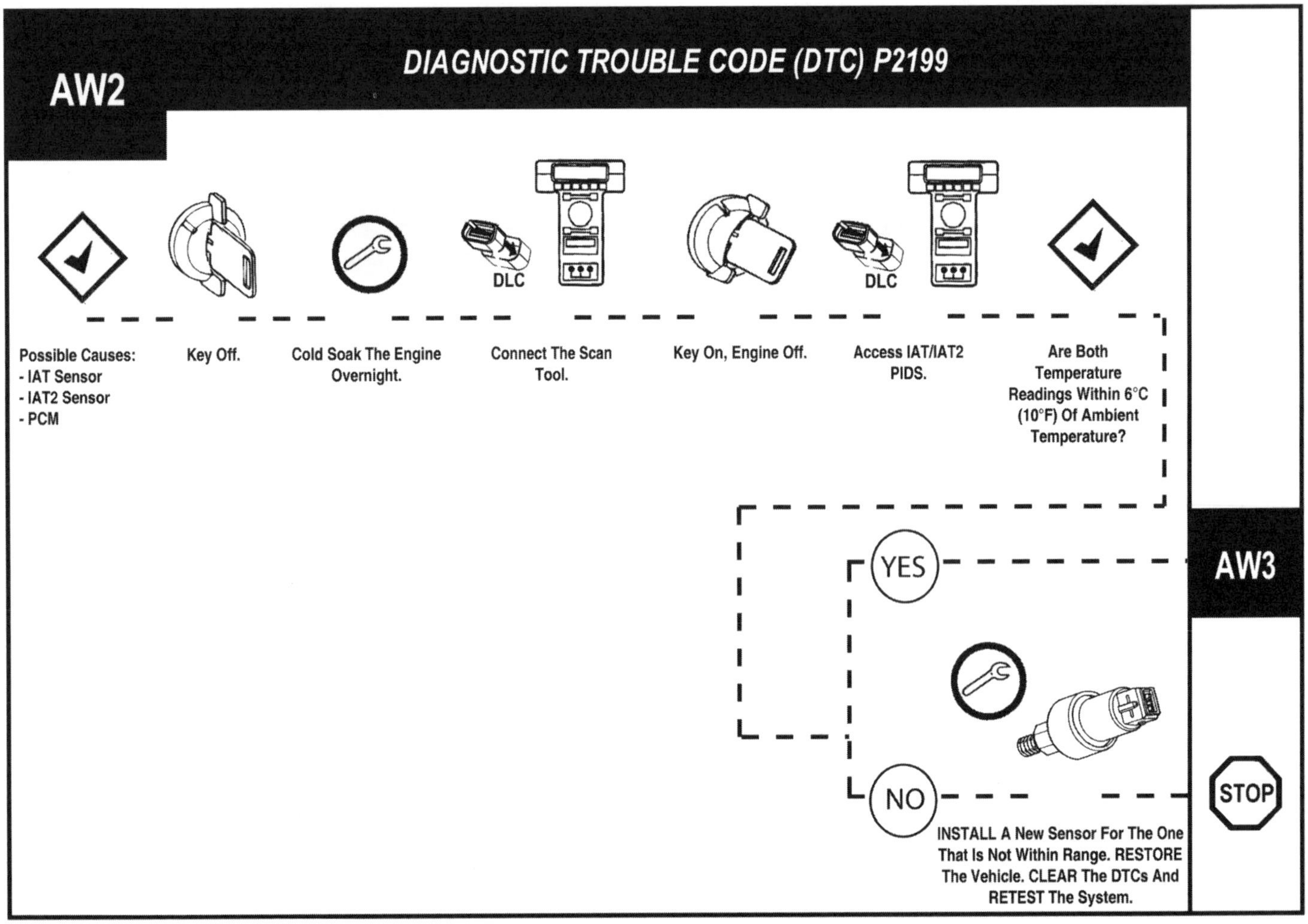
AW2
DIAGNOSTIC TROUBLE CODE (DTC) P2199
DLC
DLC
Possible Causes:
- IAT Sensor
- IAT2 Sensor
- PCM
Key Off.
Cold Soak The Engine Overnight.
Connect The Scan Tool.
Key On, Engine Off.
Access IAT/IAT2 PIDS.
Are Both Temperature Readings Within 6°C (10°F) Of Ambient Temperature?
YES
AW3
NO
INSTALL A New Sensor For The One That Is Not Within Range. RESTORE The Vehicle. CLEAR The DTCs And RETEST The System.
STOP

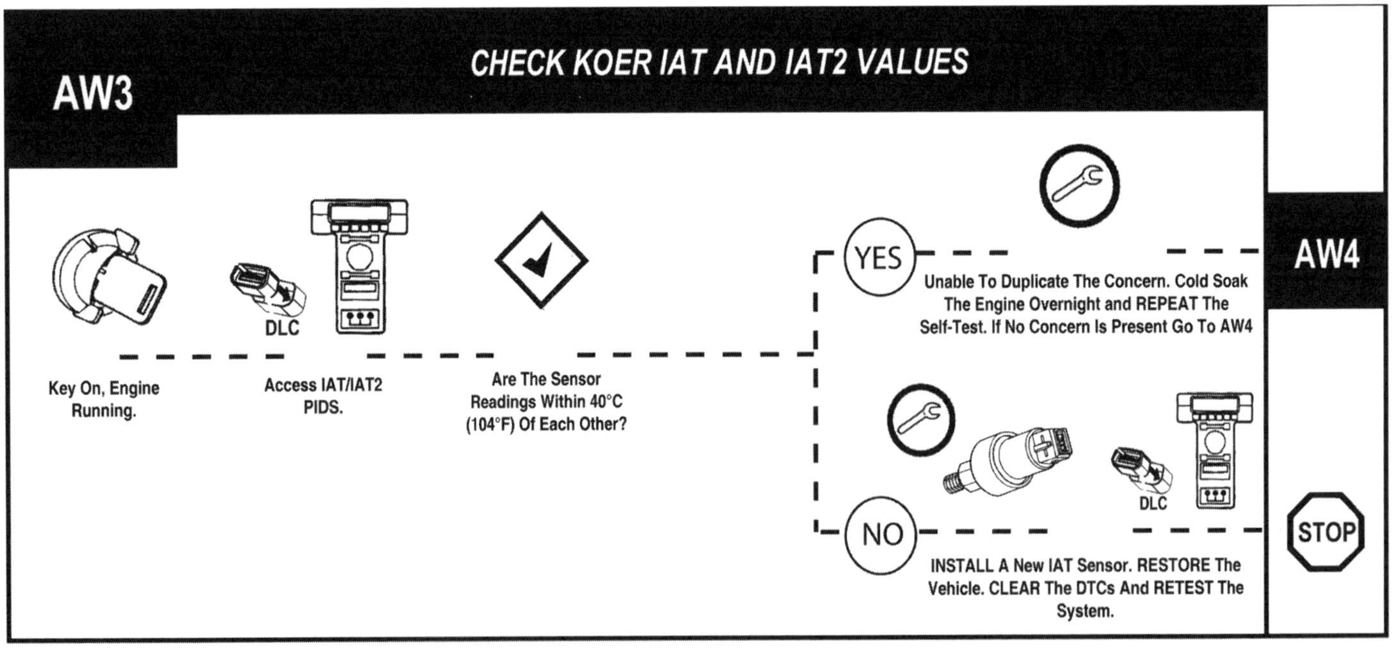
AW3
CHECK KOER IAT AND IAT2 VALUES
DLC
Key On, Engine Running.
Access IAT/IAT2 PIDS.
Are The Sensor Readings Within 40°C (104°F) Of Each Other?
YES
Unable To Duplicate The Concern. Cold Soak The Engine Overnight and REPEAT The Self-Test. If No Concern Is Present Go To AW4
AW4
DLC
NO
INSTALL A New IAT Sensor. RESTORE The Vehicle. CLEAR The DTCs And RETEST The System.
STOP

AW4 CHECK FOR CORRECT PCM OPERATION

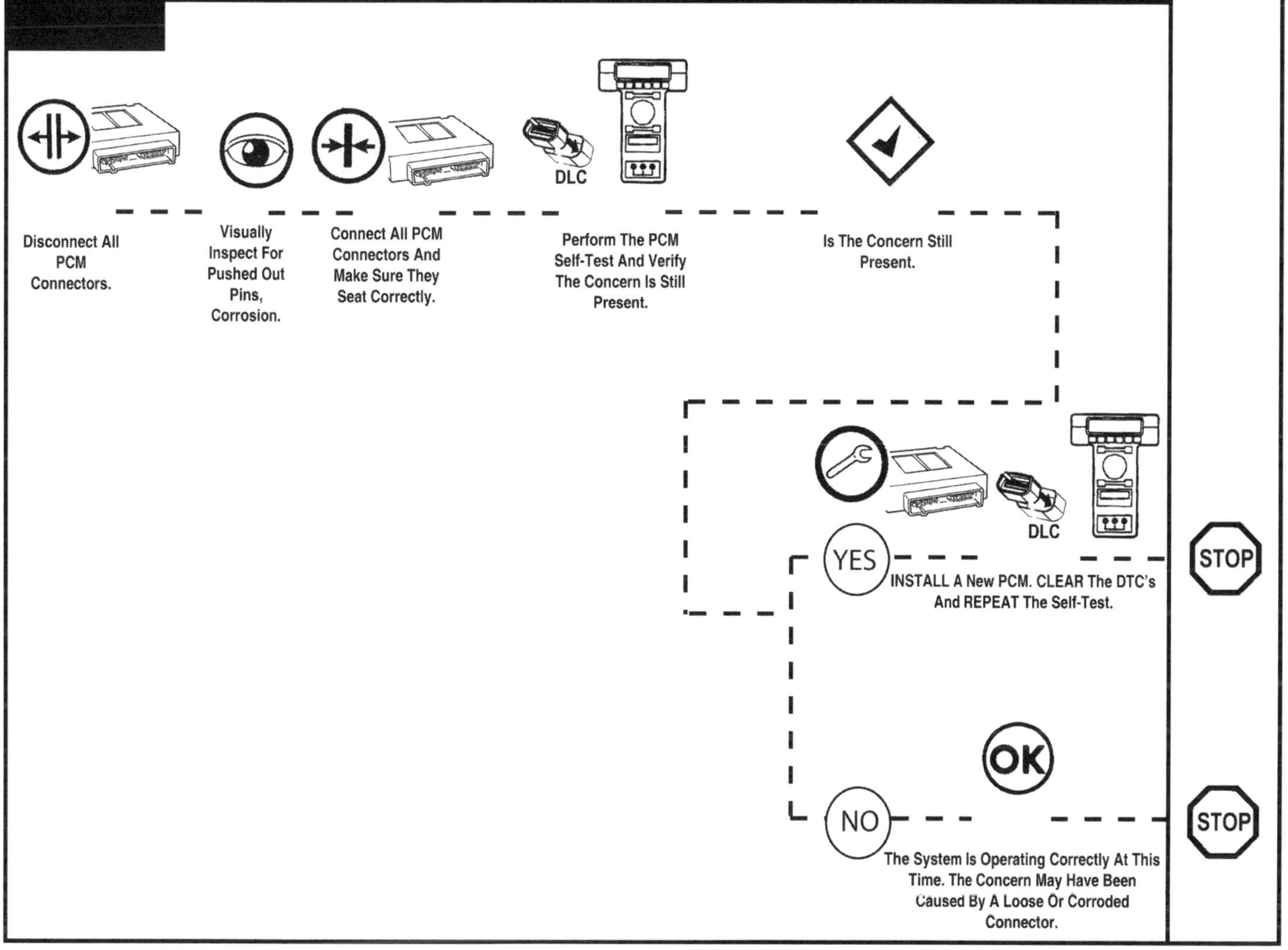

NOTES

Trouble Code (s)		Test
P0703	**Problem: Brake Pressure Applied (BPA) Switch**	AY 2006-07

REFERENCE LIST

This pinpoint test is intended to diagnose the following:

- BPA switch
- BPA harness circuits BPS and B+
- Powertrain control module (PCM) (12A650)

DIAGNOSTIC TROUBLE CODE INDEX			
Codes	KOEO	KOER	CM
P0703	AY1	AY1	AY1

P0703 = Brake Switch B Input Circuit

Signal Functions: The BPA switch provides a backup signal for the brake pedal position (BPP) switch. If the BPP switch signal is lost, the BPA switch provides the brakes-applied signal to deactivate the speed control system.

CAUTION!:
The powertrain control module (PCM) harness connectors must be properly seated and the connector latch properly attached to eliminate possible driveability concerns or a no start condition. Installing the PCM connectors on an angle may cause an improper connection, misdiagnosis and damaged components. Install the connector until the lever pivots and seats itself. Apply light pressure to get the connector into position on the PCM and then fully seat the connector.

Note: Visually inspect the harness connectors for corrosion, damage, proper mating and correct pin tension.

Note: When the PCM is disconnected additional DTCs will be set. Clear all DTCs after restoring the vehicle.

PINPOINT TEST SCHEMATICS AND CONNECTORS

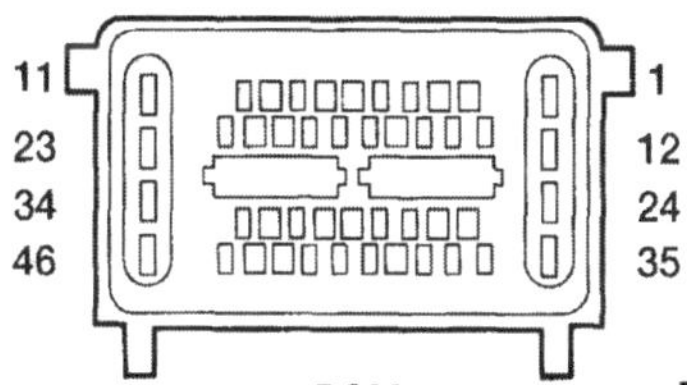

PCM
BODY HARNESS
CONNECTOR

Pin	Circuit
B28	BPS (Brake Pedal Switch)

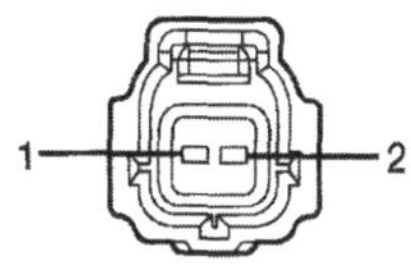

Vehicle	Connector	Pin	Circuit
E-Series	A	1	BPS
		2	B+
F-Super Duty	A	2	BPS
		1	B+

AY1 CHECK THE BPA SIGNAL

Note: Full Brake Pedal Application May Be Necessary To Correctly Operate The BPA Switch.

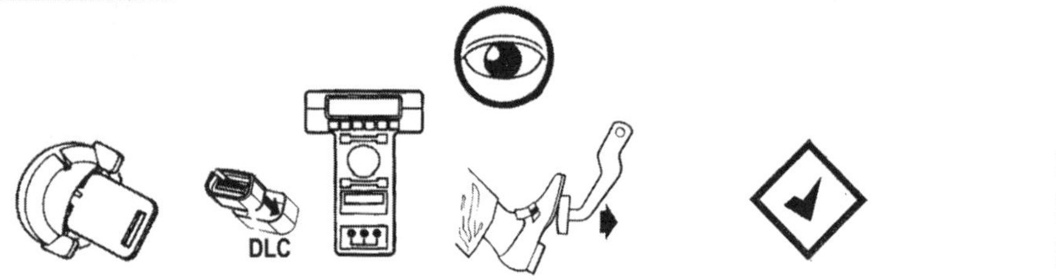

Key On, Engine Off.

Access The BPA PID.

Monitor The BPA PID While Applying And Releasing The Brake Pedal.

Does The BPA PID State Change When The Brake Pedal Is Fully Applied And Released?

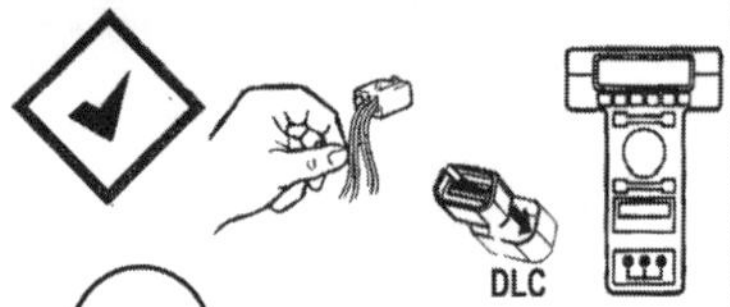

YES → STOP

UNABLE To Duplicate Condition. CHECK For Loose Connections, Damaged Or Corroded Pins. REPAIR As Necessary. REPEAT The Self-Test. REFER To Symptom Charts If A Drivability Concern Exists.

NO → AY2

AY2 CHECK THE BPA SIGNAL VOLTAGE

Note: Full Brake Pedal Application May Be Necessary To Correctly Operate The BPA Switch.

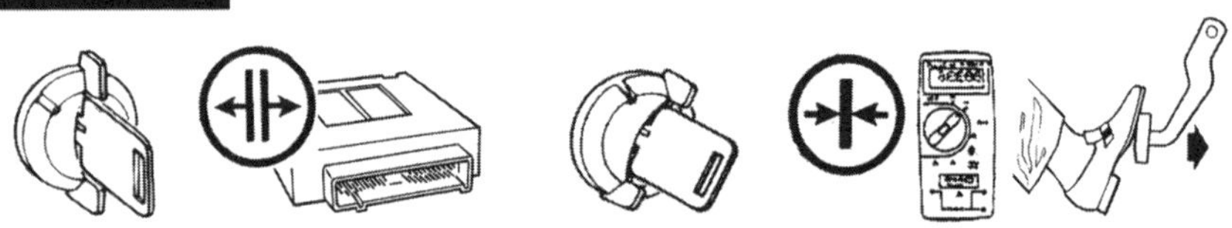

Key Off.

Disconnect The PCM.

Key On, Engine Off.

Measure the voltage between:	
(+) PCM Connector, Harness Side	(-)
BPS - Pin B28	Ground
Fully apply and release the brake pedal while measuring the voltage.	

10 Volts With Pedal Released

1 Volt With The Pedal Fully Applied

Is The Voltage Greater Than 10 Volts With The Pedal Released And Is The Voltage Less Than 1 Volt With The Pedal Fully Applied?

YES → AY5

NO → AY3

AY3 CHECK THE BPA SENSOR VOLTAGE

Key Off.

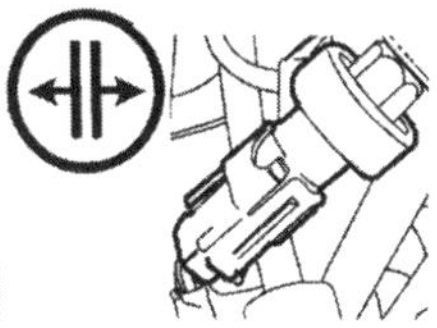

Disconnect The BPA Switch.

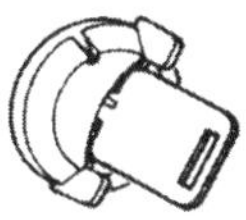

Key On, Engine Off.

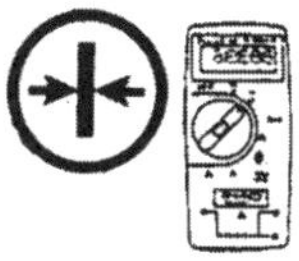

Measure the voltage between:	
(+) BPA Switch Connector, Harness Side	(-)
F-SERIES PIN 1	Ground
E-SERIES PIN 2	Ground

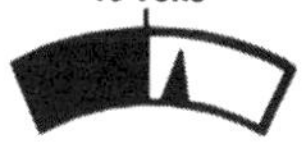

Is The Voltage Greater Than 10 Volts?

YES → AY4

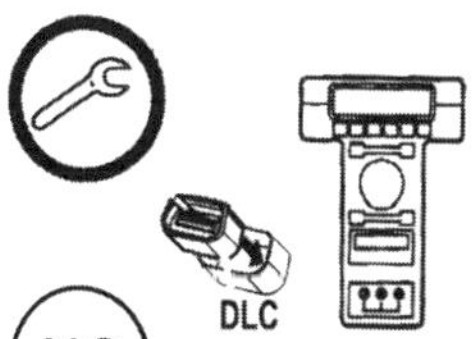

NO → REPAIR The Circuit. CLEAR The DTC's. Repeat The Self-Test. → STOP

AY4 CHECK THE BPA SWITCH

Note: Full Brake Application May Be Necessary To Correctly Operate The BPA Switch.

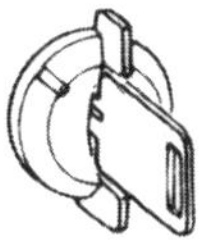

Key Off.

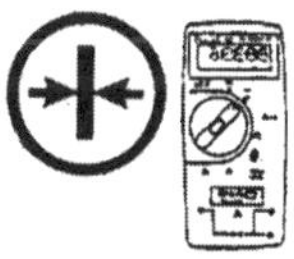

Measure the resistance between:	
(+) BPA Switch Connector, Component Side	(-) BPA Switch Connector, Component Side
B+	BPS
Fully apply and release the brake pedal while measuring the resistance.	

Is The Resistance Less Than 5 Ohms With The Pedal Released And Is The Resistance Greater Than 10,000 Ohms With The Pedal Fully Applied?

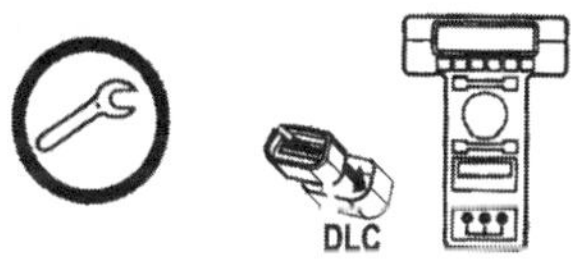

YES → REPAIR The Signal Circuit. CLEAR The DTC's. Repeat The Self-Test. → STOP

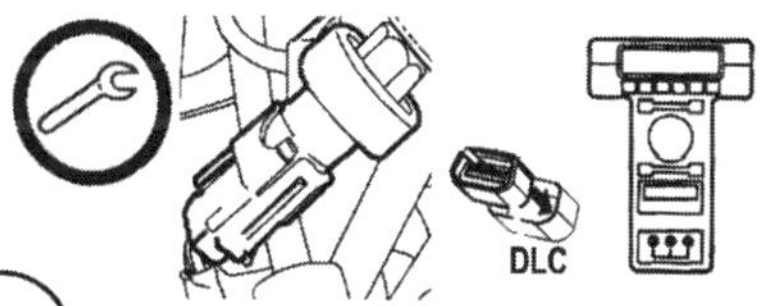

NO → INSTALL A New BPA Switch. CLEAR The DTC's. Repeat The Self-Test. → STOP

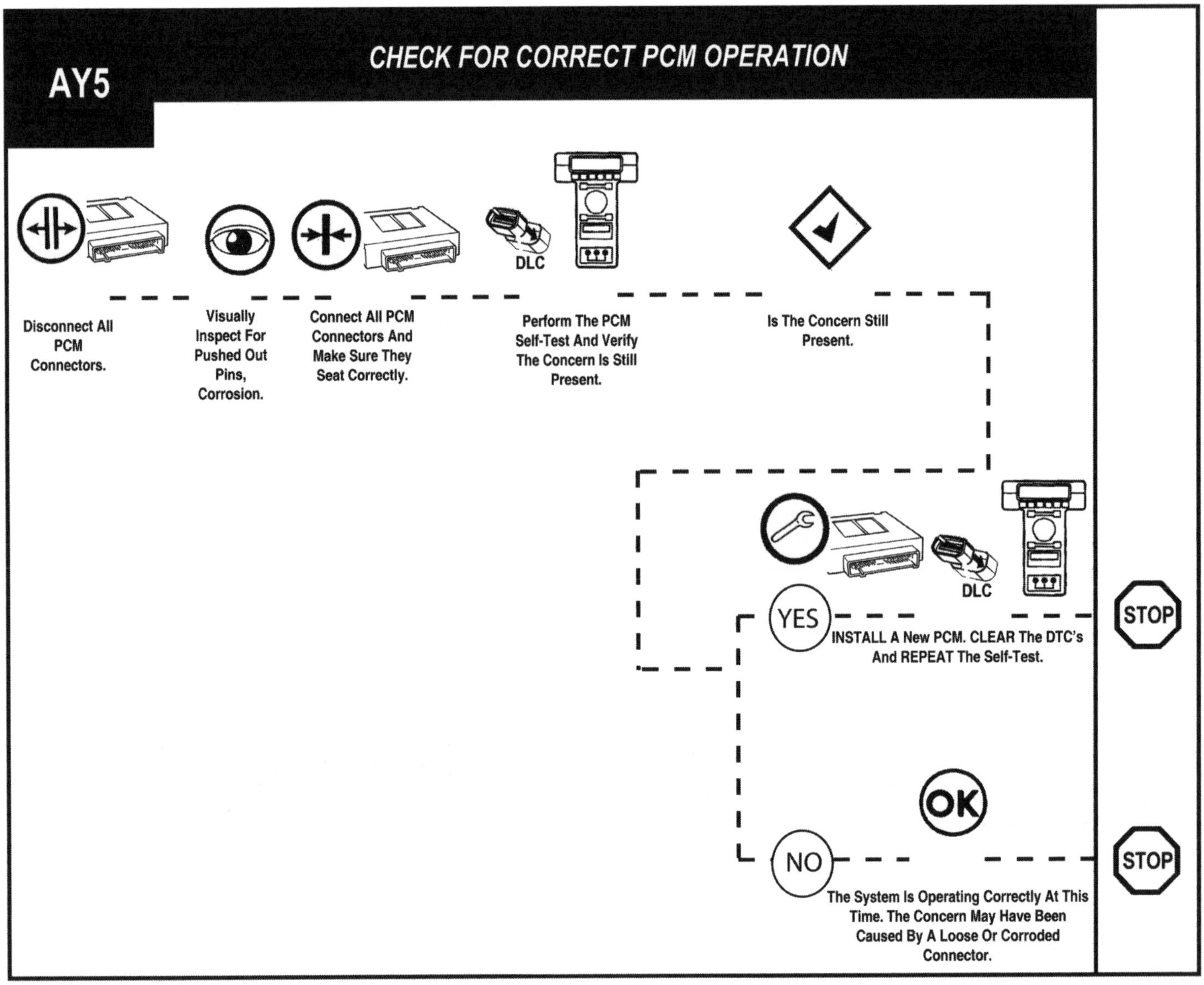
AY5
CHECK FOR CORRECT PCM OPERATION
Disconnect All PCM Connectors.
Visually Inspect For Pushed Out Pins, Corrosion.
Connect All PCM Connectors And Make Sure They Seat Correctly.
DLC
Perform The PCM Self-Test And Verify The Concern Is Still Present.
Is The Concern Still Present.
DLC
YES
INSTALL A New PCM. CLEAR The DTC's And REPEAT The Self-Test.
STOP
OK
NO
The System Is Operating Correctly At This Time. The Concern May Have Been Caused By A Loose Or Corroded Connector.
STOP

Trouble Code (s)	Problem: EGR Cooler System Performance	Test
P2457		AZ 2006-07

REFERENCE LIST

This pinpoint test is intended to diagnose the following:

- Engine overheating condition
- Restricted EGR cooler
- Charge air cooler (CAC) system

DIAGNOSTIC TROUBLE CODE INDEX			
Codes	KOEO	KOER	CM
P2457	AZ1	AZ1	AZ1

P2457 =Exhaust Gas Recirculation (EGR) Cooler System Performance

Fault Detection:

The low idle engine speed is between 600 and 750 RPM. The exhaust gas recirculation (EGR) valve position is greater than 0.08 volt. The mass fuel desired (MFDES) is between 4 and 16 mg/stk, at low idle for a minimum of 30 seconds. No exhaust pressure (EP), barometric pressure (BAP), manifold absolute pressure (MAP), pressure signal, intake air temperature (IAT), IAT2 temperature signal, visctronic drive fan (VDF) fan P0480, or EGR related faults. Fault condition: IAT2 exceeds a calibrated temperature for more than 30 seconds, based on IAT temperature.

AZ1 PRELIMINARY DIAGNOSIS

Carry Out A Visual Inspection.

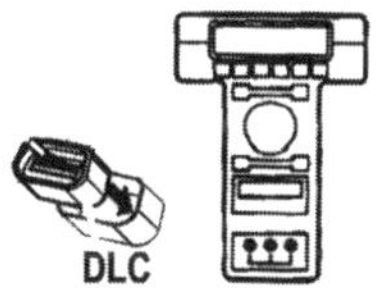

Retrieve And Record All DTCs. Record The Freeze Frame Data.

Are Any DTCs Other Than DTC P2457 Retrieved?

YES → DTC INDEX

NO → AZ2

AZ2 CHECK FOR ENGINE OVERHEATING CONDITION

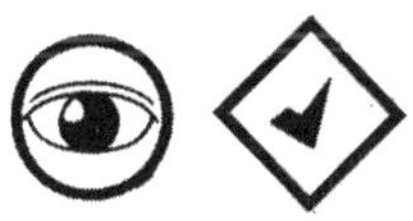

Check For Engine Overheating Condition Or Loss Of Coolant Condition.

Is The Engine Overheating Or Losing Coolant?

YES →

Diagnose And Repair Engine Cooling Stem. CLEAR The DTC's. REPEAT The Self-Test. → STOP

NO → AZ3

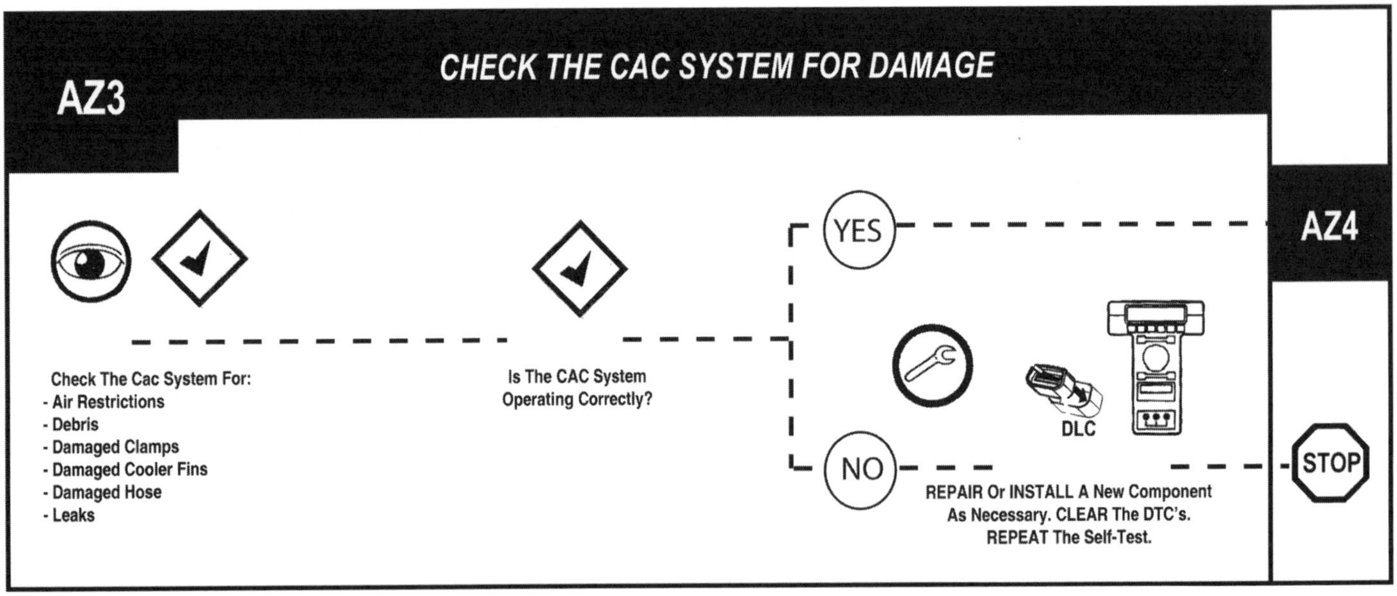
AZ3
CHECK THE CAC SYSTEM FOR DAMAGE
Check The Cac System For:
- Air Restrictions
- Debris
- Damaged Clamps
- Damaged Cooler Fins
- Damaged Hose
- Leaks
Is The CAC System Operating Correctly?
YES
AZ4
NO
DLC
REPAIR Or INSTALL A New Component As Necessary. CLEAR The DTC's. REPEAT The Self-Test.
STOP

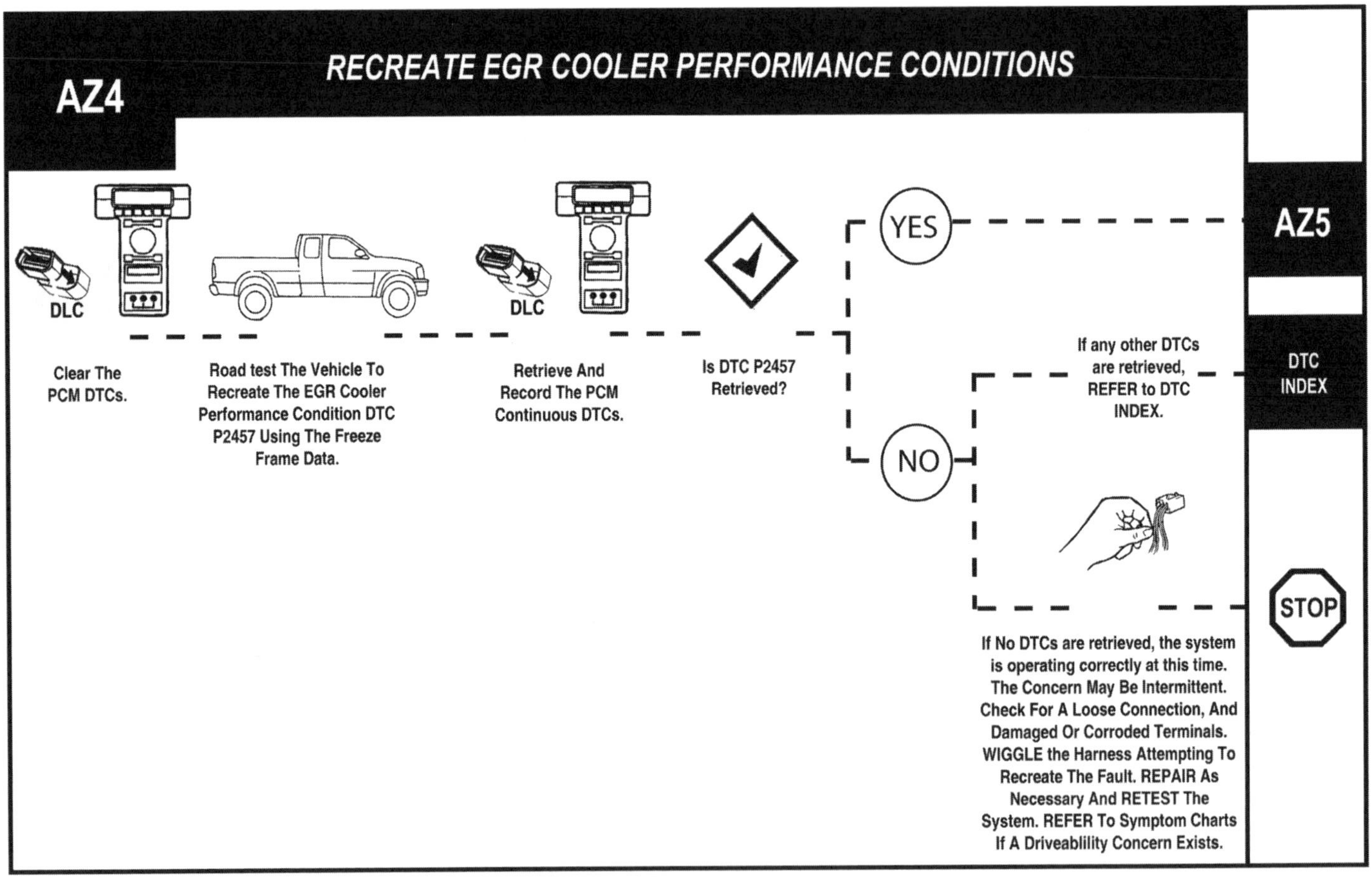
AZ4
RECREATE EGR COOLER PERFORMANCE CONDITIONS
DLC
Clear The PCM DTCs.
Road test The Vehicle To Recreate The EGR Cooler Performance Condition DTC P2457 Using The Freeze Frame Data.
DLC
Retrieve And Record The PCM Continuous DTCs.
Is DTC P2457 Retrieved?
YES
AZ5
NO
If any other DTCs are retrieved, REFER to DTC INDEX.
DTC INDEX
STOP
If No DTCs are retrieved, the system is operating correctly at this time. The Concern May Be Intermittent. Check For A Loose Connection, And Damaged Or Corroded Terminals. WIGGLE the Harness Attempting To Recreate The Fault. REPAIR As Necessary And RETEST The System. REFER To Symptom Charts If A Driveablility Concern Exists.

AZ5 CHECK THE EGR COOLER FOR RESTRICTIONS

Remove The EGR Cooler. Check the EGR Cooler Exhaust Port, Coolant Port And Coolant Hose For Restrictions.

Is The EGR Cooler Exhaust Port Restricted?

YES — DLC — REPAIR The Restricted EGR Cooler Exhaust Port Or INSTALL A New EGR Cooler As Necessary. CLEAR The DTCs. REPEAT The Self-Test. — STOP

NO — DLC — REPAIR The Restricted EGR Cooler Coolant Port Or Coolant Hose As Necessary. INSTALL A New EGR Cooler Or Coolant Hose As Necessary. CLEAR The DTCs. REPEAT The Self-Test. — STOP

NOTES

Trouble Code (s)		Test
NA	**Problem: Reference Voltage (VREF)**	B 2006-07

REFERENCE LIST

This pinpoint test is intended to diagnose only the following:

- Harness circuits: VREF and SIGRTN
- Accelerator pedal position (APP) sensor (9F836)
- Barometric pressure (BARO) sensor
- Exhaust gas recirculation (EGR) actuator
- Exhaust pressure (EP) sensor
- Injection control pressure (ICP) sensor
- Manifold absolute pressure (MAP) sensor (9F479)
- PCM

Note: Enter this pinpoint test only when check for VREF has failed in the sensor pinpoint tests for 3-wire sensors and actuators. Refer to Wiring Diagrams Manual for the specific sensors and pin locations connected to VREF.

Note: A VREF short to ground may cause a no-start condition. No diagnostic trouble codes (DTCs) may be present due to the inability of the powertrain control module (PCM) to communicate with the scan tool.

CAUTION!:
The powertrain control module (PCM) harness connectors must be properly seated and the connector latch properly attached to eliminate possible driveability concerns or a no start condition. Installing the PCM connectors on an angle may cause an improper connection, misdiagnosis and damaged components. Install the connector until the lever pivots and seats itself. Apply light pressure to get the connector into position on the PCM and then fully seat the connector.

Note: Visually inspect the harness connectors for corrosion, damage, proper mating and correct pin tension.

PINPOINT TEST SCHEMATICS AND CONNECTORS

Accelerator Pedal Position (APP) Sensor Connector

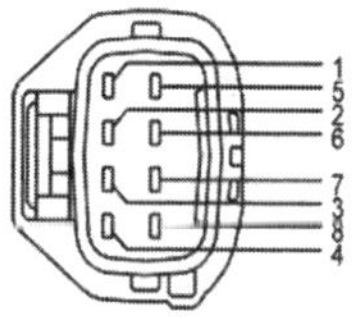

Pin	Circuit
1, 3	SIGRTN (Signal Return)
6, 7	VREF (Reference Voltage)

Barometric Pressure (BARO) Sensor Connector

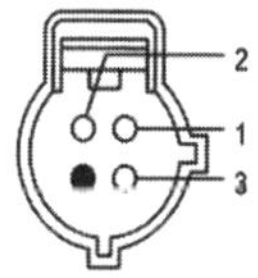

Pin	Circuit
1	SIGRTN (Signal Return)
2	VREF (Reference Voltage)

EGR Actuator Connector

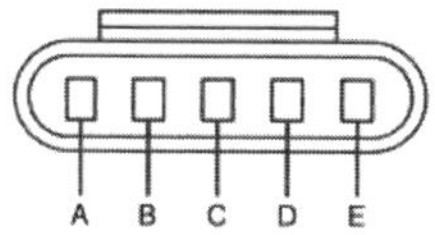

Pin	Circuit
B	SIGRTN (Signal Return)
D	VREF (Reference Voltage)

Exhaust Pressure (EP) Sensor Connector

Pin	Circuit
1	SIGRTN (Signal Return)
2	VREF (Reference Voltage)

Injection Control Pressure (ICP) Sensor Connector

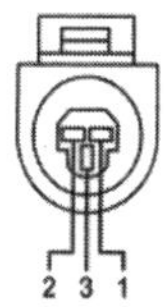

Pin	Circuit
1	SIGRTN (Signal Return)
2	VREF (Reference Voltage)

Manifold Absolute Pressure (MAP) Sensor Connector

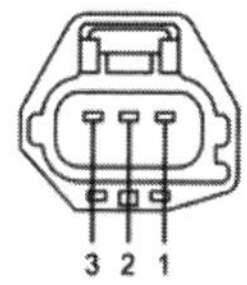

Pin	Circuit
3	SIGRTN (Signal Return)
1	VREF (Reference Voltage)

Trouble Code (s)	Problem: Reference Voltage (VREF)	Test
NA		B 2006-07

PINPOINT TEST SCHEMATICS AND CONNECTORS (CONTINUED)

Powertrain Control Module (PCM) Connector

Pin	Circuit
B20, B33, E25	SIGRTN (Signal Return)
B34, B46	VPWR (Vehicle Power)
B29, B45, E36	VREF (Reference Voltage)

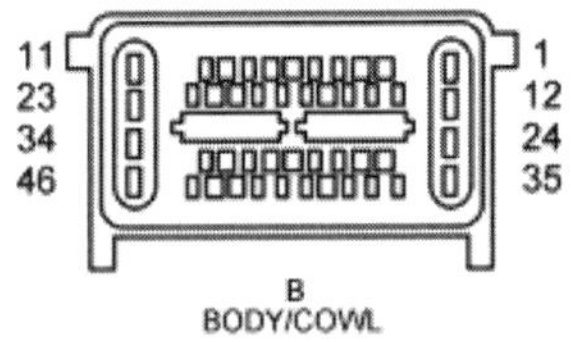

B1 CHECK FOR VOLTAGE BETWEEN THE VREF AND SIGRTN CIRCUITS

Note: More Than One VREF Circuit May Be Present, All VREF Circuits Must Be Checked.

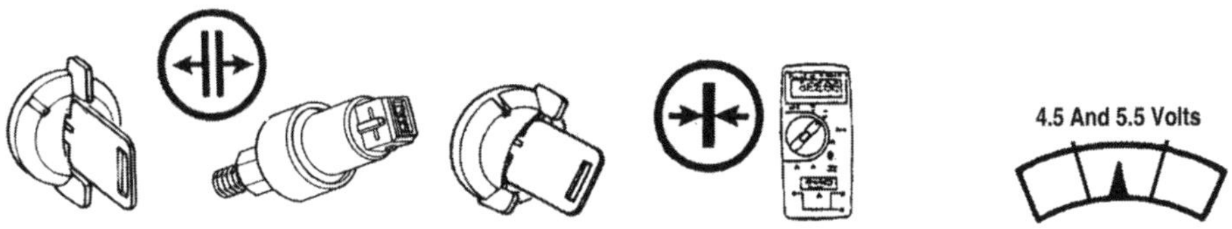

Key Off.

Disconnect The Suspect Sensor.

Key On. Engine Off.

Measure the voltage between:	
(+) Suspect Sensor Connector, Harness Side	(-) Suspect Sensor Connector, Harness Side
VREF	SIGRTN

Is The Voltage Between 4.5 And 5.5 Volts?

YES → B15

NO → B2

B2 — CHECK THE VREF CIRCUIT FOR VOLTAGE

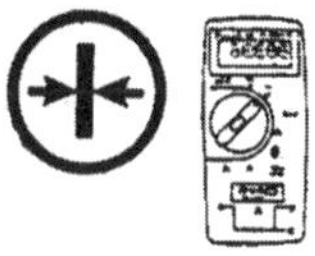

4.5 - 5.5 V

Is The Voltage Between 4.5 - 5.5 V?

Measure the voltage between:	
(+) Suspect Sensor Connector, Harness Side	(-)
VREF	Ground

YES — B14

NO — B3

B3 — CHECK THE VREF CIRCUIT VOLTAGE WITH ALL SENSORS DISCONNECTED

Note: More Than One VREF Circuit May Be Present, All VREF Circuits Must Be Checked.

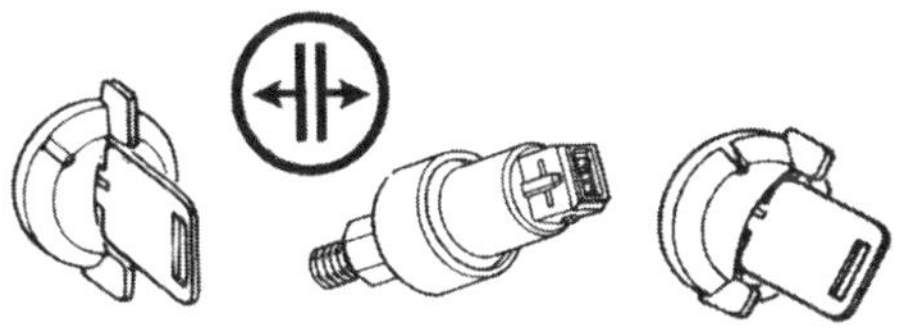

Key Off.

Disconnect All Sensors Connected To The VREF Circuit.

Key On. Engine Off.

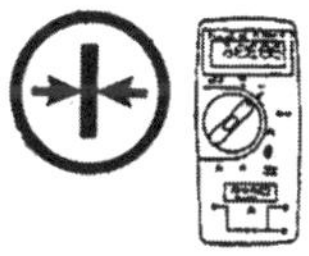

Measure the voltage at the sensor disconnected in step B1.

Measure the voltage between:	
(+) Suspect Sensor Connector, Harness Side	(-)
VREF	Ground

4.5 And 5.5 Volts

Is The Voltage Between 4.5 And 5.5 Volts?

YES — B8

NO — B4

B4 CHECK THE VREF CIRCUIT FOR AN OPEN

Note: When measuring the resistance of the VREF circuit, only measure the VREF circuit that corresponds to the suspect sensor.

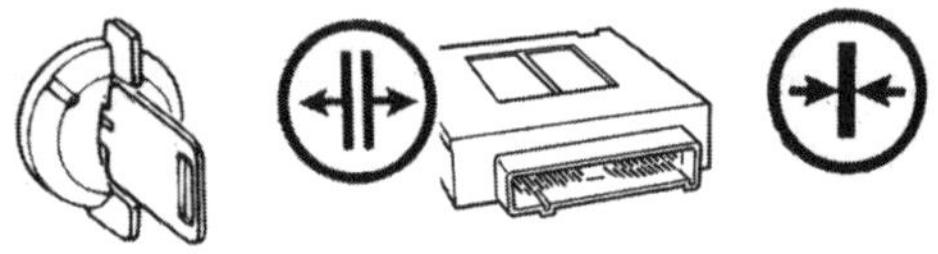

Key Off.

Disconnect The PCM Harness Connector.

Measure the resistance between:	
(+) PCM Connector, Harness Side	(-) Suspect Sensor Connector, Harness Side
VREF - Pin B29, B45, E36	VREF

Is The Resistance Less Than 5.0 Ohms?

YES → B5

NO → STOP
REPAIR The Circuit. CLEAR The DTC's. Repeat The Self-Test.

B5 CHECK THE VREF CIRCUIT FOR A SHORT TO GROUND

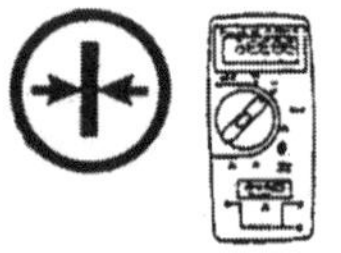

Measure the resistance between:	
(+) Suspect Sensor Connector, Harness Side	(-)
VREF	Ground

Measure the resistance between:	
(+) Suspect Sensor Connector, Harness Side	(-) Suspect Sensor Connector, Harness Side
VREF	SIGRTN

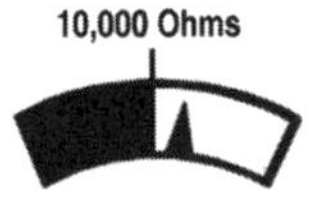

Is The Resistance Greater Than 10,000 Ohms?

YES → B6

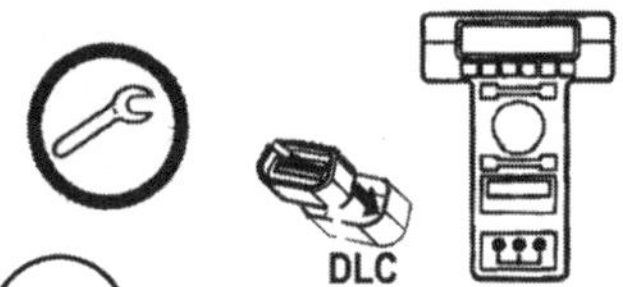

NO → STOP
REPAIR The Circuit. Clear The DTCs. Repeat The Self-Test.

B6 CHECK THE VREF CIRCUIT FOR A SHORT TO VPWR

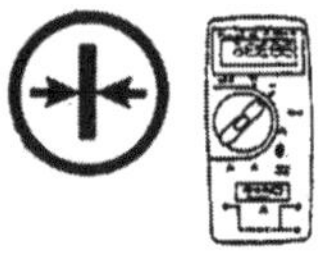

Measure the resistance between:	
(+) PCM Connector, Harness Side	(-) PCM Connector, Harness Side
VREF - Pin B29, B45, E36	VPWR - Pin B34, B46

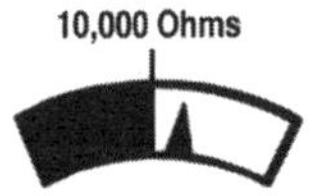

Is The Resistance Greater Than 10,000 Ohms?

YES → B7

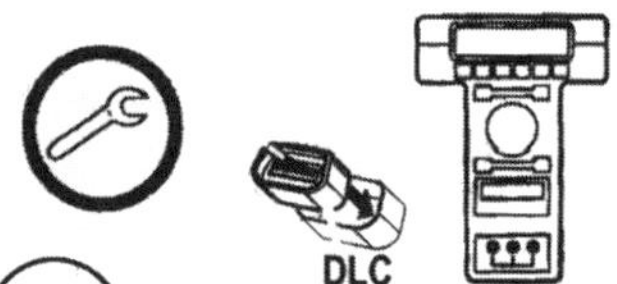

NO → STOP
REPAIR The Circuit. Clear The DTCs. Repeat The Self-Test.

B7 CHECK THE VREF CIRCUIT FOR A SHORT TO VOLTAGE

Key On.
Engine Off.

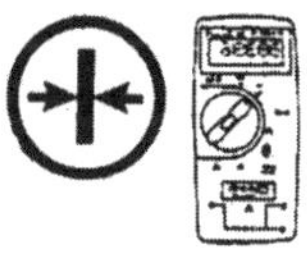

Measure the voltage between:	
(+) Suspect Sensor Connector, Harness Side	(-)
VREF	Ground

Is Any Voltage indicated?

YES → 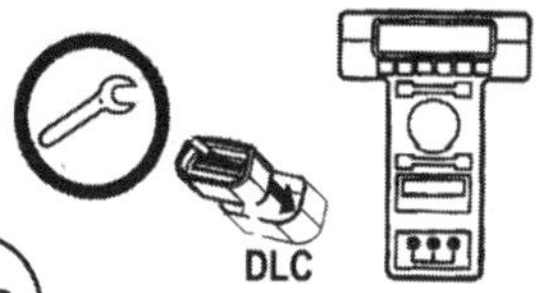

REPAIR The Circuit. CLEAR The DTC's. Repeat The Self-Test. → STOP

NO → B16

B8 CHECK THE VREF VOLTAGE WITH THE MAP SENSOR CONNECTED

Note: If this sensor was used for the VREF measurement in Step B3, GO to B9 .

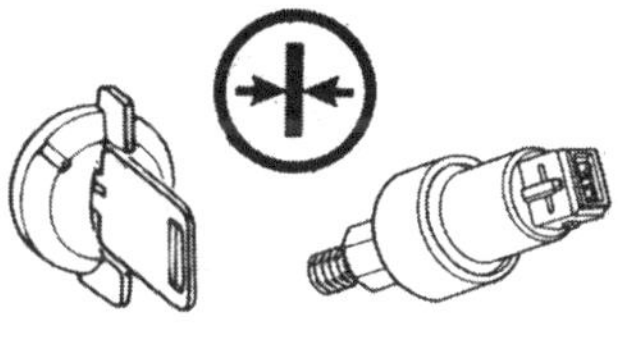

Key Off.

MAP Sensor Connector Connected.

Key On.
Engine Off.

Measure the voltage between:	
(+) Suspect Sensor Connector, Harness Side	( )
VREF	Ground

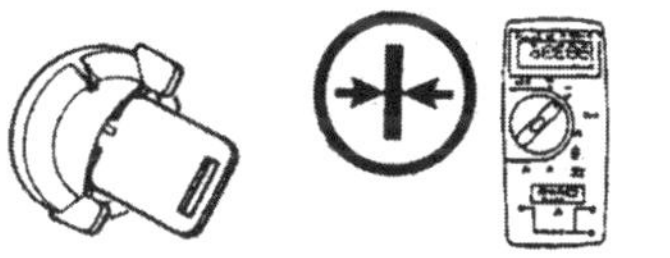

Is The Voltage Between 4.5 And 5.5 Volts?

YES → B9

NO → 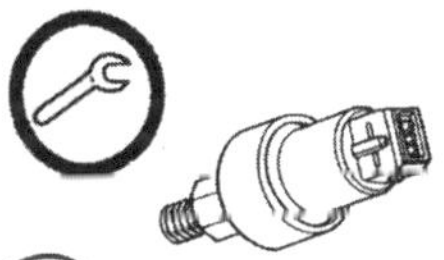

INSTALL A New MAP Sensor. CLEAR The DTC's. REPEAT The Self-Test. → STOP

B9 CHECK THE VREF VOLTAGE WITH THE BARO SENSOR CONNECTED

Note: If this sensor was used for the VREF measurement in Step B3, GO to B10 .

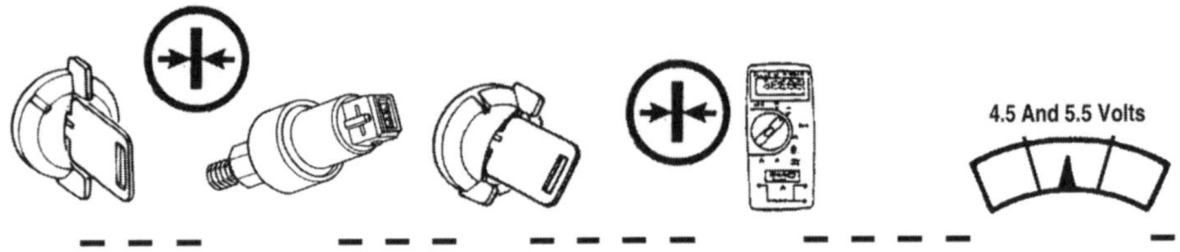

Key Off.

BARO Sensor Connector Connected.

Key On. Engine Off.

Measure the voltage between:	
(+) Suspect Sensor Connector, Harness Side	(-)
VREF	Ground

Is The Voltage Between 4.5 And 5.5 Volts?

YES → B10

NO →

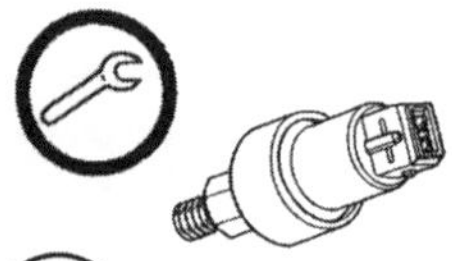

INSTALL A New BARO Sensor. CLEAR The DTC's. REPEAT The Self-Test.

STOP

B10 CHECK THE VREF VOLTAGE WITH THE EGR ACTUATOR CONNECTED

Note: If this sensor was used for the VREF measurement in Step B3, GO to B11 .

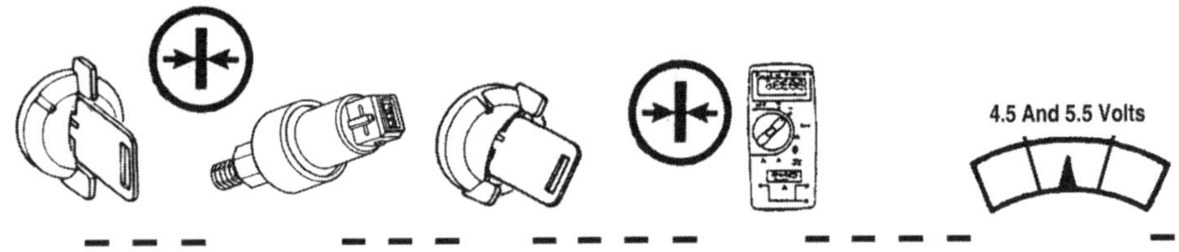

Key Off.

EGR Actuator Connector Connected.

Key On. Engine Off.

Measure the voltage between:	
(+) Suspect Sensor Connector, Harness Side	(-)
VREF	Ground

Is The Voltage Between 4.5 And 5.5 Volts?

YES → B11

NO →

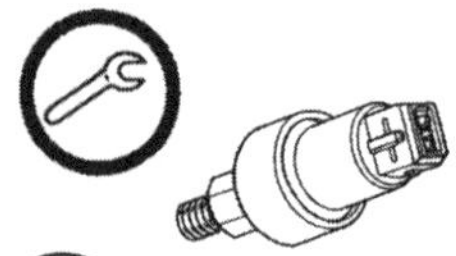

INSTALL A New EGR Actuator . CLEAR The DTC's. REPEAT The Self-Test.

STOP

B11 CHECK THE VREF VOLTAGE WITH THE ICP SENSOR CONNECTED

Note: If this sensor was used for the VREF measurement in Step B3, GO to B12 .

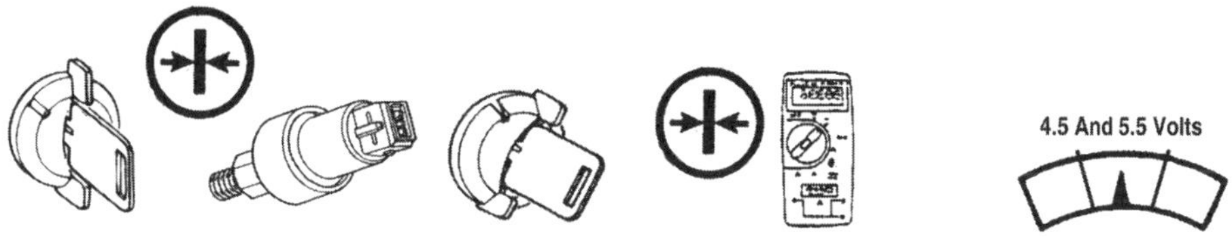

Key Off.

ICP Sensor Connector Connected.

Key On. Engine Off.

Measure the voltage between:	
(+) Suspect Sensor Connector, Harness Side	(-)
VREF	Ground

Is The Voltage Between 4.5 And 5.5 Volts?

YES → B12

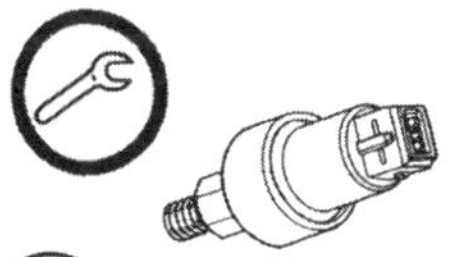

NO → INSTALL A New ICP Sensor. CLEAR The DTC's. REPEAT The Self-Test. STOP

B12 CHECK THE VREF VOLTAGE WITH THE APP SENSOR CONNECTED

Note: If this sensor was used for the VREF measurement in Step B3, GO to B13

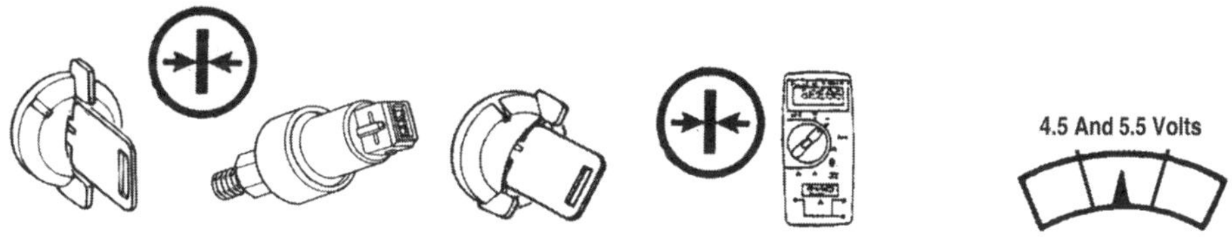

Key Off.

APP Sensor Connector Connected.

Key On. Engine Off.

Measure the voltage between:	
(+) Suspect Sensor Connector, Harness Side	(-)
VREF	Ground

Is The Voltage Between 4.5 And 5.5 Volts?

YES → B13

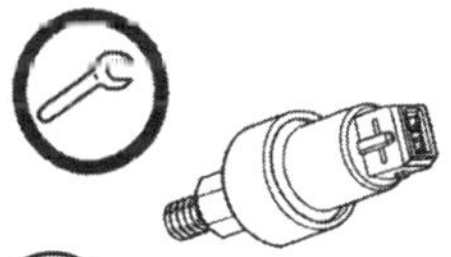

NO → INSTALL A New APP Sensor. CLEAR The DTC's. REPEAT The Self-Test. STOP

B13 CHECK THE VREF VOLTAGE WITH THE EP SENSOR CONNECTED

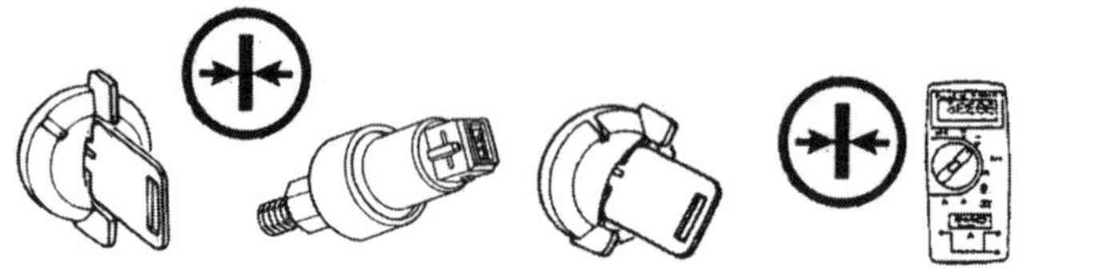

Key Off.

EP Sensor Connector Connected.

Key On. Engine Off.

Measure the voltage between:	
(+) Suspect Sensor Connector, Harness Side	(-)
VREF	Ground

4.5 And 5.5 Volts

Is The Voltage Between 4.5 And 5.5 Volts?

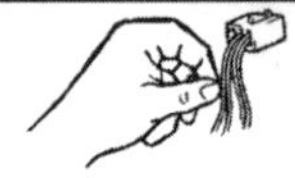

YES → STOP

UNABLE To Duplicate Concern. Check For A Loose Connection, And Damaged Or Corroded Terminals. WIGGLE the Harness Attempting To Recreate The Fault. REPAIR As Necessary And RETEST The System. REFER To Symptom Charts If A Driveablility Concern Exists.

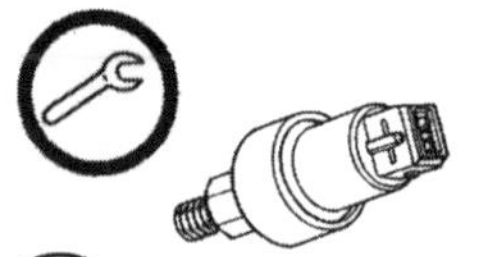

NO → STOP

INSTALL A New EP Sensor. CLEAR The DTC's. REPEAT The Self-Test.

B14 CHECK THE SIG RTN CIRCUIT FOR AN OPEN

NOTE: When measuring the resistance of the SIGRTN circuit, only measure the SIGRTN circuit that corresponds to the suspect sensor.

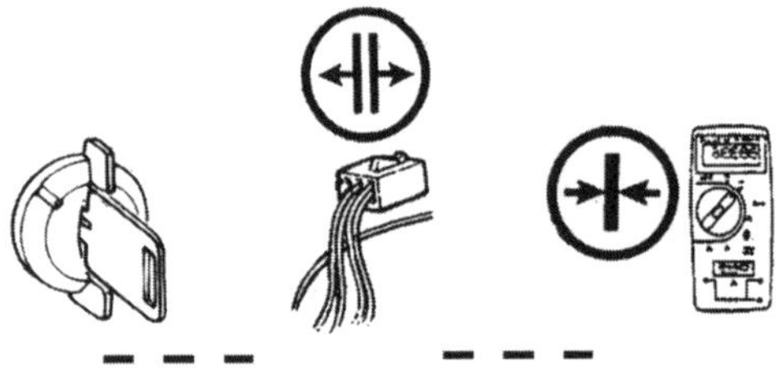

Key Off.

Disconnect The PCM Harness Connector.

Measure the resistance between:	
(+) Suspect Sensor Connector, Harness Side	(-) PCM Connector, Harness Side
SIGRTN	SIGRTN - Pin B20, B33, E25

5 Ohms

Is The Resistance Less Than 5 Ohms?

YES → B16

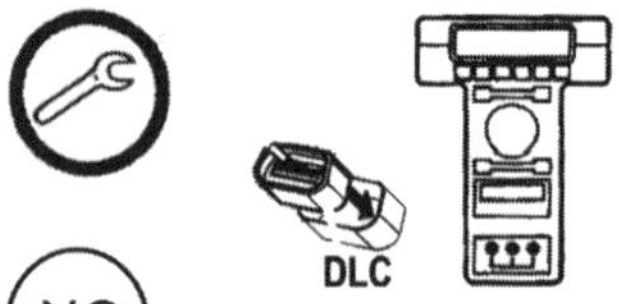

NO → STOP

REPAIR The Circuit. CLEAR DTCs And RPEAT The Self-Test.

B15 CHECK THE SUSPECT SENSOR FOR AN INTERNAL SHORT

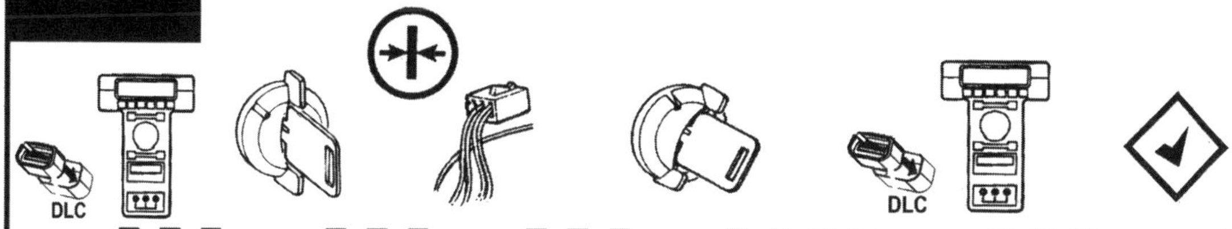

Clear The KOEO, KOER, And Continuous DTCs.

Key OFF.

Suspect Sensor Connector Connected.

Key ON. Engine OFF.

Carry Out The PCM Self-Test.

Is The Concern Still Present?

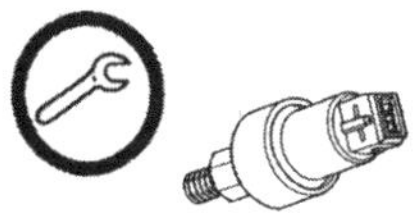

YES — INSTALL A New Sensor For The Sensor In Question. CLEAR The DTC's. REPEAT The Self-Test. STOP

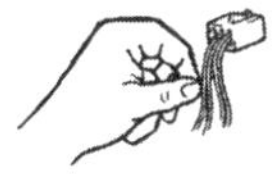

NO — UNABLE To Duplicate Concern. Check For A Loose Connection, And Damaged Or Corroded Terminals. WIGGLE the Harness Attempting To Recreate The Fault. REPAIR As Necessary And RETEST The System. REFER To Symptom Charts If A Driveablility Concern Exists. STOP

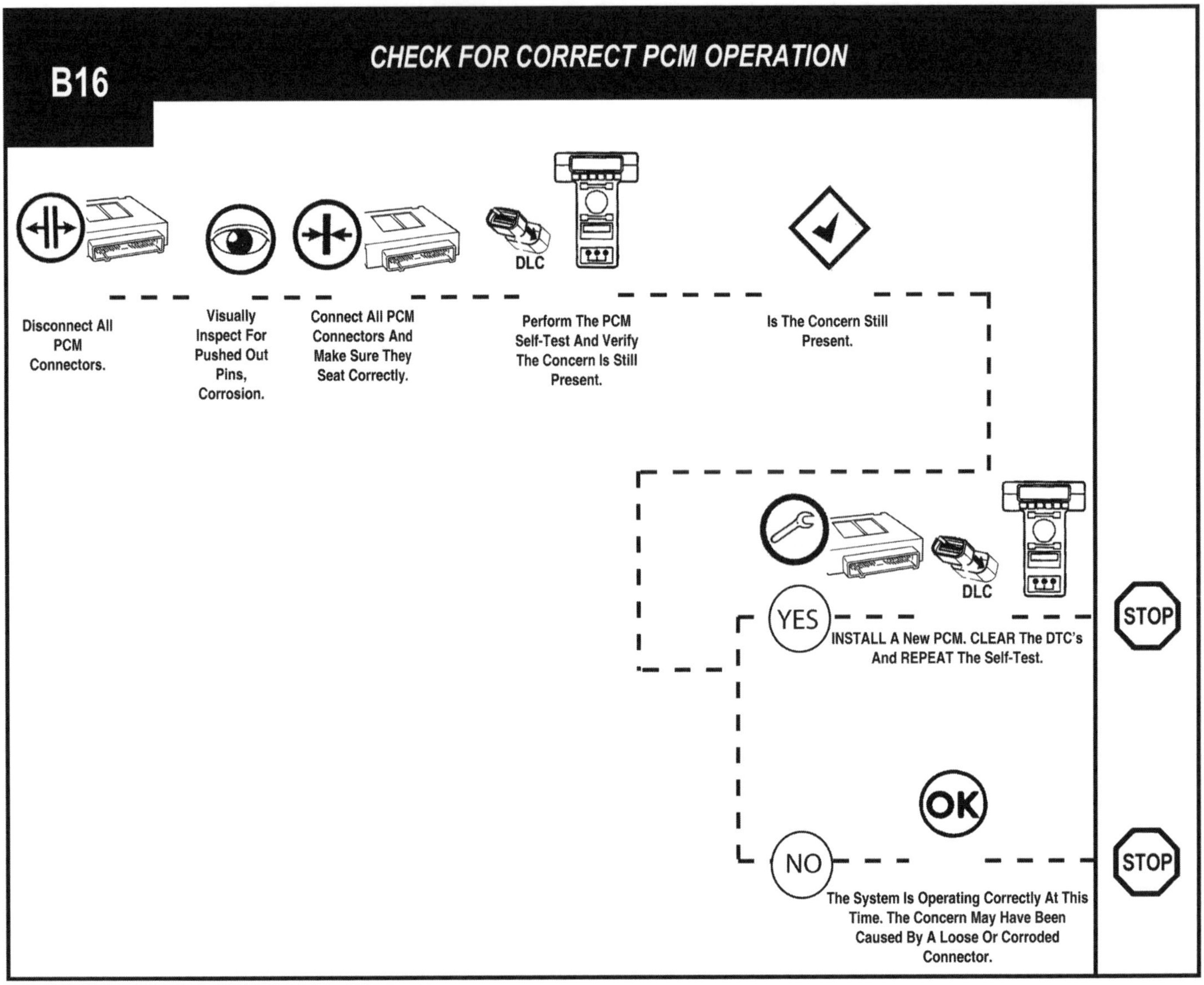
B16
CHECK FOR CORRECT PCM OPERATION
Disconnect All PCM Connectors.
Visually Inspect For Pushed Out Pins, Corrosion.
Connect All PCM Connectors And Make Sure They Seat Correctly.
DLC
Perform The PCM Self-Test And Verify The Concern Is Still Present.
Is The Concern Still Present.
DLC
YES
INSTALL A New PCM. CLEAR The DTC's And REPEAT The Self-Test.
STOP
OK
NO
The System Is Operating Correctly At This Time. The Concern May Have Been Caused By A Loose Or Corroded Connector.
STOP

Trouble Code (s)	Problem: Clutch Pedal Position (CPP) Switch	Test
P0830, P0833		C 2006-07

REFERENCE LIST

This pinpoint test is intended to diagnose the following:

- Damaged CPP top of travel switch
- Damaged CPP top of travel switch
- CPP harness circuits: CPP top of travel, CPP bottom of travel, VPWR and GND
- Powertrain control module (PCM) (12A650)

DIAGNOSTIC TROUBLE CODE INDEX			
Codes	KOEO	KOER	CM
P0830	C1	C1	C1
P0833	C1	C1	C1

P0830 = Clutch Switch A Circuit
P0833 = Clutch Switch B Circuit

Signal Functions:
The CPP switches indicate the state of the clutch pedal. CPP switch A is a normally open switch that indicates clutch disengagement (bottom of travel). CPP switch B is a normally closed switch that indicates clutch engagement (top of travel). The application ad release of the clutch pedal disables the speed control system, the cold idle kicker, or the power take-off (PTO) system, if enabled.

CAUTION!:
The powertrain control module (PCM) harness connectors must be properly seated and the connector latch properly attached to eliminate possible driveability concerns or a no start condition. Installing the PCM connectors on an angle may cause an improper connection, misdiagnosis and damaged components. Install the connector until the lever pivots and seats itself. Apply light pressure to get the connector into position on the PCM and then fully seat the connector.

Note: Visually inspect the harness connectors for corrosion, damage, proper mating and correct pin tension.

PINPOINT TEST SCHEMATICS AND CONNECTORS

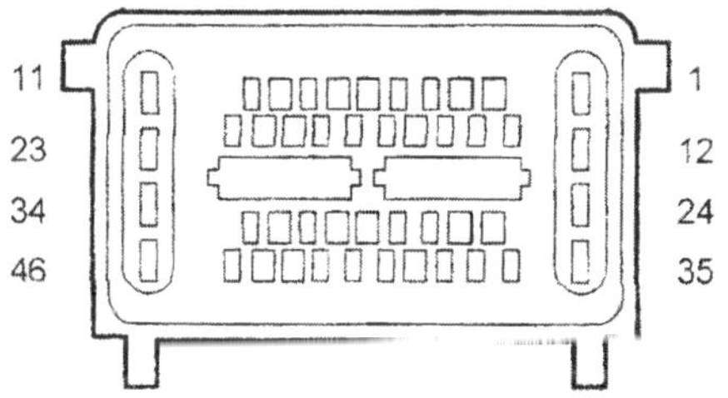

PCM
BODY HARNESS
CONNECTOR

Pin	Circuit
B36	CPP-TT (Clutch Pedal Position - Top of Travel)
B27	CPP-BT (Clutch Pedal Position - Bottom of Travel)

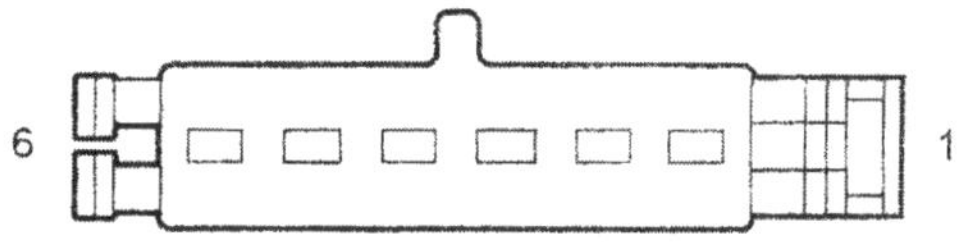

CLUTCH PEDAL POSITION
SWITCH

Pin	Circuit
3	VPWR (Vehicle Power)
4	CPP-TT (Clutch Pedal Position - Top of Travel)
5	CPP-BT (Clutch Pedal Position - Bottom of Travel)
6	GND (Ground)

C1 — PRELIMINARY DIAGNOSIS FOR DTCS P0830 OR P0833

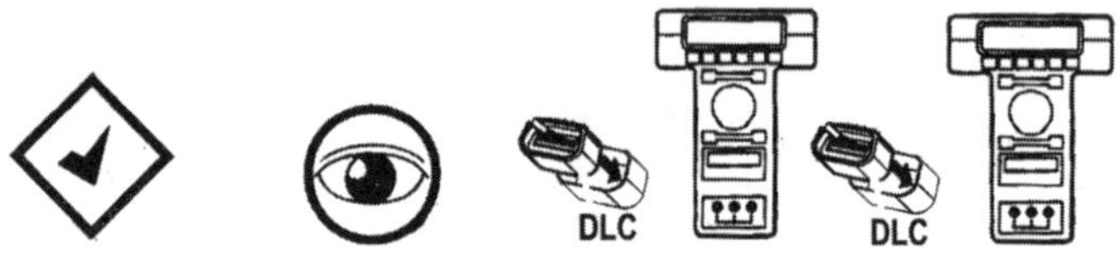

DTC's P0830 And P0833 Are Set If The PCM Does Not Detect A CPP Switch Change Of State During The Key On Engine Running (KOER) Self-test.
Possible Causes:
- Clutch Pedal Not Fully Applied And Released During The Seft Test.
- Signal Circuit Open Or Shorted
  - CPP switch
- PCM

Carry Out A Visual Inspection.

Retrive And Record Any Continuous And On-demand DTC's.

Clear All DTC's.

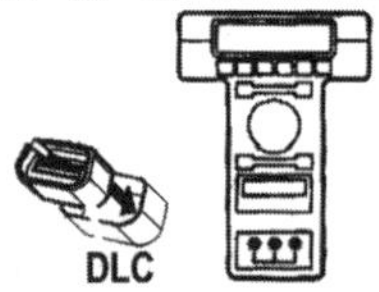

Carry Out The KOER Self-Test.

Is DTC P0830 Or P0833 Present?

YES — DTC P0830 → C2

YES — DTC P0833 → C7

NO → STOP

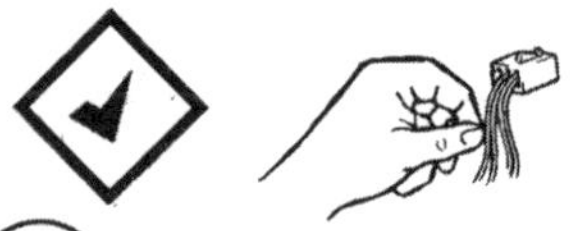

UNABLE To Duplicate Concern. ROAD TEST Vehicle, Check For A Loose Connection, And Damaged Or Corroded Terminals. WIGGLE the Harness Attempting To Recreate The Fault. REPAIR As Necessary And RETEST The System. REFER To Symptom Charts If A Driveability Concern Exists.

C2 — CHECK THE CPP SWITCH FOR VOLTAGE

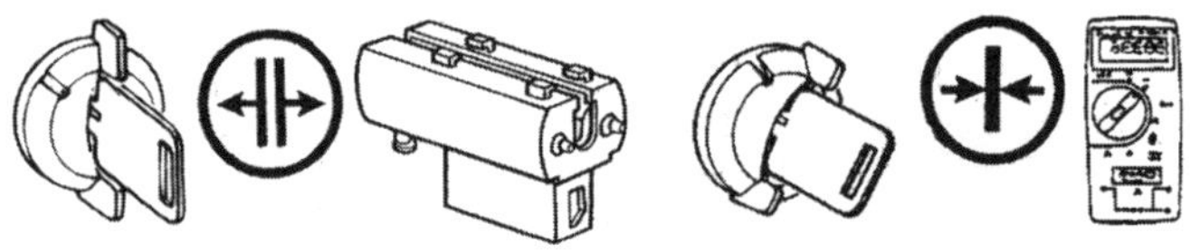

Key Off.

Disconnect The CPP Switch .

Key On, Engine Off.

Measure the voltage between:	
(+) CPP Switch Connector, Harness Side	(-)
CPP-BT - Pin 5	Ground

4.5 And 5.5 Volts

Is The Voltage Between 4.5 And 5.5 Volts.

YES → C3

NO → C4

C3 CHECK THE CPP SWITCH GROUND CIRCUIT FOR AN OPEN

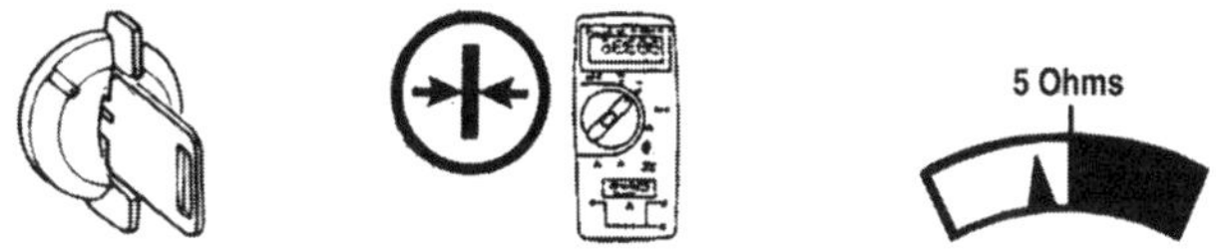

Key Off.

Measure the resistance between:	
(+) CPP Switch Connector, Harness Side	(-)
GND - Pin 6	Ground

Is The Resistance Less Than 5 Ohms?

YES → 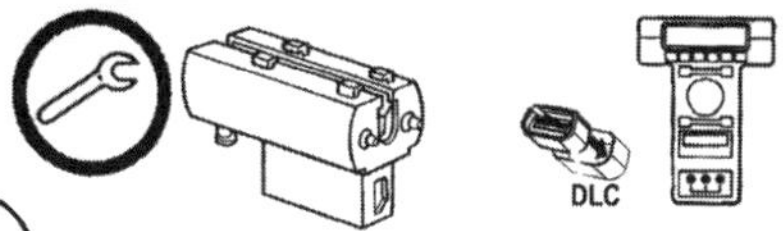

INSTALL A New CPP Switch. CLEAR The DTC's. REPEAT The Self-Test. → STOP

NO → 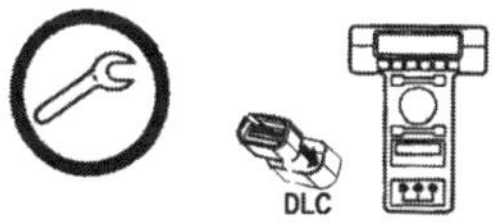

REPAIR The Circuit. CLEAR The DTC's. REPEAT The Self-Test. → STOP

C4 CHECK THE CPP SIGNAL CIRCUIT FOR AN OPEN

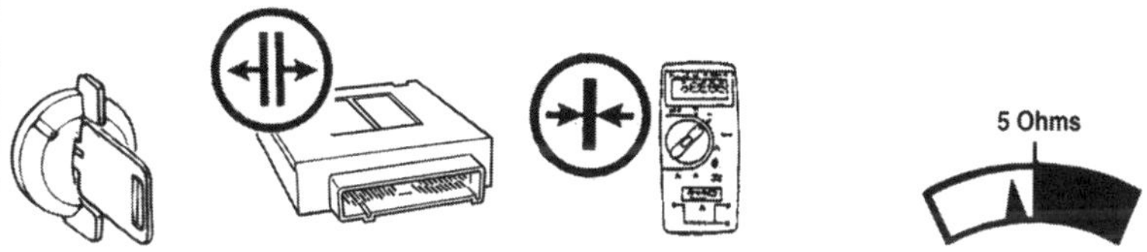

Key Off.

Disconnect The PCM Body Harness Connector.

Measure the resistance between:	
(+) PCM Connector, Harness Side	(-) CPP Switch Connector, Harness Side
CPP BT - Pin B27	CPP-BT - Pin 5

Is The Resistance Less Than 5 Ohms?

YES → C5

NO → 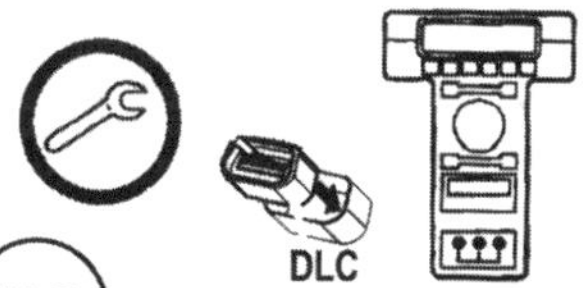

REPAIR The Circuit. CLEAR DTCs. REPEAT The Self-Test. → STOP

C5 CHECK THE CPP SIGNAL CIRCUIT FOR A SHORT TO GROUND

Measure the resistance between:	
(+) PCM Connector, Harness Side	(-)
CPP-BT - Pin B27	Ground

Is The Resistance Greater Than 10,000 Ohms.

YES → C6

NO → 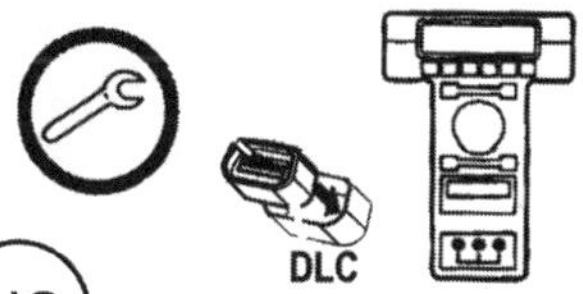

REPAIR The Circuit. CLEAR DTCs. REPEAT The Self-Test. → STOP

C6 CHECK THE CPP SIGNAL CIRCUIT FOR A SHORT TO VOLTAGE

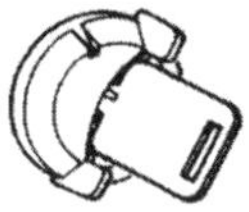

Key On, Engine Off

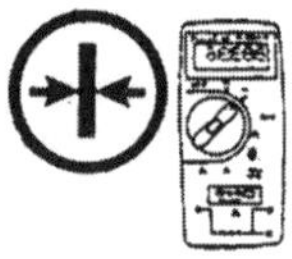

Measure the voltage between:	
(+) PCM Connector, Harness Side	(-)
CPP-BT - Pin B27	Ground

Is Any Voltage Indicated?

YES — REPAIR The Circuit. CLEAR DTCs. REPEAT The Self-Test. STOP

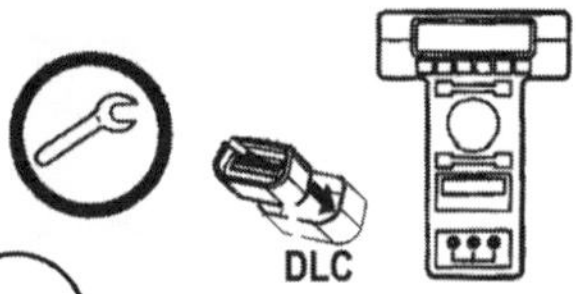

NO — C11

C7 CHECK THE CPP SWITCH FOR VOLTAGE

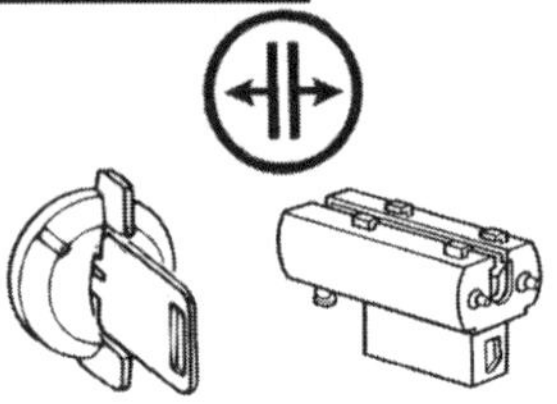

Key Off.

Remove CPP Switch.

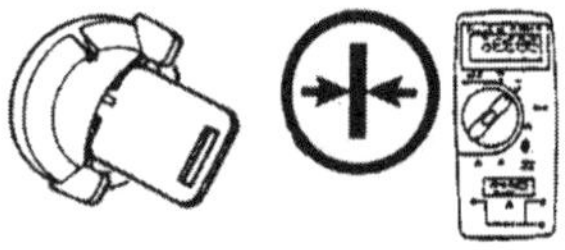

Key On, Engine Off.

Measure the voltage between:	
(+) CPP Switch Connector, Harness Side	(-)
VPWR - Pin 3	Ground

Is The Voltage Greater Than 10.5 Volts?

YES — C8

NO — Repair The Circuit. CLEAR DTCs. REPEAT The Self-Test. STOP

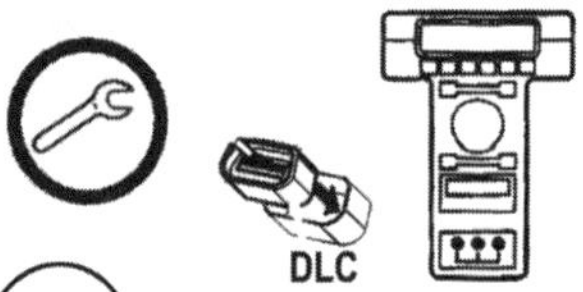

C8 CHECK THE CPP SWITCH

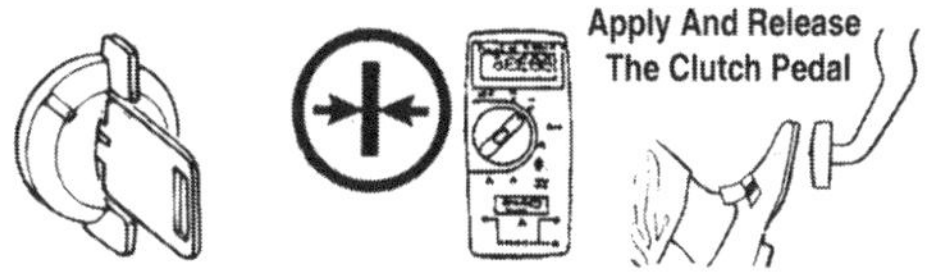

Key Off.

Measure the resistance between:	
(+) CPP Switch Connector, Component Side	(-) CPP Switch Connector, Component Side
VPWR - Pin 3	CPP-TT - Pin 4
Apply and release the clutch pedal.	

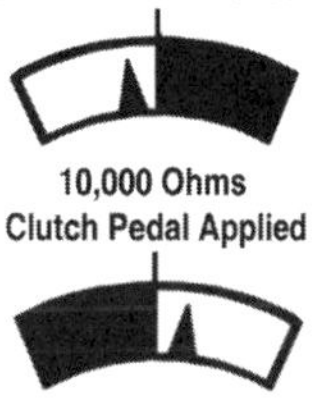

Is The Resistance Less Than 5 Ohms With The Clutch Pedal Released And Greater Than 10,000 Ohms With The Clutch Pedal Applied?

YES → C9

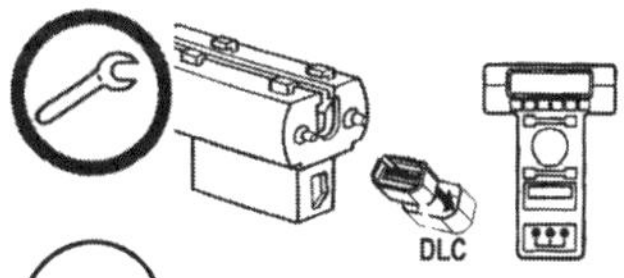

NO → STOP

INSTALL A New CPP Switch. CLEAR DTCs. REPEAT The Self-Test.

C9 CHECK THE CPP SIGNAL CIRCUIT FOR AN OPEN

Disconnect The PCM Body Harness Connector.

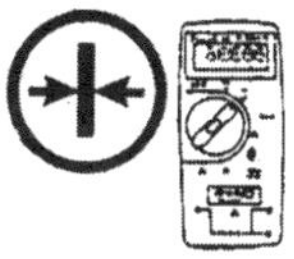

Measure the resistance between:	
(+) PCM Connector, Harness Side	(-) CPP Switch Connector, Harness Side
CPP-TT - Pin B36	CPP-TT - Pin 4

Is The Resistance Less Than 5 Ohms?

YES → C10

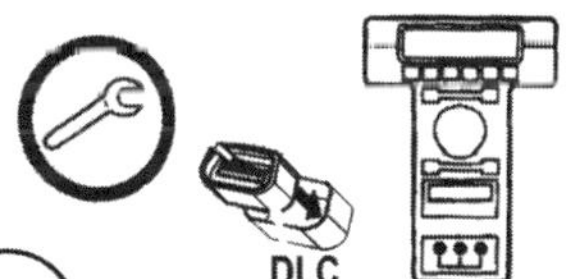

NO → STOP

REPAIR The Circuit. CLEAR DTCs. REPEAT The Self-Test.

C10 CHECK THE CPP SIGNAL CIRCUIT FOR A SHORT TO VOLTAGE

Key On, Engine Off

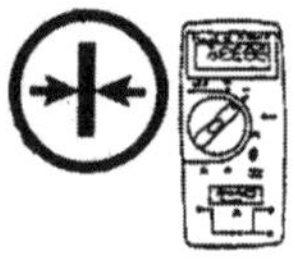

Measure the voltage between:	
(+) PCM Connector, Harness Side	(-)
CPP-TT - Pin B36	Ground

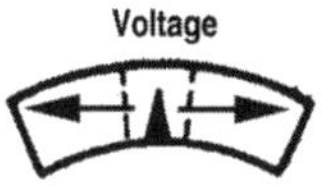

Is Any Voltage Indicated?

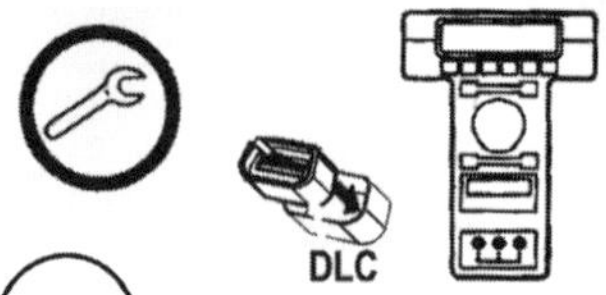

YES — REPAIR The Circuit. CLEAR DTCs. REPEAT The Self-Test. — STOP

NO — C11

C11 CHECK FOR CORRECT PCM OPERATION

Disconnect All PCM Connectors.

Visually Inspect For Pushed Out Pins, Corrosion.

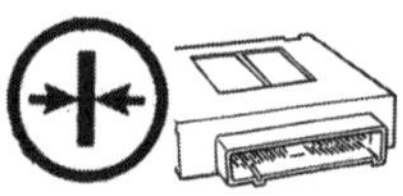

Connect All PCM Connectors And Make Sure They Seat Correctly.

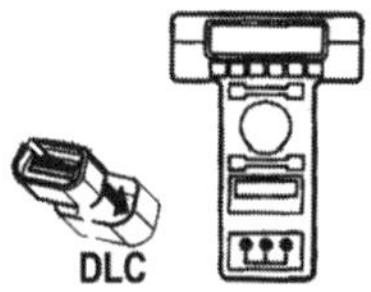

Perform The PCM Self-Test And Verify The Concern Is Still Present.

Is The Concern Still Present.

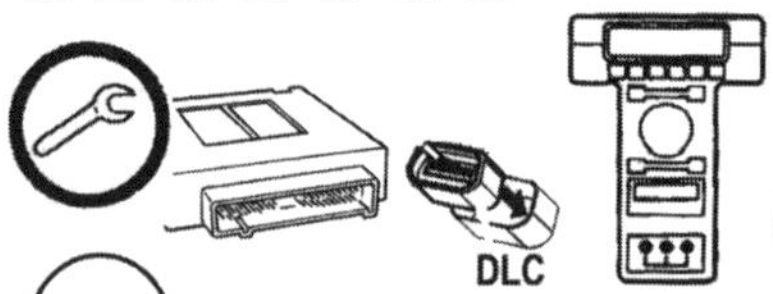

YES — INSTALL A New PCM. CLEAR The DTC's And REPEAT The Self-Test. — STOP

NO — The System Is Operating Correctly At This Time. The Concern May Have Been Caused By A Loose Or Corroded Connector. — STOP

Trouble Code (s)	Problem: Crankshaft Position (CKP) Sensor	Test
P0335, P0336, P2617		D 2006-07

REFERENCE LIST

This pinpoint test is intended to diagnose the following:
- **CKP sensor**
- **Harness circuits: CKP (+), CKP (-), CKP out and CKP shield**
- **PCM**

DIAGNOSTIC TROUBLE CODE INDEX			
Codes	KOEO	KOER	CM
P0335	D1	D1	D1
P0336	D1	D1	D1
P2617	D1	D1	D1

P0335 = Crankshaft Position Sensor Circuit A
P0336 = Crankshaft Position Sensor Circuit A Range/Performance
P2617 = Crankshaft Position Out Fault

Circuit Function:
The CKP signal source is a variable reluctance sensor mounted in the right front side of the engine block. The sensor reacts to a target wheel on the crankshaft. The target wheel is a 60 minus 2 tooth steel disk with 58 evenly spaced teeth and a SYNC gap (a minus 2 slot wide tooth). The sensor produces pulses for each tooth edge that passes it. The crankshaft speed is derived from the frequency of the CKP sensor signal. The crankshaft position is determined by synchronizing the SYNC tooth with the SYNC gap signals from the target wheel. Diagnostic information on the CKP input signal is obtained by carrying out accuracy checks on the frequency and duty cycle with software strategies.

The powertrain control module (PCM) uses the CKP and camshaft position (CMP) signals to calculate the engine speed and piston position. The CKP creates a signal used by the PCM to indicate cylinder identification in a particular bank. The CKP contains a permanent magnet that creates a magnetic field. The signal is created when the target wheel rotates and breaks the magnetic field created by the sensor. The engine will not operate without a CKP signal.

Engine Speed - Is determined by counting the 15 windows on the crankshaft gear each crankshaft revolution.

Injection Control Pressure - Engine speed is one of the controlling variables in the calculation of desired injection control pressure.

Exhaust Pressure - Exhaust pressure control is a function of engine speed and load.

Fuel Quantity ControllTorque Limiting - Engine torque and fuel is controlled and is dependent on engine speed. Fuel quantity is determined by engine speed.

CAUTION!:
The powertrain control module (PCM) harness connectors must be properly seated and the connector latch properly attached to eliminate possible driveability concerns or a no start condition. Installing the PCM connectors on an angle may cause an improper connection, misdiagnosis and damaged components. Install the connector until the lever pivots and seats itself. Apply light pressure to get the connector into position on the PCM and then fully seat the connector.

Note: Visually inspect the harness connectors for corrosion, damage, proper mating and correct pin tension.

Note: When the PCM is disconnected additional DTCs will be set. Clear all DTCs after restoring the vehicle.

PINPOINT TEST SCHEMATICS AND CONNECTORS

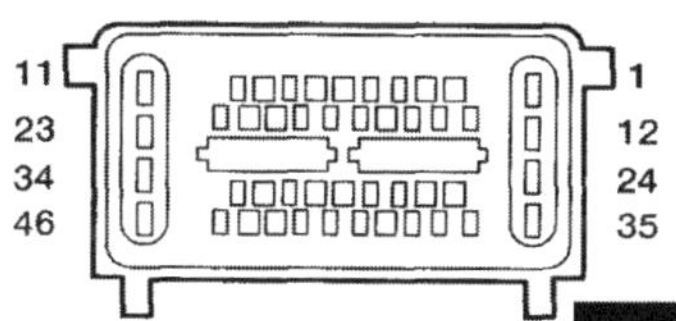

PCM
ENGINE HARNESS
CONNECTOR

Pin	Circuit
E19	CKPO (Crankshaft Position Output)
E30	CKP+ (Crankshaft Position Sensor Positive)
E41	CKP- (Crankshaft Position Sensor Negative)
E42	GND (Ground)

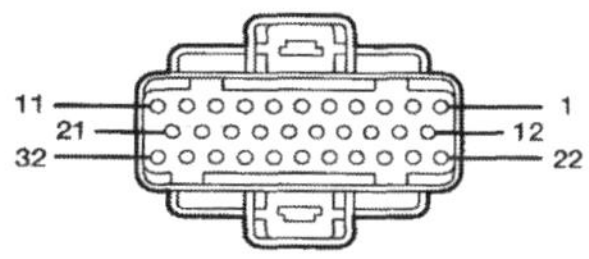

FICM
HARNESS
CONNECTOR

Pin	Circuit
5	CKPO (Crankshaft Position Output)

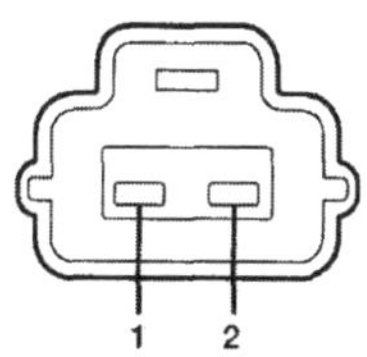

Pin	Circuit
1	CKP+ (Crankshaft Position Sensor Positive)
2	CKP- (Crankshaft Position Sensor Negative)

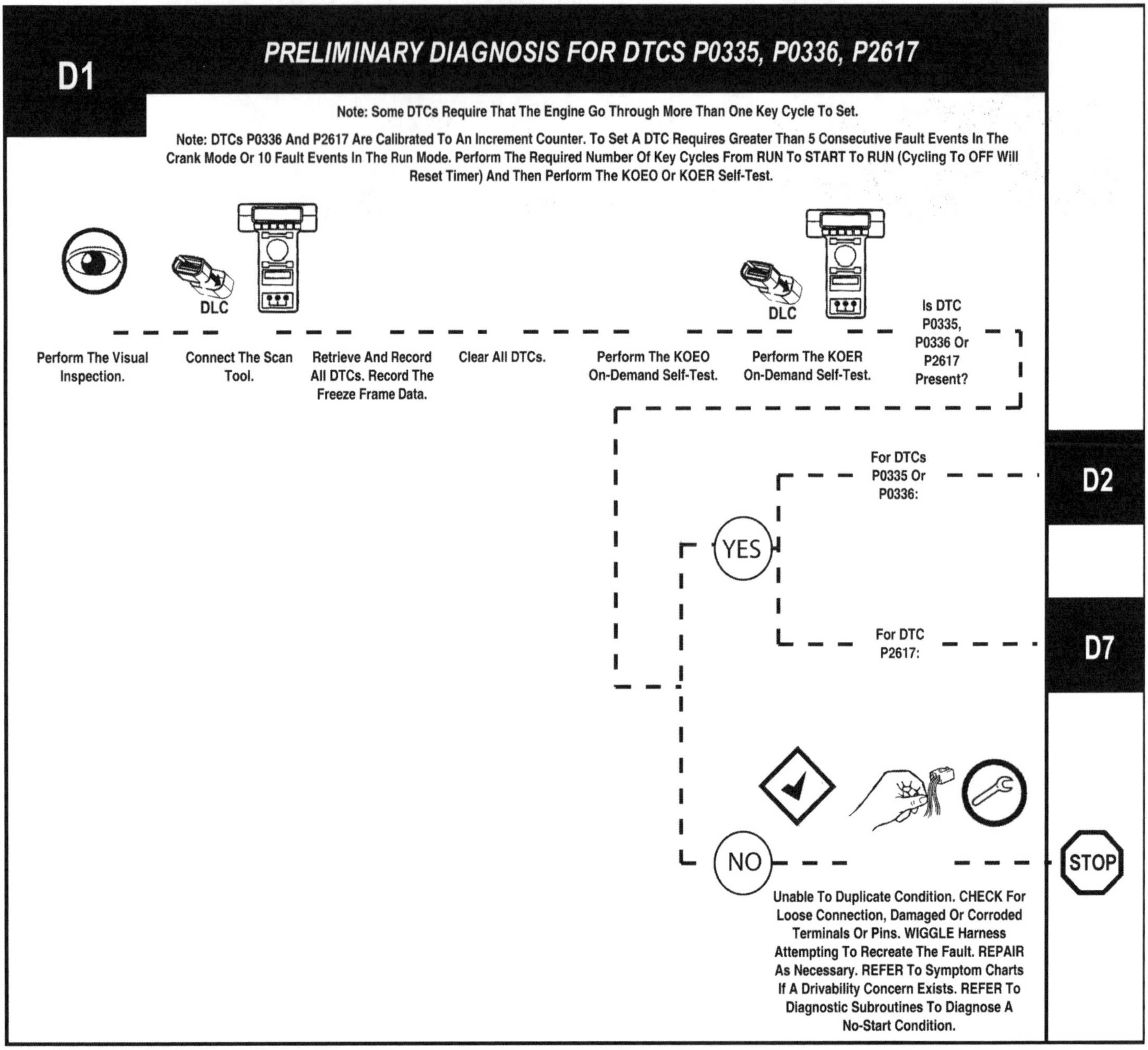
D1
PRELIMINARY DIAGNOSIS FOR DTCS P0335, P0336, P2617
Note: Some DTCs Require That The Engine Go Through More Than One Key Cycle To Set.
Note: DTCs P0336 And P2617 Are Calibrated To An Increment Counter. To Set A DTC Requires Greater Than 5 Consecutive Fault Events In The Crank Mode Or 10 Fault Events In The Run Mode. Perform The Required Number Of Key Cycles From RUN To START To RUN (Cycling To OFF Will Reset Timer) And Then Perform The KOEO Or KOER Self-Test.
DLC
DLC
Perform The Visual Inspection.
Connect The Scan Tool.
Retrieve And Record All DTCs. Record The Freeze Frame Data.
Clear All DTCs.
Perform The KOEO On-Demand Self-Test.
Perform The KOER On-Demand Self-Test.
Is DTC P0335, P0336 Or P2617 Present?
YES
For DTCs P0335 Or P0336:
D2
For DTC P2617:
D7
NO
STOP
Unable To Duplicate Condition. CHECK For Loose Connection, Damaged Or Corroded Terminals Or Pins. WIGGLE Harness Attempting To Recreate The Fault. REPAIR As Necessary. REFER To Symptom Charts If A Drivability Concern Exists. REFER To Diagnostic Subroutines To Diagnose A No-Start Condition.

D2 — DIAGNOSTIC TROUBLE CODE (DTC) P0335 OR P0336

Note: Refer To The PCM And Component Connector Pin Numbers At The Beginning Of This Pinpoint Test.

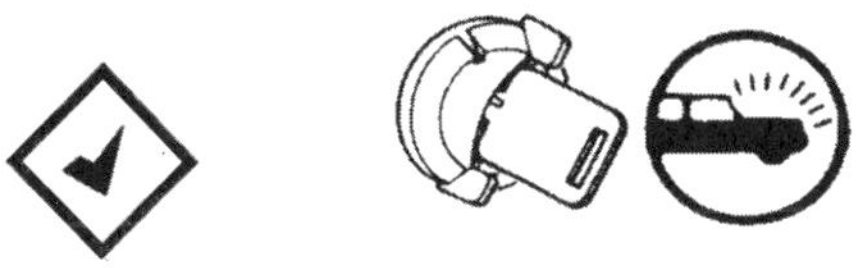

Possible Causes:
- One Or Both Sensor Outputs Open
- One Of Both Sensor Outputs Shorted To Ground
- Sensor Outputs Shorted Together
- One Output Open And The Other Output Shorted To Ground
- Sensor Output Shorted To Voltage
- CKP Sensor
- PCM

Does The Engine Start?

YES → D3

NO → STOP

DIAGNOSTIC SUBROUTINES

REFER To Diagnostic Subroutines To Diagnose No-Start Condition.

D3 — CHECK THE RESISTANCE OF THE CKP SENSOR

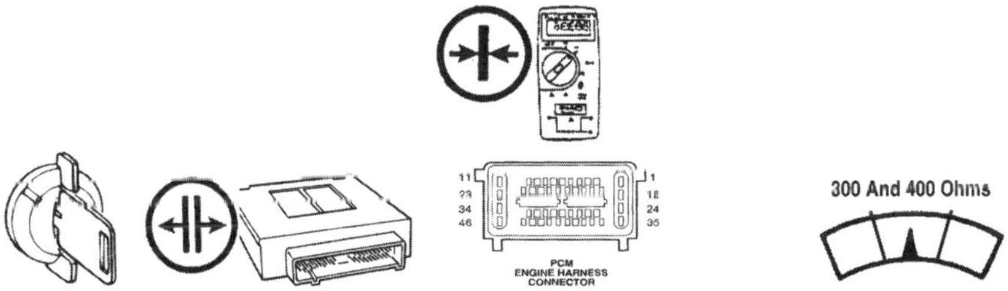

Key Off.

Disconnect The PCM Engine Harness Connector.

Measure the resistance between:	
(+) PCM Connector, Harness Side	(-) PCM Connector, Harness Side
CKP+ - Pin E30	CKP- - Pin E41

Is The Resistance Between 300 And 400 Ohms?

YES → D4

NO → D5

D4 — CONFIRM PCM FAULT

Note: Some DTCs Require That The Engine Go Through More Than One Key Cycle To Set.

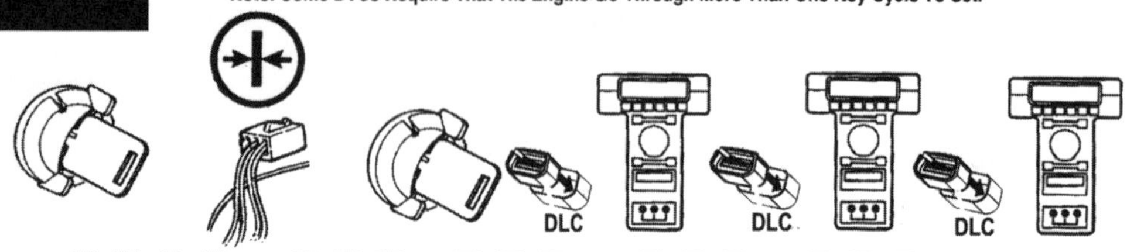

- Key Off.
- Reconnect All Electrical Connectors.
- Key On. Engine Off.
- Clear All DTCs.
- Perform The KOEO On-Demand Self-Test.
- Perform The KOER On-Demand Self-Test.

Is DTC P0335, P0336 Or P2617 Present?

YES → D11

NO →

Unable To Duplicate The Condition. CHECK For Loose Connection, Damaged Or Corroded Terminals Or Pins. WIGGLE Harness Attempting To Recreate The Fault. REFER To Symptom Charts If A Drivability Concern Exists.

STOP

D5 — CHECK CKP CIRCUITS FOR A SHORT TO GROUND

Disconnect The CKP Sensor.

Inspect The CKP Sensor For Damaged Or Corroded Pins.

Measure the resistance between:	
(+) PCM Connector, Harness Side	(-)
CKP+ - Pin E30	Ground
CKP- - Pin E41	Ground

Measure the resistance between:	
(+) PCM Connector, Harness Side	(-) PCM Connector, Harness Side
CKP+ - Pin E30	GND - Pin E42
CKP- - Pin E41	GND - Pin E42

Are The Resistances Greater Than 10,000 Ohms?

YES → D6

NO → STOP

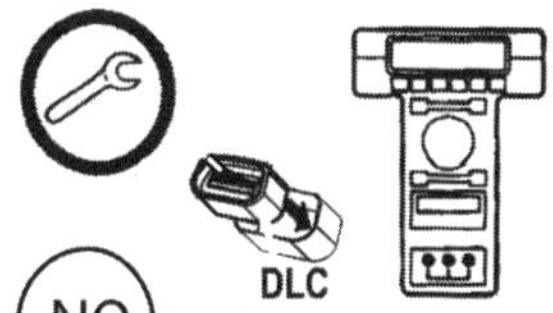

REPAIR Short In CKP Sensor Circuit. RESTORE The Vehicle. CLEAR DTCs And RETEST The System.

D6 — CHECK CKP SENSOR CIRCUITS FOR AN OPEN

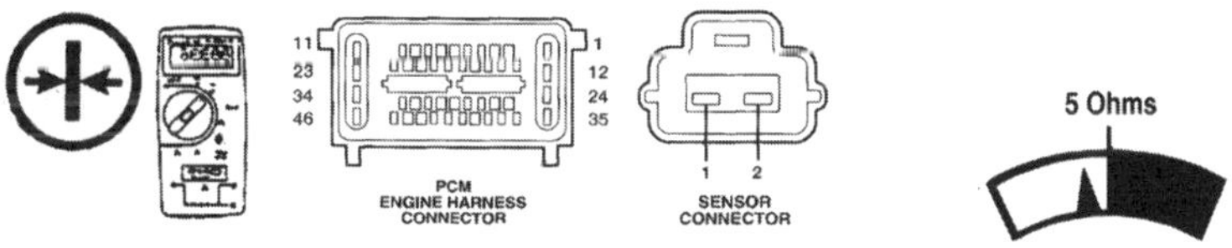

Measure the resistance between:	
(+) PCM Connector, Harness Side	(-) CKP Sensor Connector, Harness Side
CKP+ - Pin E30	CKP+ - Pin 1
CKP- - Pin E41	CKP- - Pin 2

Are The Resistances Less Than 5 Ohms?

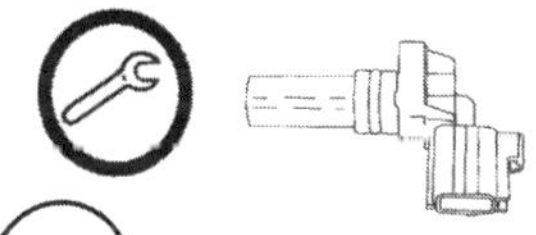

YES → STOP

INSTALL A New CKP Sensor. RESTORE The Vehicle. CLEAR DTCs And RETEST The System.

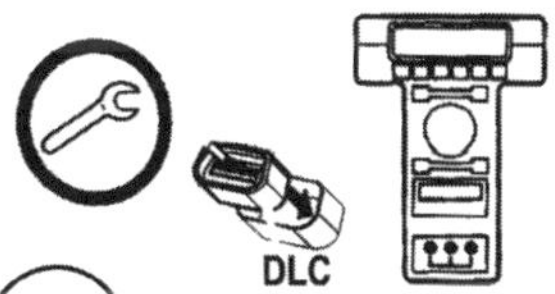

NO → STOP

REPAIR Open Circuit. RESTORE The Vehicle. CLEAR DTCs And RETEST The System.

D7 — DIAGNOSTIC TROUBLE CODE (DTC) P2617

Note: Refer to the PCM and FICM connector pin numbers at the beginning of this pinpoint test.
Note: Some DTCs require that the engine goes through more than one key cycle to set.
Note: DTCs P2617 and P2614 may set simultaneously after an engine stalling event.
Note: DTCs P0336 and P2617 are calibrated to an increment counter. To set a DTC, greater than 10 consecutive fault events in the crank or run mode are required. Carry out the required number of key cycles from RUN to START to RUN (cycling to OFF will reset the timer) and then carry out the KOEO or KOER self-test.

Possible Causes:
- Signal Shorted To Voltage
- Signal Shorted To Ground
- Signal Circuit Open
- PCM

Perform The KOEO On-Demand Self-Test.

Perform The KOER On-Demand Self-Test.

Is DTC P2617 Present?

YES → D8

NO → STOP

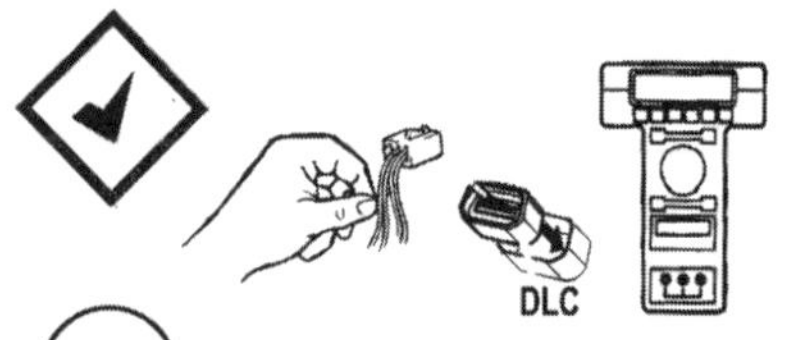

Unable To Duplicate Condition. CHECK For Loose Connections, Damaged Or Corroded Pins. WIGGLE Harness Attempting To Recreate Fault. REPAIR As Necessary. RETEST The System. REFER To Symptom Charts If A Drivability Concern Exists.

D8 — CHECK CRANKSHAFT POSITION SIGNAL CIRCUIT FOR A SHORT TO VOLTAGE

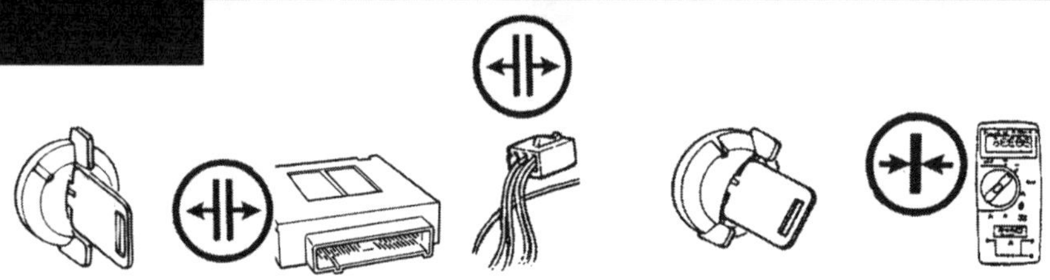

Key Off.

Disconnect The PCM Engine Harness Connector.

Disconnect The FICM Harness Connector C.

Key On. Engine Off.

Measure the voltage between:	
(+) FICM-C Connector, Harness Side	(-)
CKPO - Pin 5	Ground

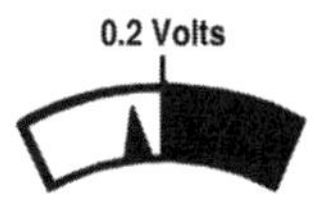

Is The Voltage Less Than 0.2 Volts?

YES → D9

NO → STOP

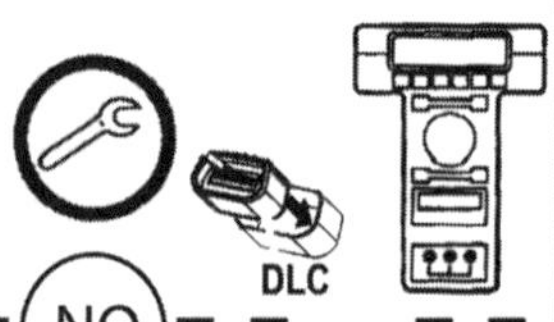

REPAIR Short To Voltage. RESTORE The Vehicle. CLEAR DTCs And RETEST The System.

D9 CHECK CRANKSHAFT POSITION SIGNAL FOR A SHORT TO GROUND

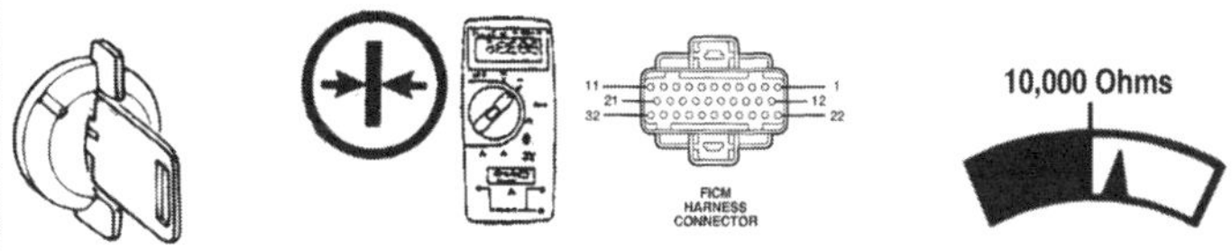

Key Off.

Measure the resistance between:	
(+) FICM-C Connector, Harness Side	(-)
CKPO - Pin 5	Ground

Is The Resistance Greater Than 10,000 Ohms?

YES → D10

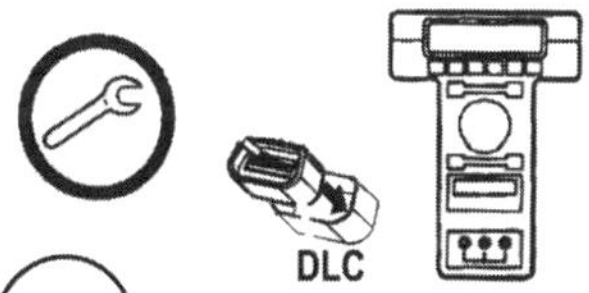

NO → STOP

REPAIR Short To Ground In Signal Output Circuit. RESTORE The Vehicle. CLEAR DTCs And RETEST The System.

D10 CHECK CRANKSHAFT POSITION SIGNAL CIRCUIT FOR AN OPEN

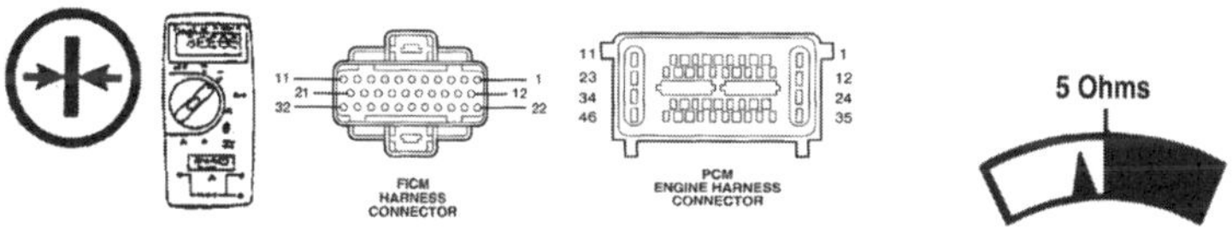

Measure the resistance between:	
(+) FICM-C Connector, Harness Side	(-) PCM Connector, Harness Side
CKPO - Pin 5	CKPO - Pin E19

Is The Resistance Less Than 5 Ohms?

YES → D11

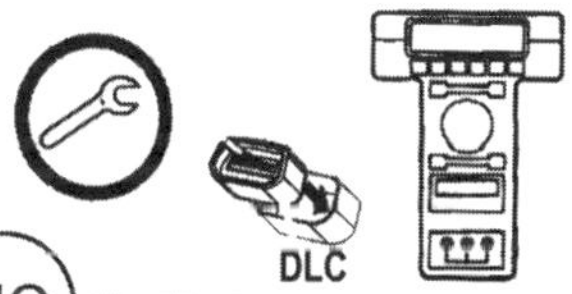

NO → STOP

REPAIR Open In Output Signal Circuit. CLEAR DTCs And RETEST The System.

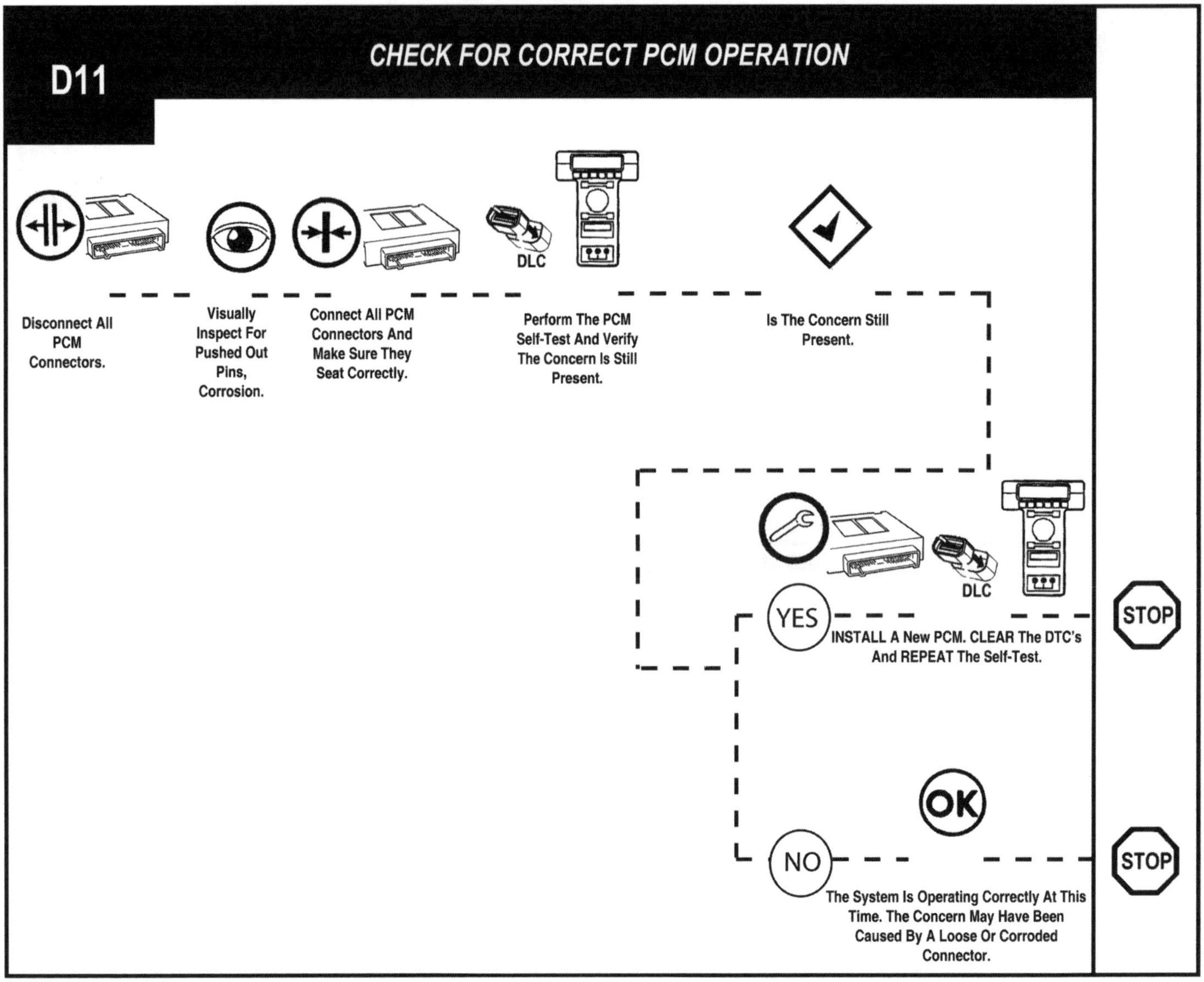
D11
CHECK FOR CORRECT PCM OPERATION
DLC
Disconnect All PCM Connectors.
Visually Inspect For Pushed Out Pins, Corrosion.
Connect All PCM Connectors And Make Sure They Seat Correctly.
Perform The PCM Self-Test And Verify The Concern Is Still Present.
Is The Concern Still Present.
DLC
YES
INSTALL A New PCM. CLEAR The DTC's And REPEAT The Self-Test.
STOP
OK
NO
The System Is Operating Correctly At This Time. The Concern May Have Been Caused By A Loose Or Corroded Connector.
STOP

Trouble Code (s)	Problem: Vehicle Speed Sensor (VSS)	Test
P0500, P1501		DF 2006-07

REFERENCE LIST

This pinpoint test is intended to diagnose the following:

- Harness circuit vehicle speed circuit VSC
- Anti-lock brake system (ABS) module
- Powertrain control module (PCM)

DIAGNOSTIC TROUBLE CODE INDEX			
Codes	KOEO	KOER	CM
P0500	DF1	DF1	DF1
P1501	DF1	DF1	DF1

P0500 = Vehicle Speed Sensor A
P1501 = Vehicle Speed Sensor Out Of Self-Test Range

Signal Function:
The anti-lock brake system (ABS) module provides the vehicle speed signal to the powertrain control module (PCM). The vehicle speed signal is used by the PCM to calculate speed control and power take-off control strategies.

CAUTION!:
The powertrain control module (PCM) harness connectors must be properly seated and the connector latch properly attached to eliminate possible driveability concerns or a no start condition. Installing the PCM connectors on an angle may cause an improper connection, misdiagnosis and damaged components. Install the connector until the lever pivots and seats itself. Apply light pressure to get the connector into position on the PCM and then fully seat the connector.

Note: Visually inspect the harness connectors for corrosion, damage, proper mating and correct pin tension.

Note: When the PCM is disconnected additional DTCs will be set. Clear all DTCs after restoring the vehicle.

PINPOINT TEST SCHEMATICS AND CONNECTORS

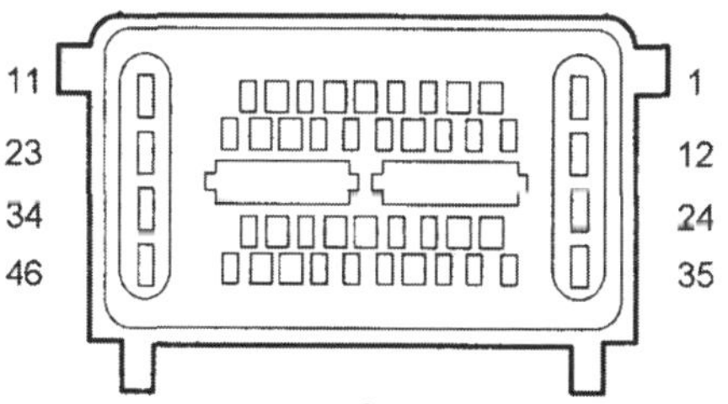

PCM
BODY HARNESS
CONNECTOR

Pin	Circuit
B35	VSC (Vehicle Speed Circuit)

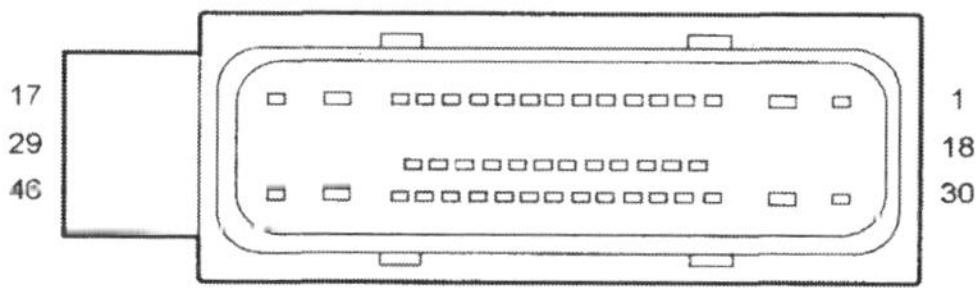

ABS
MODULE CONNECTOR

Pin	Circuit
21	VSC (Vehicle Speed Circuit)

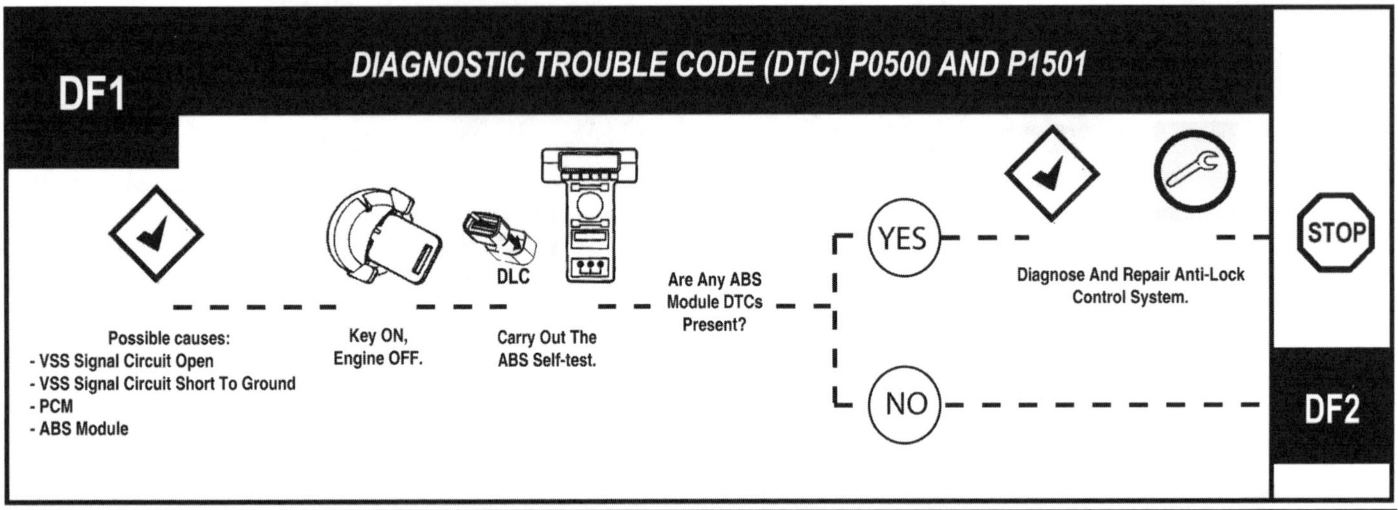
DF1
DIAGNOSTIC TROUBLE CODE (DTC) P0500 AND P1501
Possible causes:
- VSS Signal Circuit Open
- VSS Signal Circuit Short To Ground
- PCM
- ABS Module
Key ON, Engine OFF.
DLC
Carry Out The ABS Self-test.
Are Any ABS Module DTCs Present?
YES
Diagnose And Repair Anti-Lock Control System.
STOP
NO
DF2

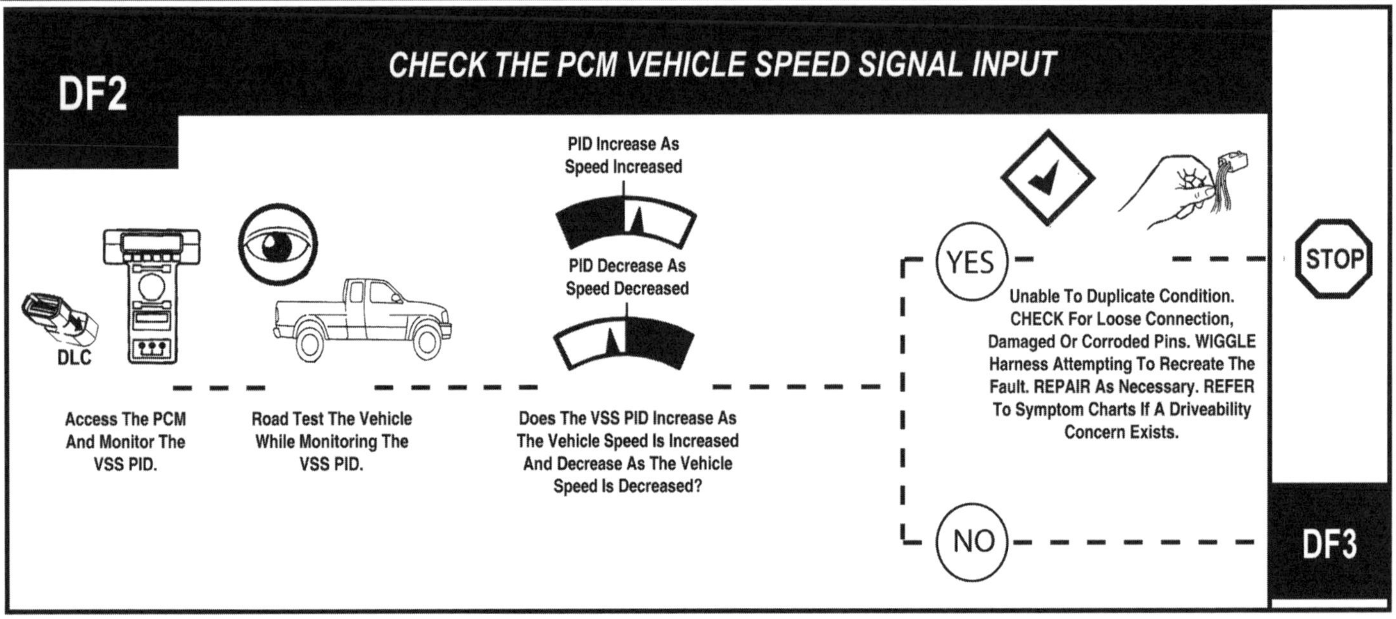
DF2
CHECK THE PCM VEHICLE SPEED SIGNAL INPUT
DLC
Access The PCM And Monitor The VSS PID.
Road Test The Vehicle While Monitoring The VSS PID.
PID Increase As Speed Increased
PID Decrease As Speed Decreased
Does The VSS PID Increase As The Vehicle Speed Is Increased And Decrease As The Vehicle Speed Is Decreased?
YES
Unable To Duplicate Condition. CHECK For Loose Connection, Damaged Or Corroded Pins. WIGGLE Harness Attempting To Recreate The Fault. REPAIR As Necessary. REFER To Symptom Charts If A Driveability Concern Exists.
STOP
NO
DF3

DF3 — CHECK THE VEHICLE SPEED SIGNAL CIRCUIT FOR AN OPEN

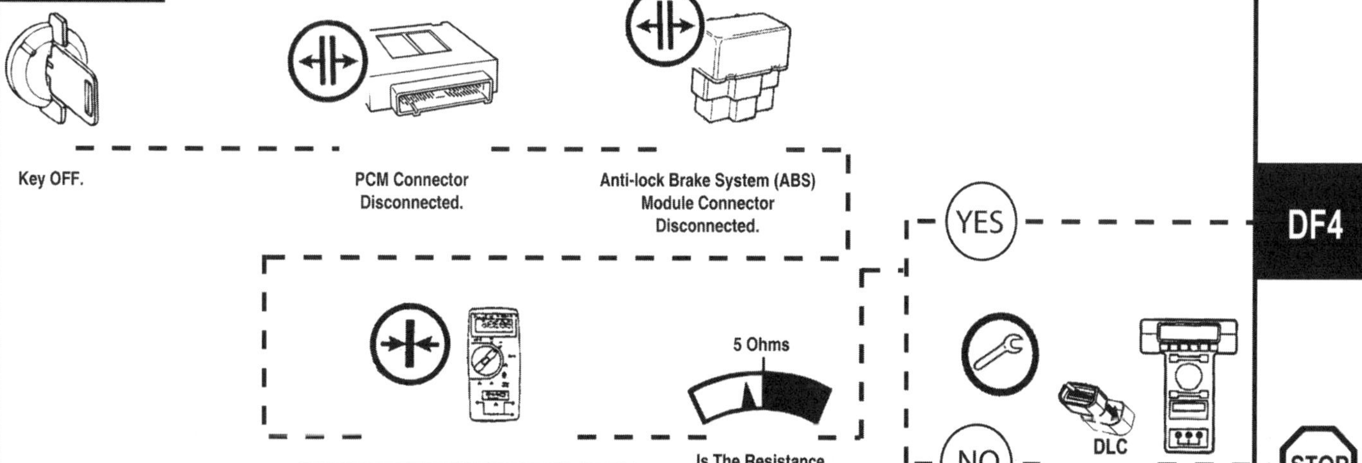

Measure the resistance between:	
(+) PCM Connector, Harness Side	(-) Anti-lock Brake System (ABS) Module Connector, Harness Side
VSC - Pin B35	VSC - Pin 21

DF4 — CHECK THE VEHICLE SPEED SIGNAL CIRCUIT FOR A SHORT TO GROUND

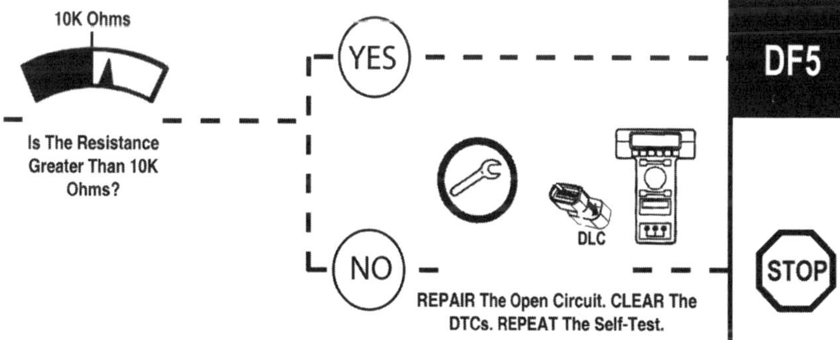

Measure the resistance between:	
(+) PCM Connector, Harness Side	(-)
VSC - Pin B35	Ground

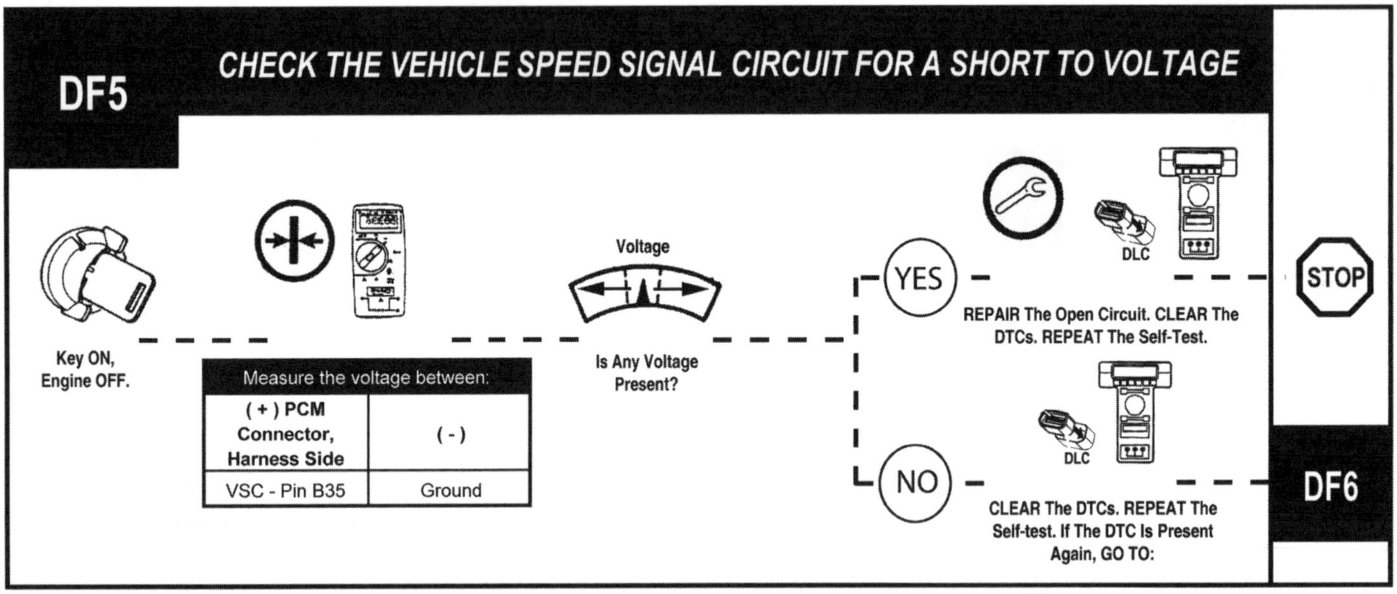

Measure the voltage between:	
(+) PCM Connector, Harness Side	(-)
VSC - Pin B35	Ground

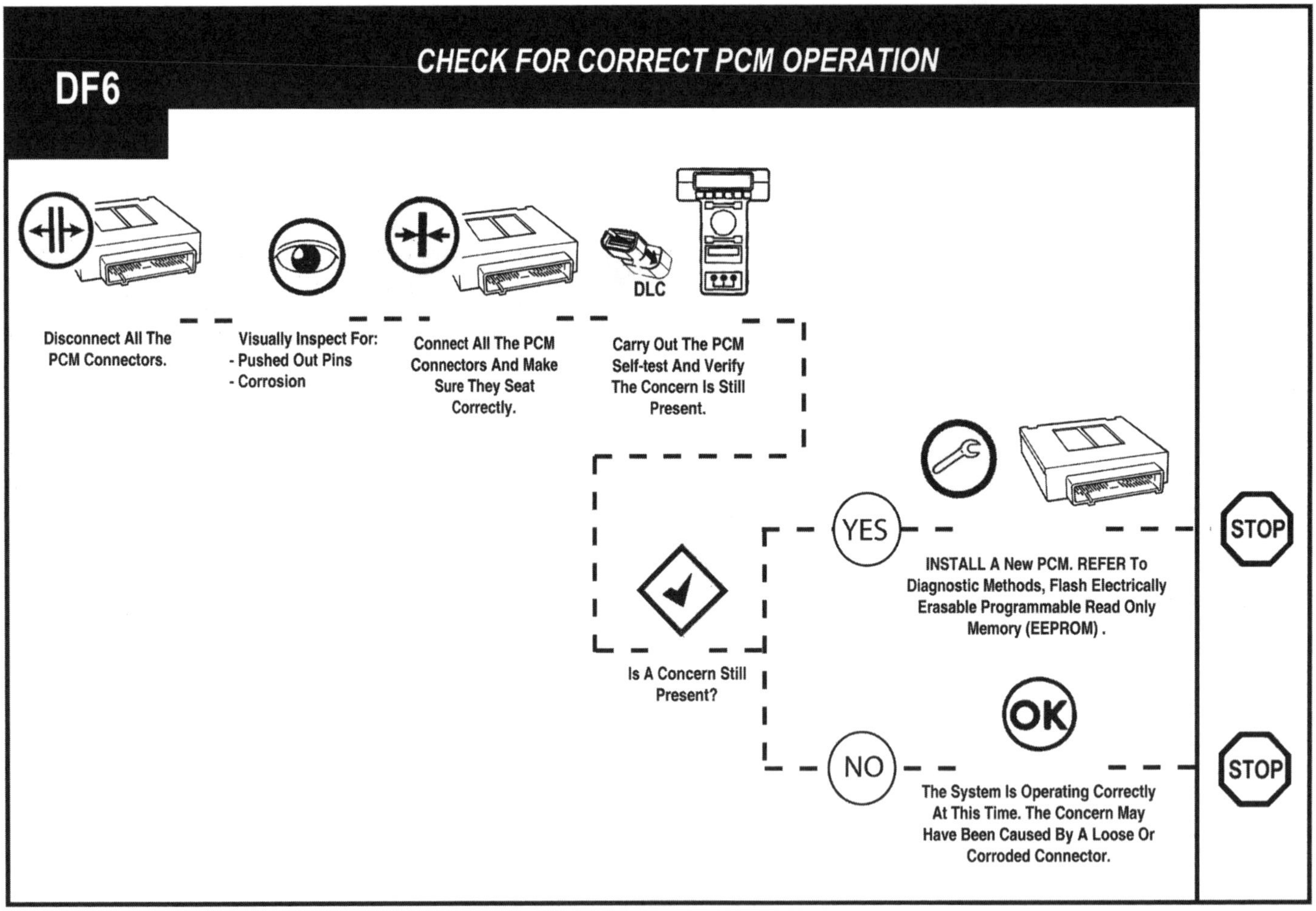

Trouble Code (s)	Problem: Manifold Absolute Pressure (MAP) Sensor	Test
P0236, P0237, P0238		E 2006-07

REFERENCE LIST

This pinpoint test is intended to diagnose the following:

- Harness circuits: MAP, SIGRTN, and VREF
- MAP sensor
- Powertrain control module (PCM) (12A650)

MAP Sensor		
Volts	kPa	psi
0.02	20	2.9
0.5	50	7.25
0.92	76.6	11.1
3.8	255.9	37.1
4.49	299	43.4
5	300	43.5

DIAGNOSTIC TROUBLE CODE INDEX			
Codes	KOEO	KOER	CM
P0236	E1	E1	E1
P0237	E1	E1	E1
P0238	E1	E1	E1

P0236 = Turbo Boost Sensor A Circuit Performance
P0237 = Turbo Boost Sensor A Circuit Low Input
P0238 = Turbo Boost Sensor A Circuit High Input

Note: ± 0.3 volt from expected voltage reading is allowed.

Circuit Function:
The manifold absolute pressure (MAP) sensor is a variable capacitance sensor that, when supplied with a 5-volt reference signal from the powertrain control module (PCM), produces an analog voltage signal that indicates pressure.

Smoke Control - The MAP signal is used to control smoke by limiting fuel quantity during acceleration until a specified boost pressure is obtained.

Dynamic Injection Timing - Optimizes injection timing for boost pressure measured.

Fault Detection/Management:
A MAP signal that is detected by the PCM to be out of range or at an incorrect value for specific conditions will cause the PCM to ignore the MAP signal and operate the engine from an inferred boost pressure signal.

CAUTION!:
The powertrain control module (PCM) harness connectors must be properly seated and the connector latch properly attached to eliminate possible driveability concerns or a no start condition. Installing the PCM connectors on an angle may cause an improper connection, misdiagnosis and damaged components. Install the connector until the lever pivots and seats itself. Apply light pressure to get the connector into position on the PCM and then fully seat the connector.

Note: Visually inspect the harness connectors for corrosion, damage, proper mating and correct pin tension.

Note: When the PCM is disconnected additional DTCs will be set. Clear all DTCs after restoring the vehicle.

PINPOINT TEST SCHEMATICS AND CONNECTORS

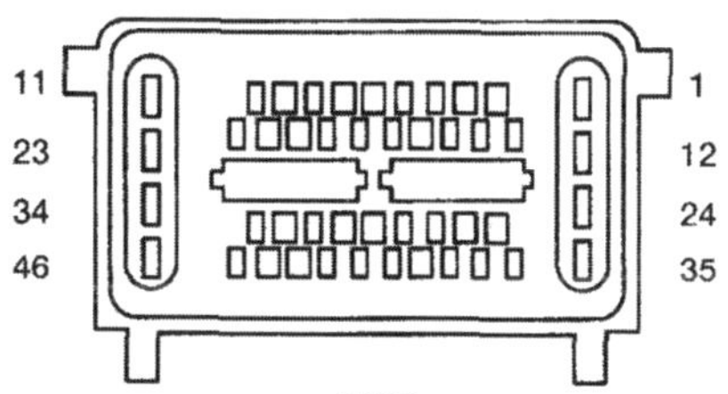

PCM
BODY HARNESS
CONNECTOR

Pin	Circuit
B33	SIGRTN (Signal Return)
B41	MAP (Manifold Absolute Pressure)

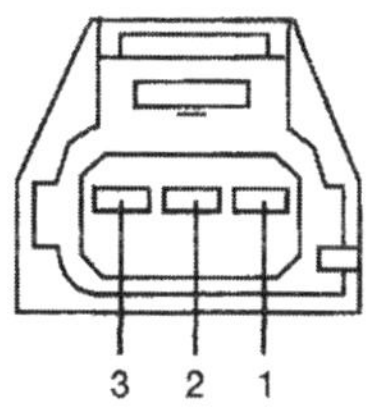

SENSOR
CONNECTOR

Pin	Circuit
1	VREF (Reference Voltage)
2	MAP (Manifold Absolute Pressure)
3	SIGRTN (Signal Return)

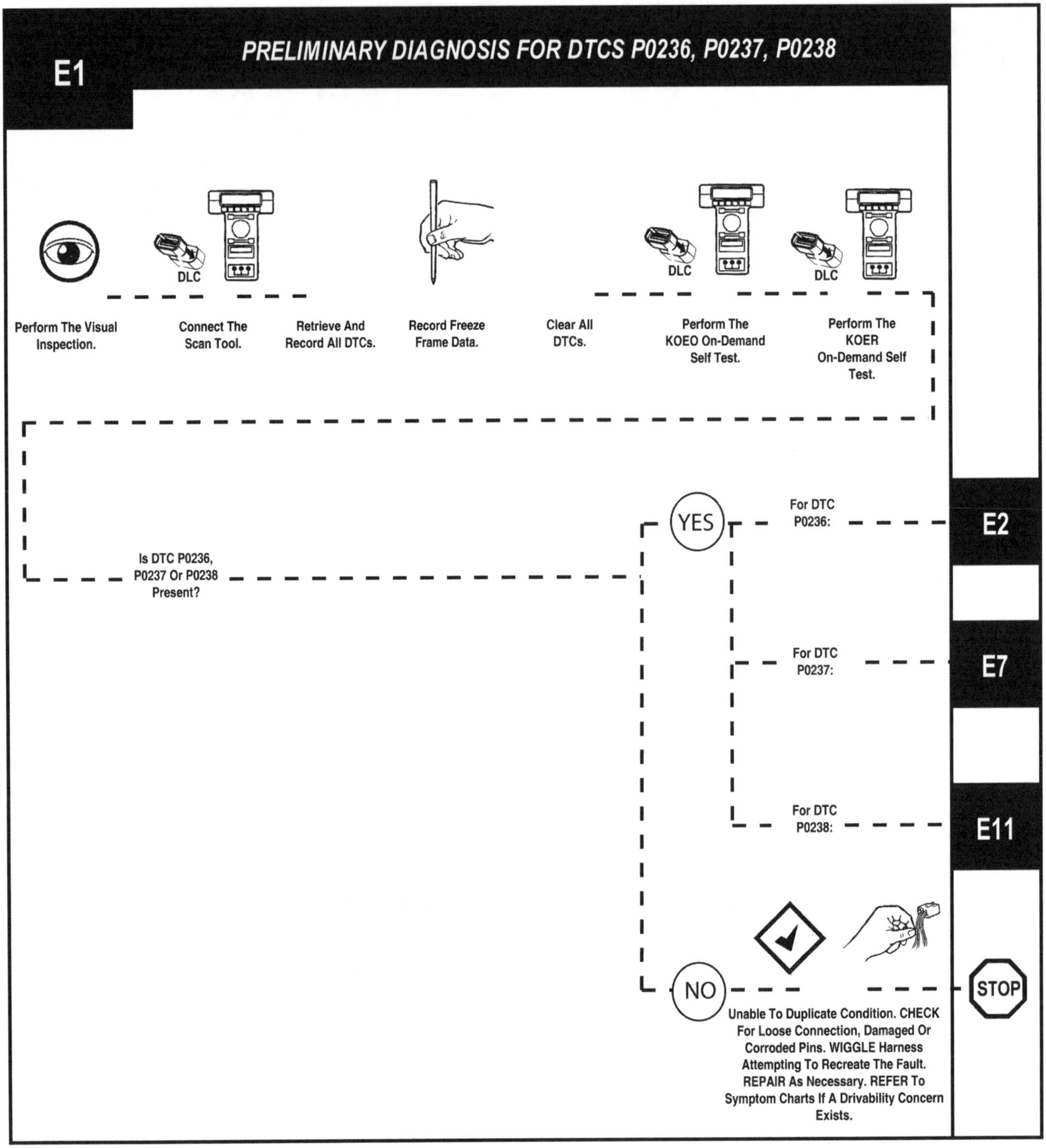
E1
PRELIMINARY DIAGNOSIS FOR DTCS P0236, P0237, P0238
DLC
DLC
DLC
Perform The Visual Inspection.
Connect The Scan Tool.
Retrieve And Record All DTCs.
Record Freeze Frame Data.
Clear All DTCs.
Perform The KOEO On-Demand Self Test.
Perform The KOER On-Demand Self Test.
Is DTC P0236, P0237 Or P0238 Present?
YES
For DTC P0236:
E2
For DTC P0237:
E7
For DTC P0238:
E11
NO
Unable To Duplicate Condition. CHECK For Loose Connection, Damaged Or Corroded Pins. WIGGLE Harness Attempting To Recreate The Fault. REPAIR As Necessary. REFER To Symptom Charts If A Drivability Concern Exists.
STOP

E2 — DIAGNOSTIC TROUBLE CODE (DTC) P0236

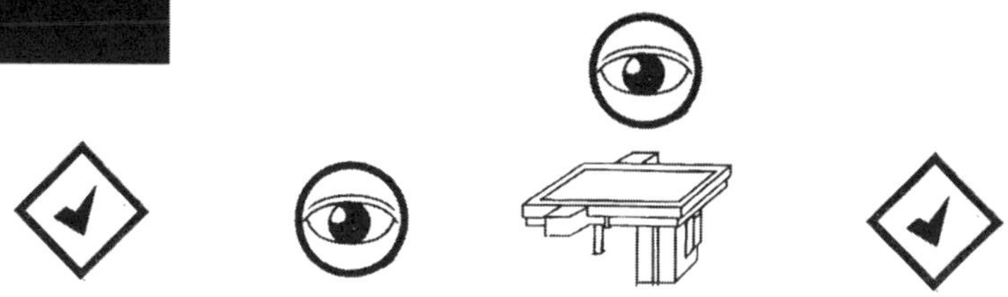

Possible Causes:
- Restricted MAP Hose
- Low Turbo Boost
- Intake Manifold
- Restricted MAP Sensor
- PCM

Perform Visual Inspection.

Inspect MAP Sensor Hose And Manifolds For Damage, Leaks, Restrictions And Misrouting.

Are Manifolds And MAP Hose OK And Free From Damage?

YES → E3

NO → STOP

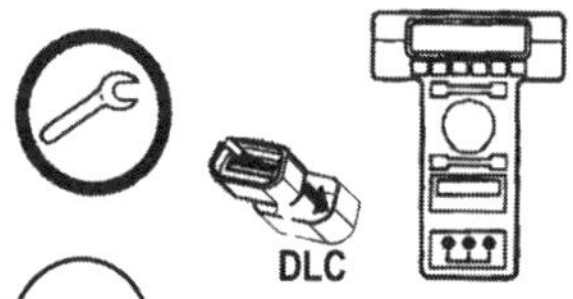

REPAIR Leaks As Necessary. RESTORE The Vehicle. CLEAR DTCs And RETEST The System.

E3 — CHECK MAP SENSOR

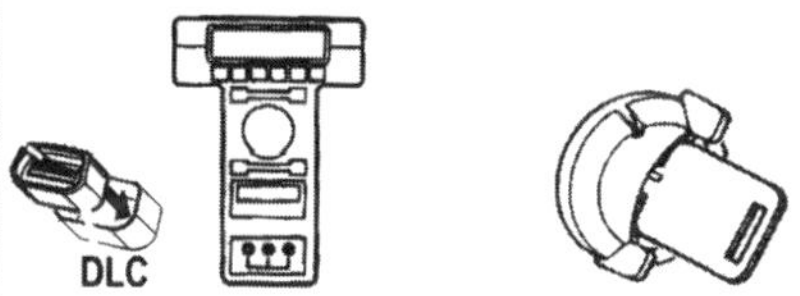

MAP Sensor		
Volts	kPa	psi
0.02	20	2.9
0.5	50	7.25
0.92	76.6	11.1
3.8	255.9	37.1
4.49	299	43.4
5	300	43.5

Note: ± 0.3 volt from expected voltage reading is allowed.

Access MAP VOLT PID.

Key On.

Is The Voltage ± 0.5 Volts For Given Altitude?

YES → E4

NO → STOP

INSTALL A New MAP Sensor. RESTORE The Vehicle. CLEAR DTCs And RETEST The System.

E4 MAP SENSOR VOLTAGE CHECK

Note: DTC P0069 MAP/Barometric Pressure Correlation will set.

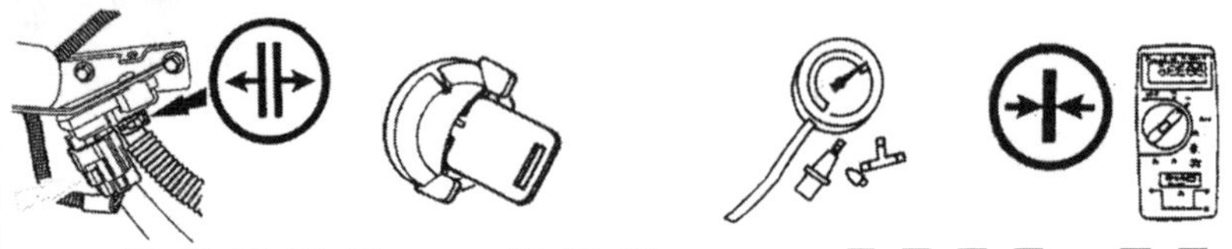

Disconnect Pressure Hose From MAP Sensor.

Key On, Engine Off.

Using Pressure Adapter Kit 014-00761 Or Equivalent (Gauge Bar), Apply 69 KPa (10 Psi) Of Pressure To The MAP Sensor.

Measure the voltage between:	
(+) MAP Sensor Connector, Harness Side	(-)
MAP - Pin 2	Ground

2.5 - 3.1 V

Is the voltage between 2.5 - 3.1 V?

YES → E5

NO

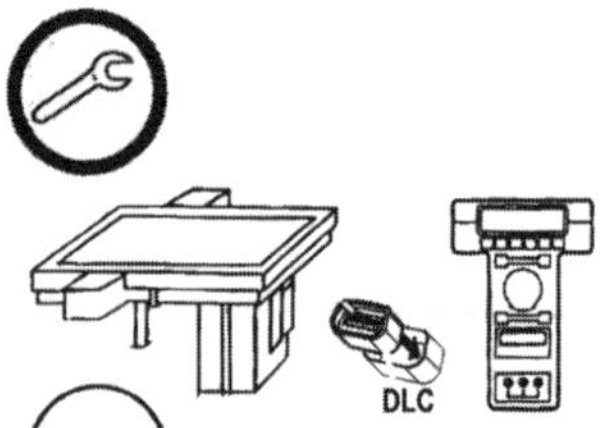

INSTALL A New MAP Sensor. RESTORE The Vehicle. CLEAR DTCs And RETEST The System.

E5 CHECK THE MAP SENSOR PERFORMANCE

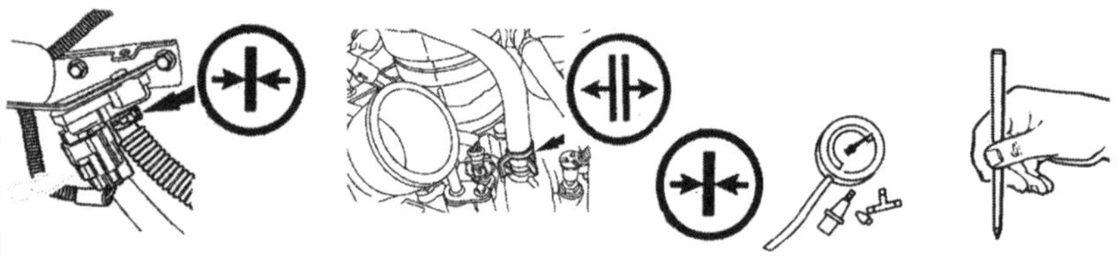

Connect Pressure Hose To MAP Sensor.

Disconnect MAP Sensor Pressure Hose From Intake Manifold And Install A Pressure (Boost) Gauge.

Record The Gauge Pressure.

Achieve Full Boost

Road Test The Vehicle And Accelerate Vehicle To Achieve Full Boost.

Does The Boost Pressure Exceed 82 KPa (12 Psi) At Full Acceleration/Boost And Return To The Recorded Gauge Pressure At Idle?

YES → E6

NO

INSPECT Intake Manifolds For Leaks. CHECK Turbo Condition. REPAIR As Necessary. CLEAR The DTC's. REPEAT The Self-Test.

E6 — CHECK FOR THE PRESENCE OF DTC P0236

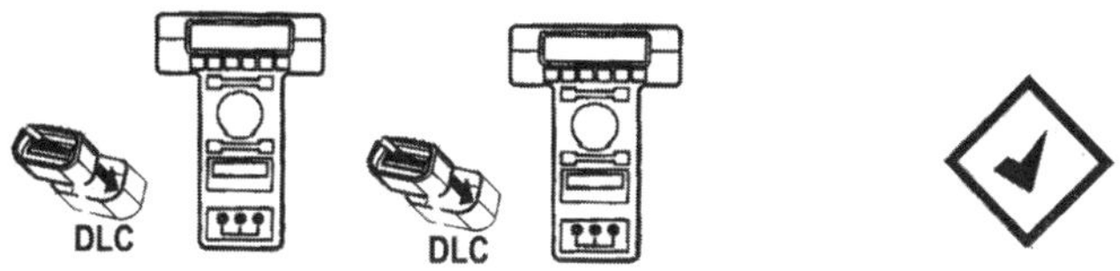

Clear The DTCs.

Carry Out The KOEO And KOER Self-Test.

Is DTC P0236 Present?

YES → E14

NO → OK — The Test Is Complete. STOP

E7 — DIAGNOSTIC TROUBLE CODE (DTC) P0237

Note: Refer To The PCM And Component Connector Pin Numbers At The Beginning Of This Pinpoint Test.

Possible Causes: MAP Signal Circuit Open Or Shorted To Ground
- Open VREF Circuit
- MAP Sensor
- PCM

Key Off.

Disconnect The MAP Sensor.

Key On.

Measure the voltage between:

(+) MAP Sensor Connector, Harness Side	(-) MAP Sensor Connector, Harness Side
VREF - Pin 1	SIGRTN - Pin 3

4.5 And 5.5 Volts

Is The Voltage Between 4.5 And 5.5 Volts?

YES → E8

NO → REPAIR Open In VREF Or Signal Return Circuit. RESTORE The Vehicle. CLEAR DTCs And RETEST The System. STOP

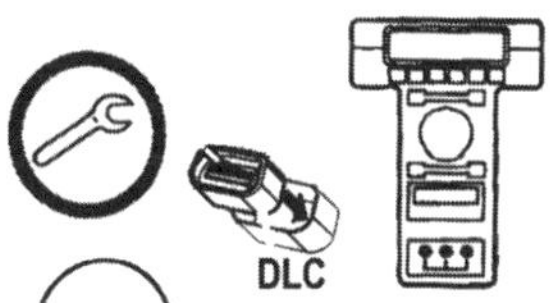

E8 — CHECK SIGNAL CIRCUIT FOR A SHORT TO GROUND

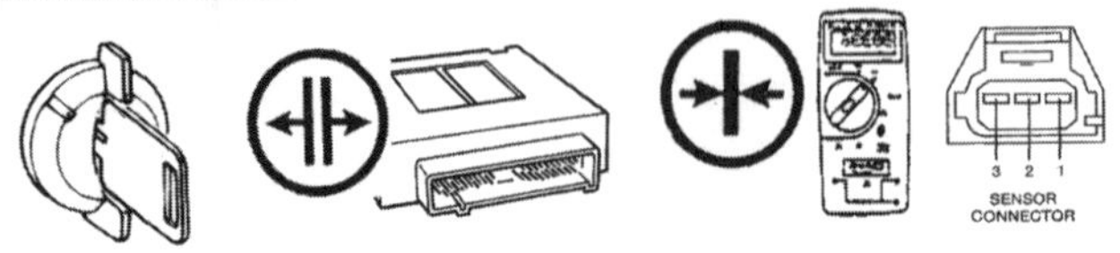

Key Off.

Disconnect The PCM Body Harness Electrical Connector.

Measure the resistance between:	
(+) MAP Sensor Connector, Harness Side	(-)
MAP - Pin 2	Ground

10,000 Ohms

Is The Resistance Greater Than 10,000 Ohms?

YES → E9

NO → STOP

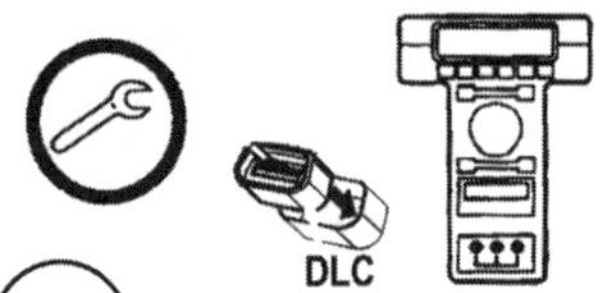

REPAIR Short To Ground In The Signal Circuit. RESTORE The Vehicle. CLEAR DTCs And RETEST The System.

E9 — CHECK THE MAP SIGNAL CIRCUIT FOR AN OPEN

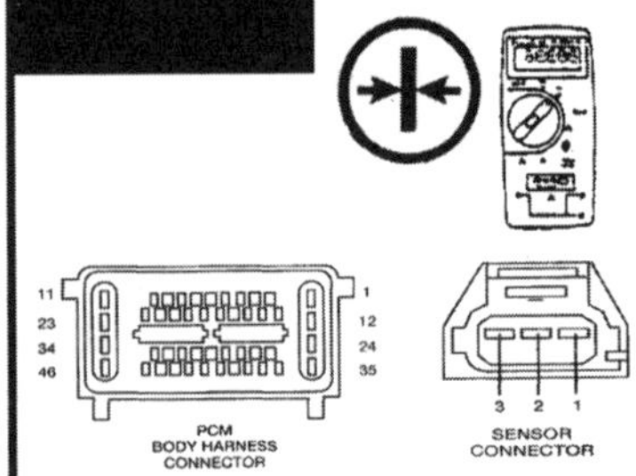

Measure the resistance between:	
(+) MAP Sensor Connector, Harness Side	(-) PCM Connector, Harness Side
MAP - Pin 2	MAP - Pin B41

5 Ohms

Is The Resistance Less Than 5 Ohms?

YES → E10

NO → STOP

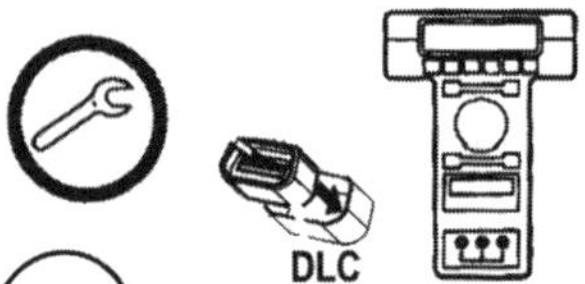

REPAIR Open In Signal Circuit. RESTORE The Vehicle. CLEAR DTCs And RETEST The System.

E10 CHECK MAP SENSOR CIRCUIT

Note: This Step May Cause DTC P0238 To Set.

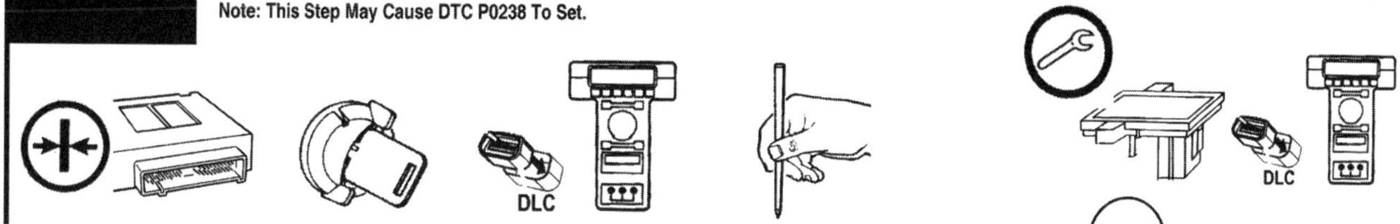

Reconnect The PCM Body Harness Connector.

Key On, Engine Off.

Access MAP VOLT PID.

Record The MAP VOLT PID Reading.

Connect a 5 amp fused jumper wire between the following:	
Point A MAP Sensor Connector, Harness Side	Point B MAP Sensor Connector, Harness Side
VREF - Pin 1	MAP - Pin 2

Is The MAP PID Voltage Reading Approximately 4.5 To 5.5 Volts?

YES — INSTALL A New MAP Sensor. RESTORE The Vehicle. CLEAR DTCs And RETEST The System. STOP

NO — E14

E11 DIAGNOSTIC TROUBLE CODE (DTC) P0238

Note: Refer To The PCM And Component Connector Pin Numbers At The Beginning Of This Pinpoint Test.
Note: Visually Inspect Vehicle For Aftermarket Accessories And Performance Modifications (Exhaust System, Turbocharger, Performance Chip Etc.)

Possible Causes:
- Signal Circuit Shorted To Power
- Open Ground Circuit
- MAP Sensor
- PCM

Induce Opposite DTC.

Install The Scan Tool.

Clear All DTCs.

Disconnect The MAP Sensor Electrical Connector.

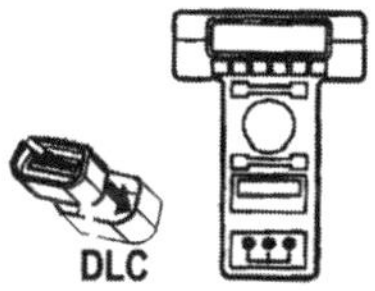

Carry Out The KOEO And KOER Self-test.

Is DTC P0237 Set?

YES — INSTALL A New MAP Sensor. RESTORE The Vehicle. CLEAR DTCs And RETEST The System. STOP

NO — E12

E12 CHECK MAP SENSOR SIGNAL CIRCUIT FOR A SHORT TO VOLTAGE

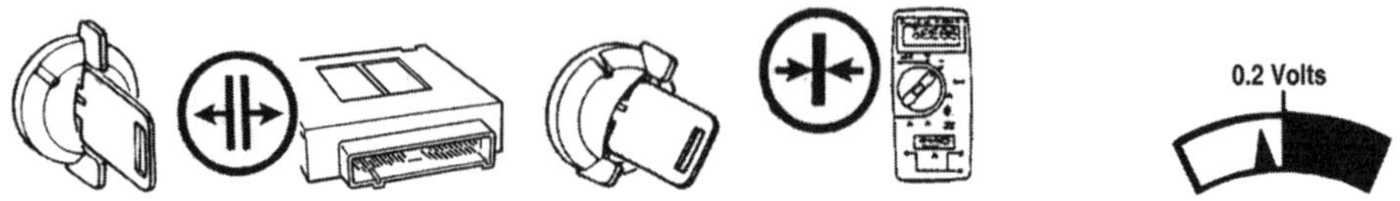

Key Off.

Disconnect The PCM Body Harness Connector.

Key On.

Measure the voltage between:	
(+) MAP Sensor Connector, Harness Side	(-)
MAP - Pin 2	Ground

Is The Voltage Less Than 0.2 Volts?

YES → E13

NO → STOP

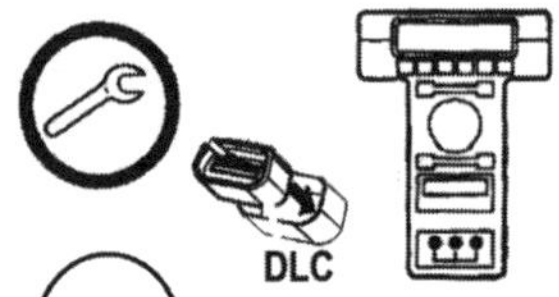

REPAIR Short To Voltage In Signal Circuit. RESTORE The Vehicle. CLEAR DTCs And RETEST The System.

E13 CHECK MAP SENSOR SIGNAL RETURN FOR AN OPEN

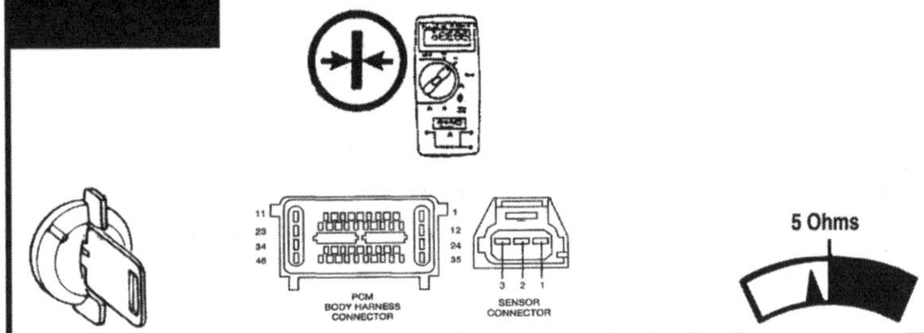

Key Off.

Measure the resistance between:	
(+) MAP Sensor Connector, Harness Side	(-) PCM Connector, Harness Side
SIGRTN - Pin 3	SIGRTN - Pin B33

Is The Resistance Less Than 5 Ohms?

YES → E14

NO → STOP

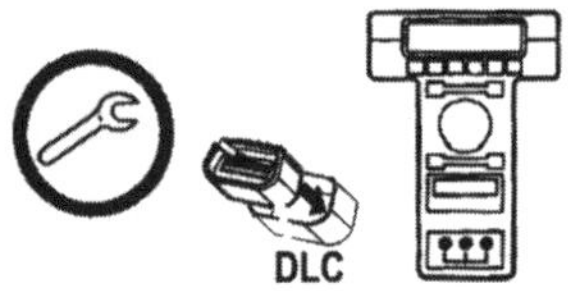

REPAIR Open In Signal Return Circuit. RESTORE The Vehicle. CLEAR DTCs And RETEST The System.

E14 CHECK FOR CORRECT PCM OPERATION

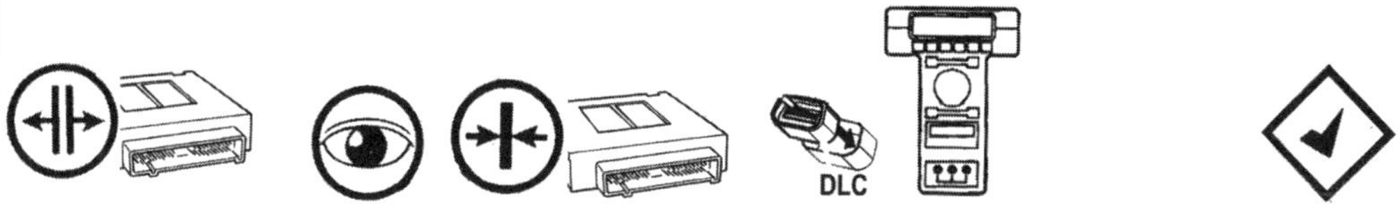

Disconnect All PCM Connectors.

Visually Inspect For Pushed Out Pins, Corrosion.

Connect All PCM Connectors And Make Sure They Seat Correctly.

Perform The PCM Self-Test And Verify The Concern Is Still Present.

Is The Concern Still Present.

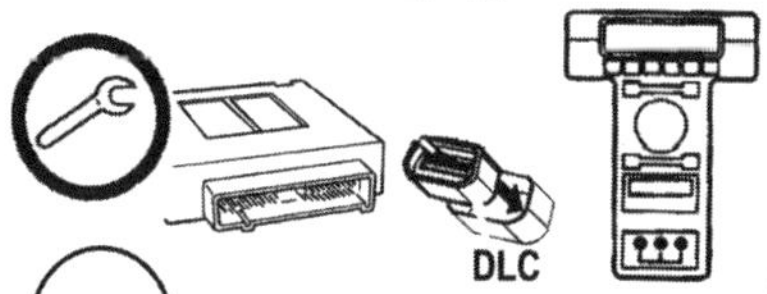

YES — INSTALL A New PCM. CLEAR The DTC's And REPEAT The Self-Test. STOP

NO — The System Is Operating Correctly At This Time. The Concern May Have Been Caused By A Loose Or Corroded Connector. STOP

NOTES

Trouble Code (s)	Problem: Intake Air Temperature (IAT) Sensor	Test
P0112, P0113		F 2006-07

REFERENCE LIST

This pinpoint test is intended to diagnose the following:

- Harness circuits: IAT, and SIGRTN
- IAT sensor
- Powertrain control module (PCM) (12A650)

DIAGNOSTIC TROUBLE CODE INDEX			
Codes	KOEO	KOER	CM
P0112	F1	F1	F1
P0113	F1	F1	F1

P0112 = IAT Sensor Circuit High Input
P0113 = IAT Sensor Circuit Low Input

Circuit Function:
The intake air temperature (IAT) sensor, located inside the mass air flow housing, is a thermistor-type sensor with a variable resistance that is inversely proportional to temperature changes. When interfaced with the powertrain control module (PCM), it produces a 0-5 volt analog signal that will measure temperature.

The IAT sensor's primary function is to measure intake air temperature to control timing and fuel rate when cold-starting. Continuous monitoring by the IAT sensor limits smoke emission.

Fault Detection/Management:
An IA T signal that is detected out of range high or low by the PCM will cause the engine to ignore the IAT signal and assume an ambient temperature of 25°C (77°F).

CAUTION!:
The powertrain control module (PCM) harness connectors must be properly seated and the connector latch properly attached to eliminate possible driveability concerns or a no start condition. Installing the PCM connectors on an angle may cause an improper connection, misdiagnosis and damaged components. Install the connector until the lever pivots and seats itself. Apply light pressure to get the connector into position on the PCM and then fully seat the connector.

Note: Visually inspect the harness connectors for corrosion, damage, proper mating and correct pin tension.

PINPOINT TEST SCHEMATICS AND CONNECTORS

A

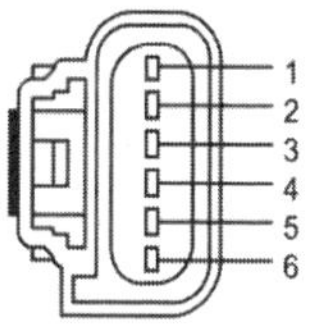

C

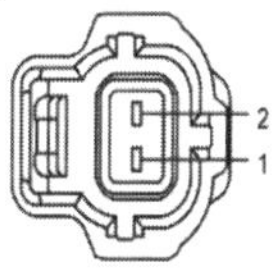

B

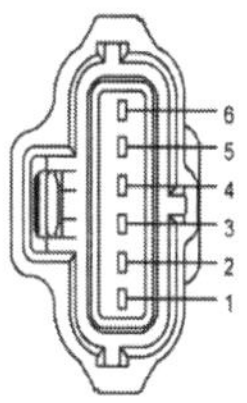

Vehicle	Connector	Pin	Circuit
E-Series	A	2	SIGRTN
		1	IAT
F-Super Duty vehicles with an integrated MAF/IAT sensor	B	1	SIGRTN
		6	IAT
F-Super Duty vehicles with a stand alone IAT sensor	C	1	SIGRTN
		2	IAT

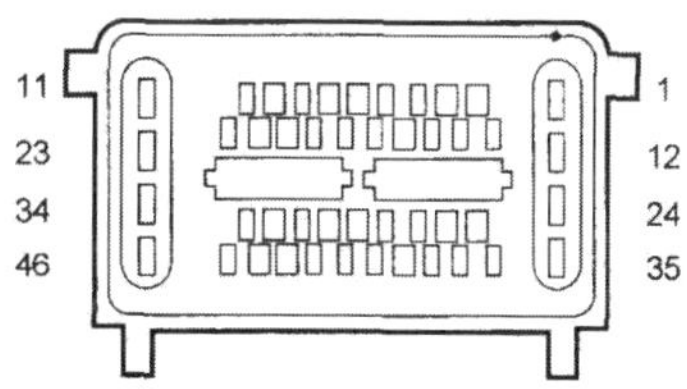

PCM
BODY HARNESS
CONNECTOR

Pin	Circuit
B20, B33, E25	SIGRTN (Signal Return)
B43	IAT (Intake Air Temperature)

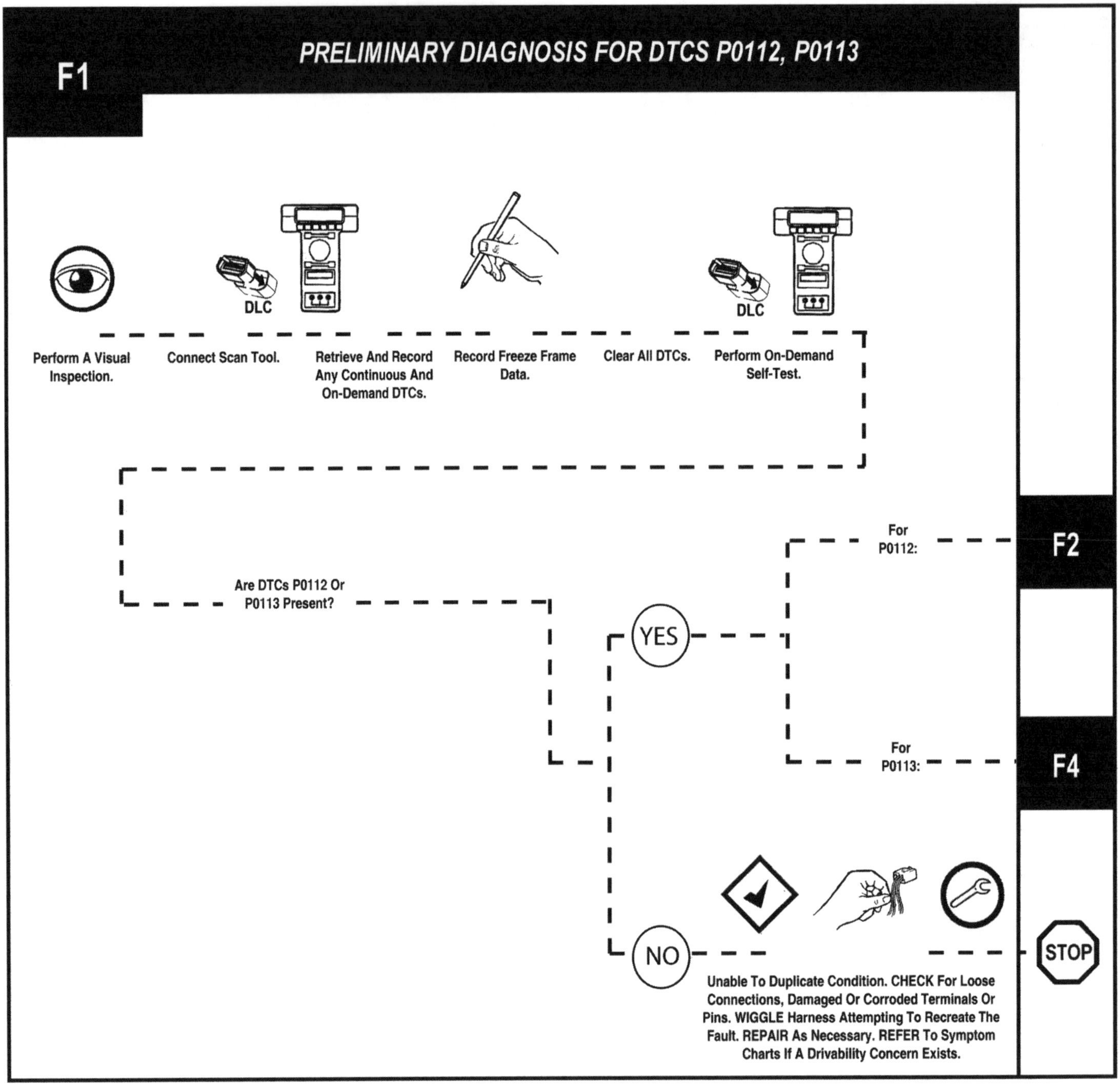

F1
PRELIMINARY DIAGNOSIS FOR DTCS P0112, P0113
DLC
DLC
Perform A Visual Inspection.
Connect Scan Tool.
Retrieve And Record Any Continuous And On-Demand DTCs.
Record Freeze Frame Data.
Clear All DTCs.
Perform On-Demand Self-Test.
Are DTCs P0112 Or P0113 Present?
YES
For P0112:
F2
For P0113:
F4
NO
Unable To Duplicate Condition. CHECK For Loose Connections, Damaged Or Corroded Terminals Or Pins. WIGGLE Harness Attempting To Recreate The Fault. REPAIR As Necessary. REFER To Symptom Charts If A Drivability Concern Exists.
STOP

F2 — DIAGNOSTIC TROUBLE CODE (DTC) P0112

Note: Refer To The PCM And Component Connector At The Beginning Of This Pinpoint Test.

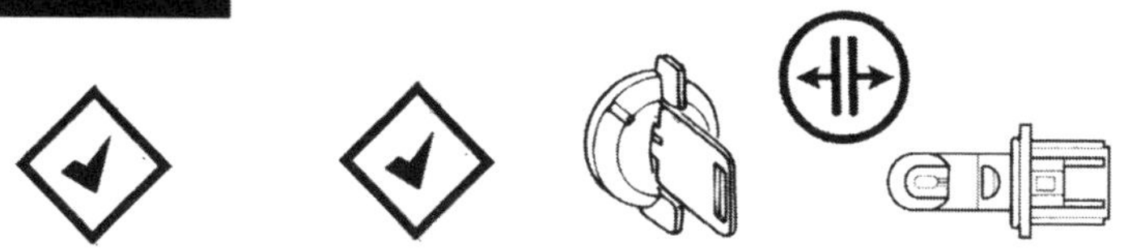

Possible Causes:
- Circuit Shorted To Ground
- IAT Sensor
- PCM

Check For Opposite Failure.

Key Off.

Disconnect The IAT Sensor.

Key On. Engine Off.

Perform On-Demand Self-Test.

Is DTC P0113 Present?

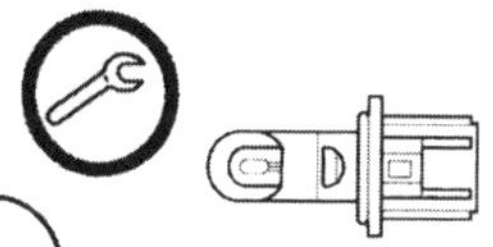

YES — INSTALL A New IAT Sensor. RESTORE The Vehicle. CLEAR DTCs And RETEST The System. STOP

NO — F3

F3 — CHECK THE IAT SIGNAL CIRCUIT FOR A SHORT TO GROUND

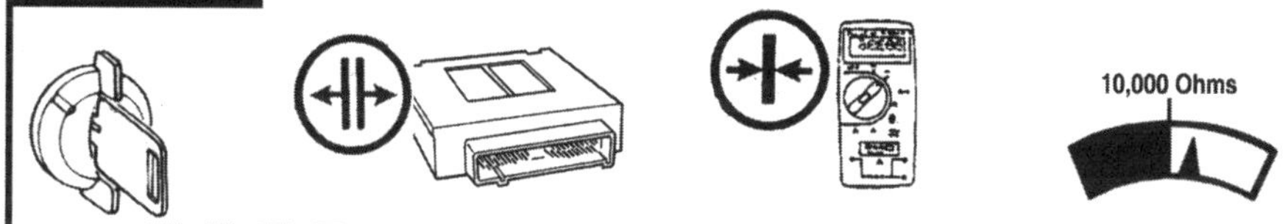

Key Off

Disconnect The PCM Body Harness Connector.

Measure the resistance between:	
(+) IAT Sensor Connector, Harness Side	(-)
E-Series: Pin 1	Ground
F-Super Duty vehicles with an integrated MAF/IAT sensor: Pin 6	Ground
F-Super Duty vehicles with a stand alone IAT sensor: Pin 2	Ground

Is The Resistance Greater Than 10,000 Ohms?

YES — F8

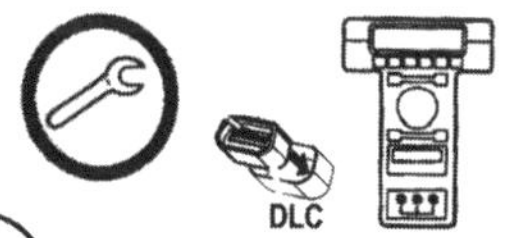

NO — REPAIR Short To Ground In The IAT Signal Circuit. RESTORE The Vehicle. CLEAR DTCs And RETEST The System. STOP

F4 DIAGNOSTIC TROUBLE CODE P0113

Note: Refer To The PCM And Component Connector At The Beginning Of This Pinpoint Test.

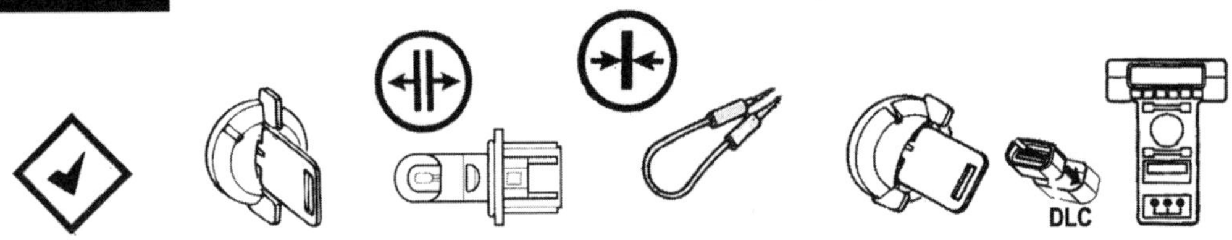

Possible Causes:
- Signal Circuit Open Or Shorted To Voltage
- Open In Signal Return Circuit
- IAT Sensor
- PCM

Key Off.

Disconnect The IAT Sensor.

Connect a 5 amp fused jumper wire between the following:

	Point A MAF/IAT Sensor Connector, Harness Side	Point B MAF/IAT Sensor Connector, Harness Side
E-Series	Pin 1	Pin 2
F-Super Duty vehicles with an integrated MAF/IAT sensor	Pin 6	Pin 1
F-Super Duty vehicles with a stand alone IAT sensor	Pin 2	Pin 1

Key On. Engine Off.

Perform On-Demand Self-Test.

Is DTC P0112 Present?

YES → 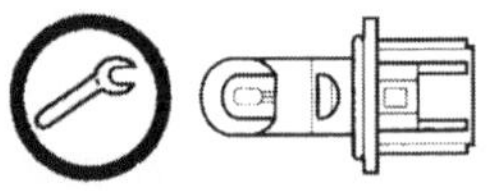

INSTALL A New IAT Sensor. RESTORE The Vehicle. CLEAR DTCs And RETEST The System. STOP

NO → F5

F5 CHECK THE SIGNAL CIRCUIT FOR AN OPEN

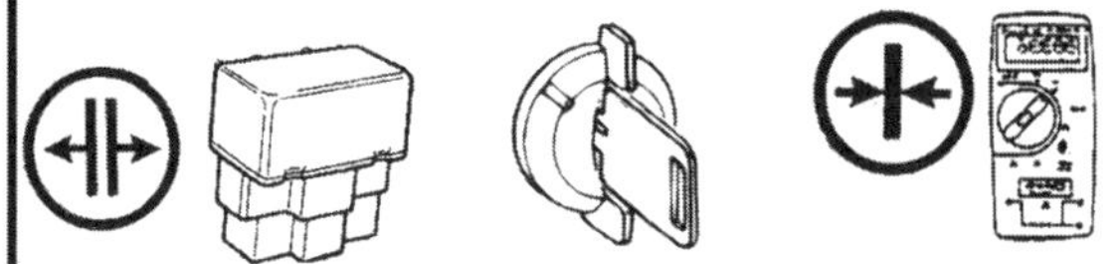

PCM Power Relay Connector Disconnected.

Key Off.

Measure the resistance between:	
(+) PCM Connector, Harness Side	(-) MAF/IAT Sensor Connector, Harness Side
IAT - Pin B43	E-Series: Pin 1
IAT - Pin B43	F-Super Duty vehicles with an integrated MAF/IAT sensor : Pin 6
IAT - Pin B43	F-Super Duty vehicles with a stand alone IAT sensor: Pin 2

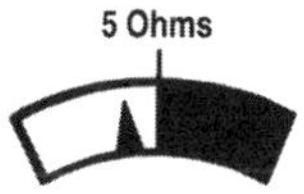

Is The Resistance Less Than 5 Ohms?

YES → F6

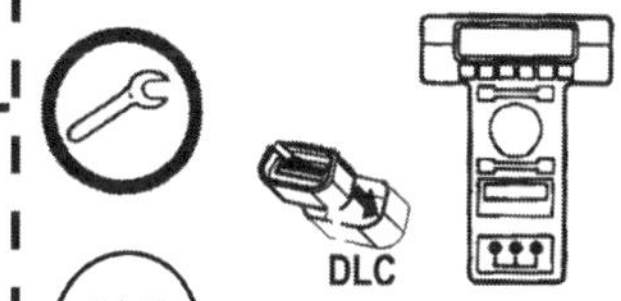

NO → STOP

REPAIR Open In The IAT Signal Circuit. CLEAR The DTCs. REPEAT The Self-Test.

F6 CHECK THE SIGNAL RETURN CIRCUIT FOR AN OPEN

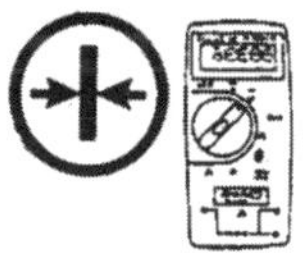

	Measure the resistance between:	
	(+) MAF/IAT Sensor Connector, Harness Side	(-) PCM Connector, Harness Side
E-Series	Pin 2	SIGRTN - Pin B20, B33, E25
F-Super Duty vehicles with an integrated MAF/IAT sensor	Pin 1	SIGRTN - Pin B20, B33, E25
F-Super Duty vehicles with a stand alone IAT sensor	Pin 1	SIGRTN - Pin B20, B33, E25

Is The Resistance Less Than 5.0 Ohms?

YES → F7

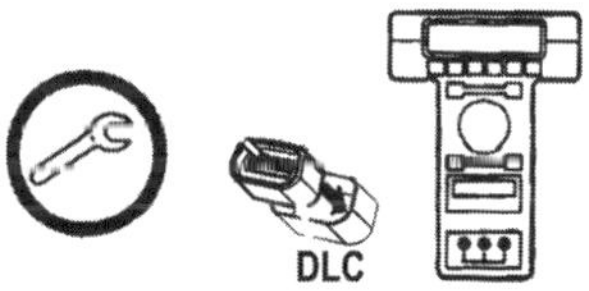

NO → STOP

REPAIR The Open In The IAT Sensor Signal Return Circuit. CLEAR The DTCs. REPEAT The Self-Test.

F7 CHECK THE IAT SIGNAL CIRCUIT FOR A SHORT TO VOLTAGE

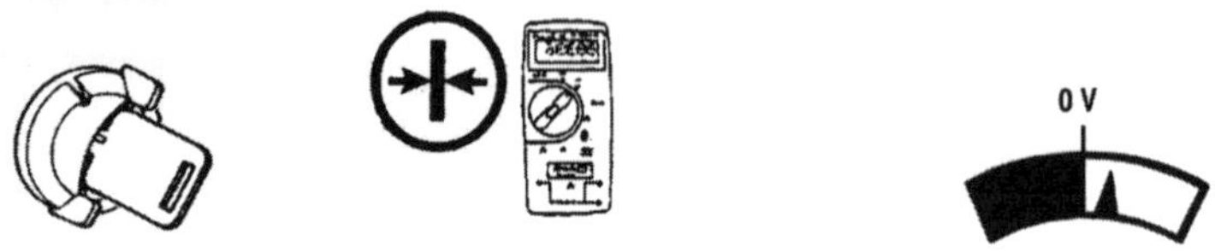

Key On.
Engine Off.

Measure the voltage between:	(+) MAF/IAT Sensor Connector, Harness Side	(-)
E-Series	Pin 1	Ground
F-Super Duty vehicles with an integrated MAF/IAT sensor	Pin 6	Ground
F-Super Duty vehicles with a stand alone IAT sensor	Pin 2	Ground

Is The Voltage Greater Than 0 V?

YES — REPAIR The Short Circuit To PWR. CLEAR The DTCs. REPEAT The Self-Test. STOP

NO — F8

F8 CHECK FOR CORRECT PCM OPERATION

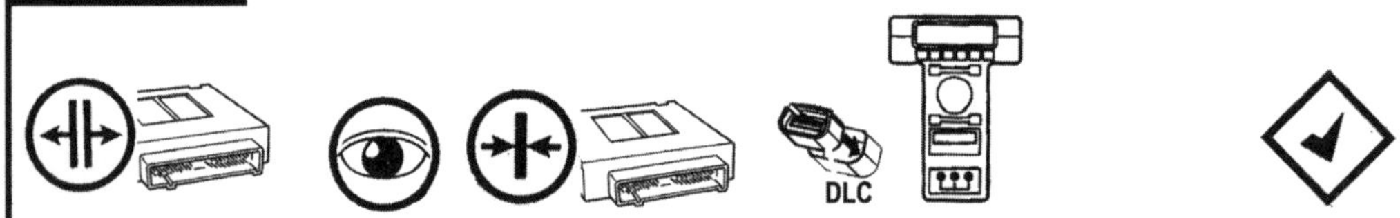

Disconnect All PCM Connectors.

Visually Inspect For Pushed Out Pins, Corrosion.

Connect All PCM Connectors And Make Sure They Seat Correctly.

Perform The PCM Self-Test And Verify The Concern Is Still Present.

Is The Concern Still Present.

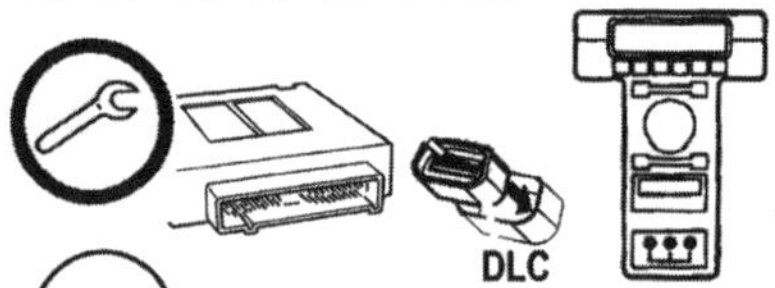

YES — INSTALL A New PCM. CLEAR The DTC's And REPEAT The Self-Test. STOP

OK

NO — The System Is Operating Correctly At This Time. The Concern May Have Been Caused By A Loose Or Corroded Connector. STOP

Trouble Code (s)	Problem: Intake Air Temperature 2 (IAT2) Sensor	Test
P0096, P0097, P0098		G 2006-07

REFERENCE LIST

This pinpoint test is intended to diagnose the following:

- Harness circuits: IAT2 and SIGRTN
- IAT2 sensor
- Powertrain control module (PCM) (12A650)

DIAGNOSTIC TROUBLE CODE INDEX			
Codes	KOEO	KOER	CM
P0096	G1	G1	G1
P0097	G1	G1	G1
P0098	G1	G1	G1

P0096 = IAT2 Sensor 2 Circuit Range/Performance
P0097 = IAT2 Sensor 2 Circuit Low Input
P0098 = IAT2 Sensor 2 Circuit High Input

Circuit Function:
The intake air temperature 2 (IAT2) sensor is a thermistor-type sensor with a variable resistance that is inversely proportional to temperature changes. When interfaced with the powertrain control module (PCM), it produces a 0-5 volt analog signal that will measure temperature.

The primary function of the IAT2 sensor is to provide a feedback signal to the PCM indicating manifold air temperature.

Fault Detection/Management:
The PCM continuously monitors the IAT2 sensor. If the PCM detects the signal voltage is higher or lower than expected, a DTC will be set.

CAUTION!:
The powertrain control module (PCM) harness connectors must be properly seated and the connector latch properly attached to eliminate possible driveability concerns or a no start condition. Installing the PCM connectors on an angle may cause an improper connection, misdiagnosis and damaged components. Install the connector until the lever pivots and seats itself. Apply light pressure to get the connector into position on the PCM and then fully seat the connector.

Note: Visually inspect the harness connectors for corrosion, damage, proper mating and correct pin tension.

Note: When the PCM is disconnected additional DTCs will be set. Clear all DTCs after restoring the vehicle.

PINPOINT TEST SCHEMATICS AND CONNECTORS

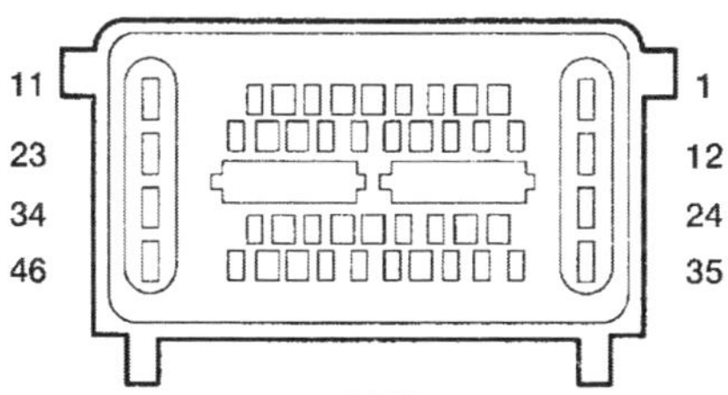

PCM
ENGINE HARNESS
CONNECTOR

Pin	Circuit
E25	SIGRTN (Signal Return)
E45	IAT2 (Intake Air Temperature 2)

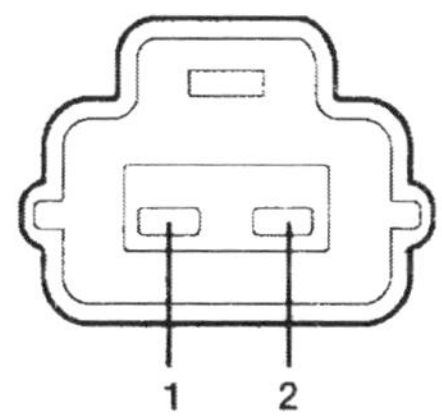

SENSOR
CONNECTOR

Pin	Circuit
1	SIGRTN (Signal Return)
2	IAT2 (Intake Air Temperature 2)

G1 PRELIMINARY DIAGNOSIS FOR DTCS P0096, P0097, P0098

Perform The Visual Inspection.

Connect The Scan Tool. (DLC)

Retrieve And Record Any Continuous And On-Demand DTCs. Record Freeze Frame Data

Clear All DTCs.

Perform The KOEO And KOER Self-Test. (DLC)

Are DTCs P0096, P0097 Or P0098 Present?

- YES — For P0096: **G2**
- YES — For P0097: **G5**
- YES — For P0098: **G7**
- NO — Unable To Duplicate Condition. CHECK For Loose Connection, Damaged Or Corroded Terminals Or Pins. WIGGLE Harness Attempting To Recreate The Fault. REPAIR As Necessary. REFER To Symptom Charts If A Drivability Concern Exists. **STOP**

G2 DIAGNOSTIC TROUBLE CODE P0096

Possible Causes:
- IAT2 Sensor Circuitry
- IAT2 Sensor
- PCM

Inspect IAT2 Sensor Circuitry At Sensor And PCM.

Are All Connectors And Terminals OK?

- YES — **G3**
- NO — REPAIR As Necessary. RESTORE The Vehicle. CLEAR DTCs And RETEST The System. (DLC) **STOP**

G3 — CHECK FOR IAT2 SENSOR DTCS

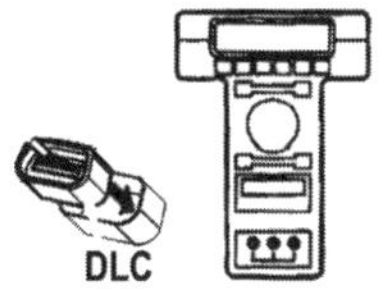

Carry Out The KOEO And KOER Self-Test.

Are Other DTCs Present?

- YES — Refer To Diagnostic Trouble Code (DTC) Index. → DTC INDEX
- NO — Only P0096 Present: → G4

G4 — CHECK IAT2 SENSOR OPERATION

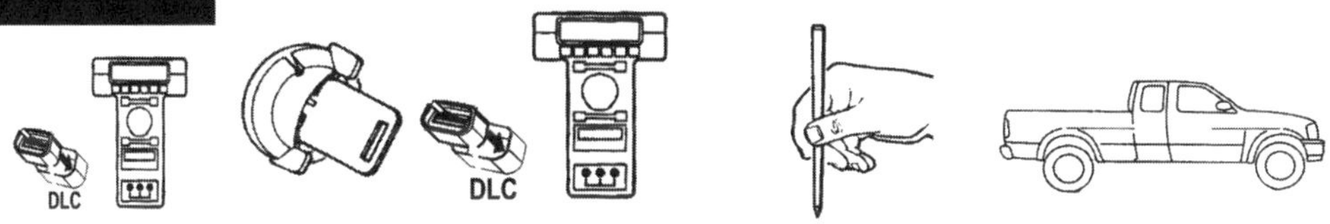

Clear Continuous DTCs.

Key On. Engine Off.

Using Scan Tool, Monitor IAT2 And Engine Oil Temperature PIDS.

Record IAT2 Sensor Temperature Reading.

Drive Vehicle While Monitoring IAT2 Sensor Reading.

Operate Vehicle Until It Reaches Normal Operating Temperature, EOT Above 176°F.

Idle Engine For Ten Minutes.

Did IAT2 Readings Change From Value Recorded At Any Time?

- YES — Unable To Duplicate The Condition. CHECK For Loose Connections, Damaged Or Corroded Terminals. REPAIR As Necessary. REFER To Symptom Charts If Drivability Concern Exists. → STOP

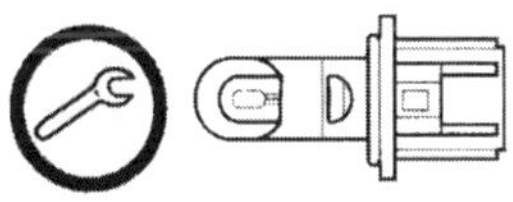

- NO — INSTALL A New IAT2 Sensor. CLEAR DTCs . REPEAT The Self-Test. → STOP

G5 DIAGNOSTIC TROUBLE CODE P0097

Note: Refer To The PCM And Component Connector At The Beginning Of This Pinpoint Test.

Possible Causes:
- Circuit Shorted To Ground
- IAT2 Sensor
- PCM

Check For Opposite Failure.

Key Off.

Disconnect The IAT2 Sensor.

DLC

Perform The KOEO And KOER Self-Test.

Is DTC P0098 Present?

YES — INSTALL A New IAT2 Sensor. CLEAR DTCs. REPEAT The Self-Test. STOP

NO — G6

G6 CHECK SIGNAL CIRCUIT FOR A SHORT TO GROUND

Key Off.

Disconnect The PCM Engine Harness Connector.

Measure the resistance between:	
(+) IAT2 Sensor Connector, Harness Side	(-)
IAT2 - Pin 2	Ground

1 2
SENSOR CONNECTOR

10,000 Ohms

Is The Resistance Greater Than 10,000 Ohms?

YES — G11

NO — DLC — REPAIR Short To Ground In The IAT2 Sensor Signal Circuit. CLEAR DTCs. REPEAT The Self Test.P — STOP

G7 — DIAGNOSTIC TROUBLE CODE P0098

Note: Refer To The PCM And Component Connector At The Beginning Of This Pinpoint Test.

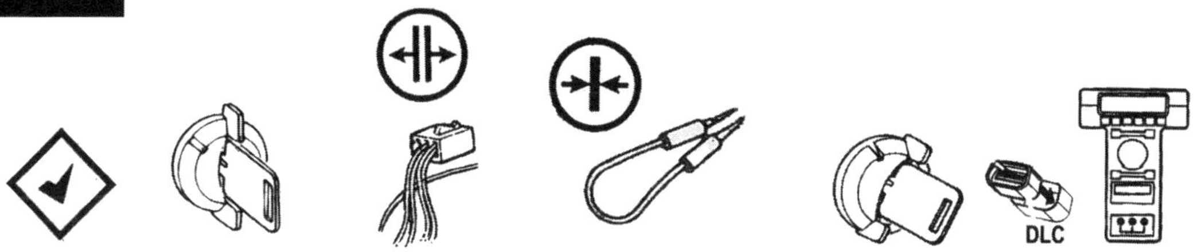

Possible Causes:
- Circuit Shorted To Voltage
- Open In Signal Return Circuit
- IAT2 Sensor
- PCM

Key Off.

Disconnect The IAT2 Sensor.

Connect a 5 amp fused jumper wire between the following:	
Point A IAT2 Sensor Connector, Harness Side	Point B IAT2 Sensor Connector, Harness Side
SIGRTN - Pin 1	IAT2 - Pin 2

Key On. Engine Off.

Perform The KOEO And KOER Self-Test.

Is DTC P0097 Present?

YES → INSTALL A New IAT2 Sensor. CLEAR The DTCs. REPEAT The Self-Test. STOP

NO → G8

G8 — CHECK THE SIGNAL CIRCUIT FOR AN OPEN

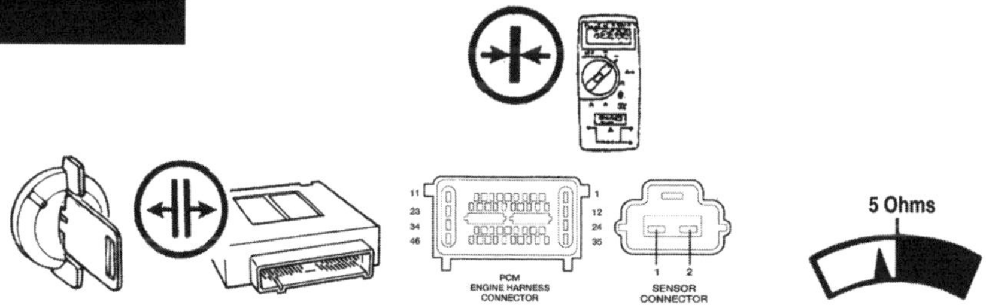

Key Off.

Disconnect The PCM Engine Harness Connector.

Measure the resistance between:	
(+) PCM Connector, Harness Side	(-) IAT2 Sensor Connector, Harness Side
IAT2 - Pin E45	IAT2 - Pin 2

Is The Resistance Less Than 5 Ohms?

YES → G9

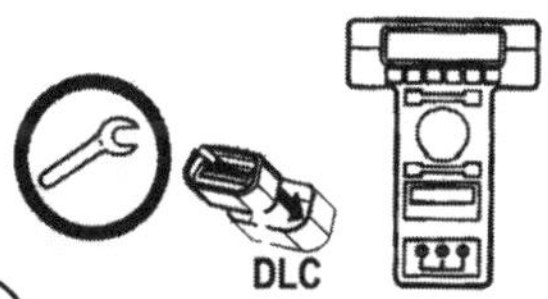

NO → REPAIR Open In The IAT2 Signal Circuit. CLEAR The DTCs. REPEAT The Self Test. STOP

G9 — CHECK THE SIGNAL RETURN CIRCUIT FOR AN OPEN

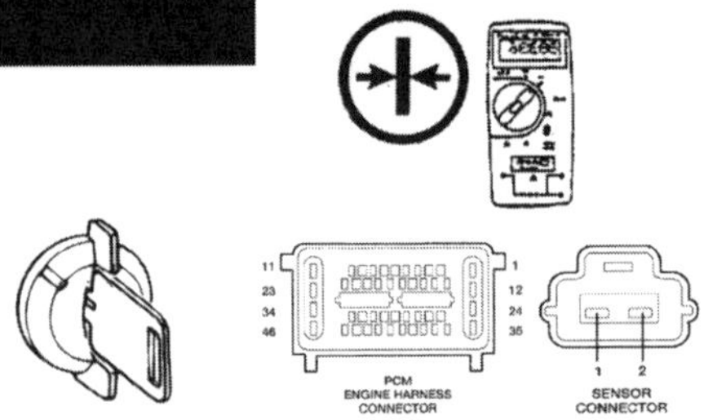

Key Off.

Measure the resistance between:	
(+) IAT2 Sensor Connector, Harness Side	(-) PCM Connector, Harness Side
SIGRTN - Pin 1	SIGRTN - Pin E25

Is The Resistance Less Than 5 Ohms?

YES — G10

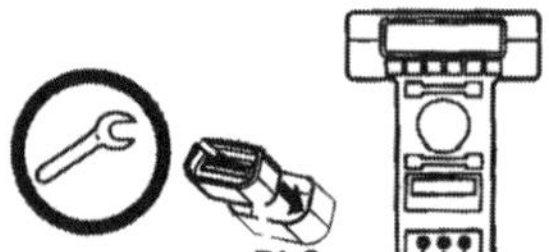

NO — REPAIR Open In The IAT2 Signal Circuit. CLEAR The DTCs. REPEAT The Self Test — STOP

G10 — CHECK THE SIGNAL CIRCUIT FOR A SHORT TO VOLTAGE

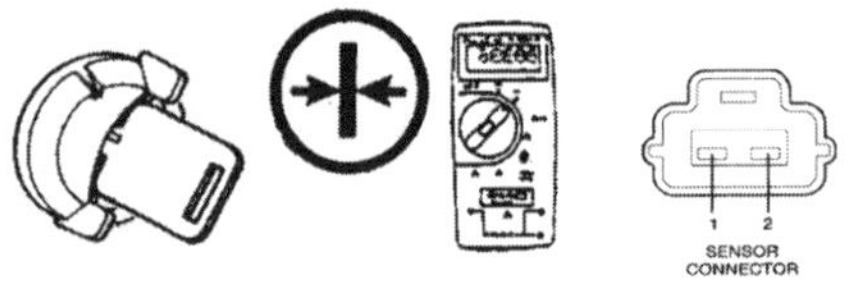

Key On. Engine Off.

Measure the voltage between:	
(+) IAT2 Sensor Connector, Harness Side	(-)
IAT2 - Pin 2	Ground

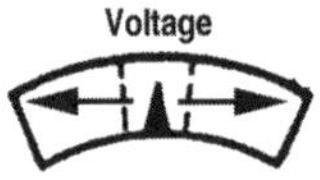

Is Any Voltage Indicated?

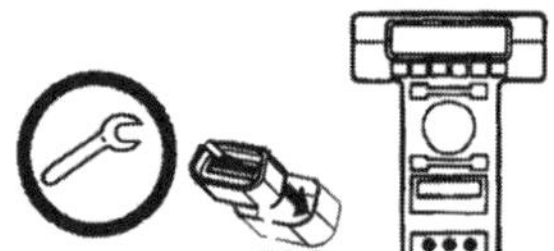

YES — REPAIR The Short To Voltage In The IAT2 Sensor Signal Circuit. CLEAR The DTCs. REPEAT The Self-Test. — STOP

NO — G11

G11 CHECK FOR CORRECT PCM OPERATION

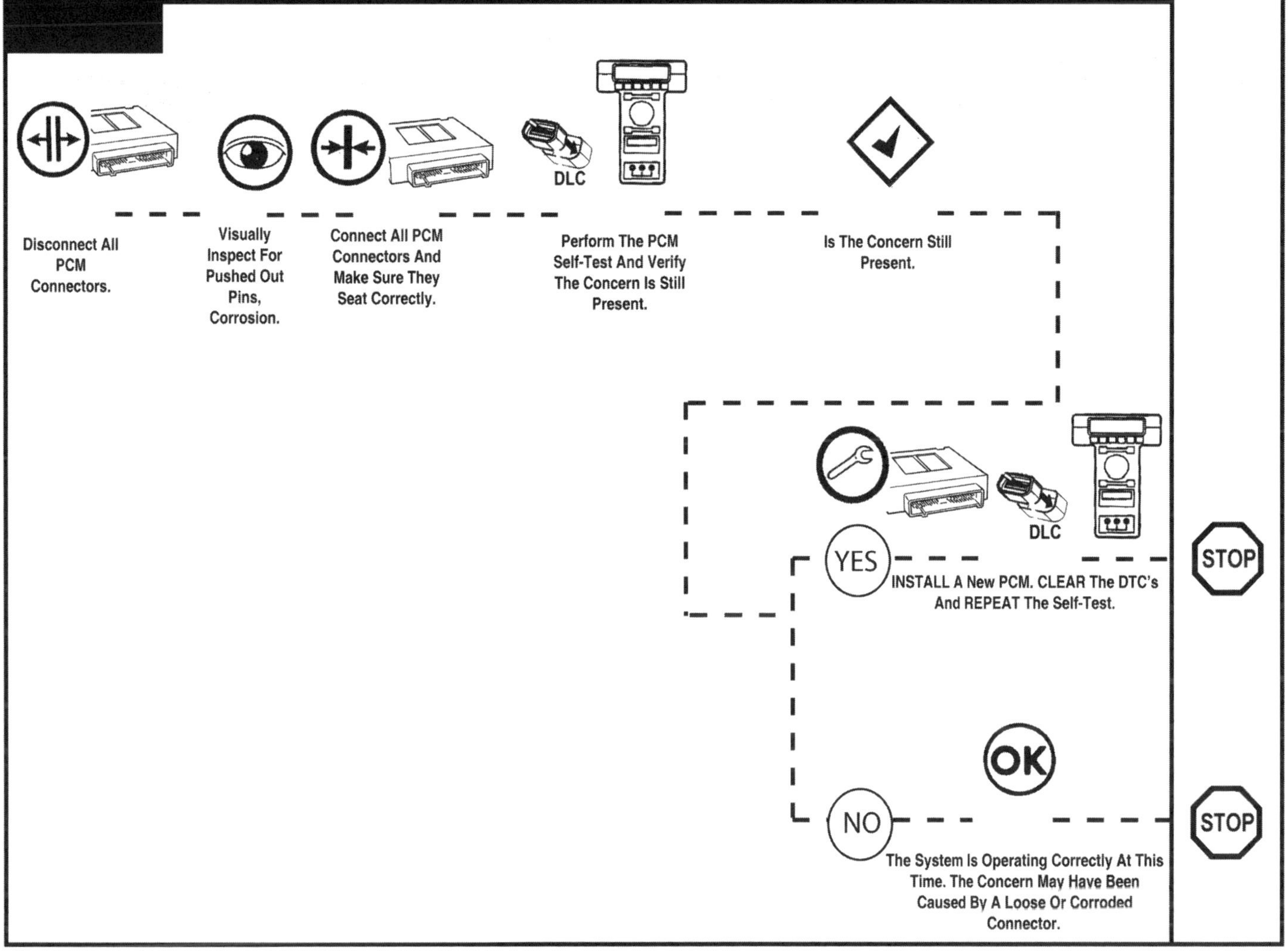

NOTES

Trouble Code (s)	Problem: Barometric Pressure (BARO) Sensor	Test
P0107, P0108		H 2006-07

REFERENCE LIST

This pinpoint test is intended to diagnose the following:

- Harness circuits: BARO, SIGRTN, and VREF
- BARO sensor
- Powertrain control module (PCM) (12A650)

DIAGNOSTIC TROUBLE CODE INDEX			
Codes	KOEO	KOER	CM
P0107	H1	H1	H1
P0108	H1	H1	H1

P0107 = BARO Sensor Low Input
P0108 = BARO Sensor High Input

BARO Sensor

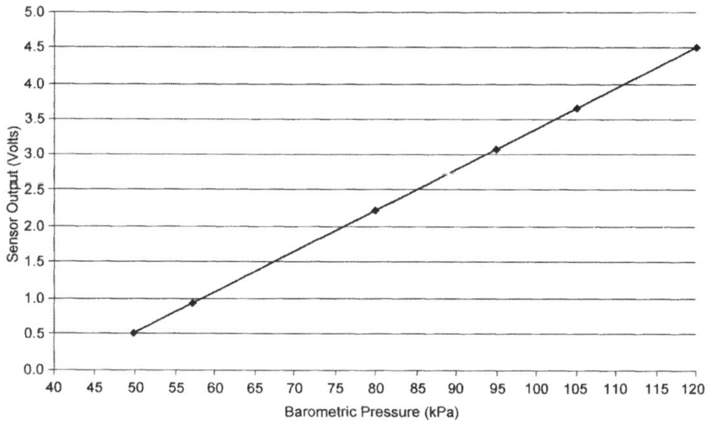

Circuit Function:
The barometric pressure (BARO) sensor is a cab-mounted variable capacitance sensor used to determine altitude. The BARO signal affects injection timing and fuel quantity to optimize engine operation and control smoke throughout all altitude conditions. The BARO signal is one of the variables used to calculate glow plug ON time. At higher altitudes, glow plug ON time is increased to reduce start-up smoke.

Fault Detection/Management:
A BARO signal that is detected out of range high or low will cause the powertrain control module (PCM) to ignore the BARO signal and use the manifold absolute pressure (MAP) signal generated at low idle as an indication of barometric pressure.

CAUTION!:
The powertrain control module (PCM) harness connectors must be properly seated and the connector latch properly attached to eliminate possible driveability concerns or a no start condition. Installing the PCM connectors on an angle may cause an improper connection, misdiagnosis and damaged components. Install the connector until the lever pivots and seats itself. Apply light pressure to get the connector into position on the PCM and then fully seat the connector.

Note: Visually inspect the harness connectors for corrosion, damage, proper mating and correct pin tension.

PINPOINT TEST SCHEMATICS AND CONNECTORS

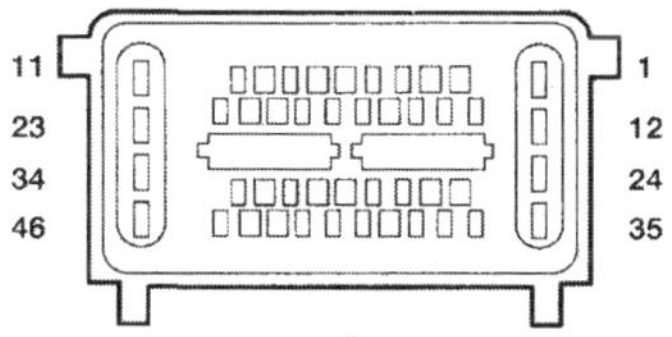

B
BODY/COWL

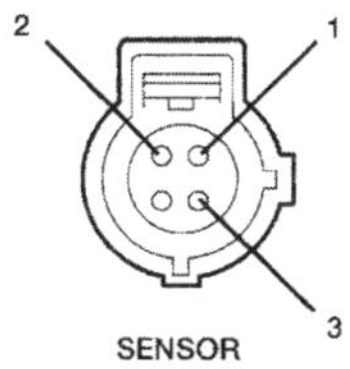

SENSOR CONNECTOR

Pin	Circuit
B20, B33, E25	SIGRTN (Signal Return)
B38	BARO (Barometric Pressure)

Pin	Circuit
3	BARO (Barometric Pressure)
1	SIGRTN (Signal Return)
2	VREF (Reference Voltage)

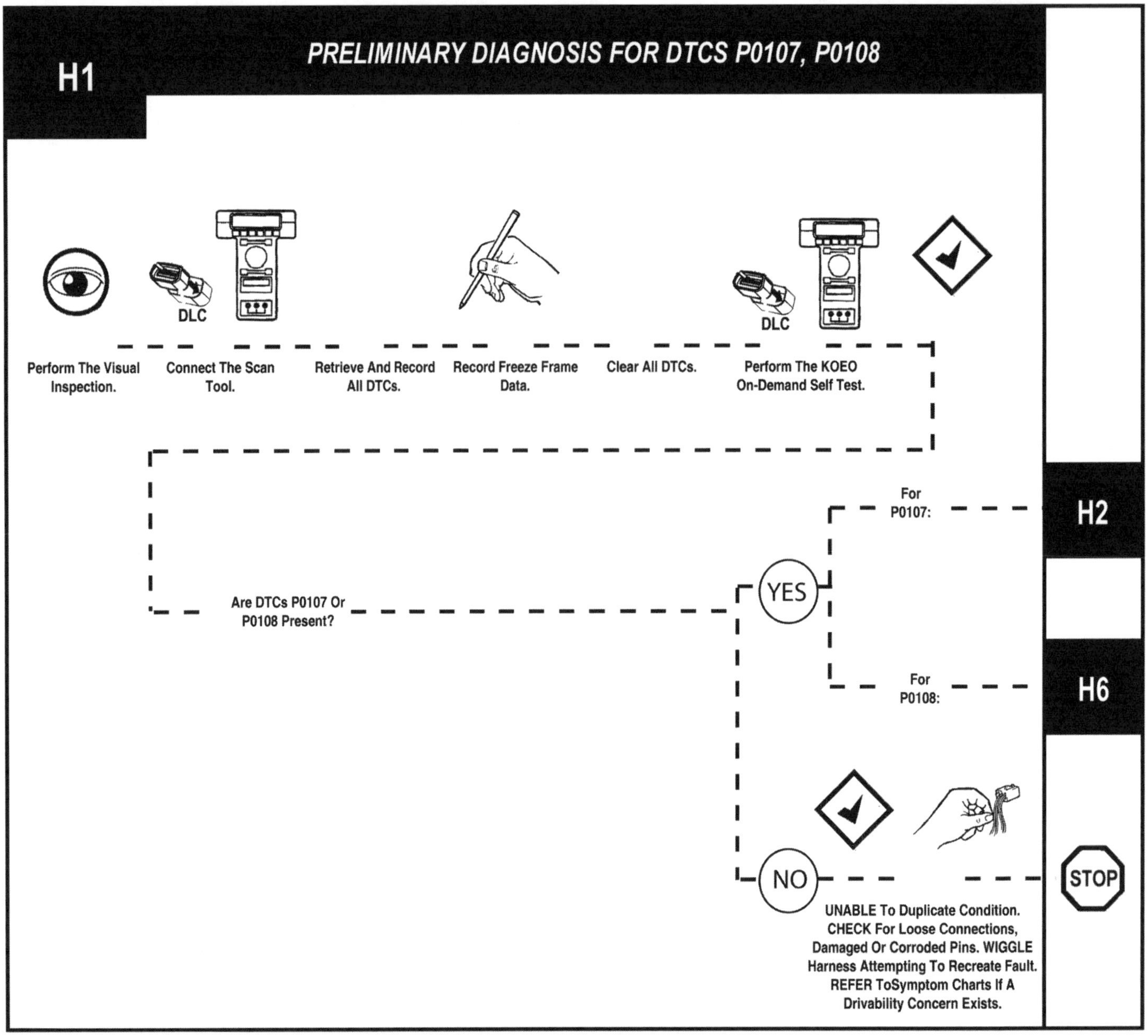
H1
PRELIMINARY DIAGNOSIS FOR DTCS P0107, P0108
DLC
DLC
Perform The Visual Inspection.
Connect The Scan Tool.
Retrieve And Record All DTCs.
Record Freeze Frame Data.
Clear All DTCs.
Perform The KOEO On-Demand Self Test.
Are DTCs P0107 Or P0108 Present?
YES
For P0107:
H2
For P0108:
H6
NO
UNABLE To Duplicate Condition. CHECK For Loose Connections, Damaged Or Corroded Pins. WIGGLE Harness Attempting To Recreate Fault. REFER ToSymptom Charts If A Drivability Concern Exists.
STOP

H2 — DIAGNOSTIC TROUBLE CODE (DTC) P0107

Note: Refer To The PCM And Component Connector Pin Numbers At The Beginning Of This Pinpoint Test.

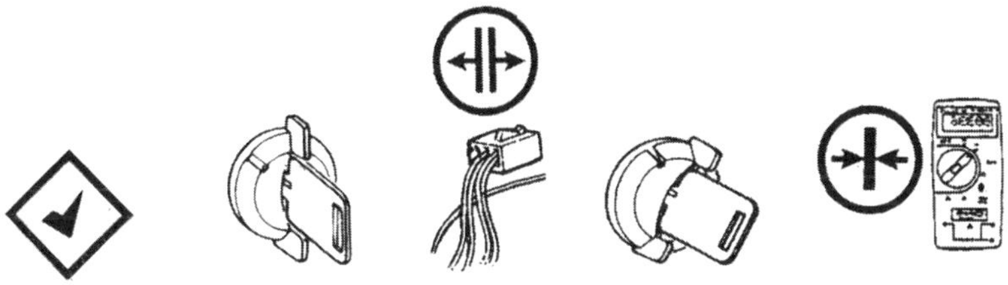

Possible Causes:
- Signal Circuit Short To Ground
- VREF Circuit Open
- BARO Sensor
- PCM

Key Off.

Disconnect BARO Sensor Electrical Connector.

Key On. Engine Off.

Measure the voltage between:	
(+) BARO Connector, Harness Side	(-) BARO Connector, Harness Side
VREF - Pin 2	SIGRTN - Pin 1

Is The VREF Voltage Between 4.5 And 5.5 Volts?

YES → H3

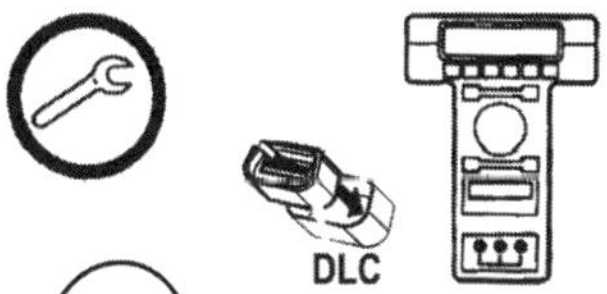

NO → STOP

REPAIR The Open In The Reference Voltage Or Signal Return Circuit. CLEAR The DTCs. REPEAT The Self-Test.

H3 — SIMULATE THE OPPOSITE DTC

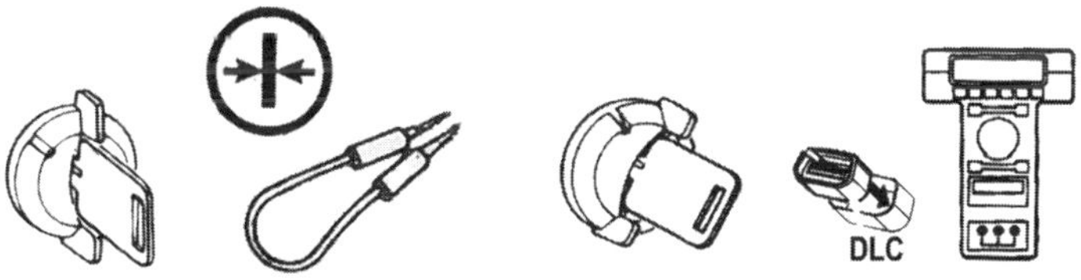

Key Off.

Connect a 5 amp fused jumper wire between the following:	
Point A BARO Connector, Harness Side	Point B BARO Connector, Harness Side
VREF - Pin 2	BARO - Pin 3

Key On. Engine Off.

Perform The KOEO On-Demand Self Test.

Is DTC P0108 Present?

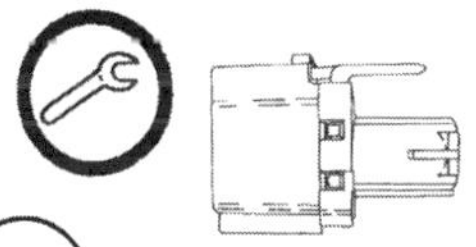

YES → STOP

INSTALL A New BARO Sensor. CLEAR DTCs And REPEAT The Self-Test.

NO → H4

H4 CHECK THE SIGNAL CIRCUIT FOR A SHORT TO GROUND

Key Off.

Disconnect The PCM Body Harness Electrical Connector.

Measure the resistance between:	
(+) BARO Connector, Harness Side	(-)
BARO - Pin 3	Ground

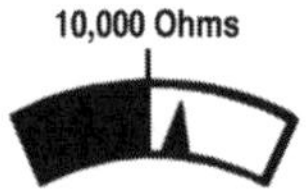

Is The Resistance Greater Than 10,000 Ohms?

YES → H5

NO → REPAIR The Short To Ground. CLEAR DTCs. REPEAT The Self-Test. STOP

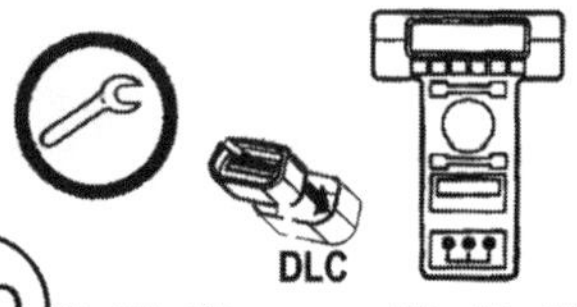

H5 CHECK SIGNAL CIRCUIT FOR AN OPEN

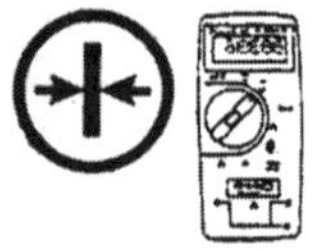

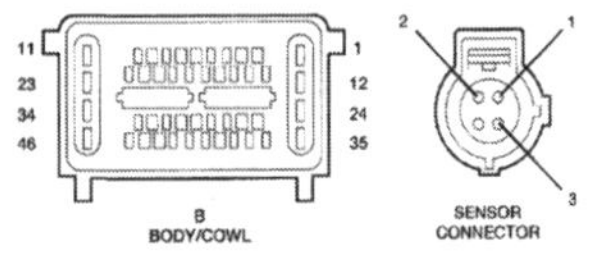

Measure the resistance between:	
(+) BARO Connector, Harness Side	(-) PCM Connector, Harness Side
BARO - Pin 3	BARO - Pin B38

Is The Resistance Less Than 5 Ohms?

YES → H9

NO → REPAIR The Open Circuit. CLEAR The DTCs. REPEAT The Self-Test STOP

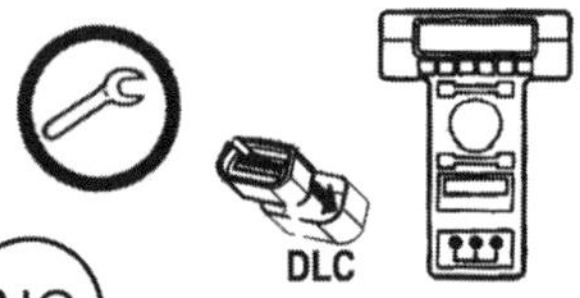

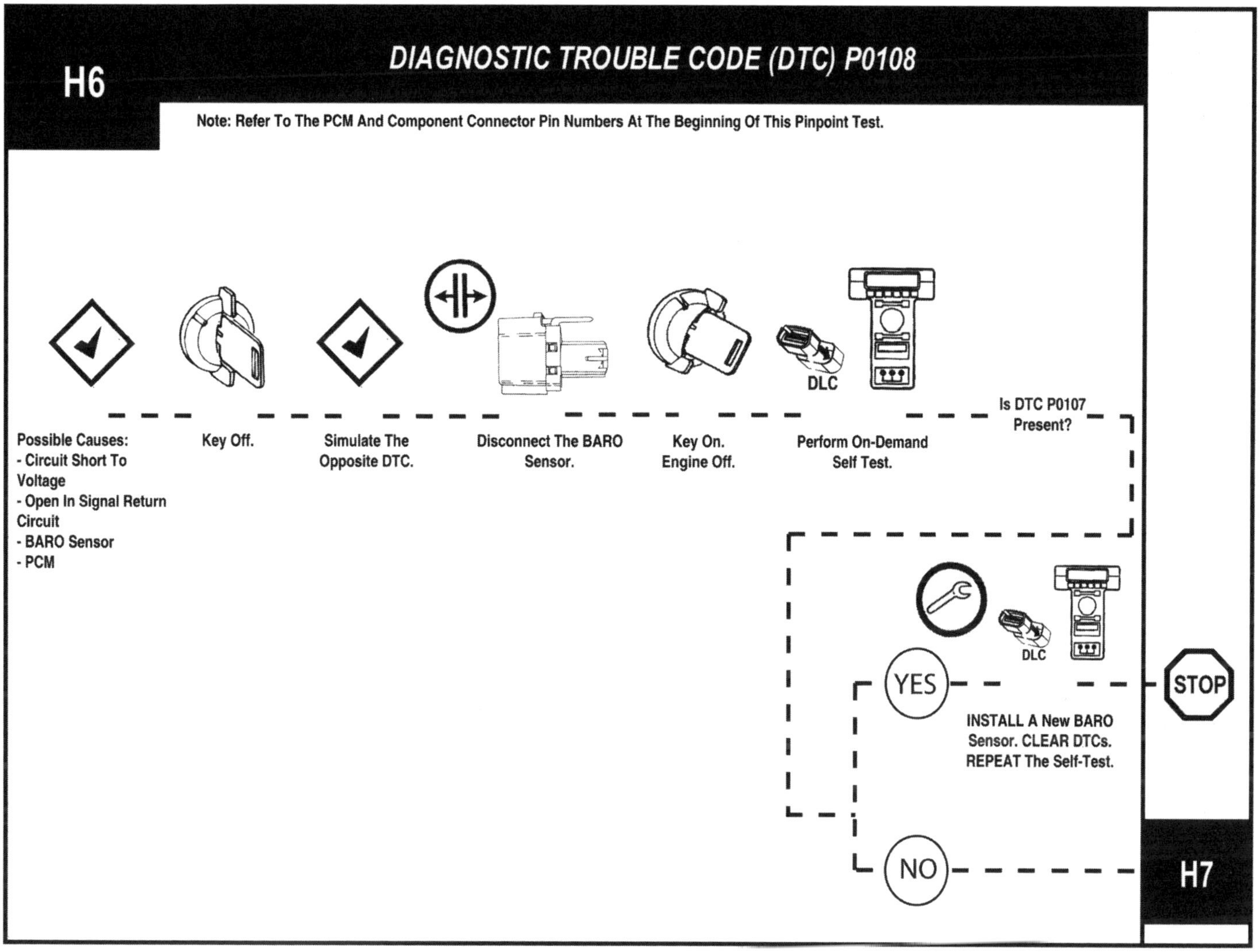
H6
DIAGNOSTIC TROUBLE CODE (DTC) P0108
Note: Refer To The PCM And Component Connector Pin Numbers At The Beginning Of This Pinpoint Test.
DLC
Possible Causes:
- Circuit Short To Voltage
- Open In Signal Return Circuit
- BARO Sensor
- PCM
Key Off.
Simulate The Opposite DTC.
Disconnect The BARO Sensor.
Key On. Engine Off.
Perform On-Demand Self Test.
Is DTC P0107 Present?
DLC
YES
STOP
INSTALL A New BARO Sensor. CLEAR DTCs. REPEAT The Self-Test.
NO
H7

H7 CHECK THE SIGNAL CIRCUIT FOR A SHORT TO VOLTAGE

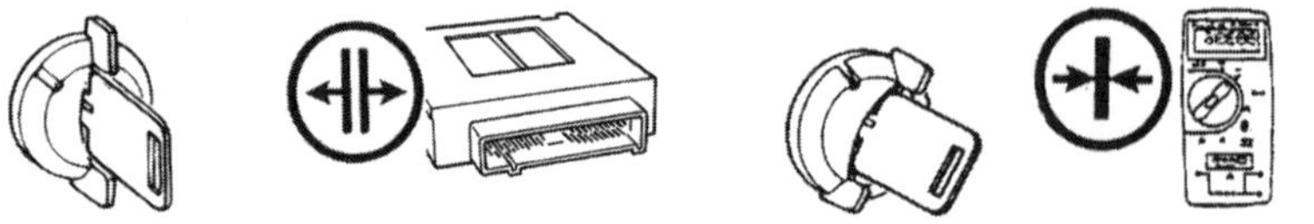

Key Off.

Disconnect The PCM Body Harness Electrical Connector.

Key On. Engine Off.

Measure the voltage between:	
(+) BARO Connector, Harness Side	(-)
BARO - Pin 3	Ground

Is Any Voltage Indicated?

YES

REPAIR The Short To Voltage. CLEAR The DTCs. REPEAT The Self-Test

NO → H8

H8 CHECK THE SIGNAL RETURN CIRCUIT FOR AN OPEN

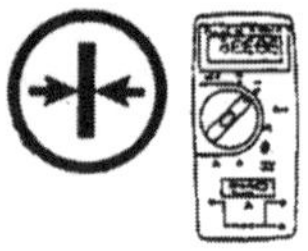

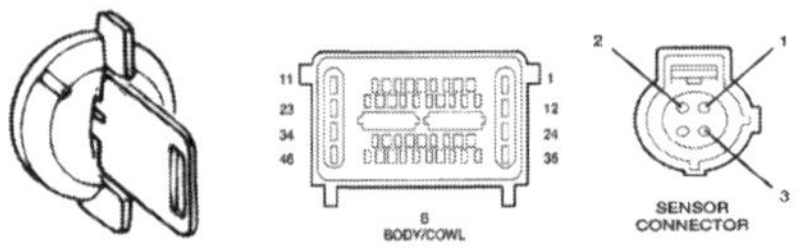

Key Off.

Measure the resistance between:	
(+) BARO Connector, Harness Side	(-) PCM Connector, Harness Side
SIGRTN - Pin 1	SIGRTN - Pin B20, B33, E25

Is The Resistance Less Than 5 Ohms?

YES → H9

NO

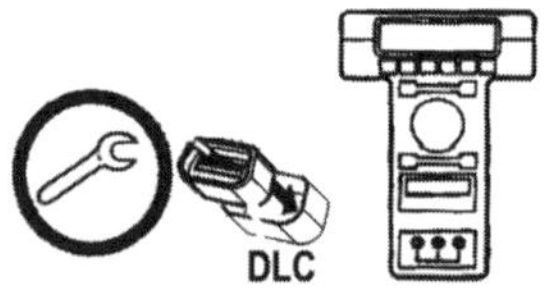

REPAIR The Open Circuit. CLEAR The DTCs. REPEAT The Self-Test.

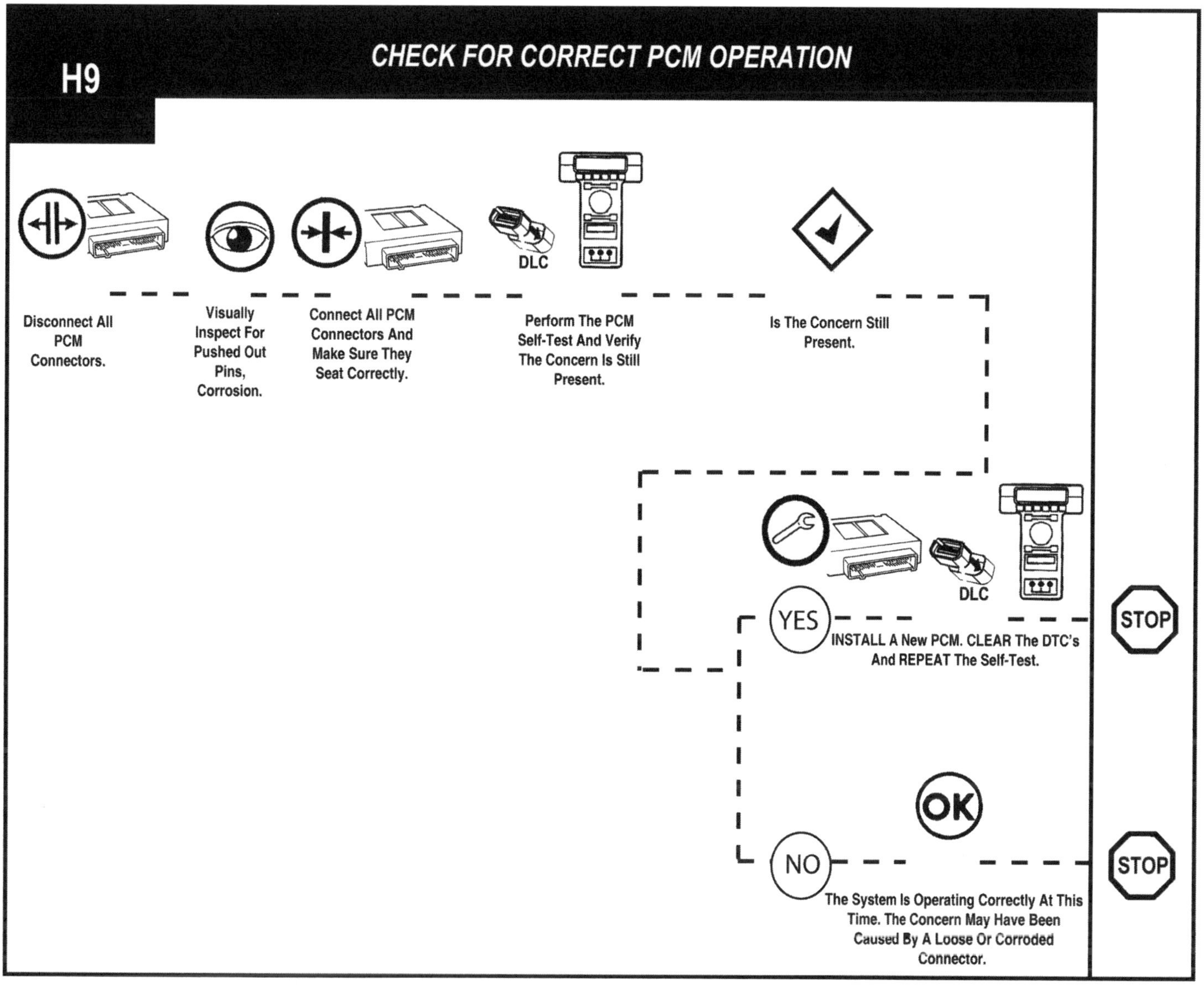
H9
CHECK FOR CORRECT PCM OPERATION
Disconnect All PCM Connectors.
Visually Inspect For Pushed Out Pins, Corrosion.
Connect All PCM Connectors And Make Sure They Seat Correctly.
DLC
Perform The PCM Self-Test And Verify The Concern Is Still Present.
Is The Concern Still Present.
DLC
YES
INSTALL A New PCM. CLEAR The DTC's And REPEAT The Self-Test.
STOP
OK
NO
The System Is Operating Correctly At This Time. The Concern May Have Been Caused By A Loose Or Corroded Connector.
STOP

NOTES

Trouble Code (s)	Problem: Parking Brake Applied (PBA) Switch	Test
P1536		I 2006-07

REFERENCE LIST

This pinpoint test is intended to diagnose the following:

- PBA switch
- Parking brake signal circuit short to ground
- Parking brake signal circuit open
- Powertrain control module (PCM) (12A650)

DIAGNOSTIC TROUBLE CODE INDEX			
Codes	KOEO	KOER	CM
P1536	I1	I1	I1

P1536 = Parking Brake Applied Fail

Signal Function:
The parking brake switch provides an input signal to the powertrain control module (PCM) indicating the current status of the parking brake (applied or released). The parking brake signal circuit is pulled to ground with the application of the parking brake.

CAUTION!:
The powertrain control module (PCM) harness connectors must be properly seated and the connector latch properly attached to eliminate possible driveability concerns or a no start condition. Installing the PCM connectors on an angle may cause an improper connection, misdiagnosis and damaged components. Install the connector until the lever pivots and seats itself. Apply light pressure to get the connector into position on the PCM and then fully seat the connector.

Note: Visually inspect the harness connectors for corrosion, damage, proper mating and correct pin tension.

Note: When the PCM is disconnected additional DTCs will be set. Clear all DTCs after restoring the vehicle.

PINPOINT TEST SCHEMATICS AND CONNECTORS

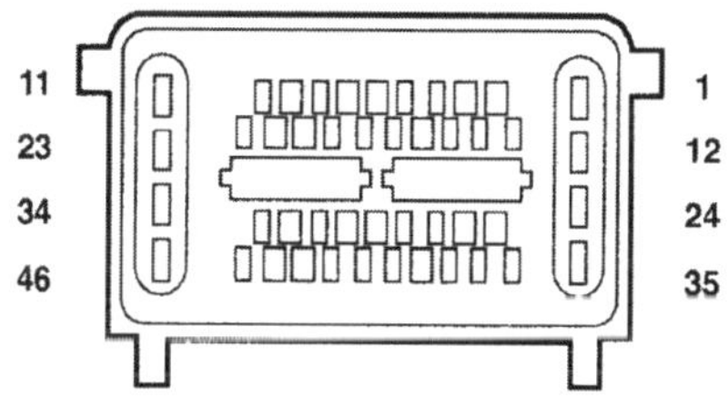

PCM BODY HARNESS CONNECTOR

Pin	Circuit
B17	PBA (Parking Brake Applied)

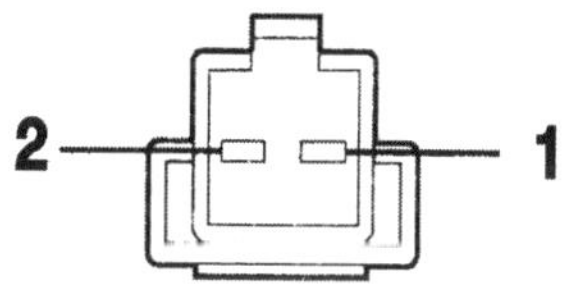

PARKING BRAKE SWITCH

Pin	Circuit
2	GND (Ground)
1	PBA (Parking Brake Applied)

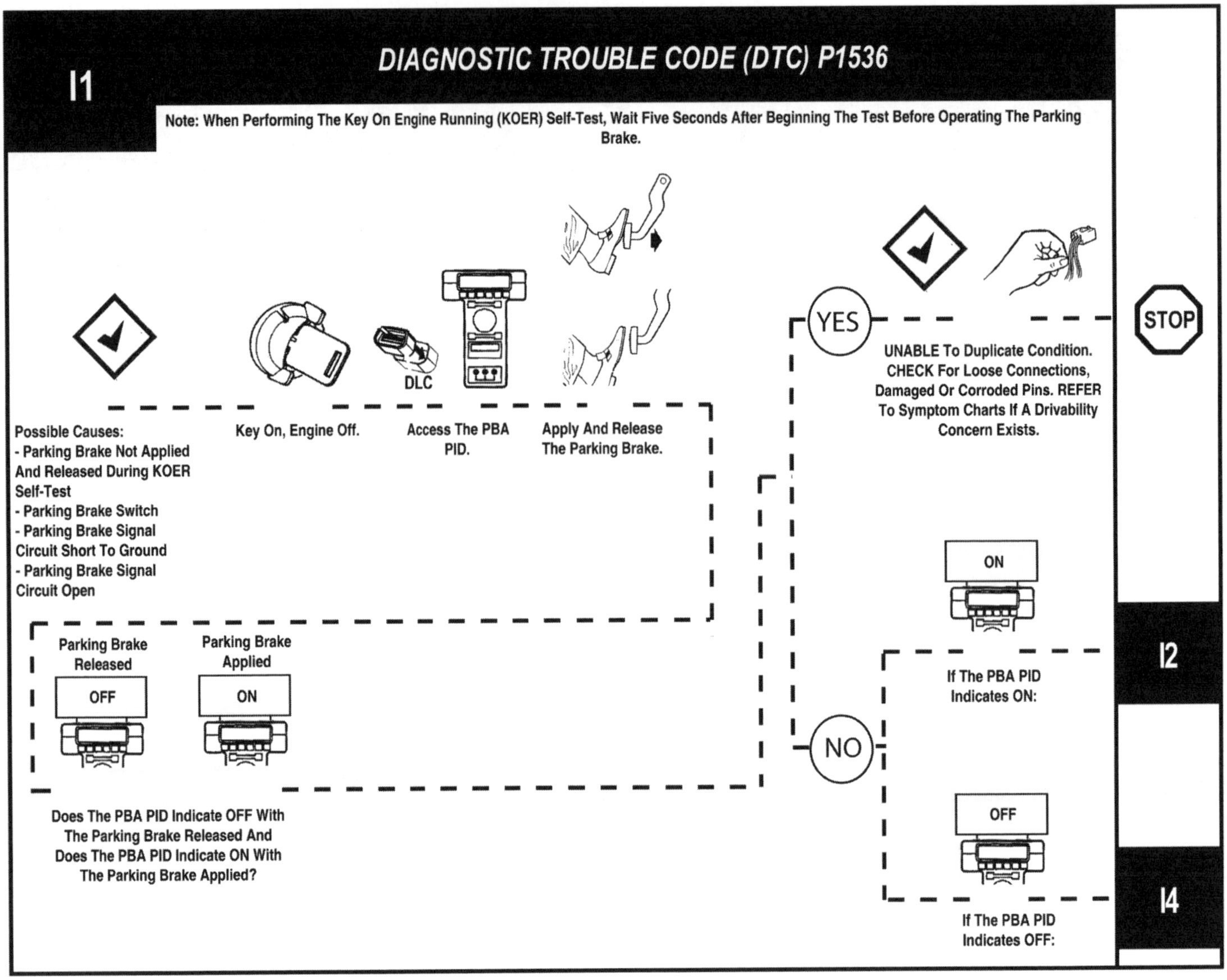
I1
DIAGNOSTIC TROUBLE CODE (DTC) P1536
Note: When Performing The Key On Engine Running (KOER) Self-Test, Wait Five Seconds After Beginning The Test Before Operating The Parking Brake.
DLC
Possible Causes:
- Parking Brake Not Applied And Released During KOER Self-Test
- Parking Brake Switch
- Parking Brake Signal Circuit Short To Ground
- Parking Brake Signal Circuit Open
Key On, Engine Off.
Access The PBA PID.
Apply And Release The Parking Brake.
Parking Brake Released
OFF
Parking Brake Applied
ON
Does The PBA PID Indicate OFF With The Parking Brake Released And Does The PBA PID Indicate ON With The Parking Brake Applied?
YES
UNABLE To Duplicate Condition. CHECK For Loose Connections, Damaged Or Corroded Pins. REFER To Symptom Charts If A Drivability Concern Exists.
STOP
NO
ON
If The PBA PID Indicates ON:
I2
OFF
If The PBA PID Indicates OFF:
I4

I2 CHECK THE PBA BRAKE SWITCH

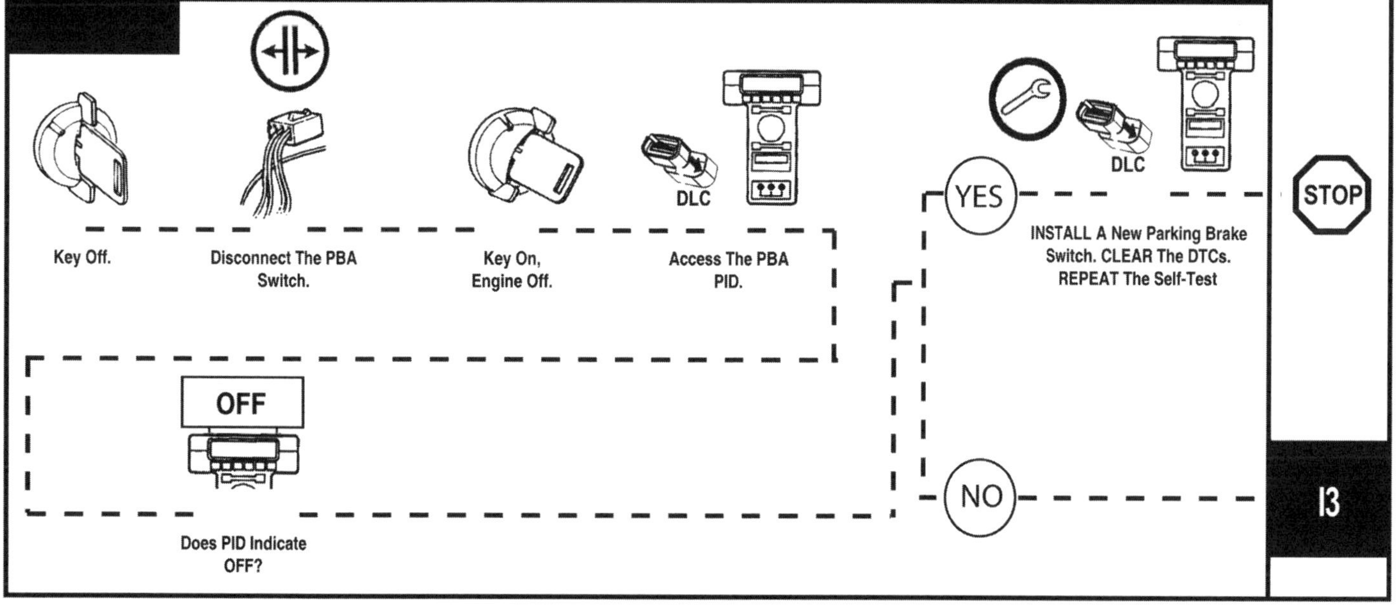

I3 CHECK THE PARKING BRAKE SIGNAL CIRCUIT FOR A SHORT TO GROUND

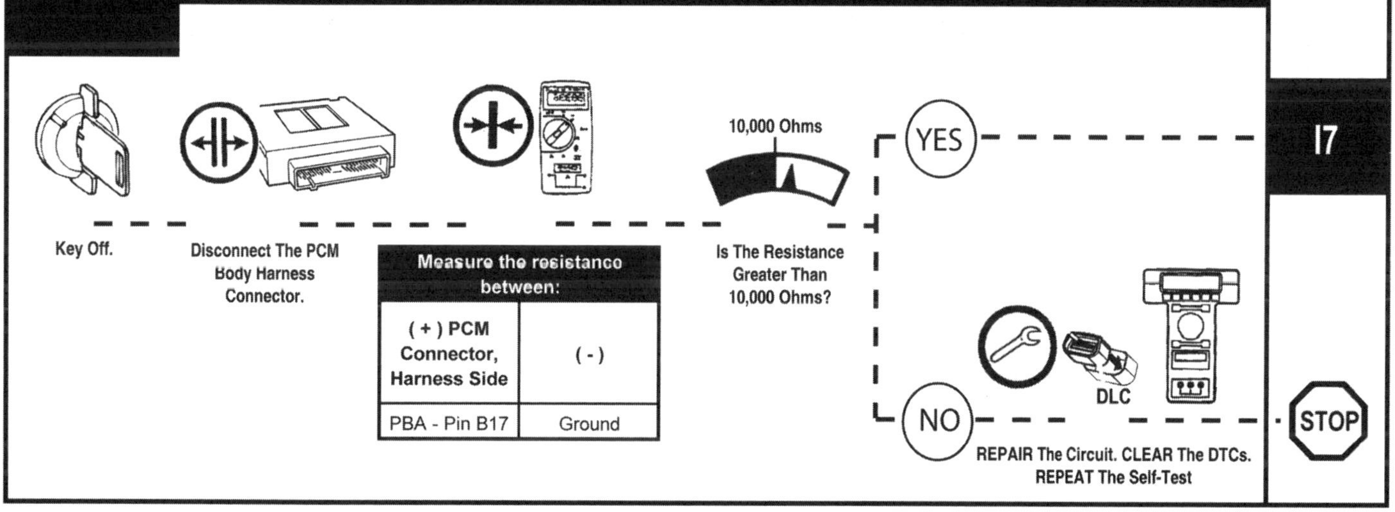

Measure the resistance between:	
(+) PCM Connector, Harness Side	(-)
PBA - Pin B17	Ground

I4 CHECK THE PARKING BRAKE SIGNAL CIRCUIT VOLTAGE

Key Off.

Disconnect The PBA Switch.

Key On, Engine Off.

Measure the voltage between:	
(+) PBA Switch Connector, Harness Side	(-)
PBA - Pin 1	Ground

10 Volts

Is The Voltage Greater Than 10 Volts?

YES → I6

NO → I5

I5 CHECK THE PARKING BRAKE CIRCUIT FOR AN OPEN

Key Off.

Disconnect The PCM Body Harness Connector.

Measure the resistance between:	
(+) PBA Switch Connector, Harness Side	(-) PCM Connector, Harness Side
PBA - Pin 1	PBA - Pin B17

5 Ohms

Is The Resistance Less Than 5 Ohms?

YES → I7

NO → DLC → STOP

REPAIR The Circuit. CLEAR The DTCs. REPEAT The Self-Test.

I6 CHECK THE PARKING BRAKE GROUND CIRCUIT FOR AN OPEN

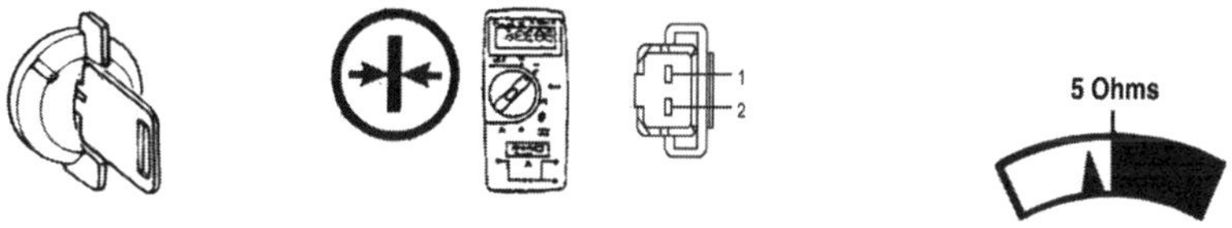

Key Off.

Measure the resistance between:	
(+) PBA Switch Connector, Harness Side	(-)
GND - Pin 2	Ground

Is The Resistance Less Than 5 Ohms?

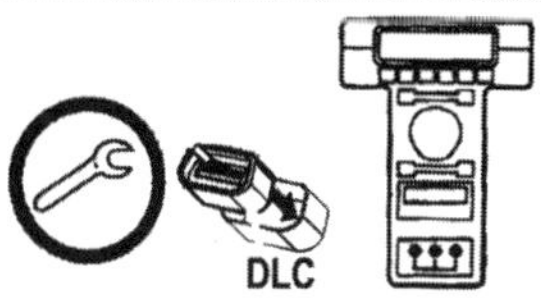

YES — INSTALL A New Parking Brake Switch. CLEAR The DTCs. REPEAT The Self-Test. — STOP

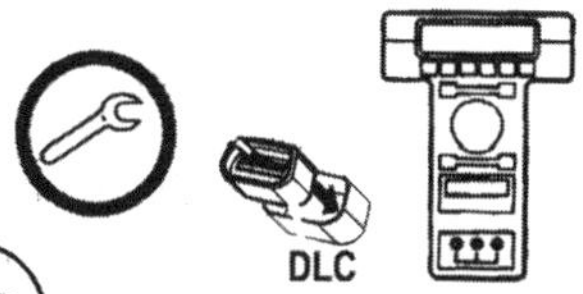

NO — REPAIR The Circuit. CLEAR The DTCS. REPEAT The Self-Test. — STOP

I7 CHECK FOR CORRECT PCM OPERATION

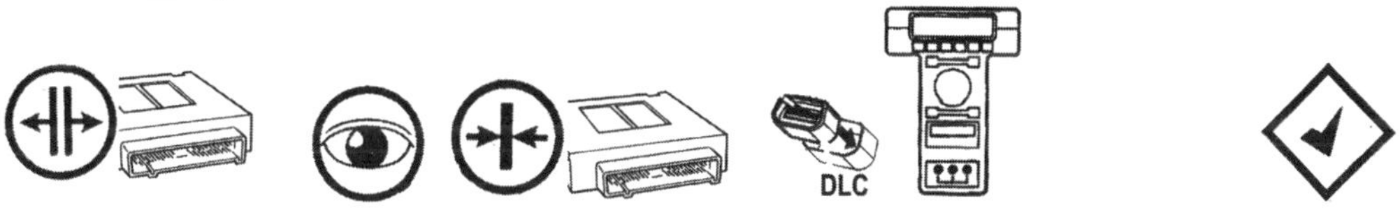

Disconnect All PCM Connectors.

Visually Inspect For Pushed Out Pins, Corrosion.

Connect All PCM Connectors And Make Sure They Seat Correctly.

Perform The PCM Self-Test And Verify The Concern Is Still Present.

Is The Concern Still Present.

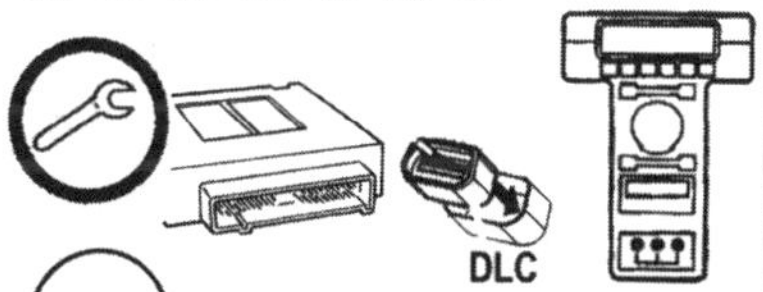

YES — INSTALL A New PCM. CLEAR The DTC's And REPEAT The Self-Test. — STOP

NO — The System Is Operating Correctly At This Time. The Concern May Have Been Caused By A Loose Or Corroded Connector. — STOP

NOTES

Trouble Code (s)		Test
P0101, P0103, P1102	**Problem: Mass Air Flow (MAF) Sensor**	J 2006-07

REFERENCE LIST

This pinpoint test is intended to diagnose the following:

- MAF sensor
- harness circuits: MAF SIG, MAF RTN, vehicle power (VPWR), and power ground (PWR GND)
- PCM

DIAGNOSTIC TROUBLE CODE INDEX			
Codes	KOEO	KOER	CM
P0101	J1	J1	J1
P0103	J1	J1	J1
P1102	J1	J1	J1

P0101 = Mass Or Volume Air Flow Circuit Range/Performance
P0103 = Mass Or Volume Air Flow Circuit High Input
P1102 = Mass Air Flow Sensor In range but Lower Than Expected

Circuit Function:

The mass airflow (MAF) sensor measures the airflow rate into the engine and provides an analog signal to the powertrain control module (PCM) which is used to calculate air flow for the EGR monitor.

Current PCM implementation contains a mass airflow interface, which consists of a two pole, differential input, low pass filter. This filter is designed to pass an analog signal, originating from the mass air sensor, while rejecting unwanted ignition noise, high frequency electrical interference, and ground offsets.

The input to the interface is the analog voltage signal difference between MAF+ (mass airflow signal) and MAF- (mass airflow signal return). The MAF sensor outputs an analog voltage signal calibrated to 0.5 V at the lower limit of the flow range and 4.75 V at the upper limit of the flow range.

The intake air temperature (IAT) is also integrated into the MAF sensor. The PCM uses the IAT signal to control timing and fuel rate during cold starts.

CAUTION!:

The powertrain control module (PCM) harness connectors must be properly seated and the connector latch properly attached to eliminate possible driveability concerns or a no start condition. Installing the PCM connectors on an angle may cause an improper connection, misdiagnosis and damaged components. Install the connector until the lever pivots and seats itself. Apply light pressure to get the connector into position on the PCM and then fully seat the connector.

Note: Visually inspect the harness connectors for corrosion, damage, proper mating and correct pin tension.

PINPOINT TEST SCHEMATICS AND CONNECTORS

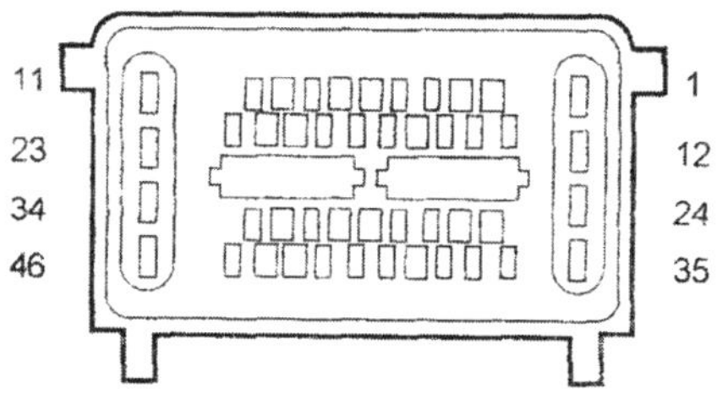

PCM
BODY HARNESS
CONNECTOR

Pin	Circuit
B21	MAF RTN (Mass Air Flow Return)
B42	MAF (Mass Air Flow)

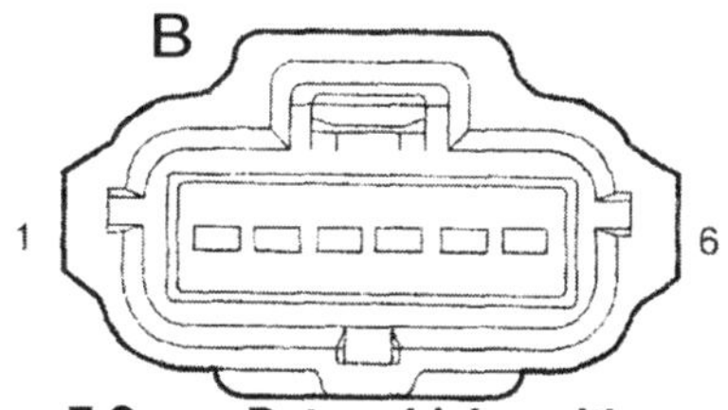

F-Super Duty vehicles with an integrated MAF/IAT sensor

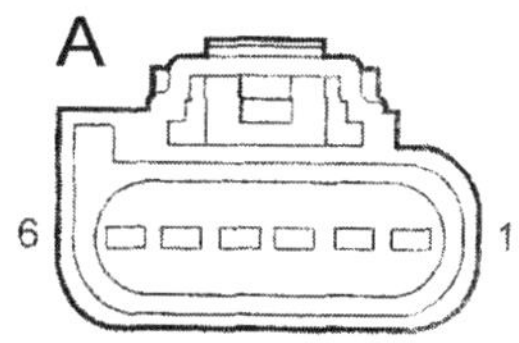

E-Series

Vehicle	Connector	Pin	Circuit
E-Series	A	3	MAF
		4	MAF RTN
		5	PWRGND
		6	VPWR
F-Super Duty vehicles with an integrated MAF/IAT sensor	B	5	MAF
		4	MAF RTN
		3	PWRGND
		2	VPWR

J1 PRELIMINARY DIAGNOSIS FOR DTCS P0101, P0103, P1102

DLC

DLC

DLC

Perform The Visual Inspection. → Connect The Scan Tool. → Retrieve And Record All DTCs. → Record Freeze Frame Data. → Clear All DTCs. → Perform The KOEO On-Demand Self-Test. → Perform The KOER On-Demand Self-Test.

Are DTCs P0101, P0103 Or P1102 Present?

YES

- For P0101: → J2
- For P1102: → J4
- For P0103: → J9

NO → STOP

Unable To Duplicate Condition. CHECK For Restriction In Air Flow And Water Intrusion. CHECK For Loose Connection, Damaged Or Corroded Terminals Or Pins. WIGGLE Harness Attempting To Recreate The Fault. REFER To Symptom Charts If Drivability Concern Exists.

J2 — DIAGNOSTIC TROUBLE CODE (DTC) P0101

Note: Visually inspect the vehicle for aftermarket accessories and performance modifications (exhaust system, turbocharger, performance chip). Refer to Diesel Electronic Engine Control (EEC) System, Modifications to OBD Vehicles.

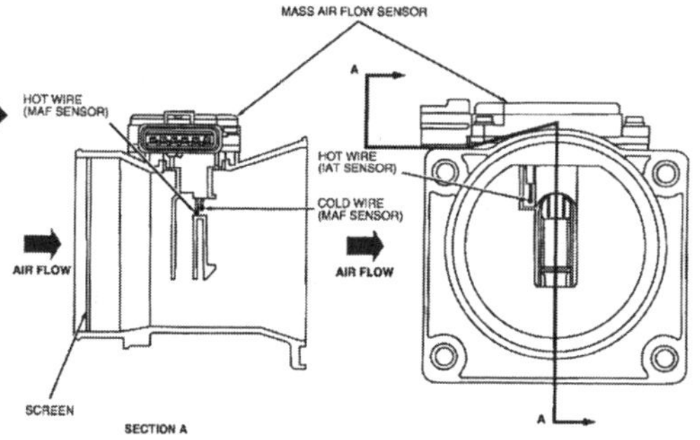

Possible Causes:
- MAF Sensor
- Contaminated Sensor
- Water Intrusion
- PCM

Check MAF Sensor For Contamination.

Is Any Restriction Or Contamination Present?

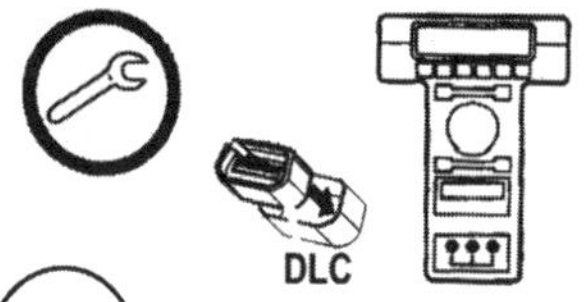

YES → REMOVE Restriction And REPAIR System As Necessary. CLEAR DTCs And RETEST The System. → STOP

NO → J3

J3 — CHECK MAF OUTPUT SIGNAL TO PCM

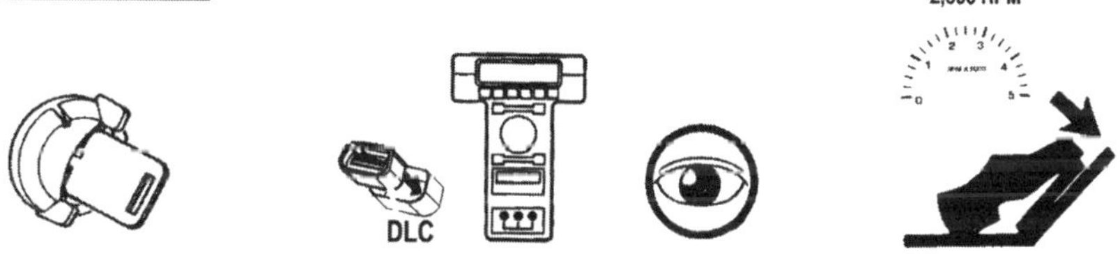

Key ON, Engine OFF.

Access MAF Volt PIDs.

Monitor MAF PID Volts At Idle.

Increase Idle To Approximately 2,500 Rpm While Monitoring PID. Record The Value.

1.2 To 1.8 Volts with Engine Idling

2.4 To 3.0 Volts At 2,500 Rpm

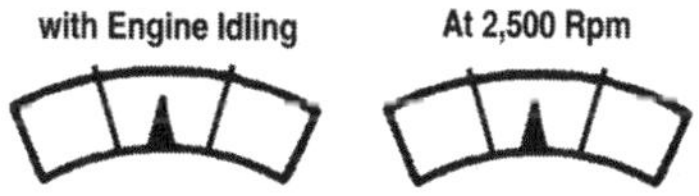

Is MAF PID Value 1.2 To 1.8 Volts With Engine Idling And 2.4 To 3.0 Volts At 2,500 Rpm?

YES → J14

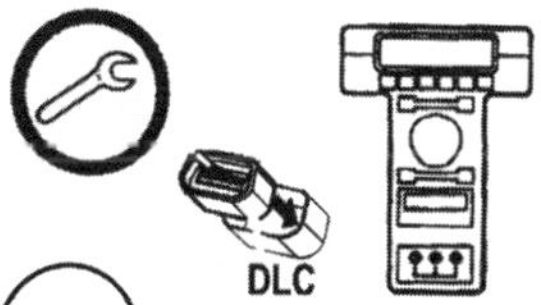

NO → INSTALL A New MAF Sensor. RESTORE The Vehicle CLEAR DTCs And RETEST System. → STOP

J4 DIAGNOSTIC TROUBLE CODE (DTC) P1102

Note: Refer To The PCM Connector Pin Numbers At The Beginning Of This Pinpoint Test.

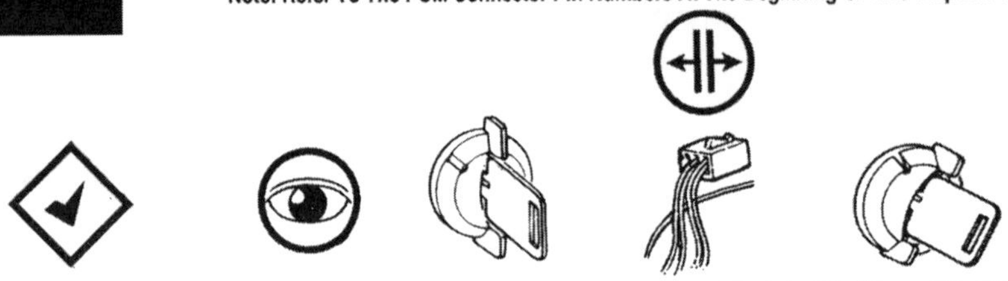

Possible Causes:
- Restricted Air Flow
- MAF Sensor Disconnected
- Unmetered Air Flow Into Engine
- VPWR Circuit Open To MAF
- PWR GND Circuit Open To MAF
- MAF RTN Signal Open
- MAF Signal Shorted To Ground
- Mechanical Failure
- MAF Sensor
- PCM

Inspect Air Intake For Restrictions

Key Off.

Disconnect The MAF Sensor Harness Connector.

Key On. Engine Off.

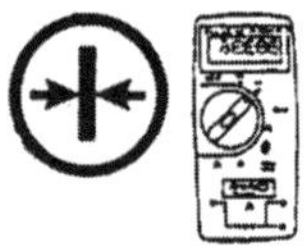

Measure the voltage between:	(+) MAF/IAT Sensor Connector, Harness Side	(-) MAF/IAT Sensor Connector, Harness Side
	VPWR	PWRGND
E-Series	Pin 6	Pin 5
F-Super Duty	Pin 2	Pin 3

Is The Voltage Greater Than 10.5 Volts Between Pins 2 And 3 (F-Super Duty/Excursion) Or Pins 6 And 5 (E-Series)?

YES → J5

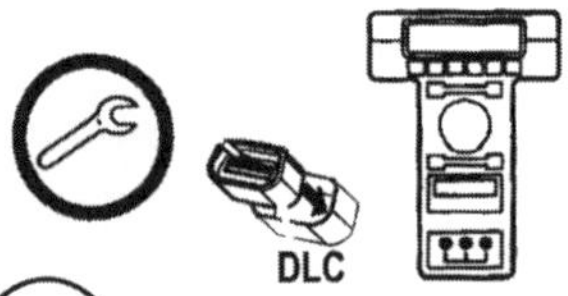

NO → REPAIR Open In The Vehicle Power Or Vehicle Ground Circuit. RESTORE The Vehicle. CLEAR DTCs And RETEST The System.

J5 CHECK MAF RTN AND MAF PWRGND FOR AN OPEN

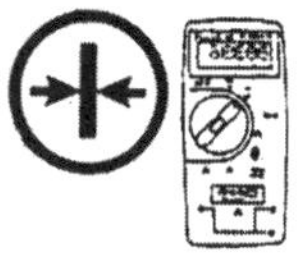

	Measure the voltage between:	
	(+) MAF/IAT Sensor Connector, Harness Side	(-) MAF/IAT Sensor Connector, Harness Side
	MAF RTN	PWRGND
E-Series	Pin 4	Pin 5
F-Super Duty	Pin 4	Pin 3

Is The Voltage Between 5.5V And 6.5V?

YES → J6

NO → STOP

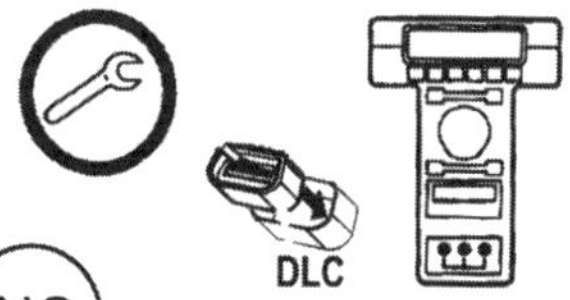

REPAIR Open Circuit. RESTORE The Vehicle. CLEAR DTCs And REPEAT The Self-Test.

J6 CHECK MAF SENSOR SIGNAL CIRCUIT FOR A SHORT TO GROUND

Key Off.

Disconnect The PCM Body Harness Electrical Connector.

	Measure the resistance between:	
	(+) MAF/IAT Sensor Connector, Harness Side	(-)
	MAF	Ground
E-Series	Pin 3	Ground
F-Super Duty	Pin 5	Ground

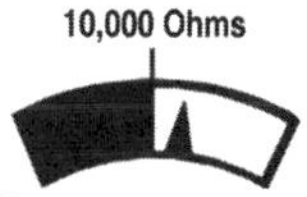

Is The Resistance Greater Than 10,000 Ohms?

YES → J7

NO → STOP

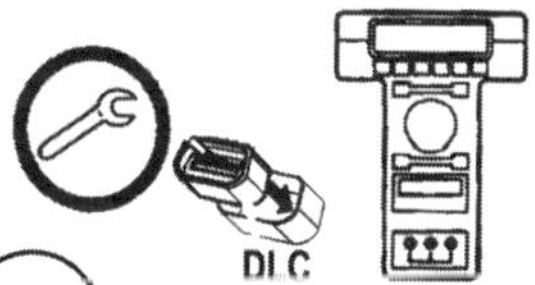

REPAIR Short To Ground In Signal Circuit. RESTORE The Vehicle. CLEAR DTCs And RETEST The System.

J7 CHECK MAF SENSOR SIGNAL CIRCUIT FOR AN OPEN

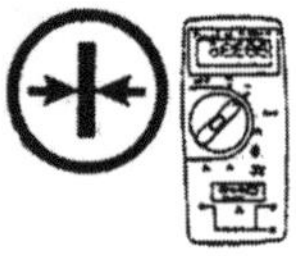

	Measure the resistance between:	
	(+) MAF/IAT Sensor Connector, Harness Side	(-) PCM Connector, Harness Side
	MAF	MAF - Pin B42
E-Series	Pin 3	MAF - Pin B42
F-Super Duty	Pin 5	MAF - Pin B42

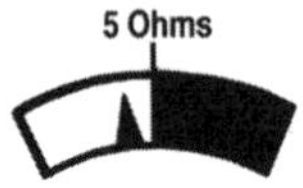

Is The Resistance Less Than 5 Ohms?

YES → J8

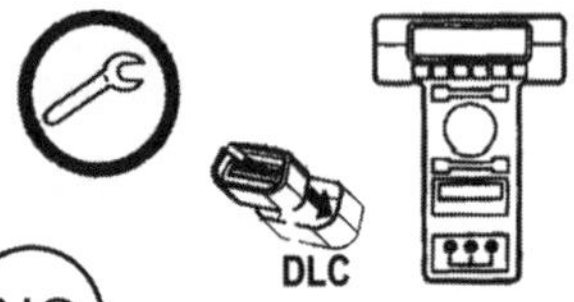

NO → STOP

REPAIR Open In Signal Circuit. RESTORE The Vehicle. CLEAR DTCs And RETEST The System.

J8 CHECK THE MAF SENSOR SIGNAL

Note: This Step May Cause DTC P0103 To Set.

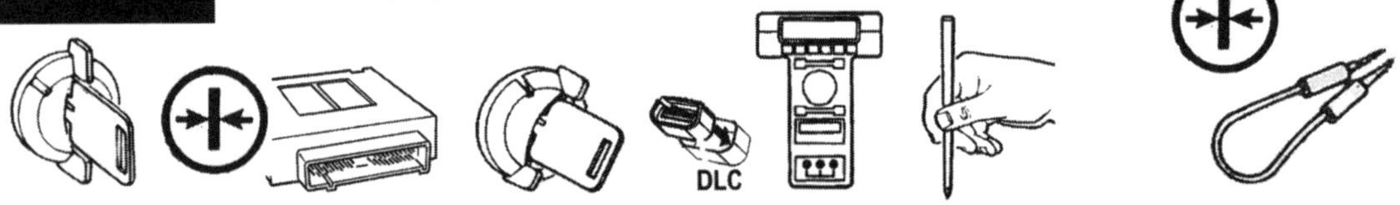

Key Off. → Reconnect The PCM Body Harness Electrical Connector. → Key On, Engine Off. → Access MAF VOLT PID. → Record MAF VOLT PID Reading.

	Connect a 5 amp fused jumper wire between the following:	
	Point A MAF/IAT Sensor Connector, Harness Side	Point B MAF/IAT Sensor Connector, Harness Side
	MAF	VPWR
E-Series	Pin 3	Pin 6
F-Super Duty	Pin 5	Pin 2

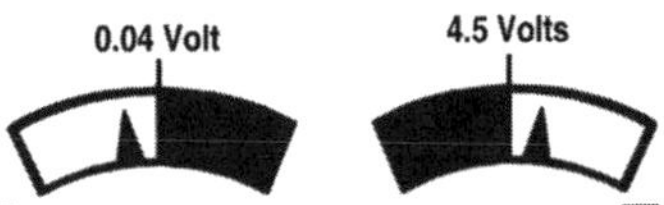

Does The MAF PID Change From Less Than 0.04 Volt To Greater Than 4.5 Volts?

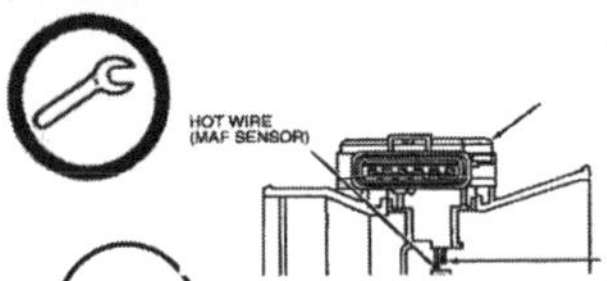

YES → STOP

INSTALL A New MAF Sensor. RESTORE The Vehicle. CLEAR DTCs And RETEST The System.

NO → J14

J9 — DIAGNOSTIC TROUBLE CODE (DTC) P0103

Note: Refer To The PCM And Component Connector Pin Numbers At The Beginning Of This Pinpoint Test.

Possible Causes:
- Restricted Inlet Air Flow
- Water Intrusion
- MAF Signal Circuit Shorted To Power
- Open Ground Circuit
- MAF Sensor
- PCM

Disconnect Air Inlet And Check For Restrictions And Water Intrusion.

Is A Concern Present?

- **YES** → STOP. REPAIR As Necessary. RESTORE The Vehicle. CLEAR DTCs And RETEST The System.
- **NO** → J10

J10 — SIMULATE THE OPPOSITE SIGNAL

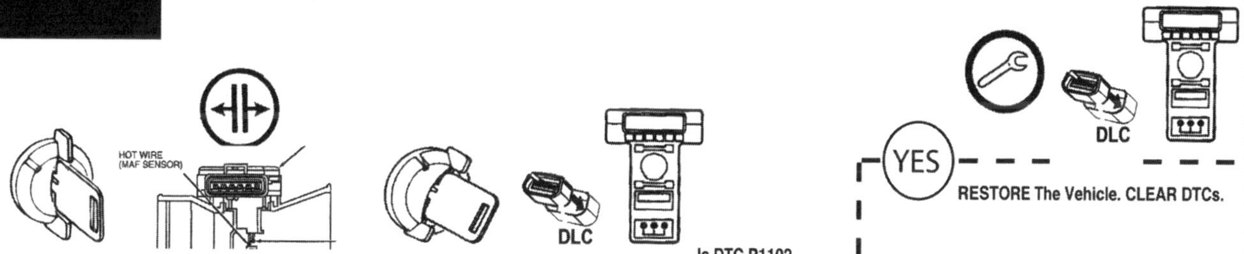

Key Off.

Disconnect The MAF Sensor.

Key On. Engine Off

Perform The KOER On-Demand Self-Test.

Is DTC P1102 Present?

- **YES** → J13. RESTORE The Vehicle. CLEAR DTCs.
- **NO** → J11

J11 — CHECK MAF SENSOR SIGNAL CIRCUIT FOR A SHORT TO VOLTAGE

Key Off.

Disconnect The PCM Body Harness Electrical Connector.

Key On. Engine Off.

	Measure the voltage between:	
	(+) MAF/IAT Sensor Connector, Harness Side	(-)
	MAF	Ground
E-Series	Pin 3	Ground
F-Super Duty	Pin 5	Ground

Is The Voltage Less Than 0.2 Volt?

- **YES** → J12
- **NO** → STOP. REPAIR Short To Voltage In Signal Circuit. RESTORE The Vehicle. CLEAR DTCs And RETEST The System.

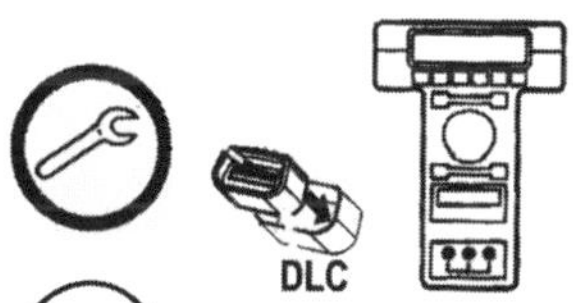

J12 CHECK MAF RETURN CIRCUITS FOR AN OPEN

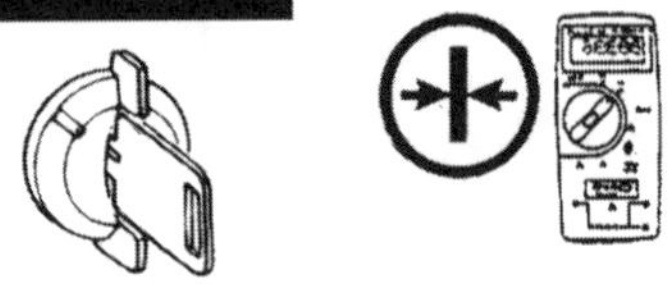

Key Off.

Measure the resistance between:

	(+) MAF/IAT Sensor Connector, Harness Side	(-) PCM Connector, Harness Side
	MAF RTN	MAF RTN - Pin B21
E-Series	Pin 4	MAF RTN - Pin B21
F-Super Duty	Pin 4	MAF RTN - Pin B21

5 Ohms

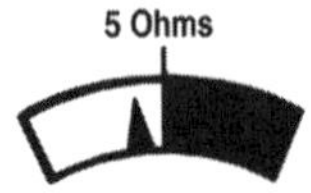

Is The Resistance Less Than 5 Ohms?

YES → J14

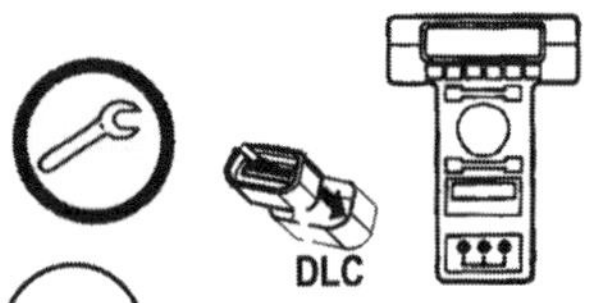

NO → REPAIR Open In Signal Return Circuit. RESTORE The Vehicle. CLEAR DTCs And RETEST The System. STOP

J13 CHECK THE MAF SIGNAL TO THE PCM

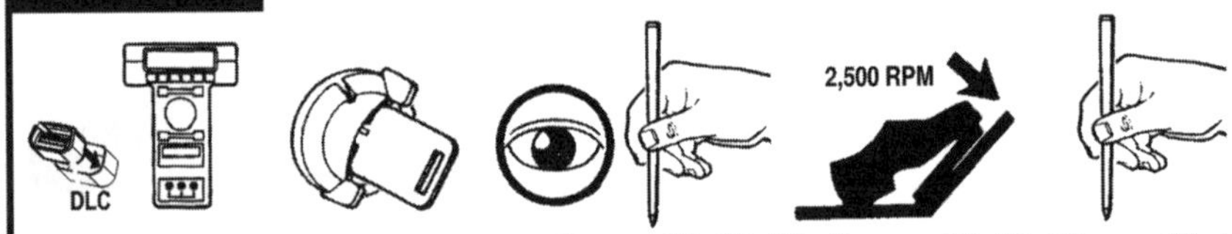

Access The MAF PID.

Key On, Engine Running.

Monitor And Record The MAF PID Voltage At Idle.

Increase The Engine Speed To 2,500 RPM.

Record The MAF PID Voltage.

1.2 To 1.8 Volts At Idle

2.4 To 3.0 Volts At 2,500 RPM

Is The MAF PID Voltage Between 1.2 To 1.8 Volts At Idle And 2.4 To 3.0 Volts At 2,500 RPM?

YES → UNABLE To Duplicate Condition. Concern May Have Set Due To Water Intrusion. REFER To Symptom Charts If A Drivability Concern Exists. STOP

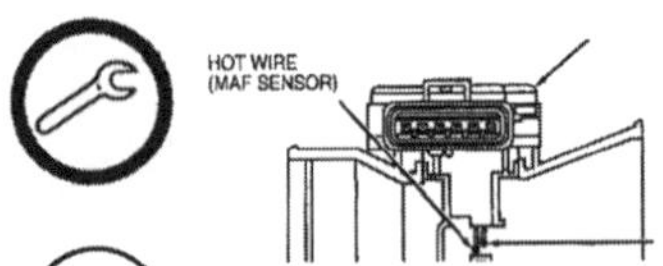

NO → INSTALL A New MAF Sensor. RESTORE The Vehicle. CLEAR The DTCs And RETEST The System. STOP

J14 CHECK FOR CORRECT PCM OPERATION

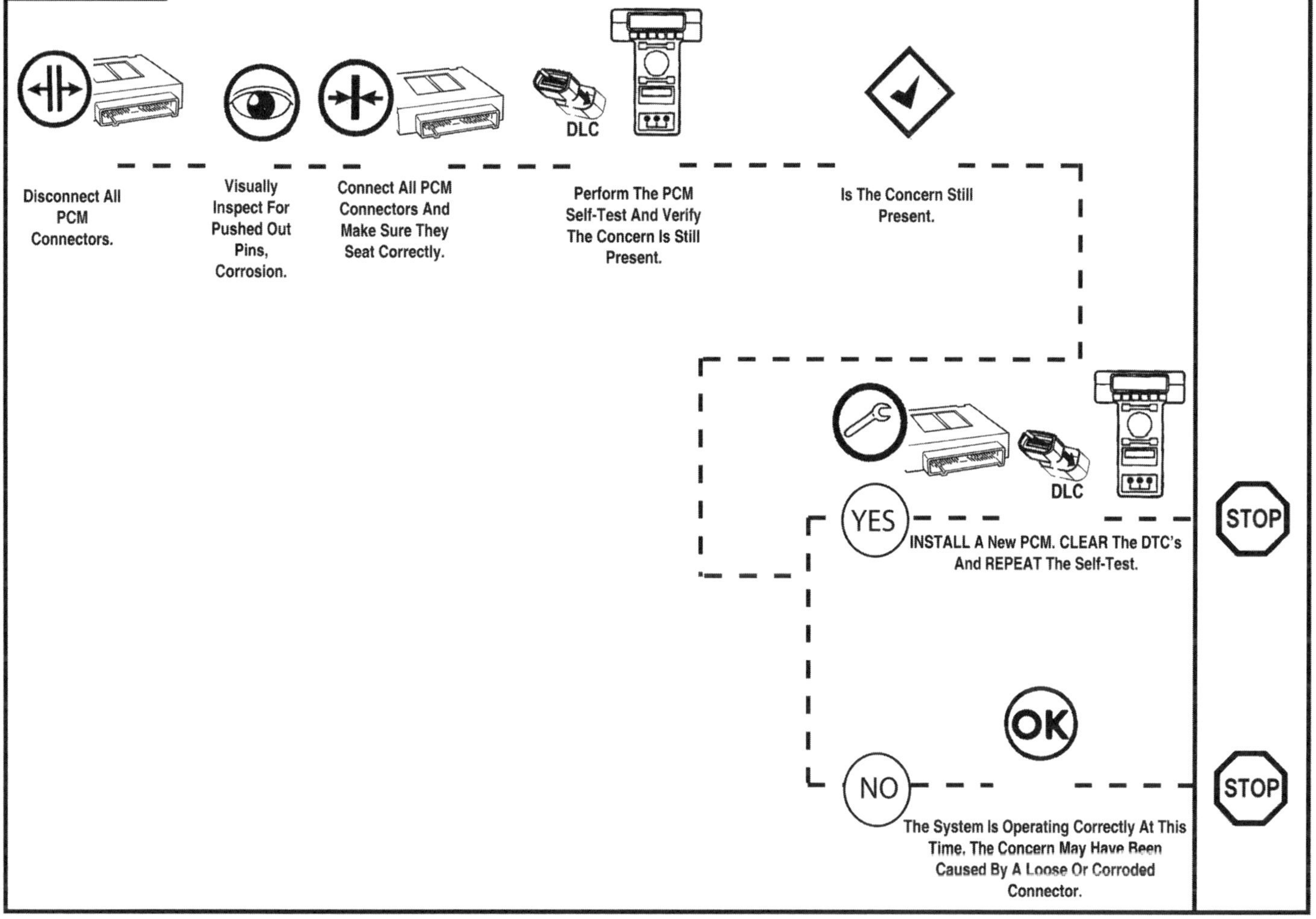

NOTES

Trouble Code (s)	Problem: Engine Coolant Temperature (ECT) Sensor	Test
P0117, P0118		K 2006-07

REFERENCE LIST

This pinpoint test is intended to diagnose the following:

- ECT sensor (12A648)
- Harness circuits: ECT and SIGRTN
- Powertrain control module (PCM) (12A650)

TEMPERATURE VS. RESISTANCE VALUES (APPROXIMATE)		
Degrees C	Degrees F	Ohms
100	212	2,080
90	194	2,803
80	176	3,836
70	158	5,337
60	140	7,556
50	122	10,908
45	113	13,216
40	104	16,092
35	95	19,696
30	86	24,329
25	77	30,000
20	68	37,352
15	59	46,797
10	50	59,016
5	41	79,940
0	32	95,851
-5	23	124,485
-10	14	160,313
-15	5	209,816
-20	-4	276,959
-30	-22	496,051
-40	-40	925,021

DIAGNOSTIC TROUBLE CODE INDEX			
Codes	KOEO	KOER	CM
P0117	K1	K1	K1
P0118	K1	K1	K1

P0117 = ECT Sensor Circuit Low Input
P0118 = ECT Sensor Circuit High Input

Circuit Function:
The engine coolant temperature (ECT) sensor is a thermistor device in which resistance changes with temperature. The electrical resistance of a thermistor decreases as the temperature increases, and increases as the temperature decreases. The varying resistance affects the voltage drop across the sensor terminals and provides electrical signals to the powertrain control module (PCM) corresponding to temperature.

The engine coolant sensor is used as the primary input to the electronic control system to enable adaptive cooling. This provides a means of providing adequate cooling in severe engine temperature conditions. The PCM will limit the fueling rate of the engine to provide cooling protection and prevent engine damage due to overheating.

Fault Detection/Management:
The ECT sensor is located on the left side of the front cover. The PCM supplies a five volt reference signal that the ECT sensor uses to produce an analog voltage, indicating temperature. The PCM continuously monitors the signal of the ECT sensor to determine if the signal is within an expected range. If the PCM detects an out of range high or low, the PCM will ignore the ECT signal and substitute the EOT signal. If both ECT and EOT signals are out of range, the PCM will assume an engine coolant temperature of -34°C (-29°F) for starting and 82°C (180'T) for engine running conditions.

CAUTION!:
The powertrain control module (PCM) harness connectors must be properly seated and the connector latch properly attached to eliminate possible driveability concerns or a no start condition. Installing the PCM connectors on an angle may cause an improper connection, misdiagnosis and damaged components. Install the connector until the lever pivots and seats itself. Apply light pressure to get the connector into position on the PCM and then fully seat the connector.

Note: Visually inspect the harness connectors for corrosion, damage, proper mating and correct pin tension.

Note: When the PCM is disconnected additional DTCs will be set. Clear all DTCs after restoring the vehicle.

PINPOINT TEST SCHEMATICS AND CONNECTORS

Pin	Circuit
E25	SIGRTN (Signal Return)
E32	ECT (Engine Coolant Temperature)

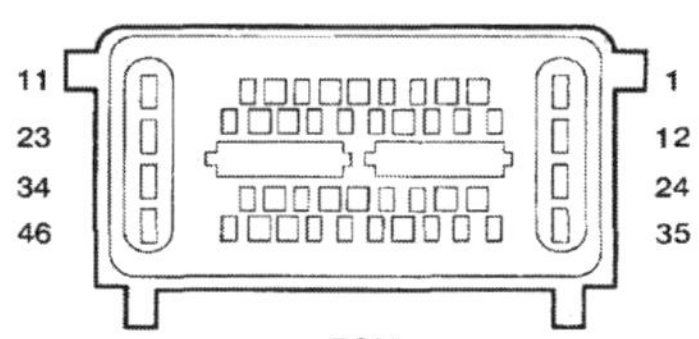

PCM
ENGINE HARNESS
CONNECTOR

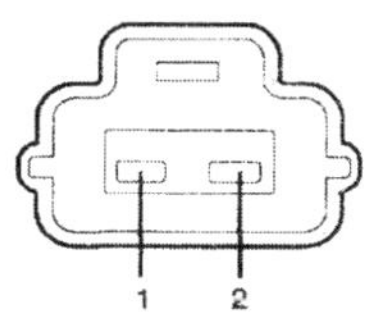

SENSOR
CONNECTOR

Pin	Circuit
1	SIGRTN (Signal Return)
2	ECT (Engine Coolant Temperature)

K1 PRELIMINARY DIAGNOSIS FOR DTCS P0117, P0118

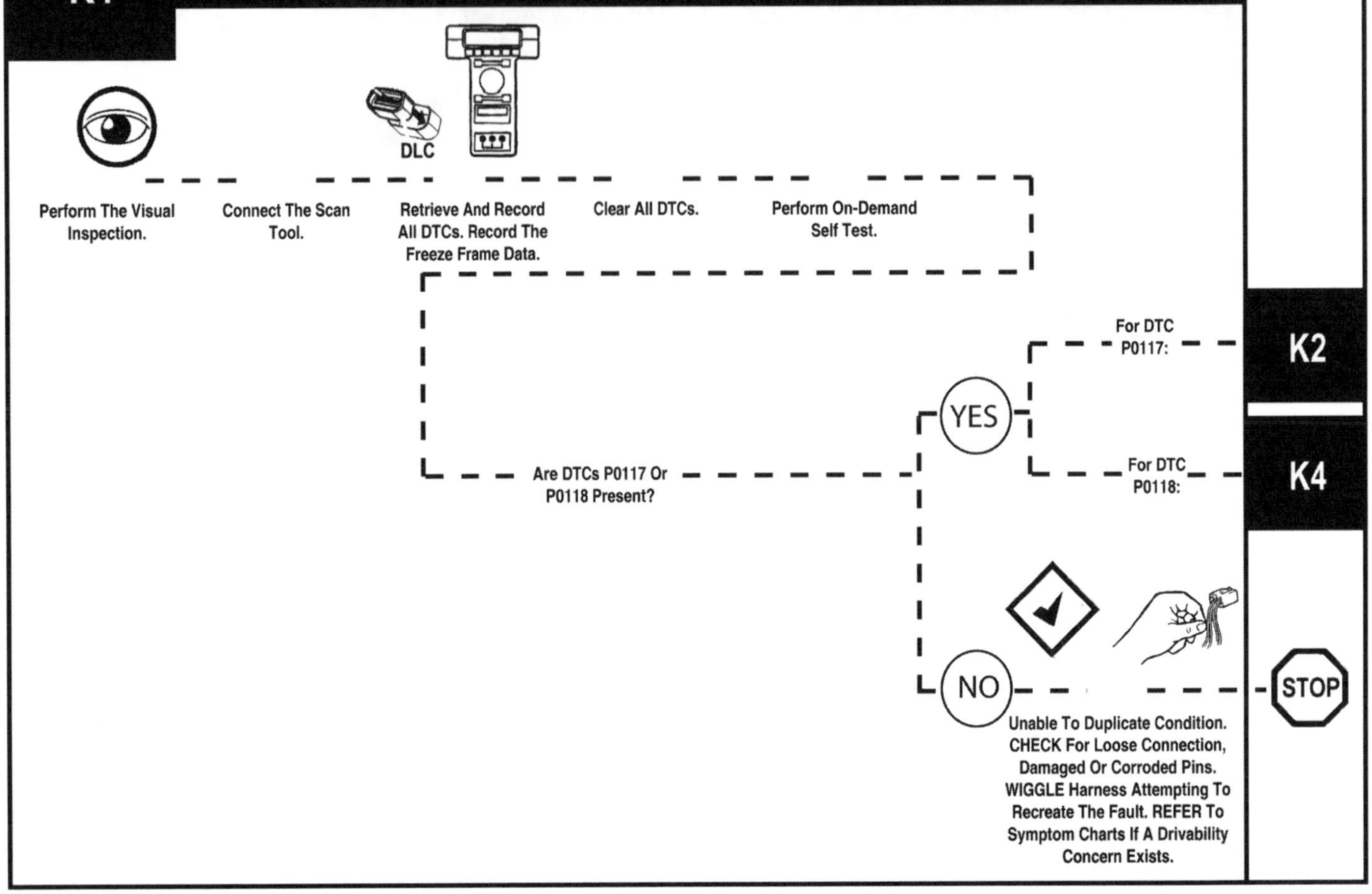

K2 DIAGNOSTIC TROUBLE CODE (DTC) P0117

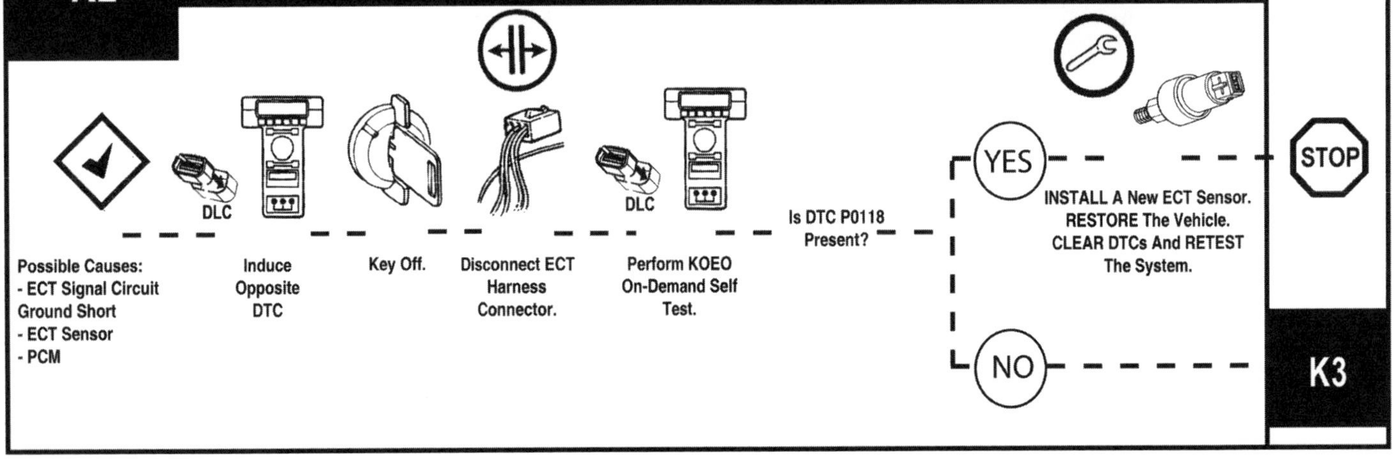

K3 CHECK THE ECT SIGNAL CIRCUIT FOR A SHORT TO GROUND

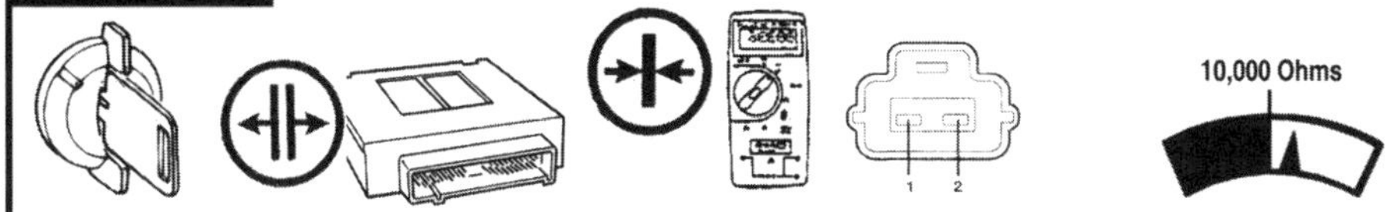

Key Off.

Disconnect The PCM Engine Harness Connector.

Measure the resistance between:	
(+) ECT Sensor Connector, Harness Side	(-)
ECT - Pin 2	Ground

Is The Resistance Greater Than 10,000 Ohms?

YES — K8

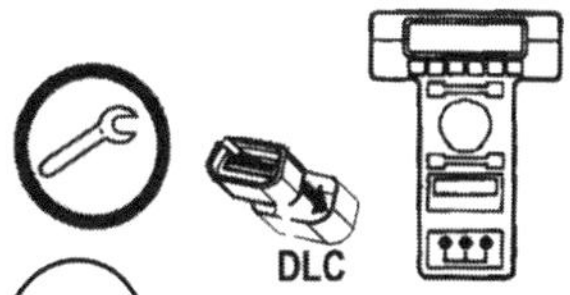

NO — STOP

REPAIR Short To Ground In The ECT Signal Circuit. CLEAR The DTCs And REPEAT The Self-Test.

K4 DIAGNOSTIC TROUBLE CODE (DTC) P0118

Possible Causes:
- Short To Voltage In The ECT Signal Circuit
- ECT Sensor
- PCM

Simulate Opposite DTC.

Key Off.

Connect a 5 amp fused jumper wire between the following:	
Point A ECT Sensor Connector, Harness Side	Point B ECT Sensor Connector, Harness Side
ECT - Pin 2	SIGRTN - Pin 1

Key On, Engine Off.

Carry Out The KOEO On-Demand Self-Test.

Is DTC P0117 Present?

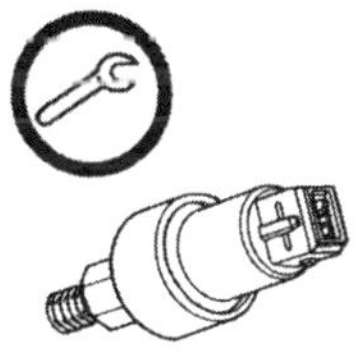

YES — STOP

INSTALL A New ECT Sensor. CLEAR The DTCs. REPEAT The Self-Test

NO — K5

K5 — CHECK THE ECT SIGNAL CIRCUIT FOR AN OPEN

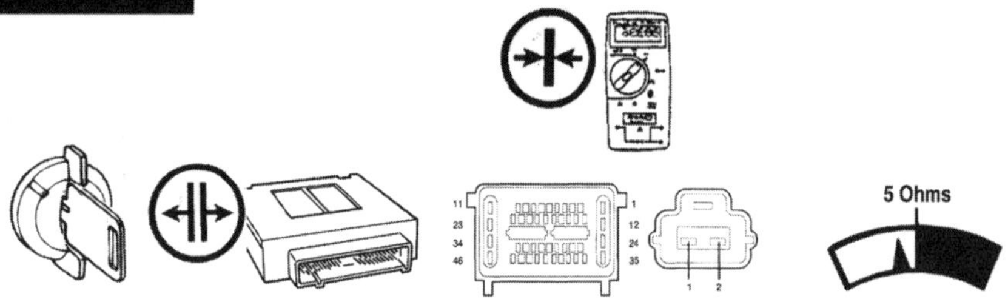

Key Off.

Disconnect The PCM Engine Connector.

Measure the resistance between:	
(+) PCM Connector, Harness Side	(-) ECT Sensor Connector, Harness Side
ECT - Pin E32	ECT - Pin 2

Is The Resistance Less Than 5 Ohms?

YES → K6

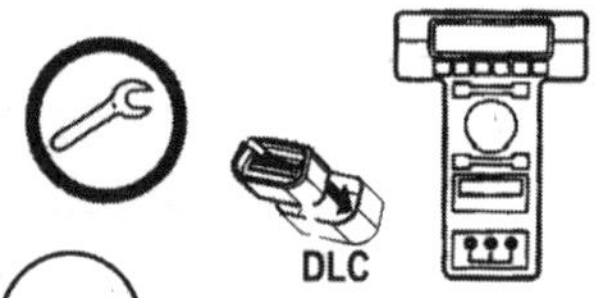

NO → REPAIR The Circuit. CLEAR The DTCs. REPEAT The Self-Test. STOP

K6 — CHECK ECT SIGNAL RETURN CIRCUIT FOR AN OPEN

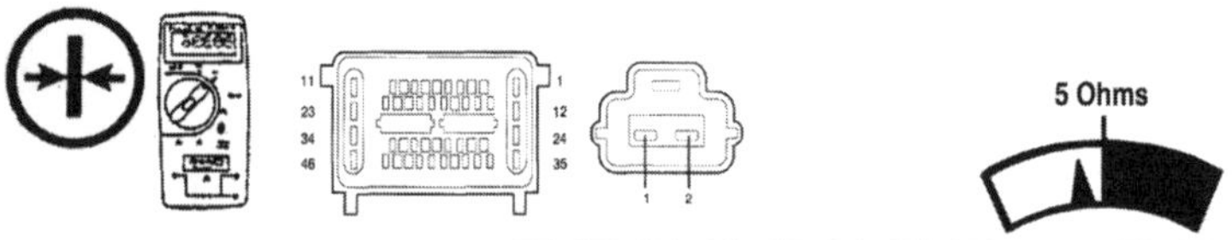

Measure the resistance between:	
(+) PCM Connector, Harness Side	(-) ECT Sensor Connector, Harness Side
SIGRTN - Pin E25	SIGRTN - Pin 1

Is The Resistance Less Than 5 Ohms?

YES → K7

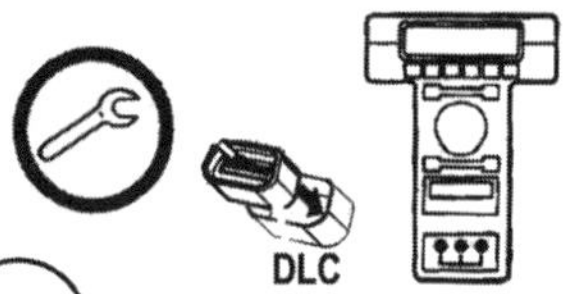

NO → REPAIR The Circuit. CLEAR The DTCs. REPEAT The Self-Test. STOP

K7 — CHECK ECT SIGNAL CIRCUIT FOR A SHORT TO VOLTAGE

Key On, Engine Off.

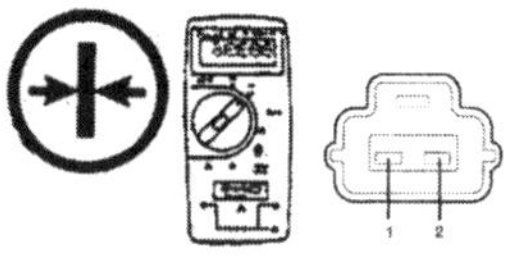

Measure the voltage between:	
(+) ECT Sensor Connector, Harness Side	(-)
ECT - Pin 2	Ground

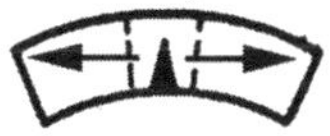

Is Any Voltage Indicated?

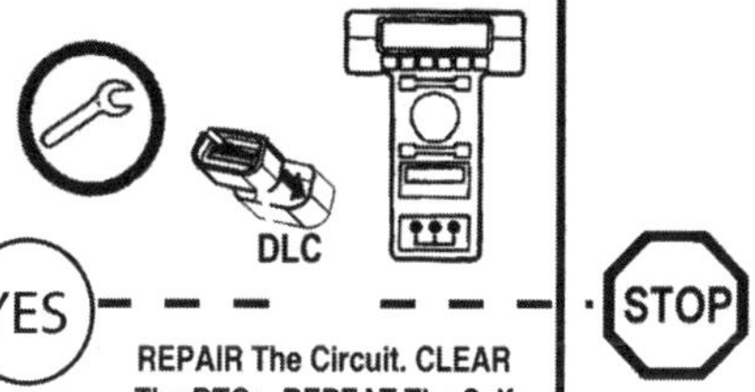

YES — REPAIR The Circuit. CLEAR The DTCs. REPEAT The Self Test. STOP

NO — K8

K8 — CHECK FOR CORRECT PCM OPERATION

Disconnect All PCM Connectors.

Visually Inspect For Pushed Out Pins, Corrosion.

Connect All PCM Connectors And Make Sure They Seat Correctly.

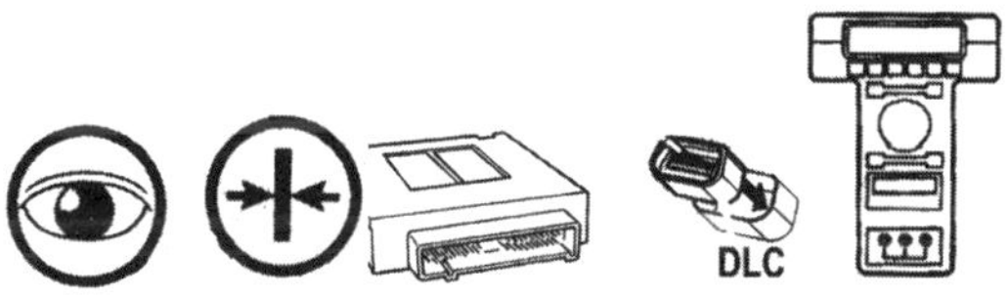

Perform The PCM Self-Test And Verify The Concern Is Still Present.

Is The Concern Still Present.

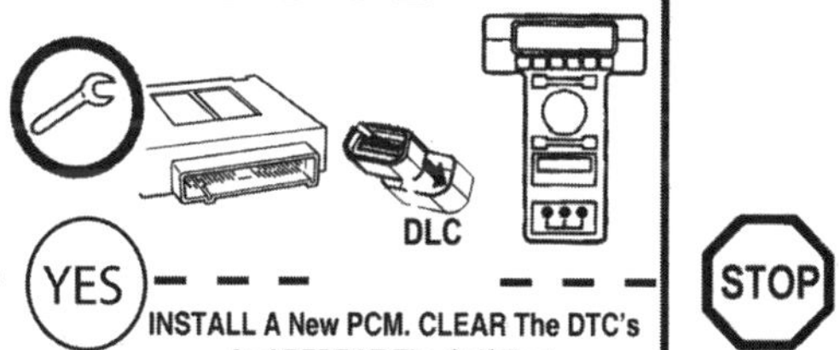

YES — INSTALL A New PCM. CLEAR The DTC's And REPEAT The Self-Test. STOP

NO — The System Is Operating Correctly At This Time. The Concern May Have Been Caused By A Loose Or Corroded Connector. STOP

NOTES

Trouble Code (s)		Test
P0299, P0478, P2262, P2263	**Problem: Turbocharger System Performance**	KA 2006-07

REFERENCE LIST

This pinpoint test is intended to diagnose the following:

- Turbocharger actuator solenoid
- Harness circuit: PWRGND
- Powertrain control module (PCM) (12A650)

DIAGNOSTIC TROUBLE CODE INDEX

Codes	KOEO	KOER	CM
P0299	KA1	KA1	KA1
P0478	KA1	KA1	KA1
P2262	KA1	KA1	KA1
P2263	KA1	KA1	KA1

P0299 = Turbo/Super Charger Underboost
P0478 = Exhaust Pressure Control Valve High Input
P2262 = Turbo/Super Charger boost Pressure Not Detected
P2263 = Turbo/Super Charger System Boost Performance

Signal Functions:
The variable geometry turbocharger (VGT) is controlled by the powertrain control module (PCM). The PCM utilizes an exhaust pressure (EP) sensor to monitor the pressure and adjust the VGT solenoid duty cycle. The VGT solenoid receives a pulse width modulated (PWM) signal from the PCM that controls the solenoid on/off time. The VGT solenoid directs oil to a piston within the actuator housing. The direction of oil flow to the piston increases or decreases the exhaust pressure.

Sensor Bias:
The VGT solenoid control is based on input sensors. The input sensors are used to calculate engine speed, desired fuel quantity, altitude, and exhaust pressure. The amount of voltage the sensor deviates from a calculated reference value (sensor bias) may cause a commanded versus actual pressure calculation error.

Detection/Management:
The PCM monitors the exhaust pressure. A DTC is set when the difference between the commanded and the actual exhaust pressure is not within the calibrated limits.

CAUTION!:
The powertrain control module (PCM) harness connectors must be properly seated and the connector latch properly attached to eliminate possible driveability concerns or a no start condition. Installing the PCM connectors on an angle may cause an improper connection, misdiagnosis and damaged components. Install the connector until the lever pivots and seats itself. Apply light pressure to get the connector into position on the PCM and then fully seat the connector.

Note: Visually inspect the harness connectors for corrosion, damage, proper mating and correct pin tension.

Note: When the PCM is disconnected additional DTCs will be set. Clear all DTCs after restoring the vehicle.

KA1 PRELIMINARY DIAGNOSIS

Key Off.

Perform A Visual Inspection.

Connect The Scan Tool.

Key On, Engine Off (KOEO)

Record The Freeze Frame Data. Retrieve The Continuous Memory DTC's. Carry Out The KOEO And KOER Self-Test

Are Any DTCs Present?

- YES:
  - For DTC P0299, P2262, P2263 Or P0404: **KA2**
  - For DTC P0478: **KA6**
  - For All Other DTCs, REFER To The Diagnostic Trouble Code (DTC) Charts. **DTC INDEX**
- NO: **KA2**

KA2 CHECK THE MANIFOLD ABSOLUTE PRESSURE (MAP) SIGNAL ACCURACY

Note: Route the pressure gauge hose to the passenger compartment so it is not crimped and does not come in contact with any hot surface.

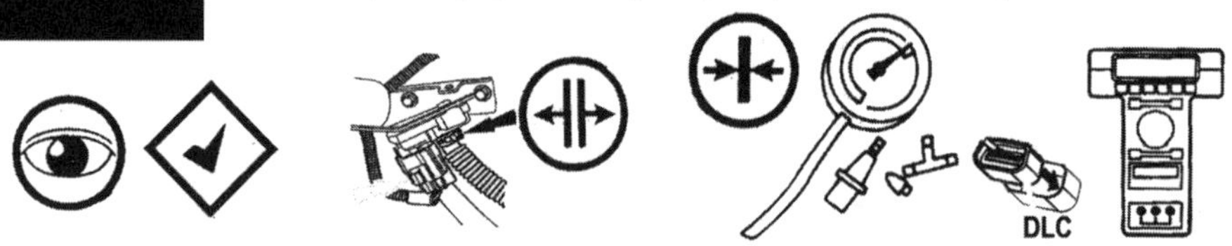

Inspect The MAP Sensor Hose And Manifolds For Damage, Leaks, Restrictions, And Correct Routing.

Disconnect The MAP Sensor Pressure Hose From The Sensor.

Install The Pressure Test Adapter Kit 014-00761 0-30 Psi Gauge Between The MAP Sensor And The Pressure Hose.

Access The PCM And Monitor The MGP And EOT PIDs.

Allow The EOT To Reach 70°C (158°F) And Road Test The Vehicle.

Does The MGP PID Match The Actual Gauge Pressure?

- YES: **KA3**
- NO: GO TO PINPOINT TEST: E **PINPOINT TEST E**

KA3 — CHECK THE ICP SENSOR FOR BIAS

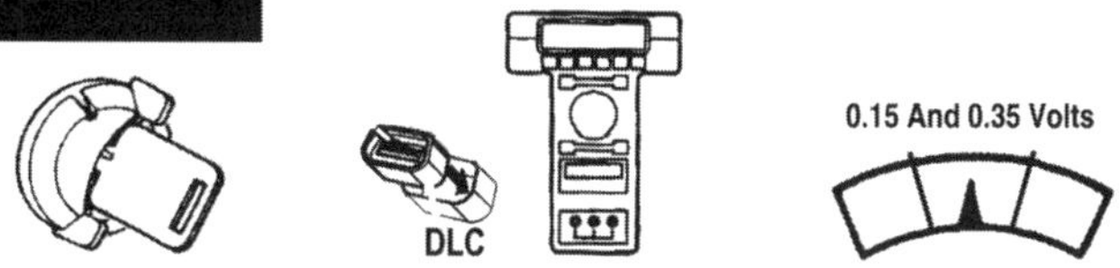

Key On, Engine Off.

Access The ICP PID.

Is The ICP PID Voltage Between 0.18 And 0.24 Volts?

YES → KA4

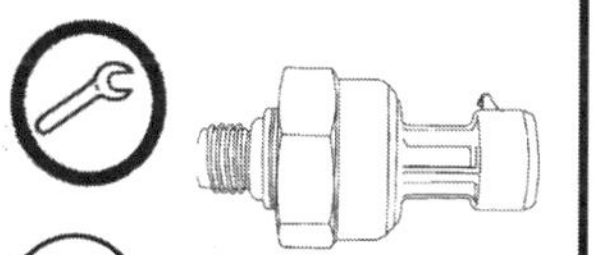

NO → INSTALL A New ICP Sensor. CLEAR The DTC's. Repeat The Self-Test. → STOP

KA4 — CHECK FOR INPUT SENSOR BIAS

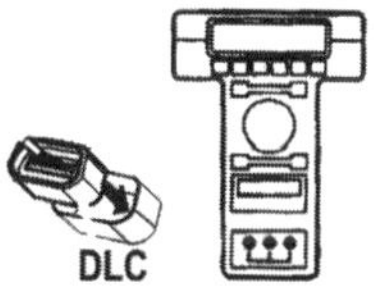

Reference Values

Access The PCM And Monitor The BARO, EGR_DC, EBP, MAF V And MAP PIDs.

Refer To Reference Values For Normal Operating Values.

Are The BARO, EGR_DC, EBP, MAF, And MAP PIDs within Specifications And Is The EBP PID Value Within 10.34 kPa (1.5 psi) Of The MAP And BARO PID Values?

YES → KA5

NO → REFER To The Appropriate Pinpoint Test To Continue Sensor Diagnostics. → STOP

KA5 — CHECK THE EGR VALVE OPERATION

NOTE: Incorrect EGR valve operation may affect turbocharger diagnostics.

Diagnostic Subroutines

Check The EGR Valve Operation. Refer To Diagnostic Subroutines.

Is A Concern Present?

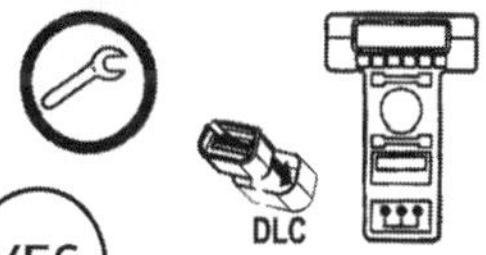

YES → REPAIR As Necessary. CLEAR The DTC's. REPEAT The Self-Test. → STOP

NO → KA6

KA6

CHECK THE TURBOCHARGER OPERATION

Note: Additional DTCs may set during the use of output state control (OSC). Clear all PCM and transmission control module (TCM) DTCs after the diagnostics have been completed. Repeat the self-test to verify all DTCs have been cleared.

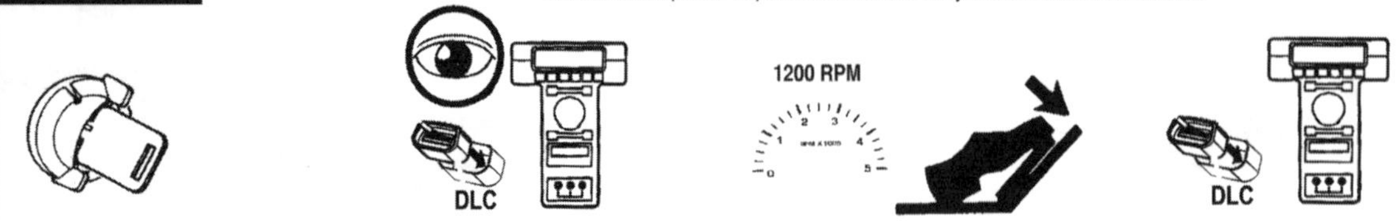

Key On, Engine Running.

Access The PCM And Monitor The RPM, MGP, EGR_DC And VGTDC PIDs.

Access The PCM And Control The EGRDC, RPM_DSD, And VGTDC PIDs.

Increase The Commanded Engine Speed To 1,200 Rpm Using Output State Control. The Calibration May Limit The Actual RPM Between 1,150 And 1,200 RPM.

Decrease the EGR duty cycle to 0%.
Decrease the VGT duty cycle to 0%.
Record the MGP PID value.
Increase the VGT duty cycle to greater than 85%.
Record the MGP PID value.
Decrease the VGT duty cycle to 0%.
Record the MGP PID value.

Is The MGP PID Below 3 kPa (0.44 PSI) At 0% VGT Duty Cycle And Above 6 kPa (.87 PSI) At 85% VGT Duty Cycle?

- YES → KA7
- NO → KA8

KA7

CHECK THE TURBOCHARGER OPERATION BACK PRESSURE

Note: Additional DTCs may set during the use of output state control (OSC). Clear all PCM and transmission control module (TCM) DTCs after the diagnostics have been completed. Repeat the self-test to verify all DTCs have been cleared.

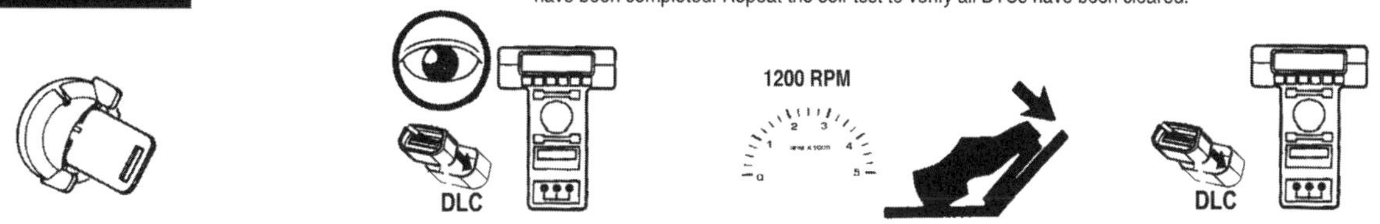

Key On, Engine Running.

Access The PCM And Monitor The RPM, EBP_G, EGR_DC And VGTDC PIDs.

Access The PCM And Control The EGRDC, RPM_DSD, And VGTDC PIDs.

Increase the Commanded Engine Speed To 1,200 RPM Using Output State Control. The Calibration May Limit The Actual RPM Between 1,150 And 1,200 RPM.

Decrease the EGR duty cycle to 0%.
Decrease the VGT duty cycle to 0%.
Record the EBP_G PID value.
Increase the VGT duty cycle to greater than 85%.
Record the EBP_G PID value.
Decrease the VGT duty cycle to 0%.
Record the EBP_G PID value.

Is The EBP_G PID Below 5 kPa (0.73 PSI) At 0% VGT Duty Cycle And Above 50 kPa (7.3 PSI) At 85% VGT Duty Cycle?

YES → KA9

NO → KA8

KA8

CHECK THE TURBOCHARGER OPERATION BACK PRESSURE

Note: Additional DTCs may set during the use of output state control (OSC). Clear all PCM and transmission control module (TCM) DTCs after the diagnostics have been completed. Repeat the self-test to verify all DTCs have been cleared.

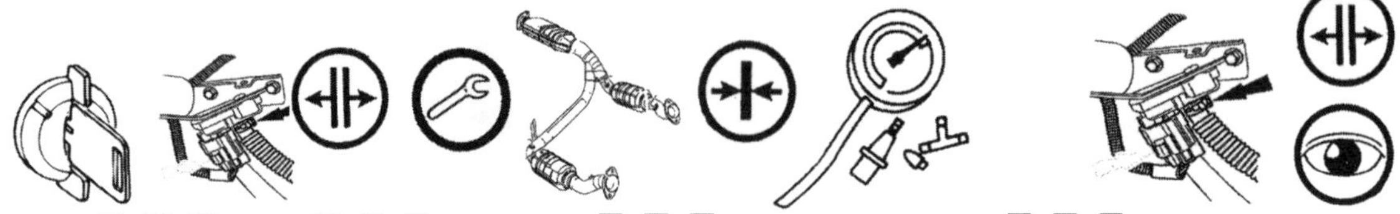

Key OFF.

Disconnect the Hose From The EP Sensor.

Plug The Exhaust Pipe.

Connect The Rotunda Smoke Machine, Fuel Evaporative Emission System Tester 218-00001 (522) Or Equivalent To The EP Hose. Fill the Exhaust System With Smoke. Pressurize The Exhaust System With 20 PSI Regulated Shop Air.

Disconnect The Hose From The MAP Sensor. Inspect The MAP Sensor Hose And Manifolds For Damage, Leaks, Restrictions, And Correct Routing.

Connect The Rotunda Smoke Machine, Fuel Evaporative Emission System Tester 218-00001 (522) Or Equivalent To The Map Hose. Fill The Air Intake System With Smoke. Pressurize The Air Intake System With 20 PSI Regulated Shop Air.

Check Air Intake And Exhaust Systems For Leaks.

Are Any Leaks Present?

- YES → REPAIR The Leaks. REFER Performance Diagnostic Procedures If A Driveability Concern Exists. → STOP
- NO → KA10

KA9 CHECK THE TURBOCHARGER OPERATION BACK PRESSURE

Note: Additional DTCs may set during the use of output state control (OSC). Clear all PCM and transmission control module (TCM) DTCs after the diagnostics have been completed. Repeat the self-test to verify all DTCs have been cleared.

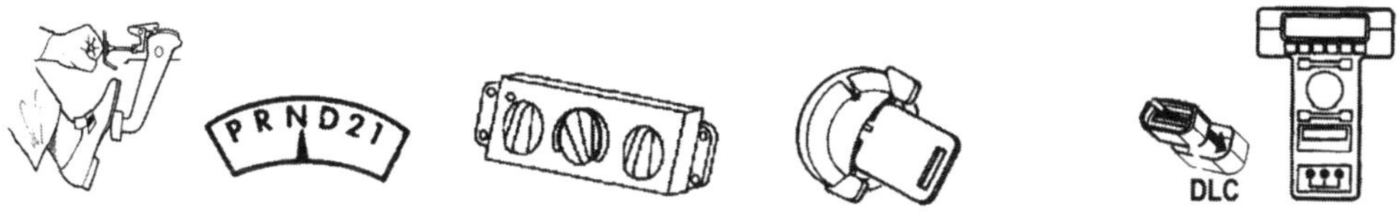

Apply The Parking Brake.

Place The Transmission In PARK Or NEUTRAL.

Turn The A/C And Defrost OFF.

Key ON, Engine Running.

Clear The DTCs.
Access The PCM And Monitor the EBP, MGP, EGR_DC, VGTDC And IPR PIDs.
Access The PCM And Control The EGRDC PID.
Decrease The EGR duty cycle to 0%.
Allow The IPR To Stabilize.

Slowly press the accelerator pedal and hold for 5 seconds at each of the following RPMs:
1,500 RPM
1,800 RPM
2,000 RPM
2,200 RPM
2,500 RPM
2,800 RPM

Do The VGTDC, MGP, Or EBP PIDs Remain Steady At All Of The RPMs?

YES

The Turbocharger System Is Operating Correctly.
REFER To Section 4, Performance Diagnostic Procedures .
CLEAR The DTCs. REPEAT The Self-test.

NO → AK7

KA10

CHECK THE OPERATION OF THE VGT ACTUATOR

CAUTION: The VGT actuator valve is sensitive to contamination. All work areas must be clean before starting this procedure. Do not allow contamination to enter the valve ports, cam follower, or turbocharger housing. Handle the VGT actuator by the solenoid body only. Do not attempt to clean or wipe oil off the valve at this time. Do not let the valve come in contact with materials that could contaminate the valve mechanism.

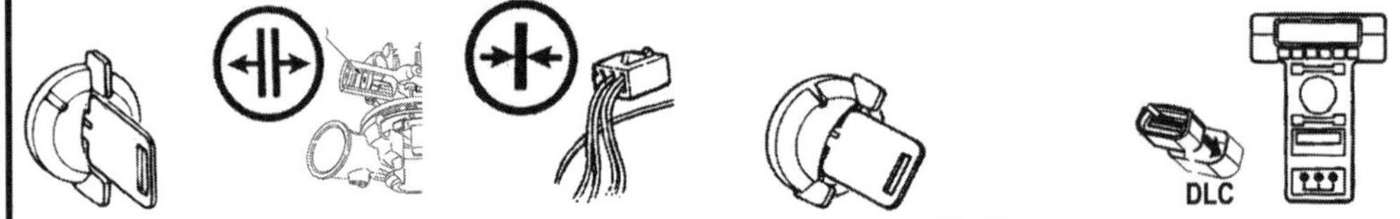

Key OFF

VGT Actuator Connector Disconnected. Remove The VGT Actuator From The Turbocharger Assembly.

VGT Actuator Connector Connected.

Key ON, Engine OFF.

Access The PCM And Control The VGTDC PID. Apply Light Pressure To The Cam Follower (Tip Of The Actuator), While Commanding The VGT Duty Cycle And Check For The Internal Valve Movement. Increase The VGT Duty Cycle To Greater Than 70%

Is Any Internal Valve Cam Follower Movement Present When The VGT Duty Cycle Is Increased?

YES → KA11

NO → AK2

KA11

CHECK THE VGT VANE OPERATION

NOTE: Inoperative vane movement may be an intermittent condition. Access the PCM and record the freeze frame data. The freeze frame data contains information which can be used to recreate original conditions that caused the symptom.

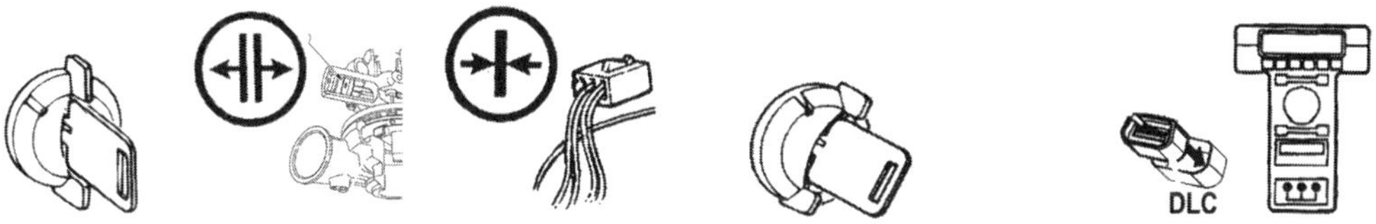

Key OFF

Install A VGT Actuator In The Turbocharger Assembly.

Remove The Pipe Plug From The Top Of The VGT Actuator Housing, Located Near The Oil Supply Tube. Apply An Index Mark On The Internal Valve Cam Follower (Tip Of The Actuator).

Key ON, Engine RUNNING.

Access the PCM and control the VGTDC PID. Increase The VGT Duty Cycle From 20% To Greater Than 85% In 10% Steps. Check The Internal Valve Cam Follower For Movement At Each Step While Increasing The Duty Cycle.

Is Any Internal Valve Cam Follower Movement Present In Each Step When The VGT Duty Cycle Is Increased?

YES

The Turbocharger System Is Operating Correctly. REFER To Performance Diagnostic Procedures. CLEAR The DTCs. REPEAT The Self-test.

STOP

NO

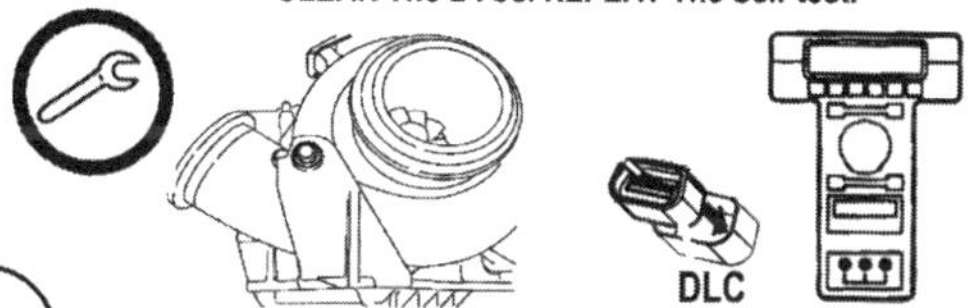

INSTALL A New Turbocharger Assembly. CLEAR The DTCs. REPEAT The Self-test.

STOP

NOTES

Trouble Code (s)	Problem: Engine Oil Temperature (EOT) Sensor	Test
P0196, P0197, P0198, P0298, P1184		L 2006-07

REFERENCE LIST

TEMPERATURE VS. RESISTANCE VALUES (APPROXIMATE)		
°C	°F	Ohms
100	212	2,080
90	194	2,083
90	176	2,836
70	158	5,337
60	140	7,556
50	122	10,908
45	113	13,216
40	104	16,092
35	95	19,696
30	86	24,329
25	77	30,000
20	68	37,352
15	59	46,797
10	50	59,016
5	41	79,940
0	32	95,851
-5	23	124,485
-10	14	160,313
-15	5	209,816
-20	-4	276,959
-30	-22	496,051
-40	-40	925,021

This pinpoint test is intended to diagnose the following:

- EOT sensor (12A648)
- Harness circuits: EOT and SIGRTN
- Powertrain control module (PCM) (12A650)

DIAGNOSTIC TROUBLE CODE INDEX			
CODES	KOEO	KOER	CM
P0196	L1	L1	L1
P0197	L1	L1	L1
P0198	L1	L1	L1
P0298	L1	L1	L1
P1184	L1	L1	L1

P0196 = ENGINE OIL TEMPERATURE SENSOR CIRCUIT RANGE/PERFORMANCE
P0197 = ENGINE OIL TEMPERATURE SENSOR CIRCUIT LOW INPUT
P0198 = ENGINE OIL TEMPERATURE SENSOR CIRCUIT HIGH INPUT
P0298 = ENGINE OIL OVERTEMPERATURE CONDITION
P1184 = ENGINE OIL TEMPERATURE SENSOR OUT OF SELF TEST RANGE

WARNING: BEWARE OF MOVING VEHICLE COMPONENTS AND HEAT.

Circuit Functions:
The engine oil temperature (EOT) sensor is a thermistor type sensor that has a variable resistance that changes when exposed to different temperatures. When interfaced with the powertrain control module (PCM), it produces a 0 to 5 volt analog signal that will deduce temperature.

Cranking Fuel Quantity/Timing Control - The EOT sensor signal is used to determine the timing and quantity of fuel required to optimize starting over all temperature conditions.

Idle Speed - At oil temperatures below 70°C (158°F) low idle is incrementally increased to a maximum of 950 rpm.

Temperature Compensation - Fuel quantity and timing is controlled throughout the total operating range to ensure adequate torque and power is available.

Glow Plug Control -The Glow Plug Control Module (GPCM) and lamp ON times are controlled by engine oil temperature.

Detection/Management :
An EOT sensor signal that is detected out of range (high or low) by the PCM will cause the PCM to ignore the EOT sensor signal and assume an engine oil temperature of -20°C (-4°F) for starting and a temperature of 100°C (212°F) for engine-running conditions. The CHECK ENGINE light will also be illuminated as long as the condition exists.

CAUTION!:
The powertrain control module (PCM) harness connectors must be properly seated and the connector latch properly attached to eliminate possible driveability concerns or a no start condition. Installing the PCM connectors on an angle may cause an improper connection, misdiagnosis and damaged components. Install the connector until the lever pivots and seats itself. Apply light pressure to get the connector into position on the PCM and then fully seat the connector.

Note: Visually inspect the harness connectors for corrosion, damage, proper mating and correct pin tension.

Note: When the PCM is disconnected additional DTCs will be set. Clear all DTCs after restoring the vehicle.

Trouble Code (s)		Test
P0196, P0197, P0198, P0298, P1184	Problem: Engine Oil Temperature (EOT) Sensor (Continued)	L 2006-07

PINPOINT TEST SCHEMATICS AND CONNECTORS

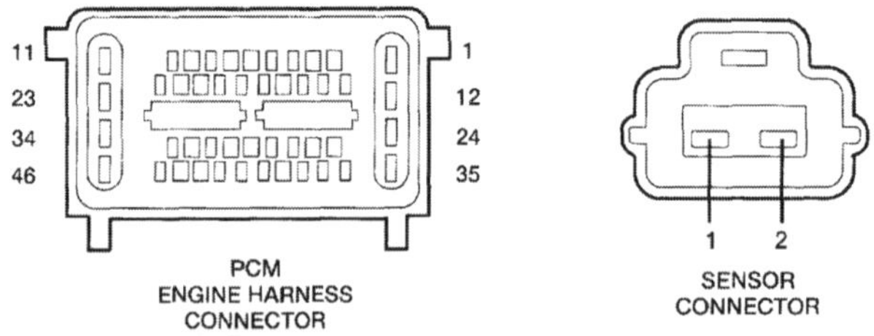

Pin	Circuit
E44	EOT (Engine Oil Temperature)
E25	SIGRTN (Signal Return)

Pin	Circuit
1	SIGRTN (Signal Return)
2	EOT (Engine Oil Temperature)

L1 PRELIMINARY DIAGNOSIS FOR DTCS P0196, P0197, P0198, P1184, P0298

Perform Visual Inspection.

DLC

Connect The Scan Tool.

Retrieve And Record Any Continuous And On-Demand DTCs.

Record The Freeze Frame Data Clear All DTCs.

DLC

Perform The KOEO On-Demand Self-Test.

DLC

Perform The KOER On-Demand Self-Test.

Are DTCs P0196, P0197, P0198, P1184, P0298 Present?

YES

For P0196: L2

For P0197: L7

For P0198: L9

For P1184: L13

For P0298: L18

NO

Unable To Duplicate Condition. CHECK For Loose Connection, Damaged Or Corroded Pins. WIGGLE Harness Attempting To Recreate The Fault. REFER To Symptom Charts If A Drivability Concern Exists.

STOP

L2 — DIAGNOSTIC TROUBLE CODE (DTC) P0196

DTC P0196 Indicates That An EOT Sensor Range/Performance Fault Has Been Detected. P0196 Is Set When The Vehicle Has Been Driven Above 1 ,250 Rpm And 12 Mg/Stroke MFDES Without EOT Sensor Signal Increasing Above 50°C (122°F). The Time To Set The Fault Is Dependent On EOT And IAT Temperatures, And Can Vary Between Approx. 15 To 45 Minutes.

Possible Causes:

- EOT Sensor
- EOT Sensor Circuitry
- Thermostat
- PCM

Inspect EOT Sensor Circuitry At Sensor And PCM. Inspect For Damaged, Loose Or Pushed-Out Pins, Loose Or Poorly Crimped Wires.

Are All Connectors And Terminals OK?

- YES → L3
- NO → DLC. REPAIR As Necessary. RESTORE The Vehicle. CLEAR DTCs And RETEST The System. → STOP

L3 — CHECK FOR DTCS

DLC

Perform KOEO On-Demand Self-Test.

Are Other DTCs Present?

- YES → DLC. REPAIR Other DTCs Before Proceeding. RESTORE Vehicle And RETEST The System. → DTC INDEX
- NO → Only P0196 Present: → L4

L4 CHECK EOT SENSOR OPERATION

Note: Verify No Accessories Are In Use (Engine Block Or Oil Heaters)

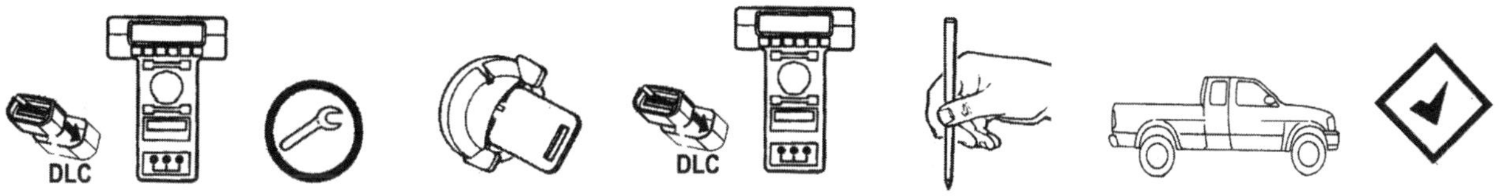

1. Clear All DTCs.
2. Soak Vehicle At Ambient Temperature For At Least 10 Hours.
3. Key On, Engine Off.
4. Using Scan Tool, Monitor EOT, RPM And MFDES PIDs.
5. Record EOT Sensor Temperature Reading.
6. Drive Vehicle Above 1,800 Rpm And 15 Mg/Stroke MFDES For At Least 15 Minutes. Select Appropriate Gear To Achieve Operating Conditions.
7. Does EOT Temperature Reading Change By At Least 6°C (10°F) Of Key On Value Recorded?

YES → L5

NO →

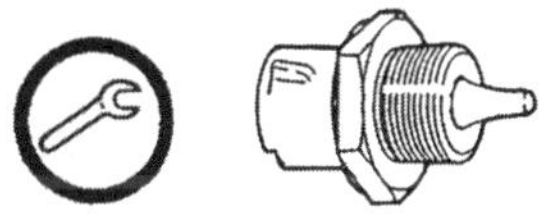

INSTALL A New EOT Sensor. RESTORE The Vehicle. CLEAR DTCs And RETEST The System. → STOP

L5 COOLING SYSTEM CHECK

Above 1,800 Rpm

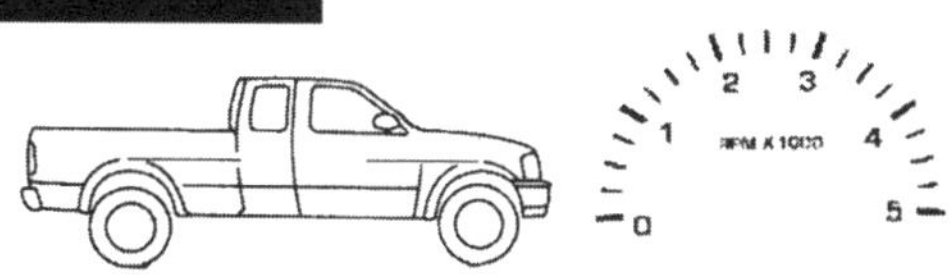

50°C (122°F)

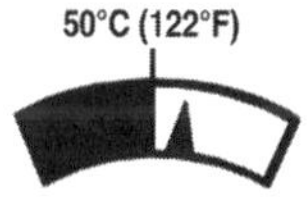

Drive Vehicle An Additional 15 Minutes Above 1,800 Rpm And 15 Mg/Stroke.

Does EOT Temperature Reading Increase Above 50°C (122°F)?

YES → L6

NO →

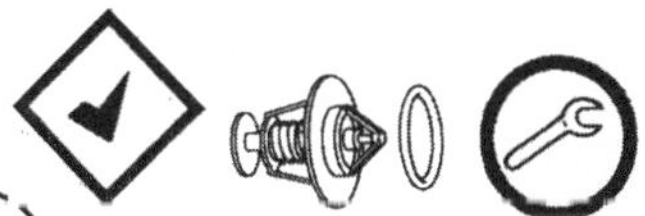

CHECK Thermostat For Correct Operation. If Fault Is Indicated, INSTALL A New Thermostat. If Thermostat Fault Is Not Present, INSTALL A New EOT Sensor. RESTORE The Vehicle. CLEAR DTCs And RETEST The System. → STOP

L6 CHECK THE PCM FOR DTCS

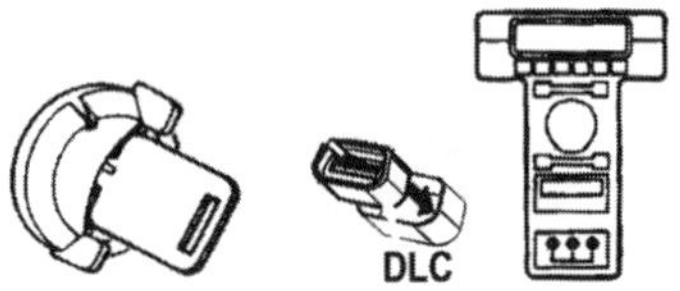

Key On, Engine Off.

Check Continuous Memory DTCs.

Is P0196 Present?

YES → L23

NO →

Unable To Duplicate Condition. CHECK For Loose Connections, Damaged Or Corroded Pins. WIGGLE Harness Attempting To Recreate Fault. REPAIR As Necessary. REFER To Symptom Charts If A Driveability Concern Exists. → STOP

L7 DIAGNOSTIC TROUBLE CODE (DTC) P0197

Note: Refer To The PCM And Component Connector Pin Numbers At The Beginning Of This Pinpoint Test.

Possible Causes: - EOT Signal Circuit Short To Ground
- EOT Sensor
- PCM
- EOT Electrical Connections

Key Off.

Disconnect EOT Sensor.

Check For Damaged, Pushed-Out Or Corroded Pins.

Check For Opposite Failure With EOT Disconnected.

Perform The KOEO On-Demand Self-Test.

Is DTC P0198 Present?

YES →

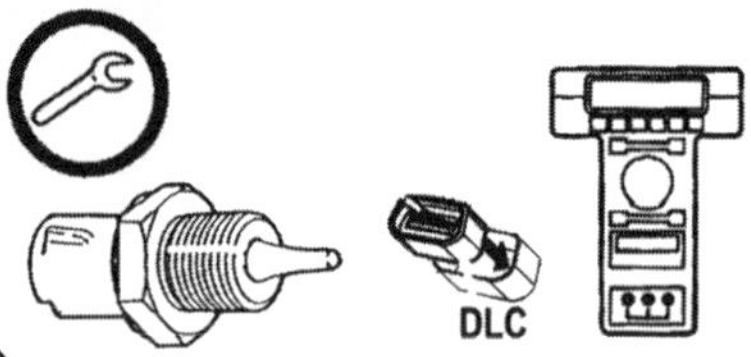

INSTALL A New EOT Sensor. RESTORE The Vehicle. CLEAR DTCs And RETEST The System. → STOP

NO → L8

L8 CHECK EOT SENSOR SIGNAL CIRCUIT FOR A SHORT TO GROUND

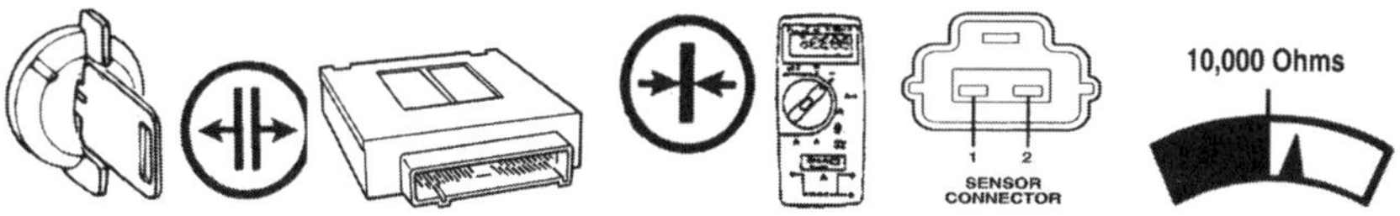

Key Off.

Disconnect The PCM Engine Harness Connector.

Measure the resistance between:	
(+) EOT Sensor Connector, Harness Side	(-)
EOT - Pin 2	Ground

Is The Resistance Greater Than 10,000 Ohms?

YES → L23

NO → STOP

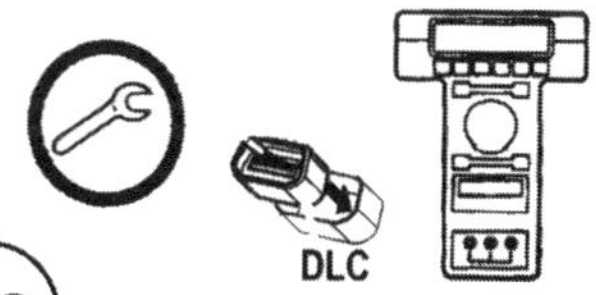

REPAIR Open In Signal Return EOT Signal Circuit. RESTORE The Vehicle. CLEAR DTCs And RETEST The System.

L9 DIAGNOSTIC TROUBLE CODE (DTC) P0198

Note: Refer To The PCM And Component Connector Pin Numbers At The Beginning Of This Pinpoint Test.

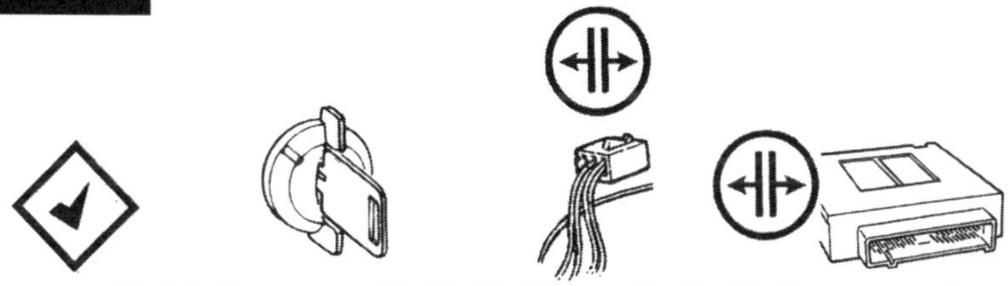

Possible Causes:
- Open Or Short To Voltage In EOT Signal Circuit
- Open In Signal Return Circuit
- Poor Connections
- EOT Sensor
- PCM

Key Off.

Disconnect EOT Sensor Harness Connector.

Disconnect The PCM Engine Harness Connector.

Is The Resistance Less Than 5 Ohms?

Measure the resistance between:	
(+) EOT Sensor Connector, Harness Side	(-) PCM Connector, Harness Side
SIGRTN - Pin 1	SIGRTN - Pin E25

YES → L10

NO → STOP

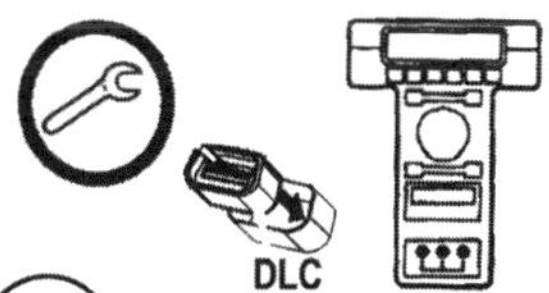

REPAIR Open In Signal Return Circuit. RESTORE The Vehicle. CLEAR DTCs And RETEST The System.

L10 — *INDUCE OPPOSITE FAILURE*

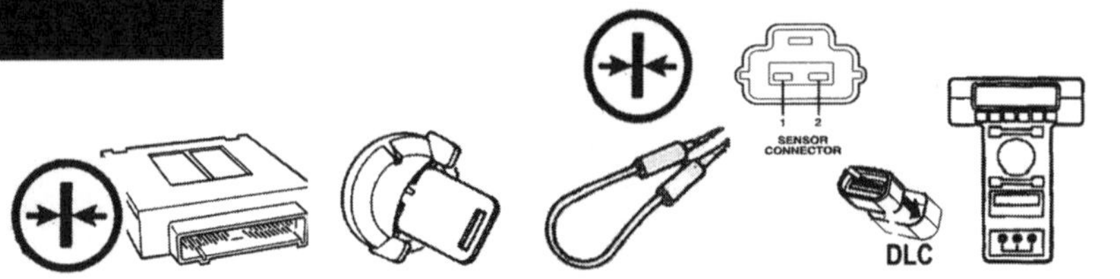

Connect The PCM Engine Harness Connector

Key On, Engine Off.

Connect a 5 amp fused jumper wire between the following:	
Point A EOT Sensor Connector, Harness Side	Point B EOT Sensor Connector, Harness Side
SIGRTN - Pin 1	EOT - Pin 2

Perform The On-Demand Self-Test.

Is DTC P0197 Present?

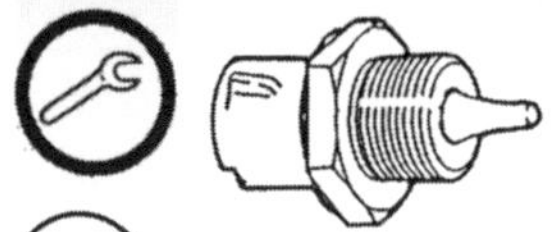

YES — INSTALL A New EOT Sensor. RESTORE The Vehicle. CLEAR DTCs And RETEST The System. — STOP

NO — L11

L11 — *CHECK SIGNAL CIRCUIT FOR A SHORT TO VOLTAGE*

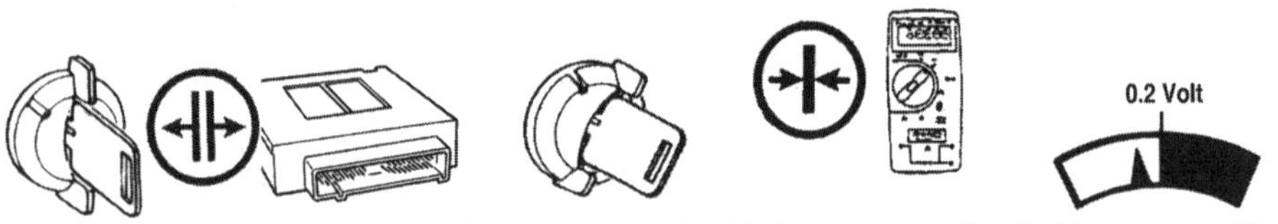

Key Off.

Disconnect The PCM Engine Harness Electrical Connector.

Key On, Engine Off.

Measure the voltage between:	
(+) EOT Sensor Connector, Harness Side	(-)
EOT - Pin 2	Ground

Is The Voltage Less Than 0.2 Volt?

YES — L12

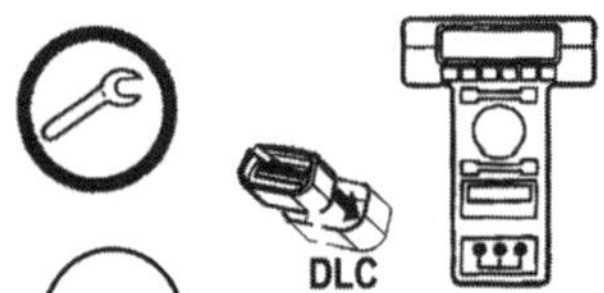

NO — REPAIR Short To Voltage In Signal Circuit. RESTORE The Vehicle. CLEAR DTCs And RETEST The System. — STOP

L12 — *CHECK SIGNAL CIRCUIT FOR AN OPEN*

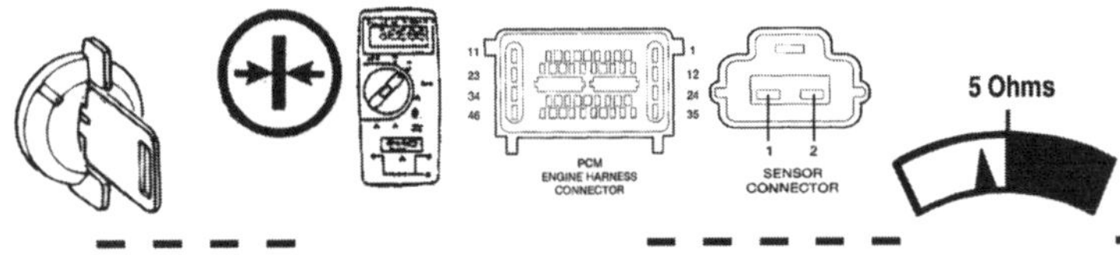

Key Off.

Measure the resistance between:	
(+) PCM Connector, Harness Side	(-) EOT Sensor Connector, Harness Side
EOT - Pin E44	EOT - Pin 2

Is The Resistance Less Than 5 Ohms?

YES — L23

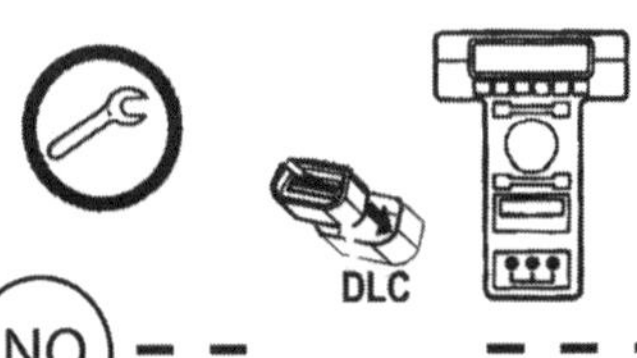

NO — REPAIR Open Signal Circuit. RESTORE The Vehicle. CLEAR DTCs And RETEST The System. — STOP

L13 DIAGNOSTIC TROUBLE CODE (DTC) P1184

DTC P1184 Indicates That The Engine Oil Temperature Is Not Within The Required Range To Perform The KOER Cylinder Contribution Self-Test. The Engine Oil Temperature Must Be Between 20°C (68°F) And 117°C (242°F).

Possible Causes:
- Engine Not Fully Warmed Up
- Low Oil Level
- Cooling System Failure
- EOT Sensor
- Thermostat
- EOT Sensor Circuit

Verify No KOEO DTCs Are Present.

Drive Vehicle Until Thermostat Opens.

Fully Warm Engine.

Check That Upper Radiator Hose Is Hot And Pressurized.

Rerun The KOER Cylinder Contribution Self-Test.

Is DTC P1184 Present?

YES → L14

NO → DISREGARD The Current DTC At This Time. DIAGNOSE The Next DTC. → DTC INDEX

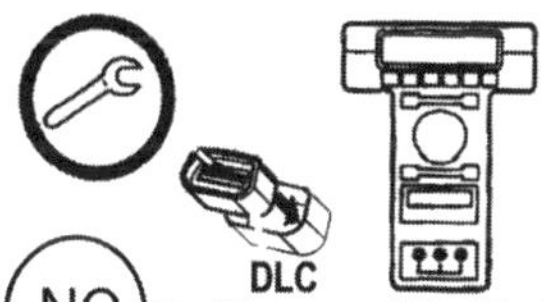

L14 EOT SENSOR CHECK

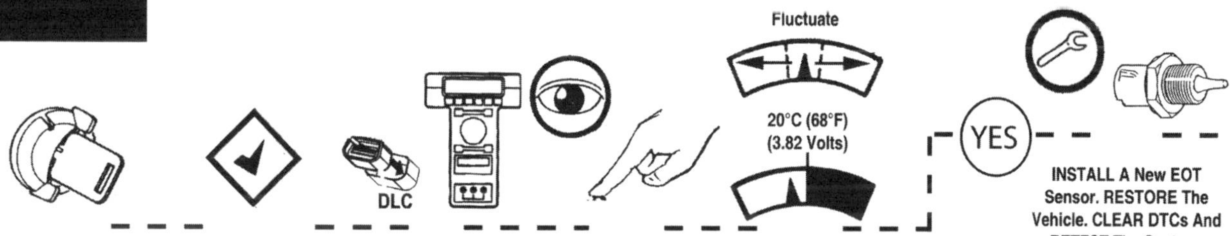

Key On, Engine Off.

Engine At Normal Operating Temperature.

Access EOT PID.

Observe EOT PID While Tapping On EOT Sensor.

Does The EOT Value Fluctuate Or Go Below 20°C (68°F) (3.82 Volts)?

YES → INSTALL A New EOT Sensor. RESTORE The Vehicle. CLEAR DTCs And RETEST The System. → STOP

NO → L15

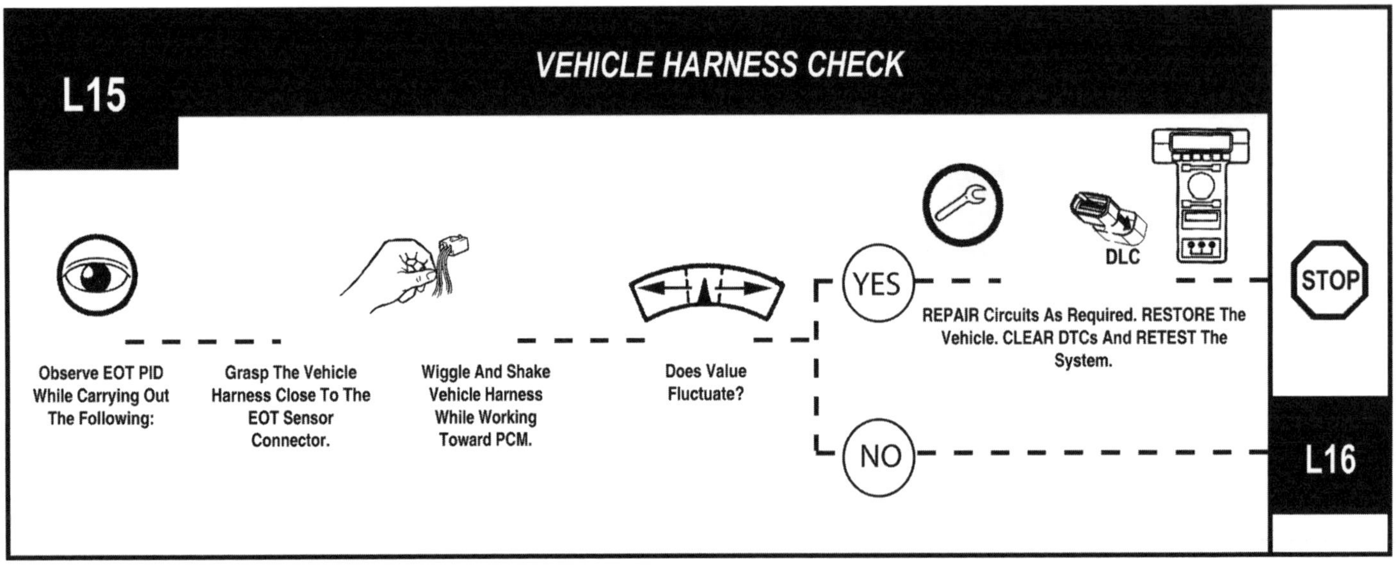

L15
VEHICLE HARNESS CHECK
Observe EOT PID While Carrying Out The Following:
Grasp The Vehicle Harness Close To The EOT Sensor Connector.
Wiggle And Shake Vehicle Harness While Working Toward PCM.
Does Value Fluctuate?
YES
DLC
REPAIR Circuits As Required. RESTORE The Vehicle. CLEAR DTCs And RETEST The System.
STOP
NO
L16

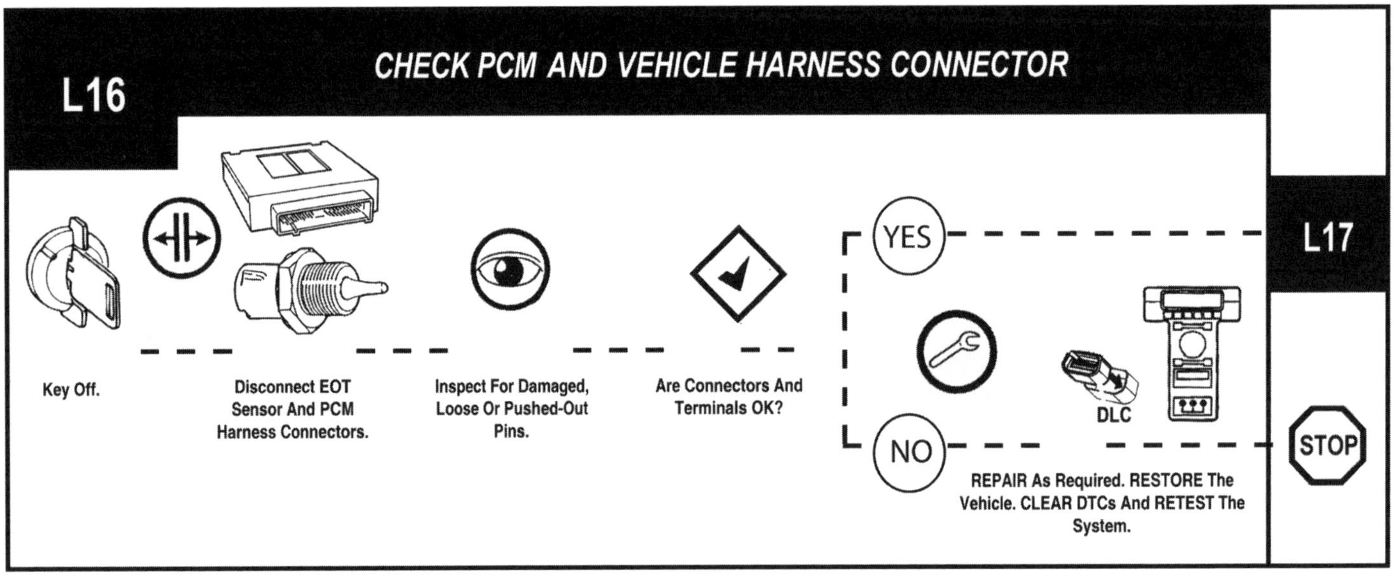

L16
CHECK PCM AND VEHICLE HARNESS CONNECTOR
Key Off.
Disconnect EOT Sensor And PCM Harness Connectors.
Inspect For Damaged, Loose Or Pushed-Out Pins.
Are Connectors And Terminals OK?
YES
L17
DLC
NO
REPAIR As Required. RESTORE The Vehicle. CLEAR DTCs And RETEST The System.
STOP

L17 CHECK FOR BIASED EOT SENSOR

Note: Verify No Accessories Are In Use (Engine Block Or Oil Heaters)

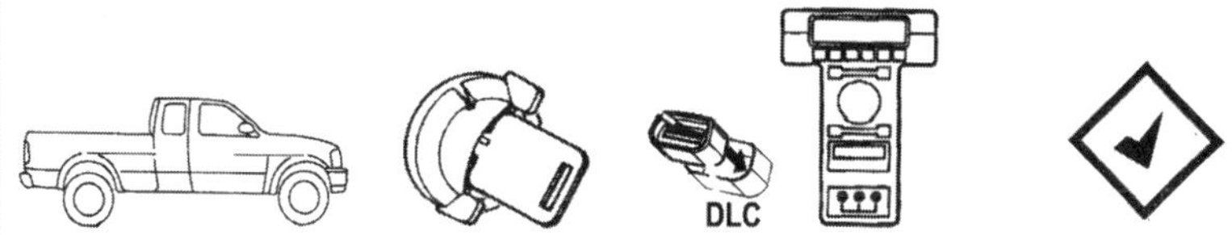

Allow Vehicle To Sit Overnight.

Key On, Engine Off.

Read EOT And IAT PIDs Using Scan Tool With KOEO.

Do The Readings Agree Within 6°C (11°F)?

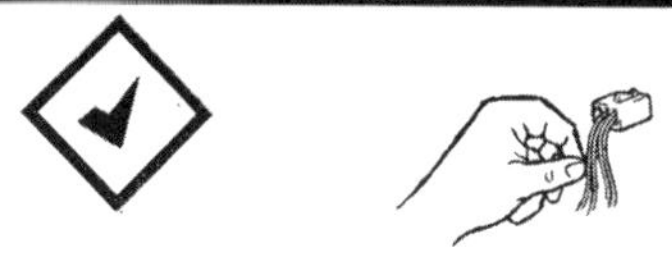

YES — UNABLE To Duplicate Concern. ROAD TEST Vehicle, Check For A Loose Connection, And Damaged Or Corroded Terminals. WIGGLE the Harness Attempting To Recreate The Fault. REPAIR As Necessary And RETEST The System. REFER To Symptom Charts If A Driveablility Concern Exists. — STOP

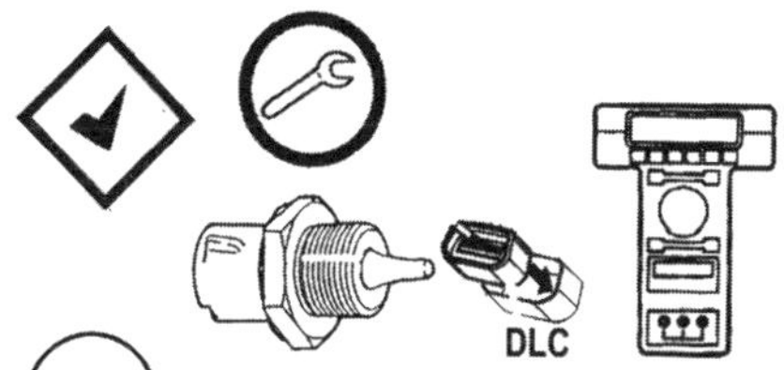

NO — CONFIRM Wiring Is OK. If OK, INSTALL A New EOT. CLEAR The DTCs. REPEAT The Self-Test. — STOP

L18 DIAGNOSTIC TROUBLE CODE (DTC) P0298

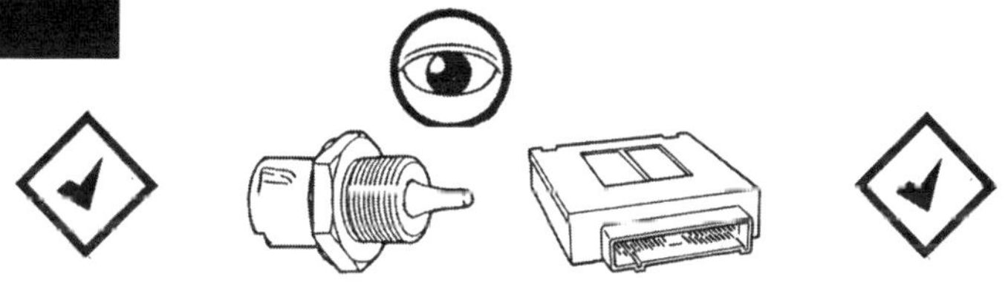

DTC P0298 Indicates That An EOT Sensor Range/Performance Fault Has Been Detected. P0298 Is Set When The Vehicle Has Been Running Below 1,000 Rpm And Less Than 20 Mg/Stroke MFDES With EOT Sensor Signal Reading Above 110°C (230°F). The Time To Set The Fault Is Dependent On EOT And IAT Temperatures, And Can Vary Between Approx. 15 To 45 Minutes.

Possible Causes:

- EOT Sensor
- EOT Sensor Circuitry
- Cooling System
- Thermostat
- PCM

Inspect EOT Sensor Circuitry At Sensor And PCM. Inspect For Damaged, Loose Or Pushed-Out Pins, Loose Or Poorly Crimped Wires.

Are All Connectors And Terminals OK?

YES — L19

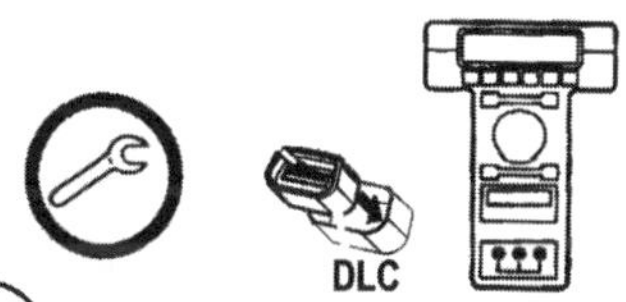

NO — REPAIR As Necessary. RESTORE The Vehicle. CLEAR DTCs And RETEST The System. — STOP

L19 — COOLING SYSTEM CHECKS

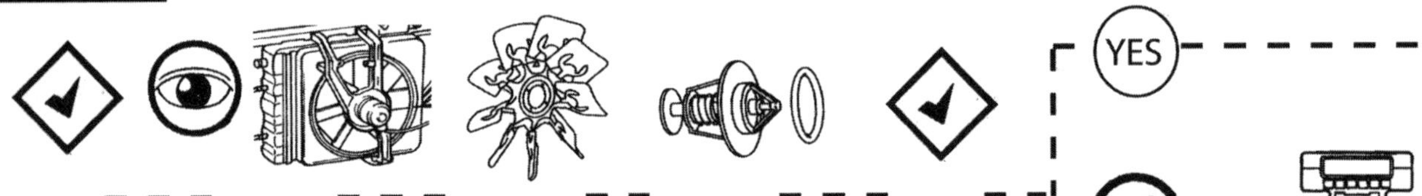

Check Coolant System For Correct Level.

Check Radiator For Correct Performance. Verify There Are No Obstructions To Airflow Across Radiator Core, And That The Radiator Core Is Not Plugged.

Verify Correct Cooling Fan/Clutch Operation.

Verify Correct Thermostat Operation.

Is Coolant System OK?

YES → L20

NO → DLC. REPAIR As Necessary. RESTORE The Vehicle. CLEAR DTCs And RETEST The System. → STOP

L20 — CHECK EOT SENSOR FAULT CODES

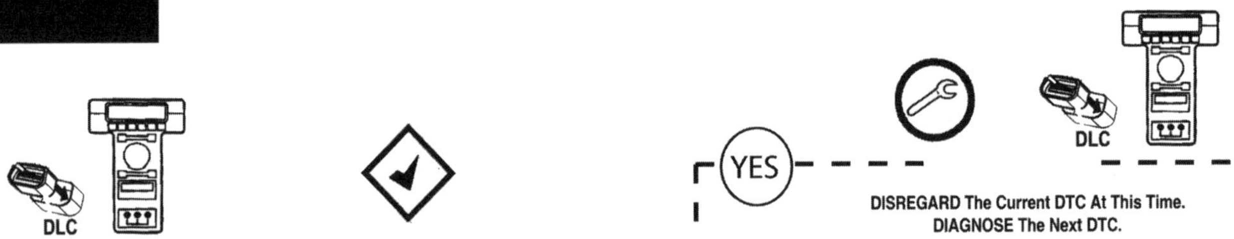

DLC. Perform KOEO On-Demand Self-Test.

Are Other DTCs Present?

YES → DLC. DISREGARD The Current DTC At This Time. DIAGNOSE The Next DTC. → DTC INDEX

NO → Only P0298 Present: → L21

L21 — CHECK EOT SENSOR OPERATION

Note: Verify No Accessories Are In Use (Engine Block Or Oil Heaters)

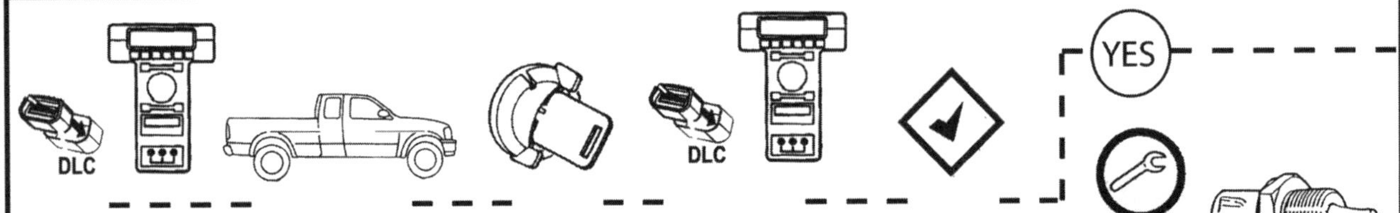

DLC. Clear Continuous DTCs.

Soak Vehicle At Ambient Temperature For At Least 8 Hours.

Key On, Engine Off.

DLC. Using Scan Tool, Monitor EOT PID.

Is EOT Sensor Equal To Ambient Temperature?

YES → L22

NO → INSTALL A New EOT Sensor. RESTORE The Vehicle. CLEAR DTCs And RETEST The System. → STOP

L22 — CONFIRM PCM FAULT

Run Engine For Approximately 15 To 20 Minutes.

Check Continuous Memory DTCs.

Is DTC P0298 Present?

YES → L23

NO → STOP

OPERATE Vehicle To VERIFY EOT Does Not Increase Above 110°C (230°F) While Running Below 1,000 Rpm. REFER To Symptom Charts If A Driveability Concern Exists.

L23 — CHECK FOR CORRECT PCM OPERATION

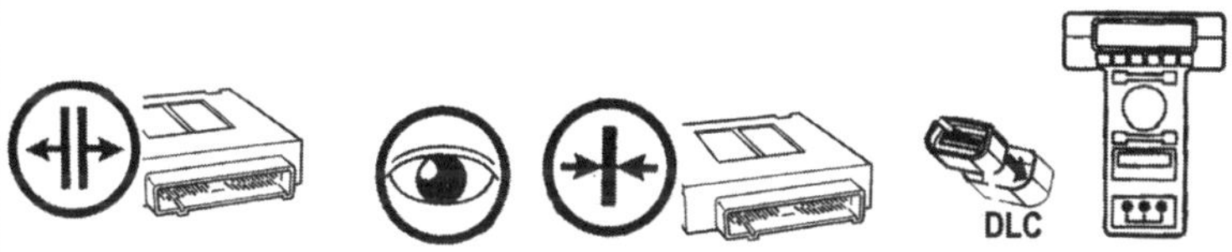

Disconnect All PCM Connectors.

Visually Inspect For Pushed Out Pins, Corrosion.

Connect All PCM Connectors And Make Sure They Seat Correctly.

Perform The PCM Self-Test And Verify The Concern Is Still Present.

Is The Concern Still Present.

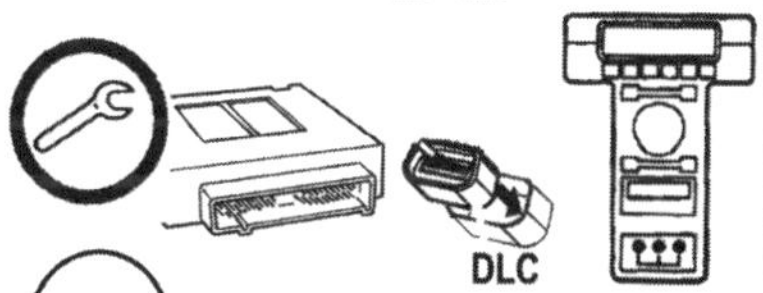

YES → STOP

INSTALL A New PCM. CLEAR The DTC's And REPEAT The Self-Test.

NO → STOP

The System Is Operating Correctly At This Time. The Concern May Have Been Caused By A Loose Or Corroded Connector.

NOTES

Trouble Code (s)	Problem: Fuel Pump Monitor/Control	Test
P0230, P0231, P0232		M 2006-07

REFERENCE LIST

This pinpoint test is intended to diagnose the following:

- Fuel pump relay
- Fuel pump
- Fuel pump circuitry
- Inertia fuel shutoff (IFS) switch
- Powertrain control module (PCM)

DIAGNOSTIC TROUBLE CODE INDEX			
CODES	KOEO	KOER	CM
P0230	M1	M1	M1
P0231	M1	M1	M1
P0232	M1	M1	M1

P0230 = Fuel Pump Primary Circuit
P0231 = Fuel Pump Secondary Circuit Low
P0232 = Fuel Pump Secondary Circuit High

Circuit Function:

The fuel pump relay coil ground is controlled by the powertrain control module (PCM) with vehicle power (VPWR) supplied to the relay coil by the PCM power relay. Energizing the fuel pump relay closes the internal switch contacts and supplies B+ voltage through the inertia switch to the electric fuel pump. Fuel pump voltage is monitored by the PCM through the fuel pump monitor (FPM) circuit which is downstream of the inertia switch. When the ignition switch is turned on, the electric fuel pump will operate for approximately 20 seconds and will then be commanded off by the PCM if an RPM signal is not detected.

CAUTION!:

The powertrain control module (PCM) harness connectors must be properly seated and the connector latch properly attached to eliminate possible driveability concerns or a no start condition. Installing the PCM connectors on an angle may cause an improper connection, misdiagnosis and damaged components. Install the connector until the lever pivots and seats itself. Apply light pressure to get the connector into position on the PCM and then fully seat the connector.

Note: Visually inspect the harness connectors for corrosion, damage, proper mating and correct pin tension.

Note: When the PCM is disconnected additional DTCs will be set. Clear all DTCs after restoring the vehicle.

PINPOINT TEST SCHEMATICS AND CONNECTORS

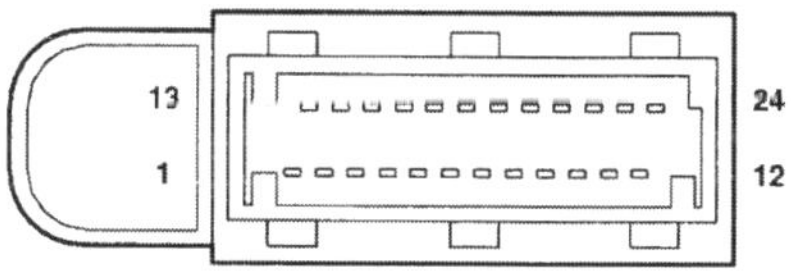

CENTRAL JUNCTION BOX (CJB) C270a

Pin	Circuit
22	IS (Inertia Switch)

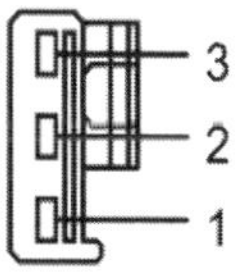

Inertia Fuel Shutoff (IFS) Switch Connector

Pin	Circuit
1	IS (Inertia Switch)
2	B+ (Battery Positive Voltage)

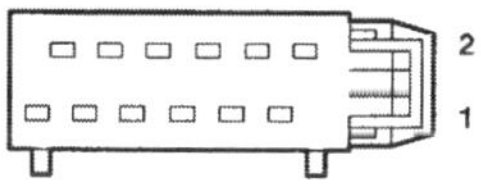

CENTRAL JUNCTION BOX (CJB) C270F

Pin	Circuit
10	FP (Fuel Pump)

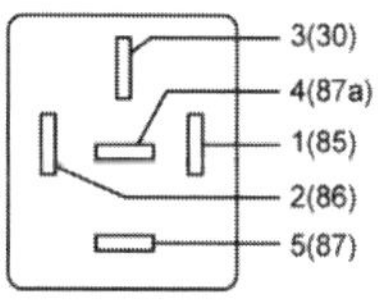

Fuel Pump (FP) Relay Connector

Pin	Circuit
1	FP (Fuel Pump)
2	VPWR (Vehicle Power)
3	B+ (Battery Positive Voltage)

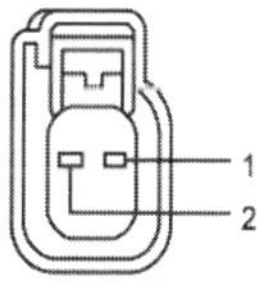

Fuel Pump (FP) Connector

Pin	Circuit
1	FPPWR (Fuel Pump Power)
2	FPGND (Fuel Pump Ground)

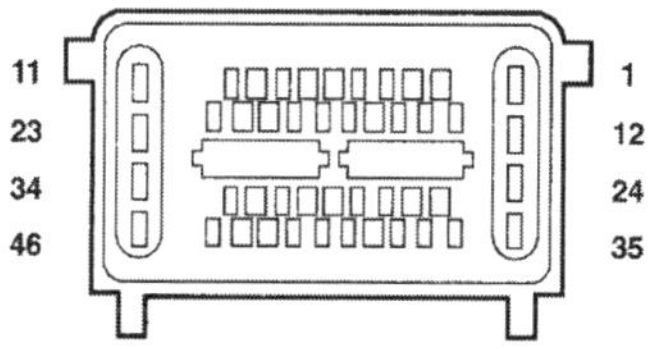

PCM BODY HARNESS CONNECTOR

Pin	Circuit
B19	IS (Inertia Switch)
B5	FP (Fuel Pump)

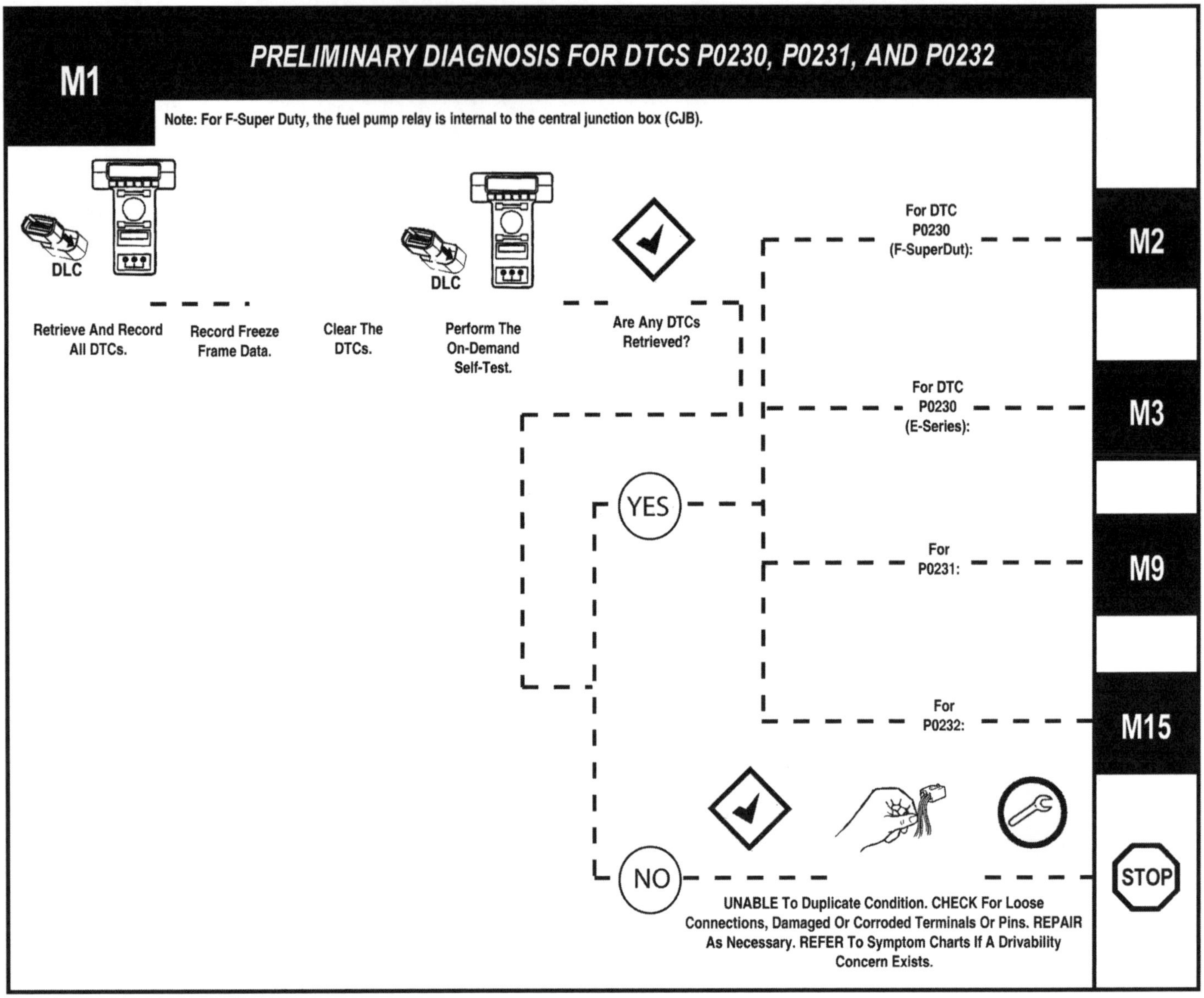

M1
PRELIMINARY DIAGNOSIS FOR DTCS P0230, P0231, AND P0232
Note: For F-Super Duty, the fuel pump relay is internal to the central junction box (CJB).
DLC
DLC
Retrieve And Record All DTCs.
Record Freeze Frame Data.
Clear The DTCs.
Perform The On-Demand Self-Test.
Are Any DTCs Retrieved?
YES
For DTC P0230 (F-SuperDut):
M2
For DTC P0230 (E-Series):
M3
For P0231:
M9
For P0232:
M15
NO
UNABLE To Duplicate Condition. CHECK For Loose Connections, Damaged Or Corroded Terminals Or Pins. REPAIR As Necessary. REFER To Symptom Charts If A Drivability Concern Exists.
STOP

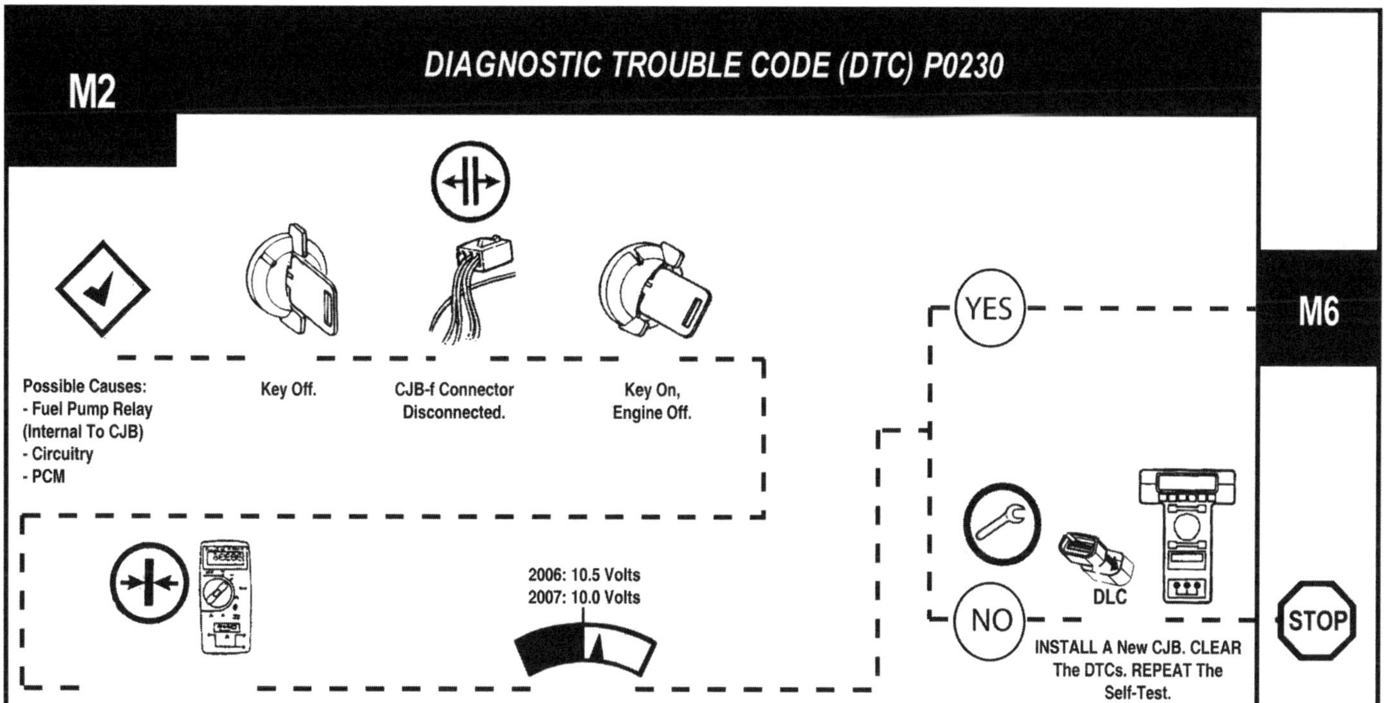

Measure the voltage between:	
(+) CJB-f Connector, Component Side	(-)
FP - Pin 10	Ground

M3 — CHECK THE FUEL PUMP RELAY VOLTAGE CIRCUIT

Note: Refer To The Wiring Diagrams Manual For Fuel Pump Relay Coil Pin Locations.

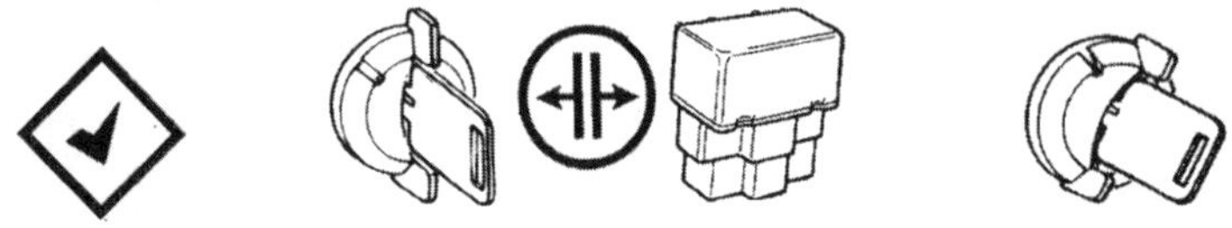

Possible Causes:
- Fuel Pump Relay
- Circuitry
- PCM

Key Off.

Disconnect The Fuel Pump Relay.

Key On, Engine Off.

2006: 10.5 Volts
2007: 10.0 Volts

Measure the voltage between:	
(+) FP Relay Connector, Harness Side	(-)
VPWR - Pin 2	Ground

Is The Voltage Greater Than 2006: 10.5 Volts 2007: 10.0 Volts?

YES — M4

NO — STOP

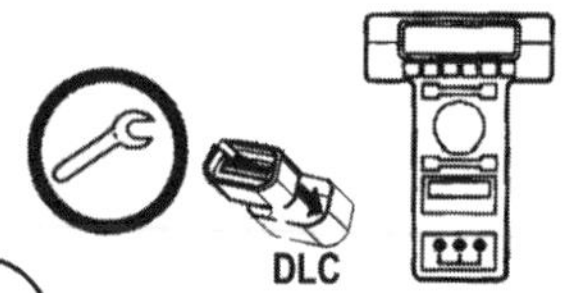

REPAIR The Circuit. CLEAR The DTCs. REPEAT The Self-Test

M4 — CHECK THE FUEL PUMP RELAY

Perform The Fuel Pump Relay Component Test.

Does The Fuel Pump Relay Pass The Component Test?

YES — E-SERIES — M5

YES — F-SUPERDUTY — M6

NO — STOP

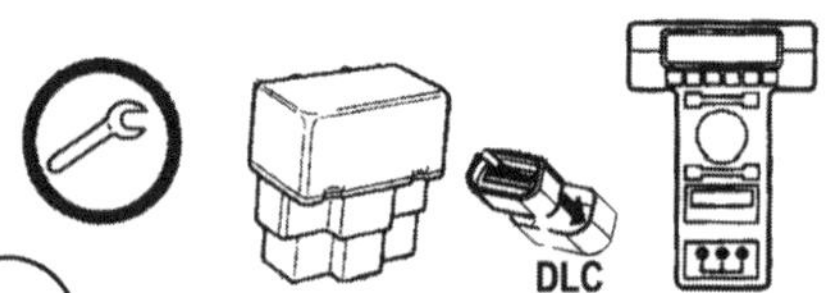

INSTALL A New Fuel Pump Relay. CLEAR The DTCs. REPEAT The Self-Test.

M5 CHECK THE FUEL PUMP RELAY CONTROL CIRCUIT FOR AN OPEN

Note: Refer to the Wiring Diagrams Manual for the fuel pump relay coil pin locations.

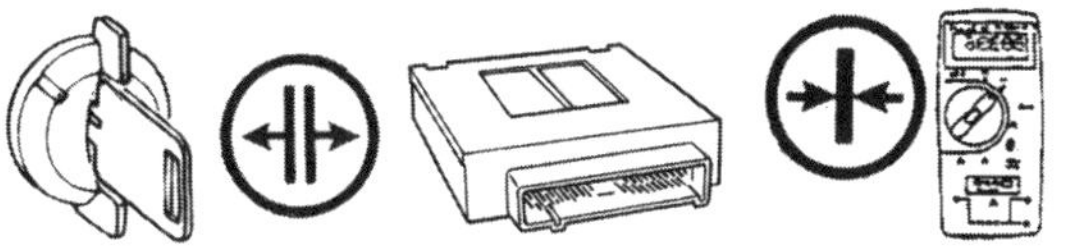

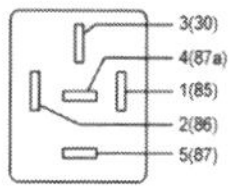

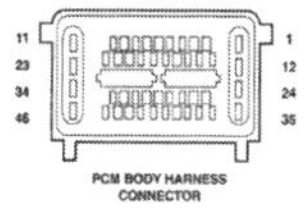

Key Off.

Disconnect The PCM Body Harness Connector.

Measure the resistance between:	
(+) FP Relay Connector, Harness Side	(-) PCM Connector, Harness Side
FP - Pin 1	FP - Pin B5

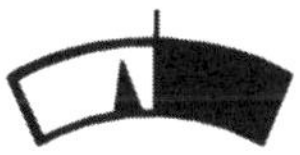

Is The Resistance Less Than 5 Ohms?

YES → M7

NO → REPAIR The Circuit. CLEAR The DTCs. REPEAT The Self-Test. STOP

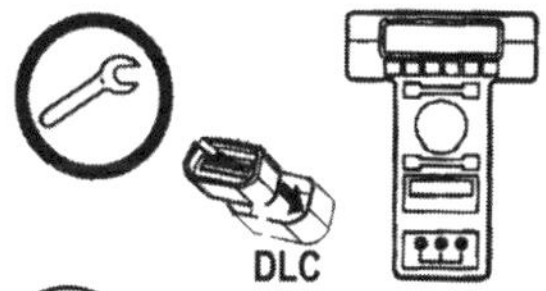

M6 CHECK THE FUEL PUMP RELAY CONTROL CIRCUIT FOR AN OPEN

Note: Refer to the Wiring Diagrams Manual for the fuel pump relay coil pin locations.

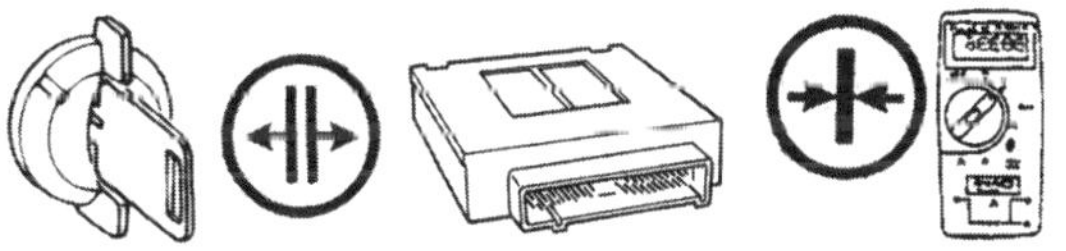

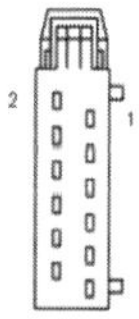

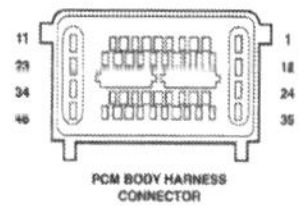

Key Off.

Disconnect The PCM Body Harness Connector.

Measure the resistance between:	
(+) CJB-f Connector, Harness Side	(-) PCM Connector, Harness Side
FP - Pin 10	FP - Pin B5

Is The Resistance Less Than 5 Ohms?

YES → M7

NO → REPAIR The Circuit. CLEAR The DTCs. REPEAT The Self-Test. STOP

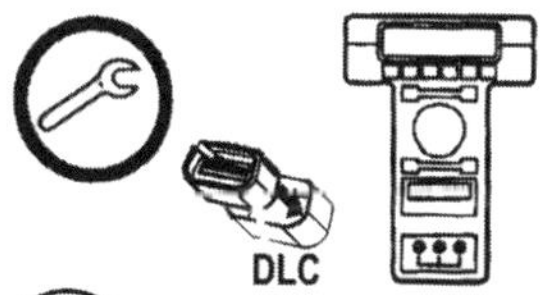

M7 — CHECK THE FUEL PUMP RELAY CONTROL CIRCUIT FOR A SHORT TO GROUND

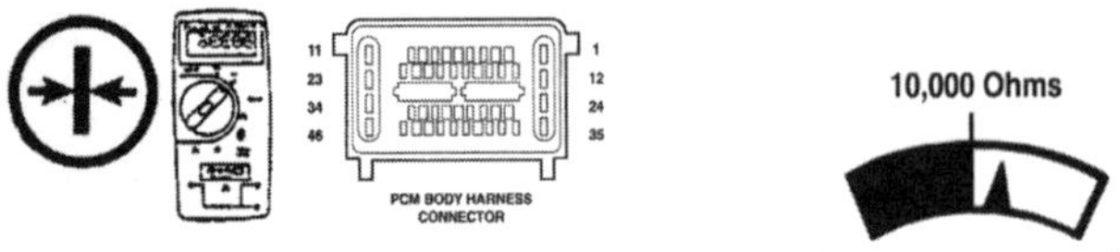

Measure the resistance between:	
(+) PCM Connector, Harness Side	(-)
FP - Pin B5	Ground

Is The Resistance Greater Than 10,000 Ohms?

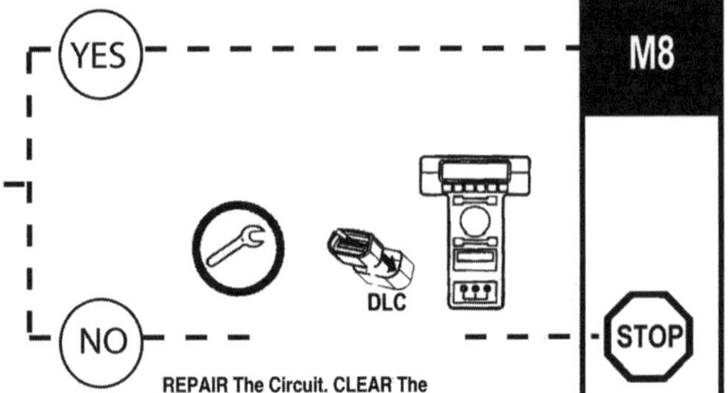

YES → M8

NO → REPAIR The Circuit. CLEAR The DTCs. REPEAT The Self-Test. STOP

M8 — CHECK THE FUEL PUMP RELAY CONTROL CIRCUIT FOR A SHORT TO VOLTAGE

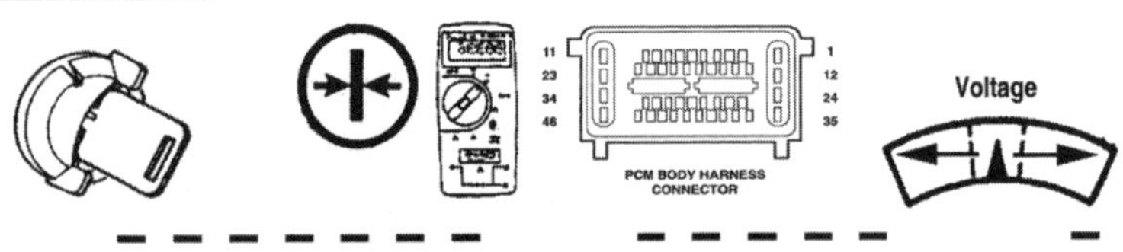

Key On, Engine Off.

Measure the resistance between:	
(+) PCM Connector, Harness Side	(-)
FP - Pin B5	Ground

Is Any Voltage Indicated?

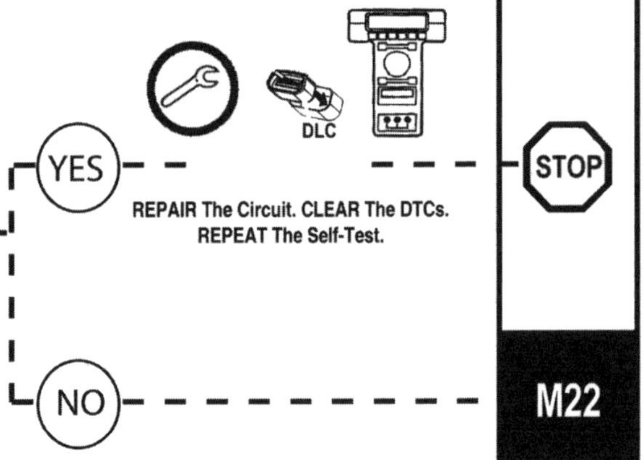

YES → REPAIR The Circuit. CLEAR The DTCs. REPEAT The Self-Test. STOP

NO → M22

M9 DIAGNOSTIC TROUBLE CODE (DTC) P0231

Note: Diagnose And Repair DTC P0230 Before Addressing DTC P0231.

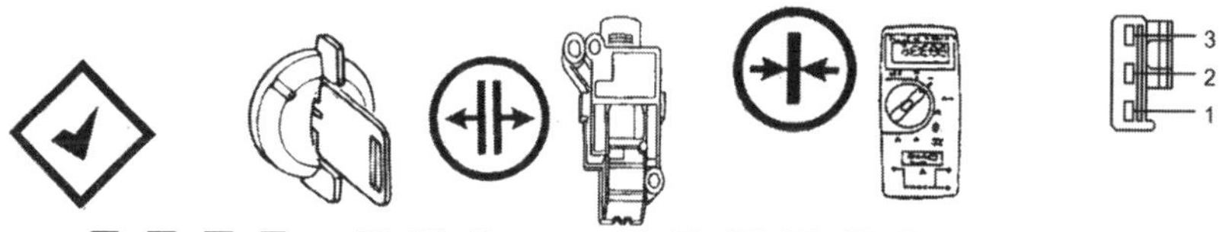

Possible Causes:
- Inertia Switch
- Fuel Pump Monitor Circuit Open
- Fuel Pump Relay

Key Off.

Disconnect The Inertia Switch.

Measure the resistance between:	
(+) IFS Switch Connector, Component Side	(-) IFS Switch Connector, Component Side
B+ - Pin 2	IS - Pin 1

5 Ohms

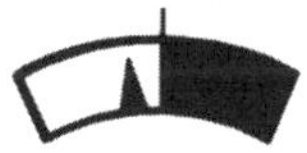

Is The Resistance Less Than 5 Ohms?

YES → M10

NO → STOP

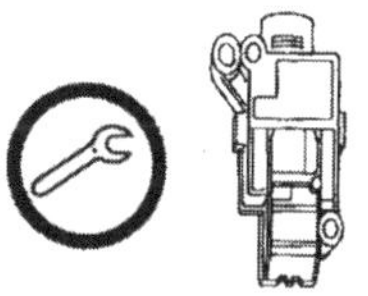

INSTALL A New Inertia Switch. CLEAR The DTCs. REPEAT The Self-Test.

M10 CHECK THE FUEL PUMP VOLTAGE

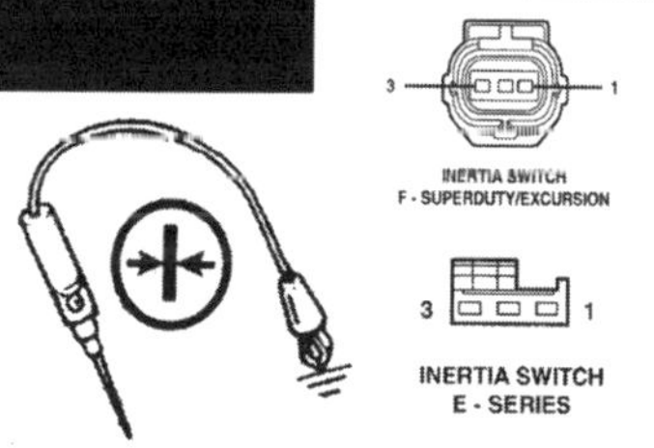

Connect a non-powered test lamp between:	
(+) IFS Switch Connector, Harness Side	(-)
B+ - Pin 2	Ground

Key On, Engine Off.

20 Seconds

Is The Test Lamp Illuminated For Approximately 20 Seconds?

YES → M14

NO → For F-SuperDuty → M11

For E-Series: → M12

M11 CHECK THE FUEL PUMP VOLTAGE CIRCUIT FOR AN OPEN

Key Off.

CJB-a Connector Disconnected.

Measure the resistance between:	
(+) CJB-a Connector, Harness Side	(-) IFS Switch Connector, Harness Side
IS - Pin 22	B+ - Pin 2

Is The Resistance Less Than 5 Ohms?

YES → CHECK fuse. If OK, INSTALL A New CJB. CLEAR The DTCs. REPEAT The Self-test. **STOP**

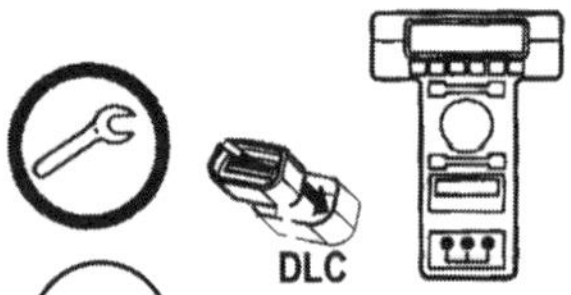

NO → REPAIR The Circuit. CLEAR The DTC's. Repeat The Self-Test. **STOP**

M12 CHECK THE FUEL PUMP RELAY POWER CIRCUIT FOR VOLTAGE

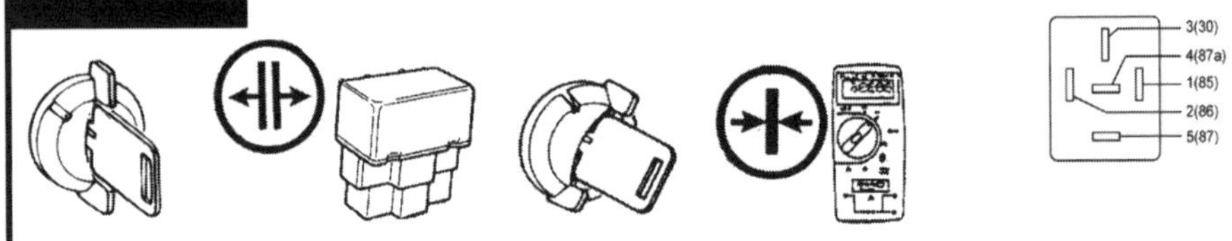

Key Off.

Disconnect The Fuel Pump Relay.

Key On, Engine Off.

Measure the voltage between:	
(+) FP Relay Connector, Harness Side	(-)
B+ - Pin 3	Ground

Is The Voltage Greater Than 2006: 10.5 Volts 2007: 10.0 Volts?

YES → **M13**

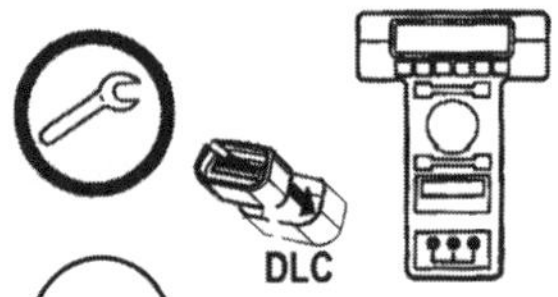

NO → REPAIR The Circuit. CLEAR The DTCs. REPEAT The Self-Test. **STOP**

M13 CHECK THE FUEL PUMP RELAY

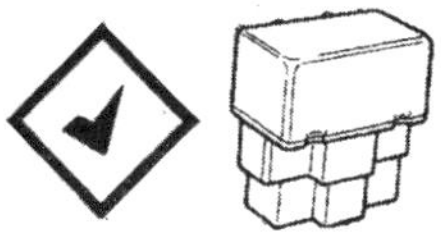

Perform The Fuel Pump Relay Component Test.

Does The Fuel Pump Relay Pass The Component Test?

YES

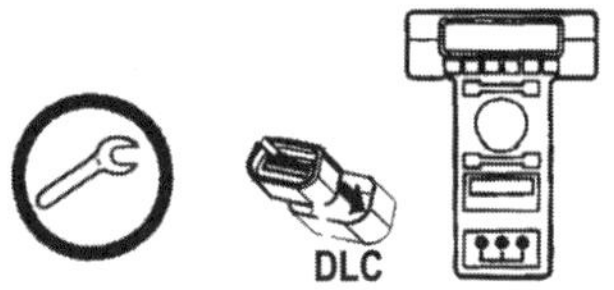

REPAIR The Circuit. CLEAR The DTC's. Repeat The Self-Test.

STOP

NO

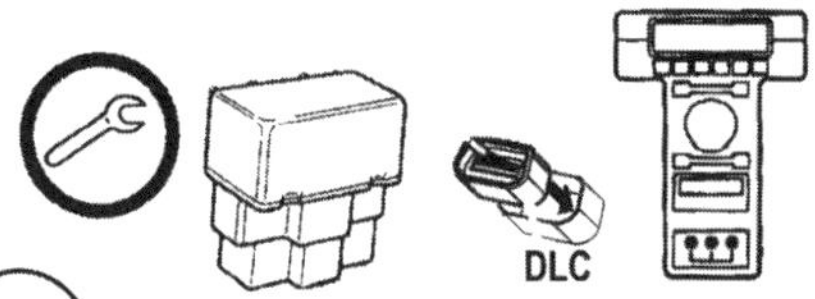

INSTALL A New Fuel Pump Relay. CLEAR The DTC's. Repeat The Self-Test.

STOP

M14 CHECK THE FUEL PUMP MONITOR CIRCUIT FOR AN OPEN

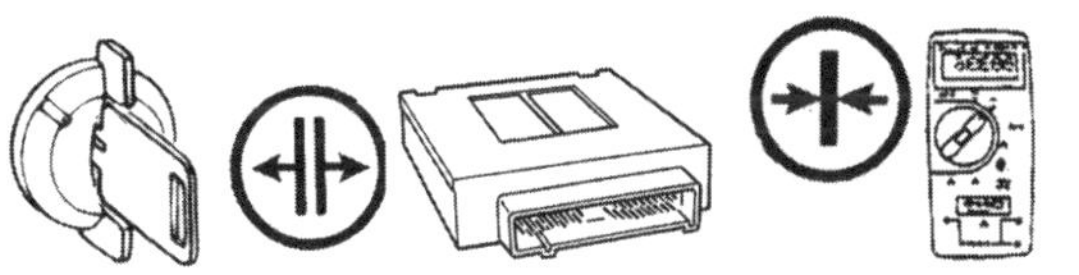

Key Off.

Disconnect The PCM.

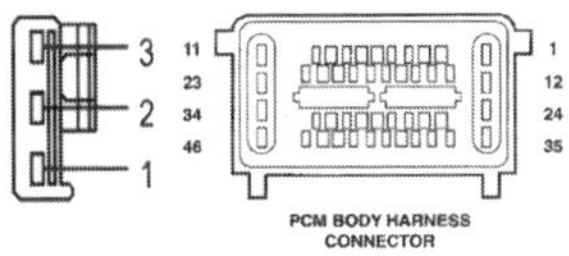

Measure the resistance between:	
(+) IFS Switch Connector, Harness Side	(-) PCM Connector, Harness Side
IS - Pin 1	IS - Pin B19

5 Ohms

Is The Resistance Less Than 5 Ohms?

YES → M22

NO

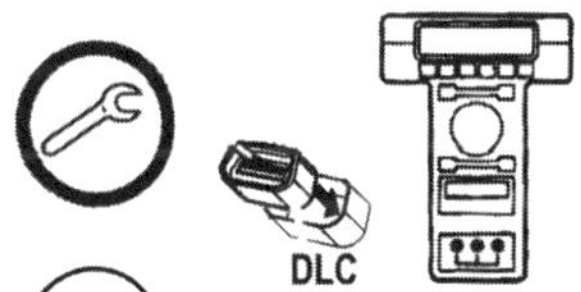

REPAIR The Circuit. CLEAR The DTC's. Repeat The Self-Test.

STOP

M15 — *DIAGNOSTIC TROUBLE CODE (DTC) P0232*

Note: Diagnose And Repair DTC P0230 Before Addressing DTC P0232.

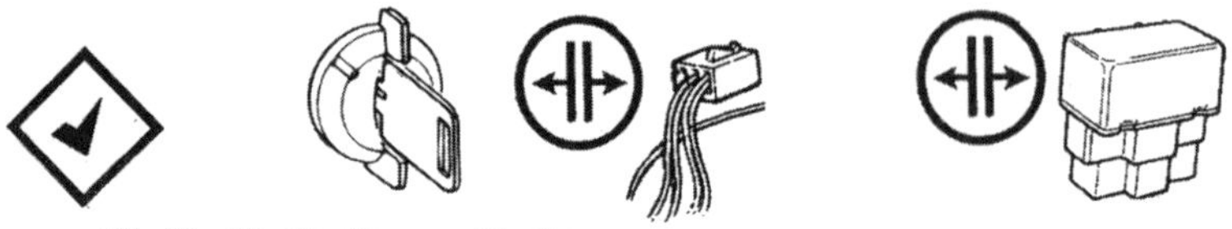

Possible Causes:
- Fuel Pump Relay
- Fuel Pump Monitor Circuit Short To Voltage
- Fuel Ground Circuit Open Fuel Pump

Key Off.

Disconnect CJB-a (F-SuperDuty).

Disconnect The Fuel Pump Relay (E-Series).

Key On, Engine Off.

Is The Fuel Pump ON?

YES — M16

NO — M17

M16 — *CHECK THE FUEL PUMP MONITOR CIRCUIT A SHORT TO VOLTAGE*

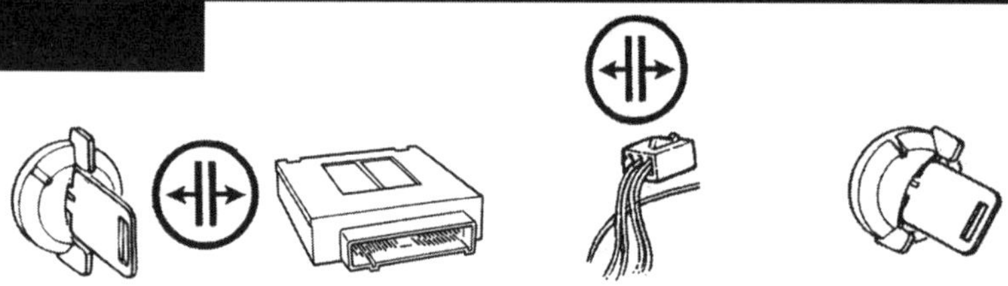

Key Off.

Disconnect The PCM Body Harness Connector.

Disconnect The Fuel Pump.

Key On, Engine Off.

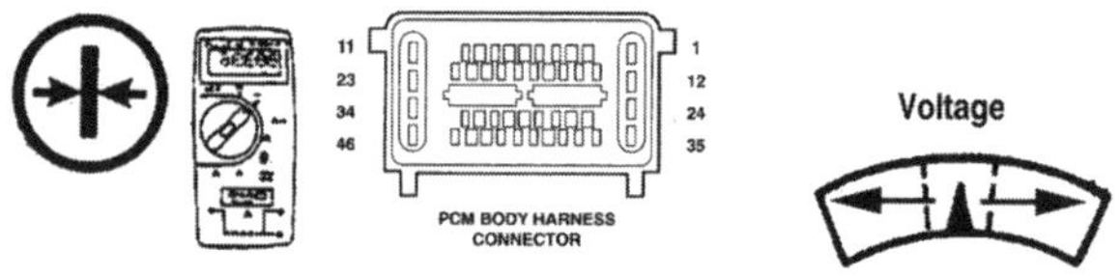

Is Any Voltage Indicated?

Measure the voltage between:	
(+) PCM Connector, Harness Side	(-)
IS - Pin B19	Ground

YES — STOP

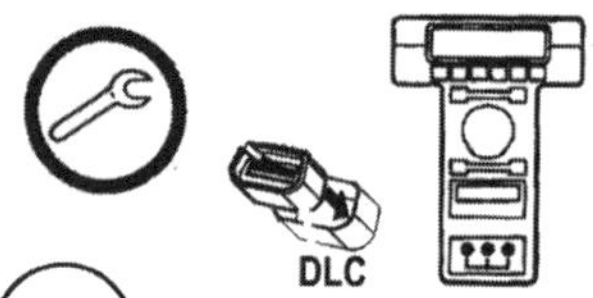

REPAIR The Circuit. CLEAR The DTCs. REPEAT The Self-Test.

NO — M22

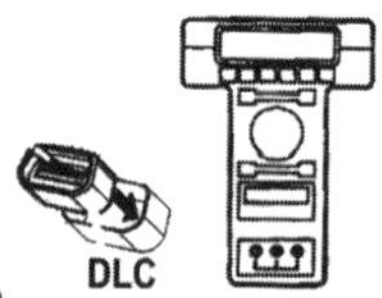

CLEAR The DTCs. REPEAT The Self- Test. If DTC P0232 Is Present Again, GO TO M22

M17 CHECK THE FUEL PUMP GROUND FOR AN OPEN

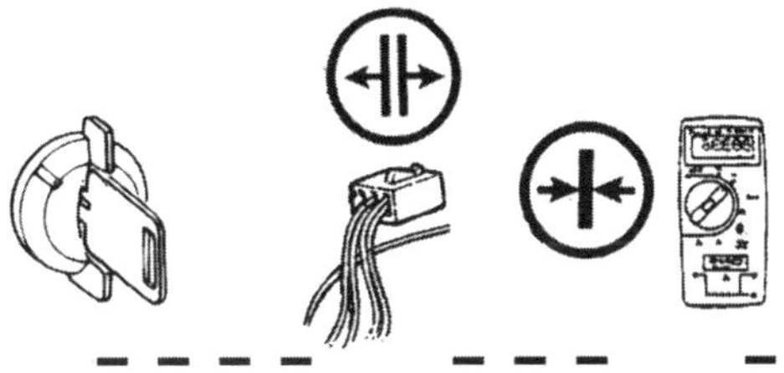

Key Off.

Disconnect The Fuel Pump.

Measure the resistance between:	
(+) FP Connector, Harness Side	(-)
FPGND - Pin 2	Ground

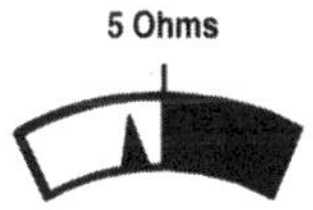

Is The Resistance Less Than 5 Ohms?

YES → M18

NO → STOP

REPAIR The Circuit. CLEAR The DTC's. Repeat The Self-Test.

M18 CHECK THE FUEL PUMP

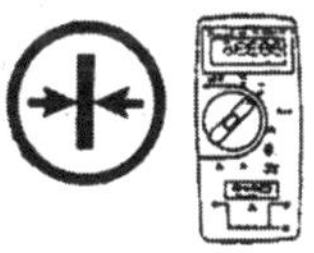

Measure the resistance between:	
(+) FP Connector, Component Side	(-) FP Connector, Component Side
FPPWR - Pin 1	FPGND - Pin 2

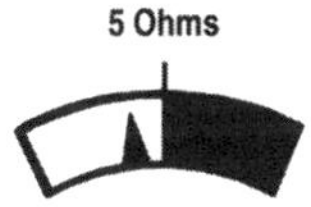

Is The Resistance Less Than 5 Ohms?

YES → M19

NO → STOP

INSTALL A New Fuel Pump. CLEAR The DTC's. Repeat The Self-Test.

M19 CHECK THE FUEL PUMP MONITOR CIRCUIT FOR AN OPEN

Disconnect The PCM.

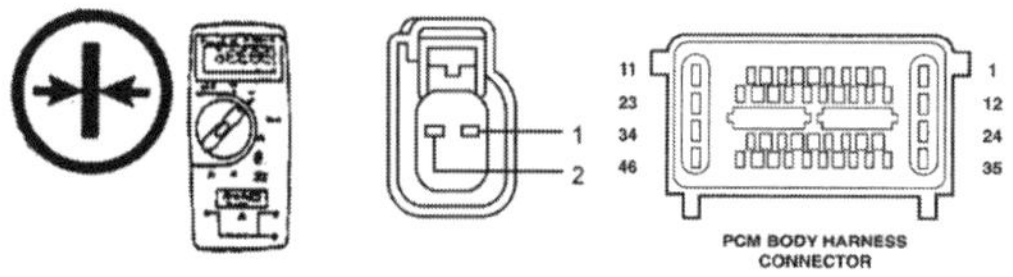

Measure the resistance between:	
(+) PCM Connector, Harness Side	(-) FP Connector, Harness Side
IS - Pin B19	FPPWR - Pin 1

Is The Resistance Less Than 5 Ohms?

YES — For E-Series: M20

For F-Super Duty: M21

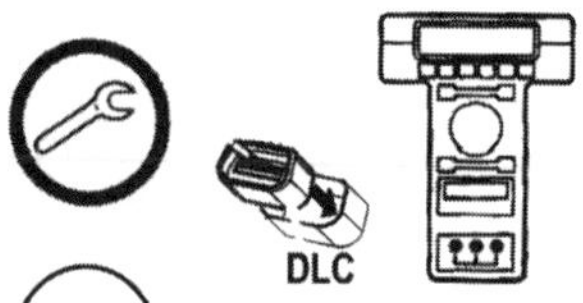

NO — REPAIR The Circuit. CLEAR The DTC's. Repeat The Self-Test. STOP

M20 CHECK THE FUEL PUMP RELAY

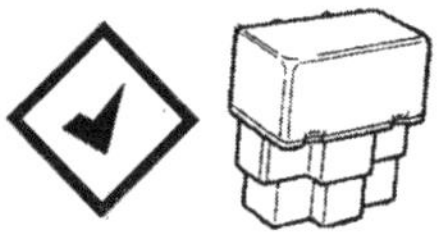

Perform The Fuel Pump Relay Component Test.

Does The Fuel Pump Relay Pass The Component Test?

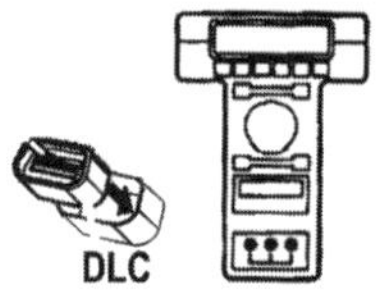

YES — CLEAR The DTC's And RETEST The System. If DTC P0232 Is Present Again GO TO M22

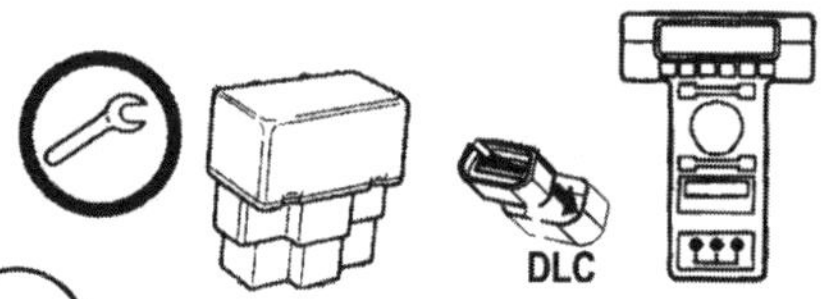

NO — INSTALL A New Fuel Pump Relay. CLEAR The DTC's. Repeat The Self-Test. STOP

M21 CHECK THE CJB FUEL PUMP RELAY

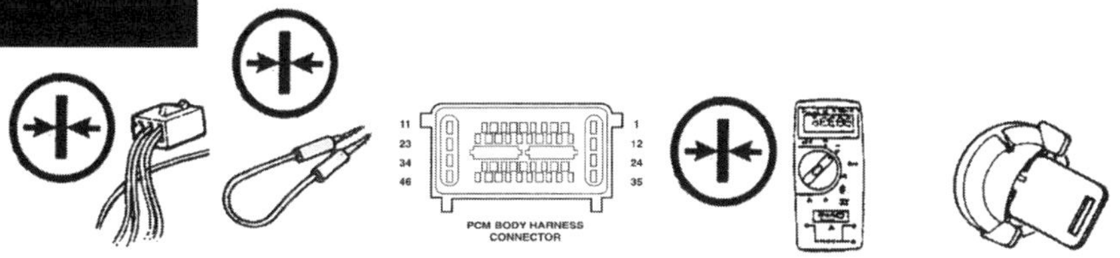

CJB-a Connector Connected.

Connect a 5 amp fused jumper wire between the following:	
Point A PCM Connector, Harness Side	**Point B**
FP - Pin B5	Ground

Measure the voltage between:	
(+) PCM Connector, Harness Side	**(-)**
IS - Pin B19	Ground

Key On. Engine Off.

1 Volt With The Jumper Removed

10 Volts With The Jumper Installed

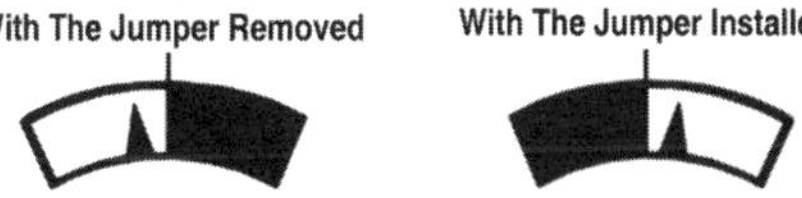

Is The Voltage Greater Than 10 Volts With The Jumper Installed And Less Than 1 Volt With The Jumper Removed?

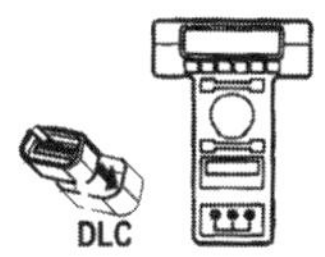

YES — CLEAR The DTC's And RETEST The System. If DTC P0232 Is Present Again GO TO M22

M22

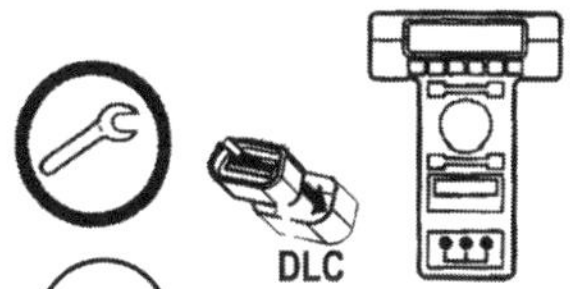

NO — INSTALL A New CJB. CLEAR The DTC's. Repeat The Self-Test.

STOP

M22 CHECK FOR CORRECT PCM OPERATION

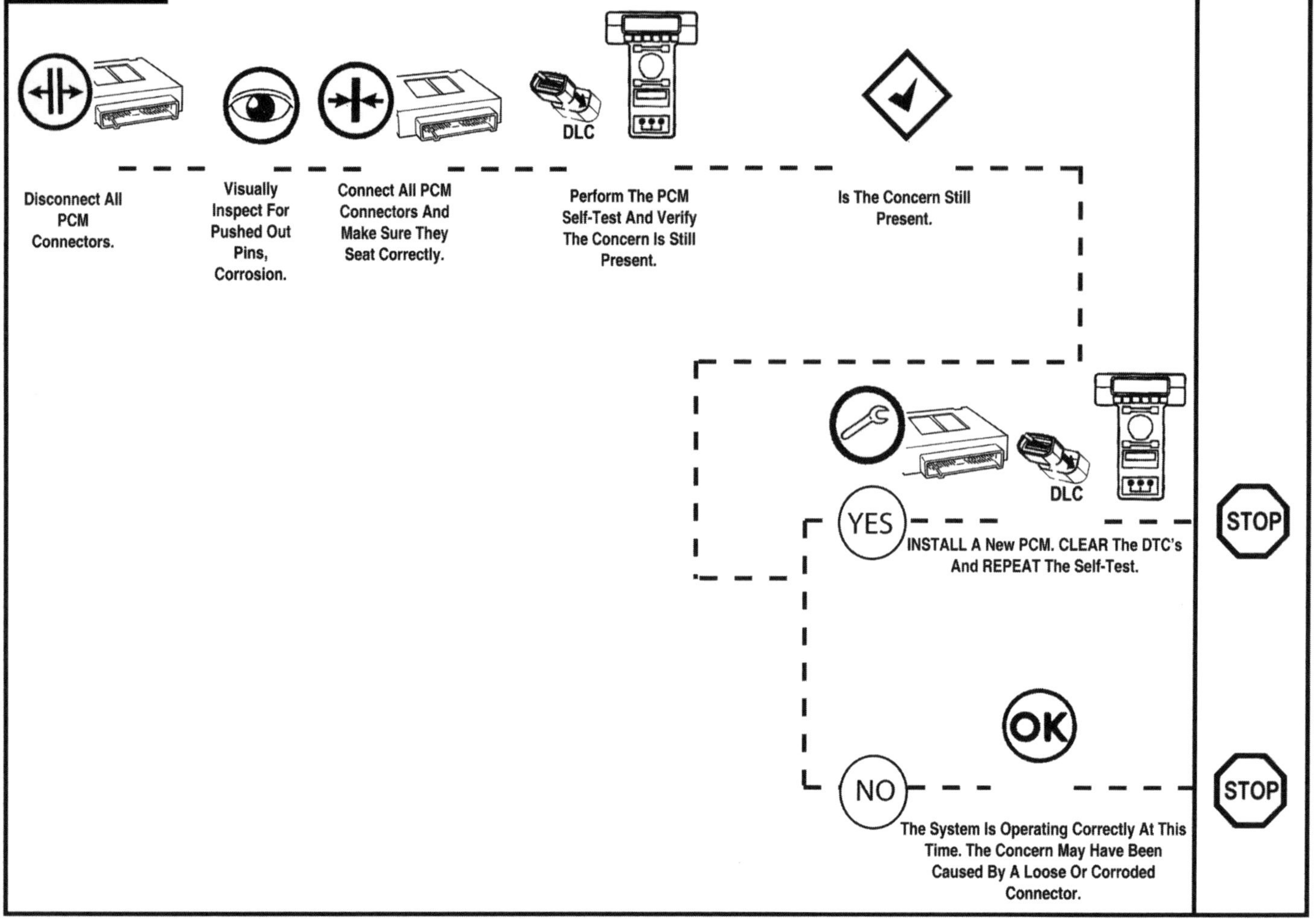

Trouble Code (s)	Problem: Water In Fuel (WIF) Sensor	Test
P2269		O 2006-07

REFERENCE LIST

This pinpoint test is intended to diagnose the following:
- Fuel system contamination
- Circuitry
- WIF sensor

DIAGNOSTIC TROUBLE CODE INDEX			
Codes	KOEO	KOER	CM
P2269	01	01	01

P2269 = Water In Fuel Condition

Circuit Functions:
The water in fuel (WIF) sensor is used to detect water in the fuel system and is located in the fuel control module. The WIF sensor is monitored by the powertrain control module (PCM). If the PCM detects water for more than 5 seconds, it will set continuous DTC P2269 and turn on the WATER IN FUEL indicator lamp (WIFIL). Route a hose from the fuel drain line to a 1-qt clear container. Open the fuel control module drain valve. Close the valve when you have filled the container. Inspect fuel in the container. If no water or contaminants are in the container, you may have a circuit fault.

CAUTION!:
The powertrain control module (PCM) harness connectors must be properly seated and the connector latch properly attached to eliminate possible driveability concerns or a no start condition. Installing the PCM connectors on an angle may cause an improper connection, misdiagnosis and damaged components. Install the connector until the lever pivots and seats itself. Apply light pressure to get the connector into position on the PCM and then fully seat the connector.

Note: Visually inspect the harness connectors for corrosion, damage, proper mating and correct pin tension.

PINPOINT TEST SCHEMATICS AND CONNECTORS

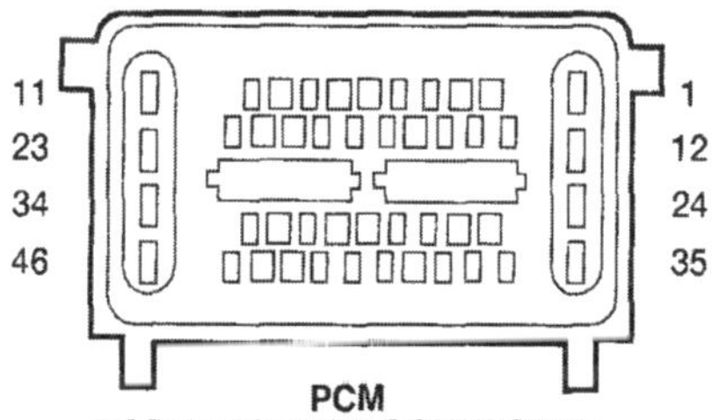

PCM
BODY HARNESS CONNECTOR

Pin	Circuit
B15	WFS (Water In Fuel Sensor)

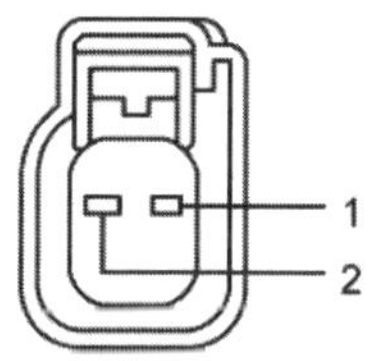

Pin	Circuit
2	SIGRTN (Signal Return)
1	WFS (Water In Fuel Sensor)

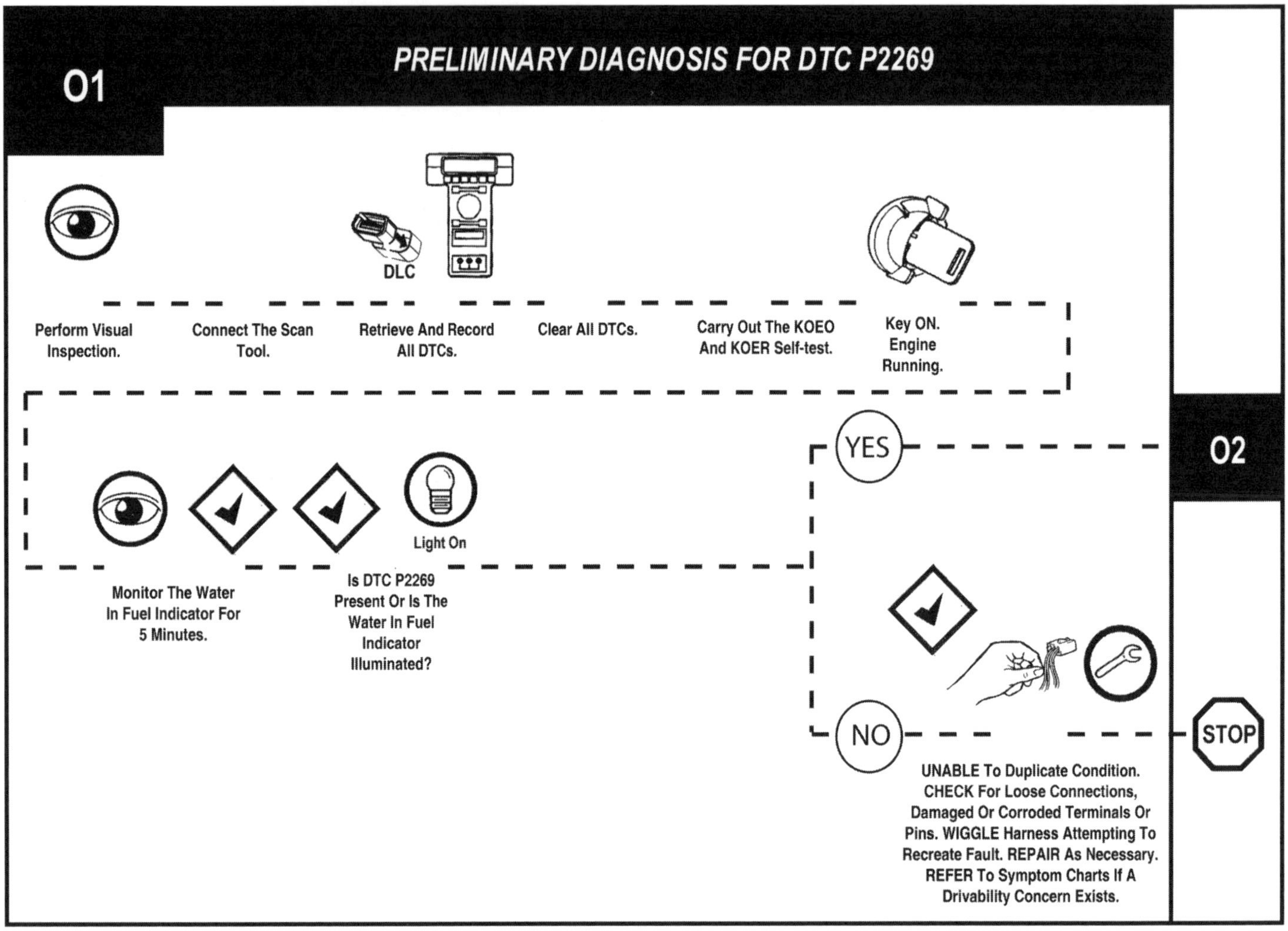
01
PRELIMINARY DIAGNOSIS FOR DTC P2269
DLC
Perform Visual Inspection.
Connect The Scan Tool.
Retrieve And Record All DTCs.
Clear All DTCs.
Carry Out The KOEO And KOER Self-test.
Key ON. Engine Running.
Light On
Monitor The Water In Fuel Indicator For 5 Minutes.
Is DTC P2269 Present Or Is The Water In Fuel Indicator Illuminated?
YES
02
NO
STOP
UNABLE To Duplicate Condition. CHECK For Loose Connections, Damaged Or Corroded Terminals Or Pins. WIGGLE Harness Attempting To Recreate Fault. REPAIR As Necessary. REFER To Symptom Charts If A Drivability Concern Exists.

O2 DIAGNOSTIC TROUBLE CODE (DTC) P2269

Note: If the PCM detects water for greater than 4 minutes, continuous DTC P2269 is set and the PCM sends a message to the instrument cluster to illuminate the WATER IN FUEL indicator.

Note: The fuel sample should fill the 0.95 L (1 quart) clear container half full. Flow out of the fuel drain should be a steady stream. Insufficient flow may indicate a fuel drain restriction or contaminants blocking the fuel drain. Repair as necessary.

Note: Inspect the fuel sample in the clear container. The fuel sample should be clear, not cloudy. Some water and contaminants may be present in the fuel sample if the fuel filter has not been replaced recently. If water in the fuel is present, the water will separate from the fuel and condense in the bottom of the clear container. Contaminants may appear in the form of sludge, dirt or metallic corrosion particles in the fuel sample. If the fuel sample is cloudy, the fuel is aerated or emulsified with water. If the fuel sample is aerated, refer to Performance Diagnostic Procedures and carry out the Fuel Aeration Test. If the fuel sample is emulsified, replace the fuel filters.

Possible Causes:
- Water In The Fuel Filter Housing
- Shorted Circuit
- Damaged Connection
- WIF Sensor
- PCM

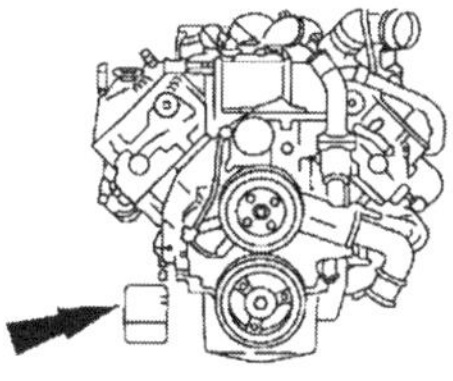

Route A Hose From The Fuel Drain Line, Located On The Fuel Control Module, To A 1-Qt Clear Container.

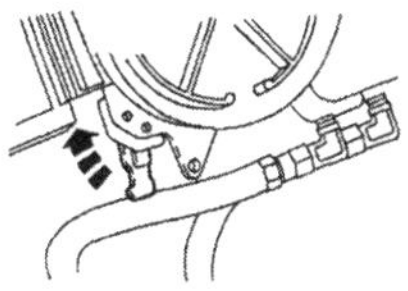

Open The Fuel Control Module Drain Valve And Close The Drain Valve When You Have Filled The Container.

Does The Container Have Any Water Or Contaminants?

- YES → O3
- NO → O4

03 CHECK 2 ADDITIONAL FUEL SAMPLES FOR WATER IN THE FUEL

Note: Checking 2 additional fuel samples for water will determine if the fuel system has water contamination in the fuel tank.

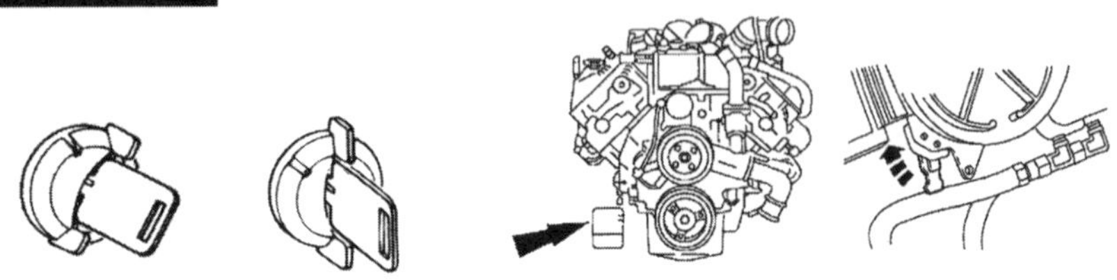

Start The Engine.

Key OFF.

Route A Hose From The Fuel Drain Line, Located On The Fuel Control Module, To A 1-Qt Clear Container.

Open The Fuel Control Module Drain Valve And Close The Drain Valve When You Have Filled The Container.

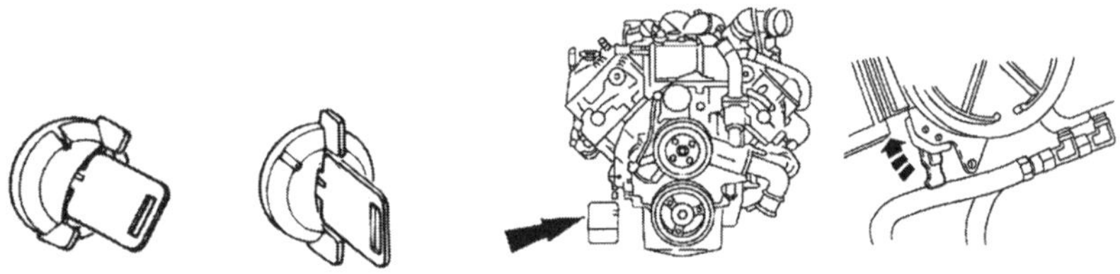

Start The Engine.

Key OFF.

Route A Hose From The Fuel Drain Line, Located On The Fuel Control Module, To A 1-Qt Clear Container.

Open The Fuel Control Module Drain Valve And Close The Drain Valve When You Have Filled The Container.

Does The Container Have Any Water Or Contaminants?

YES — FLUSH And CLEAN The Fuel Tank And Fuel System. — 07

NO — 04

O4 CHECK WIF SENSOR

WIF Sensor Connector Disconnected.

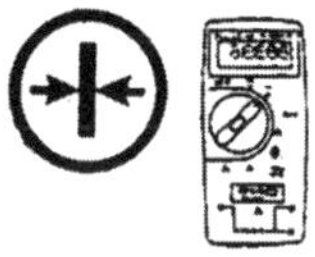

Measure the resistance between:	
(+) WIF Sensor Connector, Component Side	(-) WIF Sensor Connector, Component Side
WFS - Pin 1	SIGRTN - Pin 2

Is The Resistance Greater Than 10K Ohms?

YES → O5

NO → O6

O5 CHECK FOR SHORT TO GROUND IN THE SIGNAL CIRCUIT

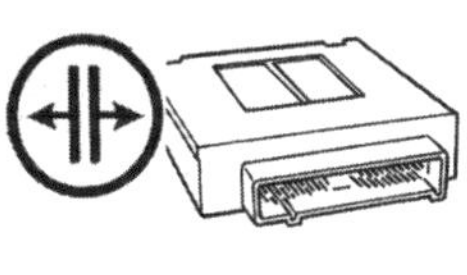

Disconnect PCM Body Harness Connector.

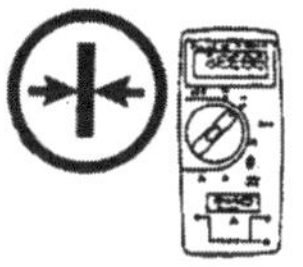

Measure the resistance between:	
(+) WIF Sensor Connector, Harness Side	(-)
WFS - Pin 1	Ground

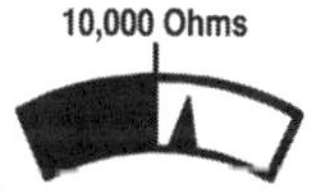

Is The Resistance Greater Than 10,000 Ohms?

YES → O7

NO → O7

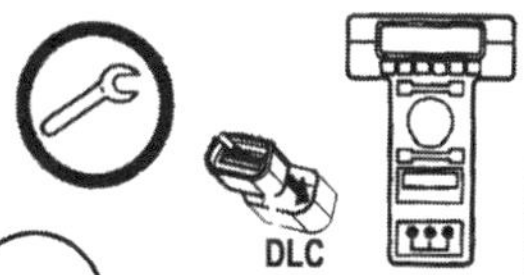

REPAIR Short To Ground. CLEAR The DTC's. REPEAT The Self-Test.

06 CHECK THE WIF SENSOR FOR CONTAMINATION

Note: Any contamination or metallic particles on the WIF sensor pins may cause a short between the pins.

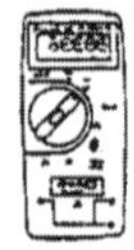

Remove The Fuel Conditioning Module Manifold.

Check The WIF Sensor Pins And The Fuel Conditioning Module Manifold And Reservoir For Contamination.

Clean The WIF Sensor Pins And The Fuel Conditioning Module Manifold And Reservoir As Necessary.

Measure the resistance between:	
(+) WIF Sensor Connector, Component Side	(-) WIF Sensor Connector, Component Side
WFS - Pin 1	SIGRTN - Pin 2

10K Ohms

Is The Resistance Greater Than 10K Ohms?

YES — INSTALL The Fuel Conditioning Module Manifold. → 07

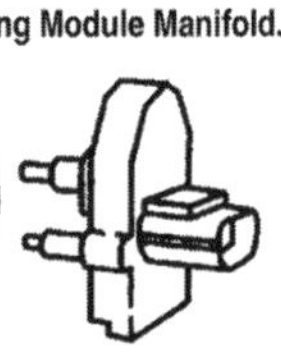

NO — INSTALL A New WIF Sensor. → 07

07 CHECK WIF SENSOR

Note: To check the correct operation of the Water in Fuel indicator, a period of time greater than 4 minutes is required for the PCM to complete the diagnostic routine.

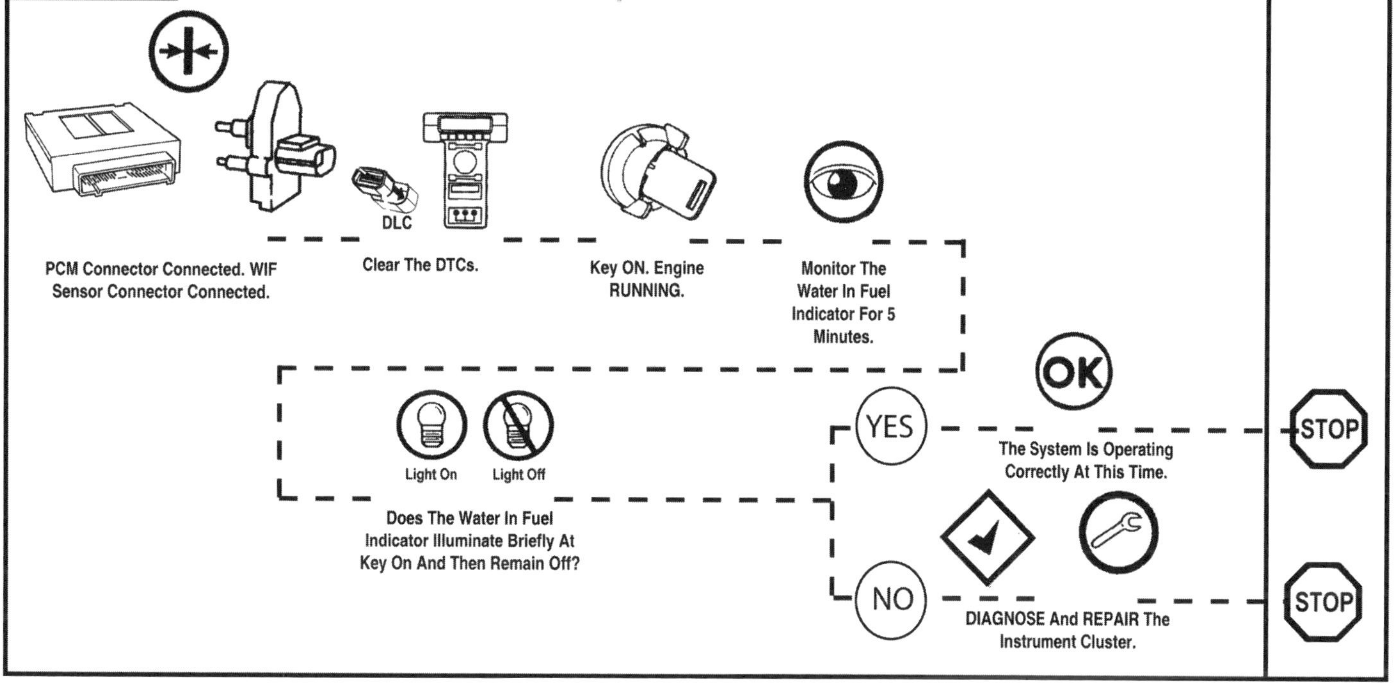

NOTES

Trouble Code (s)	Problem: Misfire/Injector Driver Circuit Operation	Test
SEE DTC INDEX IN REFERENCE LIST BELOW		P 2006-07

REFERENCE LIST

DTC	DTC DESCRIPTION
P0261 =	CYLINDER #1 INJECTOR CIRCUIT LOW
P0262 =	CYLINDER #1 INJECTOR CIRCUIT HIGH
P0263 =	CYLINDER #1 CONTRIBUTION/BALANCE
P0264 =	CYLINDER #2 INJECTOR CIRCUIT LOW
P0265 =	CYLINDER# 2 INJECTOR CIRCUIT HIGH
P0266 =	CYLINDER# 2 CONTRIBUTION/BALANCE
P0267 =	CYLINDER# 3 INJECTOR CIRCUIT LOW
P0268 =	CYLINDER# 3 INJECTOR CIRCUIT HIGH
P0269 =	CYLINDER# 3 CONTRIBUTION/BALANCE
P0270 =	CYLINDER# 4 INJECTOR CIRCUIT LOW
P0271 =	CYLINDER# 4 INJECTOR CIRCUIT HIGH
P0272 =	CYLINDER# 4 CONTRIBUTION/BALANCE
P0273 =	CYLINDER# 5 INJECTOR CIRCUIT LOW
P0274 =	CYLINDER# 5 INJECTOR CIRCUIT HIGH
P0275 =	CYLINDER# 5 CONTRIBUTION/BALANCE
P0276 =	CYLINDER# 6 INJECTOR CIRCUIT LOW
P0277 =	CYLINDER# 6 INJECTOR CIRCUIT HIGH
P0278 =	CYLINDER# 6 CONTRIBUTION/BALANCE
P0279 =	CYLINDER# 7 INJECTOR CIRCUIT LOW
P0280 =	CYLINDER# 7 INJECTOR CIRCUIT HIGH
P0281 =	CYLINDER# 7 CONTRIBUTION/BALANCE
P0282 =	CYLINDER# 8 INJECTOR CIRCUIT LOW
P0283 =	CYLINDER# 8 INJECTOR CIRCUIT HIGH
P0284 =	CYLINDER# 8 CONTRIBUTION/BALANCE
P0301 =	CYLINDER #1 MISFIRE DETECTED
P0302 =	CYLINDER #2 MISFIRE DETECTED
P0303 =	CYLINDER #3 MISFIRE DETECTED
P0304 =	CYLINDER #4 MISFIRE DETECTED
P0305 =	CYLINDER #5 MISFIRE DETECTED
P0306 =	CYLINDER #6 MISFIRE DETECTED
P0307 =	CYLINDER #7 MISFIRE DETECTED
P0308 =	CYLINDER #8 MISFIRE DETECTED

DTC	KOEO	KEOR	CM
P0261	P1	P1	P1
P0262	P1	P1	P1
P0263	P1	P1	P1
P0264	P1	P1	P1
P0265	P1	P1	P1
P0266	P1	P1	P1
P0267	P1	P1	P1
P0268	P1	P1	P1
P0269	P1	P1	P1
P0270	P1	P1	P1
P0271	P1	P1	P1
P0272	P1	P1	P1
P0273	P1	P1	P1
P0274	P1	P1	P1
P0275	P1	P1	P1

DTC	KOEO	KEOR	CM
P0276	P1	P1	P1
P0277	P1	P1	P1
P0278	P1	P1	P1
P0279	P1	P1	P1
P0280	P1	P1	P1
P0281	P1	P1	P1
P0282	P1	P1	P1
P0283	P1	P1	P1
P0284	P1	P1	P1
P0301	P1	P1	P1
P0302	P1	P1	P1
P0303	P1	P1	P1
P0304	P1	P1	P1
P0305	P1	P1	P1
P0306	P1	P1	P1
P0307	P1	P1	P1
P0308	P1	P1	P1

Output Functions:

Four High Side Driver Outputs (2 Right and 2 Left Bank) - The high side driver output function is to distribute energy to the correct bank based on a camshaft position output (CMPO) signal and provide regulated current to the injectors.The powertrain control module (PCM) commands the fuel quantity and the fuel injection control module (FICM) controls the duration of the injection event.

Low Side Driver Outputs - The low side driver outputs control the sequencing (firing order) of the engine.

CAUTION! The PCM harness connectors must be properly seated and the connector latch properly attached to eliminate possible driveability concerns or a no-start condition. Installing PCM connectors on an angle may cause an improper connection, misdiagnosis, and damaged components. Install the connector until the lever pivots and seats itself. Apply light pressure to get the connector into position on the PCM and then fully seat the connector.

Note: Inspect the FICM harness for damage, chafing and correct routing within the engine compartment and around engine components. Inspect the harness connectors for corrosion, damage, proper mating and correct pin tension.

Trouble Code (s)	Problem: Misfire/Injector Driver Circuit Operation (Continued)	Test
SEE DTC INDEX IN REFERENCE LIST		P 2006-07

REFERENCE LIST (CONTINUED)

CAUTION!: Do not pierce engine electrical wires or damage to the harness can occur.

Detection/Management:
- Key on.
- Record the DTCs stored in KOEO On-Demand Self-Test and KOEO Injector Electrical Self-Test.
- Access the Continuous DTCs.
- Record the DTCs displayed.
- CLEAR the DTCs.
- DTCs are now cleared from memory. Repeat the KOEO On-Demand Self-Test and KOEO Injector Electrical Self-Test. Any DTCs that reappear are hard faults. If no DTCs reappear, then the fault is intermittent.

The fuel injection control module (FICM) is capable of detecting individual injector open and shorts to ground or voltage while the engine is running. The FICM is also capable of detecting right or left bank high side opens or shorts to ground. An on-demand injector self-test will allow the operator to verify operation of all injector solenoids with the engine off. FICM detected DTCs will not be transmitted if the intermodule (PCM to FICM) controller area network (CAN) is inoperative.

PINPOINT TEST SCHEMATICS AND CONNECTORS

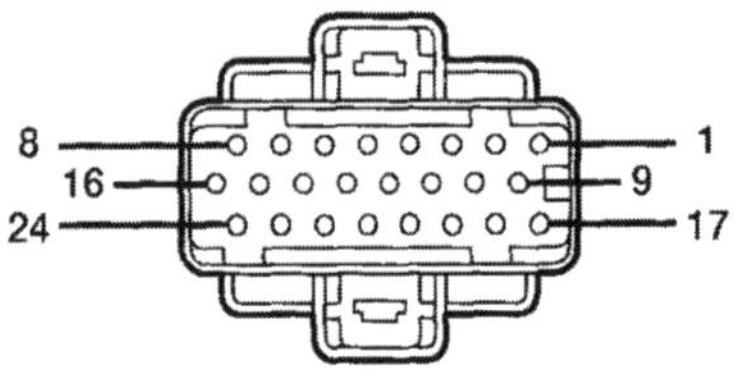

FICM HARNESS CONNECTOR

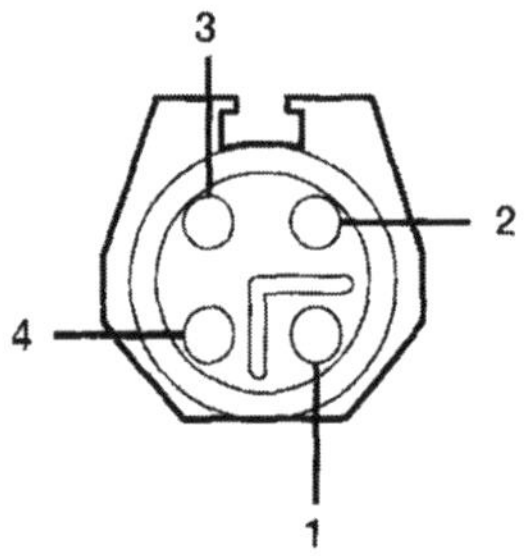

FUEL INJECTOR

FICM-A Connector	
Pin	Circuit
20	Fuel Injector 1, Close Coil, Power
5	Fuel Injector 1, Close Coil, Ground
2	Fuel Injector 1, Open Coil, Ground
19	Fuel Injector 1, Open Coil, Power
6	Fuel Injector 4, Close Coil, Ground
24	Fuel Injector 4, Close Coil, Power
1	Fuel Injector 4, Open Coil, Ground
23	Fuel Injector 4, Open Coil, Power
7	Fuel Injector 6, Close Coil, Ground
22	Fuel Injector 6, Close Coil, Power
4	Fuel Injector 6, Open Coil, Ground
21	Fuel Injector 6, Open Coil, Power
8	Fuel Injector 7, Close Coil, Ground
18	Fuel Injector 7, Close Coil, Power
3	Fuel Injector 7, Open Coil, Ground
17	Fuel Injector 7, Open Coil, Power

FICM-B Connector	
Pin	Circuit
6	Fuel Injector 2, Close Coil, Ground
18	Fuel Injector 2, Close Coil, Power
2	Fuel Injector 2, Open Coil, Ground
17	Fuel Injector 2, Open Coil, Power
8	Fuel Injector 3, Close Coil, Ground
20	Fuel Injector 3, Close Coil, Power
4	Fuel Injector 3, Open Coil, Ground
19	Fuel Injector 3, Open Coil, Power
5	Fuel Injector 5, Close Coil, Ground
22	Fuel Injector 5, Close Coil, Power
1	Fuel Injector 5, Open Coil, Ground
21	Fuel Injector 5, Open Coil, Power
7	Fuel Injector 8, Close Coil, Ground
24	Fuel Injector 8, Close Coil, Power
3	Fuel Injector 8, Open Coil, Ground
23	Fuel Injector 8, Open Coil, Power

Pin	Circuit
1	Fuel Injector, Open Coil, Power
2	Fuel Injector, Open Coil, Ground
3	Fuel Injector, Close Coil, Power
4	Fuel Injector, Close Coil, Ground

P1 PRELIMINARY DIAGNOSIS

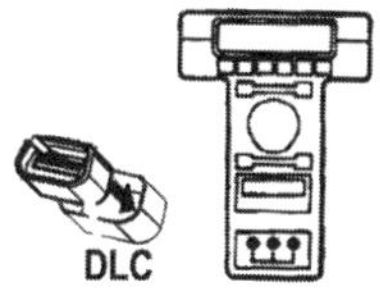

Perform A Visual Inspection.

Connect The Scan Tool.

Retrieve And Record Any Continuous And On-Demand DTCs.

Record The Freeze Frame Data. Clear The DTCs.

Perform The Injector Electrical Self-Test (Click Test). Carry Out The KOER Self-Test.

Are Any DTCs Retrieved?

YES

- For DTCs P0261, P0264, P0267, P0270, P0273, P0276, P0279 Or P0282: **P2**
- For DTCs P0262, P0265, P0268, P0271, P0274. P0277, P0280 Or P0283: **P5**
- For DTCs P0301, P0302, P0303, P0304, P0305, P0306, P0307 Or P0308: **P8**
- For DTCs P0263, P0266, P0269, P0272, P0275, P0278, P0281 Or P0284: **P16**

NO

Unable To Duplicate Condition. CHECK For Loose Connections, Damaged Or Corroded Pins. WIGGLE Harness Attempting To Recreate Fault. REFER To Symptom Charts If A Drivability Concern Exists. **STOP**

P2

DIAGNOSTIC TROUBLE CODES (DTCS) P0261, P0264, P0267, P0270, P0273, P0276, P0279 OR P0282

Note: Disconnect the suspect fuel injector harness connector. Only the suspect injector needs to be diagnosed.

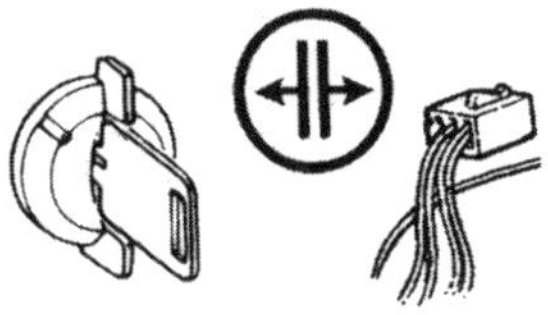

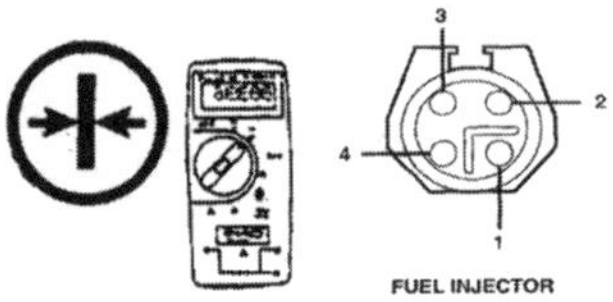

Possible Causes:
- Open Circuit
- Coil To Coil Short
- FICM
- FICM Connector Loose
- Fuel Injector

Key Off.

Disconnect The Suspect Fuel Injector.

Visually Inspect The Suspect Injector Pins.

Measure the resistance between:	
(+) Fuel Injector Connector, Component Side	(-) Fuel Injector Connector, Component Side
Fuel Injector, Open Coil, Power - Pin 1	Fuel Injector, Open Coil, Ground - Pin 2
Fuel Injector, Close Coil, Power - Pin 3	Fuel Injector, Close Coil, Ground - Pin 4

Refer to the injector coil resistance table for resistance

INJECTOR COIL RESISTANCE		
Temperature	Minimum Resistance (Ohms)	Maximum Resistance (Ohms)
Ambient	0.4	0.6
Normal Operating	0.4	1.5

Is The Resistance Within Specification?

YES → P3

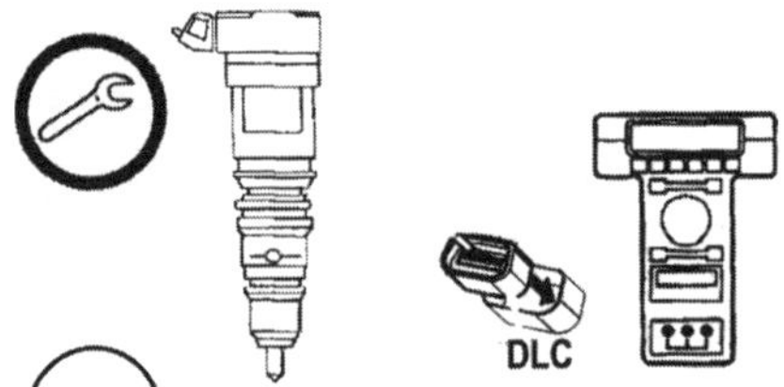

NO → INSTALL A New Fuel Injector. CLEAR The DTCs. REPEAT The Self-Test. STOP

P3 CHECK FOR SHORTS IN THE INJECTOR

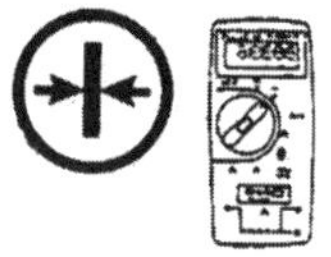

10,000 Ohms

Measure the resistance between:	
(+) Fuel Injector Connector, Component Side	(-) Fuel Injector Connector, Component Side
Fuel Injector, Open Coil, Power - Pin 1	Fuel Injector, Close Coil, Power - Pin 3
Fuel Injector, Open Coil, Power - Pin 1	Fuel Injector, Close Coil, Ground - Pin 4
Fuel Injector, Open Coil, Ground - Pin 2	Fuel Injector, Close Coil, Power - Pin 3
Fuel Injector, Open Coil, Ground - Pin 2	Fuel Injector, Close Coil, Ground - Pin 4

Resistances Greater Than 10,000 Ohms.

YES — P4

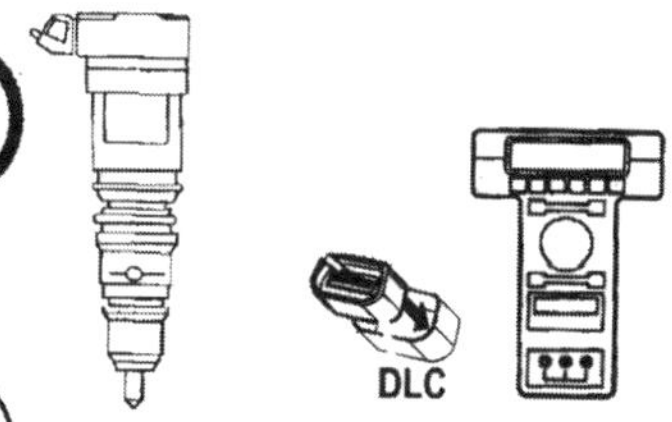

NO — INSTALL A New Fuel Injector. CLEAR The DTCs. REPEAT The Self-Test. STOP

P4 CHECK THE FUEL INJECTOR CIRCUITS FOR AN OPEN

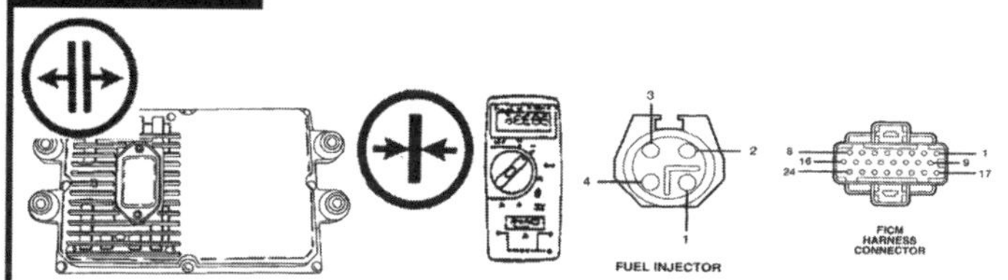

Disconnect The Suspect FICM.

Measure the resistance between:	
(+) Fuel Injector Connector, Harness Side	(-) Suspect FICM Connector, Harness Side
Fuel Injector, Open Coil, Power - Pin 1	Fuel Injector, Open Coil, Power
Fuel Injector, Close Coil, Power - Pin 3	Fuel Injector, Close Coil, Power
Fuel Injector, Open Coil, Ground - Pin 2	Fuel Injector, Open Coil, Ground
Fuel Injector, Open Coil, Power - Pin 1	Fuel Injector, Open Coil, Power

5 Ohm

Are The Resistances Less Than 5 Ohms?

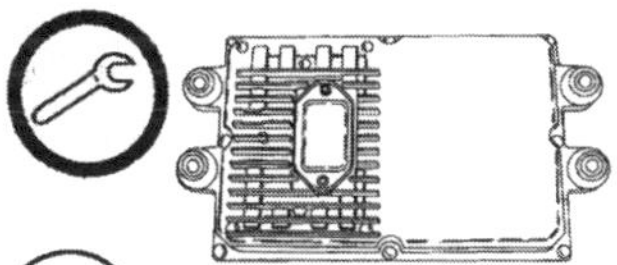

YES — INSTALL A New FICM. CLEAR The DTC's. Repeat The Self-Test. STOP

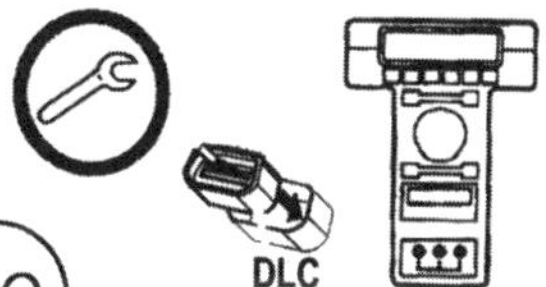

NO — REPAIR The Circuit. CLEAR The DTC's. Repeat The Self-Test. STOP

P5

DIAGNOSTIC TROUBLE CODES (DTCS) P0262, P0265, P0268, P0271, P0274, P0277, P0280 OR P0283

Note: Disconnect the suspect fuel injector harness connector. Only the suspect injector needs to be diagnosed.

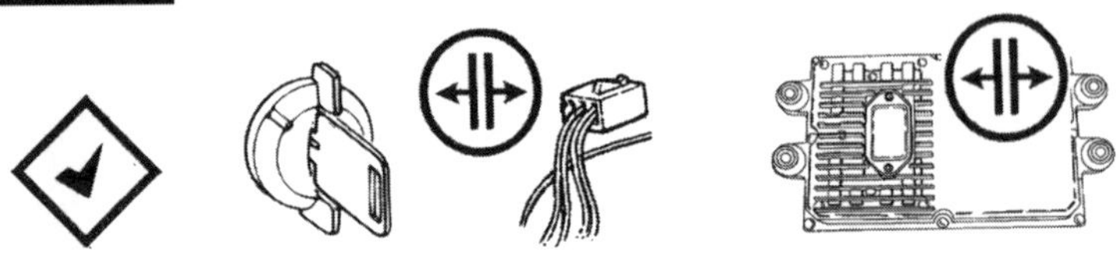

Possible Causes:
- Circuit Short To Voltage
- Circuit Short To Ground
- Fuel Injector
- FICM

Key Off.

Fuel Injector Connector Disconnected.

Disconnect The Suspect FICM.

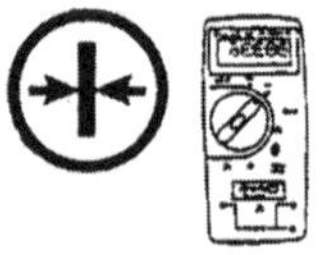

Measure the resistance between:	
(+) Fuel Injector Connector, Harness Side	**(-) Fuel Injector Connector, Harness Side**
Fuel Injector, Close Coil, Ground - Pin 4	Fuel Injector, Close Coil, Power - Pin 3
Fuel Injector, Close Coil, Ground - Pin 4	Fuel Injector, Open Coil, Ground - Pin 2
Fuel Injector, Close Coil, Ground - Pin 4	Fuel Injector, Open Coil, Power - Pin 1
Fuel Injector, Close Coil, Power - Pin 3	Fuel Injector, Open Coil, Ground - Pin 2
Fuel Injector, Close Coil, Power - Pin 3	Fuel Injector, Open Coil, Power - Pin 1
Fuel Injector, Open Coil, Ground - Pin 2	Fuel Injector, Open Coil, Power - Pin 1

Are The Resistances Greater Than 10K Ohms?

YES → **P6**

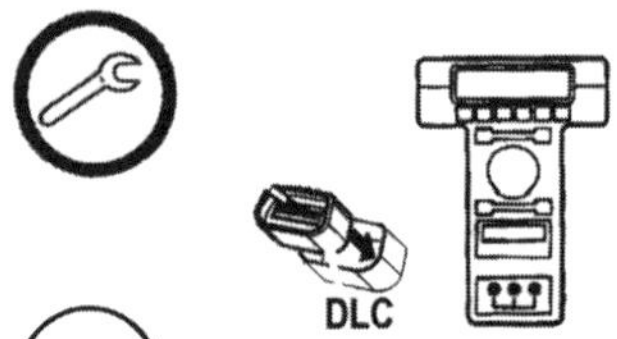

NO → STOP

REPAIR The Short Circuit. CLEAR The DTCs. REPEAT The Self-Test.

P6 CHECK THE FUEL INJECTOR CIRCUITS FOR A SHORT TO GROUND

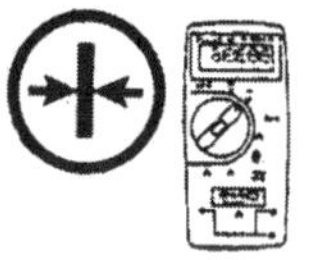

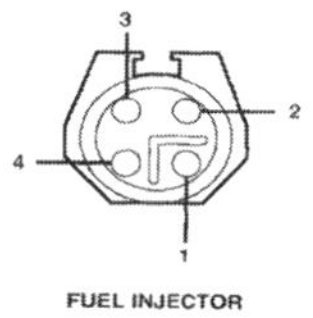

Measure the resistance between:	
(+) Fuel Injector Connector, Harness Side	(-)
Fuel Injector, Close Coil, Ground - Pin 4	Ground
Fuel Injector, Close Coil, Power - Pin 3	Ground
Fuel Injector, Open Coil, Ground - Pin 2	Ground
Fuel Injector, Open Coil, Power - Pin 1	Ground

Are The Resistances Greater Than 10,000 Ohms?

YES → P7

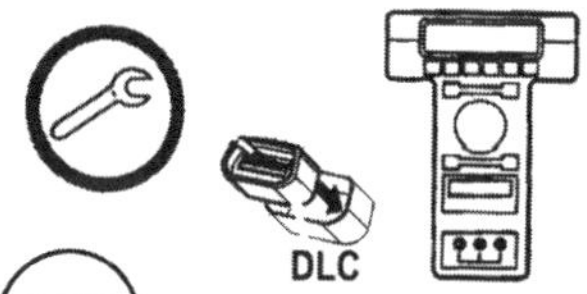

NO → STOP

REPAIR The Circuit. CLEAR The DTC's. Repeat The Self-Test.

P7 — CHECK THE FUEL INJECTOR CIRCUITS FOR A SHORT TO VOLTAGE

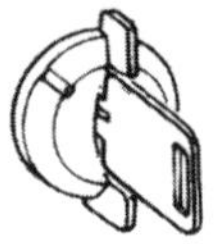

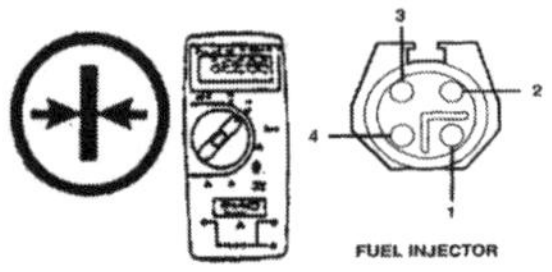

Key On, Engine Off.

Measure the voltage between:	
(+) Fuel Injector Connector, Harness Side	(-)
Fuel Injector, Open Coil, Power - Pin 1	Ground
Fuel Injector, Open Coil, Ground - Pin 2	Ground
Fuel Injector, Close Coil, Power - Pin 3	Ground
Fuel Injector, Close Coil, Ground - Pin 4	Ground

Is Any Voltage Indicated?

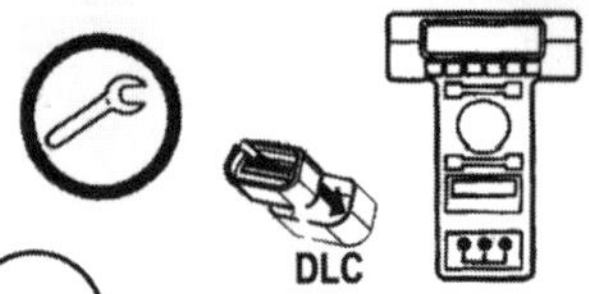

YES — REPAIR The Circuit In Question. CLEAR The DTCs. REPEAT The Self-Test. → STOP

NO → P2

P8 — CHECK FOR CONTINUOUS MEMORY FAULT DTCS

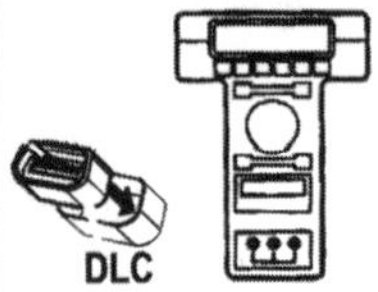

Check For Continuous Memory DTCs.

Are DTCs P0301, P0302, P0303, P0304, P0305, P0306, P0307, P0308 Present?

YES — Visually Inspect For Valve Train, Piston Or Any Other Base Engine Concerns. REPAIR As Necessary. If OK, REFER To The Performance Diagnostics Procedure, To Continue Diagnostics. → STOP

NO → P9

P9 PERFORM KOEO SELF-TEST AND INJECTOR ELECTRICAL SELF-TEST

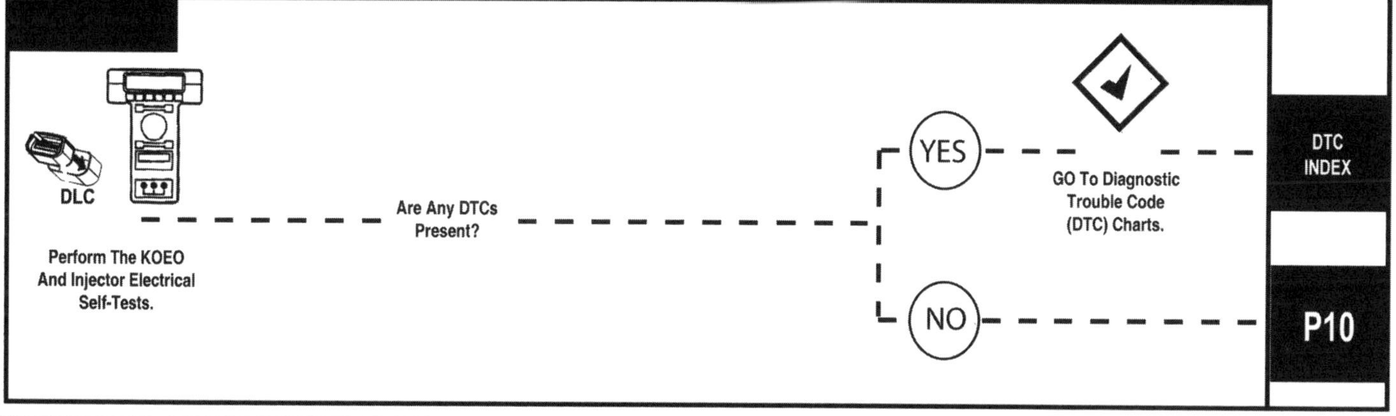

P10 CHECK FOR KOER CODES

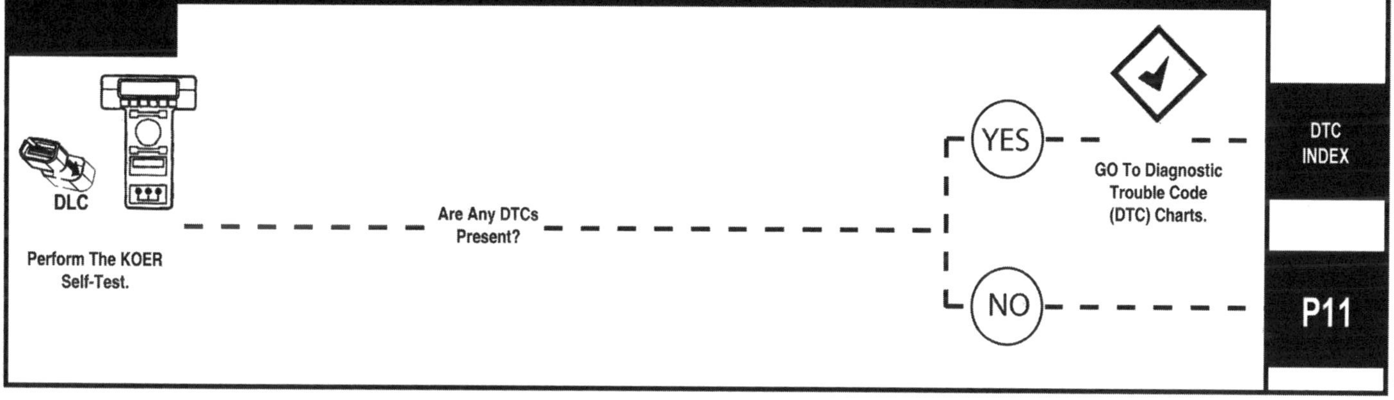

P11 CHECK FUEL SUPPLY PRESSURE

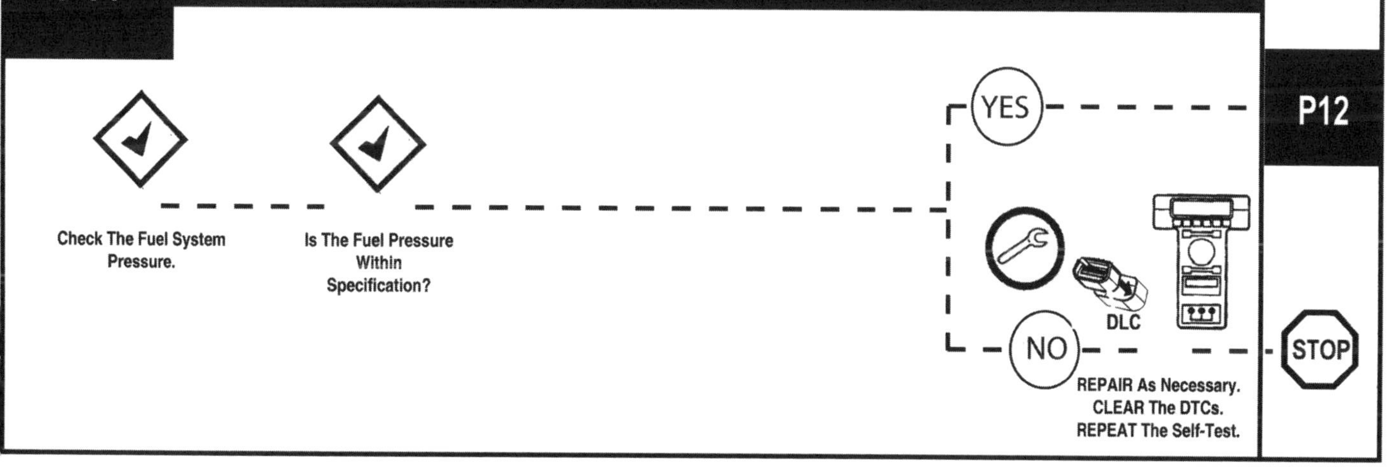

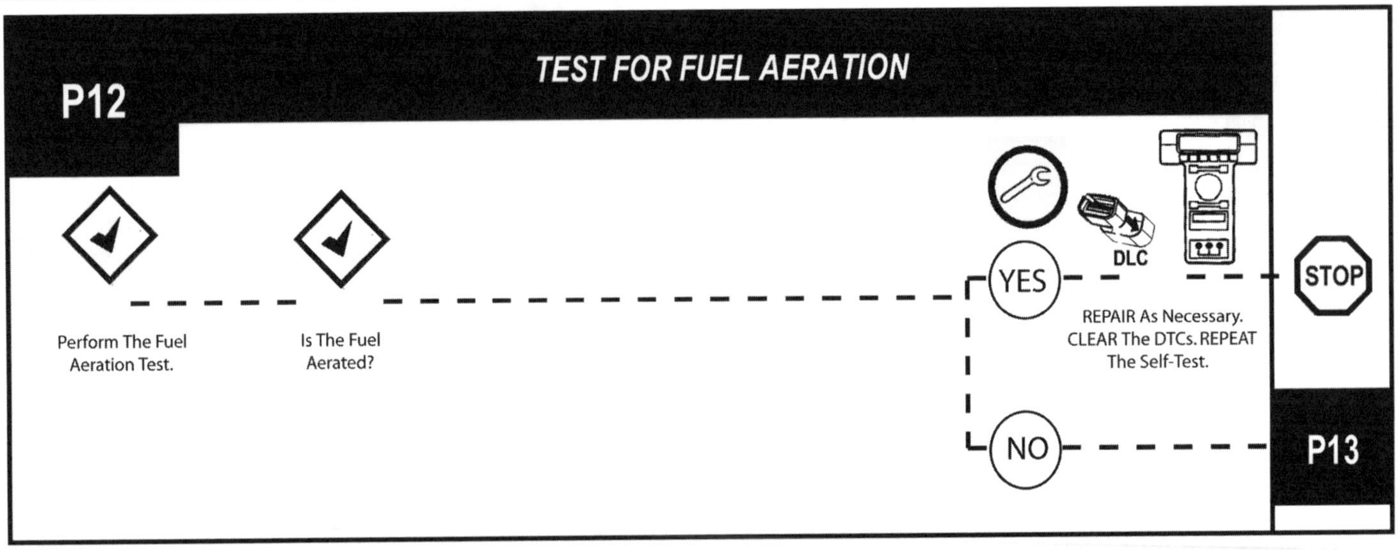
P12
TEST FOR FUEL AERATION
Perform The Fuel Aeration Test.
Is The Fuel Aerated?
YES
DLC
REPAIR As Necessary. CLEAR The DTCs. REPEAT The Self-Test.
STOP
NO
P13

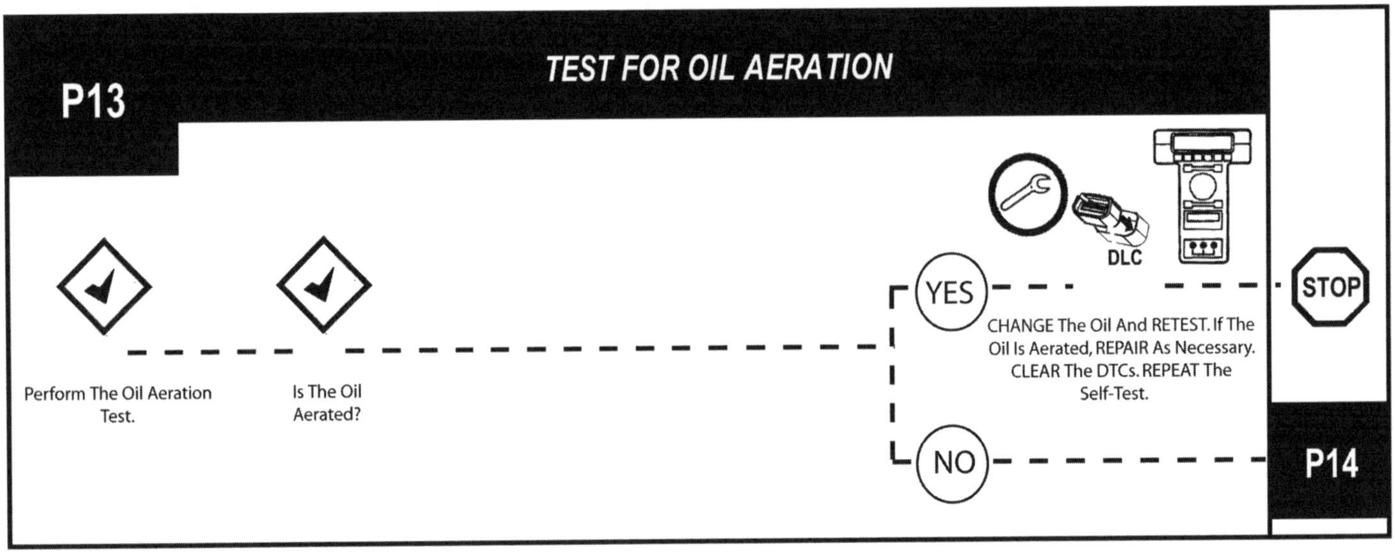
P13
TEST FOR OIL AERATION
Perform The Oil Aeration Test.
Is The Oil Aerated?
YES
DLC
CHANGE The Oil And RETEST. If The Oil Is Aerated, REPAIR As Necessary. CLEAR The DTCs. REPEAT The Self-Test.
STOP
NO
P14

P14 — *PERFORM THE INJECTOR ENHANCED DIAGNOSTICS*

Perform Injector Interrupt For Enhanced Diagnostics.

Is A Weak Cylinder Identified?

YES

If All Four Cylinders On One Bank Are Identified, CHECK The Following:
- Oil Rail Check Valve For Restriction
- Fuel Inlet Check Valve
- FICM
- Compression/Head Gaskets
- Fuel Contamination

STOP

If Multiple Or Individual Cylinders Are Identified, CHECK The Following:
- Dual Mass Flywheel

STOP

If Individual Cylinders Are Identified, CHECK The Following:
- Wiring To Injector
- Injector Hold-Down Bolt Tightness
- Injector Copper Gasket Presence
- Faulty Injector

STOP

If The Two Rear Cylinders Are Identified On Either Bank As Weak, CHECK The Following:
- AWA Fitting In Oil Rail
- Compression/Head Gasket

STOP

Any Individual Cylinder Concern Can Be Seen In The Conditions Listed Above.

STOP

If Any Two Adjacent Cylinders Are Identified As Weak, CHECK The Following:
- Compression/Head Gasket

STOP

NO → P15

P15 CHECK CRANKCASE PRESSURE

Perform The Crankcase Pressure Test.

Is The Crankcase Pressure Above Specification?

YES — DLC — REPAIR As Necessary. CLEAR The DTCs. REPEAT The Self-Test. — STOP

NO — P20

P16 DTCS P0263, P0266, P0269, P0272, P0275, P0278, P0281 OR P0284

Check The Fuel System Pressure.

Is The Fuel Pressure Within Specification?

YES — P17

NO — DLC — REPAIR As Necessary. CLEAR The DTCs. REPEAT The Self-Test. — STOP

P17 CYLINDER POWER BALANCE CHECK

Carry Out The Power Balance Test. Referperformance Diagnostic Procedures .

Are All Cylinders Contributing Correctly During The Power Balance Test?

YES — P19

NO — P18

P18 CRANKCASE AND COMPRESSION TESTS

Perform The Crankcase Pressure Test.

Perform A Cylinder Compression Test On The Suspect Cylinder.

Are Any Concerns Present?

YES:

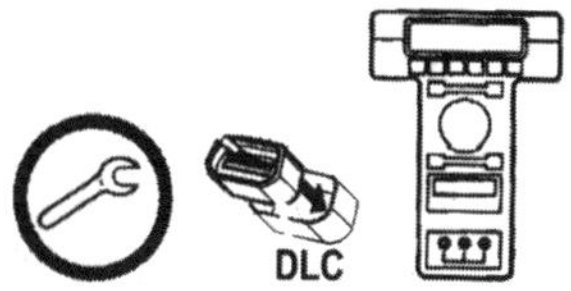

REPAIR As Necessary. CLEAR The DTCs. REPEAT The Self-Test. STOP

NO: P19

P19 CARRY OUT A VISUAL INSPECTION

Perform A Visual Inspection For The Following Mechanical Concerns:

- Push Rods
- Rocker Arms
- Valves
- Valve Springs

Are Any Of The Above Concerns Present?

YES:

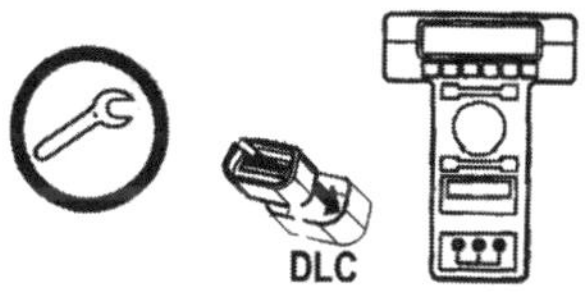

REPAIR As Necessary. CLEAR The DTCs. REPEAT The Self-Test. STOP

NO:

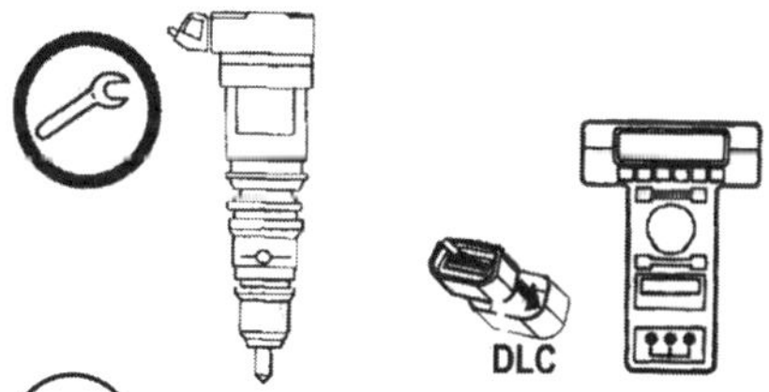

INSPECT The Fuel Injector Copper Compression Washer And O-rings. REPAIR As Necessary. If The Fuel Injector Copper Compression Washer And O-rings Are OK, INSTALL A New Fuel Injector And Copper Compression Washer. CLEAR The DTC'Ss. REPEAT The Self-Test. STOP

P20 — CHECK FOR CORRECT PCM OPERATION

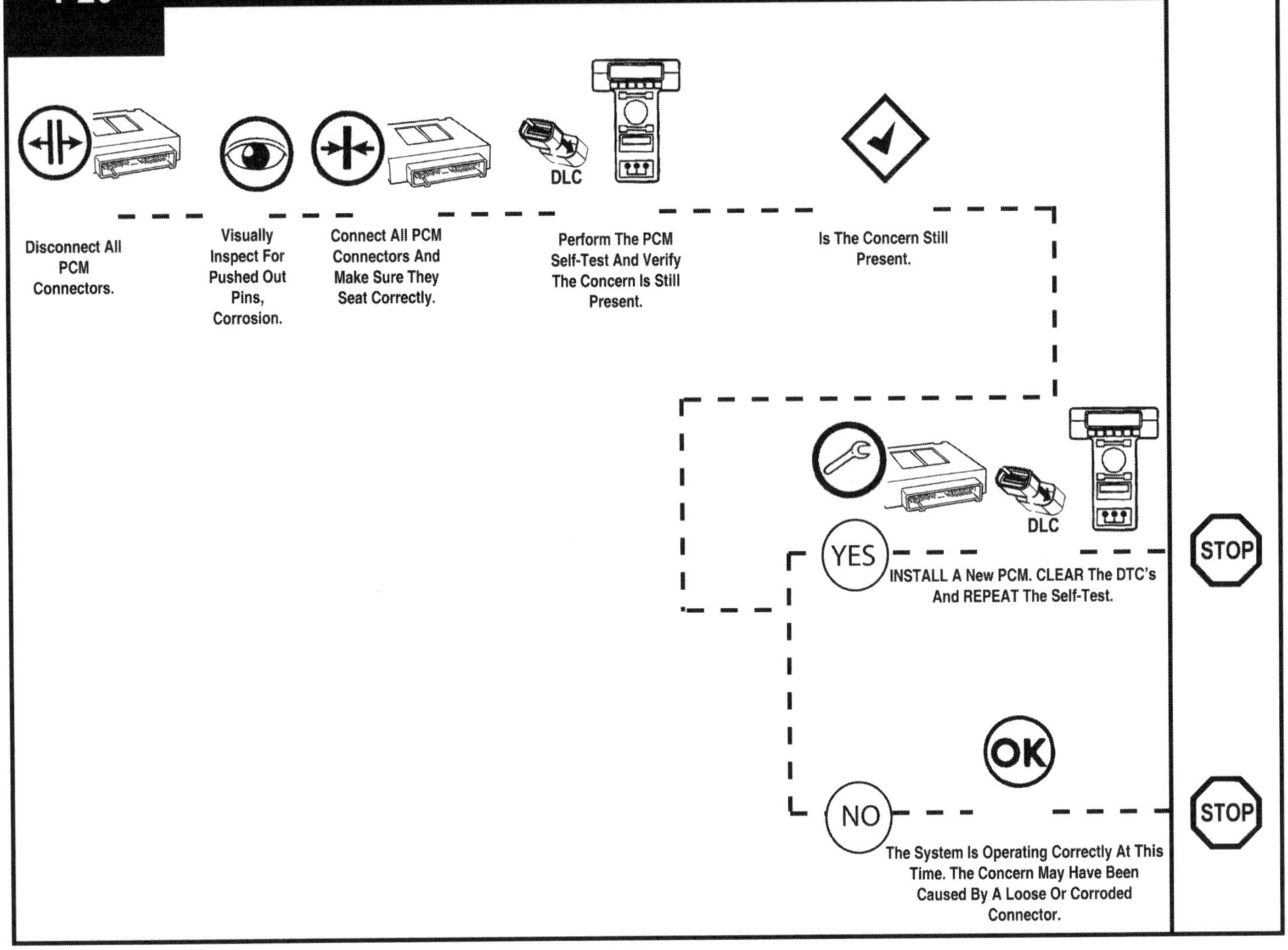

Trouble Code (s)	Problem: Injection Control Pressure (ICP) Sensor	Test
P2285, P2286		Q 2006-07

REFERENCE LIST

This pinpoint test is intended to diagnose the following:

- ICP circuit open in the harness
- ICP circuit short to ground in the harness
- ICP circuit short to voltage in the harness
- ICP sensor
- Powertrain control module (PCM)

DIAGNOSTIC TROUBLE CODE INDEX			
Codes	KOEO	KOER	CM
P2285	Q1	Q1	Q1
P2286	Q1	Q1	Q1

P2285 = ICP Circuit Out Of Range Low
P2286 = ICP Out Of Range High

Circuit Function:
The ICP sensor is a variable capacitance sensor that, when supplied with a 5-volt reference signal from the powertrain control module (PCM), produces a linear analog voltage signal that indicates pressure.

The primary function of the ICP sensor is to provide a feedback signal to indicate the rail pressure so that the PCM can command the correct injector timing, pulse width, and the correct injection control pressure for proper fuel delivery at all speed and load conditions.

Injection Control Pressure (ICP) Sensor Bias:
The ICP sensor bias varies between sensors. The amount of voltage the sensor deviates from a calculated reference value (sensor bias) may cause an injection control pressure calculation error.

Fault Detection/Management:
If the PCM detects an out of range high or low ICP sensor, the malfunction indicator lamp (MIL) is illuminated and the PCM functions from an estimated injection control pressure (open loop control of injection control pressure).

CAUTION!:
The powertrain control module (PCM) harness connectors must be properly seated and the connector latch properly attached to eliminate possible driveability concerns or a no start condition. Installing the PCM connectors on an angle may cause an improper connection, misdiagnosis and damaged components. Install the connector until the lever pivots and seats itself. Apply light pressure to get the connector into position on the PCM and then fully seat the connector.

Note: Visually inspect the harness connectors for corrosion, damage, proper mating and correct pin tension.

Note: When the PCM is disconnected additional DTCs will be set. Clear all DTCs after restoring the vehicle.

PINPOINT TEST SCHEMATICS AND CONNECTORS

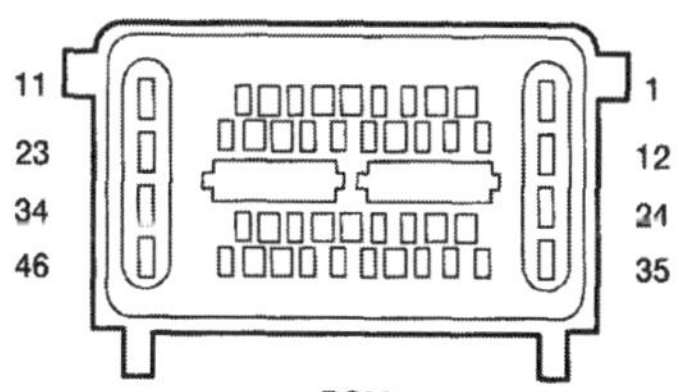

PCM
ENGINE HARNESS
CONNECTOR

SENSOR
CONNECTOR

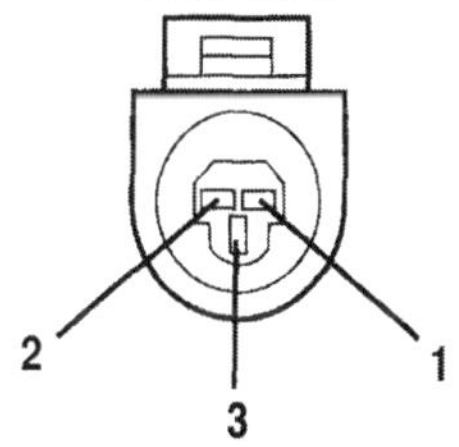

Pin	Circuit
E25	SIGRTN (Signal Return)
E29	ICP (Injection Control Pressure)
E36	VREF (Reference Voltage)

Pin	Circuit
1	SIGRTN (Signal Return)
2	VREF (Reference Voltage)
3	ICP (Injection Control Pressure)

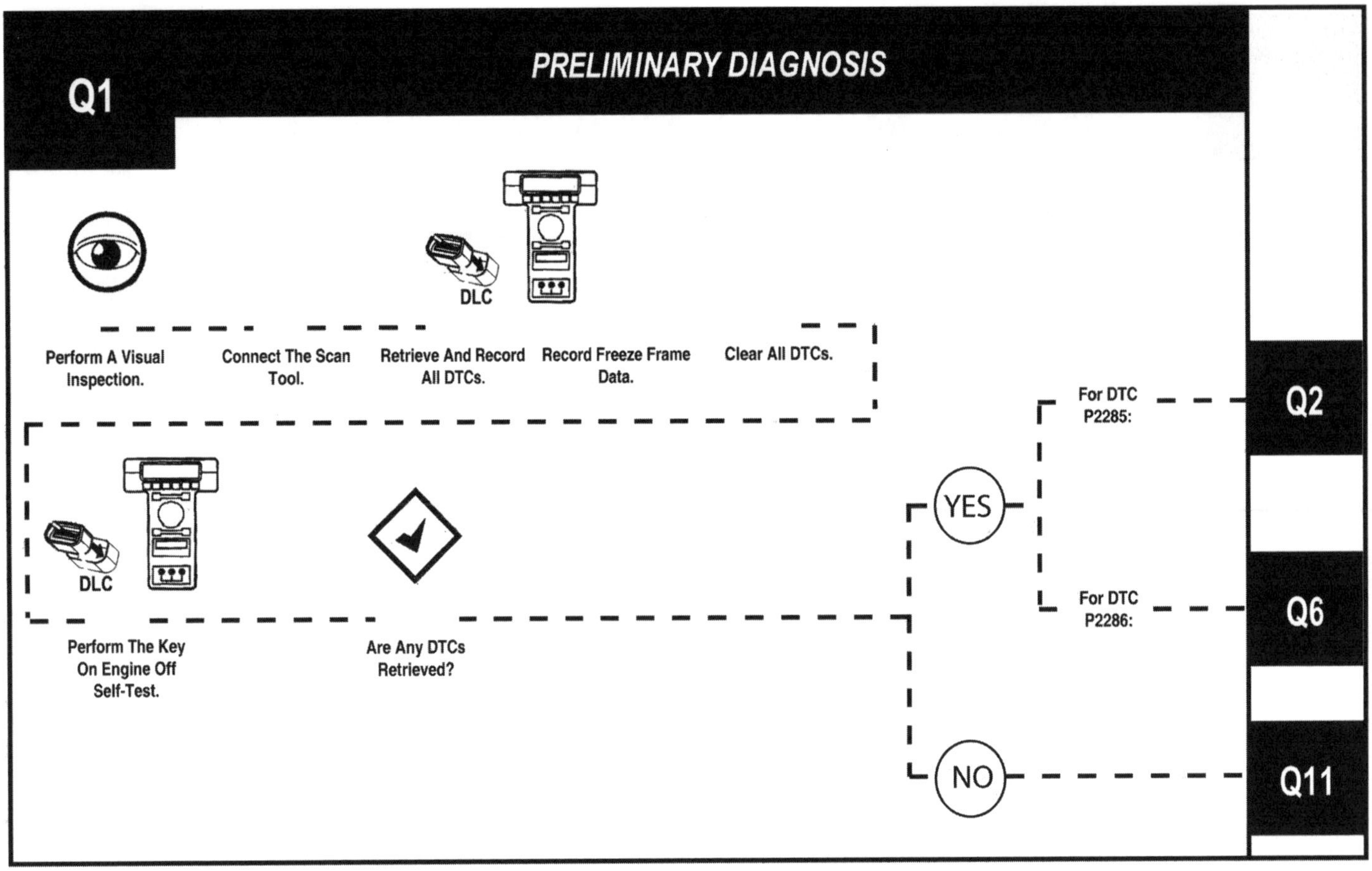

Q1
PRELIMINARY DIAGNOSIS
DLC
Perform A Visual Inspection.
Connect The Scan Tool.
Retrieve And Record All DTCs.
Record Freeze Frame Data.
Clear All DTCs.
DLC
Perform The Key On Engine Off Self-Test.
Are Any DTCs Retrieved?
YES
For DTC P2285:
Q2
For DTC P2286:
Q6
NO
Q11

Q2 DIAGNOSTIC TROUBLE CODE (DTC) P2285

Note: DTC P2285 Indicates The Injector Control Signal Circuit Out Of Range Low Was Detected During The KOEO Self-Test Or During Continuous Diagnostic Monitoring.

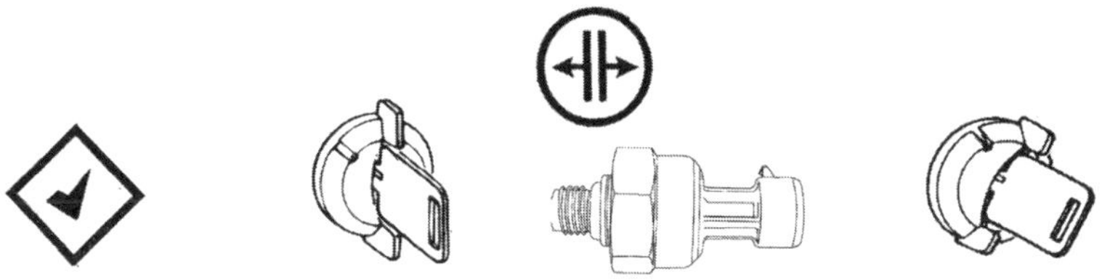

Possible Causes:
- ICP Signal Circuit Open
- ICP Disconnected
- ICP Signal Circuit Shorted To Ground
- VREF Circuit Open
- Biased ICP Sensor
- ICP Sensor - PCM

Key Off.

Disconnect The ICP Sensor.

Key On. Engine Off.

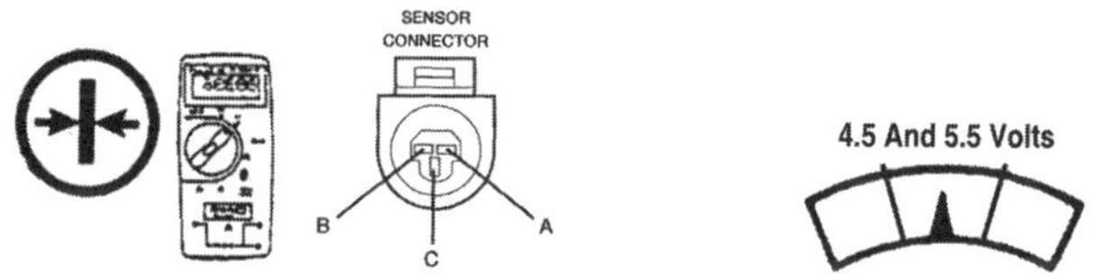

Measure the voltage between:	
(+) ICP Sensor Connector, Harness Side	(-)
VREF - Pin 2	Ground

Is The Voltage Between 4.5 And 5.5 Volts?

YES → Q3

NO → Q9

Q3 CHECK THE SIGNAL CIRCUIT FOR A SHORT TO GROUND

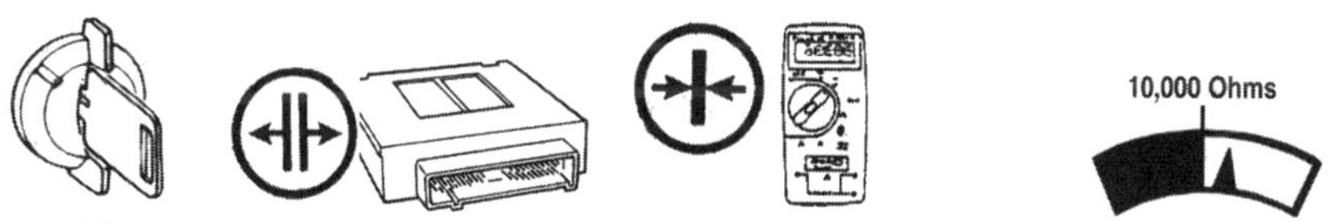

Key Off.

Disconnect The PCM.

Measure the resistance between:	
(+) ICP Sensor Connector, Harness Side	(-)
ICP - Pin 3	Ground

Is The Resistance Greater Than 10,000 Ohms?

YES → Q4

NO → STOP

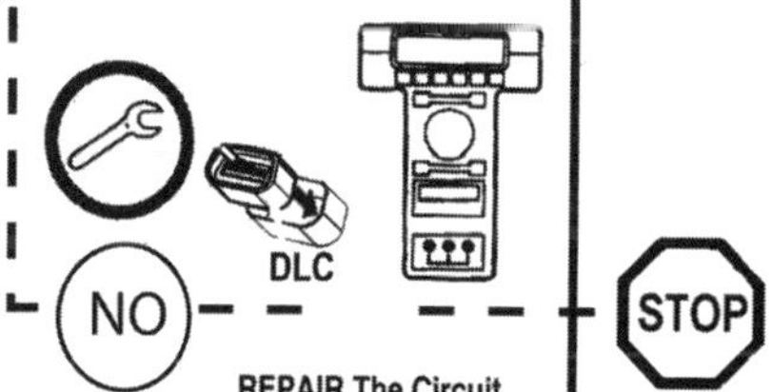

REPAIR The Circuit.
CLEAR The DTCs.
REPEAT The Self-Test.

Q4 CHECK THE SIGNAL CIRCUIT FOR AN OPEN

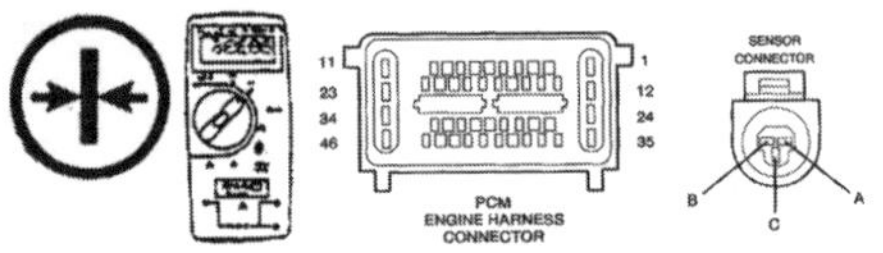

Measure the resistance between:	
(+) ICP Sensor Connector, Harness Side	(-) PCM Connector, Harness Side
ICP - Pin 3	ICP - Pin E29

Is The Resistance Less Than 5 Ohms?

YES → Q5

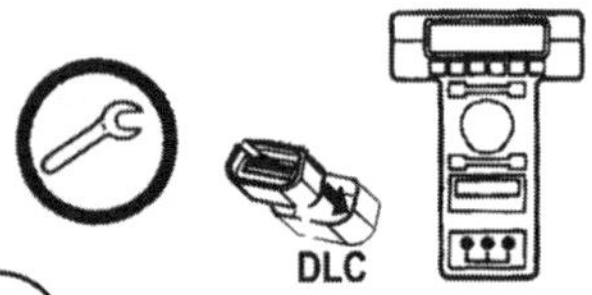

NO → STOP

REPAIR The Circuit. CLEAR The DTCs. REPEAT The Self-Test.

Q5 CHECK THE ICP SENSOR

Note: This Step May Set DTC P2286.

Reconnect The PCM.

Key On, Engine Off.

Connect a 5 amp fused jumper wire between the following:	
Point A ICP Sensor Connector, Harness Side	Point B ICP Sensor Connector, Harness Side
ICP - Pin 3	VREF - Pin 2

Access The ICP PID.

Is The ICP PID Voltage Reading Approximately 4.0 To 5.5 Volts?

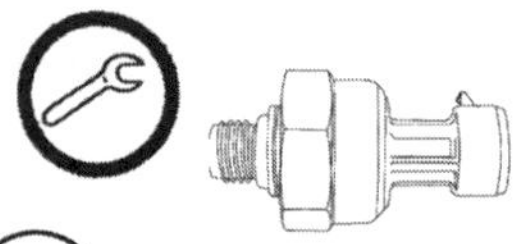

YES → STOP

INSTALL A New ICP Sensor. CLEAR The DTCs. REPEAT The Self-Test.

NO → Q12

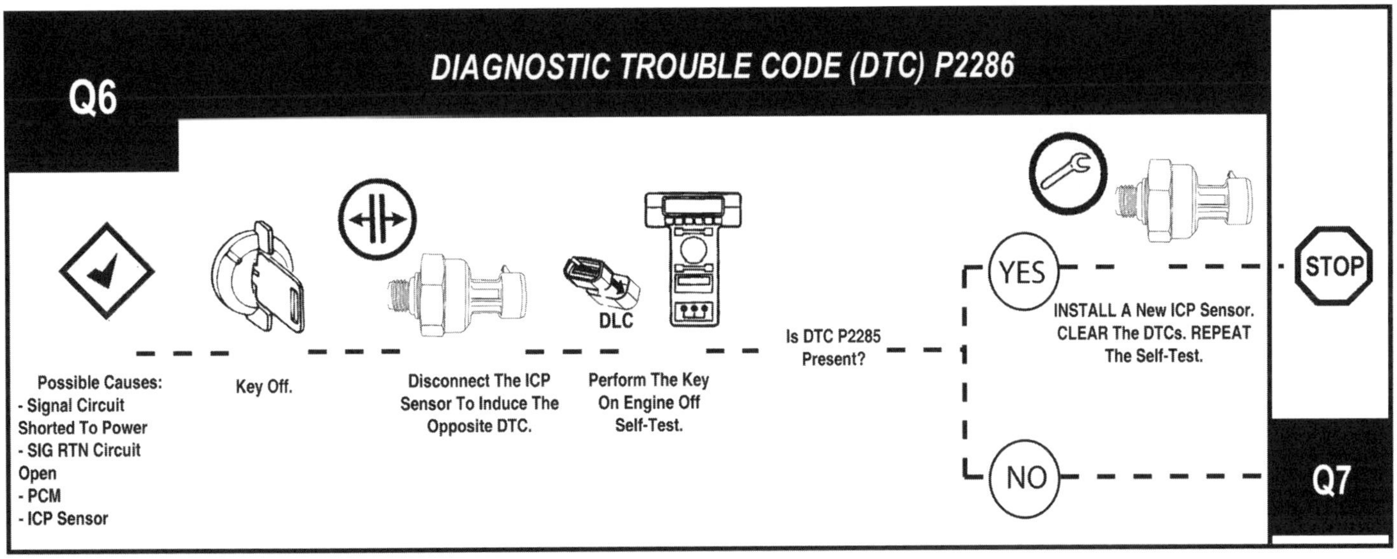

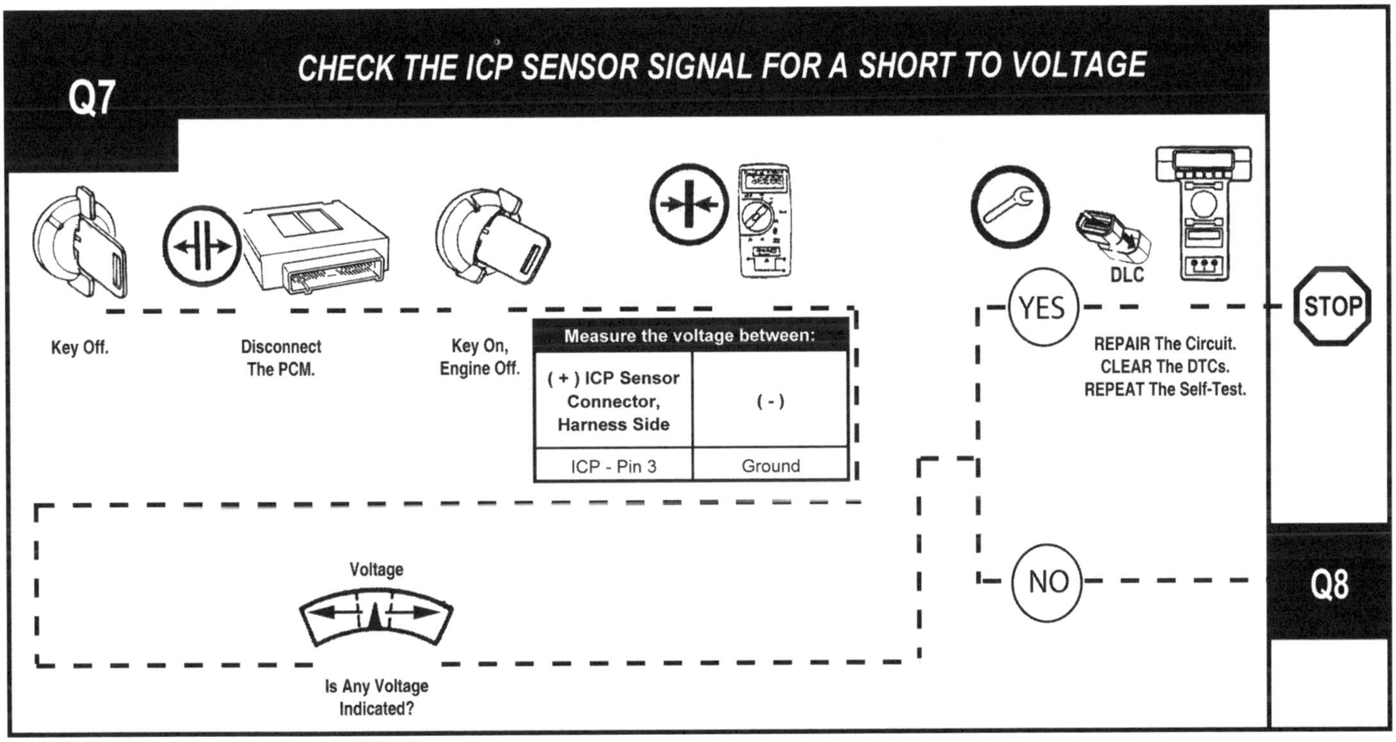

Measure the voltage between:	
(+) ICP Sensor Connector, Harness Side	(-)
ICP - Pin 3	Ground

Q8 CHECK THE ICP SIG RTN CIRCUIT FOR AN OPEN

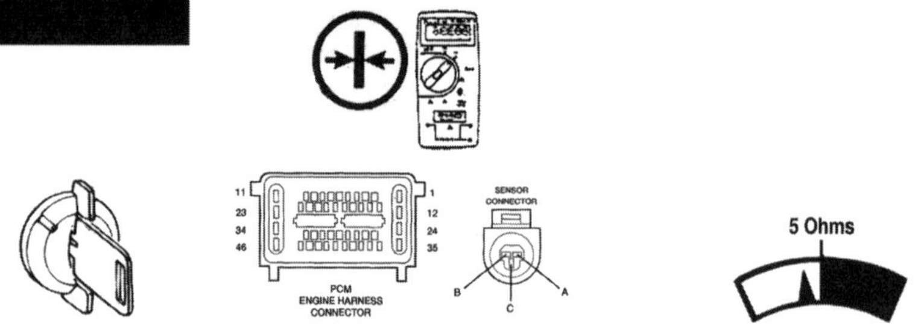

Key Off.

Measure the resistance between:	
(+) ICP Sensor Connector, Harness Side	(-) PCM Connector, Harness Side
SIGRTN - Pin 1	SIGRTN - Pin E25

Is The Resistance Less Than 5 Ohms?

YES → Q12

NO → STOP

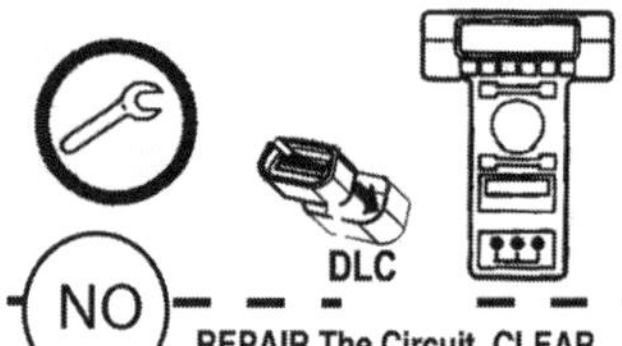

REPAIR The Circuit. CLEAR The DTCs. REPEAT The Self-Test.

Q9 CHECK THE VREF CIRCUIT FOR AN OPEN

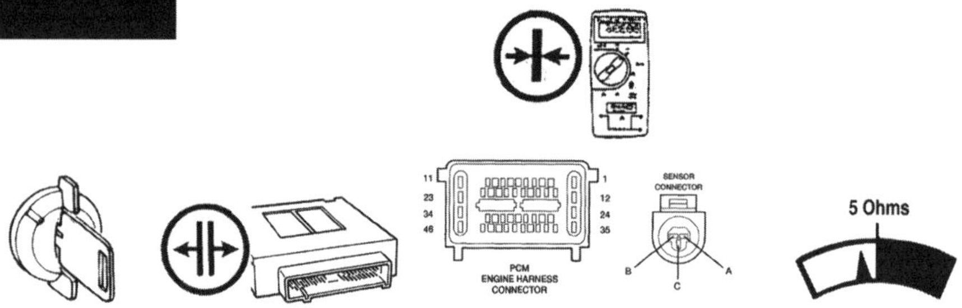

Key Off.

Disconnect The PCM.

Measure the resistance between:	
(+) ICP Sensor Connector, Harness Side	(-) PCM Connector, Harness Side
VREF - Pin 2	VREF - Pin E36

Is Resistance Less Than 5 Ohms?

YES → Q10

NO → STOP

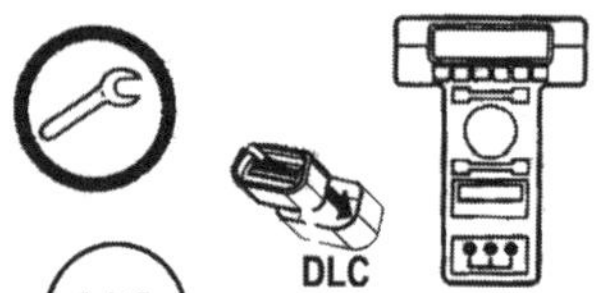

REPAIR The Circuit. CLEAR The DTCs. REPEAT The Self-Test.

Q10 CHECK THE VREF CIRCUIT FOR A SHORT TO GROUND

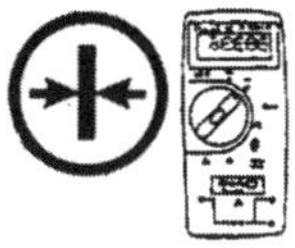

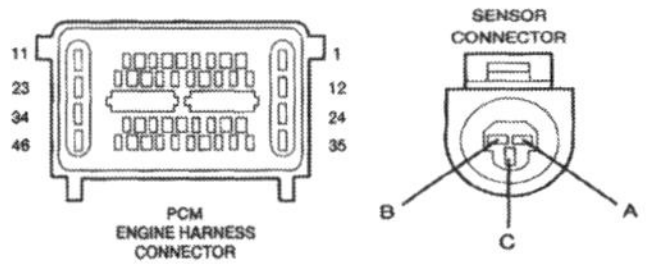

Measure the resistance between:	
(+) ICP Sensor Connector, Harness Side	(-)
VREF - Pin 2	Ground

10, 000 Ohms

Is Resistance Greater Than 10, 000 Ohms?

YES → Q12

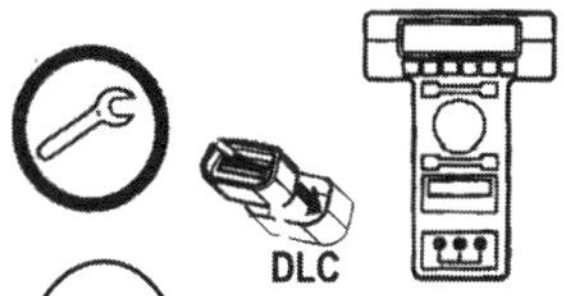

NO → STOP: REPAIR The Circuit. CLEAR The DTCs. REPEAT The Self-Test.

Q11 ICP SENSOR BIAS CHECK

Note: Verify the engine oil temperature (EOT) is greater than 82 C (180 F).

Key On, Engine Off.

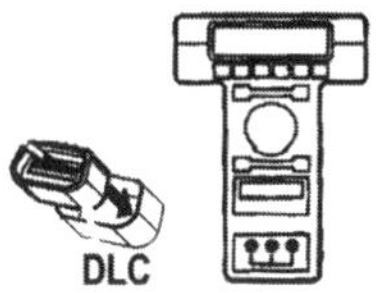

Access The PCM And Monitor The ICP_V PID.

0.15 And 0.35 Volts

Is The ICP PID Voltage Between 0.15 And 0.35 Volts?

YES → STOP: UNABLE To Duplicate Concern. ROAD TEST Vehicle, Check For A Loose Connection, And Damaged Or Corroded Terminals. WIGGLE the Harness Attempting To Recreate The Fault. REPAIR As Necessary And RETEST The System. REFER To Symptom Charts If A Driveablility Concern Exists.

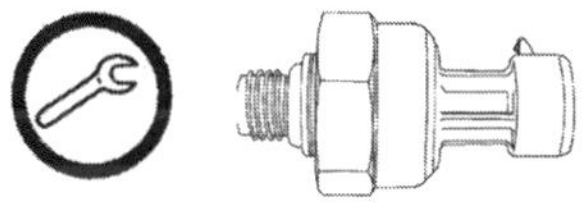

NO → STOP: INSTALL A New ICP Sensor. CLEAR The DTCs. REPEAT The Self-Test.

Q12 CHECK FOR CORRECT PCM OPERATION

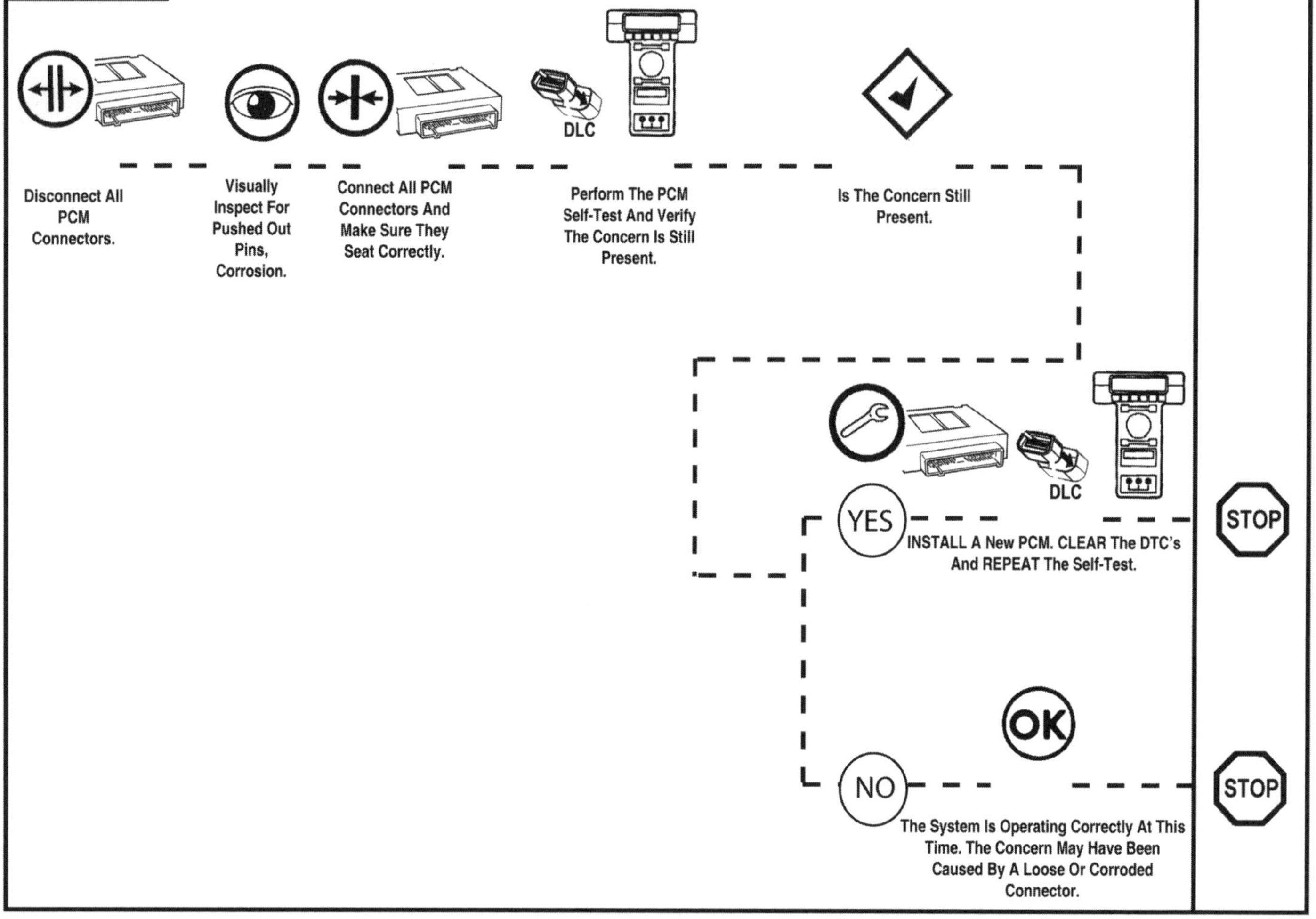

Q13 ICP SENSOR INTERMITTENT CHECK

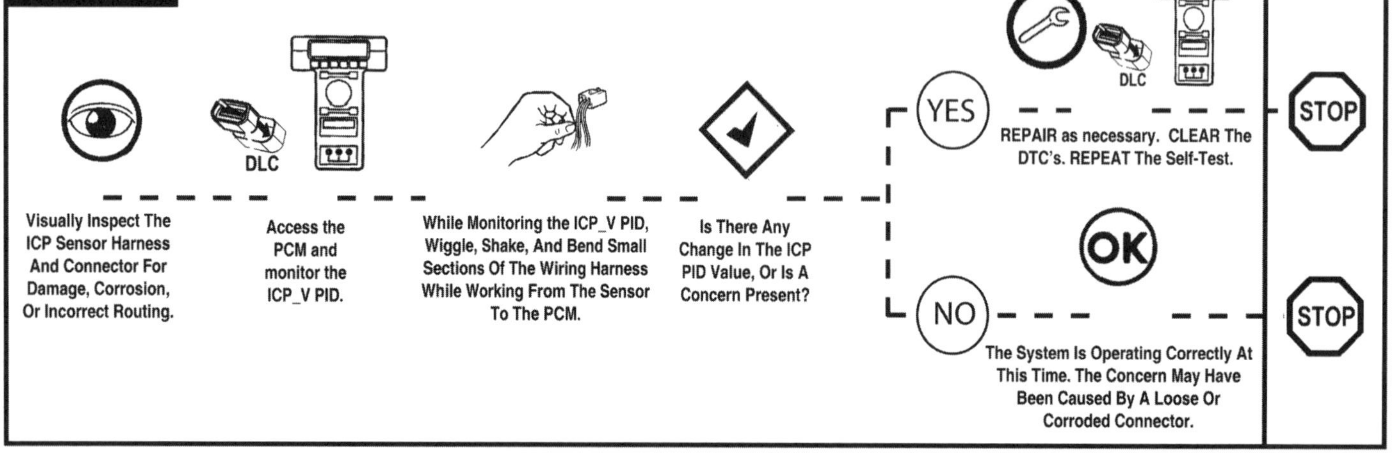

Trouble Code (s)		Test
P0460, P0462, P0463, P115A, P2067, P2068	**Problem: Fuel Level Input (FLI)**	QH 2006-07

REFERENCE LIST

This pinpoint test is intended to diagnose the following:
- Fuel tank control module
- Fuel control module circuitry
- Instrument cluster
- Powertrain control module (PCM)

DIAGNOSTIC TROUBLE CODE INDEX			
CODES	KOEO	KOER	CM
P0460	QH1	QH1	QH1
P0462	QH1	QH1	QH1
P0463	QH1	QH1	QH1
P115A	QH1	QH1	QH1
P2067	QH1	QH1	QH1
P2068	QH1	QH1	QH1

P0460 = Fuel Level Sensor A Circuit
P0462 = Fuel Level Sensor A Circuit Low Input
P0463 = Fuel Level Sensor A Circuit High Input
P115A = Low Fuel Level - Forced Limited Power
P2067 = Fuel Level Sensor B Circuit Low Input
P2068 = Fuel Level Sensor B Circuit High Input

Circuit Function:
The powertrain control module (PCM) receives the FLI message through the controller area network (CAN) from the instrument cluster. The FLI message is one of several required inputs used by the PCM to enable the misfire detection monitor.

QH1 *PRELIMINARY DIAGNOSIS*

Note: Diagnostic trouble code (DTC) P115A may set with a low fuel level input. If an excessively low fuel level input message is received by the PCM from the instrument cluster, the PCM limits the injection control pressure (ICP). This may cause a lack of power, rough idle, or other driveability concerns. If no additional DTCs are present, the concern may have been caused by the vehicle running too low on fuel.

Possible Causes:
Fuel Tank Empty
Fuel Tank Overfilled
Fuel Control Module Circuitry
Fuel Control Module
Instrument Cluster
Pcm

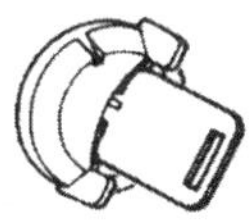

Key ON, Engine OFF.

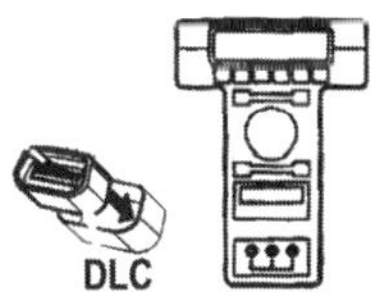

Carry Out The Instrument Cluster Self-Test.

Carry Out The PCM Self-Test.

Are Any Instrument Cluster DTCs Present Or Any Additional PCM DTCs Present?

YES — For instrument cluster DTCs, DIAGNOSE And REPAIR Instrument Cluster. → STOP

YES — For Additional PCM DTCs, REFER to DTC INDEX. → DTC INDEX

NO → QH2

QH2 — CHECK THE INSTRUMENT CLUSTER FLI PID

DLC

Access The Instrument Cluster And Monitor The FLI PID.

Record The FLI PID Value.

Compare The Instrument Cluster FLI PID Recorded Value To The Instrument Cluster Fuel Gauge.

Does The Instrument Cluster FLI PID Correspond To The Fuel Gauge Level?

YES — QH3

NO — DIAGNOSE And REPAIR The Instrument Cluster. — STOP

QH3 — CHECK THE PCM FLI PID

DLC

Access The PCM And Monitor The FLI PID.

Record The FLI PID Value.

Compare The PCM FLI PID Recorded Value To The Instrument Cluster FLI PID Recorded Value.

Does The PCM FLI PID Correspond To The Instrument Cluster FLI PID?

YES — UNABLE To Duplicate The Condition. CHECK For A Loose Connection, And Damaged Or Corroded Terminals. WIGGLE The Harness Attempting To Recreate The Fault. REPAIR As Necessary. REFER To Symptom Charts, If A Driveability Concern Exists. — STOP

NO — QH4

QH4 — CHECK IF A DTC IS STILL PRESENT

DLC

Clear The DTCs.

DLC

Carry Out The Self-test.

Are DTCs P0460, P0462, P0463, P115A, P2067 Or P2068 Present?

YES — QH5

NO — The Test Is Complete. — STOP

QH5

CHECK FOR CORRECT PCM OPERATION

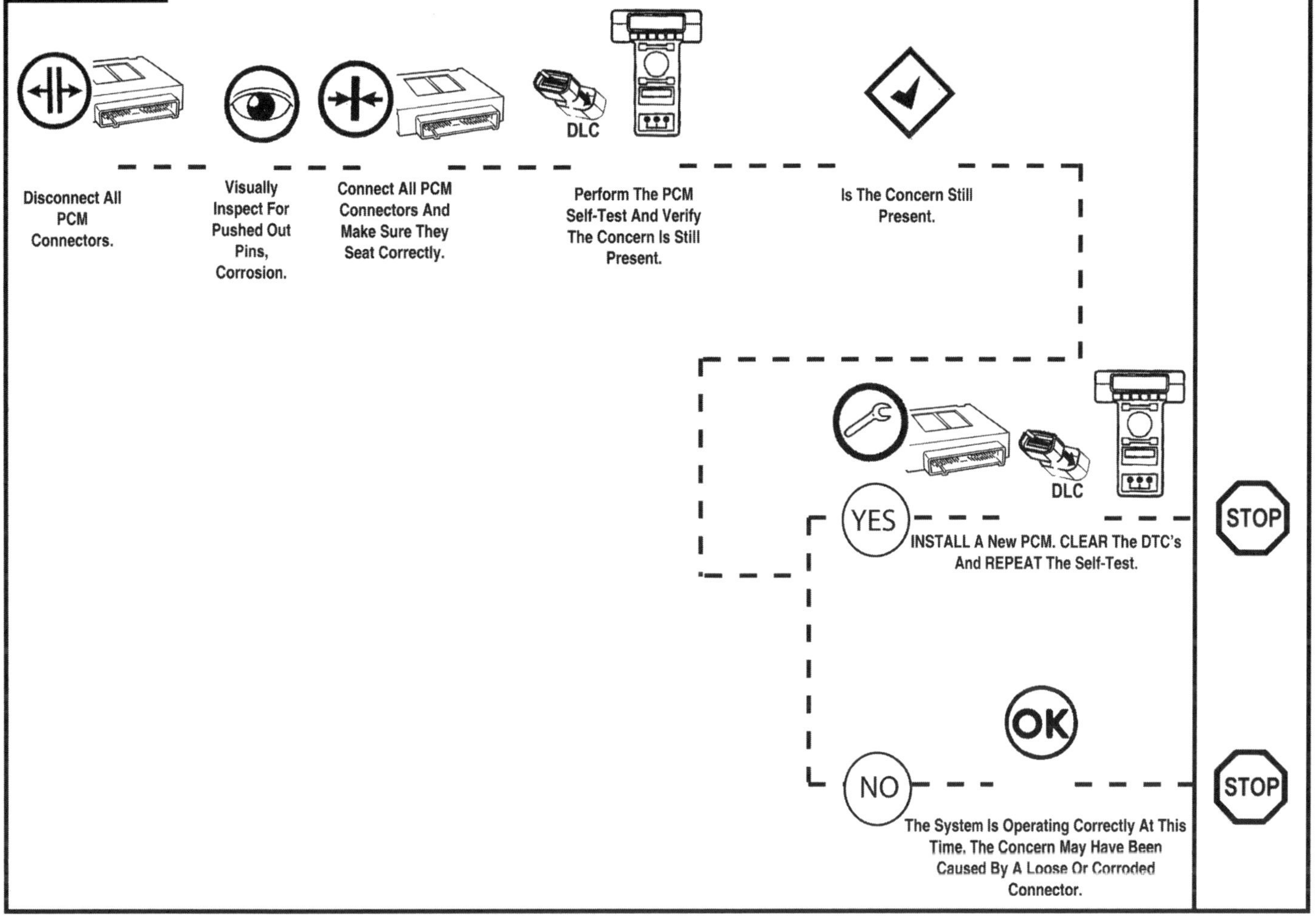

NOTES

Trouble Code (s)		Test
NA	**Problem: Powertrain Control Module (PCM) Quick Test**	QT 2006-07

QT1 — CARRY OUT A VISUAL INSPECTION

Note: Visually inspect the vehicle for aftermarket accessories and performance modifications (air filter, exhaust system, intake system, performance chip, and turbocharger). Refer to Diesel Electronic Engine Control (EEC) System, Modifications to OBD Vehicles.

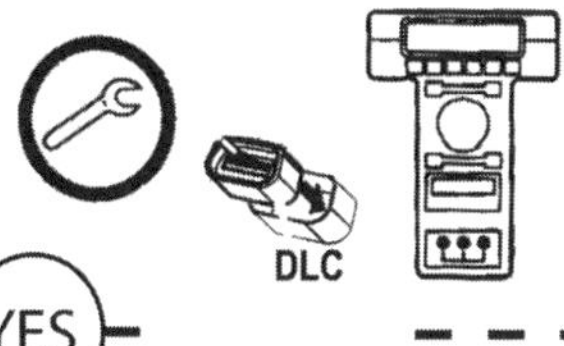

Access Any Related OASIS Or TSB Information.

1.) Inspect the air cleaner, inlet ducting, tubes and clamps.
2.) Inspect all hoses for damage, leaks, cracks, blockage, and proper routing.
3.) Inspect the fuel system including the fuel tank and the fuel lines for kinks, bends and/or leakage.
4.) Inspect the wiring harness for correct routing and chafing.
5.) Inspect the exhaust system.
6.) Check for the correct engine oil level and quality.
7.) Check the engine coolant for proper level.

Is Any Concern Present?

- YES — REPAIR As Necessary. CLEAR The DTC's. REPEAT The Self-Test. → STOP
- NO → QT2

QT2 — VEHICLE PREPARATION

Note: Visually inspect the vehicle for aftermarket accessories and performance modifications (air filter, exhaust system, intake system, performance chip, and turbocharger). Refer to Diesel Electronic Engine Control (EEC) System, Modifications to OBD Vehicles.

PARK (Automatic Transmission)
NEUTRAL (Manual Transmission)

Place The Transmission Gear Selector Into PARK (Automatic Transmission) Or NEUTRAL (Manual Transmission)

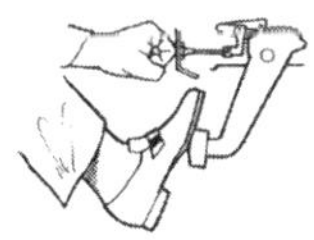

Apply The Parking Brake. Block The Drive Wheels.

OFF

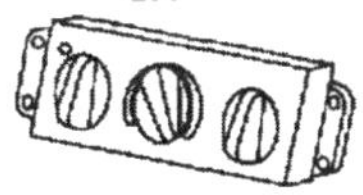

Turn OFF All Electrical Loads (Radios, Lights, A/c, Blower, And Fans).

Has The Vehicle Preparation Been Completed?

- YES → QT3
- NO → QT2

DTC(S): NA

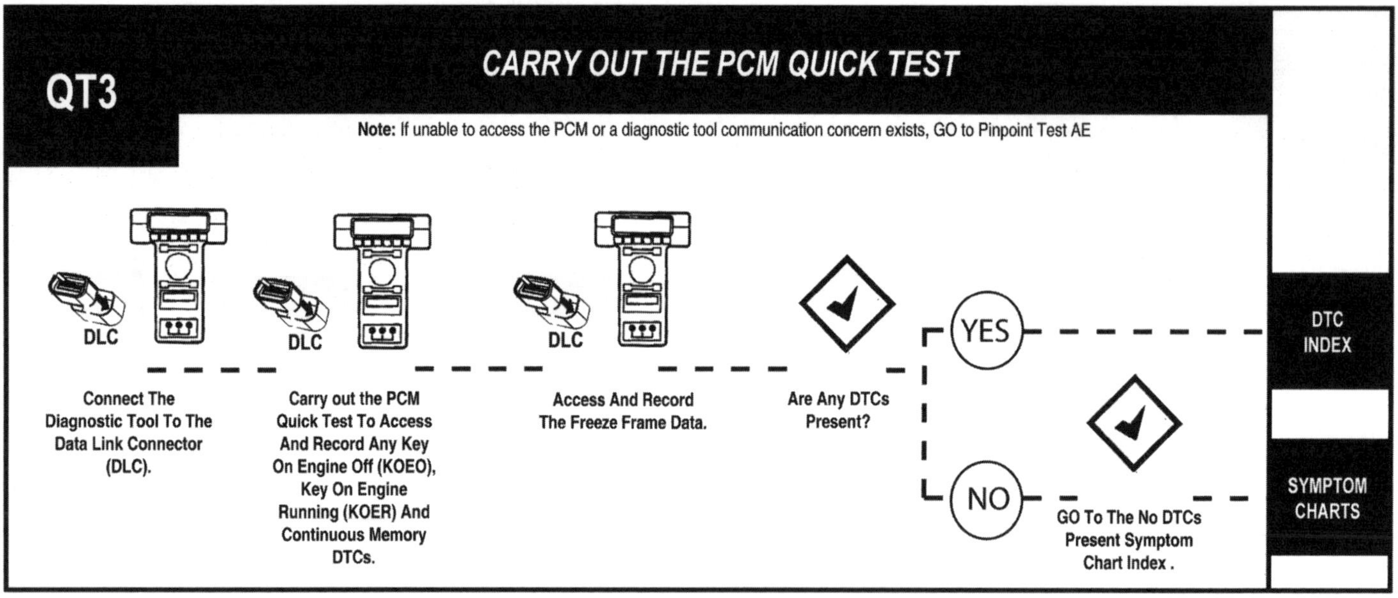

QT3
CARRY OUT THE PCM QUICK TEST
Note: If unable to access the PCM or a diagnostic tool communication concern exists, GO to Pinpoint Test AE
DLC
DLC
DLC
Connect The Diagnostic Tool To The Data Link Connector (DLC).
Carry out the PCM Quick Test To Access And Record Any Key On Engine Off (KOEO), Key On Engine Running (KOER) And Continuous Memory DTCs.
Access And Record The Freeze Frame Data.
Are Any DTCs Present?
YES
NO
GO To The No DTCs Present Symptom Chart Index .
DTC INDEX
SYMPTOM CHARTS

Trouble Code (s)		Test
P2623	**Problem: Injection Pressure Regulator (IPR)**	R 2006-07

REFERENCE LIST

This pinpoint test is intended to diagnose the following:

- Harness circuits: IPR and VPWR
- IPR
- Powertrain control module (PCM)

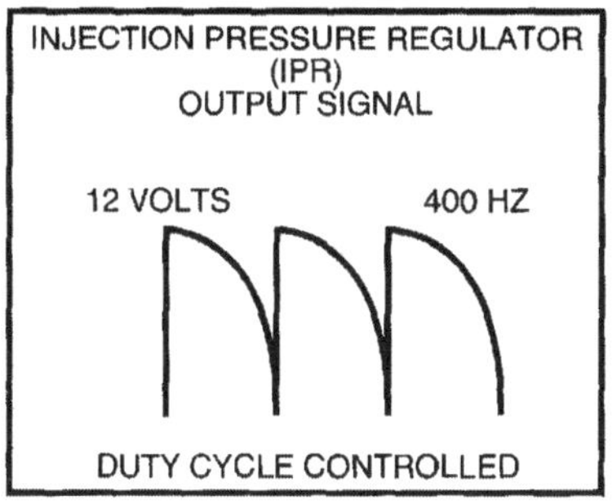

DIAGNOSTIC TROUBLE CODE INDEX			
Codes	KOEO	KOER	CM
P2623	R1	R1	R1

P2623 = Injector Control Pressure Regulator Circuit

Circuit Functions:
The IPR is a pulse width modulated (PWM) variable position valve that regulates the pressure in the injection control pressure (ICP) system. The IPR restricts the return flow path for oil from the high pressure pump. As the duty cycle increases, the IPR restriction to the drain increases, thus increasing the ICP. The IPR duty cycle is controlled by the powertrain control module (PCM) and is modulated from 0 to 65% depending upon the desired injection control pressure.

Detection/Management:
An open or short to ground in the IPR control circuit can be detected by an on-demand output circuit check carried out during the key on, engine off (KOEO) self-test. An injection control pressure step test, in which the PCM commands and then measures specific preprogrammed pressures is carried out during the key on, engine running (KOER) self-test.

The PCM is also capable of detecting whether the desired injection control pressure is equal to measured injection control pressure while the engine is running. If the measured injection control pressure does not reasonably compare to the desired injection control pressure, the PCM ignores the measured ICP sensor signal and attempts to control the engine with the desired value. (If the problem is in the ICP sensor circuit this strategy causes minimal performance deterioration. If the problem is in the IPR control circuit, engine performance may be unsatisfactory).

CAUTION!:
The powertrain control module (PCM) harness connectors must be properly seated and the connector latch properly attached to eliminate possible driveability concerns or a no start condition. Installing the PCM connectors on an angle may cause an improper connection, misdiagnosis and damaged components. Install the connector until the lever pivots and seats itself. Apply light pressure to get the connector into position on the PCM and then fully seat the connector.

Note: Visually inspect the harness connectors for corrosion, damage, proper mating and correct pin tension.

Note: When the PCM is disconnected additional DTCs will be set. Clear all DTCs after restoring the vehicle.

Note: The engine will not operate with an IPR circuit that is not functioning.

PINPOINT TEST SCHEMATICS AND CONNECTORS

Pin	Circuit
E2	IPR (Injection Pressure Regulator)

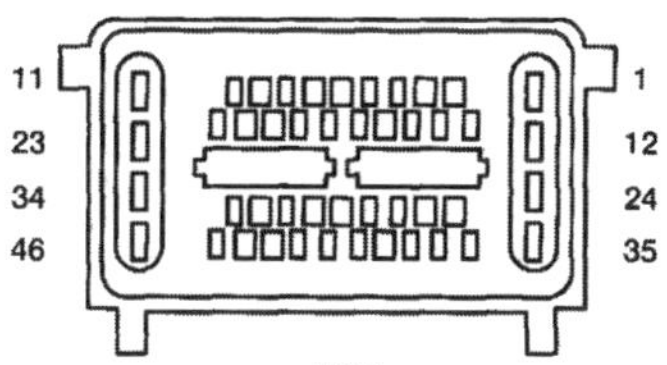

PCM ENGINE HARNESS CONNECTOR

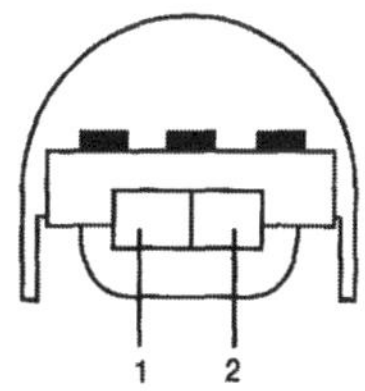

ACTUATOR CONNECTOR

Pin	Circuit
1	VPWR (Vehicle Power)
2	IPR (Injection Pressure Regulator)

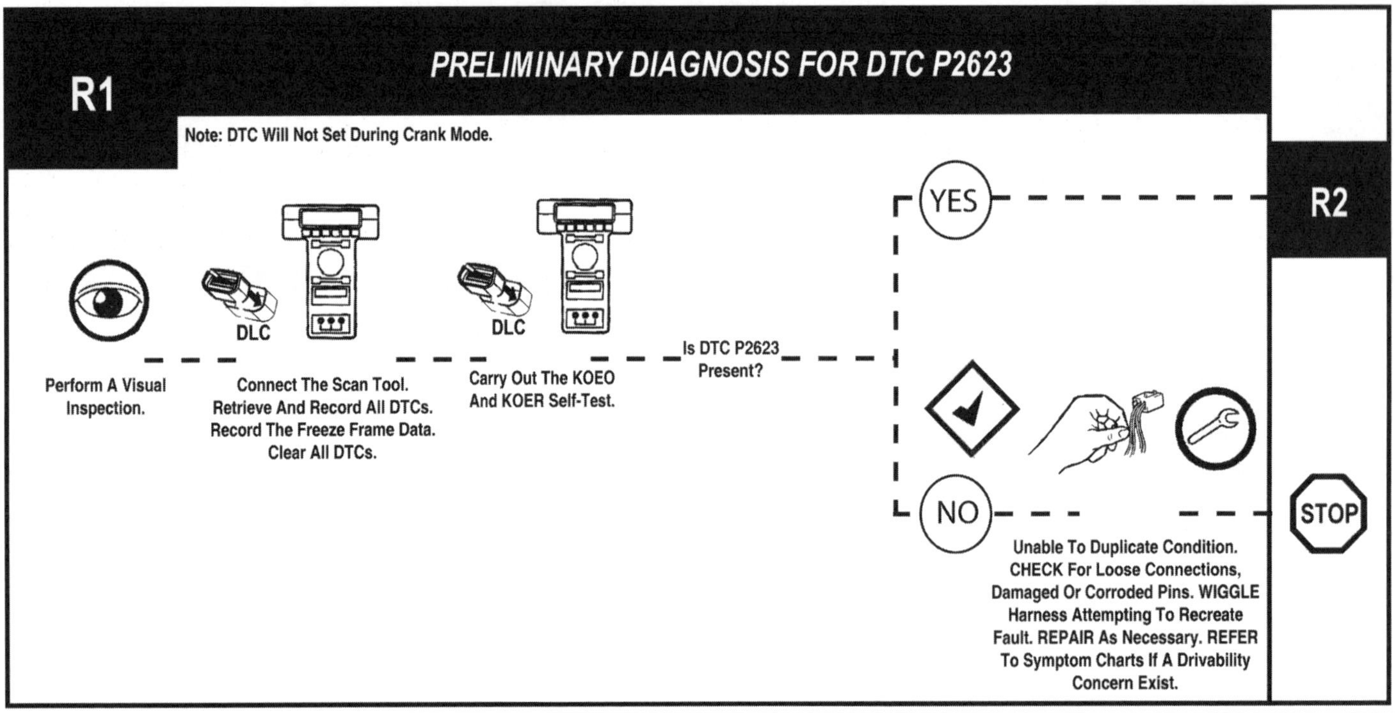
R1
PRELIMINARY DIAGNOSIS FOR DTC P2623
Note: DTC Will Not Set During Crank Mode.
DLC
DLC
Perform A Visual Inspection.
Connect The Scan Tool. Retrieve And Record All DTCs. Record The Freeze Frame Data. Clear All DTCs.
Carry Out The KOEO And KOER Self-Test.
Is DTC P2623 Present?
YES
R2
NO
STOP
Unable To Duplicate Condition. CHECK For Loose Connections, Damaged Or Corroded Pins. WIGGLE Harness Attempting To Recreate Fault. REPAIR As Necessary. REFER To Symptom Charts If A Drivability Concern Exist.

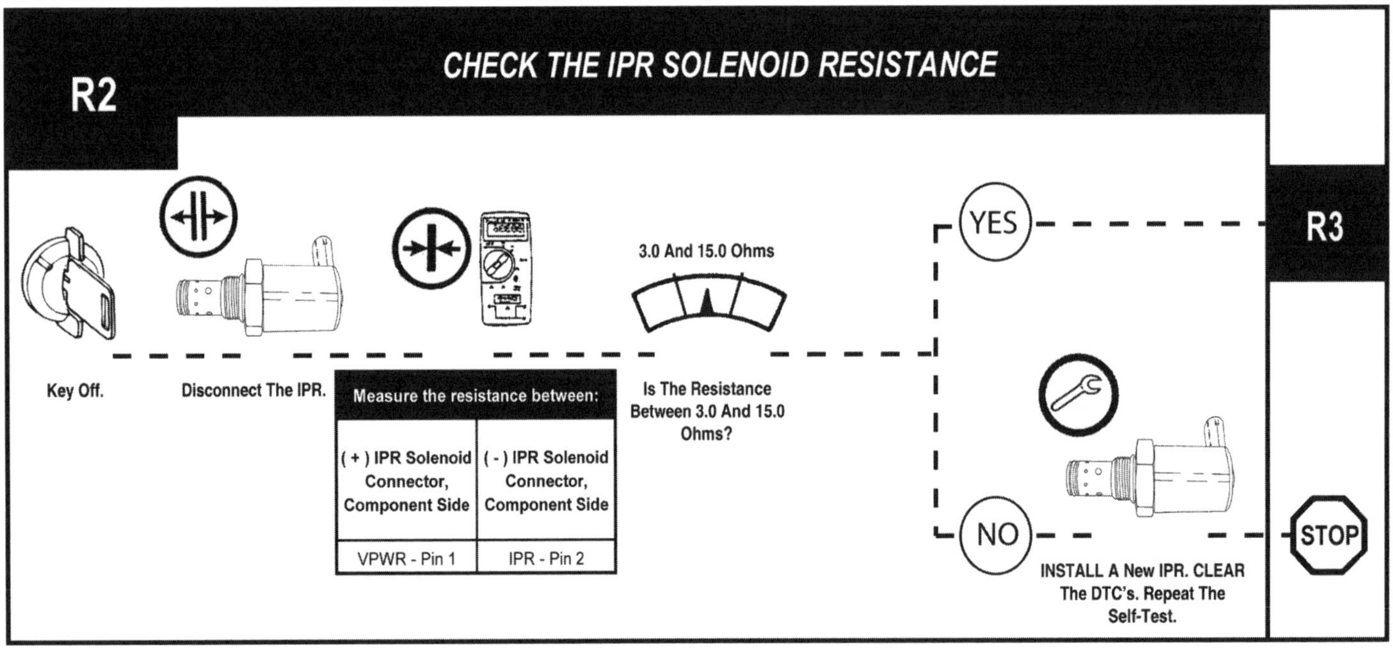
R2
CHECK THE IPR SOLENOID RESISTANCE
Key Off.
Disconnect The IPR.
Measure the resistance between:
(+) IPR Solenoid Connector, Component Side
(-) IPR Solenoid Connector, Component Side
VPWR - Pin 1
IPR - Pin 2
3.0 And 15.0 Ohms
Is The Resistance Between 3.0 And 15.0 Ohms?
YES
R3
NO
STOP
INSTALL A New IPR. CLEAR The DTC's. Repeat The Self-Test.

R3 CHECK THE IPR SOLENOID VOLTAGE

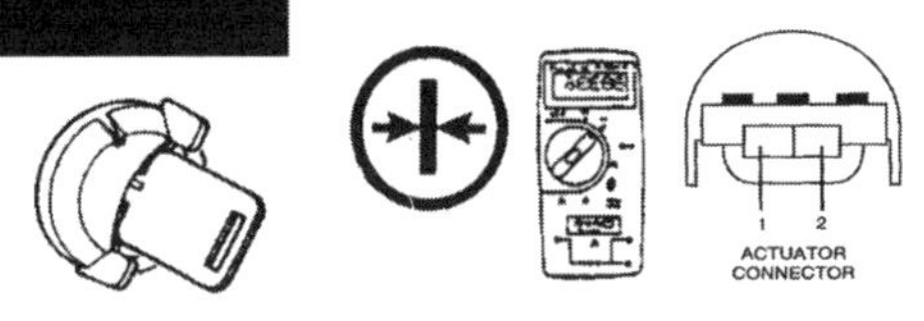

Key On, Engine Off.

Measure the voltage between:	
(+) IPR Solenoid Connector, Harness Side	(-)
VPWR - Pin 1	Ground

Is The Voltage Greater Than 10.5 Volts?

YES → R4

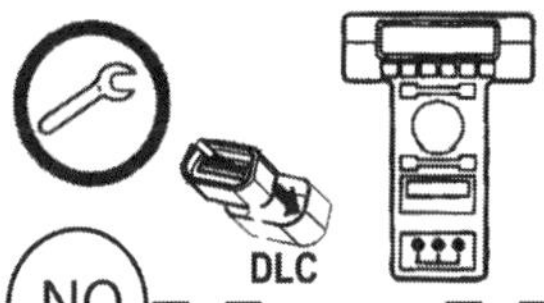

NO → REPAIR The Circuit. CLEAR The DTC's. Repeat The Self-Test. → STOP

R4 CHECK THE IPR CONTROL CIRCUIT FOR A SHORT TO POWER

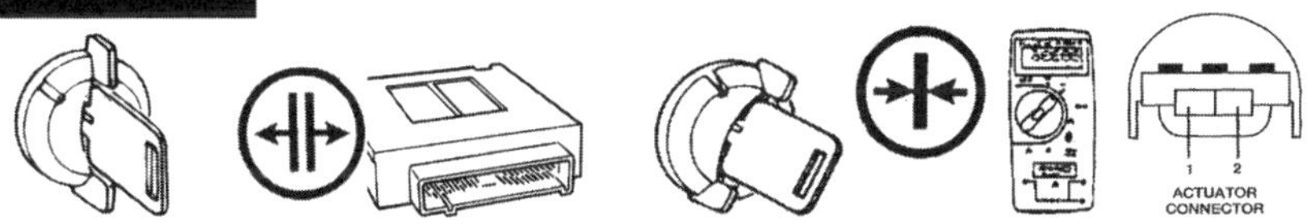

Key Off.

Disconnect The PCM Engine Harness Connector.

Key On. Engine Off.

Measure the voltage between:	
(+) IPR Solenoid Connector, Harness Side	(-)
IPR - Pin 2	Ground

Voltage

Is Any Voltage Indicated?

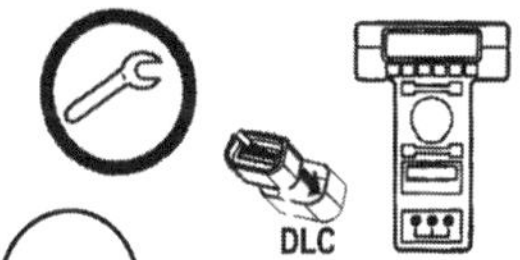

YES → REPAIR The Circuit. CLEAR The DTC's. Repeat The Self-Test. → STOP

NO → R5

R5 — CHECK THE IPR CONTROL CIRCUIT FOR A SHORT TO GROUND

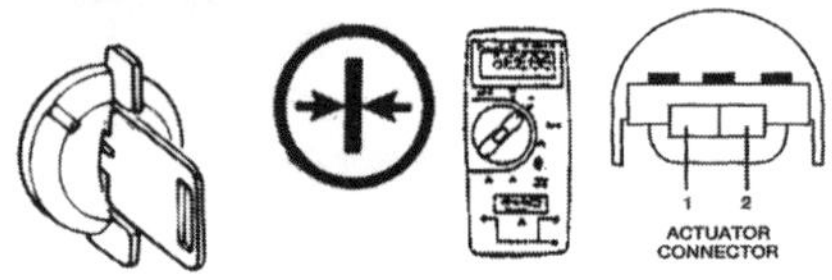

Key Off.

Measure the resistance between:	
(+) IPR Solenoid Connector, Harness Side	(-)
IPR - Pin 2	Ground

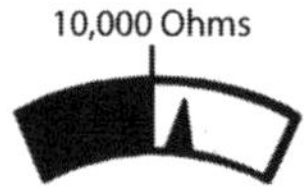

10,000 Ohms

Is The Resistance Greater Than 10,000 Ohms?

YES → R6

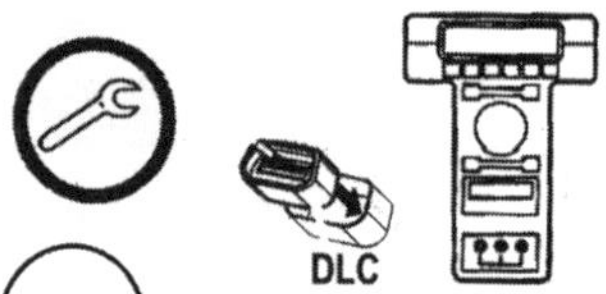

NO → STOP: REPAIR The Circuit. CLEAR The DTC's. Repeat The Self-Test.

R6 — CHECK THE IPR CONTROL CIRCUIT FOR AN OPEN

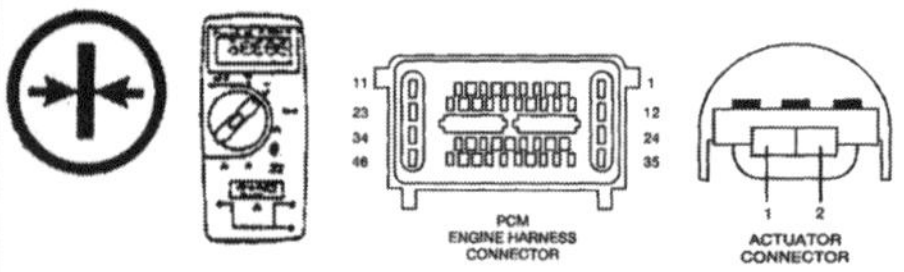

Measure the resistance between:	
(+) PCM Connector, Harness Side	(-) IPR Solenoid Connector, Harness Side
IPR - Pin E2	IPR - Pin 2

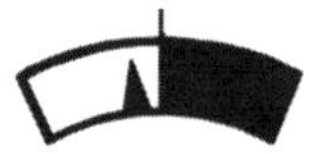

5 Ohms

Is The Resistance Less Than 5 Ohms?

YES → R7

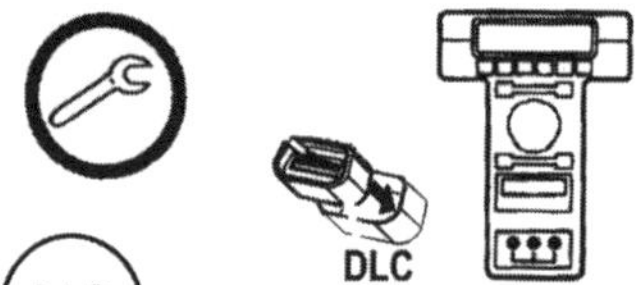

NO → STOP: REPAIR The Circuit. CLEAR The DTC's. Repeat The Self-Test.

R7 CHECK FOR CORRECT PCM OPERATION

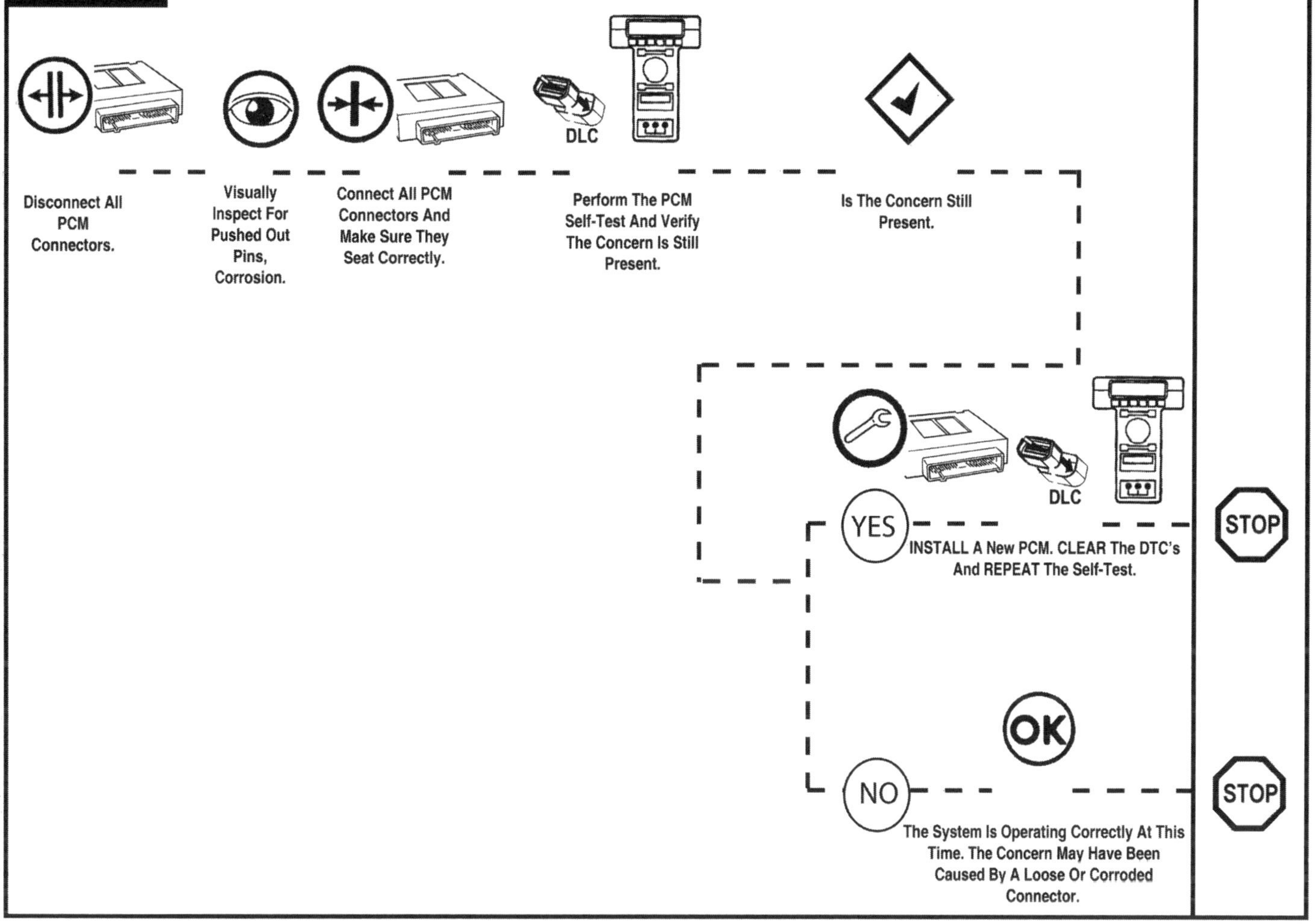

NOTES

Trouble Code (s)	Problem	Test
P0611 P1378 P1379	**Problem: Fuel Injector Control Module (FICM) Low Voltage**	S 2006-07

REFERENCE LIST

This pinpoint test is intended to diagnose the following:

- FICM
- FICM power relay
- Harness circuits: B+, FICM control, FICM vehicle power

DIAGNOSTIC TROUBLE CODE INDEX			
CODES	KOEO	KOER	CM
P0611	S1	S1	S1
P1378	S1	S1	S1
P1379	S1	S1	S1

P0611 = FICM Performance
P1378 = FICM Circuit Low Voltage
P1379 = FICM Circuit High Voltage

Circuit Function:
The FICM requires a 12-volt power source. The FICM receives power from the vehicle batteries through the FICM relay contacts each time the key is turned to the ON position. As the key is turned to the ON position, the FICM provides an internal ground to the coil side of the FICM relay. This closes the relay contacts and provides the FICM with the necessary power.

Fault/Management:
When the FICM continuously receives less than 7 volts or more than 16 volts, diagnostic trouble codes (DTC) P1378 or P1379 are set. DTCs P1378 or P1379 may also cause DTCs P2552 and P0148 to set.

CAUTION!:
The powertrain control module (PCM) harness connectors must be properly seated and the connector latch properly attached to eliminate possible driveability concerns or a no start condition. Installing the PCM connectors on an angle may cause an improper connection, misdiagnosis and damaged components. Install the connector until the lever pivots and seats itself. Apply light pressure to get the connector into position on the PCM and then fully seat the connector.

Note: Visually inspect the harness connectors for corrosion, damage, proper mating and correct pin tension.

PINPOINT TEST SCHEMATICS AND CONNECTORS

2006	
Pin	Circuit
23, 24, 25, 4	FICM Vehicle Power
27	FICM Control
8	FICM Logic Power

2007	
Pin	Circuit
8	FICM Vehicle Power
27	FICM Control

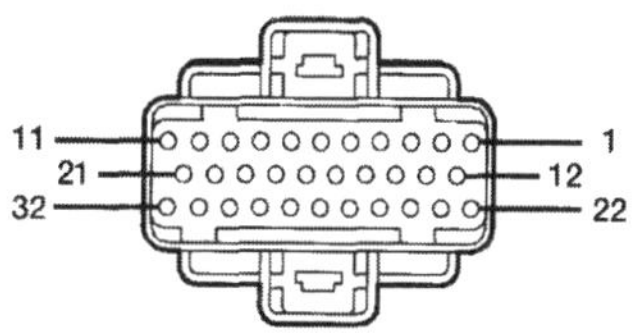

FICM HARNESS CONNECTOR

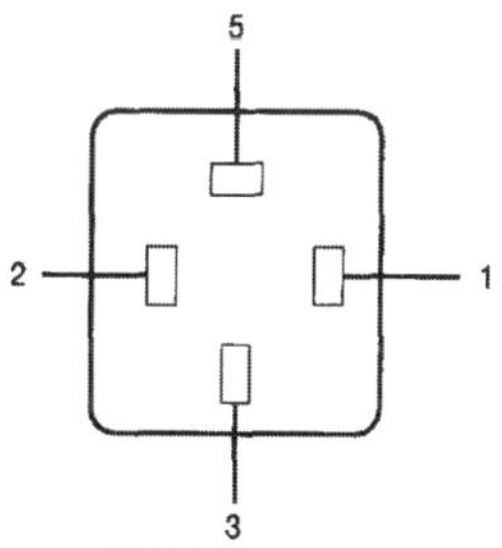

FICM RELAY

Pin	Circuit
1	FICM Control
2	B+ (coil)
3	FICM Vehicle Power
5	B+ (contacts)

2006-07 TEST S: FUEL INJECTOR CONTROL MODULE (FICM) LOW VOLTAGE

DTC(S): P0611, P1378, P1379

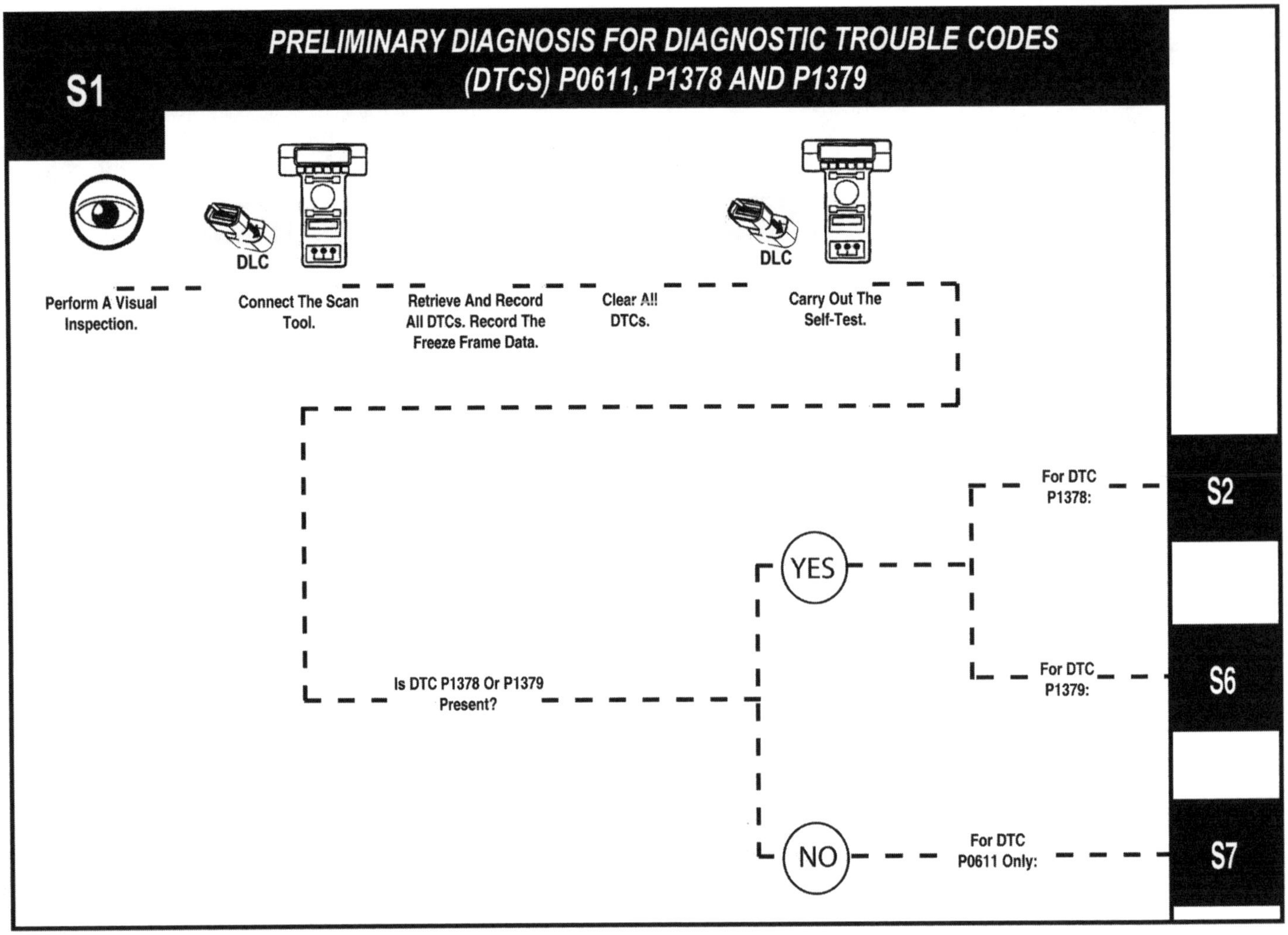

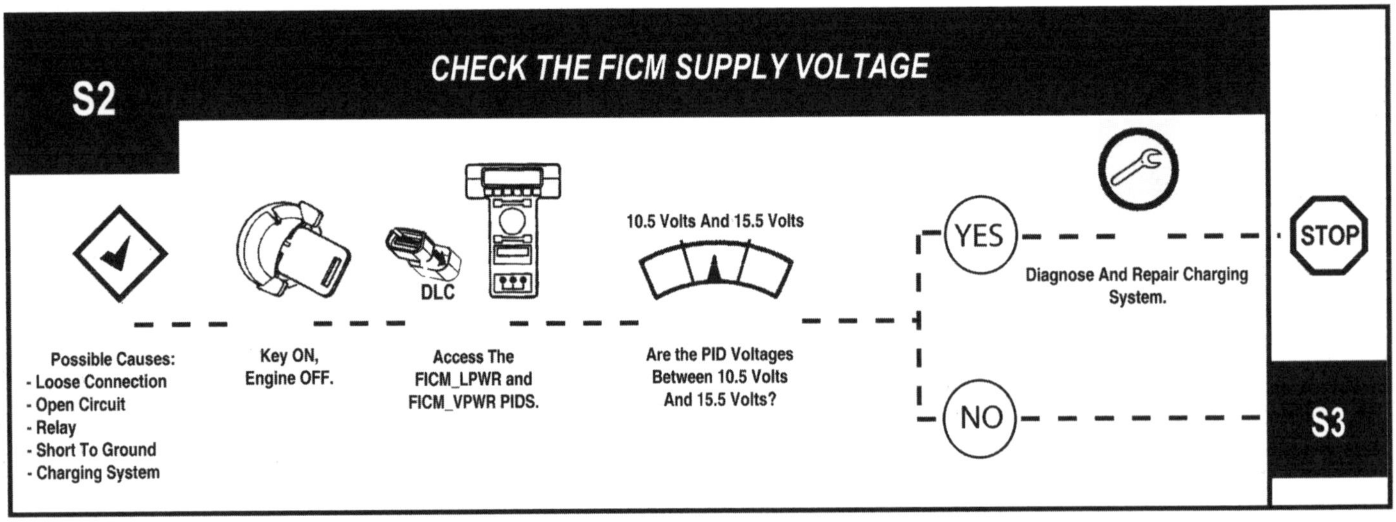

S3 — CHECK THE FICM VOLTAGE CIRCUIT FOR AN OPEN

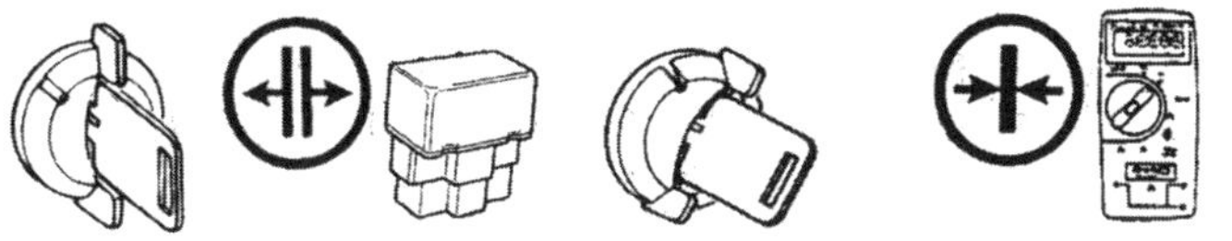

Key Off.

Remove The FICM Relay.

Key On. Engine Off.

Measure the voltage between:	
(+) FICM Relay Connector, Harness Side	(-)
B+ (coil) - Pin 2	Ground
B+ (contacts) - Pin 5	Ground

Are The Voltages Greater Than 10.5 Volts?

YES → S4

NO →

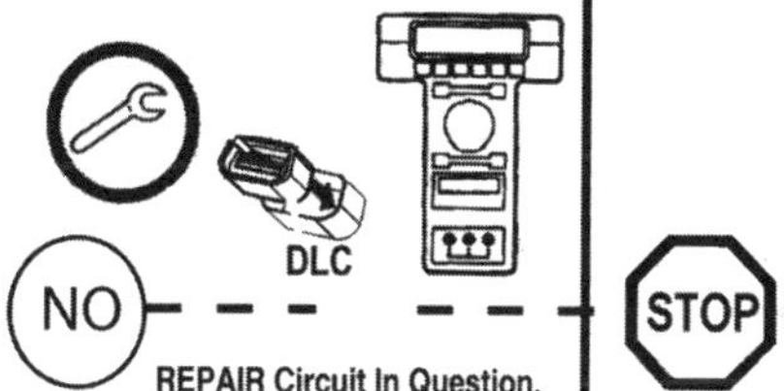

REPAIR Circuit In Question. CLEAR DTCs And REPEAT The Self-Test.

S4 — CHECK THE FICM RELAY CIRCUIT FOR AN OPEN

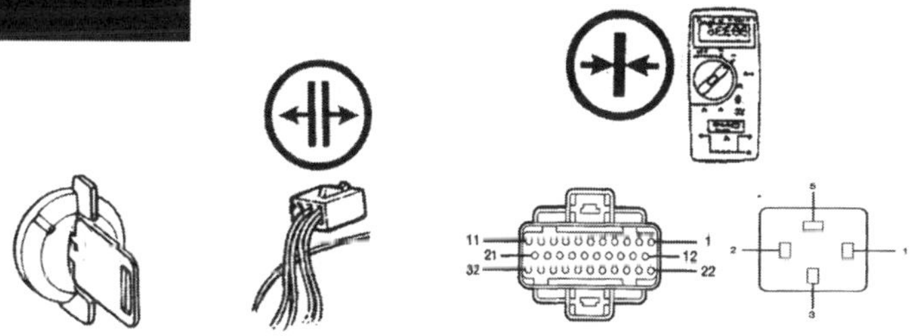

Key Off.

Disconnect The FICM Connector C.

Measure the resistance between:	
(+) FICM Relay Connector, Harness Side	(-) FICM-C Connector, Harness Side
FICM Vehicle Power - Pin 3	FICM Logic Power Pin 8

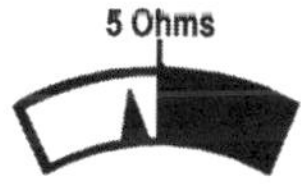

Is The Resistance Less Than 5 Ohms?

YES → S5

NO →

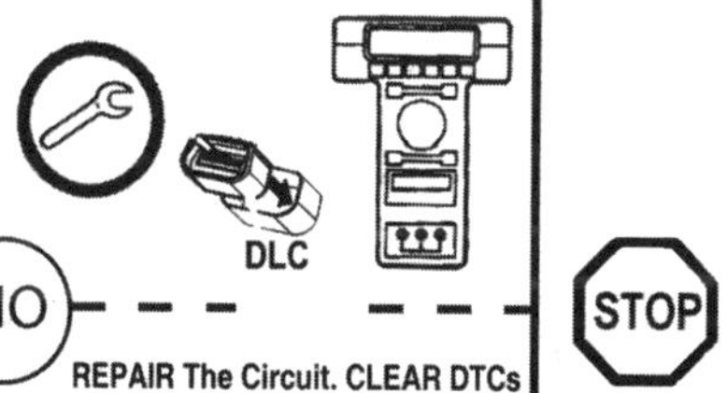

REPAIR The Circuit. CLEAR DTCs And REPEAT The Self-Test.

S5 CHECK FICM RELAY CONTROL CIRCUIT FOR AN OPEN

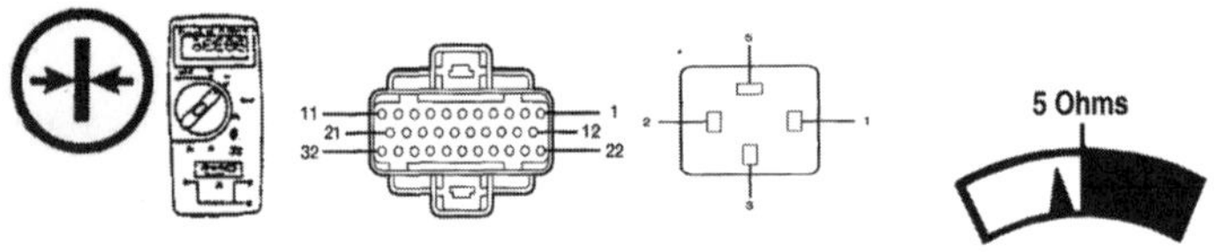

Measure the resistance between:

(+) FICM Relay Connector, Harness Side	(-) FICM-C Connector, Harness Side
FICM Control - Pin 1	FICM Control - Pin 27

Is The Resistance Less Than 5 Ohms?

YES

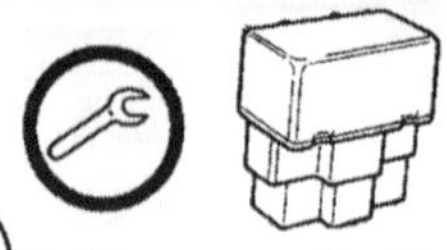

INSTALL A New FICM Relay. CLEAR The DTCs And REPEAT The Self-Test.

STOP

NO

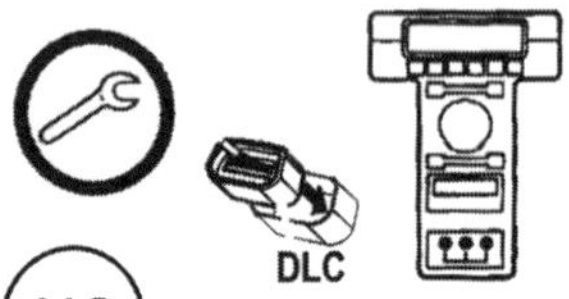

REPAIR The Circuit. CLEAR The DTCs And REPEAT The Self-Test.

STOP

S6 DIAGNOSTIC TROUBLE CODE (DTC) P1379

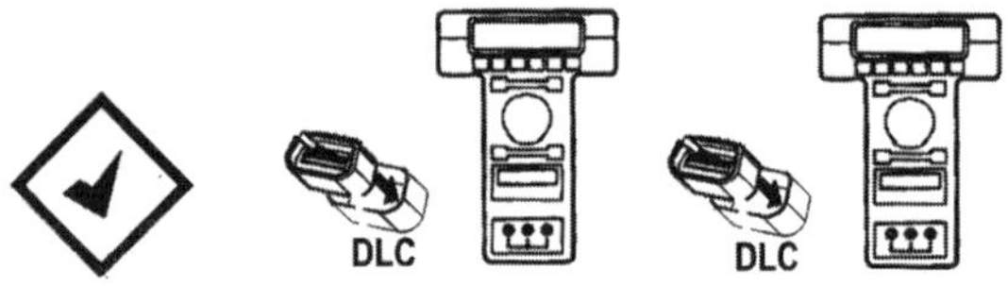

DTC P1379 Indicates The FICM Supply Voltage High.
Possible Causes:
- Charging System
- Alternator

Clear All DTCs.

Carry Out The Self-Test.

Is DTC P1379 Present?

YES

Diagnose And Repair Charging System.

STOP

NO

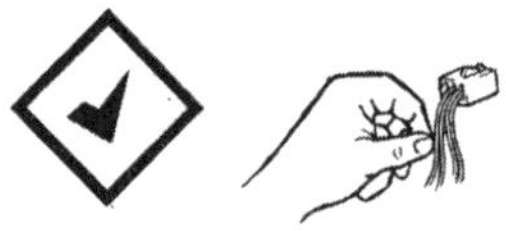

Unable To Duplicate Condition. CHECK For Loose Connections, Damaged Or Corroded Pins. WIGGLE Harness Attempting To Recreate Fault. REFER To Symptom Charts If A Driveability Concern Exists.

STOP

S7 CHECK THE FICM POWER RELAY

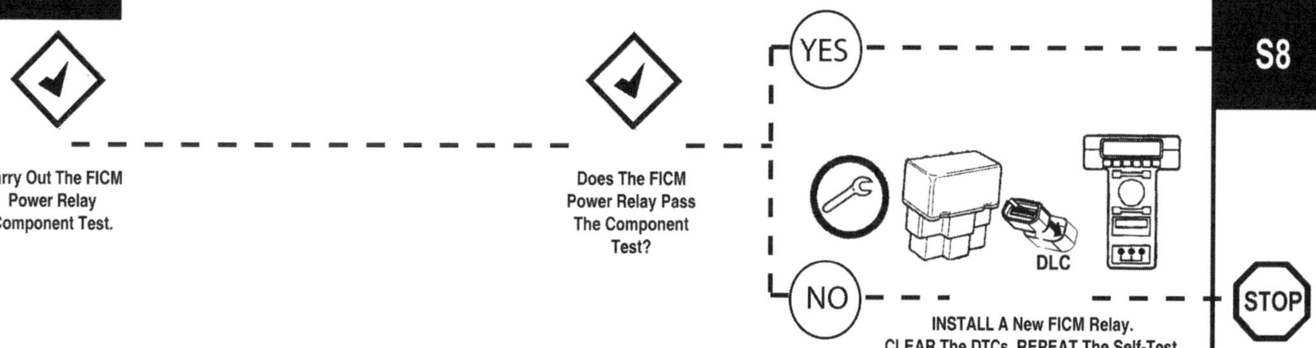

Carry Out The FICM Power Relay Component Test.

Does The FICM Power Relay Pass The Component Test?

YES → S8

NO → INSTALL A New FICM Relay. CLEAR The DTCs. REPEAT The Self-Test. STOP

S8 CHECK THE FICM SUPPLY VOLTAGE

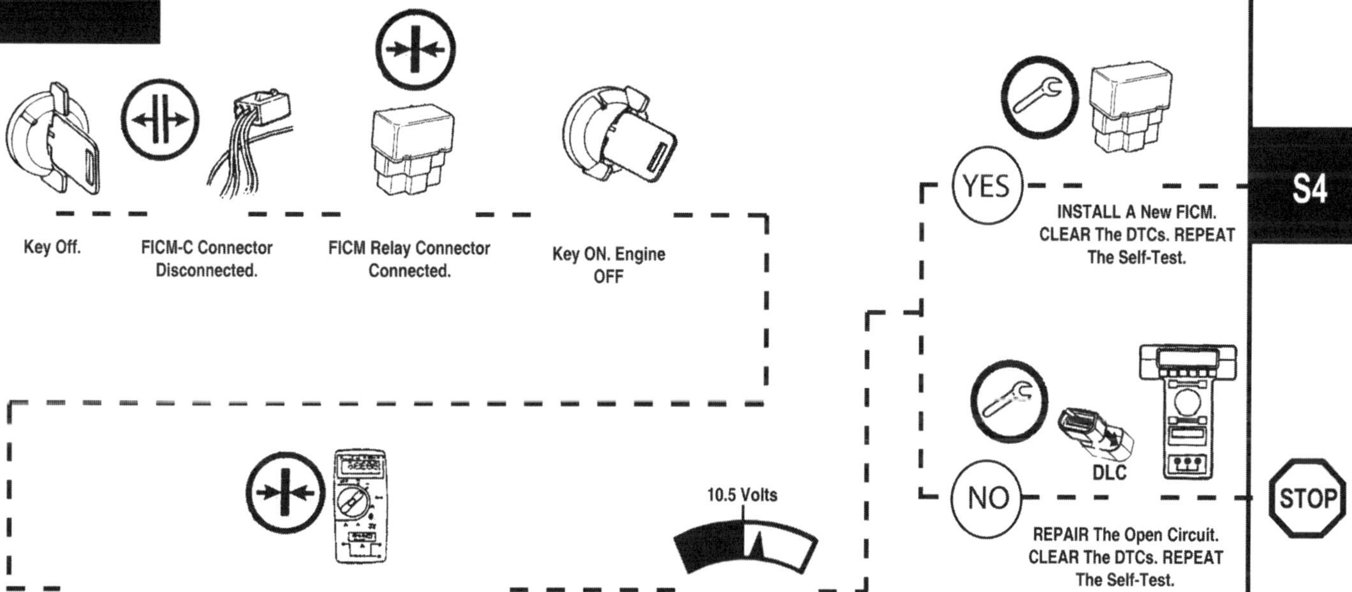

Key Off.

FICM-C Connector Disconnected.

FICM Relay Connector Connected.

Key ON. Engine OFF

2006	
Measure the voltage between:	
(+) FICM-C Connector, Harness Side	(-)
FICM Vehicle Power - Pin 23, 24, 25, 4	Ground

2007	
Measure the voltage between:	
(+) FICM-C Connector, Harness Side	(-)
FICM Vehicle Power - Pin 8	Ground

Are The Voltages Greater Than 10.5 Volts?

YES → INSTALL A New FICM. CLEAR The DTCs. REPEAT The Self-Test. → S4

NO → REPAIR The Open Circuit. CLEAR The DTCs. REPEAT The Self-Test. STOP

NOTES

Trouble Code (s)	Problem: Camshaft Position (CMP) Sensor	Test
P0340, P0341, P2614		V 2006-07

REFERENCE LIST

This pinpoint test is intended to diagnose the following:

- CMP sensor (12K073)
- Harness circuits: CMP+, CMP-, and CMPO
- Fuel injector control module (FICM)
- Powertrain control module (PCM) (12A650)

DIAGNOSTIC TROUBLE CODE INDEX			
CODES	KOEO	KOER	CM
P0340	V1	V1	V1
P0341	V1	V1	V1
P2614	V1	V1	V1

P0340 = CMP Sensor Circuit Malfunction
P0341 = CMP Sensor Circuit Performance
P2614 = Camshaft Position Out

Circuit Function:
The CMP signal source is a magnetic pickup sensor mounted on the left front side of the engine block. The sensor reacts to a peg on a target wheel that is positioned on the camshaft. The peg passes the sensor once per camshaft revolution and produces a single pulse correspondingly. The camshaft speed is calculated from the frequency of the CMP sensor signal. Diagnostic information on the CMP input signal is obtained by carrying out accuracy checks on signal levels, frequency, and duty cycle with software strategies.

The powertrain control module (PCM) requires a crankshaft position (CKP) sensor and CMP signal to calculate engine speed and piston position. The CMP sensor creates a signal that the PCM uses to indicate the left or right bank. The engine does not operate without a CMP signal.

Detection/Management:
An inactive CMP signal during cranking is detectable by the PCM. An inactive CMP signal causes a no-start condition. Electrical noise can also be detected by the PCM. If the level is sufficient to affect engine operation, a corresponding diagnostic trouble code (DTC) is set.

CAUTION!:
The powertrain control module (PCM) harness connectors must be properly seated and the connector latch properly attached to eliminate possible driveability concerns or a no start condition. Installing the PCM connectors on an angle may cause an improper connection, misdiagnosis and damaged components. Install the connector until the lever pivots and seats itself. Apply light pressure to get the connector into position on the PCM and then fully seat the connector.

Note: Visually inspect the harness connectors for corrosion, damage, proper mating and correct pin tension.

Note: When the PCM is disconnected additional DTCs will be set. Clear all DTCs after restoring the vehicle.

PINPOINT TEST SCHEMATICS AND CONNECTORS

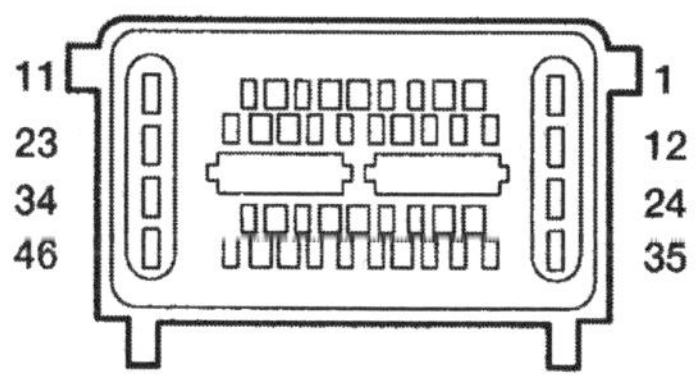

PCM ENGINE HARNESS CONNECTOR

Pin	Circuit
E20	CMPO (Camshaft Position Output)
E31	CMP+ (Camshaft Position +)
E43	CMP- (Camshaft Position -)

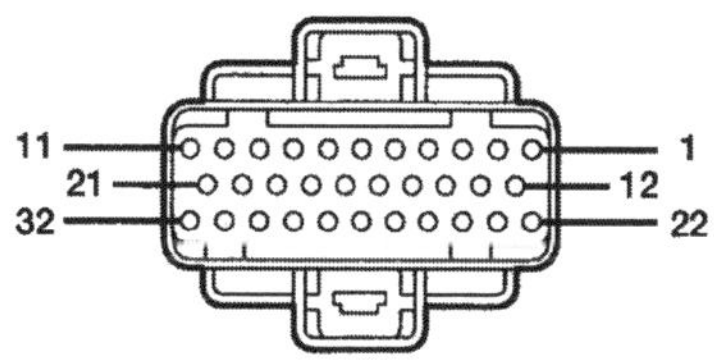

FICM HARNESS CONNECTOR

Pin	Circuit
10	CMPO (Camshaft Position Output)

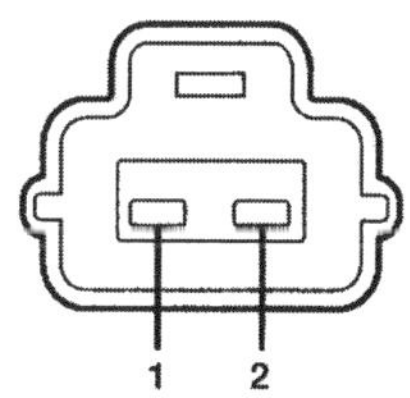

SENSOR CONNECTOR

Pin	Circuit
1	CMP+ (Camshaft Position +)
2	CMP- (Camshaft Position -)

V1 — PRELIMINARY DIAGNOSIS FOR DTCS P0340, P0341, P2614

Note: Some DTCs require that the engine goes through more than one key cycle to set.
Note: DTCs P2617 and P2614 may set simultaneously after an engine stalling event.
Note: DTC P2614 is calibrated to an increment counter. To set the DTC requires more than 10 consecutive fault events in the crank or the run mode. Carry out the required number of key cycles from ON to START to ON (cycling to OFF resets the timer) and then carry out the key on engine off (KOEO) or key on engine running (KOER) self-test.

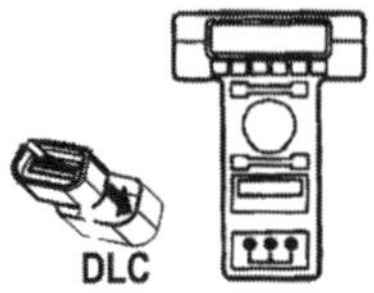

Perform A Visual Inspection. — Connect The Scan Tool. — Retrieve And Record All DTCs. — Record Freeze Frame Data. — Clear All DTCs. — Perform The On-Demand Self-Test.

Are DTCs P0340, P0341 Or P2614 Present?

- YES → V2
- NO → STOP

UNABLE To Duplicate Condition. CHECK For Loose Connections, Damaged Or Corroded Pins. WIGGLE Harness Attempting To Recreate Fault. REFER To Symptom Charts If A Drivability Concern Exists. REFER To Diagnostic Subroutines To Diagnose A No-Start Condition.

V2 — DIAGNOSTIC TROUBLE CODES (DTCS) P0340, P0341 OR P2614

Note: Refer To The PCM And Component Connector Pin Numbers At The Beginning Of This Pinpoint Test.

Possible Causes:
- CMP Circuit Open
- CMP Circuit Short To Ground
- CMP Circuit Short To Power
- Signal Circuit Open
- Signal Circuit Shorted
- PCM
- CMP Sensor

Does The Engine Start?

- YES → V3
- NO → V4

V3 — *ATTEMPT TO RECREATE THE DTCS*

Greater Than 1,500 RPM

1. Clear All DTCs.
2. Start The Engine.
3. Increase RPM To Greater Than 1,500 RPM For 10 Seconds. Repeat Two Times.
4. Key On, Engine Off.
5. Perform On-Demand Self-Test.

Is DTC P0340, P0341 Or P2614 Present?

- **YES** → V4
- **NO** → STOP

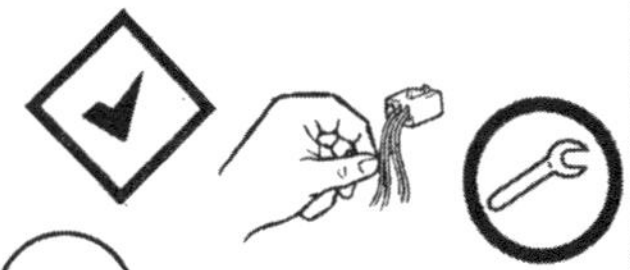

CHECK For A Loose Connection. REPAIR As Necessary. CHECK Pins For Damage Or Corrosion. CLEAR DTCs And RETEST The System.

V4 — *CHECK THE RESISTANCE OF THE CMP SENSOR CIRCUITS*

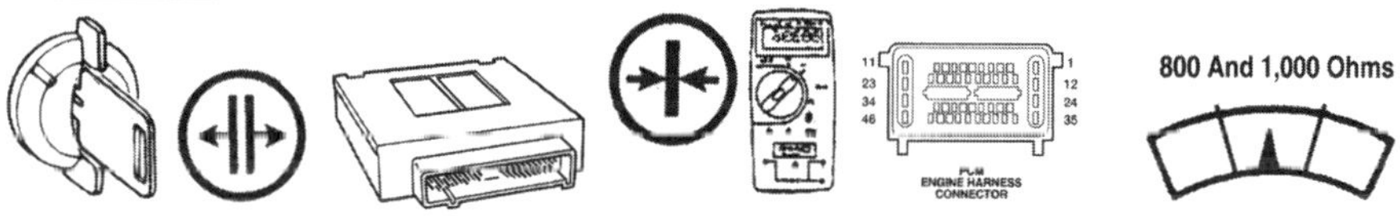

800 And 1,000 Ohms

1. Key Off.
2. Disconnect The PCM Engine Harness Connector.

Measure the resistance between:	
(+) PCM Connector, Harness Side	**(-) PCM Connector, Harness Side**
CMP+ - Pin E31	CMP- - Pin E43

Is The Resistance Between 800 And 1,000 Ohms?

- **YES** → V5
- **NO** → V7

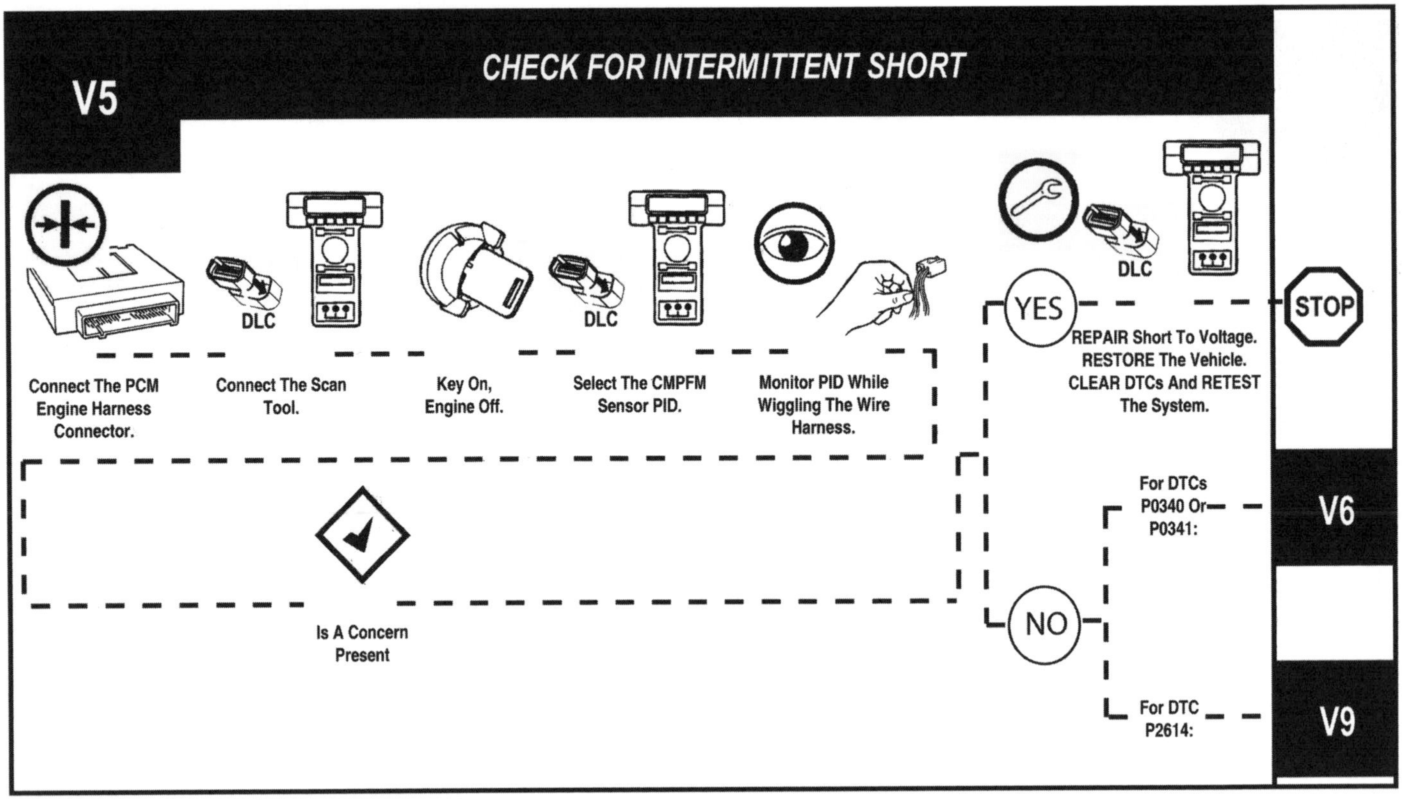
V5
CHECK FOR INTERMITTENT SHORT
DLC
DLC
DLC
Connect The PCM Engine Harness Connector.
Connect The Scan Tool.
Key On, Engine Off.
Select The CMPFM Sensor PID.
Monitor PID While Wiggling The Wire Harness.
Is A Concern Present
YES
REPAIR Short To Voltage. RESTORE The Vehicle. CLEAR DTCs And RETEST The System.
STOP
NO
For DTCs P0340 Or P0341:
V6
For DTC P2614:
V9

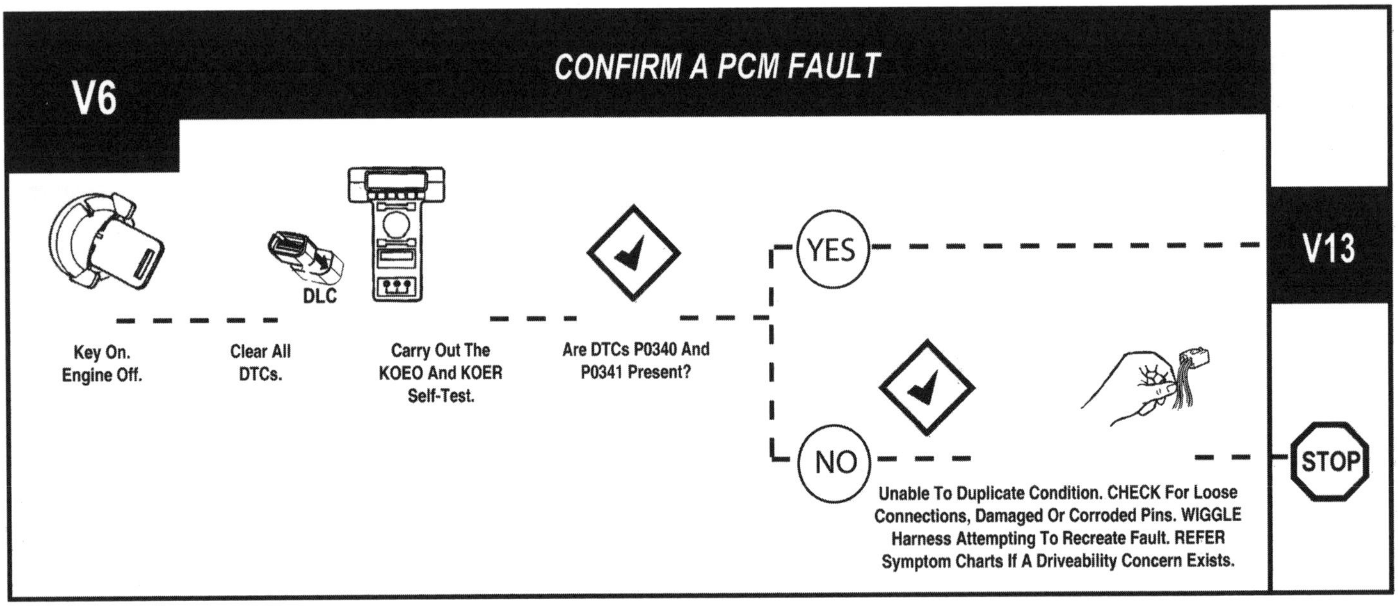
V6
CONFIRM A PCM FAULT
DLC
Key On. Engine Off.
Clear All DTCs.
Carry Out The KOEO And KOER Self-Test.
Are DTCs P0340 And P0341 Present?
YES
V13
NO
Unable To Duplicate Condition. CHECK For Loose Connections, Damaged Or Corroded Pins. WIGGLE Harness Attempting To Recreate Fault. REFER Symptom Charts If A Driveability Concern Exists.
STOP

V7 — CHECK CMP SENSOR CIRCUITS FOR A SHORT TO GROUND

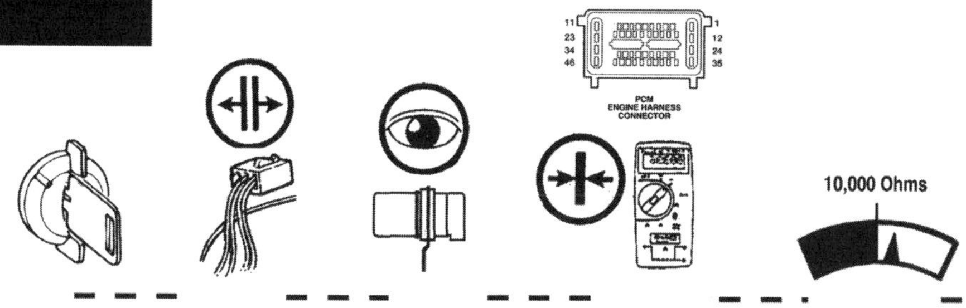

Key Off.

Disconnect The CMP Sensor.

Inspect The CMP Sensor Connector For Damaged Or Corroded Pins.

Measure the resistance between:

(+) PCM Connector, Harness Side	(-)
CMP+ - Pin E31	Ground
CMP- - Pin E43	Ground

Is The Resistance Greater Than 10,000 Ohms?

YES → V8

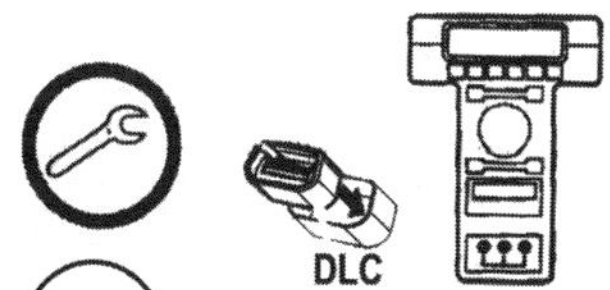

NO → REPAIR Short In CMP Sensor Circuits. CLEAR DTCs And RETEST The System. STOP

V8 — CHECK CMP CIRCUITS FOR AN OPEN

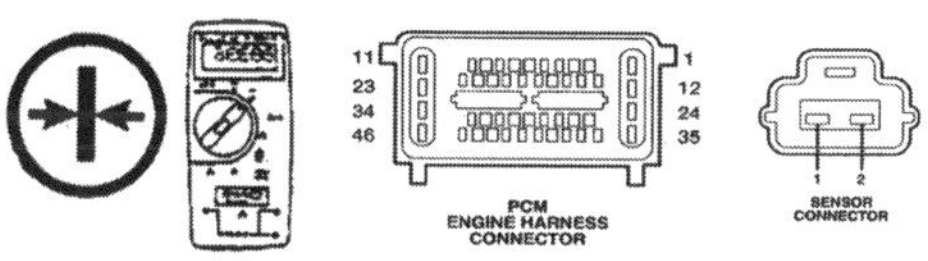

Measure the resistance between:

(+) PCM Connector, Harness Side	(-) CMP Sensor Connector, Harness Side
CMP+ - Pin E31	CMP+ - Pin 1
CMP- - Pin E43	CMP- - Pin 2

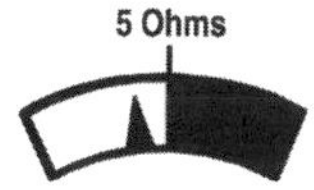

Is The Resistance Less Than 5 Ohms?

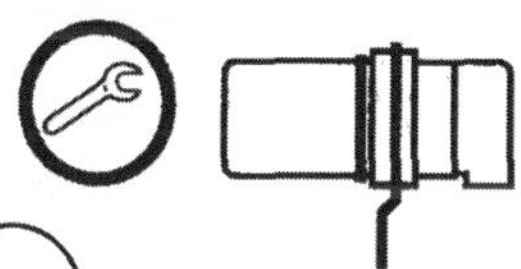

YES → INSTALL A New CMP Sensor. CLEAR DTCs And RETEST The System. STOP

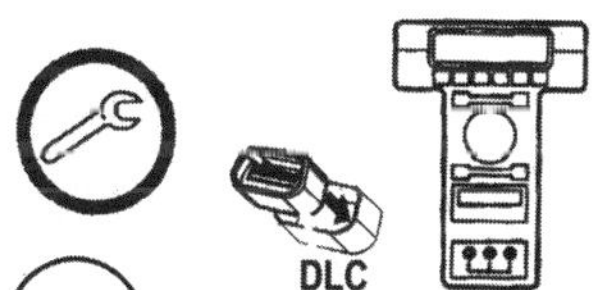

NO → REPAIR Open Circuit. CLEAR DTCs And RETEST The System. STOP

V9 CHECK CAMSHAFT POSITION SIGNAL FOR A SHORT TO VOLTAGE

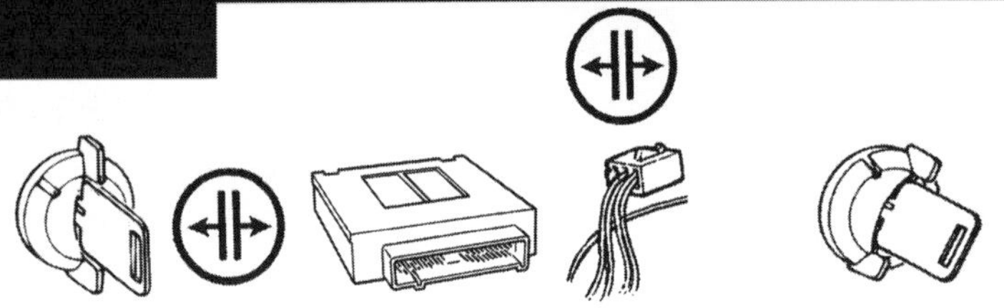

Key Off. → Disconnect The PCM Engine Harness Connector. → Disconnect The FICM Harness Connector C. → Key On, Engine Off.

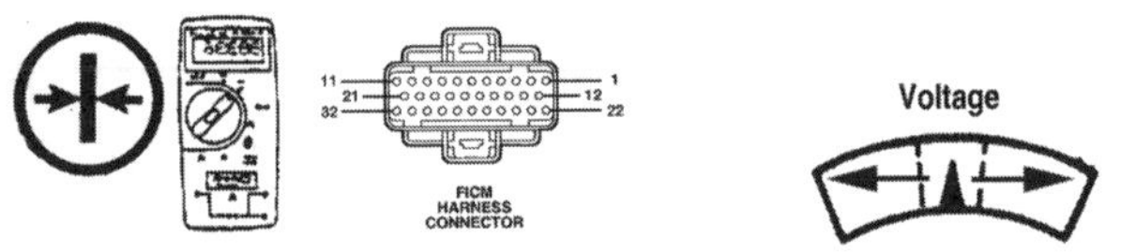

Measure the voltage between:	
(+) FICM-C Connector, Harness Side	(-)
CMPO - Pin 10	Ground

Is Any Voltage Indicated?

YES:

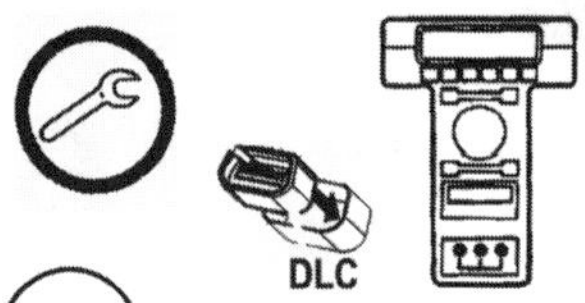

REPAIR Short To Voltage. CLEAR DTCs And RETEST The System. → STOP

NO: → V10

V10 CHECK CAMSHAFT POSITION SIGNAL FOR A SHORT TO GROUND

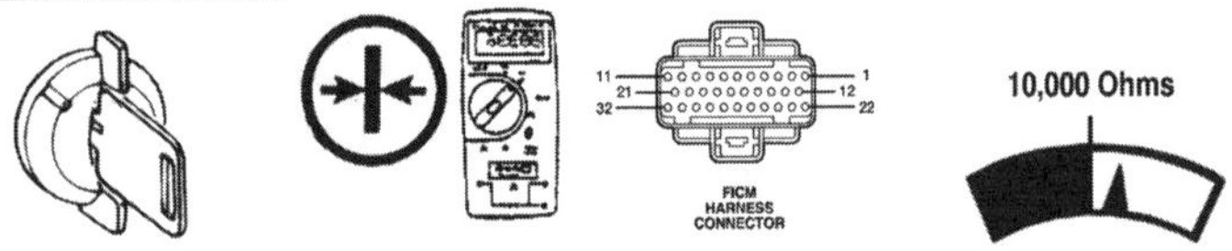

Key Off.

Measure the resistance between:	
(+) FICM-C Connector, Harness Side	(-)
CMPO - Pin 10	Ground

Is The Resistance Greater Than 10,000 Ohms?

YES: → V11

NO:

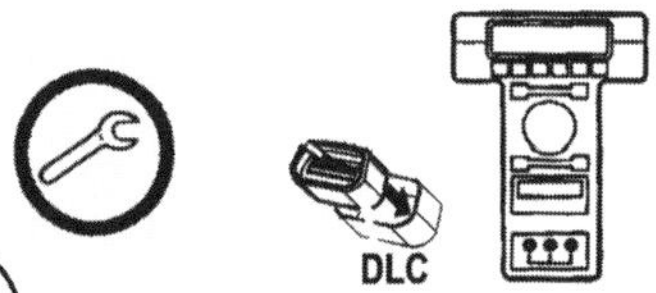

REPAIR Short To Ground In The Camshaft Output Signal Circuit. RESTORE The Vehicle. CLEAR DTCs And RETEST The System. → STOP

V11 CHECK CAMSHAFT POSITION SIGNAL FOR AN OPEN

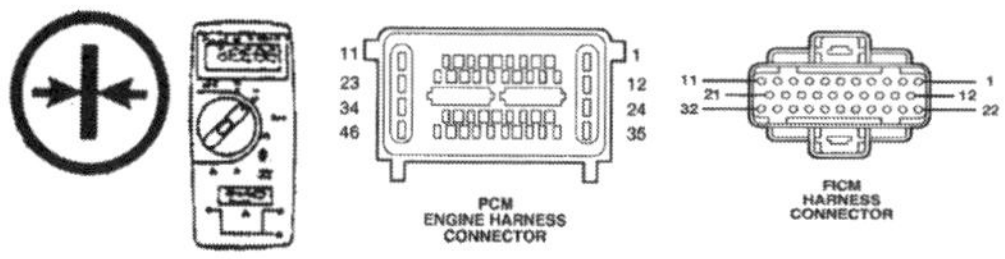

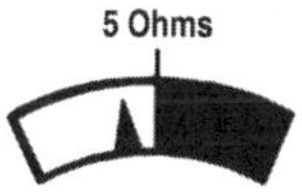

Measure the resistance between:	
(+) FICM-C Connector, Harness Side	**(-) PCM Connector, Harness Side**
CMPO - Pin 10	CMPO - Pin E20

Is The Resistance Less Than 5 Ohms?

YES — V12

NO —

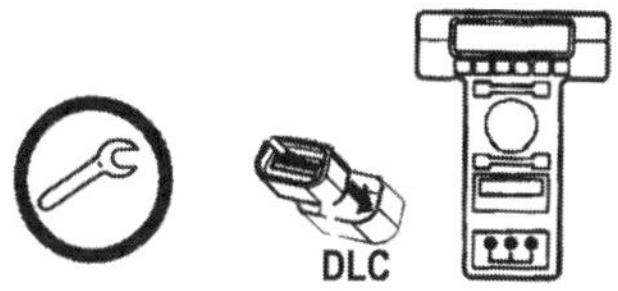

REPAIR Open In The Camshaft Signal Output Circuit. RESTORE The Vehicle. CLEAR DTCs And RETEST The System. — STOP

V12 CHECK THE CMP SENSOR

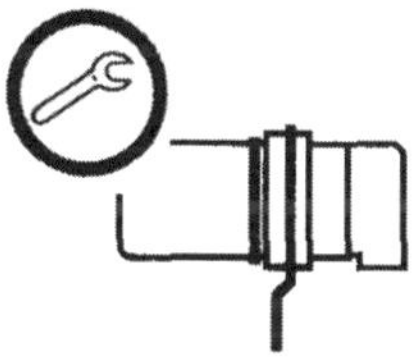

Remove The CMP Sensor.

Visually Inspect The Sensor For Damage, Misalignment And Improper Installation.

Is A Concern Present?

YES —

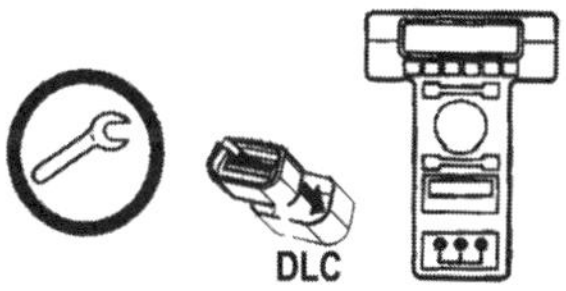

REPAIR As Necessary. CLEAR The DTCs And RETEST The System. — STOP

NO — V13

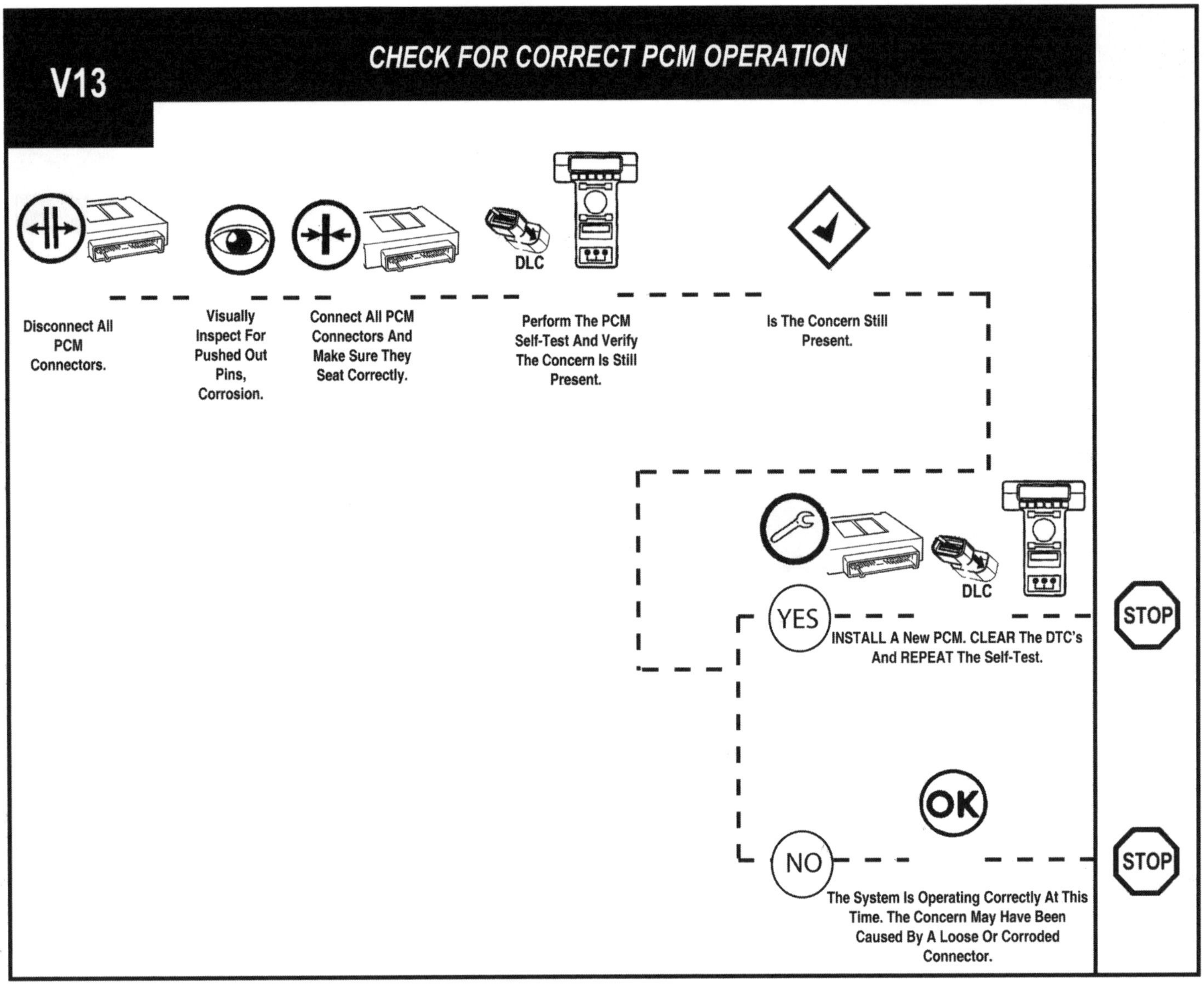
V13
CHECK FOR CORRECT PCM OPERATION
DLC
Disconnect All PCM Connectors.
Visually Inspect For Pushed Out Pins, Corrosion.
Connect All PCM Connectors And Make Sure They Seat Correctly.
Perform The PCM Self-Test And Verify The Concern Is Still Present.
Is The Concern Still Present.
DLC
YES
INSTALL A New PCM. CLEAR The DTC's And REPEAT The Self-Test.
STOP
OK
NO
The System Is Operating Correctly At This Time. The Concern May Have Been Caused By A Loose Or Corroded Connector.
STOP

Trouble Code (s)	Problem: Exhaust Gas Recirculation (EGR)	Test
P0401, P0402, P0403, P0404, P0405, P0406, P1335		W 2006

REFERENCE LIST

This pinpoint test is intended to diagnose the following:

- Harness circuits: VPWR, VREF, EGR, EGRMC and SIGRTN
- EGR actuator
- Powertrain control module (PCM)

DIAGNOSTIC TROUBLE CODE INDEX			
Codes	KOEO	KOER	CM
P0401	W1	W1	W1
P0402	W1	W1	W1
P0403	W1	W1	W1
P0404	W1	W1	W1
P0405	W1	W1	W1
P0406	W1	W1	W1
P1335	W1	W1	W1

P0401 = EGR Flow Insufficient Detected
P0402 = EGR Flow Excessive Detected
P0403 = EGR Control Circuits
P0404 = EGR Control Circuit Range/Performance
P0405 = EGR Sensor A Circuit Low
P0406 = EGR Sensor A Circuit High
P1335 = EGR Position Sensor Minimum Stop Performance

Circuit Function:
The EGR actuator is a variable position valve that controls the amount of exhaust that enters the intake manifold. The EGR is controlled by the powertrain control module (PCM) using a pulse width modulated signal that varies from 0-100%. The EGR actuator consists of 2 components, a valve with an actuator (solenoid), and a position sensor to monitor valve movement. The EGR sensor is a potentiometer sensor. When the EGR receives a 5-volt reference signal and a ground from the PCM, a linear analog voltage signal from the sensor indicates the position of the EGR valve.

Fault Detection/Management:
The PCM detects an open or short in the EGR control circuit after the key on engine off (KOEO) test has run and set a diagnostic trouble code (DTC). The EGR actuator can be tested for a stuck or sticking valve by carrying out the output state low test while monitoring the exhaust gas recirculation position (EGRP) percentage.

CAUTION!:
The powertrain control module (PCM) harness connectors must be properly seated and the connector latch properly attached to eliminate possible driveability concerns or a no start condition. Installing the PCM connectors on an angle may cause an improper connection, misdiagnosis and damaged components. Install the connector until the lever pivots and seats itself. Apply light pressure to get the connector into position on the PCM and then fully seat the connector.

Note: Visually inspect the harness connectors for corrosion, damage, proper mating and correct pin tension.

PINPOINT TEST SCHEMATICS AND CONNECTORS

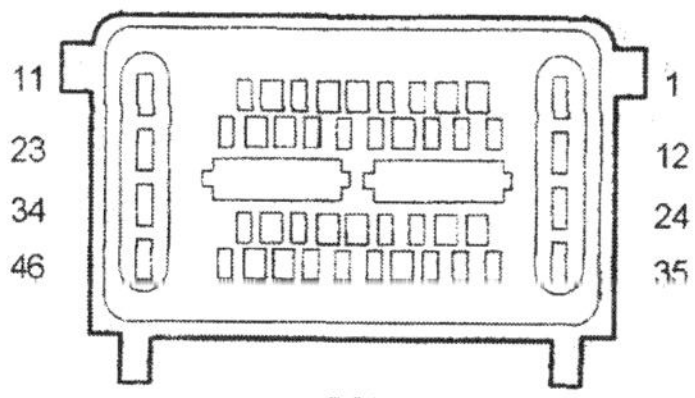

PCM
ENGINE HARNESS
CONNECTOR

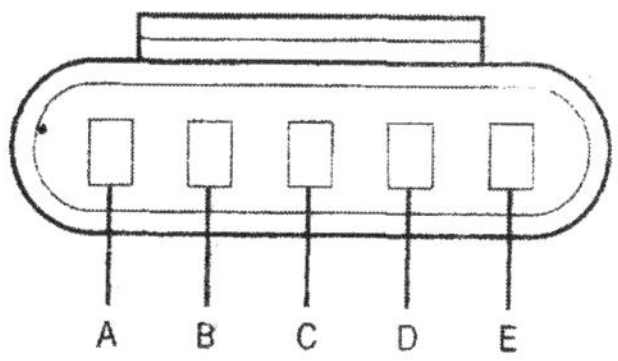

EGR
VALVE ACTUATOR
CONNECTOR

Pin	Circuit
B20, B33, E25	SIGRTN (Signal Return)
E33	EGR (Exhaust Gas Recirculation)
E23	EGRMC (EGR Motor Control)

Pin	Circuit
A	EGRMC (EGR Motor Control)
B	SIGRTN (Signal Return)
C	EGR (Exhaust Gas Recirculation)
D	VREF (Reference Voltage)
E	VPWR (Vehicle Power)

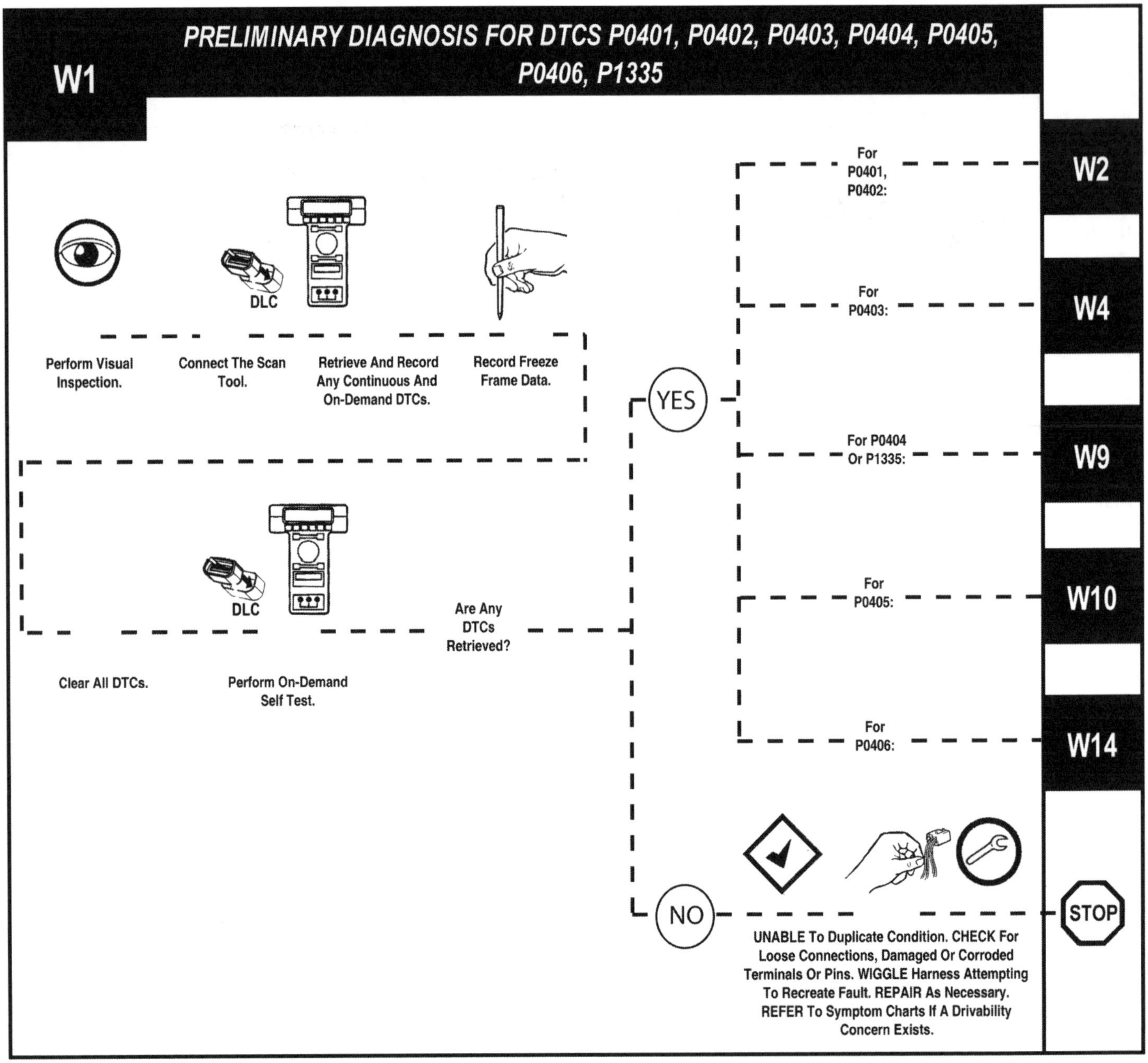

W1
PRELIMINARY DIAGNOSIS FOR DTCS P0401, P0402, P0403, P0404, P0405, P0406, P1335
DLC
Perform Visual Inspection.
Connect The Scan Tool.
Retrieve And Record Any Continuous And On-Demand DTCs.
Record Freeze Frame Data.
DLC
Clear All DTCs.
Perform On-Demand Self Test.
Are Any DTCs Retrieved?
YES
For P0401, P0402:
W2
For P0403:
W4
For P0404 Or P1335:
W9
For P0405:
W10
For P0406:
W14
NO
UNABLE To Duplicate Condition. CHECK For Loose Connections, Damaged Or Corroded Terminals Or Pins. WIGGLE Harness Attempting To Recreate Fault. REPAIR As Necessary. REFER To Symptom Charts If A Drivability Concern Exists.
STOP

W2

DIAGNOSTIC TROUBLE CODE (DTC) P0401, P0402

Note: Visually Inspect The Vehicle For Aftermarket Accessories And Performance Modifications (Air Filter, Exhaust Sstem Intake System, Performance Chip, Turbocharger). Refer To Diesel Electronic Engine Control (EEC) System Modifications To OBD Vehicles.

Possible Causes:
- Restricted Air Flow (Intake Or Exhaust)
- Charge Air Cooler (CAC) System Leaks (Hoses/Cooler)
- EGR Valve Stuck Or Sticking EGR Valve O-Rings
- Restricted EGR Cooler
- EGR Valve Position Sensor Bias
- EP Sensor Bias

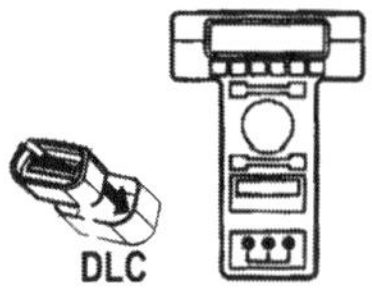

Perform On-Demand Self Test.

Are Other DTCs Present?

YES → DTC INDEX

NO → W3

W3 CHECK EGR VALVE OPERATION

Note: Additional DTCs may set during the use of output state control (OSC). Clear all PCM and transmission control module (TCM) DTCs after the diagnostics are completed. Repeat the self-test to verify all DTCs are cleared.

Note: EGR valve override control with the engine running is 75% maximum duty cycle. With the KOEO, the duty cycle can be controlled to 100%.

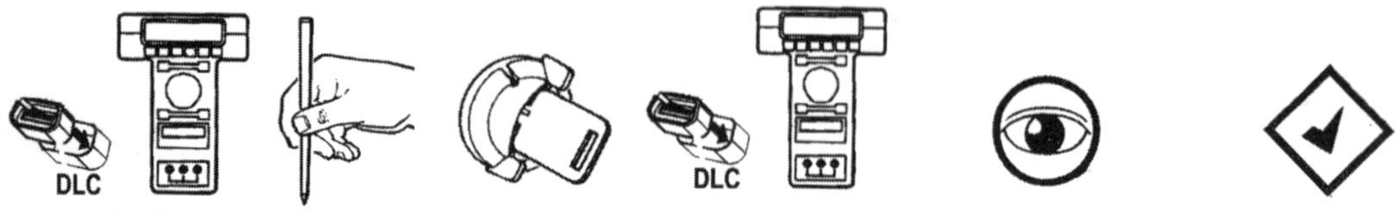

Access EP, EGRVPDES, EGRDC, EGRVP PIDS.

Record Readings.

Start Engine.

Access The PCM And Control The EGR DC PID.

Monitor The EP PID While Increasing And Decreasing The EGRDC.

Does EP Reading Indicate EGR Valve Movement?

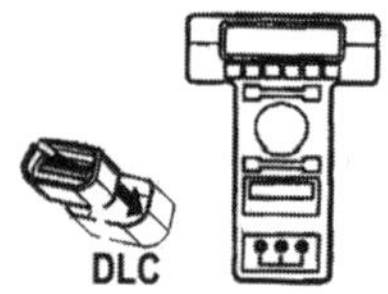

YES — CLEAR DTCs And RETEST The System. If Other DTCs Are Present, GO To The Appropriate Pinpoint Test. If A Drivability Concern Is Present, REFER To Diagnostic Subroutines, Performance Testing Of EGR System. STOP

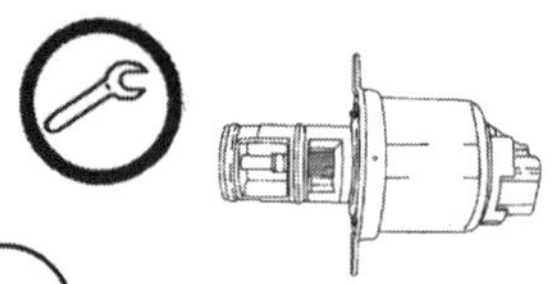

NO — INSTALL A New EGR Valve. RESTORE The Vehicle. CLEAR DTCs And RETEST The System. STOP

W4 — DIAGNOSTIC TROUBLE CODE (DTC) P0403

Note: Refer To The PCM Connector Pin Numbers At The Beginning Of This Pinpoint Test.

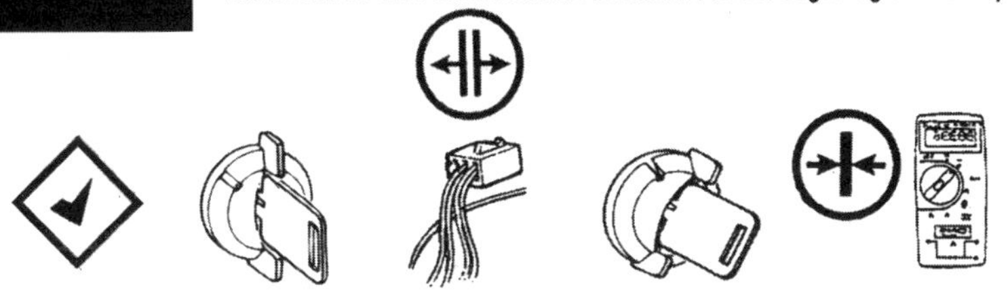

Possible Causes:
- Open In Control Circuit
- Control Circuit Shorted To Voltage
- Control Circuit Shorted To Ground
- EGR Actuator
- PCM

Key Off.

Disconnect The EGR Actuator Electrical Harness Connector.

Key On, Engine Off.

Measure the voltage between:	
(+) EGR Actuator Connector, Harness Side	(-)
VPWR - Pin E	Ground

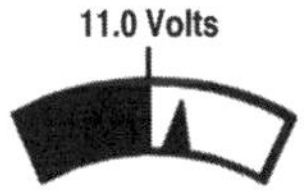

Is The Voltage Above 11.0 Volts?

YES → W5

NO →

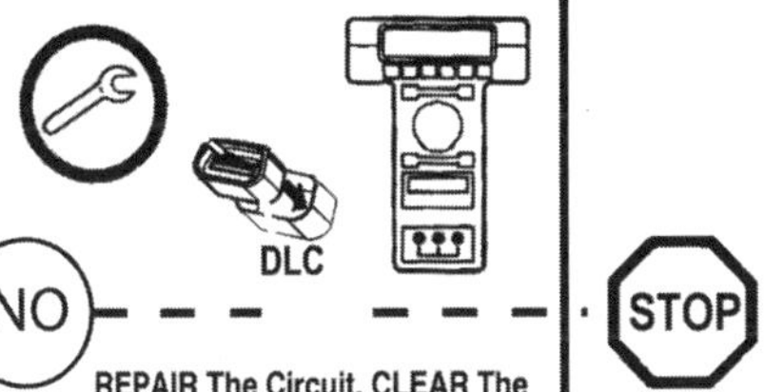

REPAIR The Circuit. CLEAR The DTCs. REPEAT The Self-Test.

W5 — CHECK EGR ACTUATOR CIRCUIT FOR A SHORT TO VOLTAGE

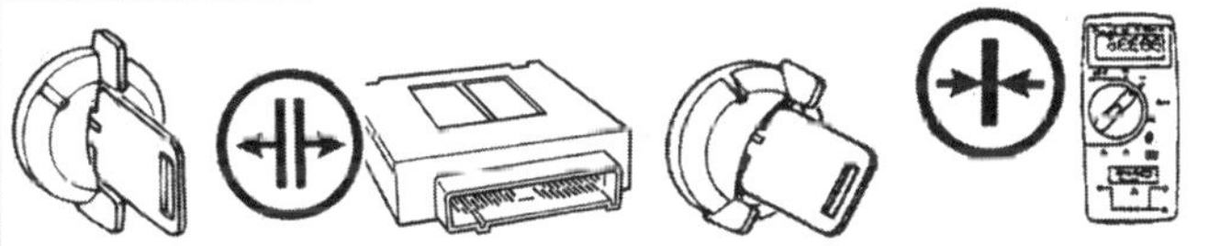

Key Off.

Disconnect The PCM Engine Harness Connector.

Key On. Engine Off.

Measure the voltage between:	
(+) EGR Actuator Connector, Harness Side	(-)
EGRMC - Pin A	Ground

Is The Voltage Greater Than 0.2 Volts?

YES → W6

NO →

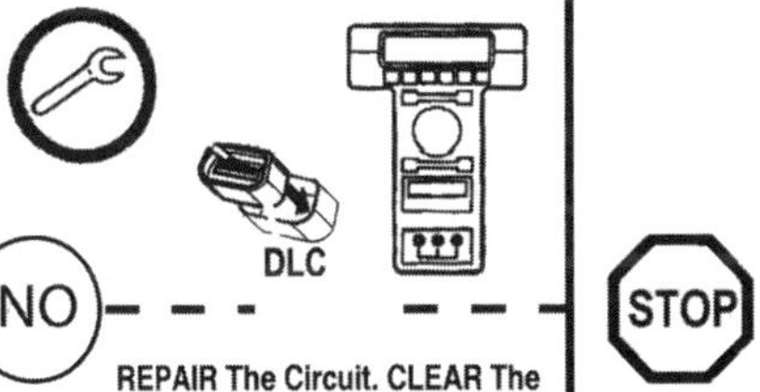

REPAIR The Circuit. CLEAR The DTCs. REPEAT The Self-Test.

W6 CHECK THE EGR ACTUATOR CIRCUIT FOR A SHORT TO GROUND

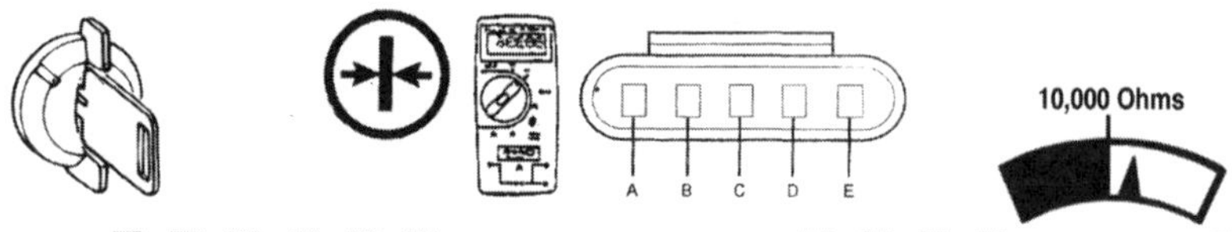

Key Off.

Measure the resistance between:	
(+) EGR Actuator Connector, Harness Side	(-)
EGRMC - Pin A	Ground

Is The Resistance Greater Than 10,000 Ohms?

YES → W7

NO → REPAIR The Circuit. CLEAR The DTCs. REPEAT The Self-Test. STOP

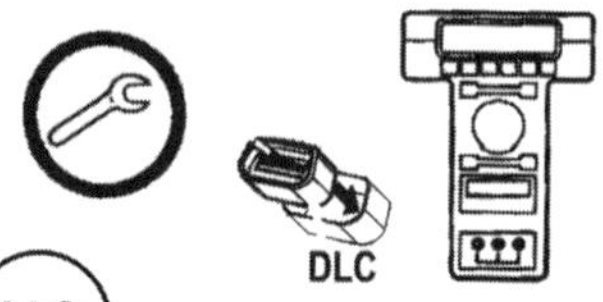

W7 CHECK EGR CONTROL CIRCUIT FOR AN OPEN

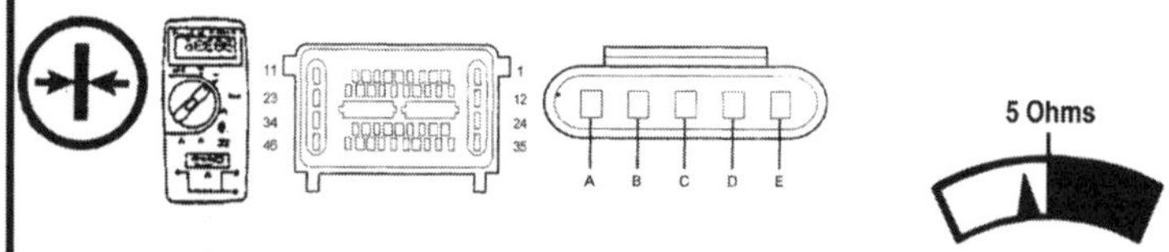

Measure the resistance between:	
(+) EGR Actuator Connector, Harness Side	(-) PCM Connector, Harness Side
EGRMC - Pin A	EGRMC - Pin E23

Is The Resistance Less Than 5 Ohms?

YES → W8

NO → REPAIR The Circuit. CLEAR The DTCs. REPEAT The Self-Test. STOP

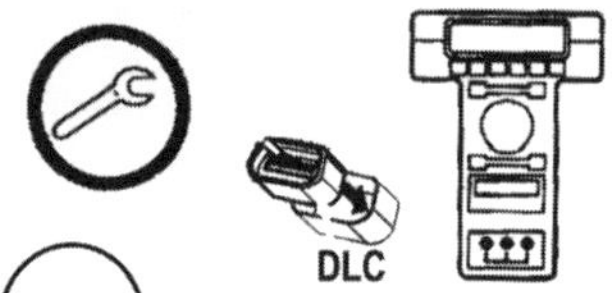

W8 CHECK EGR ACTUATOR

Measure the resistance between:	
(+) EGR Actuator Connector, Component Side	(-) EGR Actuator Connector, Component Side
EGRMC - Pin A	VPWR - Pin E

Is The Resistance Between 4.0 And 10.0 Ohms?

YES → W17

NO → INSTALL A New EGR Actuator. CLEAR The DTCs. REPEAT The Self-Test. STOP

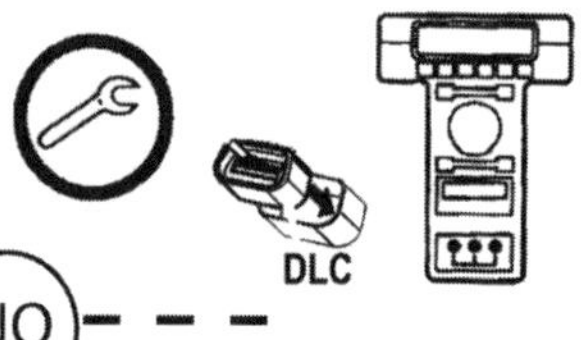

W9

DIAGNOSTIC TROUBLE CODE (DTC) P0404 OR P1335

Note: DTC P1335 Is Set When The EGR Closed Position Voltage Is Greater Than 1.20 Volts At Initial Key On.

Possible Causes:
- EGR Valve
- PCM

Are Other DTCs Present?

YES

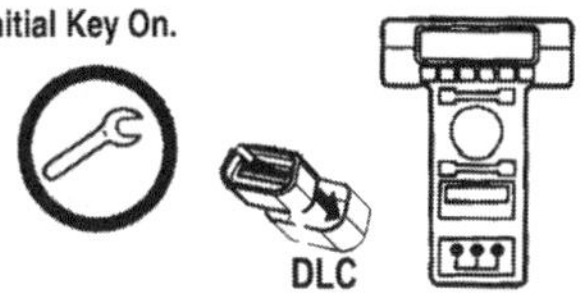

REPAIR Other DTCS. REFER To DTC Index.

DTC INDEX

NO

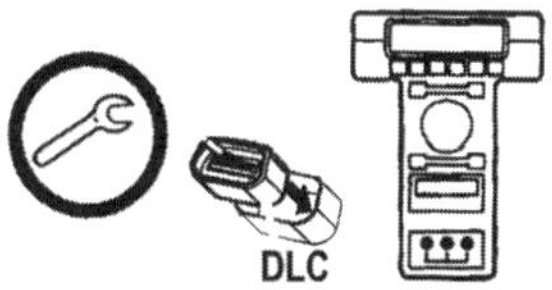

REFER To Diagnostic Subroutines, Performance Diagnostic Procedures. PERFORM The EGR Position Test. RESTORE The Vehicle. CLEAR DTCs. REPEAT The Self-Test.

Diagnostic Subroutines

W10

DIAGNOSTIC TROUBLE CODE (DTC) P0405

Note: Refer To The PCM And Component Connector Pin Numbers At The Beginning Of This Pinpoint Test.

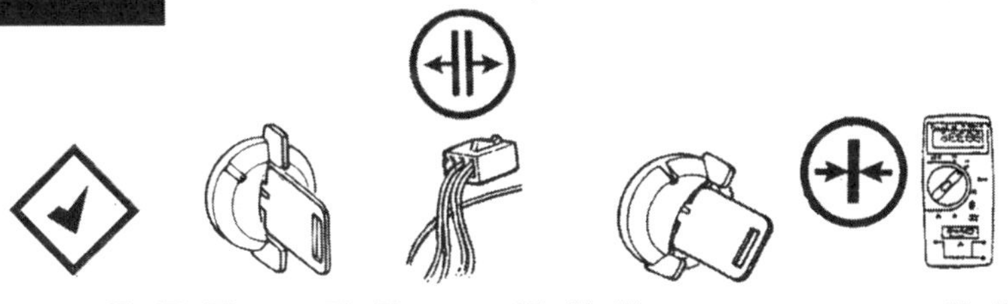

Possible Causes:
- Position Signal Circuit Shorted To Ground
- Position Signal Circuit Open
- Actuator
- PCM

Key Off.

Disconnect The EGR Actuator Valve Electrical Connector.

Key On. Engine Off.

Measure the voltage between:	
(+) EGR Actuator Connector, Harness Side	(-) EGR Actuator Connector, Harness Side
VREF - Pin D	SIGRTN - Pin B

4.5-5.5 Volts

Is The Voltage Between 4.5-5.5 Volts?

YES

W11

NO

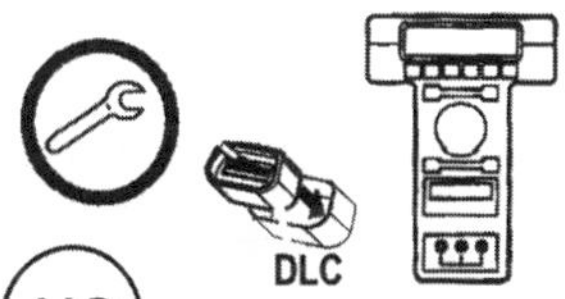

REPAIR The Circuit. CLEAR The DTCs. REPEAT The Self-Test.

STOP

W11 CHECK EGR POSITION SIGNAL CIRCUIT FOR A SHORT TO GROUND

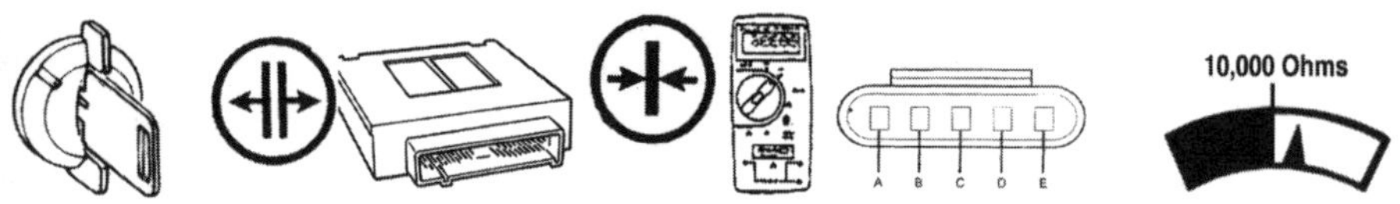

Key Off.

Disconnect The PCM Engine Harness Connector.

Measure the resistance between:	
(+) EGR Actuator Connector, Harness Side	(-)
EGR - Pin C	Ground

Is The Resistance Greater Than 10,000 Ohms?

YES → W12

NO → STOP

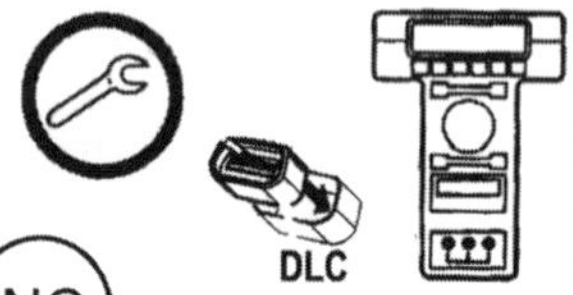

REPAIR The Circuit. CLEAR The DTCs. REPEAT The Self-Test.

W12 CHECK EGR POSITION SIGNAL CIRCUIT FOR AN OPEN

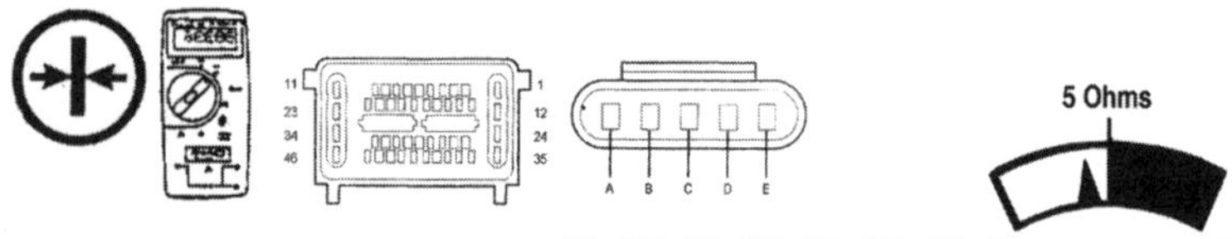

Measure the resistance between:	
(+) PCM Connector, Harness Side	(-) EGR Actuator Connector, Harness Side
EGR - Pin E33	EGR - Pin C

Is The Resistance Less Than 5 Ohms?

YES → W13

NO → STOP

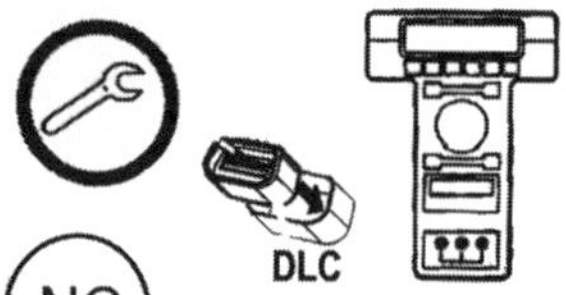

REPAIR The Circuit. CLEAR The DTCs. REPEAT The Self-Test.

W13 CHECK EGR ACTUATOR

Measure the resistance between the EGR actuator pins, component side as follows:

EGR Actuator Connector			
Pins	Pins	Minimum Resistance	Maximum Resistance
B	C	50 Ohms	2,500 Ohms
C	D	2,000 Ohms	7,000 Ohms
B	D	3,000 Ohms	7,000 Ohms

Are The Resistances Within Specifications?

YES → W17

NO → STOP

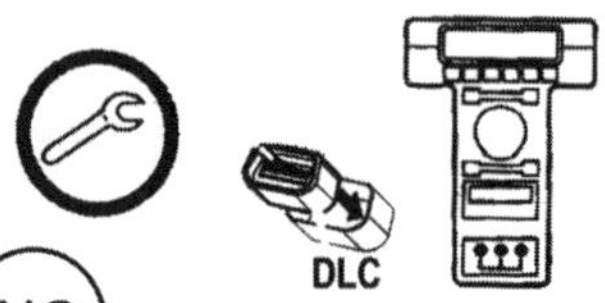

INSTALL A New EGR Actuator. CLEAR The DTCs. REPEAT The Self-Test.

W14 DIAGNOSTIC TROUBLE CODE (DTC) P0406

Note: Refer To The PCM And Component Connector Pin Numbers At The Beginning Of This Pinpoint Test.

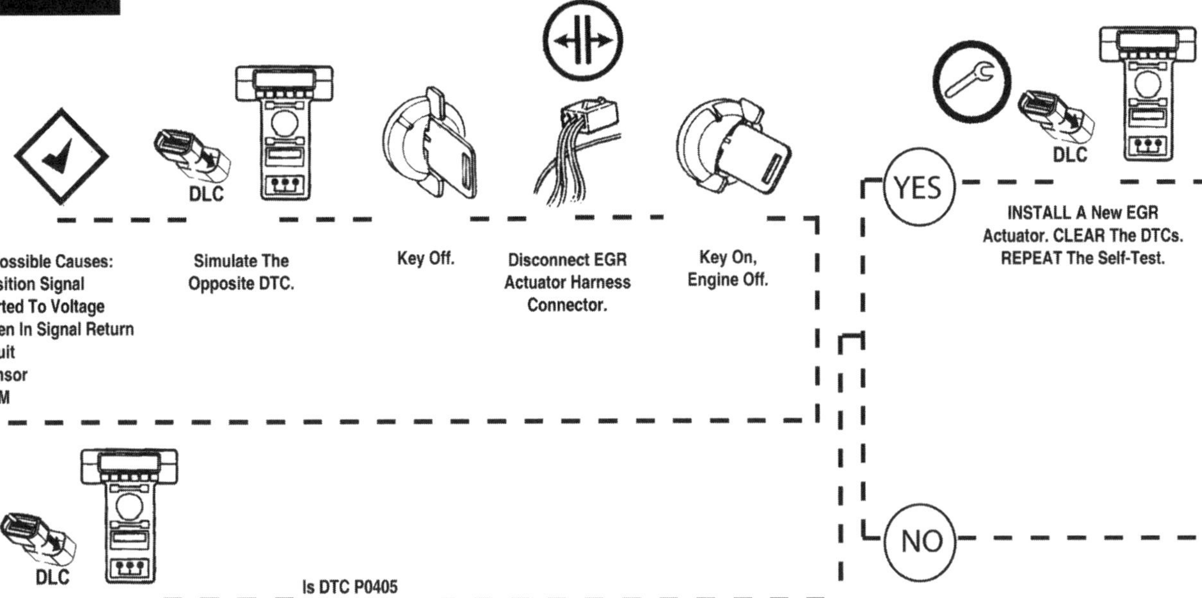

Possible Causes:
- Position Signal Shorted To Voltage
- Open In Signal Return Circuit
- Sensor
- PCM

Simulate The Opposite DTC.

Key Off.

Disconnect EGR Actuator Harness Connector.

Key On, Engine Off.

Perform On-Demand Self Test.

Is DTC P0405 Present?

YES — INSTALL A New EGR Actuator. CLEAR The DTCs. REPEAT The Self-Test. STOP

NO — W15

W15 CHECK EGR POSITION SIGNAL CIRCUIT FOR A SHORT TO VOLTAGE

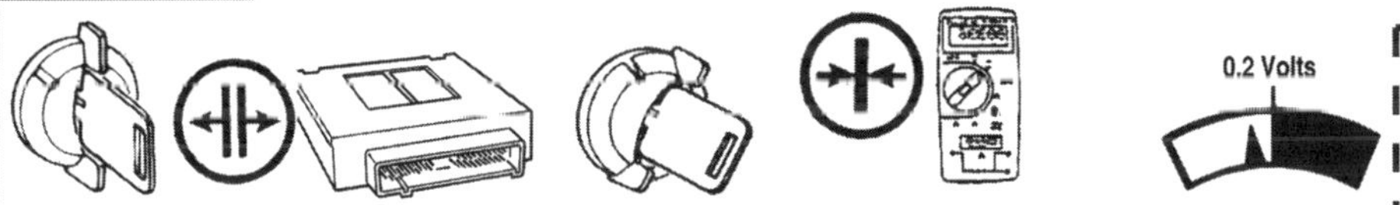

Key Off.

Disconnect The PCM Engine Harness Connector.

Key On, Engine Off.

Measure the voltage between:	
(+) EGR Actuator Connector, Harness Side	(-)
EGR - Pin C	Ground

Is The Voltage Less Than 0.2 Volts?

YES — W16

NO — REPAIR The Circuit. CLEAR The DTCs. REPEAT The Self-Test. STOP

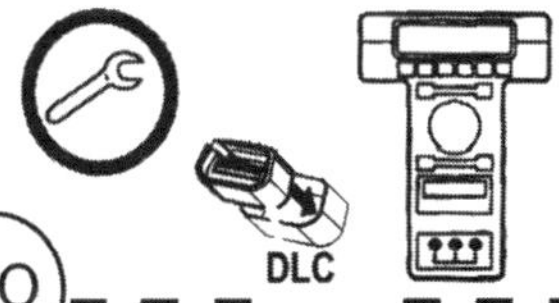

W16 CHECK EGR POSITION SIGNAL CIRCUIT SIGNAL RETURN FOR AN OPEN

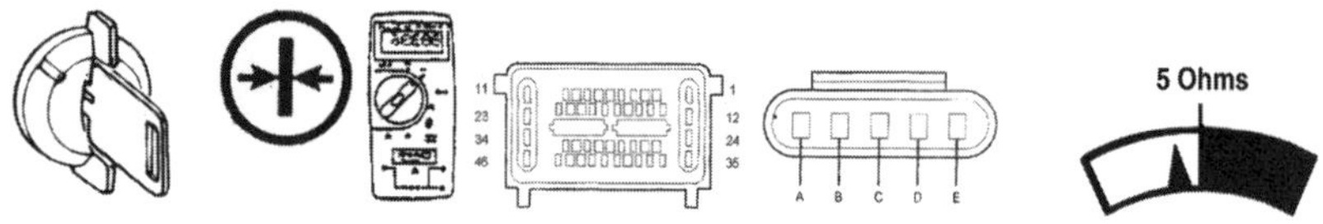

Key Off.

Measure the resistance between:	
(+) EGR Actuator Connector, Harness Side	(-) PCM Connector, Harness Side
SIGRTN - Pin B	SIGRTN - Pin B20, B33, E25

Is The Resistance Less Than 5 Ohms?

YES → W17

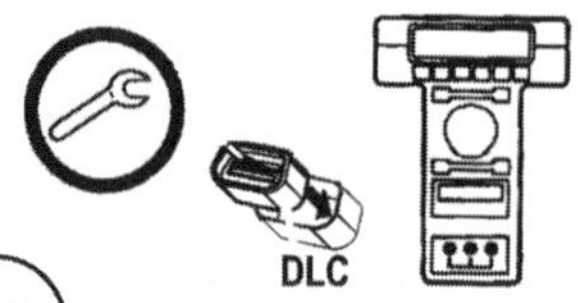

NO → STOP
REPAIR The Circuit. CLEAR The DTCs. REPEAT The Self-Test.

W17 CHECK FOR CORRECT PCM OPERATION

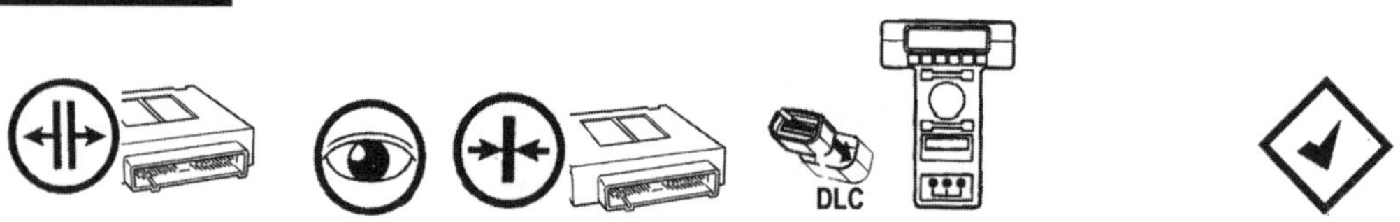

Disconnect All PCM Connectors.

Visually Inspect For Pushed Out Pins, Corrosion.

Connect All PCM Connectors And Make Sure They Seat Correctly.

Perform The PCM Self-Test And Verify The Concern Is Still Present.

Is The Concern Still Present.

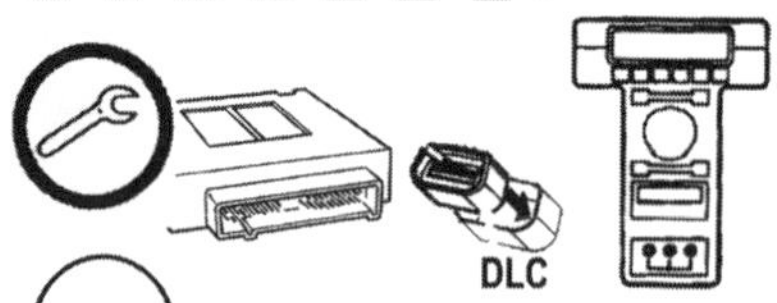

YES → STOP
INSTALL A New PCM. CLEAR The DTC's And REPEAT The Self-Test.

OK

NO → STOP
The System Is Operating Correctly At This Time. The Concern May Have Been Caused By A Loose Or Corroded Connector.

Trouble Code (s)	Problem: Exhaust Gas Recirculation (EGR)	Test
P0401, P0402, P0403, P0405, P0406, P1335		W 2007

REFERENCE LIST

This pinpoint test is intended to diagnose the following:

- Harness circuits: VPWR, VREF, EGR, EGRMC and SIGRTN
- EGR actuator
- Powertrain control module (PCM)

DIAGNOSTIC TROUBLE CODE INDEX			
Codes	KOEO	KOER	CM
P0401	W1	W1	W1
P0402	W1	W1	W1
P0403	W1	W1	W1
P0404	W1	W1	W1
P0405	W1	W1	W1
P0406	W1	W1	W1
P1335	W1	W1	W1

P0401 = EGR Flow Insufficient Detected
P0402 = EGR Flow Excessive Detected
P0403 = EGR Control Circuits
P0405 = EGR Sensor A Circuit Low
P0406 = EGR Sensor A Circuit High
P1335 = EGR Position Sensor Minimum Stop Performance

Circuit Function:
The EGR actuator is a variable position valve that controls the amount of exhaust that enters the intake manifold. The EGR is controlled by the powertrain control module (PCM) using a pulse width modulated signal that varies from 0-100%. The EGR actuator consists of 2 components, a valve with an actuator (solenoid), and a position sensor to monitor valve movement. The EGR sensor is a potentiometer sensor. When the EGR receives a 5-volt reference signal and a ground from the PCM, a linear analog voltage signal from the sensor indicates the position of the EGR valve.

Fault Detection/Management:
The PCM detects an open or short in the EGR control circuit after the key on engine off (KOEO) test has run and set a diagnostic trouble code (DTC). The EGR actuator can be tested for a stuck or sticking valve by carrying out the output state low test while monitoring the exhaust gas recirculation position (EGRP) percentage.

CAUTION!:
The powertrain control module (PCM) harness connectors must be properly seated and the connector latch properly attached to eliminate possible driveability concerns or a no start condition. Installing the PCM connectors on an angle may cause an improper connection, misdiagnosis and damaged components. Install the connector until the lever pivots and seats itself. Apply light pressure to get the connector into position on the PCM and then fully seat the connector.

Note: Visually inspect the harness connectors for corrosion, damage, proper mating and correct pin tension.

PINPOINT TEST SCHEMATICS AND CONNECTORS

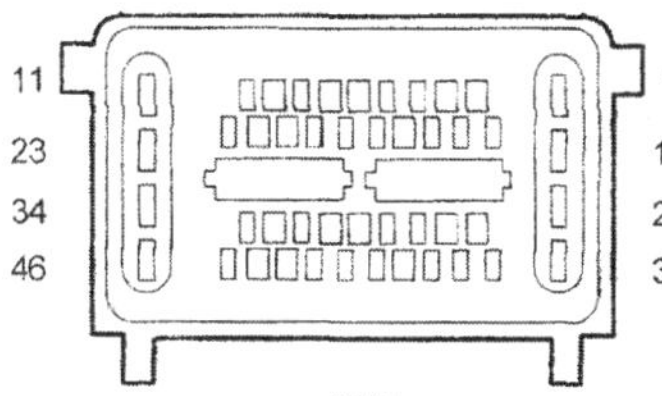

PCM
ENGINE HARNESS
CONNECTOR

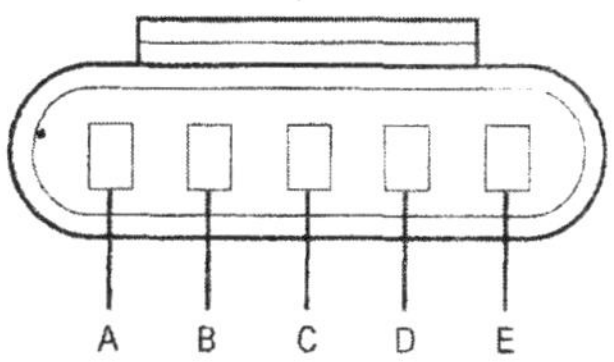

EGR
VALVE ACTUATOR
CONNECTOR

Pin	Circuit
B20, B33, E25	SIGRTN (Signal Return)
E33	EGR (Exhaust Gas Recirculation)
E23	EGRMC (EGR Motor Control)

Pin	Circuit
A	EGRMC (EGR Motor Control)
B	SIGRTN (Signal Return)
C	EGR (Exhaust Gas Recirculation)
D	VREF (Reference Voltage)
E	VPWR (Vehicle Power)

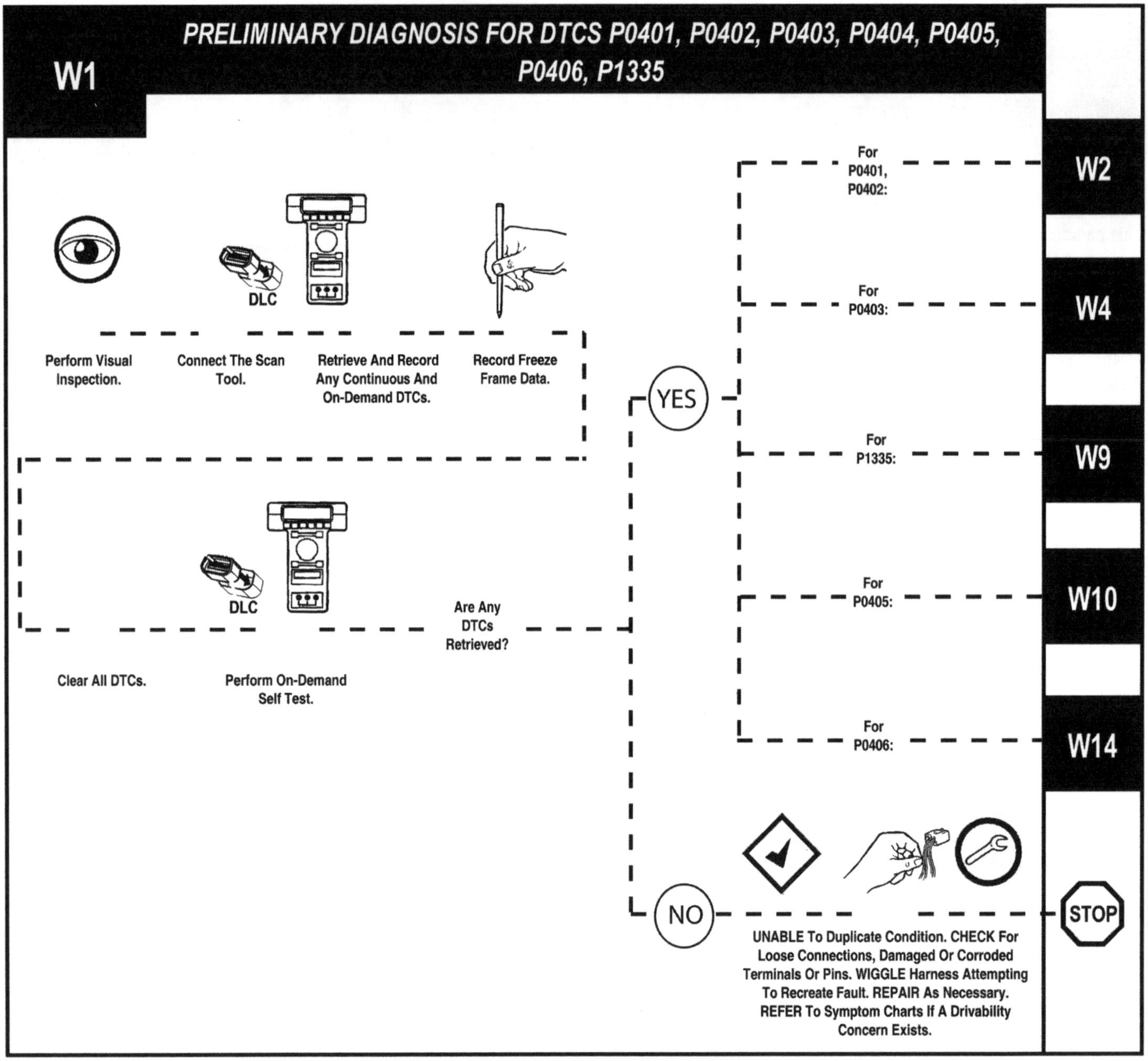
W1
PRELIMINARY DIAGNOSIS FOR DTCS P0401, P0402, P0403, P0404, P0405, P0406, P1335
DLC
Perform Visual Inspection.
Connect The Scan Tool.
Retrieve And Record Any Continuous And On-Demand DTCs.
Record Freeze Frame Data.
DLC
Clear All DTCs.
Perform On-Demand Self Test.
Are Any DTCs Retrieved?
YES
For P0401, P0402:
W2
For P0403:
W4
For P1335:
W9
For P0405:
W10
For P0406:
W14
NO
UNABLE To Duplicate Condition. CHECK For Loose Connections, Damaged Or Corroded Terminals Or Pins. WIGGLE Harness Attempting To Recreate Fault. REPAIR As Necessary. REFER To Symptom Charts If A Drivability Concern Exists.
STOP

W2

DTC'S P0401, P0402: CHECK FOR OTHER DTC'S

Note: Visually Inspect The Vehicle For Aftermarket Accessories And Performance Modifications (Air Filter, Exhaust Sstem Intake System, Performance Chip, Turbocharger). Refer To Diesel Electronic Engine Control (EEC) System Modifications To OBD Vehicles.

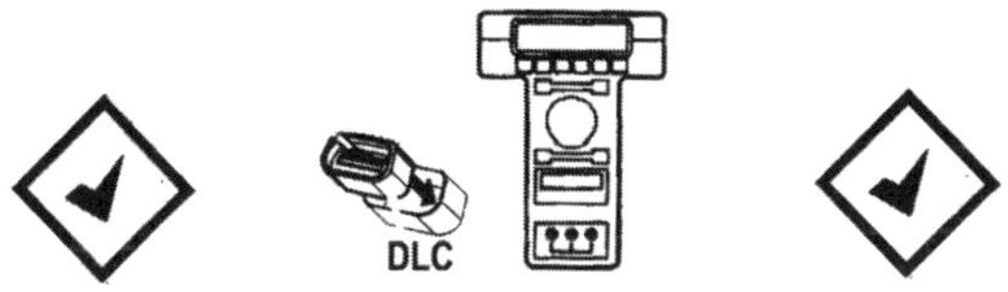

Possible Causes:
- Restricted Air Flow (Intake Or Exhaust)
- Charge Air Cooler (CAC) System Leaks (Hoses/Cooler)
- EGR Valve Stuck Or Sticking EGR Valve O-Rings
- Restricted EGR Cooler
- EGR Valve Position Sensor Bias
- EP Sensor Bias

Perform On-Demand Self Test.

Are Other DTCs Present?

YES → DTC INDEX

NO → W3

W3 CHECK THE EGR ACTUATOR OPERATION

Note: Additional DTCs may set during the use of output state control (OSC). Clear all PCM and transmission control module (TCM) DTCs after the diagnostics are completed. Repeat the self-test to verify all DTCs are cleared.

Note: EGR valve override control with the engine running is 75% maximum duty cycle. With the KOEO, the duty cycle can be controlled to 100%.

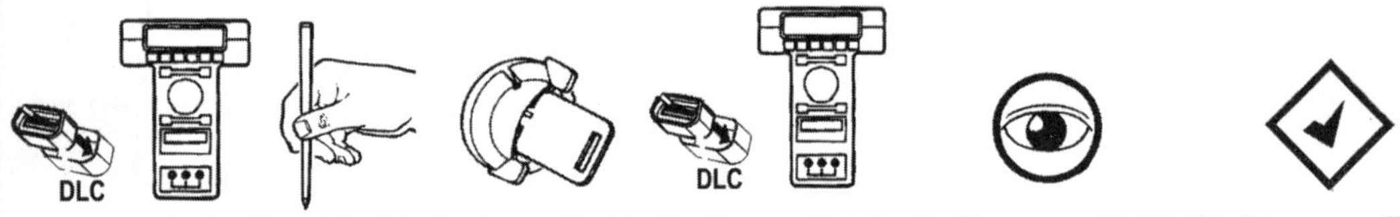

Access EP, EGRVPDES, EGRDC, EGRVP PIDS.

Record Readings.

Start Engine.

Access The PCM And Control The EGR DC PID.

Monitor The EP PID While Increasing And Decreasing The EGRDC.

Does EP Reading Indicate EGR Valve Movement?

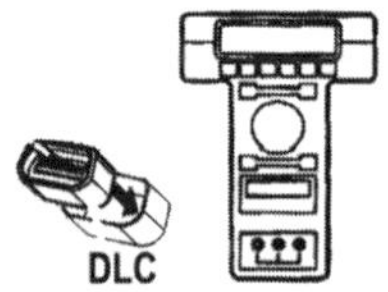

YES — CLEAR DTCs And RETEST The System. If Other DTCs Are Present, GO To The Appropriate Pinpoint Test. If A Drivability Concern Is Present, REFER To Diagnostic Subroutines, Performance Testing Of EGR System. STOP

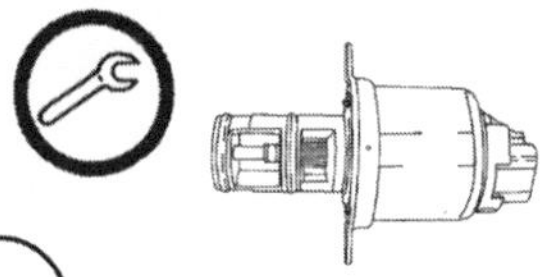

NO — INSTALL A New EGR Valve. RESTORE The Vehicle. CLEAR DTCs And RETEST The System. RESET The Keep Alive Memory (KAM). STOP

W4 DTC P0403: CHECK THE VPWR CIRCUIT FOR VOLTAGE

Note: Refer To The PCM Connector Pin Numbers At The Beginning Of This Pinpoint Test.

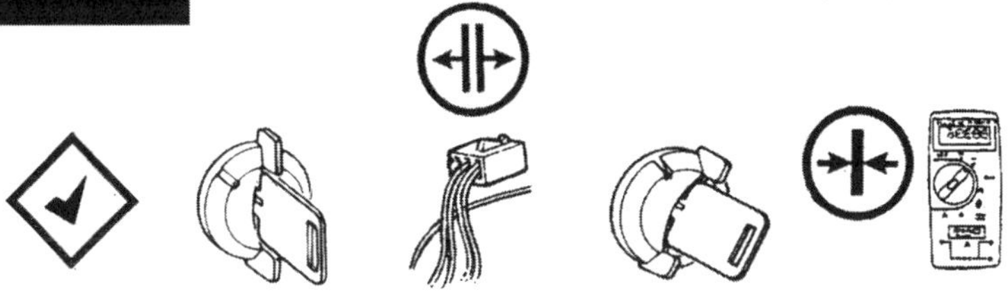

Possible Causes:
- Open In Control Circuit
- Control Circuit Shorted To Voltage
- Control Circuit Shorted To Ground
- EGR Actuator
- PCM

Key Off.

Disconnect The EGR Actuator Electrical Harness Connector.

Key On, Engine Off.

Measure the voltage between:	
(+) EGR Actuator Connector, Harness Side	(-)
VPWR - Pin E	Ground

Is The Voltage Above 11.0 Volts?

YES → W5

NO →

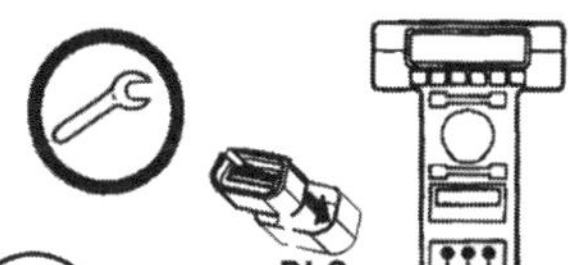

REPAIR The Circuit. CLEAR The DTCs. REPEAT The Self-Test.

W5 CHECK EGRMC CIRCUIT FOR A SHORT TO VOLTAGE

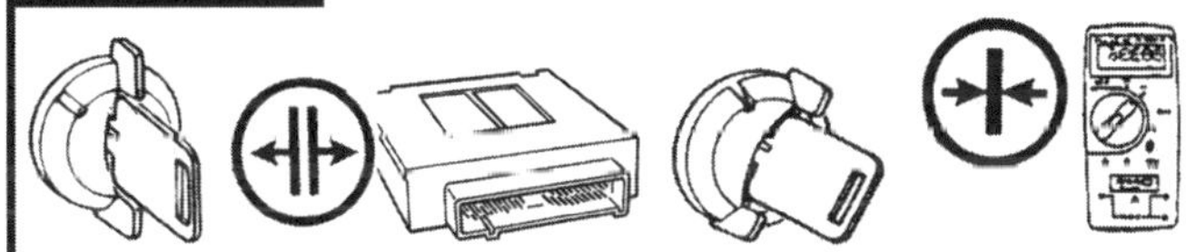

Key Off.

Disconnect The PCM Engine Harness Connector.

Key On. Engine Off.

Measure the voltage between:	
(+) EGR Actuator Connector, Harness Side	(-)
EGRMC - Pin A	Ground

Is The Voltage Greater Than 0.2 Volts?

YES → W6

NO →

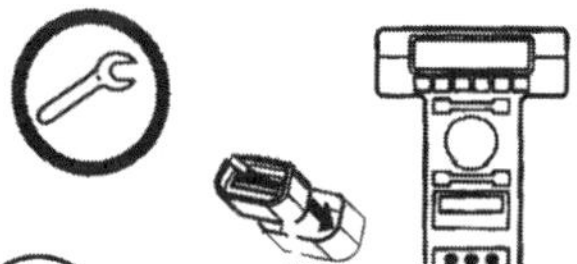

REPAIR The Circuit. CLEAR The DTCs. REPEAT The Self-Test.

W6 CHECK THE EGRMC CIRCUIT FOR A SHORT TO GROUND

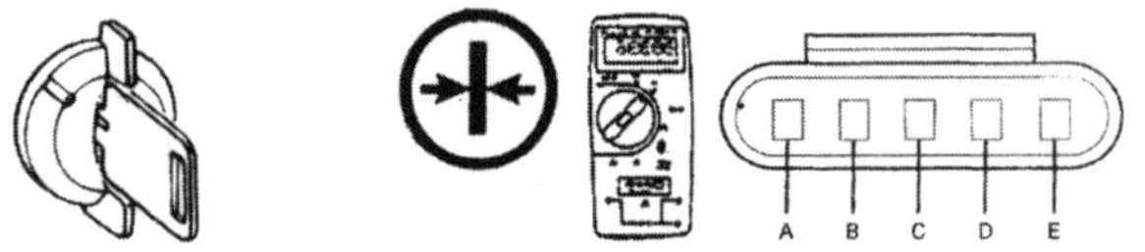

Key Off.

Measure the resistance between:	
(+) EGR Actuator Connector, Harness Side	(-)
EGRMC - Pin A	Ground

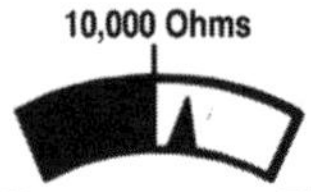

Is The Resistance Greater Than 10,000 Ohms?

YES → W7

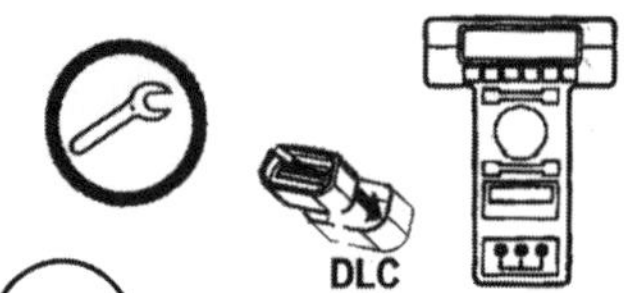

NO → REPAIR The Circuit. CLEAR The DTCs. REPEAT The Self-Test. STOP

W7 CHECK EGRMC CIRCUIT FOR AN OPEN

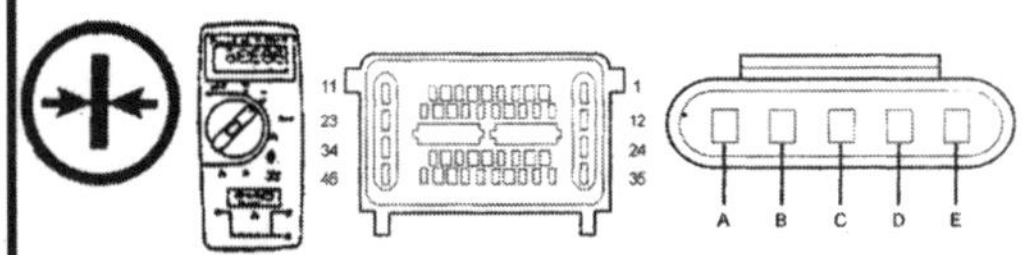

Measure the resistance between:	
(+) EGR Actuator Connector, Harness Side	(-) PCM Connector, Harness Side
EGRMC - Pin A	EGRMC - Pin E23

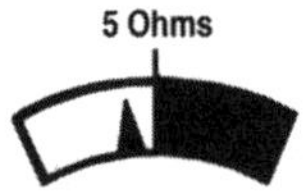

Is The Resistance Less Than 5 Ohms?

YES → W8

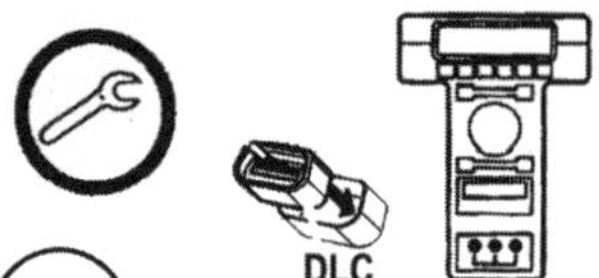

NO → REPAIR The Circuit. CLEAR The DTCs. REPEAT The Self-Test. STOP

W8 CHECK EGR ACTUATOR

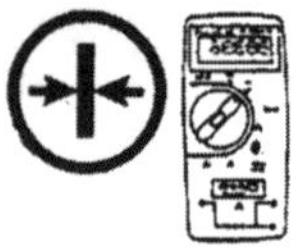

Measure the resistance between:	
(+) EGR Actuator Connector, Component Side	(-) EGR Actuator Connector, Component Side
EGRMC - Pin A	VPWR - Pin E

Is The Resistance Between 4.0 And 10.0 Ohms?

YES → W17

NO → INSTALL A New EGR Actuator. RESET The Keep Alive Memory (KAM). CLEAR The DTCs. REPEAT The Self-Test. STOP

W9 — DTC P1335: CHECK THE EGRVP PID VOLTAGE

Note: DTC P1335 Is Set When The EGR Closed Position Voltage Is Greater Than 1.20 Volts At Initial Key On.

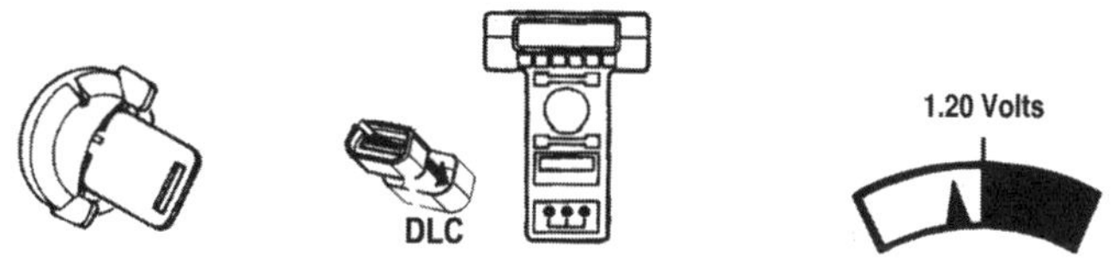

Key ON, Engine OFF.

Access The PCM And Monitor The EGRVPID

Is The EGRVP PID Voltage Under 1.20 Volts?

YES

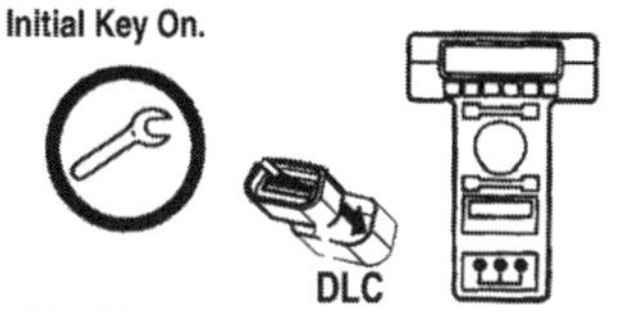

The System Is Operating Correctly At This Time. The Concern May Have Been Caused By Carbon Buildup On The EGR Valve That Has Since Been Removed. Clear The DTC's. REPEAT The Self Test. — STOP

NO

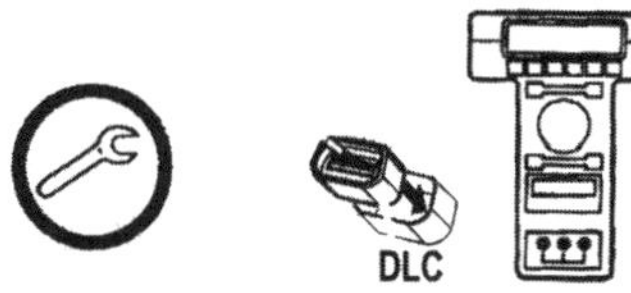

INSTALL A New EGR Actuator. RESET The Keep Alive Memory (KAM). CLEAR The DTCs. REPEAT The Self-Test. — STOP

W10 — DTC P0405: CHECK THE VREF CIRCUIT FOR VOLTAGE

Note: Refer To The PCM And Component Connector Pin Numbers At The Beginning Of This Pinpoint Test.

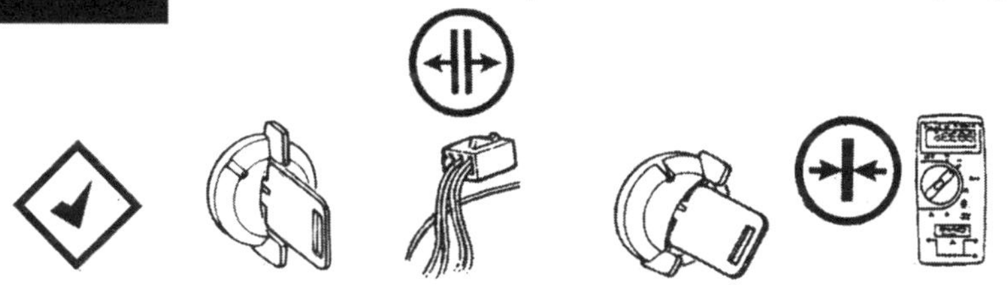

Possible Causes:
- Position Signal Circuit Shorted To Ground
- Position Signal Circuit Open
- Actuator
- PCM

Key Off.

Disconnect The EGR Actuator Valve Electrical Connector.

Key On. Engine Off.

Measure the voltage between:	
(+) EGR Actuator Connector, Harness Side	(-) EGR Actuator Connector, Harness Side
VREF - Pin D	SIGRTN - Pin B

Is The Voltage Between 4.5-5.5 Volts?

YES — W11

NO

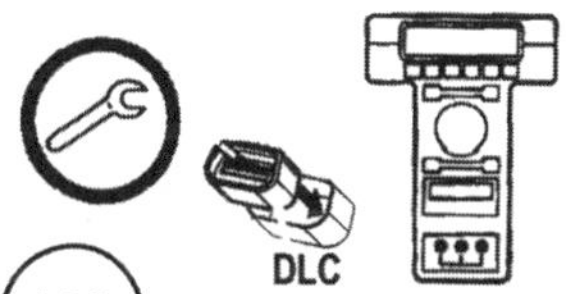

REPAIR The Circuit. CLEAR The DTCs. REPEAT The Self-Test. — STOP

W11 — CHECK THE EGR CIRCUIT FOR A SHORT TO GROUND

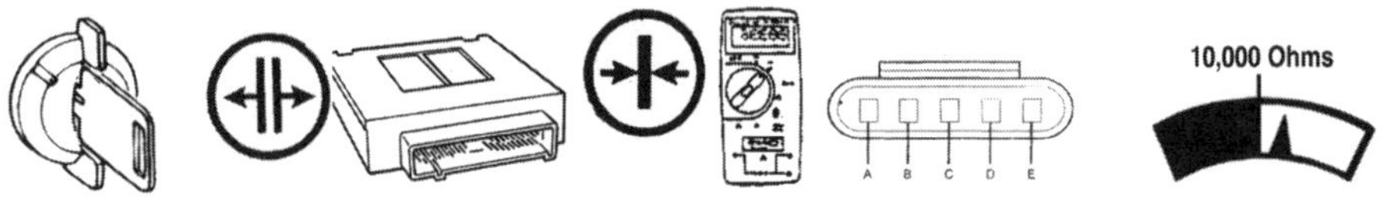

Key Off.

Disconnect The PCM Engine Harness Connector.

Measure the resistance between:	
(+) EGR Actuator Connector, Harness Side	(-)
EGR - Pin C	Ground

Is The Resistance Greater Than 10,000 Ohms?

YES → W12

NO →

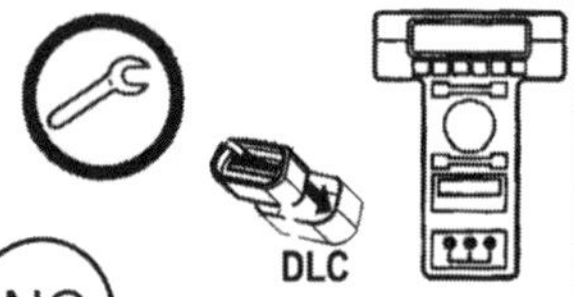

REPAIR The Circuit. CLEAR The DTCs. REPEAT The Self-Test. STOP

W12 — CHECK THE EGR CIRCUIT FOR AN OPEN

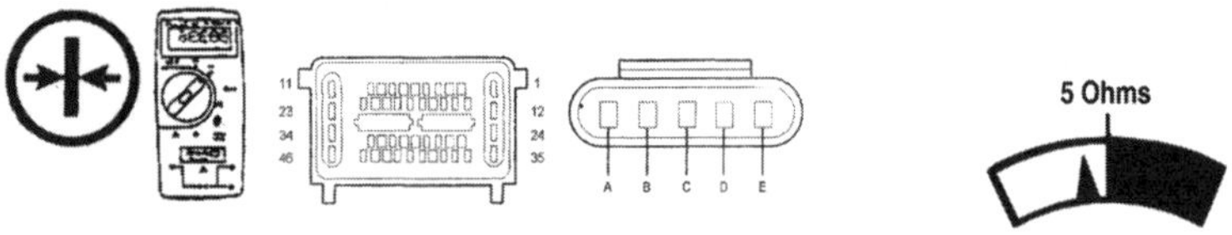

Measure the resistance between:	
(+) PCM Connector, Harness Side	(-) EGR Actuator Connector, Harness Side
EGR - Pin E33	EGR - Pin C

Is The Resistance Less Than 5 Ohms?

YES → W13

NO →

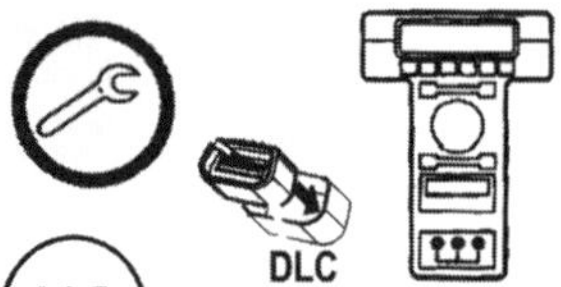

REPAIR The Circuit. CLEAR The DTCs. REPEAT The Self-Test. STOP

W13 — CHECK EGR ACTUATOR

Measure the resistance between the EGR actuator pins, component side as follows:

EGR Actuator Connector			
Pins	Pins	Minimum Resistance	Maximum Resistance
B	C	50 Ohms	2,500 Ohms
C	D	2,000 Ohms	7,000 Ohms
B	D	3,000 Ohms	7,000 Ohms

Are The Resistances Within Specifications?

YES → W17

NO →

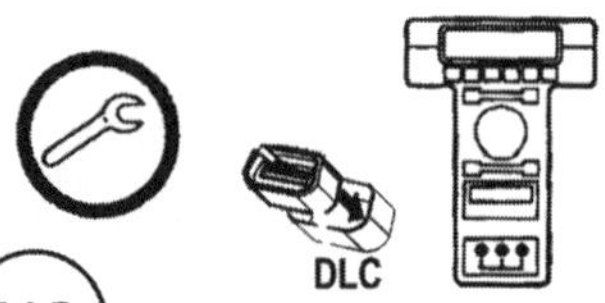

INSTALL A New EGR Actuator. RESET The Keep Alive Memory (KAM). CLEAR The DTCs. REPEAT The Self-Test. STOP

W14 — DTC P0406: CHECK FOR DTC P0405

Note: Refer To The PCM And Component Connector Pin Numbers At The Beginning Of This Pinpoint Test.

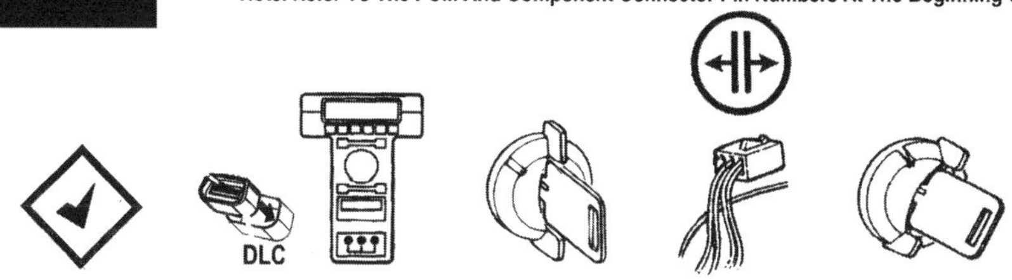

Possible Causes:
- Position Signal Shorted To Voltage
- Open In Signal Return Circuit
- Sensor
- PCM

Simulate The Opposite DTC.

Key Off.

Disconnect EGR Actuator Harness Connector.

Key On, Engine Off.

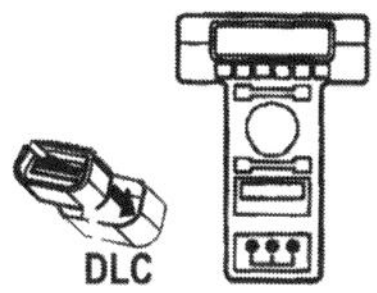

Perform On-Demand Self Test.

Is DTC P0405 Present?

YES — 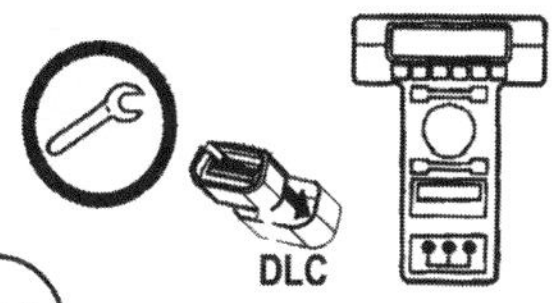

INSTALL A New EGR Actuator. RESET The Keep Alive Memory (KAM). CLEAR The DTCs. REPEAT The Self-Test. STOP

NO — W15

W15 — CHECK EGR POSITION SIGNAL CIRCUIT FOR A SHORT TO VOLTAGE

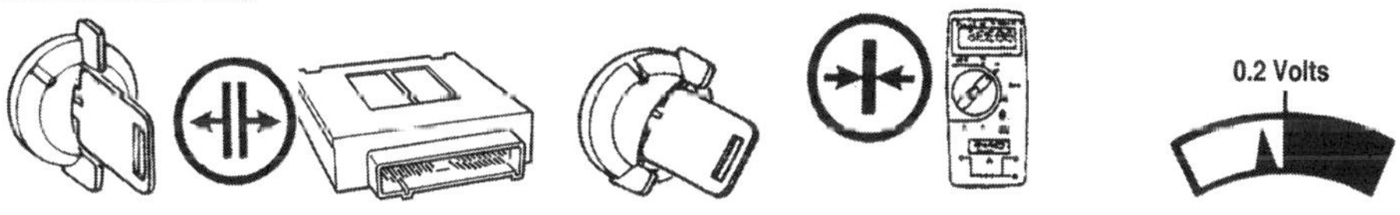

Key Off.

Disconnect The PCM Engine Harness Connector.

Key On, Engine Off.

Measure the voltage between:	
(+) EGR Actuator Connector, Harness Side	(-)
EGR - Pin C	Ground

Is The Voltage Less Than 0.2 Volts?

YES — W16

NO — 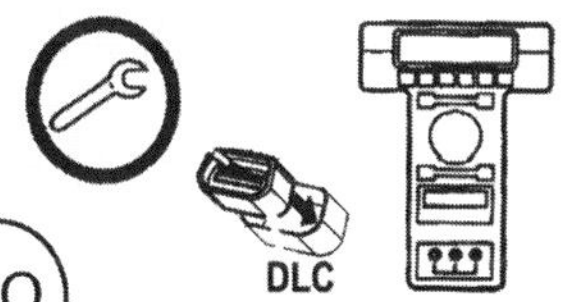

REPAIR The Circuit. CLEAR The DTCs. REPEAT The Self-Test. STOP

W16 CHECK EGR POSITION SIGNAL CIRCUIT SIGNAL RETURN FOR AN OPEN

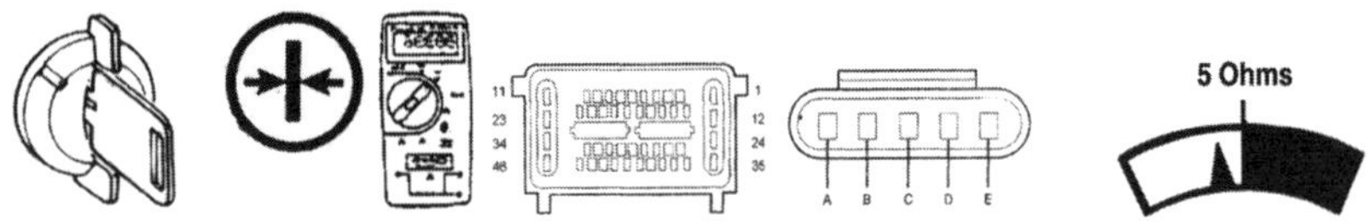

Key Off.

Measure the resistance between:	
(+) EGR Actuator Connector, Harness Side	(-) PCM Connector, Harness Side
SIGRTN - Pin B	SIGRTN - Pin B20, B33, E25

Is The Resistance Less Than 5 Ohms?

YES — W17

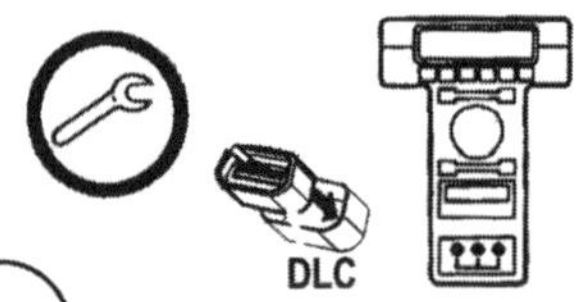

NO — REPAIR The Circuit. CLEAR The DTCs. REPEAT The Self-Test. STOP

W17 CHECK FOR CORRECT PCM OPERATION

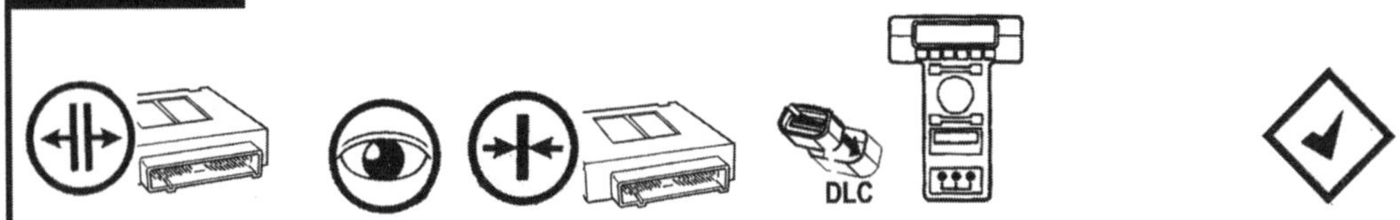

Disconnect All PCM Connectors.

Visually Inspect For Pushed Out Pins, Corrosion.

Connect All PCM Connectors And Make Sure They Seat Correctly.

Perform The PCM Self-Test And Verify The Concern Is Still Present.

Is The Concern Still Present.

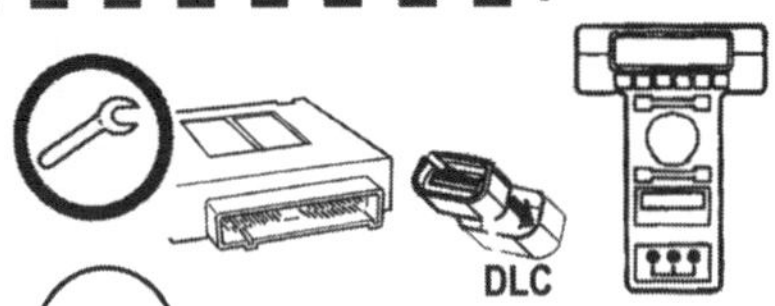

YES — INSTALL A New PCM. CLEAR The DTC's And REPEAT The Self-Test. STOP

NO — The System Is Operating Correctly At This Time. The Concern May Have Been Caused By A Loose Or Corroded Connector. STOP

Trouble Code (s)	Problem: Exhaust Pressure (EP) Sensor	Test
P0470, P0471, P0472, P0473		X 2006-07

REFERENCE LIST

This pinpoint test is intended to diagnose the following:

- Harness circuits: EP, SIGRTN, VREF
- EP sensor
- Powertrain control module (PCM) (12A650)

kPa	PSI	EP (Volts)
43.8	6.4	0.05
70.5	10.2	0.45
101.5	14.7	0.90
325.5	47.2	4.14
365.4	53	4.75

DIAGNOSTIC TROUBLE CODE INDEX			
CODES	KOEO	KOER	CM
P0470	X1	X1	X1
P0471	X1	X1	X1
P0472	X1	X1	X1
P0473	X1	X1	X1

P0470 = Exhaust Pressure Sensor
P0471 = Exhaust Pressure Sensor Range/Performance
P0472 = Exhaust Pressure Sensor Low Input
P0473 = Exhaust Pressure Sensor High Input

Circuit Function:
The exhaust pressure (EP) sensor is a variable capacitance sensor. A 5-volt reference signal from the powertrain control module (PCM) is utilized by the EP sensor to produce a linear analog voltage signal indicating exhaust pressure. The PCM uses the EP sensor input to control the variable geometry turbocharger (VGT) and the exhaust gas recirculation (EGR) systems.

Fault Detection:
The PCM will ignore an out-of-range high or low EP signal. If an out-of-range signal is detected, the EGR system will close and the PCM will default the VGT to calculated reference values.

CAUTION!:
The powertrain control module (PCM) harness connectors must be properly seated and the connector latch properly attached to eliminate possible driveability concerns or a no start condition. Installing the PCM connectors on an angle may cause an improper connection, misdiagnosis and damaged components. Install the connector until the lever pivots and seats itself. Apply light pressure to get the connector into position on the PCM and then fully seat the connector.

Note: Visually inspect the harness connectors for corrosion, damage, proper mating and correct pin tension.

Note: When the PCM is disconnected additional DTCs will be set. Clear all DTCs after restoring the vehicle.

PINPOINT TEST SCHEMATICS AND CONNECTORS

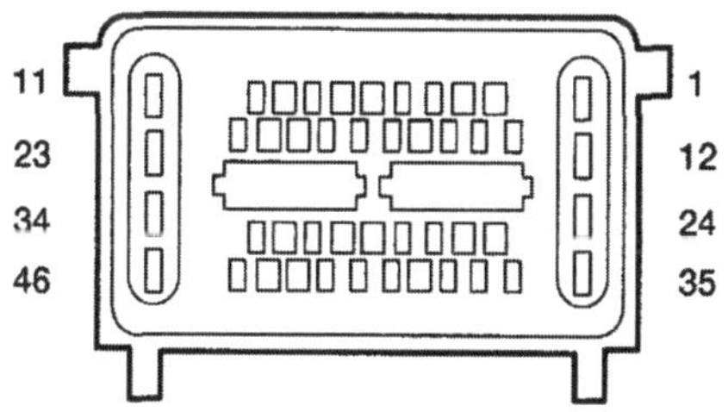

PCM
ENGINE HARNESS
CONNECTOR

Pin	Circuit
E27	EP (Exhaust Pressure)
B20, B33, E25	SIGRTN (Signal Return)

SENSOR
CONNECTOR

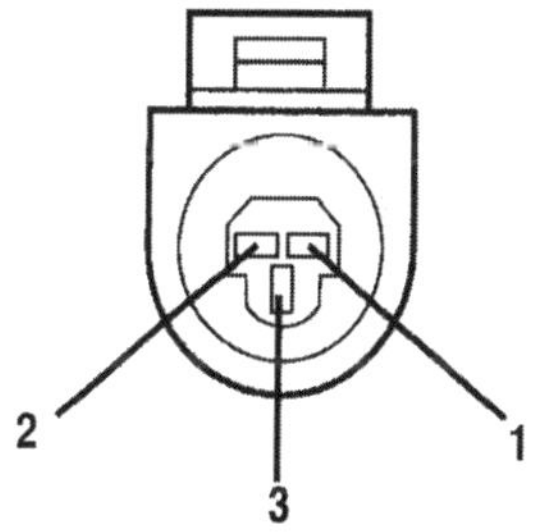

Pin	Circuit
1	SIGRTN (Signal Return)
2	VREF (Reference Voltage)
3	EP (Exhaust Pressure)

X1 — PRELIMINARY DIAGNOSIS FOR DTCS P0470, P0471, P0472 OR P0473

Note: Diagnose And Repair EP Sensor Circuit Fault DTCs Before Addressing Range/Performance Concerns.

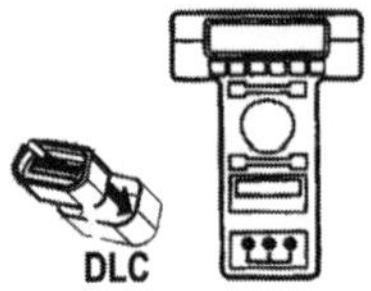

Perform A Visual Inspection.

Connect The Scan Tool.

Retrieve And Record All DTCs.

Record The Freeze Frame Data.

Clear The DTCs.

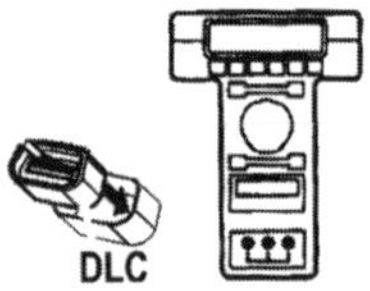

Perform The Key On Engine Off (KOEO) Self-Test.

Perform The Key On Engine Running (KOER) Self-Test.

Are DTCs P0470, P0471, P0472 Or P0473 Present?

YES:
- For KOEO DTC P0470: → X2
- For Continuous DTC P0470: → X7
- For DTC P0471: → X9
- For DTC P0472: → X13
- For DTC P0473: → X17

NO → X20

X2 — DIAGNOSTIC TROUBLE CODE (DTC) P0470

Possible Causes:
- EP Sensor
- SIG RTN Circuit Open Or Shorted To Ground
- PCM

Key On, Engine Off.

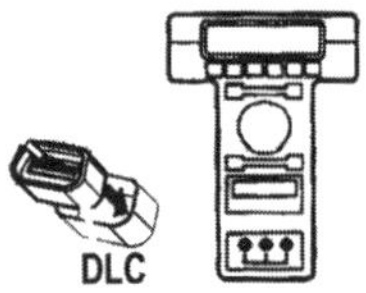

Access The EP PID.

Is The EP PID Less Than 127 KPa (18.5 Psi)?

YES → X3

NO → X4

X3 — CHECK FOR KOEO DTC P0470

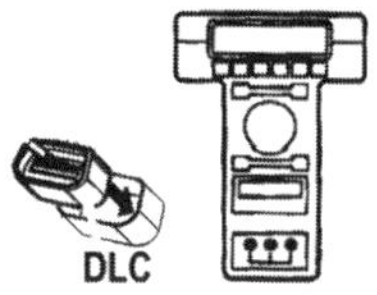

Clear The DTCs.

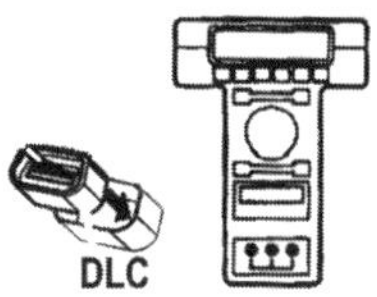

Repeat The KOEO On-Demand Self-Test.

Is DTC P0470 present?

YES → X13

NO → STOP

UNABLE To Duplicate Concern. Check For A Loose Connection, And Damaged Or Corroded Terminals. WIGGLE the Harness Attempting To Recreate The Fault. REPAIR As Necessary And RETEST The System. REFER To Symptom Charts If A Driveablility Concern Exists.

X4 — CHECK THE EP SENSOR PRESSURE

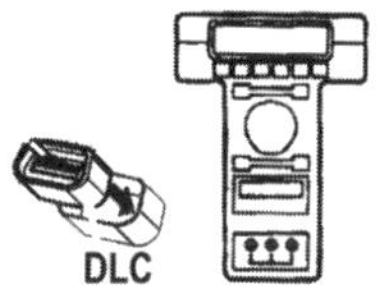

Access The EP, MAP, And BARO PIDs.

Within 10.34 kPa (1.5 psi)

Is the EP PID Value Within 10.34 kPa (1.5 psi) Of The MAP And BARO PID Values?

YES → X6

NO → X5

X5 — CHECK THE SIGNAL RETURN CIRCUIT

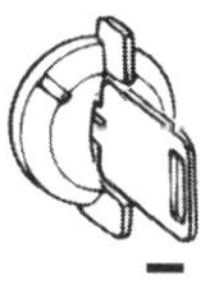

Key Off.

Disconnect The EP Sensor.

Disconnect The PCM Engine Harness Connector.

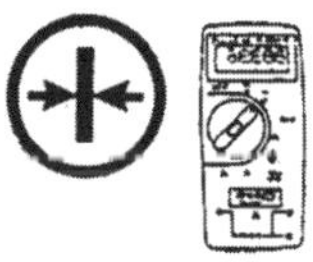

Measure the resistance between:	
(+) EP Sensor Connector, Harness Side	(-) PCM Connector, Harness Side
SIGRTN - Pin 1	SIGRTN - Pin B20, B33, E25

5 Ohms

Is The Resistance Less Than 5 Ohms?

YES → STOP

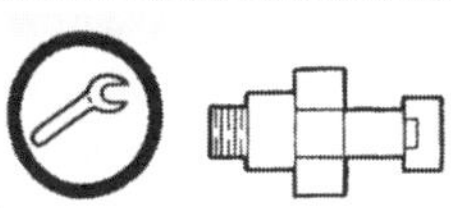

INSTALL A New EP Sensor. CLEAR The DTCs. REPEAT The Self-Test.

NO → STOP

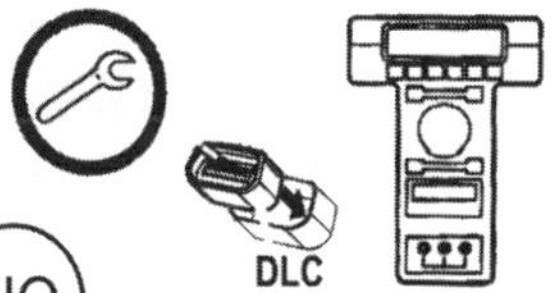

REPAIR The Circuit. CLEAR The DTCs. REPEAT The Self-Test.

X6 — CHECK THE EP SENSOR

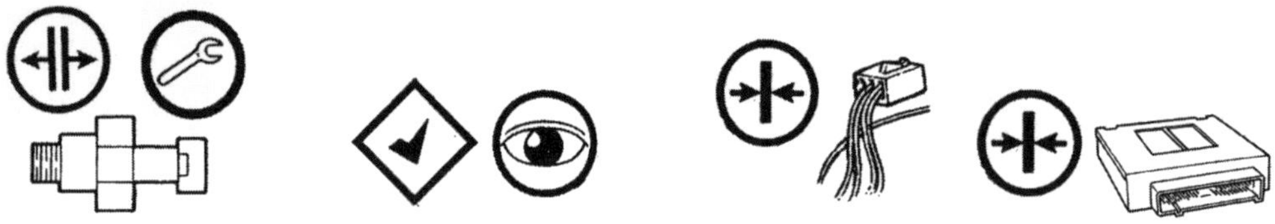

1. Remove The EP Sensor.
2. Check For Contamination, Restrictions Or Carbon Build-up In The EP Sensor Port, Mounting Bracket Or Tube. Repair As Necessary.

   Check For Moisture In The EP Sensor Port. If Moisture Is Present, Use 69 kPa (10 psi) Regulated Air Pressure To Remove The Moisture.
3. EP Sensor Connector Connected.
4. PCM Connector Connected.

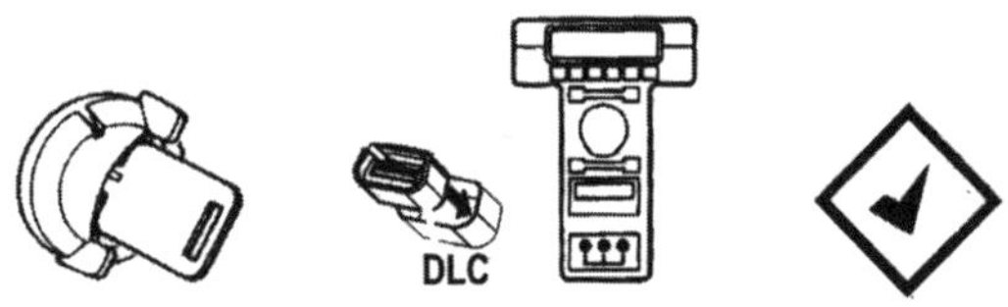

5. Key ON, Engine OFF.
6. Access the PCM and monitor the EP, MAP and BARO PIDs.
7. Is The EP PID Value Within 10.34 kPa (1.5 psi) Of The MAP And BARO PID Values?

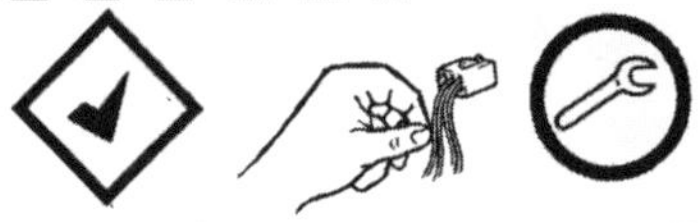

YES → UNABLE To Duplicate Concern. ROAD TEST Vehicle, Check For A Loose Connection, And Damaged Or Corroded Terminals. WIGGLE the Harness Attempting To Recreate The Fault. REPAIR As Necessary And RETEST The System. REFER To Symptom Charts If A Driveablility Concern Exists. **STOP**

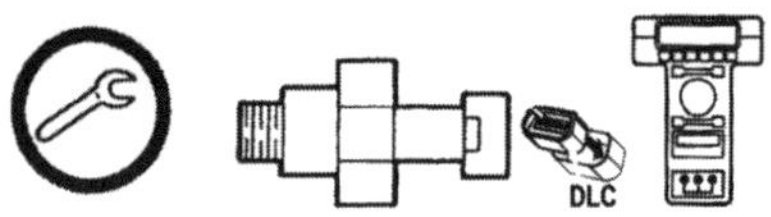

NO → INSTALL A New EP Sensor. CLEAR The DTCs. REPEAT The Self-Test. **STOP**

X7 — CONTINUOUS DIAGNOSTIC TROUBLE CODE (DTC) P0470

Possible Causes:
- EP Sensor
- PCM

1. Key On, Engine Off.
2. Access The EP PID.
3. Observe The EP PID While Carrying Out The Following:
   - Grasp The Vehicle Harness Close To The EP Sensor Connector.
   - Wiggle And Shake The Vehicle Harness While Working Toward The PCM.
4. Does The EP PID Reading Fluctuate?

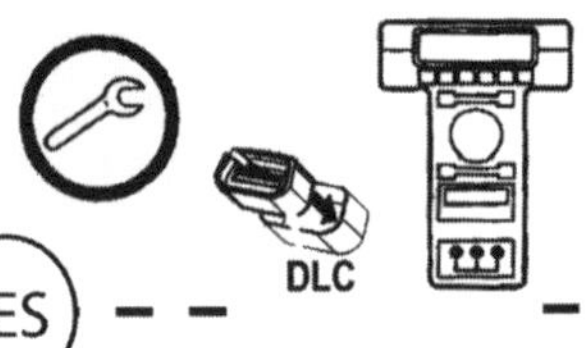

YES → ISOLATE The Intermittent Connection. REPAIR As Necessary. CLEAR The DTCs. REPEAT The Self-Test. **STOP**

NO → **X8**

X8 INSPECT THE EP SENSOR AND PCM CONNECTORS

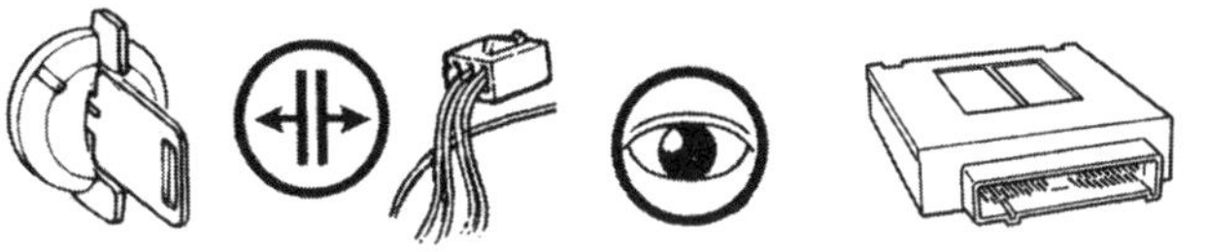

Key Off. — Disconnect The EP Sensor. — Inspect The EP Sensor Connector And Terminals. — Inspect The PCM Connectors. — Are All The Connections OK?

YES

Unable To Duplicate Condition CHECK For A Loose Connection, And Damaged Or Corroced Terminals. WIGGLE The Harness Attempting To Recreate The Fault. REPEAR As Necessary. REFER To Symptom Charts If A Driveability Concern Exists. → STOP

NO

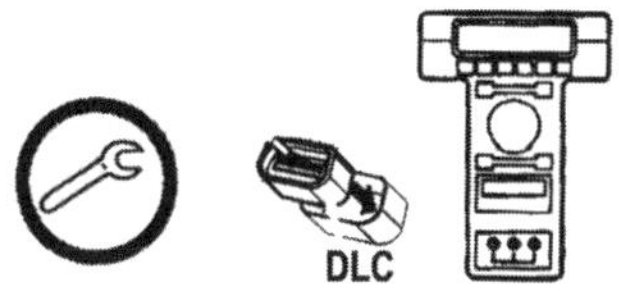

REPAIR As Necessary. CLEAR The DTCs. REPEAT The Self-Test. → STOP

X9 DIAGNOSTIC TROUBLE CODE (DTC) P0471

Note: Visually Inspect The Vehicle For Aftermarket Accessories And Performance Modifications (Exhaust System, Turbocharger, Performance Chip, Etc.). REFER Diesel Electronic Engine Control (EC) System, Modifications To OBD Vehicles.

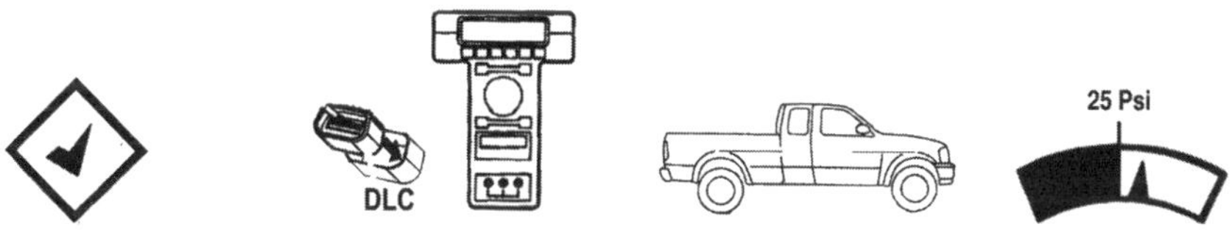

Possible Causes:
- EP Sensor
- EP Sensor Supply Tube Plugged Or Restricted
- PCM

Carry Out The KOEO Self-Test. If Other DTCs Are Present, Repair Before Continuing. — Access The EP PID. — Road Test The Vehicle Performing Hard Accelerations While Monitoring The EP PID. — Is The EP Increase Greater Than 25 Psi?

YES

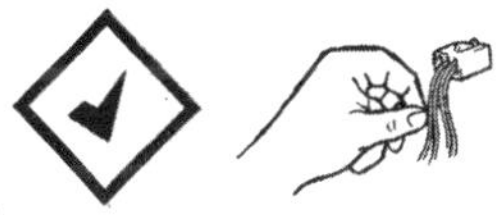

Unable To Duplicate Condition. CHECK For Loose Connections, Damaged Or Corroded Pins. WIGGLE Harness Attempting To Recreate Fault. REFER To Symptom Charts If Drivability Concerns Exist. → STOP

NO → X10

X10 INSPECT FOR EXHAUST LEAKS

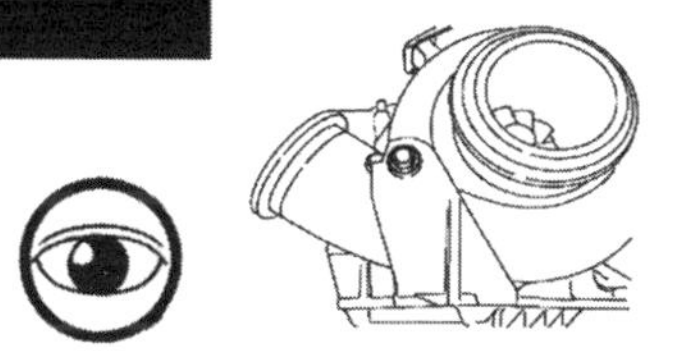

Inspect The Turbo Pipe And Exhaust Manifolds For Leaks.

Are Any Exhaust Leaks Present?

YES

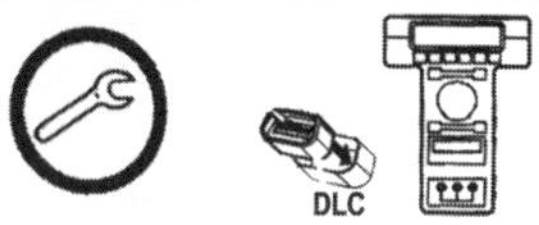

REPAIR The Exhaust Leak. CLEAR The DTCs. REPEAT The Self-Test. → STOP

NO → X11

X11 CHECK THE EP SENSOR SUPPLY TUBE

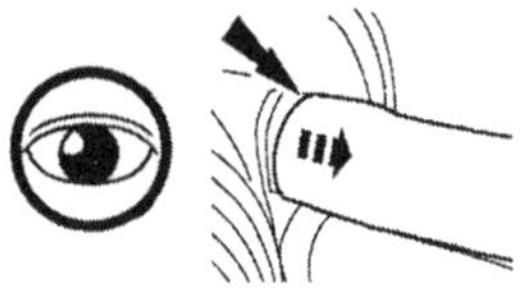

Inspect The EP Sensor Supply Tube.

Is The EP Sensor Tube Restricted Or Plugged?

YES

INSTALL A New EP Sensor Tube. CLEAR The DTCs. REPEAT The Self-Test.

STOP

NO

X12

X12 EP SENSOR CHECK

Disconnect The EP Sensor.

Install ICP/EBC Adapter Cable D94T-50-A Between	
EP Sensor	Vehicle Harness

Key On, Engine Running.

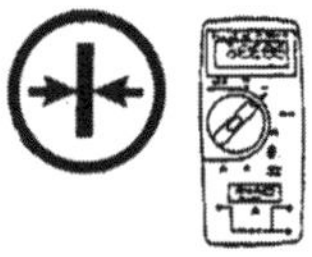

Measure Voltage Between	
EP Sensor	Chassis
Signal Circuit	Ground
ON ICP/EBC Adapter Cable D94T-50-A	

Accelerate The Engine To Wide Open Throttle (WOT) Several Times.

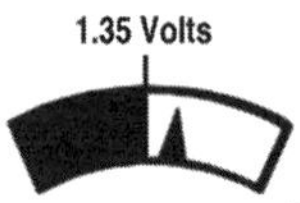

Is The Voltage Greater Than 1.35 Volts During Acceleration?

YES

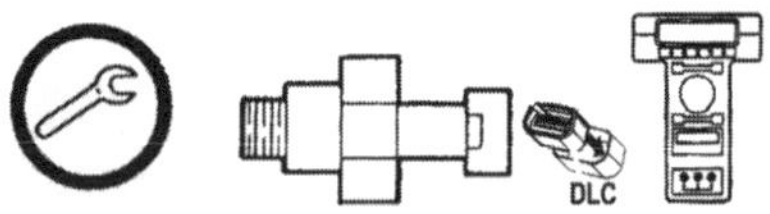

INSTALL A New EP Sensor. CLEAR The DTCs. REPEAT The Self-Test.

If DTC P0471 Is Present Again, GO To X21 .

X21

NO

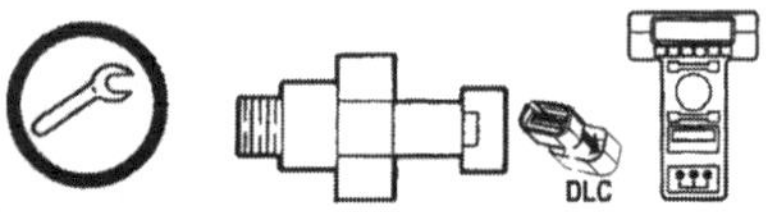

INSTALL A New EP Sensor. CLEAR The DTCs. REPEAT The Self-Test.

STOP

X13 — DIAGNOSTIC TROUBLE CODE (DTC) P0472

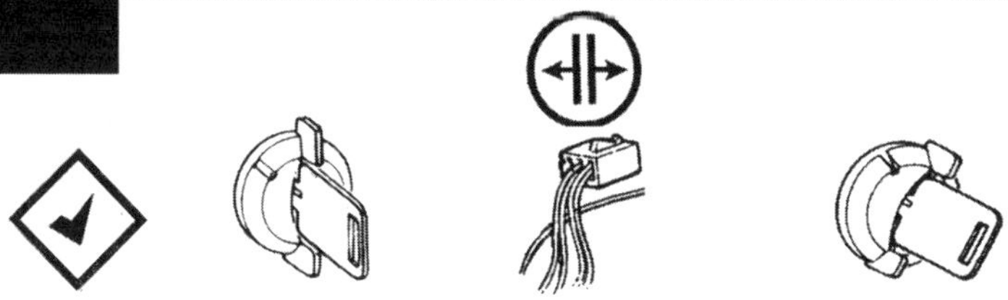

Possible Causes:
- EP Sensor Signal Circuit Open
- Biased Sensor
- EP Sensor Signal Circuit Shorted To Ground
- VREF Circuit Open
- EP Sensor
- PCM

Key Off.

Disconnect The EP Sensor.

Key On, Engine Off.

4.5 And 5.5 Volts

Measure the voltage between:	
(+) EP Sensor Connector, Harness Side	(-)
VREF - Pin 2	Ground

Is The Voltage Between 4.5 And 5.5 Volts?

YES → X14

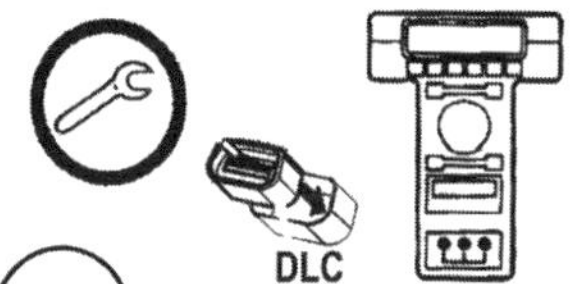

NO → STOP: REPAIR The Circuit. CLEAR The DTCs. REPEAT The Self-Test.

X14 — CHECK THE EP SENSOR SIGNAL

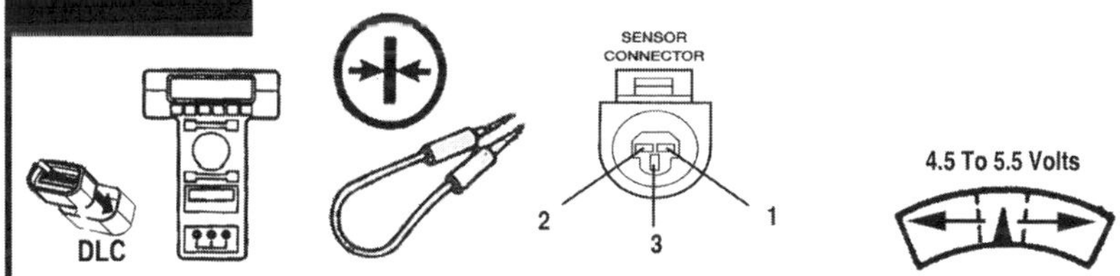

Access The EP PID.

Connect a 5 amp fused jumper wire between the following:	
Point A EP Sensor Connector, Harness Side	Point B EP Sensor Connector, Harness Side
VREF - Pin 2	EP - Pin 3

Is The EP PID Voltage Approximately 4.5 To 5.5 Volts?

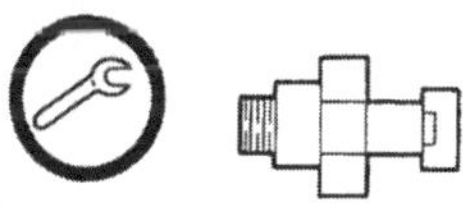

YES → STOP: INSTALL A New EP Sensor. CLEAR The DTCs. REPEAT The Self-Test.

NO → X15

X15 — CHECK THE SIGNAL CIRCUIT FOR AN OPEN

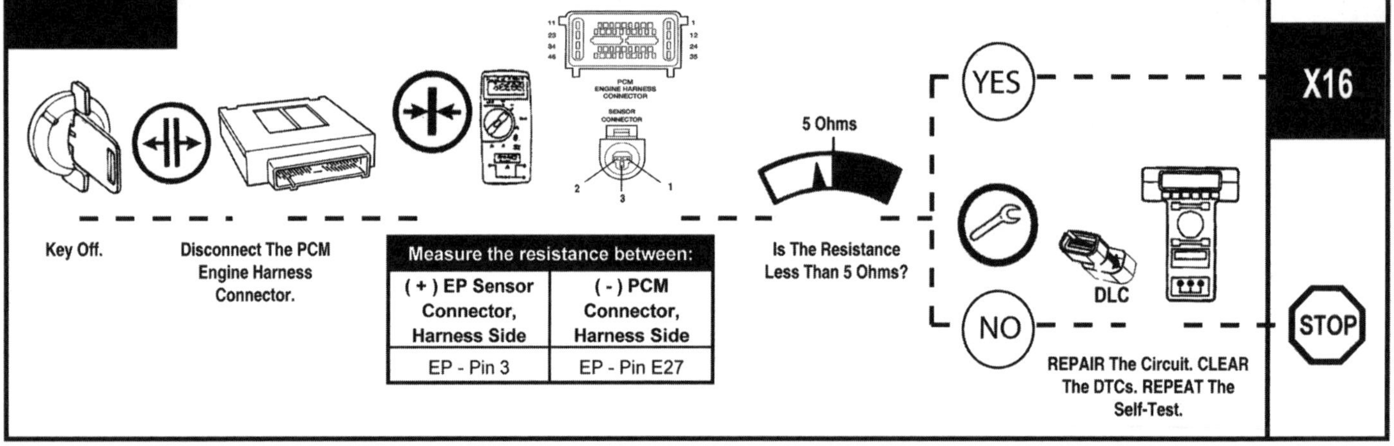

Key Off.

Disconnect The PCM Engine Harness Connector.

Measure the resistance between:	
(+) EP Sensor Connector, Harness Side	(-) PCM Connector, Harness Side
EP - Pin 3	EP - Pin E27

Is The Resistance Less Than 5 Ohms?

YES → X16

NO → REPAIR The Circuit. CLEAR The DTCs. REPEAT The Self-Test. STOP

X16 — CHECK THE SIGNAL CIRCUIT FOR A SHORT TO GROUND

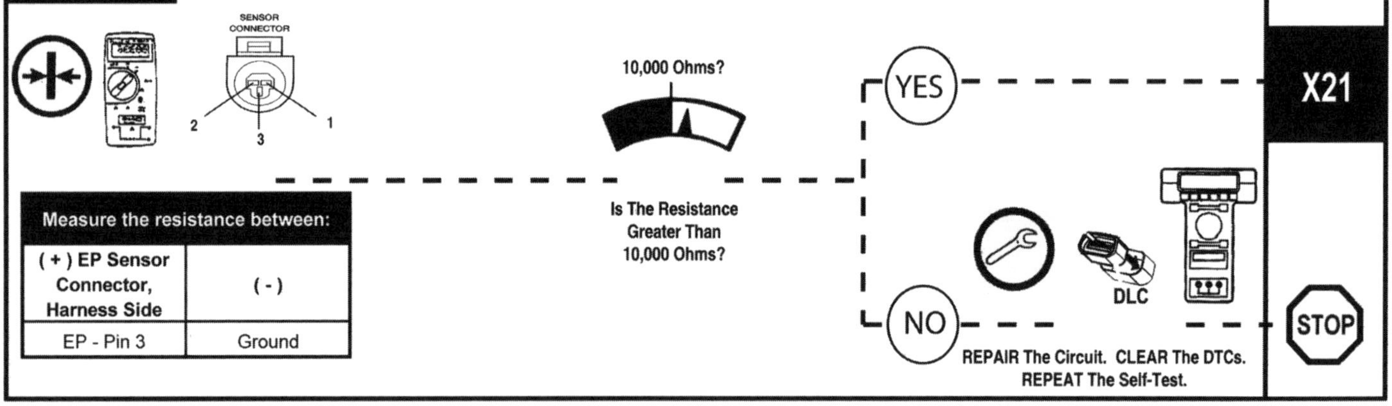

Measure the resistance between:	
(+) EP Sensor Connector, Harness Side	(-)
EP - Pin 3	Ground

Is The Resistance Greater Than 10,000 Ohms?

YES → X21

NO → REPAIR The Circuit. CLEAR The DTCs. REPEAT The Self-Test. STOP

X17 DIAGNOSTIC TROUBLE CODE (DTC) P0473

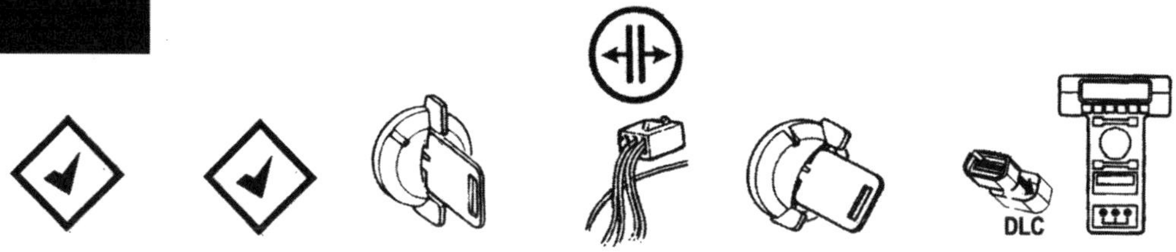

Possible Causes:
- EP Signal Shorted To Voltage
- SIG RTN Circuit Open
- EP Sensor
- PCM

Simulate The Opposite DTC.

Key OFF.

Disconnect The EP Sensor.

Key On, Engine Off.

Perform The KOEO On-Demand Self-Test.

Is DTC P0472 Present?

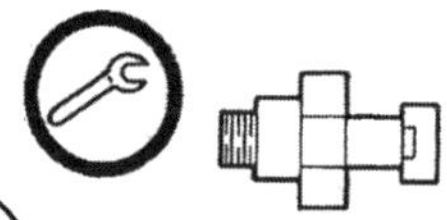

YES — INSTALL A New EP Sensor. CLEAR The DTCs. REPEAT The Self-Test. — STOP

NO — X18

X18 CHECK THE EP SENSOR SIGNAL CIRCUIT FOR A SHORT TO VOLTAGE

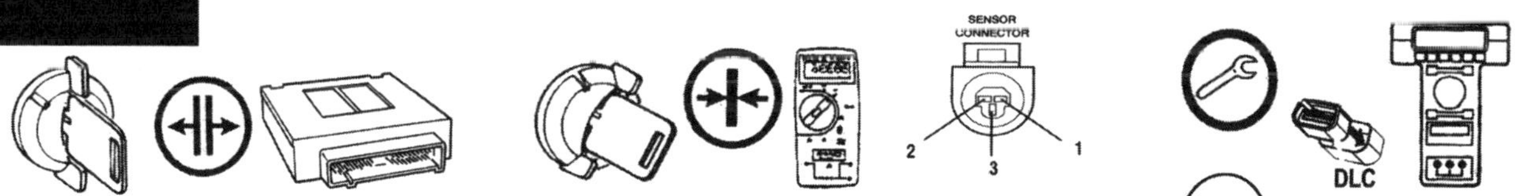

Key Off.

Disconnect The PCM Engine Harness Connector.

Key On, Engine Off.

Measure the voltage between:	
(+) EP Sensor Connector, Harness Side	(-)
EP - Pin 3	Ground

Is Any Voltage Indicated?

YES — REPAIR The Circuit. CLEAR The DTCs. REPEAT The Self-Test. — STOP

NO — X19

X19 — CHECK THE EP SENSOR SIG RTN CIRCUIT FOR AN OPEN

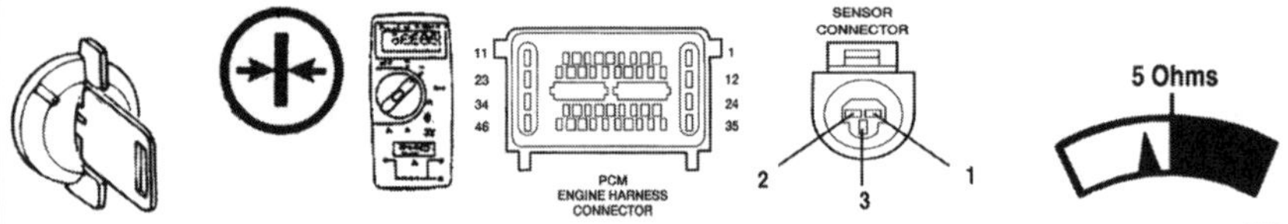

Key Off.

Measure the resistance between:	
(+) EP Sensor Connector, Harness Side	(-) PCM Connector, Harness Side
SIGRTN - Pin 1	SIGRTN - Pin E25

Is The Resistance Less Than 5 Ohms?

YES → X21

NO →

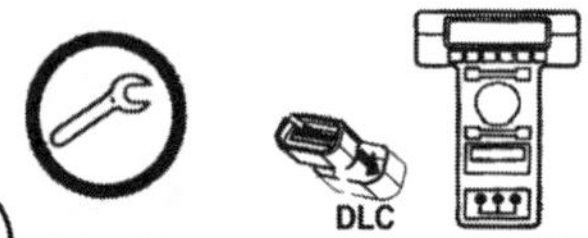

REPAIR The Circuit. CLEAR The DTCs. REPEAT The Self-Test. → STOP

X20 — CHECK THE EP SENSOR BIAS

Key On, Engine Off.

Access The MAP, BARO And EP Sensor PIDs.

Is The EP PID Value Within 10.34 kPa (1.5 psi) Of The MAP And BARO PID Values?

Refer To Reference Values For Normal Operating Values.

YES →

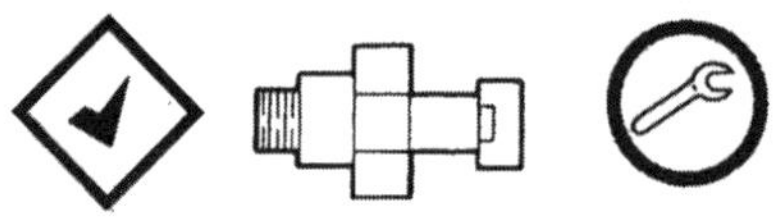

CHECK The EP Sensor And Tube For Blockage Or Restrictions. REPAIR As Necessary. REFER To Symptom Charts If A Drivability Concern Exists. → STOP

NO →

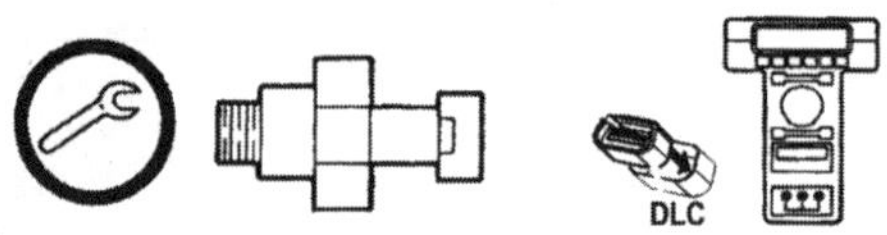

INSTALL A New EP Sensor. CLEAR The DTCs. REPEAT The Self-Test. → STOP

X21 CHECK FOR CORRECT PCM OPERATION

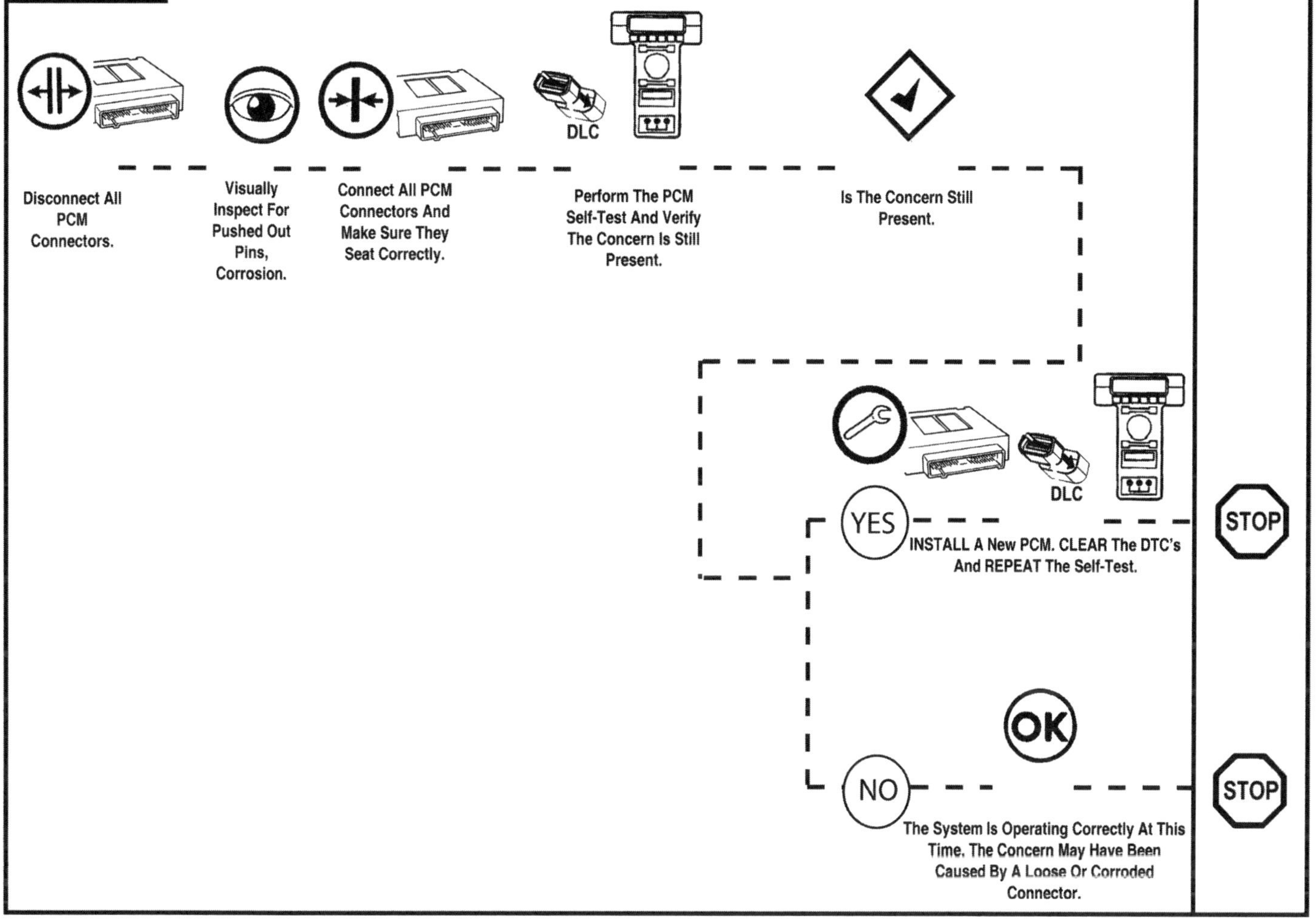

NOTES

Trouble Code (s)	Problem: Keep Alive Memory (KAM) Voltage Low	Test
P0606, P1633		Z 2006-07

REFERENCE LIST

This pinpoint test is intended to diagnose the following:

- Battery cables loose or intermittent connections
- Harness circuit: KAPWR
- Powertrain control module (PCM)

DIAGNOSTIC TROUBLE CODE INDES			
Codes	KOEO	KOER	CM
P0606	Z1	Z1	Z1
P1633	Z1	Z1	Z1

P0606 = ECM/PCM Processor
P1633 = Keep Alive Power Voltage Too low

Circuit Function:
The powertrain control module (PCM) stores the vehicle operating conditions in keep alive memory (KAM) and uses this information to compensate for component variability. The keep alive power (KAPWR) circuit supplies battery voltage (B+) to the PCM to maintain the memory contents when the ignition switch is off.

CAUTION!:
The powertrain control module (PCM) harness connectors must be properly seated and the connector latch properly attached to eliminate possible driveability concerns or a no start condition. Installing the PCM connectors on an angle may cause an improper connection, misdiagnosis and damaged components. Install the connector until the lever pivots and seats itself. Apply light pressure to get the connector into position on the PCM and then fully seat the connector.

Note: Visually inspect the harness connectors for corrosion, damage, proper mating and correct pin tension.

PINPOINT TEST SCHEMATICS AND CONNECTORS

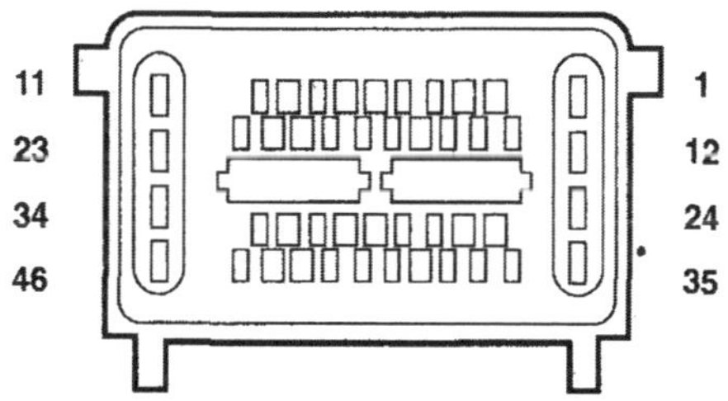

PCM BODY HARNESS CONNECTOR

Pin	Circuit
B40	KAPWR (Keep Alive Memory Power)

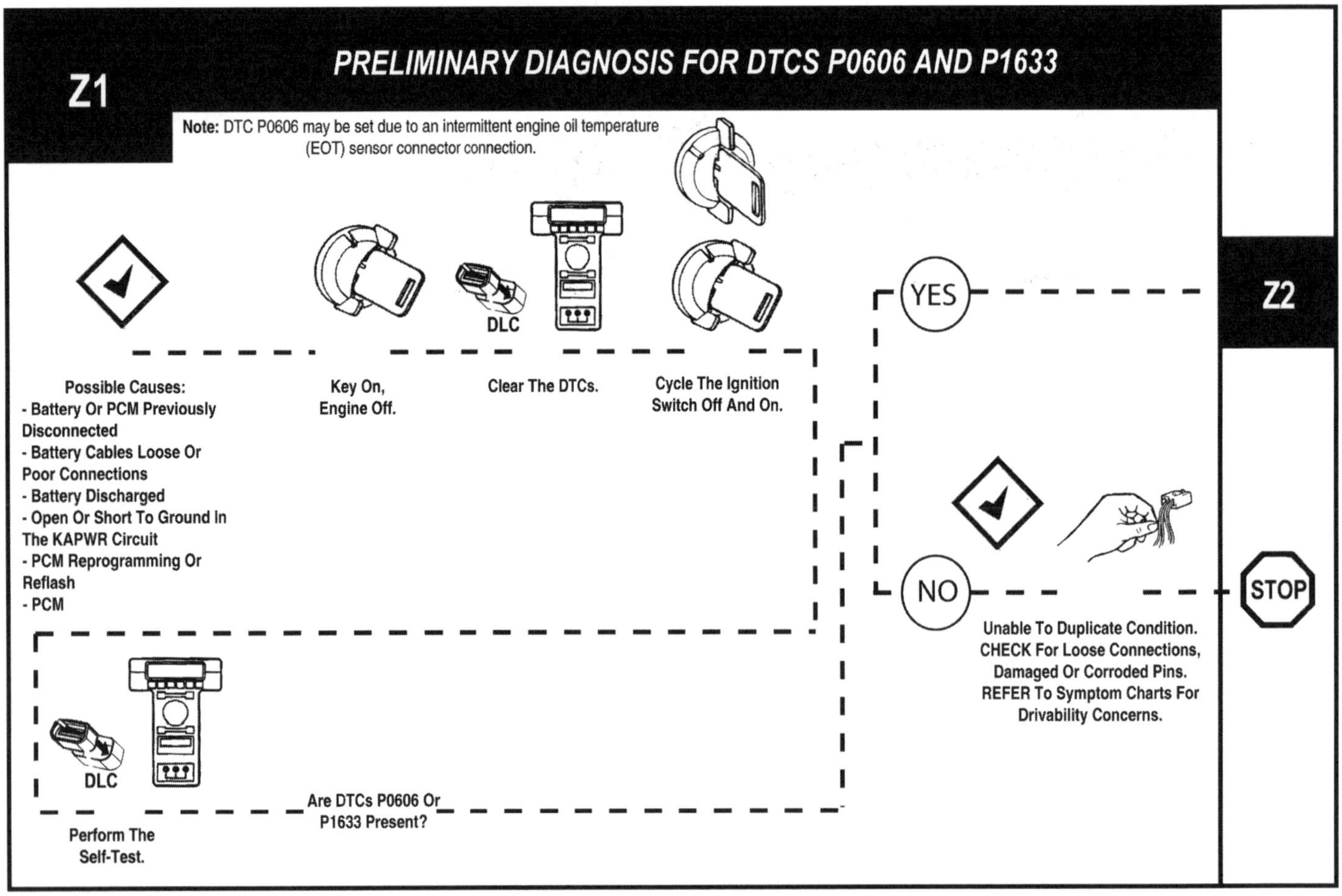

Z1
PRELIMINARY DIAGNOSIS FOR DTCS P0606 AND P1633
Note: DTC P0606 may be set due to an intermittent engine oil temperature (EOT) sensor connector connection.
DLC
Possible Causes:
- Battery Or PCM Previously Disconnected
- Battery Cables Loose Or Poor Connections
- Battery Discharged
- Open Or Short To Ground In The KAPWR Circuit
- PCM Reprogramming Or Reflash
- PCM
Key On, Engine Off.
Clear The DTCs.
Cycle The Ignition Switch Off And On.
DLC
Perform The Self-Test.
Are DTCs P0606 Or P1633 Present?
YES
Z2
NO
STOP
Unable To Duplicate Condition. CHECK For Loose Connections, Damaged Or Corroded Pins. REFER To Symptom Charts For Drivability Concerns.

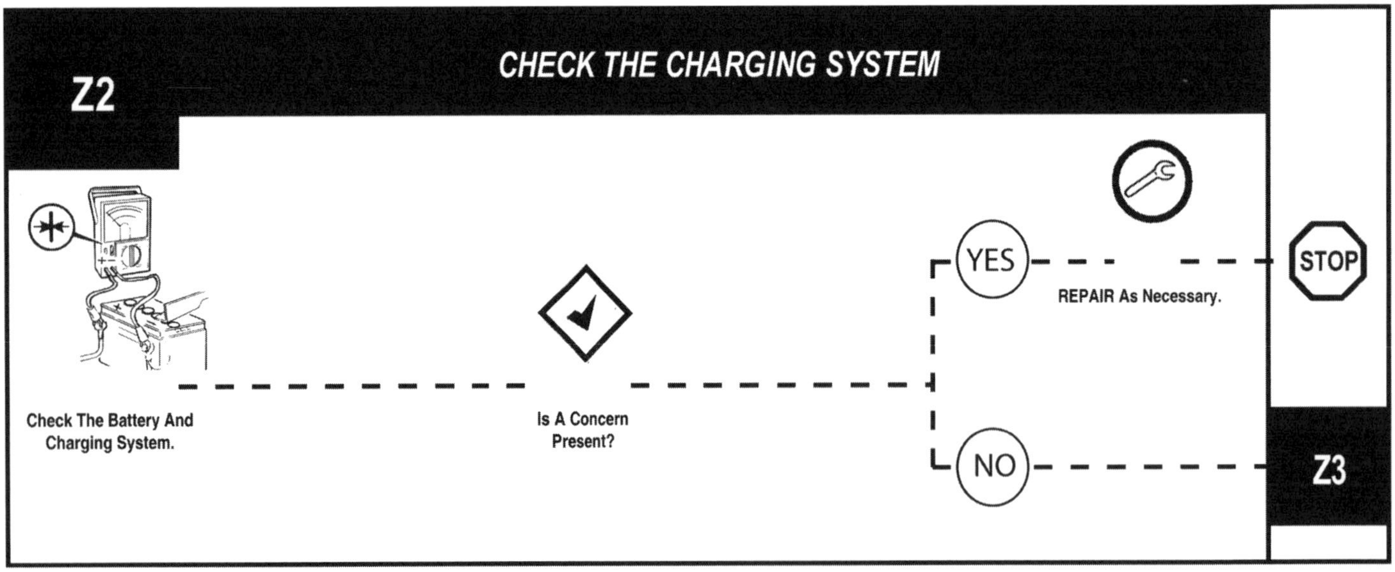

Z2
CHECK THE CHARGING SYSTEM
Check The Battery And Charging System.
Is A Concern Present?
YES
STOP
REPAIR As Necessary.
NO
Z3

Z3 CHECK THE KAPWR CIRCUIT FOR VOLTAGE

Key Off.

Disconnect The PCM Body Harness Connector.

Measure the voltage between:	
(+) PCM Connector, Harness Side	(-)
KAPWR - Pin B40	Ground

10.5 Volts

Is The Voltage Greater Than 10.5 Volts?

YES → Z4: CHECK For Loose Connections, Damaged Or Corroded Pins. WIGGLE Harness Attempting To Recreate Fault. CLEAR The DTCs And REPEAT The Self-Test. If DTC P1633 Is Present Again, INSTALL A New PCM.

DLC

NO → STOP: REPAIR The Open Circuit. CLEAR The DTCs. REPEAT The Self-test.

Z4 CHECK IF A DTC IS STILL PRESENT

Key ON. Engine OFF.

CLEAR The DTC's. (DLC)

Carry Out The Self-Test. (DLC)

Are DTCs P0606 Or P1633 Present?

YES → Z5

OK

NO → STOP: The Test Is Complete.

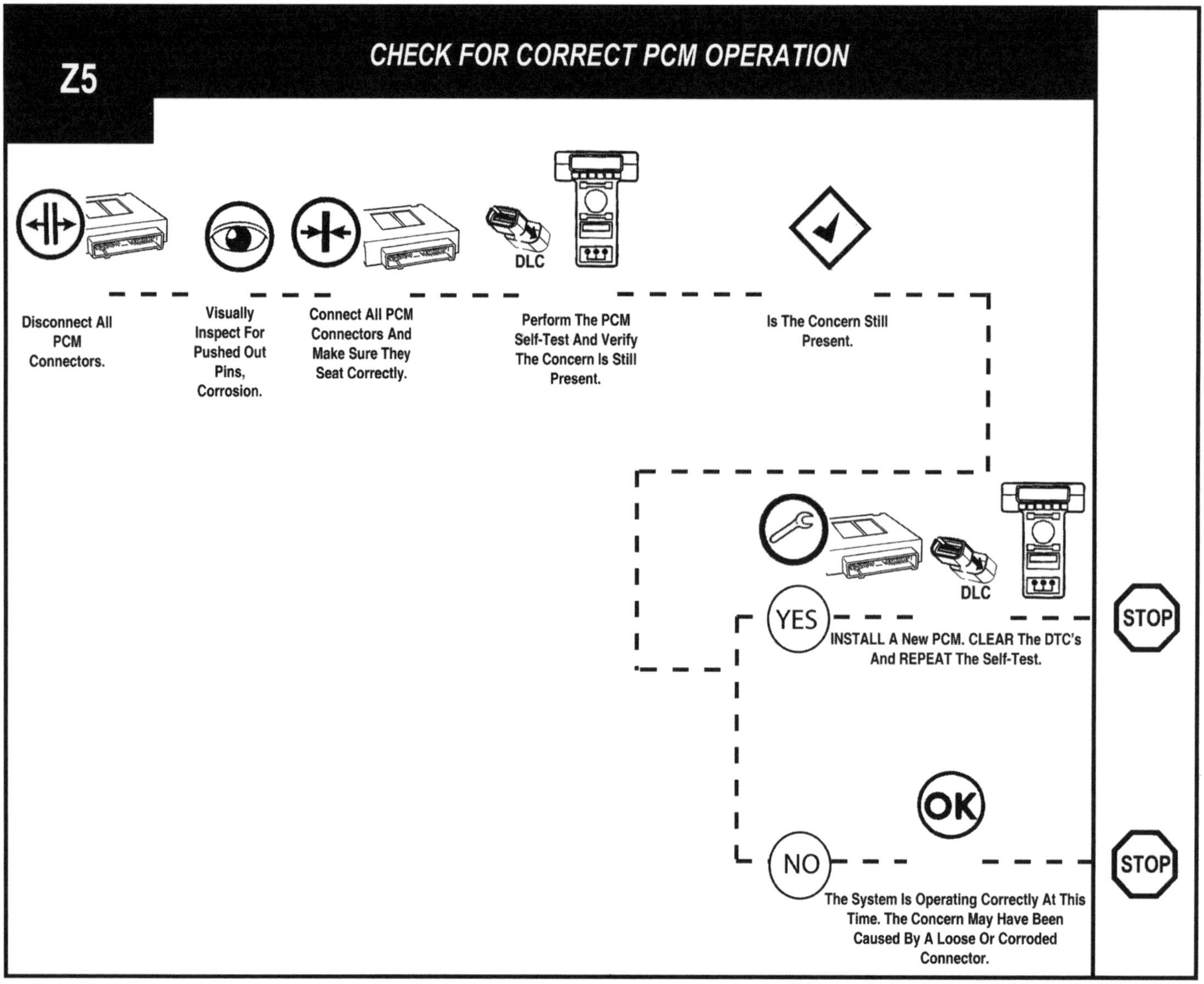
Z5
CHECK FOR CORRECT PCM OPERATION
DLC
Disconnect All PCM Connectors.
Visually Inspect For Pushed Out Pins, Corrosion.
Connect All PCM Connectors And Make Sure They Seat Correctly.
Perform The PCM Self-Test And Verify The Concern Is Still Present.
Is The Concern Still Present.
DLC
YES
INSTALL A New PCM. CLEAR The DTC's And REPEAT The Self-Test.
STOP
OK
NO
The System Is Operating Correctly At This Time. The Concern May Have Been Caused By A Loose Or Corroded Connector.
STOP

REFERENCE DATA
INDEX

2006 Parameter Identification (PID) List		
Acronym	**Description**	**Measurement Unit**
4WDL	4x4 Low by Request	Bitmap
APP	Accelerator Pedal Position Sensor	Percent
APP1	Accelerator Pedal Position #1	Volts
APP2	Accelerator Pedal Position #2	Volts
APP3	Accelerator Pedal Position #3	Volts
TSLIPRAT	Speed Ratio Across Torque Convertor	None
TSLIP	Actual Slip Across Torque Converter	RPM
ACC	Air Conditioning Clutch Command	Bitmap
ACCS	Air Conditioning Clutch Switch Compressor Cycling	Mode
GEN_F	Generator Output Fault	Bitmap Fault
GENDC	Generator Duty Cycle	Percent
ARPMDES	Ancillary RPM Desired	RPM
AXLERAT	Axle Ratio Used by PCM to Calculate Vehicle Speed	Input Revs/Output Revs
BARO	Barometric Pressure	in-Hg/kPa
BARO_V	Background BARO for Boost 2	A/D Counts
VPWR	Battery Voltage	Volts
VBAT_FF	Battery Voltage at Time of J1979 Freeze Frame	Volts
BPP1	Brake Pedal Applied	Mode
BPP2	Brake Pressure Applied	Mode
LOAD	Calculated Load Valve	Percent
ACPSW	A/C Pressure Sensor High	Mode
B+	Battery Positive Voltage	Volts
DRIVECNT	Valid Drive Counter # EOT-FM-EOT-Status	Fault
VREF	Calibration Voltage Reference Input	Volts
SYNC	Camshaft/Crankshaft Are in Sync	Mode
FICMSYNC	FICM Information for SYNC	Mode
TPLRNMIN	Closed Throttle Position	A/D Counts
CPP_A	Clutch Pedal Position A (bottom of travel)	Mode
CPP_B	Clutch Pedal Position B (top of travel)	Mode
CPP/TCS	CPP/TCS Engaged	Mode
PCA_AMP	Commanded Current for PCA	Amps
SSD_AMP	Commanded Current for SSD	Amps
TCC_AMP	Commanded Current for TCC	Amps
SSE_AMP	Commanded Current for SSE	Amps
PIP_CTR	Counter Incremented Every PIP Rising Edge	None
SHIFT_TYPE/ GEARTR	Commanded Gear and PRNDL	None
RPMDSD	Desired Idle RPM	RPM
LINEDSD	Desired Main Line Pressure	PSI
TSLIPDSD	Desired Slip Across the Torque Converter	RPM
OSS_DIR	Direction of OSS Spring	State Encoded
TR_DC	Duty Cycle of TRS	Percent
EGRVPDES	EGR Valve Position Desired	Percent
EGRVP_V	EGR Valve Position Sensor (Raw)	A/D Counts (Volts)
EGR_PCT	Commanded EGR	Percent
EGR_ERR	EGR Error From Commanded	Percent
FPDC	Modulated Fuel Pump Duty Cycle	Percent
PCA	Commanded Pressure Electronic Pressure Control A	PSI
PCA_F	Electronic Pressure Control A Fault Status	State Encoded
SSPCD_F	Shift Solenoid D Fault Status	State Encoded
TCCPC	Torque Converter Commanded Pressure	PSI
TCCPC_F	Torque Convertor Control Fault Status	State Encoded

2006 Parameter Identification (PID) List		
ECT	Engine Coolant Temperature (After Any FMEM Sub.)	Degrees F
ECT_V	Engine Coolant Temperature (Raw)	A/D Counts
EGRP	EGR Valve Position	Percent
EOT_V	Engine Oil Temperature	A/D Counts
EOT	Engine Oil Temperature, Mode	Degrees C
EP_V	Exhaust Back Pressure	A/D Counts
EP	Exhaust Back Pressure Absolute	PSI/kPa, Volts
EPDSD	Exhaust Back Pressure Desired	kPa
FANSS	Fans Speed Sensor	RPM
FANSSM	Fan Speed Sensor Monitor	Bitmap
FICMLPWR	FICM Logic Power	Volts
FICMMPWR	FICM Main Power	Volts
FICMVPWR	FICM Vehicle Power	Volts
SHFT_FLRE	Flare RPM for PID 1931	RPM
PNP	Park Neutral Switch	Bitmap
DTC	Freeze Frame DTC	None
FLI	Fuel Level Input Sensor	Percent
FLI_V	Fuel Level Input Voltage	Volts
FUELPW	Fuel Pulse Width	Time
FPMS	Fuel Pump Secondary Monitor High	Bitmap
GEAR_OSC	Gear Commanded OSC Status	State Encoded
GPCTM	Glow Plug Control Time	Seconds
GPDC	Glow Plug Control Duty Cycle	Percent
GPLTM	Glow Plug Lamp Time On	Seconds
GPIL	Glow Plug Indicator Lamp Commanded On	Bitmap
GEARMAX	Highest Gear Permitted	None
ICP	Injection Control Pressure	MPa
ICP_DES	Injection Control Pressure Desired	MPa Absolute
IPR	Injector Control Pressure Regulator	Percent
INJ_TIM	Injector Timing BTDC	Degrees BTDC
IAT2	Intake Air Temperature	Degrees F
IAT2_V	Intake Air Temperature Counts	A/D Counts
IAT_V	Intake Air Temperature (Raw)	A/D Counts
ISS_SRC	Intermediate Shaft Speed, No FMEM	RPM
SECSTAT	IPATS Security Access Status	None
HFPIP	Last PIP Up Edge to Last PIP Down Edge	Seconds
TPLRNMIN	Lowest Idle Throttle Position	A/D Counts
MAF_V	Mass Air Flow Sensor Voltage (raw)	A/D Counts
MIL	Malfunction Indicator Requested ON	Bitmap
MAP	Manifold Absolute Pressure	kPa
MAP_V	Manifold Absolute Pressure Sensor	Volts
MGP	Manifold Gauge Pressure	kPa
MAF	Mass Air Flow	Grams
MFDES	Mass Fuel Desired	Mg
T/F Delta PID	Mass Air Flow Status	MAF FM
MIS_CNT1	Misfire Count Cylinder #1 (High Byte)	# Misfires
MIS_CNT1	Misfire Count Cylinder #1 (Low Word)	# Misfires
MIS_CNT2	Misfire Count Cylinder #2 (High Byte)	# Misfires
MIS_CNT2	Misfire Count Cylinder #2 (Low Word)	# Misfires
MIS_CNT3	Misfire Count Cylinder #3 (High Byte)	# Misfires
MIS_CNT3	Misfire Count Cylinder #3 (Low Word)	# Misfires
MIS_CNT4	Misfire Count Cylinder #4 (High Byte)	# Misfires
MIS_CNT4	Misfire Count Cylinder #4 (Low Word)	# Misfires
MIS_CNT5	Misfire Count Cylinder #5 (High Byte)	# Misfires

2006 Parameter Identification (PID) List		
MIS_CNT5	Misfire Count Cylinder #5 (Low Word)	# Misfires
MIS_CNT6	Misfire Count Cylinder #6 (High Byte)	# Misfires
MIS_CNT6	Misfire Count Cylinder #6 (Low Word)	# Misfires
MIS_CNT7	Misfire Count Cylinder #7 (High Byte)	# Misfires
MIS_CNT7	Misfire Count Cylinder #7 (Low Word)	# Misfires
MIS_CNT8	Misfire Count Cylinder #8 (High Byte)	# Misfires
MIS_CNT8	Misfire Count Cylinder #8 (Low Word)	# Misfires
MIS_CNT	Misfire Count Total (High Byte)	# Misfires
MIS_CNT	Misfire Count Total (Low Word)	# Misfires
MISDEL1	Misfire Percent Delta Cylinder #1	Percent
MISDEL2	Misfire Percent Delta Cylinder #2	Percent
MISDEL3	Misfire Percent Delta Cylinder #3	Percent
MISDEL4	Misfire Percent Delta Cylinder #4	Percent
MISDEL5	Misfire Percent Delta Cylinder #5	Percent
MISDEL6	Misfire Percent Delta Cylinder #6	Percent
MISDEL7	Misfire Percent Delta Cylinder #7	Percent
MISDEL8	Misfire Percent Delta Cylinder #8	Percent
TORQUE	Net Engine Torque	Lb-Ft
DRIVECNT	Number of Completed OBD Drive Cycles	#
TRIPCNT	Number of Completed OBD Trips	#
TRIP	OBD Trip Completed	Bitmap
PATSOUT	Outcode Status	None
OSS	Output Shaft Sensor	RPM
OSS_SRC	Output Shaft Speed, No FMEM	RPM
PBPP	Park Brake Pedal Applied	Bitmap
PNP	Park/Neutral Switch in Drive	Bitmap
PSP	Power Steering Load	Bitmap
	Software ID Block Starting Address	None
TMSTRT	Time Since Engine Start	Seconds
TIREREV	Tire Size Used by PCM to Calculate VS	Tire Revs Per Mile
TCCRAT	Torque Converter Clutch Speed Ratio	None
TCC_OSC	Torque Converter Clutch OSC Status	State Encoded
DTC CNT	Total Number of Fault Codes Pending	#
DTC CNTD	Total Number of On-Demand Codes	#
TCS	Transmission Control Switch Applied	Bitmap
TFT_V	Transmission Fluid Temp. Counts (Raw)	A/D Counts
TFT	Transmission Fluid Temperature	Degrees F
TRANRAT	Transmission Gear Ratio	None
TSS	Turbine Shaft Speed	RPM
TSS_SRC	Turbine Shaft Speed, No FMEM	RPM
VGT_F	Variable Geometry Turbo Fault	Bitmap
VGTDC	Variable Geometry Turbo Duty Cycle	Percent
FANVAR	Variable Speed Fan Duty Cycle	Percent
VSS	Vehicle Speed Sensor	MPH
VS	Vehicle Speed	MPH
VFDES	Volume Fuel Desired	mm^3
WFS	Water in Fuel Condition	Bitmap
SSPCD	Commanded Pressure Shift Solenoid D	PSI
EGRDC	EGR Duty Cycle	Percent
TSS/ISS	Turbine Speed Adjusted for Chain Ratio	RPM
SSPCE	Commanded Pressure Shift Solenoid E	PSI
SSPCE_F	Shift Solenoid E Output Fault Status	State Encoded
SSPCA	Commanded Pressure Shift Solenoid A	PSI
SSPCB	Commanded Pressure Shift Solenoid B	PSI

2006 Parameter Identification (PID) List		
SSPCC	Commanded Pressure Shift Solenoid C	PSI
SSA_AMP	Commanded Current for SSA	AMPS
SSB_AMP	Commanded Current for SSB	AMPS
SSC_AMP	Commanded Current for SSC	AMPS
SSPCA_F	Shift Solenoid A Pressure Control Output Fault	State Encoded
SSPCB_F	Shift Solenoid B Pressure Control Output Fault	State Encoded
SSPCC_F	Shift Solenoid C Pressure Control Output Fault	State Encoded
IAT	Intake Air Temperature (After Any FMEM Sub.)	Degrees F
SECTIME	Security Access Time	Minute
CMPFM	Camshaft Position Failure Rate	Bitmap
DECHOKE	Dechoke Mode Flag	Bitmap
TCCFM	Converter Unlock Due to Excess Slip	Bitmap
EGRFM	EGR Failure Rate	Bitmap
MAFFM	MAF Failure Rate	Bitmap
ECTFM	ECT Failure Rate	Bitmap
IATFM	IAT Failure Rate	Bitmap
EGR EVAL	EGR System Evaluated	FP#
PTO	Power Take-Off Status	Bitmap
GEN_F	Generator Fault Detected	Bitmap
GENB_F	Generator 2 Fault Detected	Bitmap
TCILF	TCIL Fault Detected	Bitmap
TCC_OT	Overtemp Lock-Up Mode	Bitmap
TCCMACT	Torque Convertor Slip (absolute value)	RPM
TCCDC	Torque Convertor Clutch Duty Cycle	Percent
TRAN_OT	Transmission is Overtemperature	Bitmap
TRAC	Traction Assist is Available	Bitmap
VSSFM	Vehicle Speed Sensor Failure Mode	Bitmap
PCFM	Pressure Control Failure Mode	Bitmap
OSSFM	Output Shaft Speed Sensor Failure Mode	Bitmap
TFTFM	Trans Fluid Temperature Failure Mode	Bitmap
FANVAR_F	Variable Speed Fan (VSF) Output	Bitmap
PIP HIGH	PIP State is High	Bitmap
GENFI	Generator Fault Indicator Light Commanded On	Bitmap
GENFI_F	Generator Fault Indicator Light Fault	Bitmap
MIS STAT	Misfire Monitor Completed	Bitmap
MISFIRE	Currently Misfiring	Bitmap
EGR STAT	EGR Monitor Completed	Bitmap
CCM STAT	CCM Monitor Completed	Bitmap
IAT2FM	Downstream ACT FMEM Flag	Bitmap
EOTFM	Engine Oil Temperature FMEM	Bitmap
BAROFM	BARO FMEM Flag	Bitmap
MP_LRN	Misfire Correction Profile Learned	Bitmap
MFF_PNP	In Drive During Time of Misfire	Bitmap
SRC	Starter Motor Relay is Enabled	Bitmap
SPAREKEY	IPATS Spare Key Programming Enable	Bitmap
ANTISCAN	Anti-Scan Enable and Counting	Bitmap
PCM_ID	PCM ID Stored	Bitmap
PCM_VFY	PCM ID Verified	Bitmap
ENSTAT	Engine Start Enabled	Bitmap
PSA	Trans Pressure Switch A is Active	Bitmap
PSB	Trans Pressure Switch B is Active	Bitmap
PSC	Trans Pressure Switch C is Active	Bitmap
PSD	Trans Pressure Switch D is Active	Bitmap
PSE	Trans Pressure Switch E is Active	Bitmap

2006 Parameter Identification (PID) List		
HRSH_SHFT	Harsh Shift Commanded by OSC	Bitmap
WAC	WOT A/C Cutout	On/Off
INJ1	Fuel Injector Control	On/Off
INJ2	Fuel Injector Control	On/Off
INJ3	Fuel Injector Control	On/Off
INJ4	Fuel Injector Control	On/Off
INJ5	Fuel Injector Control	On/Off
INJ6	Fuel Injector Control	On/Off
INJ7	Fuel Injector Control	On/Off
INJ8	Fuel Injector Control	On/Off
ISS_F	ISS is Currently Unreliable	Bitmap
OSS_F	OSS is Currently Unreliable	Bitmap
TSS_F	TSS is Currently Unreliable	Bitmap
VSS_F	VSS is Currently Unreliable	Bitmap
TRO_N_F	Neutral Output Fault	Bitmap
TRO_P_F	Park Output Fault	Bitmap
RLC_F	Reverse Lamp Control Fault	Bitmap
TRFM	TR Failure Mode	Bitmap
TCIL	TCIL Light Commanded ON	Bitmap
IPATS	Integrated Passive Anti-Theft System Status	Bitmap
TCS FLG	Transmission Switch Applied	Bitmap
OSS_DIR	Output Shaft Speed Direction	State Encoded
TP_MODE	Throttle Mode	State Encoded
SHFT_TYP	Shift Type	State Encoded
TQ_CNTL	Torque Reduction Request	State Encoded
GEAR	Commanded Gear	State Encoded
TR	PRNDL Position	State Encoded
IAC_MODE	Idle Speed Control Operating Mode	None
TP V	Throttle Position (Before FMEM)	A/D Counts
RPM	Engine Speed	RPM
TCCMCMD	Torque Converter Clutch Desired Slip	RPM
SHIFT_ID	Shift Identification	None
SHFT_LAG	0 - 10% Shift Time	Seconds
SHFT_TM	10 - 90% Shift Time	Seconds
SHFT_DROP	RPM Overshoot For PID 1931	RPM
RUNTM	Time Since Engine Start	Seconds
MIL_DIST	Distance Traveled With MIL Activated	Km
WARM_UPS	Number of Warm-ups Since Codes Cleared	Km
TP_R	Relative Throttle Position	Percent
CLR_DIST	Distance Traveled Since Codes Cleared	Km
INJ_TIM	Injector Timing BTDC	Degrees BTDC
ICP_DES	Injection Control Pressure Desired	MPA Absolute
PIP_CTR	Counter incremented every PIP rising edge	None
PIPTIM	Last PIP time in seconds	Seconds

2006 DIAGNOSTIC TOOL — DRIVER-OPERATED CONTROLS CHECK				
PID	**Operator Setting**	**Reading**	**Voltage**	**Pinpoint Test**
APP	Accelerator Pedal Released	0%	95/100%	AG
	Accelerator Pedal Fully Applied	0%	95/100%	
BOO/BPP	Brake Applied	ON	B+	AI
	Brake Released	OFF	0V	
CPP/TCS	(ZF 6-Speed)			C
	Clutch Pedal Applied	ON	0V	
	Clutch Pedal Released	OFF	B+	
PBPP	Parking Brake ON	ON	0V	I
	Parking Brake OFF	OFF	B+	
CPP/TCS	(Torque Shift)	OFF	0V	
	Key On:			
	Push CNCL (Latch)	ON	B+	
	Push CNCL (VF Latch)	OFF	0 V	
TR	PRNDL in P	PARK		
	PRNDL in R	REV		
	PRNDL in N	NTRL		
	PRNDL in OD	OD		
	PRNDL in OD with Tow/Haul Cancel Light ON	DRIVE		
	PRNDL in L2	MAN2		
	PRNDL in L1	MAN1		
TCIL	(5R110)			
	After Key ON	OFF	0V	
	Push Tow/Haul Cancel (Latch)	ON	B+	
	Push Tow/Haul Cancel (Unlatch)	OFF	0V	
4WDL	(Transfer Case Shifter)			
	4x4H or 2H	OFF	B+	
	4x4L	ON	0 V	
ACCS	(A/C Controls)			
	MAX/NORM AC or MIX/DEF with A/C Clutch ON	ON	B+	
	Floor or Vent or A/C Clutch OFF	OFF	0 V	

2006 PCM PIN IDENTIFICATION

Engine						
Pin Number	Name	Key On	Low Idle	High Idle	Operating Range	Comments
1	GEN1C	B+	B+	B+	B+	Generator 1 Monitor 0 Fault Detected
2	IPR	10.7	10.1	9.3	0 - B+	Injection Pressure Regulator Control (Pulse Width Modulated)
3	GPE	0/B+	0/B+	0/B+	0/B+	Glow Plug Enable 0 Equals Relay ON B+ Equals Relay OFF
4	GEN2C	B+	B+	B+	B+	Generator 2 Monitor (Dual Generator Only) 0 Equals Fault Detected
5	N/A	N/A	N/A	N/A	N/A	N/A
6	FSS	0	5.4	5.1	0 - VREF	Fan Speed Signal
7	N/A	N/A	N/A	N/A	N/A	N/A
8	N/A	N/A	N/A	N/A	N/A	N/A
9	N/A	N/A	N/A	N/A	N/A	N/A
10	VGTC	12.1	7.9	10.4	0 - B+	Variable Geometry Turbo Actuator Control (Pulse Width Modulated)
11	VGTCH	B+	B+	B+	B+	Variable Geometry Turbo Actuator Voltage
12	N/A	N/A	N/A	N/A	N/A	N/A
13	N/A	N/A	N/A	N/A	N/A	N/A
14	FC-V	12.3	12.3	13.8	0 - B+	Cooling Fan Solenoid Control (Pulse Width Modulated) 0 Equals ON B+ Equals OFF
15	PTOC	0	0	0	0	PTO Control
16	BCPIL	0	0	0	0	Battery Charge Protection Indicator Lamp
17	GPD	0	0	0	0 - VPWR	Glow Plug Control Module/PCM Communication
18	N/A	N/A	N/A	N/A	N/A	N/A
19	CKPO	0	5.7V	5.7V	0 - 6000 Hz	Buffered Crankshaft Signal Output (to FICM)
20	CMPO	0	1.0V	1.2V	0.5 - 50 Hz	Buffered Camshaft Signal Out (to FICM)
21	N/A	N/A	N/A	N/A	N/A	N/A
22	TPWR GND	0	0	0	0	Cooling Fan Sensor Signal Return
23	EGRVC	12.2	11.2 - 12.2	10.3	0 - VPWR	EGR Solenoid Control (Pulse Width Modulated)
24	N/A	N/A	N/A	N/A	N/A	N/A
25	SIGRTN	0	0	0	0	Signal Return
26	CAN2L	2.1	2.1	2.2	-19	CAN- Module Communications (PCM to FICM)
27	EP	1	1.2	1.4	0.4 - 4.7	Exhaust Pressure Signal 0.9 Equals 101 kPa (14.7 psi) 4.7 Equals 365 kPa (53 psi)

2006 PCM PIN IDENTIFICATION

Engine (CONTINUED)						
Pin Number	Name	Key On	Low Idle	High Idle	Operating Range	Comments
28	FICMM	1.1	2.6	2.8	4-Jan	Intermodule Communication Signal (PCM to FICM)
29	ICP	0.2	1	1.9	0.15 - 4.7	Injection Control Pressure Sensor Signal 1.0 Equals 4.5 MPa (652 psi) 4.45 Equals 28 MPa (4061 psi)
30	CKP+	1.58V	1.60V	1.60V	0 - 6000 Hz	Crankshaft Position Signal
31	CMP+	0.25V	0.25V	0.38V	0.5 - 50 Hz	Camshaft Position Signal
32	ECT	1.9	1.9	1.1	0.3 - 4.7	Engine Coolant Temperature 4.9 Equals -40°C (-40°F) 0.47 Equals 100°C (212°F)
33	EGRVP	0.8	0.8 - 1.0	1.3	0.6 - 4.0	EGR Valve Position Signal
34	N/A	N/A	N/A	N/A	N/A	N/A
35	N/A	N/A	N/A	N/A	N/A	N/A
36	VREF	5	5	5	5 ± 0.5	VREF
37	CAN2H	2.9	2.9	2.9	-19	CAN+ Module Communications (PCM to FICM)
38	N/A	N/A	N/A	N/A	N/A	N/A
39	N/A	N/A	N/A	N/A	N/A	N/A
40	N/A	N/A	N/A	N/A	N/A	N/A
41	CKP-	1.5V	1.6V	1.6V	0 - 6000 Hz	Crankshaft Position Signal
42	CMP/CKP SHD	0	0	0	0	CMP/CKP Sensor Shield
43	CMP-	0.3V	0.3V	0.4V	0.5 - 50 Hz	Camshaft Position Signal
44	EOT	2	2.6	1.1	0.3 - 4.7	Engine Oil Temperature 4.9 Equals -40°C (-40°F) 0.47 Equals 100°C (212°F)
45	IAT2	2.7	3.7	2.9	0.2 - 4.7	Manifold Air Temperature 4.3 Equals 0°C (32°F) 2.5 Equals 50°C (122°F)
46	VBPWR	B+	B+	B+	Buffered VBAT	Cooling Fan Speed Sensor Buffered Voltage

2006 PCM PIN IDENTIFICATION

Body						
Pin Number	Name	Key On	Low Idle	High Idle	Operating Range	Comments
1	CTO	12.2	12.2	12.4	1 - VPWR	Tachometer Output (Aftermarket)
2	ACCR	B+	0/B+	0/B+	0/B+	A/C Clutch Relay Control B+ Equals A/C OFF 0 Equals A/C ON
3	DOL	0	0	0	0 - 5	Tripminder Fuel Economy Digital Signal
4	TRO PN	0	0	0	0/B+	Starter Relay Enable 0 Equals Enabled
5	FPC	0/B+	0	0	0	Fuel Pump Relay Control B+ Equals Pump OFF 0 Equals Pump ON
6	N/A	N/A	N/A	N/A	N/A	N/A
7	TPO	0	0	0	0	Throttle Position Output
8	ACCS	0	0/B+	0/B+	0/B+	A/C Cycling Pressure Switch B+ Equals A/C ON 0 Equals A/C OFF
9	BCPSW	—	—	—	—	Battery Charge Protection
10	CASEGND	0	0	0	0	Chassis Ground
11	PWR GND	0	0	0	0	Power Ground
12	PTO	0	0/B+	0/B+	0/B+	Power Take Off (Aftermarket) 0 Equals PTO Disengaged B+ Equals PTO Engaged
13	CAN1H	2.7	2.7	2.7	-19	CAN+ Module Communications
14	CAN1L	2.4	2.4	2.4	-19	CAN- Module Communications
15	WFS	4.6	4.6	4.6	> 4.4	Water In Fuel Sensor Diesel Greater Than 4.4 Water Less Than 3.3
16	ACPSW	0	0/B+	0/B+	0/B+	A/C Pressure Switch 12 Equals A/C Pressure Less Than 325 psig 0 Equals A/C Pressure Greater Than 325 psig
17	PBA	0/B+	0/B+	0/B+	0/B+	Park Brake Applied 0 Equals Brake Applied 12 Equals Brake OFF
18	BPP	0/B+	0/B+	0/B+	0/B+	Brake Pedal Position 12 Equals Brake Applied 0 Equals Brake OFF
19	FPM	0/B+	B+	B+	B+	Fuel Pump Monitor B+ Equals Fuel Pump ON
20	SIGRTN2	0	0	0	0	Accelerator Pedal Position Sensor Signal Return
21	MAFRTN	0	0	0	0	Mass Air Flow Sensor Signal Return
22	VSO	0	0	0	0 - VBAT	Vehicle Speed Out
23	PWRGND	0	0	0	0	Power Ground

2006 PCM PIN IDENTIFICATION

Body (CONTINUED)						
Pin Number	**Name**	**Key On**	**Low Idle**	**High Idle**	**Operating Range**	**Comments**
24	SCCSRTN	0	0	0	0	Speed Control Switch Signal Return
25	APP3	1	1	1.45	0.8 - 3.5	Accelerator Pedal Track 3 CT Equals 0.8 - 1.1 WOT Equals 3.1 - 3.5
26	APP1	4.1	4.1	3.5	0.7 - 4.2	Accelerator Pedal Track 1 CT Equals 3.9 - 4.2 WOT Equals 0.7 - 1.2
27	N/A	N/A	N/A	N/A	N/A	N/A
28	BPS	0/B+	0/B+	0/B+	0/B+	Brake Pressure Switch 0 Equals Brake Applied B+ Equals Brake OFF
29	VREF2	5	5	5	5 ± 0.5	Reference Voltage
30	GENIL	0	0	0	0/B+	Generator Indicator
31	SCCS	6.7	6.7	6.7	0 - 7.0	Speed Control Switches ON Equals B+ OFF Equals 0 SET+ Equals 2.8 RESUME Equals 4.7 SET- Equals 0.8 HOLD Equals 6.6
32	PTOGND	0	0	0	0	PTO Ground
33	SIGRTN	0	0	0	0	Signal Return
34	VPWR	B+	B+	B+	B+	Vehicle Power
35	VSS	8.89V	9.82V	9.85V	5.5 - 333 Hz	Vehicle Speed Signal
36	CPP/TOW	0V/B+	0V/B+	0V/B+	0V/B+	Grade Load Switch (Automatic Trans) 0 Equals Switch OFF B+ Equals Switch ON Clutch Pedal Position Switch (Manual Trans) 0 Equals Pedal APPLIED B+ Equals Pedal OFF
37	APP2	1.53	1.53	2	1.4 - 4.1	Accelerator Pedal Track 2 CT Equals 1.4 - 1.6 WOT Equals 3.6 - 4.1
38	BARO	4.6	4.6	4.6	0.5 - 4.5	Barometric Pressure
39	FEPS	0	0	0	N/A	Flash EPROM Digital Signal
40	KAPWR	B+	B+	B+	B+	Keep Alive Power
41	MAP	1.35	1.41	1.61	0.5 - 4.5	Manifold Absolute Pressure 1.35 Equals 103 kPa (14.9 psi) 4.5 Equals 300 kPa (43.5 psi)
42	MAF	0	1.36	2.65	0 - 4.7	Mass Air Flow Signal
43	IAT1	2.3	3.1	2.5	0.13 - 4.9	Intake Air Temperature (Mass Air Flow Assembly) 4.9 Equals -40°C (-40°F) 0.5 Equals 100°C (212°F)
44	PTO VREF	5	5	5	5	PTO Reference Voltage
45	VREF	5	5	5	5 ± 0.5	Reference Voltage
46	VPWR	B+	B+	B+	B+	Vehicle Power

2006 PCM PIN IDENTIFICATION

Acronym	PIDS Key On	Low Idle	High Idle	Operating Range	Measurement Units	Description
ACCS	OFF	OFF	OFF	OFF/ON	OFF/ON	AC Cycle Switch
ACP	OPEN	OPEN	OPEN	OPEN/ CLOSED	OPEN/ CLOSED	AC Pressure Switch
APP	0	0	13	0 - 17.75	PERCENT	Accelerator Pedal Position
APP1	4	4	3.4	0.7 - 4.2	VOLT	Accelerator Pedal Track 1
APP2	1.4	1.4	1.9	1.4 - 4.1	VOLT	Accelerator Pedal Track 2
APP3	0.9	0.9	1.4	0.8 - 3.5	VOLT	Accelerator Pedal Track 3
APP_MODE	CT	CT	PT	CT, PT, WOT	CT, PT, WOT	Accelerator Pedal Position
AST	0:00	0:07	1:00	-	min : sec	Time Since Start
B+	12.6	13.3	14	10.5 - 15.5	VOLT	Battery Voltage
BARO	96.5 (14)	96.5 (14)	96.5 (14)	50 - 120 (7.3 - 17.4)	kPa (PSI)	Barometric Pressure
BOO	OFF	OFF	OFF	ON/OFF	ON/OFF	Brake Pedal
BPA	OFF	OFF	OFF	ON/OFF	ON/OFF	Brake Pressure Applied
CLRDIST	926.1	926.1	926.1	-	Miles	Distance Since DTCs Cleared
CLRWRMUP	2	2	2	-	Numeric Value	Number of Warmups Since DTCs Cleared
CMPFM	NO FAULT	NO FAULT	NO FAULT	NO FAULT	NO FAULT/ YES FAULT	CMP Status
CPP/PNP	NEUTRAL	NEUTRAL	NEUTRAL	NEUTRAL	DRIVE/ NEUTRAL	Clutch Pedal Position
DECHOKE	No	No	No	-	Yes/No	Dechoke Mode
DRIVECNT	6	7	7	0 - 65535	NUMBER	Valid Drive Counter
DTCCNT	0	0	0	0 - 255	NUMBER	DTC Count
ECT	88 (190)	88 (190)	88 (190)	-40 - 121 (-40 - 250)	TEMPERA-TURE °C (°F)	Coolant Temperature
ECT_FM	NO FAULT	NO FAULT	NO FAULT	NO FAULT	NO FAULT/ YES FAULT	ECT Fault
EGR EVAL	NO	NO	NO	YES	NO/YES	EGR Evaluated
EGR DC#	0	0 - 15	32	0 - 44	PERCENT	EGR Duty Cycle
EGR_F	NO FAULT	NO FAULT	NO FAULT	0 - 44	FAULT/NO FAULT	EGR Status
EGRVP	0.8	0.8 - 1.0	1.3	0.6 - 3.5	VOLT	EGR Valve Position
EGRVPDES	0	0	9.72	0 - 14	PERCENT	EGR Desired
EOT	88 (190)	88 (190)	88 (190)	-40 - 150 (-40 - 302)	TEMPERA-TURE °C (°F)	Oil Temperature
EOTFM	NO FAULT	NO FAULT	NO FAULT	NO FAULT	NO FAULT/ YES FAULT	EOT Status
EP	103 (15)	110 (16)	110 (16)	55 - 607 (8 - 88)	kPa (PSI)	Exhaust Pressure
EP_V	0.8	1.2	1.3	0.4 - 4.7	VOLT	Exhaust Pressure

2006 PCM PIN IDENTIFICATION

PIDS (CONTINUED)

Acronym	Key On	Low Idle	High Idle	Operating Range	Measurement Units	Description
EP DSD	3.26	2.43	2.13		PSI	Exhaust Pressure Desired
FANSS	0	533	488	0 - 4000	RPM	Fan Speed
FANSSM	LOW	LOW	LOW	LOW	LOW/HIGH	Cooling Fan Driver
FAN VAR #	0	0	0	0 - 100	Percent	Fan Speed Monitor
FAN VAR_F	NO FAULT	NO FAULT	NO FAULT	NO FAULT	NO FAULT/ YES FAULT	Fan Speed Monitor
FICMSYNC	NO	YES	YES	YES	YES/NO	Synchronizatio n from FICM
FICM_LPWR	12	13.5	13.5	10.5 - 15.5	VOLTS	FICM Logic Power [1]
FICM_MPWR	47	47.5	47.5	40 - 52	VOLTS	FICM Main Power [2]
FICM_VPWR	12	14	14	10.5 - 15.5	VOLTS	FICM Vehicle Power [3]
FLI	36.1	36.1	34.1	0 - 100	PERCENT	Fuel Level Indicator
FP#	ON/OFF	ON	ON	ON	ON/OFF	Fuel Pump
FPM	ON/OFF	ON	ON	ON	ON/OFF	Fuel Pump Monitor
FUELPW	0	990	680	0 - 50000	(μ seconds)	Fuel Pulse Width
GENB_F	NO FAULT	NO FAULT	NO FAULT	NO FAULT	NO FAULT/ YES FAULT	Generator 2 Output Fault
GENFIL	OFF	OFF	OFF	OFF	OFF/ON	Generator Fault Indicator
GEN_F	NO FAULT	NO FAULT	NO FAULT	NO FAULT	NO FAULT/ YES FAULT	Generator 1 Output Fault
GPCTM	0	0	0	0 - 180	TIME (seconds)	Glow Plug Coil On Time
GPL	ON/OFF	OFF	OFF	ON/OFF	ON/OFF	Glow Plug Indicator
GPLTM	0	0	0	10-Jan	TIME (seconds)	Glow Plug Indicator On Time
IAT	98.6	98.6	96.8	-290	TEMPERA-TURE (°F)	Intake Air Temp (MAF Assembly)
IAT2	113	113	113	-290	TEMPERA-TURE °F	Intake Air Temp (Manifold Air)
IATFM	NO FAULT	NO FAULT	NO FAULT	NO FAULT	NO FAULT/ YES FAULT	IAT Status
ICP	0.2	1	1.8	0.15 - 4.7	VOLT	Injection Control Pressure
ICP_DES	0	625.5	1400 - 1450	0 - 4045	PRESSURE (PSI)	ICP Desired
INJ1_OFF #	NORMAL	NORMAL	NORMAL	NORMAL	DISABLE/ NORMAL	Injector Disabled
INJ2_OFF #	NORMAL	NORMAL	NORMAL	NORMAL	DISABLE/ NORMAL	Injector Disabled
INJ3_OFF #	NORMAL	NORMAL	NORMAL	NORMAL	DISABLE/ NORMAL	Injector Disabled
INJ4_OFF #	NORMAL	NORMAL	NORMAL	NORMAL	DISABLE/ NORMAL	Injector Disabled
INJ5_OFF #	NORMAL	NORMAL	NORMAL	NORMAL	DISABLE/ NORMAL	Injector Disabled
INJ6_OFF #	NORMAL	NORMAL	NORMAL	NORMAL	DISABLE/ NORMAL	Injector Disabled

2006 PCM PIN IDENTIFICATION

PIDS (CONTINUED)						
Acronym	Key On	Low Idle	High Idle	Operating Range	Measurement Units	Description
INJ7_OFF #	NORMAL	NORMAL	NORMAL	NORMAL	DISABLE/ NORMAL	Injector Disabled
INJ8_OFF #	NORMAL	NORMAL	NORMAL	NORMAL	DISABLE/ NORMAL	Injector Disabled
INJ_TIM	0	-8.71	3.5	-100	ANGLE (Degrees)	Injector Timing (TDC)
IPR #	14.84	24	40.6	0 - 50	PERCENT	Injection Pressure Regulator
LOAD	0	12.5	14	0 - 100	PERCENT	Calculated Load Value
MAF	0	1.44	2.65	0.5 - 4.75	VOLT	Mass Air Flow
MAF_FM	NO FAULT	NO FAULT	NO FAULT	NO FAULT	NO FAULT/ YES FAULT	Mass Air Flow Status
MAP	96 (14)	96 (14)	103 (15)	41 - 324 (6 - 47)	kPa (PSI)	Manifold Absolute Pressure
MAP_V	1.3	1.33	1.4	0.5 - 4.5	VOLT	Manifold Absolute Pressure
MFDES	0	7.5	10.7	0 - 100	MASS mg/stroke	Mass Fuel Desired
MGP	0	0.25	0.8	-40	PRESSURE (psi)	Manifold Gauge Pressure
MIL	OFF	OFF	OFF	OFF	ON/OFF	Malfunction Indicator
MIL_DIS	926.1	926.1	926.1	-	Miles	Distance Since MIL Activated
MISF_EVAL	NO	NO	YES	YES	NO/YES	Misfire Monitor
MP_LRN	YES	YES	YES	YES	NO/YES	Learned Misfire Correction
NM	0	0	0	0 - 65535	NUMBER	Number of Misfires
PATSENABL	ENABLED	ENABLED	ENABLED	ENABLED	ENABLED/ DISABLED	Passive Anti-Theft
PBA	OFF	OFF	OFF	OFF	OFF/ON	Park Brake Applied
PNP	Neutral	Neutral	Neutral	Park/Neutral	Park/Neutral	Park/Neutral Position Switch
RPM	0	628	2500	0 - 4096	RPM	Revolutions Per Minute
RPMSD	625	625	625	625	RPM	Desired Idle RPM
SCCS	NONE	NONE	NONE	OFF/ON/ RESUME/ CANCEL/ NONE/SET/ COAST/ NOT AVAILABLE	OFF/ON/ RESUME/ CANCEL/ NONE/SET/ COAST/ NOT AVAILABLE	Speed Control Switches
STRT_RLY	DISABLED	DISABLED	DISABLED	DISABLED	DISABLED/ ENABLED	Starter Relay
SYNC	NO	YES	YES	YES	NO/YES	CMP and CKP Synchronized
TORQUE	0	20-Sep	50	400	Nm	Net Engine Torque
TRIP	NO	NO	NO	YES	N0/YES	OBD Trip Completed
TRIP CNT	0	0	0	0	0 - 255	Trip Count

2006 PCM PIN IDENTIFICATION

PIDS (CONTINUED)						
Acronym	Key On	Low Idle	High Idle	Operating Range	Measurement Units	Description
VFDES	0	9	14	0 - 250	VOLUME mm^3	Volume Fuel Desired
VGTDC #	0	73	43	0 - 100	PERCENT	Turbo Duty Cycle
VGT_F	NO	NO	NO	NO	NO/YES	Turbo Fault
VREF	4.9	4.9	4.9	4.5 - 5.5	VOLTS	Reference Voltage
VSS	0	0	0	0 - 120	SPEED (MPH)	Vehicle Speed
WAC	OFF	OFF	OFF	OFF	ON/OFF	WOT A/C Cutoff
WFS	NO	NO	NO	NO	NO/YES	Water in Fuel

[1] FICM logic power from the FICM relay.

[2] FICM injector power supply from the FICM relay.

[3] FICM vehicle power from the ignition switch.

2007 Parameter Identification (PID) List		
Acronym	**Description**	**Measurement Unit**
4WDL	4x4 Low by Request	Bitmap
ACCS	Air Conditioning Cycle Switch	ON/OFF
ACP	Air Conditioning Pressure Switch	Open/Closed
ANTISCAN	Anti-Scan Enable and Counting	Bitmap
APP	Accelerator Pedal Position	Percent
APP1	Accelerator Pedal Track 1	DCV
APP2	Accelerator Pedal Track 2	DCV
APP3	Accelerator Pedal Track 3	DCV
AST	Time Since Start	Minutes/Seconds
ARPMDES	Ancillary RPM Desired	RPM
AXLERAT	Axle Ratio Used by PCM to Calculate Vehicle Speed	Input Revs/Output Revs
B+	Battery Voltage	DCV
BARO	Barometric Pressure	kPa (PSI)
BPA	Brake Pressure Applied	ON/OFF
CCM STAT	CCM Monitor Completed	Bitmap
CLRDIST	Distance Traveled Since DTCs Cleared	Miles
CLRWRMUP	Number of Warm-ups Since DTCs Cleared	Numeric Value
CMP_F	Camshaft Position Failure	Bitmap
CMPFM	CMP Status	No/Yes Fault
CPP_BOT	Clutch Pedal Position (bottom of travel)	Yes/No
CPP_TOP	Clutch Pedal Position (top of travel)	Yes/No
CPP/TCS	CPP/TCS Engaged	Mode
CPP/PNP	Clutch Pedal Position	Drive/Neutral
DECHOKE	Dechoke Mode	Yes/No
DRIVECNT	Valid Drive Counter	Numeric Value
DTCCNT	DTC Count	Numeric Value
ECT	Coolant Temperature	Degree °C (°F)
ECT_F	ECT Fault	No/Yes Fault
EGR STAT	EGR Monitor Completed	Bitmap
EGR EVAL	EGR Evaluated	No/Yes
EGR DC#	EGR Duty Cycle	Percent
EGRFM	EGR Failure Rate	Bitmap
EGR_F	EGR Status	Fault/NoFault
EGRVP	EGR Valve Position	DCV
EGRVPDES	EGR Valve Position Desired	Percent
EGR_PCT	Commanded EGR	Percent
EGR_ERR	EGR Error From Commanded	Percent
ENSTAT	Engine Start Enabled	Bitmap
EOT	Engine Oil Temperature	Degrees °C/°F
EOT_F	Engine Oil Temperature Status	No/Yes Fault
EPG	Exhaust Pressure Gauge	kPa Gauge
EPDSD	Exhaust Back Pressure Desired	kPa
EBP	Exhaust Pressure	DCV
EBP_A	Exhaust Pressure Fault	No/Yes Fault
EP_V	Exhaust Pressure	Volts
EBP_DSD	Exhaust Pressure Desired	PSI
FANSS	Fans Speed	RPM
FANSSM	Cooling Fan Driver	Low/High
FANVAR_F	Fan Speed Monitor	No/Yes Fault
FICMSYNC	Synchronization from FICM	Yes/No
FICM_LPWR	FICM Logic Power	DCV
FICM_MPWR	FICM Main Power	DCV

2007 Parameter Identification (PID) List		
FICM_VPWR	FICM Vehicle Power	DCV
FLI	Fuel Level Indicator	Percent
FP#	Fuel Pump	ON/OFF
FPM	Fuel Pump Monitor	ON/OFF
FUELPW	Fuel Pulse Width	Time
GEAR	Commanded Gear	State Encoded
GEAR_OSC	Gear Commanded OSC Status	State Encoded
GPDC	Glow Plug Control Duty Cycle	Percent
GEARMAX	Highest Gear Permitted	None
GENDC	Generator Duty Cycle	Percent
GEN2_F	Generator 2 Output Fault	No/Yes Fault
GENFIL	Generator 1 Output Fault	ON/OFF
GENFI	Generator Fault Indicator Light Commanded On	Bitmap
GEN_F(GFSF)	Generator Fault Indicator	No/Yes Fault
GPCTM	Glow Plug Coil On Time	Time (Seconds)
GPL	Glow Plug Indicator	ON/OFF
GPLTM	Glow Plug Indicator On Time	Time (Seconds)
HFPIP	Last PIP Up Edge to Last PIP Down Edge	Seconds
HRSH_SHFT	Harsh Shift Commanded by OSC	Bitmap
IAC_MODE	Idle Speed Control Operating Mode	None
IAT	Intake Air Temperature (MAF Assembly)	Degrees C° (F°)
IAT_V	Intake Air Temperature (Raw)	A/D Counts
ISS_F	ISS is Currently Unreliable	Bitmap
ISS_SRC	Intermediate Shaft Speed, No FMEM	RPM
IAT2	Intake Air Temperature (Manifold Air)	Temperature
IAT_F	IAT Status	No/Yes Fault
ICP	Injection Control Pressure	Volts
ICP_DES	Injection Control Pressure Desired	Pressure (PSI)
INJ1_OFF #	Injector Disabled	N/A
INJ2_OFF #	Injector Disabled	N/A
INJ3_OFF #	Injector Disabled	N/A
INJ4_OFF #	Injector Disabled	N/A
INJ5_OFF #	Injector Disabled	N/A
INJ6_OFF #	Injector Disabled	N/A
INJ7_OFF #	Injector Disabled	N/A
INJ8_OFF #	Injector Disabled	N/A
INJ_TIM	Injector Timing TDC	Degrees
IPATS	Integrated Passive Anti-Theft System Status	Bitmap
IPR #	Injector Pressure Regulator	Percent
LOAD	Calculated Load Valve	Percent
LINEDSD	Desired Main Line Pressure	PSI
MAF	Mass Air Flow	DCV
MAF_F	Mass Air Flow Status	No/Yes Fault
MAP	Manifold Absolute Pressure	kPa (PSI)
MAP_V	Manifold Absolute Pressure	DCV
MFF_PNP	In Drive During Time of Misfire	Bitmap
MFDES	Mass Fuel Desired	Mass Mg/Stroke
MGP	Manifold Gauge Pressure	Pressure (PSI)
MIL	Malfunction Indicator	ON/OFF
MIL_DIS	Distance Since MIL Activated	Miles
MIS STAT	Misfire Monitor Completed	Bitmap
MISF_EVAL	Misfire Monitor	No/Yes
MIS_CNT1	Misfire Count Cylinder #1 (High Byte)	# Misfires
MIS_CNT1	Misfire Count Cylinder #1 (Low Word)	# Misfires

2007 Parameter Identification (PID) List		
MIS_CNT2	Misfire Count Cylinder #2 (High Byte)	# Misfires
MIS_CNT2	Misfire Count Cylinder #2 (Low Word)	# Misfires
MIS_CNT3	Misfire Count Cylinder #3 (High Byte)	# Misfires
MIS_CNT3	Misfire Count Cylinder #3 (Low Word)	# Misfires
MIS_CNT4	Misfire Count Cylinder #4 (High Byte)	# Misfires
MIS_CNT4	Misfire Count Cylinder #4 (Low Word)	# Misfires
MIS_CNT5	Misfire Count Cylinder #5 (High Byte)	# Misfires
MIS_CNT5	Misfire Count Cylinder #5 (Low Word)	# Misfires
MIS_CNT6	Misfire Count Cylinder #6 (High Byte)	# Misfires
MIS_CNT6	Misfire Count Cylinder #6 (Low Word)	# Misfires
MIS_CNT7	Misfire Count Cylinder #7 (High Byte)	# Misfires
MIS_CNT7	Misfire Count Cylinder #7 (Low Word)	# Misfires
MIS_CNT8	Misfire Count Cylinder #8 (High Byte)	# Misfires
MIS_CNT8	Misfire Count Cylinder #8 (Low Word)	# Misfires
MIS_CNT	Misfire Count Total (High Byte)	# Misfires
MIS_CNT	Misfire Count Total (Low Word)	# Misfires
MISDEL1	Misfire Percent Delta Cylinder #1	Percent
MISDEL2	Misfire Percent Delta Cylinder #2	Percent
MISDEL3	Misfire Percent Delta Cylinder #3	Percent
MISDEL4	Misfire Percent Delta Cylinder #4	Percent
MISDEL5	Misfire Percent Delta Cylinder #5	Percent
MISDEL6	Misfire Percent Delta Cylinder #6	Percent
MISDEL7	Misfire Percent Delta Cylinder #7	Percent
MISDEL8	Misfire Percent Delta Cylinder #8	Percent
MP_LRN	Learned Misfire Correction	Number
NM	Number of Misfires	Number
OSS_DIR	Output Shaft Speed Direction	State Encoded
OSS	Output Shaft Sensor	RPM
OSS_SRC	Output Shaft Speed, No FMEM	RPM
OSSFM	Output Shaft Speed Sensor Failure Mode	Bitmap
OSS_F	OSS is Currently Unreliable	Bitmap
OSS_DIR	Direction of OSS Spring	State Encoded
PATSOUT	Outcode Status	None
PSP	Power Steering Load	Bitmap
	Software ID Block Starting Address	None
PATSENABL	Passive Anti-theft	Enabled/Disabled
PCA	Commanded Pressure Electronic Pressure Control A	PSI
PCA_F	Electronic Pressure Control A Fault Status	State Encoded
PBA	Park Brake Applied	OFF/ON
PCA_AMP	Commanded Current for PCA	Amps
PCM_ID	PCM ID Stored	Bitmap
PCM_VFY	PCM ID Verified	Bitmap
PIP_CTR	Counter Incremented Every PIP Rising Edge	None
PIP_HIGH	PIP State is High	Bitmap
PNP	Park/Neutral Position Switch	Park/Neutral
PCFM	Pressure Control Failure Mode	Bitmap
PIP_CTR	Counter incremented every PIP rising edge	None
PIPTIM	Last PIP time in seconds	Seconds
PTO	Power Take-Off Status	DCV
PSA	Trans Pressure Switch A is Active	Bitmap
PSB	Trans Pressure Switch B is Active	Bitmap
PSC	Trans Pressure Switch C is Active	Bitmap
PSD	Trans Pressure Switch D is Active	Bitmap
PSE	Trans Pressure Switch E is Active	Bitmap

2007 Parameter Identification (PID) List		
RPM	Revolutions Per Minute	RPM
RPMSD	Desired Idle RPM	RPM
RLC_F	Reverse Lamp Control Fault	Bitmap
RUNTM	Time Since Engine Start	Seconds
SCCS	Speed Control Switches	DCV
SSD_AMP	Commanded Current for SSD	Amps
SSE_AMP	Commanded Current for SSE	Amps
STRT_RLY	Starter Relay	Disabled/Enabled
SHIFT_ID	Shift Identification	None
SHIFT_TYPE/ GEARTR	Commanded Gear and PRNDL	None
SHFT_LAG	0 - 10% Shift Time	Seconds
SHFT_TM	10 - 90% Shift Time	Seconds
SHFT_DROP	RPM Overshoot For PID 1931	RPM
SSPCD_F	Shift Solenoid D Fault Status	State Encoded
SHFT_FLRE	Flare RPM for PID 1931	RPM
SYNC	Camshaft/Crankshaft Are in Sync	No/Yes
SECSTAT	IPATS Security Access Status	None
SSPCD	Commanded Pressure Shift Solenoid D	PSI
SSPCE	Commanded Pressure Shift Solenoid E	PSI
SSPCE_F	Shift Solenoid E Output Fault Status	State Encoded
SSPCA	Commanded Pressure Shift Solenoid A	PSI
SSPCB	Commanded Pressure Shift Solenoid B	PSI
SSPCC	Commanded Pressure Shift Solenoid C	PSI
SSA_AMP	Commanded Current for SSA	AMPS
SSB_AMP	Commanded Current for SSB	AMPS
SSC_AMP	Commanded Current for SSC	AMPS
SSPCA_F	Shift Solenoid A Pressure Control Output Fault	State Encoded
SSPCB_F	Shift Solenoid B Pressure Control Output Fault	State Encoded
SSPCC_F	Shift Solenoid C Pressure Control Output Fault	State Encoded
SECTIME	Security Access Time	Minute
SHFT_TYP	Shift Type	State Encoded
SRC	Starter Motor Relay is Enabled	Bitmap
SPAREKEY	IPATS Spare Key Programming Enable	Bitmap
TSS_F	TSS is Currently Unreliable	Bitmap
TRO_N_F	Neutral Output Fault	Bitmap
TRO_P_F	Park Output Fault	Bitmap
TRFM	TR Failure Mode	Bitmap
TCIL	TCIL Light Commanded ON	Bitmap
TCS FLG	Transmission Switch Applied	Bitmap
TP_MODE	Throttle Mode	State Encoded
TP_R	Relative Throttle Position	Percent
TP V	Throttle Position (Before FMEM)	A/D Counts
TCCPC	Torque Converter Commanded Pressure	PSI
TCCPC_F	Torque Convertor Control Fault Status	State Encoded
TCC_AMP	Commanded Current for TCC	Amps
TCCMCMD	Torque Converter Clutch Desired Slip	RPM
TPLRNMIN	Lowest Idle Throttle Position	A/D Counts
T/F Delta PID	Mass Air Flow Status	MAF FM
TPLRNMIN	Closed Throttle Position	A/D Counts
TSLIPDSD	Desired Slip Across the Torque Converter	RPM
TCCFM	Converter Unlock Due to Excess Slip	Bitmap
TCILF	TCIL Fault Detected	Bitmap
TCC_OT	Overtemp Lock-Up Mode	Bitmap

2007 Parameter Identification (PID) List		
TCCMACT	Torque Convertor Slip (absolute value)	RPM
TCCDC	Torque Convertor Clutch Duty Cycle	Percent
TRAN_OT	Transmission is Overtemperature	Bitmap
TRAC	Traction Assist is Available	Bitmap
TR_DC	Duty Cycle of TRS	Percent
TSLIPRAT	Speed Ratio Across Torque Convertor	None
TSLIP	Actual Slip Across Torque Converter	RPM
TQ_CNTL	Torque Reduction Request	State Encoded
TR	PRNDL Position	State Encoded
TORQUE	Net Engine Torque	Nm
TMSTRT	Time Since Engine Start	Seconds
TIREREV	Tire Size Used by PCM to Calculate VS	Tire Revs Per Mile
TCCRAT	Torque Converter Clutch Speed Ratio	None
TCC_OSC	Torque Converter Clutch OSC Status	State Encoded
TCS	Transmission Control Switch Applied	Bitmap
TFT_V	Transmission Fluid Temp. Counts (Raw)	A/D Counts
TFT	Transmission Fluid Temperature	Degrees F
TRANRAT	Transmission Gear Ratio	None
TSS/ISS	Turbine Speed Adjusted for Chain Ratio	RPM
TSS	Turbine Shaft Speed	RPM
TSS_SRC	Turbine Shaft Speed, No FMEM	RPM
TRIP	OBD Trip Completed	No/Yes
TRIP CNT	Trip Count	Numbers
TFTFM	Trans Fluid Temperature Failure Mode	Bitmap
VBAT_FF	Battery Voltage at Time of J1979 Freeze Frame	Volts
VSS_F	VSS is Currently Unreliable	Bitmap
VSSFM	Vehicle Speed Sensor Failure Mode	Bitmap
VFDES	Volume Fuel Desired	Volume/mm^3
VGTDC #	Turbo Duty Cycle	Percent
VGT_F	Variable Geometry Turbo-Fault	No/Yes Fault
VPWR	Battery Voltage	DCV
VREF	Battery Voltage	DCV
VSS	Vehicle Speed	Speed (MPH)
WAC	WOT A/C Cutoff	ON/OFF
WFS	Water in Fuel	No/Yes

2007 DIAGNOSTIC TOOL — DRIVER-OPERATED CONTROLS CHECK				
PID	**Operator Setting**	**Reading**	**Voltage**	**Pinpoint Test**
APP_MODE	Accelerator Pedal Released	0%	95/100%	AG
	Accelerator Pedal Fully Applied	0%	95/100%	
BOO/BPP	Brake Applied	ON	B+	AI
	Brake Released	OFF	0V	
CPP/TCS	(ZF 6-Speed)			C
	Clutch Pedal Applied	ON	0V	
	Clutch Pedal Released	OFF	B+	
PBPP	Parking Brake ON	ON	0V	I
	Parking Brake OFF	OFF	B+	
CPP/TCS	(Torque Shift)	OFF	0V	
	Key On:			
	Push CNCL (Latch)	ON	B+	
	Push CNCL (VF Latch)	OFF	0 V	
TR	PRNDL in P	PARK		
	PRNDL in R	REV		
	PRNDL in N	NTRL		
	PRNDL in OD	OD		
	PRNDL in OD with Tow/Haul Cancel Light ON	DRIVE		
	PRNDL in L2	MAN2		
	PRNDL in L1	MAN1		
TCIL	(5R110)			
	After Key ON	OFF	0V	
	Push Tow/Haul Cancel (Latch)	ON	B+	
	Push Tow/Haul Cancel (Unlatch)	OFF	0V	
4WDL	(Transfer Case Shifter)			
	4x4H or 2H	OFF	B+	
	4x4L	ON	0 V	
ACCS	(A/C Controls)			
	MAX/NORM AC or MIX/DEF with A/C Clutch ON	ON	B+	
	Floor or Vent or A/C Clutch OFF	OFF	0 V	

2007 PCM PIN IDENTIFICATION

Engine

Sensors/Inputs	PCM Pin/PID only	Measured/PID Values				Units Measured/PID
		KOEO	Hot Idle	48 KM/H (30 MPH)	89 KM/H (55 MPH)	
ICP	29	0.2 (A)	1.0 (A)	1.1 (A)	1.2 (A)	DCV
ECT	32	1.9 (B)	1.9 (B)	1.1 (B)	0.65 (B)	DCV
EGRVP	33	0.8	1.25	1.6	1.4	DCV
EOT	44	0.68 (B)	0.68 (B)	0.61 (B)	0.60 (B)	DCV
IAT2	45	2.7 (C)	3.7 (C)	4.31 (C)	4.31 (C)	DCV

A. 1 volt equals 4.5 MPa (652 psi) and 4.45 volts equals 28 MPa (4061 psi)

B. 4.9 volts equals -40°C (-40°F) and 0.5 volts equals 100°C (212°F)

C. 4.3 volts equals 0°C (32°F) and 2.5 volts equals 50°C (122°F)

Actuators/Outputs	PCM Pin/PID only	Measured/PID Values				Units Measured/PID
		KOEO	Hot Idle	48 KM/H (30 MPH)	89 KM/H (55 MPH)	
VGTC	10	OPEN	VARYING	VARYING	VARYING	OPEN/ VARYING/ CLOSED
PTOC	15	0	0	0	0	DCV

Body

Sensors/Inputs	PCM Pin/PID only	Measured/PID Values				Units Measured/PID
		KOEO	Hot Idle	48 KM/H (30 MPH)	89 KM/H (55 MPH)	
PTO	12	0/12.44 (D)	0/13.75 (D)	0/14 (D)	0/14.19 (D)	DCV
APP3	25	1.0 (E)	1.0 (E)	1.45 (E)	1.5 (E)	DCV
APP1	26	4.1 (F)	4.1 (F)	3.5 (F)	3.3 (F)	DCV
APP2	37	1.46 (G)	1.46 (G)	1.77 (G)	2.05 (G)	DCV
BARO	38	3.3	3.3	3.3	3.3	DCV
MAP	41	1.28 (H)	1.33 (H)	1.41 (H)	2.03 (H)	DCV
MAF	42	0	1.25	2.47	2.84	DCV
IAT	43	3.75 (B)	3.91 (B)	4.3 (B)	4.27 (B)	DCV
ACCS	PID	OFF	OFF	OFF	OFF/ON	ON/OFF
ACP	PID	OPEN (OK)	OPEN (OK)	OPEN (OK)	OPEN (OK)	OPEN/ CLOSED
APP	PID	0	0	14	18	%
APP1	PID	4	4	3.44	3.28	DCV
APP2	PID	1.5	1.5	1.9	2.1	DCV
APP3	PID	0.9	0.9	1.4	1.6	DCV
APP_MODE	PID	CT	CT	PT	PT	CT, PT, WOT
AST	PID	0:00	0:07	1:00	23:51	min : sec
B+	PID	12	14	14	14.3	DCV
BARO	PID	96.5 (14)	96.5 (14)	96.5 (14)	96.5 (14)	kPa (psi)
BOO	PID	OFF	OFF	OFF	OFF	ON/OFF
BPA	PID	OFF	OFF	OFF	OFF	ON/OFF
CLRDIST	PID	926.1	926.1	926.1	926.1	Miles
CLRWRMUP	PID	1	1	1	1	Numeric Value
CMPFM	PID	NO FAULT	NO FAULT	NO FAULT	NO FAULT	NO FAULT/ YES FAULT
CPP/PNP	PID	NEUTRAL	NEUTRAL	DRIVE	DRIVE	DRIVE/ NEUTRAL
CPP_BOT	PID	NO	NO	NO	NO	YES/NO
DECHOKE	PID	NO	NO	NO	NO	YES/NO
DRIVECNT	PID	4	3	1	1	Numeric Value
DTCCNT	PID	1	3	1	1	Numeric Value
ECT	PID	88 (190)	88 (190)	88 (190)	88 (190)	DEG °C (°F)
ECT_F	PID	NO FAULT	NO FAULT	NO FAULT	NO FAULT	NO FAULT/ YES FAULT

2007 PCM PIN IDENTIFICATION

Body (CONTINUED)						
EGR EVAL	PID	NO	NO	NO	YES	YES/NO
EGR F	PID	NO FAULT	NO FAULT	NO FAULT	NO FAULT	FAULT/NO FAULT
EGRVP	PID	0.8	1.25	1.6	1.4	DCV
EGRVPDES	PID	0	11.8	0	0	%
EOT	PID	88 (190)	88 (190)	88 (190)	92 (197)	DEG °C (°F)
EOT F	PID	NO FAULT	NO FAULT	NO FAULT	NO FAULT	NO FAULT/ YES FAULT
EBP A	PID	103 (15)	117 (17)	145 (21)	173 (25.2)	kPa (psi)
EBP	PID	0.9	1.13	1.7	1.99	DCV
EBP DSD	PID	3.26	2.46	7.5	10.25	psi
FANSS	PID	0	425	517	595	RPM
FANSSM	PID	LOW	LOW	LOW	LOW	LOW/HIGH
FANVAR F	PID	NO FAULT	NO FAULT	NO FAULT	NO FAULT	NO FAULT/ YES FAULT
FICMSYNC	PID	NO	YES	YES	YES	YES/NO
FICM LPWR	PID	12.5	13.5	14	14	DCV
FICM MPWR	PID	47.5	48	48.5	48	DCV
FICM VPWR	PID	12.5	14	14	14	DCV
FLI	PID	36.1	36.1	34.1	34.1	%
FPM	PID	ON/OFF	ON	ON	ON	ON/OFF
GEN2 F	PID	NO FAULT	NO FAULT	NO FAULT	NO FAULT	NO FAULT/ YES FAULT
GENFIL	PID	OFF	OFF	OFF	OFF	OFF/ON
GEN F (GFSF)	PID	NO FAULT	NO FAULT	NO FAULT	NO FAULT	NO FAULT/ YES FAULT
IAT	PID	98.6	98.6	96.8	96.8	DEG (°F)
IAT2	PID	113	113	113	113	DEG (°F)
IAT F	PID	NO FAULT	NO FAULT	NO FAULT	NO FAULT	NO FAULT/ YES FAULT
GPL	PID	ON/OFF	OFF	OFF	OFF	ON/OFF
LOAD	PID	0	12.5	14	31	%
MAF	PID	0	1.25	2.47	2.84	DCV
MAF F	PID	NO FAULT	NO FAULT	NO FAULT	NO FAULT	NO FAULT/ YES FAULT
MAP	PID	96 (14)	103 (15)	110 (16)	149 (21.64)	kPa (psi)
MAP V	PID	1.3	1.33	1.4	2.03	DCV
MGP	PID	0	0.18	1.26	6.77	PRESSURE (psi)
MIL	PID	OFF	OFF	OFF	OFF	ON/OFF
MIL DIS	PID	926.1	926.1	926.1	926.1	Miles
MISF EVAL	PID	NO	NO	YES	YES	YES/NO
MP LRN	PID	YES	YES	YES	YES	YES/NO
NM	PID	0	0	0	0	Numeric Value
PATSENABL	PID	ENABLED	ENABLED	ENABLED	ENABLED	ENABLED/ DISABLED
PBA	PID	OFF	OFF	OFF	OFF	ON/OFF
PNP	PID	Neutral	Neutral	Neutral	Park/Neutral	Park/Neutral
RPM	PID	0	726	1320	1751	RPM
RPMDSD	PID	4096	725	725	725	RPM
STRT RLY	PID	DISABLED	DISABLED	DISABLED	DISABLED	DISABLED/ ENABLED
SYNC	PID	NO	YES	YES	YES	YES/NO
TORQUE	PID	0	14.69	180	236	Nm
TRIP	PID	NO	NO	NO	YES	YES/NO
TRIP CNT	PID	0	0	0	0	0 - 255
VGT F	PID	NO FAULT	NO FAULT	NO FAULT	NO FAULT	NO FAULT/FAULT
WAC	PID	OFF	OFF	OFF	OFF	ON/OFF
WFS	PID	NO	NO	NO	NO	YES/NO
VREF	PID	5	5	5	5	DCV
VSS	PID	0	0	30	55	SPEED (MPH)

2007 PCM PIN IDENTIFICATION

Actuators/Outputs	PCM Pin/PID only	Measured/PID Values				Units Measured/PID
		KOEO	Hot Idle	48 KM/H (30 MPH)	89 KM/H (55 MPH)	
EGRDC	PID	0	35	38	38	%
FANVAR	PID	0	0	0	0	%
FP #	PID	ON/OFF	ON	ON	ON	ON/OFF
FUELPW	PID	0	944	880	896	(μ seconds)
GPCTM	PID	41	0	0	0	TIME (seconds)
GPLTM	PID	0	0	0	0	TIME (seconds)
ICP	PID	0.24	0.85	1.05	1.21	DCV
ICP_DES	PID	0	580	858.6	1126.6	PRESSURE (psi)
INJ_TIM	PID	0	-4	5.85	3.39	DEG
IPR #	PID	14.84	22	29.3	36.33	%
MFDES	PID	0	0.1	0.34	0.32	MASS mg/stroke
VFDES	PID	0	9	28	34	VOLUME mm^3
VGTDC #	PID	0	67	56	51	%

Other	PCM Pin/PID only	Measured/PID Values				Units Measured/PID
		KOEO	Hot Idle	48 KM/H (30 MPH)	89 KM/H (55 MPH)	
VPWR	34	12.44	13.75	14	14.19	DCV
VREF	45	5	5	5	5	DCV

A. 1 volt equals 4.5 MPa (652 psi) and 4.45 volts equals 28 MPa (4061 psi)
B. 4.9 volts equals -40°C (-40°F) and 0.5 volts equals 100°C (212°F)
C. 4.3 volts equals 0°C (32°F) and 2.5 volts equals 50°C (122°F)
D. 0 volts equals disengaged and battery voltage equals engaged
E. CT equals 0.8 - 1.1 volts and WOT equals 3.1 - 3.5 volts
F. CT equals 3.9 - 4.2 volts and WOT equals 0.7 - 1.2 volts
G. CT equals 1.4 - 1.6 volts and WOT equals 3.6 - 4.1 volts
H. 1.35 volts equals 103 kPa (14.9 psi) and 4.5 volts equals 300 kPa (43.5 psi)

DIAGNOSTIC TROUBLE CODE DESCRIPTIONS & POSSIBLE CAUSES

Diagnostic Trouble Code (DTC) Descriptions

Fault Code	Circuit Index	DTC Description	Probable Causes
P0046	VGT Solenoid	Turbo/Super Charger Boost Control Solenoid Circuit Range/Performance	Open circuit, control circuit short to voltage, control circuit short to ground, solenoid
P0069	BARO	MAP/BARO Correlation	Biased MAP or BARO sensor, circuitry
P0096	IAT2	Intake Air Temperature Sensor 2 Circuit Range/Performance	IAT2 sensor biased, circuitry
P0097	IAT2	Intake Air Temperature Sensor 2 Circuit Low Input	Short to ground, IAT2 sensor
P0098	IAT2	Intake Air Temperature Sensor 2 Circuit High Input	Short to voltage, signal circuit open, IAT2 sensor
P0101	MAF	Mass or Volume Air Flow Circuit Range/Performance	Restricted air flow, MAF sensor, contamination, intake air leak
P0103	MAF	Mass or Volume Air Flow Circuit High Input	Restriction, water intrusion, signal short to voltage, return circuit open, MAF sensor
P0107	BARO	Manifold Absolute Pressure/BARO Sensor Low Input	Open VREF, signal circuit short to ground, BARO sensor
P0108	BARO	Manifold Absolute Pressure/BARO Sensor High Input	Short to voltage, signal return circuit open, BARO sensor
P0112	IAT	Intake Air Temperature Circuit Low Input	Short to ground, IAT sensor
P0113	IAT	Intake Air Temperature Circuit High Input	Short to voltage, signal circuit open, IAT sensor
P0117	ECT	Engine Coolant Temperature Circuit Low Input	Short to ground, ECT sensor
P0118	ECT	Engine Coolant Temperature High Input	Short to voltage, signal circuit open, ECT sensor
P0196	EOT	Engine Oil Temperature Sensor Circuit Range/Performance	Circuitry, biased sensor, thermostat
P0197	EOT	Engine Oil Temperature Sensor Circuit Low Input	Short to ground, EOT sensor
P0198	EOT	Engine Oil Temperature Sensor Circuit High Input	Short to voltage, signal circuit open, EOT sensor
P0219		Engine Overspeed Condition	Incorrect downshift, CKP and CMP interference, excessive engine speed
P0230	Fuel Pump	Fuel Pump Primary Circuit	Control circuit open, fuel pump relay
P0231	Fuel Pump	Fuel Pump Secondary Circuit Low	Inertia switch, fuel pump relay, open or short circuit
P0232	Fuel Pump	Fuel Pump Secondary Circuit High	Fuel pump relay, short circuit, fuel pump
P0236	MAP	Turbo/Super Charger Boost Sensor A Circuit Range/Performance	Restricted exhaust, MAP hose, MAP sensor
P0237	MAP	Turbo/Super Charger Boost Sensor A Circuit Low	Open circuit, short to ground, MAP sensor
P0238	MAP	Turbo/Super Charger Boost Sensor A Circuit High	Circuit short to voltage, MAP sensor
P0261	INJ	Cylinder Number 1 Injector Circuit Low	Open circuit, fuel injector, FICM connector loose
P0262	INJ	Cylinder Number 1 Injector Circuit High	Circuit short to ground or voltage, fuel injector
P0263	INJ	Cylinder Number 1 Contribution/Balance	Power cylinder, valve train or injector
P0264	INJ	Cylinder Number 2 Injector Circuit Low	Open circuit, fuel injector, FICM connector loose
P0265	INJ	Cylinder Number 2 Injector Circuit High	Circuit short to ground or voltage, fuel injector
P0266	INJ	Cylinder Number 2 Contribution/Balance	Power cylinder, valve train or injector
P0267	INJ	Cylinder Number 3 Injector Circuit Low	Open circuit, fuel injector, FICM connector loose
P0268	INJ	Cylinder Number 3 Injector Circuit High	Circuit short to ground or voltage, fuel injector
P0269	INJ	Cylinder Number 3 Contribution/Balance	Power cylinder, valve train or injector
P0270	INJ	Cylinder Number 4 Injector Circuit Low	Open circuit, fuel injector, FICM connector loose
P0271	INJ	Cylinder Number 4 Injector Circuit High	Circuit short to ground or voltage, fuel injector
P0272	INJ	Cylinder Number 4 Contribution/Balance	Power cylinder, valve train or injector
P0273	INJ	Cylinder Number 5 Injector Circuit Low	Open circuit, fuel injector, FICM connector loose
P0274	INJ	Cylinder Number 5 Injector Circuit High	Circuit short to ground or voltage, fuel injector
P0275	INJ	Cylinder Number 5 Contribution/Balance	Power cylinder, valve train or injector
P0276	INJ	Cylinder Number 6 Injector Circuit Low	Open circuit, fuel injector, FICM connector loose

DIAGNOSTIC TROUBLE CODE DESCRIPTIONS & POSSIBLE CAUSES

Diagnostic Trouble Code (DTC) Descriptions

Fault Code	Circuit Index	DTC Description	Probable Causes
P0277	INJ	Cylinder Number 6 Injector Circuit High	Circuit short to ground or voltage, fuel injector
P0278	INJ	Cylinder Number 6 Contribution/Balance	Power cylinder, valve train or injector
P0279	INJ	Cylinder Number 7 Injector Circuit Low	Open circuit, fuel injector, FICM connector loose
P0280	INJ	Cylinder Number 7 Injector Circuit High	Circuit short to ground or voltage, fuel injector
P0281	INJ	Cylinder Number 7 Contribution/Balance	Power cylinder, valve train or injector
P0282	INJ	Cylinder Number 8 Injector Circuit Low	Open circuit, fuel injector, FICM connector loose
P0283	INJ	Cylinder Number 8 Injector Circuit High	Circuit short to ground or voltage, fuel injector
P0284	INJ	Cylinder Number 8 Contribution/Balance	Power cylinder, valve train or injector
P0297	INJ	Vehicle Overspeed Condition	Excessive engine speed, VSS interference
P0298	EOT	Engine Oil Overtemperature Condition	EOT sensor, circuitry, cooling system, thermostat
P0299	EP	Turbo/Super Charger Underboost	EP sensor, VGT control valve slow/lack of response
P0300	INJ	Random Misfire Detected	Oil/fuel aeration, base engine
P0301	INJ	Cylinder Number 1 Misfire Detected	Base engine, injector
P0302	INJ	Cylinder Number 2 Misfire Detected	Base engine, injector
P0303	INJ	Cylinder Number 3 Misfire Detected	Base engine, injector
P0304	INJ	Cylinder Number 4 Misfire Detected	Base engine, injector
P0305	INJ	Cylinder Number 5 Misfire Detected	Base engine, injector
P0306	INJ	Cylinder Number 6 Misfire Detected	Base engine, injector
P0307	INJ	Cylinder Number 7 Misfire Detected	Base engine, injector
P0308	INJ	Cylinder Number 8 Misfire Detected	Base engine, injector
P0335	CKP	Crankshaft Position Sensor A Circuit	Circuitry, CKP sensor, electrical noise
P0336	CKP	Crankshaft Position Sensor A Circuit Range/Performance	Circuitry, CKP sensor, electrical noise
P0340	CMP	Camshaft Position Sensor A Circuit (Bank 1 or Single Sensor)	Circuitry, CMP sensor, electrical noise
P0341	CMP	Camshaft Position Sensor A Circuit Range/Performance (Bank 1 or Single Sensor)	Circuitry, CMP sensor, electrical noise
P0381	GPL	Glow Plug/Heater Indicator Circuit	Open circuit, short to ground, indicator
P0401	EGR	Exhaust Gas Recirculation Flow Insufficient Detected	EGR valve, EGR position sensor bias, EP sensor bias
P0402	EGR	Exhaust Gas Recirculation Flow Excessive Detected	EGR valve, EGR position sensor bias, EP sensor bias
P0403	EGR	Exhaust Gas Recirculation Control Circuit	Open circuit, short to ground or voltage, EGR valve
P0404	EGR	Exhaust Gas Recirculation Control Circuit Range/Performance	EGR sensor or valve, EGR position circuit
P0405	EGR	Exhaust Gas Recirculation Sensor A Circuit Low	Open circuit, short to ground, EGR position sensor
P0406	EGR	Exhaust Gas Recirculation Sensor A Circuit High	Short to voltage, EGR position sensor
P0460	FLI	Fuel Level Sensor A Circuit	Open circuit, short circuit, fuel sending unit, instrument cluster
P0462	FLI	Fuel Level Sensor A Circuit Low Input	Open circuit, short circuit, fuel sending unit, instrument cluster
P0463	FLI	Fuel Level Sensor A Circuit High Input	Open circuit, short circuit, fuel sending unit, instrument cluster
P0470	EP	Exhaust Pressure Sensor	EP sensor, open signal return
P0471	EP	Exhaust Pressure Sensor Range/Performance	EP sensor, restricted supply tube
P0472	EP	Exhaust Pressure Sensor Low Input	Open VREF, short circuit, open circuit, EP sensor
P0473	EP	Exhaust Pressure Sensor High Input	Short to voltage, EP sensor
P0478	EP	Exhaust Pressure Control Valve High Input	EP sensor, VGT control valve slow/lack of response
P0480	VFAN	Fan 1 Control Circuit	Short to voltage, short to ground, open circuit

DIAGNOSTIC TROUBLE CODE DESCRIPTIONS & POSSIBLE CAUSES

Diagnostic Trouble Code (DTC) Descriptions

Fault Code	Circuit Index	DTC Description	Probable Causes
P0500	VSS	Vehicle Speed Sensor A	VSS sensor, circuitry
P0528	FSS	Fan Speed Sensor Circuit No Signal	Mechanical fault, short to voltage, open circuit, short to ground
P0562	PCED	System Voltage Low	Charging system fault
P0563	PCED	System Voltage High	Charging system fault
P0565	SCCS	Cruise Control ON Signal	Open circuit, short circuit, switch
P0566	SCCS	Cruise Control OFF Signal	Open circuit, short circuit, switch
P0567	SCCS	Cruise Control RESUME Signal	Open circuit, short circuit, switch
P0568	SCCS	Cruise Control SET Signal	Open circuit, short circuit, switch
P0569	SCCS	Cruise Control COAST Signal	Open circuit, short circuit, switch
P0605	PCED	Powertrain Control Module Read Only Memory (ROM) error	Internal PCM failure
P0606	PCED	ECM/PCM Processor	Internal PCM fault, Engine oil temperature (EOT) sensor connector intermittent connection
P0611	FICM	Fuel Injector Control Module Performance	Internal FICM error
P0620	ALT	Generator 1 Control Circuit	Charging system
P0623	ALT	Generator Lamp Control Circuit	Charging system
P0645	PCED	A/C Clutch Relay Control Circuit	A/C circuitry, A/C relay
P0649	PCED	Cruise Control Lamp Control Circuit	Circuitry, indicator
P0670	GPCM	Glow Plug Module Control Circuit	Short to voltage, short to ground, open circuit
P0671	GPCM	Cylinder 1 Glow Plug Circuit	Open circuit, short to ground, short to voltage, glow plug
P0672	GPCM	Cylinder 2 Glow Plug Circuit	Open circuit, short to ground, short to voltage, glow plug
P0673	GPCM	Cylinder 3 Glow Plug Circuit	Open circuit, short to ground, short to voltage, glow plug
P0674	GPCM	Cylinder 4 Glow Plug Circuit	Open circuit, short to ground, short to voltage, glow plug
P0675	GPCM	Cylinder 5 Glow Plug Circuit	Open circuit, short to ground, short to voltage, glow plug
P0676	GPCM	Cylinder 6 Glow Plug Circuit	Open circuit, short to ground, short to voltage, glow plug
P0677	GPCM	Cylinder 7 Glow Plug Circuit	Open circuit, short to ground, short to voltage, glow plug
P0678	GPCM	Cylinder 8 Glow Plug Circuit	Open circuit, short to ground, short to voltage, glow plug
P0683	GPCM	Glow Plug Control Module to PCM Communication Circuit	Open circuit, short to voltage, short to ground
P0700	TR	Transmission Control System (MIL request)	Internal transmission fault
P0703	BOO	Brake Switch B Input Circuit	Short to ground, open circuit, short to voltage, switch
P0830	CPP	Clutch Pedal Switch A Circuit	Short to ground, open circuit, short to voltage, switch
P0833	CPP	Clutch Pedal Switch B Circuit	Short to ground, open circuit, short to voltage, switch
P1000	PCED	OBD Systems Readiness Test Not Complete	OBD monitors/drive cycle incomplete
P1001	PCED	KOER Not Able To Complete, KOER Aborted	Entry conditions for KOER self-test not met
P1102	MAF	Mass Air Flow Sensor In Range But Lower Than Expected	VPWR circuit open, return circuit open, signal short to ground, sensor
P1148	ALT	Generator 2 Control Circuit	Charging system
P1149	ALT	Generator 2 Control Circuit High	Charging system
P115A	FLI	Low Fuel Level - Forced Limited Power	Low fuel level
P1184	EOT	Engine Oil Temperature Sensor Out Of Self-Test Range	Engine too cold/hot, thermostat
P1260		Theft Detected, Vehicle Immobilized	Anti-theft system fault
P1284	ICP	Aborted KOER — Injector Control Pressure regulator	IPR valve, circuit failure

DIAGNOSTIC TROUBLE CODE DESCRIPTIONS & POSSIBLE CAUSES

Diagnostic Trouble Code (DTC) Descriptions

Fault Code	Circuit Index	DTC Description	Probable Causes
P1335	EGR	EGR Position Sensor Minimum Stop Performance	EGR valve
P1378	INJ Module	FICM Supply Voltage Circuit Low	FICM connections, relay, short to ground, charging system
P1379	INJ Module	FICM Supply Voltage Circuit High	Charging system
P1397	GPCM	System Voltage Out Of Self-Test Range	Charging system under/overcharge
P1408	EGR	Exhaust Gas Recirculation Flow Out Of Self-Test Range	EGR control circuit open, short to ground or voltage
P1464	ACCS	A/C Demand Out Of Self-Test Range	A/C circuit shorted to voltage, A/C on during self-test
P1501	VSS	Vehicle Speed Sensor Out Of Self-Test Range	VSS sensor, circuitry
P1502	PCED	Invalid Test — Auxiliary Power Control Functioning	Power take-off or auxiliary idle control active during self-test
P1531	PCED	Invalid Test — Accelerator Pedal Movement	Accelerator moved during KOER self-test
P1536	PBA	Parking Brake Switch Circuit	Circuitry, switch
P1610	PCED	Install a New PCM Module	PCM flashing error
P1611	PCED	Diagnose Further	PCM flashing error
P1615	PCED	Flash Erase Error	PCM flashing error
P1616	PCED	Flash Erase Error, Low Voltage	PCM flashing error
P1617	PCED	Block Programming Error	PCM flashing error
P1618	PCED	Block Programming Error, Low Voltage	PCM flashing error
P1633	PCED	Keep Alive Power Voltage Too Low	Open circuit, short to ground
P1635	PCM	Tire/Axle Out Of Acceptable Range	PCM programming
P1639	PCM	Vehicle ID Block Corrupted, Not programmed	PCM programming
P1703	BOO	Brake Switch Out Of Self-Test Range	Switch, circuitry, switch not cycled during self-test
P1705	TR	Transmission Range Circuit Not Indicating Park/Neutral During Self-Test	TR sensor, circuitry
P1725		Insufficient Engine Speed Increase During Self-Test	System voltage out of self-test range. Repeat the self-test.
P1726		Insufficient Engine Speed Decrease During Self-Test	System voltage out of self-test range. Repeat the self-test.
P2067	FLI	Fuel Level Sensor A Circuit Low Input	Open circuit, short circuit, fuel sending unit, instrument cluster
P2068	FLI	Fuel Level Sensor A Circuit High Input	Open circuit, short circuit, fuel sending unit, instrument cluster
P2104	APP	Throttle Actuator Control System - Forced Idle	APP sensor, signal circuit shorted
P2122	APP	Throttle/Pedal Position Sensor/Switch D Circuit Low Input	APP sensor, connections, open in signal circuit, signal circuit short to ground
P2123	APP	Throttle/Pedal Position Sensor/Switch D Circuit High Input	Short to voltage, signal return open, APP sensor
P2127	APP	Throttle/Pedal Position Sensor/Switch E Circuit Low Input	APP sensor, connection, open in signal circuit, signal circuit short to ground
P2128	APP	Throttle/Pedal Position Sensor/Switch E Circuit High Input	Short to voltage, signal return open, APP sensor
P2132	APP	Throttle/Pedal Position Sensor/Switch F Circuit Low Input	APP sensor, poor connection, open in signal circuit, signal circuit short to ground
P2133	APP	Throttle/Pedal Position Sensor/Switch F Circuit High Input	Short to voltage, signal return open, APP sensor
P2138	APP	Throttle/Pedal Position Sensor/Switch D/E Voltage Correlation	Difference in sensor readings
P2139	APP	Throttle/Pedal Position Sensor/Switch D/F Voltage Correlation	Difference in sensor readings
P2140	APP	Throttle/Pedal Position Sensor/Switch E/F Voltage Correlation	Difference in sensor readings
P2199	IAT	Intake Air Temperature 1/2 Correlation	Sensor bias, open/shorted circuit
P2262	MAP	Turbo/Super Charger Boost Pressure Not Detected — Mechanical	CAC system leaks, MAP hose disconnected, exhaust system leaks, intake system leaks

DIAGNOSTIC TROUBLE CODE DESCRIPTIONS & POSSIBLE CAUSES

Diagnostic Trouble Code (DTC) Descriptions

Fault Code	Circuit Index	DTC Description	Probable Causes
P2263	MAP	Turbo/Super Charger System Performance	CAC system leaks, MAP hose disconnected, exhaust system leaks, intake system leaks
P2269	WIF	Water In Fuel Condition	Water in fuel, connections, short circuit, WIF sensor
P2284	ICP	Injector Control Pressure Sensor Circuit Range/Performance	A difference in commanded versus actual injection oil pressure was detected.
P2285	ICP	Injector Control Pressure Sensor Circuit Low	Open circuit, short to ground, ICP sensor
P2286	ICP	Injector Control Pressure Sensor Circuit High	Short to voltage, ICP sensor
P2287	ICP	Injector Control Pressure Sensor Circuit Intermittent	ICP signal is intermittent
P2288	ICP	Injector Control Pressure Too High	ICP pressure is greater than specified value
P2289	ICP	Injector Control Pressure Too High — Engine OFF	Signal ground circuit open, ICP sensor
P2290	ICP	Injector Control Pressure Too Low	Actual pressure is less than commanded for a specified time
P2291	ICP	Injector Control Pressure Too Low — Engine Cranking	ICP cranking pressure too low
P2457	EGR Cooler	EGR Cooler Range/Performance	Engine is overheating, Cooling system has a restriction, Charge air cooler restriction, EGR Cooler restriction
P2552	FICMM	FICMM Circuit — Throttle/Fuel Inhibit Circuit	Open/short circuit
P2614	CMP	Camshaft Position Output Circuit	CMPO circuit noise, connections, circuitry
P2617	CKP	Crankshaft Position Output Circuit	CKPO circuit noise, connections, circuitry
P2623	IPR	Injector Control Pressure Regulator Circuit	Open circuit, short to ground, IPR stuck
U0101	PCED	Lost Communication With ECM/PCM A	Internal communication error
U0101	PCED	Lost Communication With TCM	Internal communication error
U0105	PCED	Lost Communication With FICM	Communication error
U0155	PCED	Lost Communication With Instrument Cluster	Communication error
U0306	PCED	Software Incompatibility With Fuel Injector Control module	Communication error

NOTES

NOTES

NOTES

Acronyms and Definitions

Note: This Acronyms and Definitions listing contains technical terms applicable to the components listed in this manual. It is not intended to be an all—inclusive dictionary of components and their functions.

ABS: Anti—lock Brake System.

A/C: Air Conditioning. A vehicle accessory system that modifies the passenger compartment air by cooling and dehydrating the air.

ACC: Air Conditioning Clutch. Indicates status of the A/C clutch.

ACCS: Air Conditioning Cycling Switch. Indicates status of the A/C cycling switch.

ACD: Air Conditioning Demand. A signal input to the PCM from the Air Conditioning control panel.

ACET: Air Conditioning Evaporator Temperature (may also be referred to as Evaporator Air Discharge Temperature).

ACP: Air Conditioning Head Pressure or A/C cycling switch input state.

ACPSW: Air Conditioning Pressure Switch.

ACP V: Air Conditioning Head Pressure Volts. A voltage input to the PCM from the ACP switch or sensor.

ACR: Air Conditioning Relay. Commanded output controlled by the PCM and acts as an A/C cutout control during heavy acceleration.

A/D: Analog—to—Digital. Analog—to—Digital signal conversion.

ADC: See ATDC.

Air Diverter: Air Diverter Valve. Part of the EAIR system. Diverts fresh air to the exhaust system when the electric air pump is commanded on.

AIR EVAL: Air System Evaluated. Displays a YES or NO status indicating whether the Air System has been evaluated for OBD (On—Board Diagnostic) II purposes.

Air/Fuel Ratio: Air to fuel mixture ratio. An air/fuel mixture that is 14.7:1 is also called stoichiometry.

Ambient Air Temperature: Temperature of the air surrounding an object.

Analog (Electrical/Electronic): An electrical signal that can obtain any value within the voltage limits of the signal.

ARB: Air Resource Board.

ARPMIDES: Ancillary RPM Desired. RPM required to maintain the vehicle speed commanded by Speed Control Command Switch (SCCS) inputs.

ASCII: American Standard Code for Information Interchange.

ATDC: After Top Dead Center. The location of the piston after it has reached the top of its stroke. Measured in degrees of crankshaft rotation.

AWD: All—Wheel Drive.

BARO: Barometric Pressure.

Base Idle: Idle rpm determined by the throttle lever hardset on the throttle body with the IAC solenoid disconnected.

Base Timing: Spark advance in degrees before top dead center of the base engine without any control from the PCM or ICM.

Acronyms and Definitions (Continued)

Battery Positive Voltage (B+): The positive (+) voltage from the battery or any circuit connected directly to the battery. Compare "Vehicle Power (VPWR)."

BATTEMP: Battery Temperature.

BJB: Battery Junction Box.

BOB: Breakout Box. A test device which connects in series to the PCM and PCM harness.

BPA: Brake Pedal Applied switch. Typically located on the braking system master cylinder. Can be hydraulic or electric.

BPP: Brake Pedal Position. Indicates the position of the brake pedal, based on input from the Brake Pedal Position (BPP) switch.

BRAKE _LMP or BRKL: Brake Warning Lamp Status. Activates the Brake Warning Lamp by applying voltage to the control line.

Bus + or Bus — : Multiplex circuits that carry SCP data from module to module and to the DLC.

CAC: Charge Air Cooler. Formerly known as Intercooler. A device which lowers the temperature of pressurized intake air.

CCM: Comprehensive Component Monitor.

CCRM: Constant Control Relay Module. A relay module that provides ON—OFF control of various EEC components.

CD A through J: Coil Driver 1 through 10.

CGND or CSE GND: Case Ground. Provides a ground source for the PCM or ECU case.

CHT: Cylinder Head Temperature. Units are displayed in either degrees Fahrenheit or Centigrade.

CHTIL: Cylinder Head Temperature Indicator Lamp.

CHTV: Cylinder Head Temperature Voltage. The actual voltage drop across the CHT sensor thermistor.

CID: Cylinder Identification. PCM input signal from Camshaft Position Sensor.

CKP: Crankshaft Position. Senses the position of the crankshaft.

CKP+, CKP—: CKP+ is the Crankshaft Position (CKP) sensor signal wire. CKP— is the signal return.

CL: Closed Loop. An operating condition or mode which enables operation based on sensor feedback.

CMP: Camshaft Position. Indicates camshaft position.

CMPFM: Camshaft Position Failure Mode. Indicates when the PCM identifies a CID/CMP fault.

CO: Carbon Monoxide. A colorless, odorless and toxic gas that is a component of auto exhaust emissions.

CO_2: Carbon Dioxide. A colorless, odorless gas that is a normal by—product of the combustion of fuel.

Coil: A device consisting of windings around an iron core. In a spark ignition system, designed to increase voltage.

Cold Soak: Time given to a vehicle to sit at a low temperature (typically below 68° F / 20° C) until the temperature of external and internal components stabilize.

Acronyms and Definitions (Continued)

CONT: Continuous Memory. The portion of KAM (keep alive memory) used to store DTCs generated during Continuous Memory Self—Test.

Continuous Memory Self—Test: A continuous test of the EEC system conducted by the PCM whenever the vehicle is operating.

CPP: Clutch Pedal Position. Indicates clutch pedal position.

CPP Switch: Clutch Pedal Position Switch. Located on the clutch pedal and detects when the clutch pedal is depressed.

CSE GND: Case Ground.

CT: Closed Throttle Mode. A mode when the PCM varies the pulse width of the fuel injectors to obtain the air/fuel mixture appropriate for closed throttle operation.

Data Communications Link: A communication path between various in—vehicle electronic modules. Accessed by scan tools through the Data Link Connector (DLC).

DCL: Data Communication Link.

DI: Distributor Ignition. A system in which the ignition coil secondary circuit is sequenced by a distributor.

Digital: Controls process information by switching the current or voltage ON and OFF.

DLC: Data Link Connector. J1962 connector providing access to vehicle diagnostic information.

DOHC: Dual Overhead Cam. An engine configuration that uses two camshafts positioned above the valves.

DOL: Data Output Line. A circuit that sends certain information from the PCM to the instrument cluster.

DTM: Diagnostic Test Mode. A level of capability in an On—Board Diagnostic (OBD) system.

DTC: Diagnostic Trouble Code. An alpha/numeric identifier for a fault condition identified by the On—Board Diagnostic System.

DVOM: Digital Volt—Ohm Meter.

ECT: Engine Coolant Temperature. Displayed in either Fahrenheit or Centigrade.

ECTV: Engine Coolant Temperature Voltage. The actual voltage drop across the ECT sensor thermistor.

ECU: Electronic Control Unit. A module that handles the control strategy and monitors system inputs or outputs.

EEC: Electronic Engine Control system.

EEC—V: Fifth generation EEC system.

EFT: Engine Fuel Temperature.

EFTA: Bank 1 input. EFTA is displayed in either Fahrenheit or Centigrade.

EFTAV: Voltage drop across the EFTA (Bank 1) sensor thermistor.

EFTB: Bank 2 input. EFTB is displayed in either Fahrenheit or Centigrade.

EFTBV: Voltage drop across the EFTB (Bank 2) sensor thermistor.

Acronyms and Definitions (Continued)

EI: Integrated Electronic Ignition. An Electronic Ignition system that has the Ignition Control Module (ICM) integrated into the PCM.
EI—HDR: Electronic Ignition, High Data Rate. Formerly known as Electronic Distributorless Ignition System.

EI—LDR: Electronic Ignition, Low Data Rate. Formerly known as Distributorless Ignition System.

EMI: Electromagnetic Interference. Usually caused by ignition voltage spikes, solenoids, relay operation or noisy generator contacts.
EOL: End Of Line. A system designed specifically for use at assembly plants to make sure all new vehicles perform to design specifications.

EPA: Environmental Protection Agency (U.S. Government).

EPROM: Erasable Programmable Read—Only Memory. An electronic component in the PCM that requires the electronic storage of information.

ESOF: Electronic Shift—on—the—Fly.

EVAP: Evaporative Emissions. A system to prevent fuel vapor from escaping into the atmosphere.

EVAPCP: Evaporative Canister Purge Solenoid. Controls a solenoid which allows venting of the evaporative purge canister.
EVAPCPF: Evaporative Canister Purge Solenoid Fault. Identifies whether an electrical fault exists for the current commanded state.
EVAPCV: Evaporative Canister Vent Solenoid. Controls a solenoid which seals the EVAP system canister from atmospheric pressure during the EVAP OBD II Monitor test.
Evaporative Emissions Canister: An evaporative emission canister, containing activated charcoal which absorbs and holds fuel vapors.
EVAPPDC: Evaporative Canister Purge Duty Cycle. The duty cycle commanded to the Evap Canister Purge Solenoid by the PCM.

EVO: Electronic Variable Orifice.

Exciter Ring: A toothed or notched iron or steel disk, which is the movcable part of a wheel speed sensor.
FAN: Fan Speed. Used in conjunction with vehicles having multiple fan speed control. Displays OFF, LOW, or HIGH status.

FANSS: Fan Speed Sensor.

FC: Fan Control.

FCS: Fuel Control Solenoid.

FCIL: Fuel Cap Off Indicator Lamp. Indicates that the fuel filler cap was not properly installed.

FEAD: Front End Accessory Drive.

FEPS: Flash EEPROM Programming Signal. 18 volt DC signal sent by the scan tool to initiate PCM reprogramming.
FLI: Fuel Level Input. Used by the Evap monitor to calculate fuel tank vapor volume. Displayed as a percentage.

FLI V: Fuel Level Input Voltage.

FMEM: Failure Mode Effects Management. Operating strategy that maintains limited vehicle function in the event of a PCM or EEC component failure.

Acronyms and Definitions (Continued)

FP: 1. Fuel Pump. Indicates whether the pump has been commanded ON or OFF by the PCM. 2. Fuel Pump (Modulated). Fuel pump duty cycle percentage.

FPDM: Fuel Pump Driver Module. A module that controls the electric fuel pump.

FPF: Fuel Pump Fault. Identifies whether a fault exists in the FP circuit.

FPM: Fuel Pump Monitor. Monitors the Fuel Pump / circuits for faults.

Freeze Frame: A block of memory containing the vehicle operating conditions at a specific time.

FRP: Fuel Rail Pressure. Based on FRP V.

FRP V: Fuel Rail Pressure Voltage. A voltage input to the PCM from the Fuel Rail Pressure Sensor.

FSC: Fail—Safe Cooling.

FSV: Fuel Shut—Off Valve. A component of Natural Gas Vehicles. This valve either allows or prevents Natural Gas flow to the fuel rail.

FSVF: Fuel Shut—Off Valve Fault. Indicates if there is a fault in the FSV circuit. Displayed as YES or NO.

FSVM: Fuel Shut—Off Valve Monitor. Monitors operation of the Fuel Shut—Off Valve / circuit.

FTP: Fuel Tank Pressure. Displayed as inches of water, kPa, or volts.

FTP V: Fuel Tank Pressure Voltage. From the FTP transducer.

FUEL PR: Fuel Pressure. Measurement of the force of the fuel delivered via the fuel pump.

FUELPW: Fuel Pulse Width. Displays the commanded pulse width at time of last data update.

FUELPW1: Fuel Injector Pulse Width #1. Corresponds to injectors normally affected by O2S1 (HEGO1).

FUELPW2: Fuel Injector Pulse Width #2. Corresponds to injectors normally affected by O2S2 (HEGO2).

FUELSYS: Fuel System Status (OPEN/CLOSED Loop). Formerly known as LOOP.

Fuel Tank Vapor Valve: A valve mounted in the top of the fuel tank that vents excess vapor and pressure from the fuel tank into the Evaporative Emission Control System.

GEM: Generic Electronic Module.

GEN: Generator.

GENF: Generator output fault.

GENFDC: Generator field control output.

GFS: Generator field signal monitor.

GND: Ground.

GPM: Grams Per Mile. Also known as Gallons Per Minute.

GSS: Gear Select Solenoid.

Acronyms and Definitions (Continued)

GVW: Gross Vehicle Weight.

Hall Effect: A process where current is passed through a small slice of semi—conductor material and a magnetic field to produce a small voltage in the semi—conductor.

Hard Fault: A fault currently present in the system.

HC: 1. Hydrocarbon. A by—product of combustion and a component of auto exhaust emissions. 2. High Compression.

HCF: Hydraulic Cooling Fan.

HCFD: Hydraulic Cooling Fan Drive.

HFC: Hydraulic Cooling Fan.

HFCF: High Fan Control Fault. Identifies if there is a fault in the HFC circuit.

HFP: High Fuel Pump.

HLOS: Hardware Limited Operating Strategy. A mode of operation where the PCM replaces output commands with fixed values in response to internal PCM malfunctions.

HO: High Output.

Hot Soak: Period of time after an engine operates where localized combustion heat dissipates throughout the engine.

Hz: Hertz. Cycles per second.

IAC: Idle Air Control. Electrical control of throttle bypass air.

IAT: Intake Air Temperature.

IATV: Intake Air Temperature Voltage. Actual voltage drop across the IAT sensor.

IAT2: Intake Air Temperature 2. Displayed in either Fahrenheit or Centigrade. Used on supercharged vehicles.

IAT2V: Intake Air Temperature 2 Voltage. Actual voltage drop across the IATV2 sensor.

IC: Integrated Circuit. A small semi—conductor device capable of doing many separate circuit functions.

ICM: Ignition Control Module. The module that controls the ignition system.

IFDM: Integrated Fuel Delivery Module.

IFS: Inertia Fuel Shutoff.

IGN GND: Ignition Ground.

Ignition: System used to provide high voltage spark for internal combustion engines.

IGN_KEY (IGKY): Ignition Key status.

IGN_SW (IGSW): Ignition Switch Position.

Acronyms and Definitions (Continued)

IMRC: Intake Manifold Runner Control. Controls airflow through the high—speed runners in the intake manifold.

IMRCM: Intake Manifold Runner Control Monitor. Monitors the IMRC / circuits for faults.

IMTV: Intake Manifold Tuning Valve. Controls airflow through runners in a split intake manifold.

INJ1, INJ2, INJ3, INJ4, INJ5, INJ6, INJ7, INJ8, INJ9, INJ10: Injector number or its signal output from the PCM.

Injector: A device for delivering metered pressurized fuel to the intake system or the cylinders.

Intake Air: Air drawn through a filter and distributed to each cylinder for use in combustion.

Intercooler: See CAC.

IPATS: Integrated Passive Anti—Theft System.

ISO: International Standards Organization.

KAM: Keep Alive Memory. A portion of the memory within the PCM that must have power even when the vehicle is not operating.

KAPWR: Keep Alive Power. Dedicated, unswitched power circuit that maintains KAM.

Key On Engine Off Self—Test: A test of the EEC system conducted by the PCM with power applied and the engine at rest.

Key On Engine Running Self—Test: A test of the EEC system conducted by the PCM with the engine running and the vehicle at rest.

KEYPWR: Key Power. Battery voltage supplied when the ignition key is in the ON position.

Knock: The sharp metallic sound produced when two combustion pressure fronts collide in the combustion chamber of an engine.

KOEC: Key On Engine Continuous.

KOEO: Key On Engine Off.

KOER: Key On Engine Running.

KPA: Kilopascal. Unit of pressure. 3.386 kPa = 1 inch of mercury (Hg.).

KPH: Kilometers Per Hour.

KS: Knock Sensor. Detects engine knock.

L: Liters. The unit of volume in the metric measuring system. One liter equals 1.06 quarts.

LFC: Low Fan Control.

LFP: Low Fuel Pump. Reduced operating speed for multi—speed fuel pumps.

LOOP: Indicates OPEN or CLOSED loop status.

LPLR: Low Pressure Low Resistance fuel injector.

Acronyms and Definitions (Continued)

MAF: Mass Air Flow. Used to measure the mass (weight) of the air entering the engine.

MAF RTN: Mass Air Flow Return. A return circuit for the MAF sensor.

MAP: Manifold Absolute Pressure. The internal pressure of the intake manifold.

MFC: Medium Fan Control.

MFI: Multiport Fuel Injection. A fuel—delivery system in which each cylinder is individually fueled.

MFP: Modulated Fuel Pump.

Microprocessor: A digital processor on a chip which performs arithmetic and control logic.

MIL: Malfunction Indicator Lamp. An indicator lamp alerting the driver of an emission related malfunction. May also read "CHECK ENGINE" or "SERVICE ENGINE SOON."

MISF: Misfire. Any event in the cylinder that causes a sudden change in acceleration of the crankshaft.

Monolithic Substrate: The ceramic honeycomb structure used in the catalytic converter.

MY: Model Year.

NA: Naturally Aspirated. Engine that is not supercharged or turbocharged.

NC: Normally Closed.

NGS: New Generation STAR (Self—Test Automatic Readout) tester.

NO: Normally Open.

NO_x: Oxides of Nitrogen. Formed at high combustion temperatures.

NVH: Noise, Vibration, Harshness. A classification of vehicle concerns.

OASIS: On—line Automotive Service Information System.

OBD, OBD—II: On—Board Diagnostics, On—Board Diagnostics Second Generation. A system that monitors PCM input and output control signals.

On—Demand Test: Technician initiated "KOEO" and "KOER" tests performed by the PCM.

OEM: Original Equipment Manufacturer.

OHC: OverHead Cam. An engine configuration that uses a single camshaft positioned above the valves.

OWL: Overheat Warning Lamp or its signal output from the PCM. Turns the TEMP warning lamp ON when engine oil temperature exceeds safe limits.

Open Circuit: A circuit which does not provide a complete path for flow of current.

OL: Open Loop. An operating condition based on instructions not modified by PCM feedback.

OSC: Output State Control.

Acronyms and Definitions (Continued)

OSS: Output Shaft Speed.

Particulate: Small solid matter found in exhaust gases, especially prevalent in diesel engines.

PATS: Passive Anti—Theft System.

PATSIL: Passive Anti—Theft System Indicator Light.

PATSIN: Passive Anti—Theft System Receive Signal.

PATSOUT: Passive Anti—Theft System Transmit Signal.

PATSTRT: Passive Anti—Theft System Starter Relay Control

PCM: Powertrain Control Module. Formerly known as the EEC (Electronic Engine Control) Processor.

PF: Purge Flow. Amount of fuel vapor burned in the engine.

Photochemical: Term describing the action of light on air pollutants which results in creating smog.

PID: Parameter Identifier. Identifies an address in PCM memory which contains operating information.

Powertrain: Engine and transmission/transaxle components.

Pressure — Absolute: A pressure referenced to a perfect vacuum.

Pressure — Atmospheric: The pressure of the surrounding air at any given temperature and altitude. Sometimes called Barometric Pressure.

Pressure — Barometric: Pertaining to atmospheric pressure or the results obtained by a barometer.

Pressure — Differential: The pressure difference between two regions, such as between the intake manifold and atmospheric pressure.

Pressure — Gage: The amount by which absolute pressure exceeds the ambient atmospheric pressure.

Potentiometer: An adjustable resistance component commonly used as a sensor (Example: TP Sensor).

PPM: Parts Per Million. A measure used in emission analysis.

PROM: Programmable Read—Only Memory. Similar to ROM except without program instructions.

Protocol: A set of rules for the exchange of information on a network.

PSOM: Programmable Speedometer/Odometer Module. A module that processes vehicle speed information.

PSP: Power Steering Pressure. Indicates the pressure in the power steering system.

PSP V: Power Steering Pressure Input Voltage.

PTEC: PowerTrain Electronic Controller.

PTO: Power Take—Off.

Acronyms and Definitions (Continued)

PW: Pulse Width. The length of time an actuator, such as a fuel injector, remains energized.

PWM: Pulse Width Modulation. Controls the intensity of an output by varying the signal duty cycle.

PWR GND: Power Ground. The main ground circuit in the EEC system.

Quick Test: A series of diagnostic tests of the EEC system consisting of KOEO, KOER and Continuous Memory Self—Tests. Results are displayed as a series of DTCs.

RABS: Rear Antilock Brake System.

RAM: Random Access Memory. Memory into which information can be written as well as read.

Regulator: Controls the alternator/generator field current to maintain proper battery charge. Contained within the PCM in smart charging applications.

Relay: An electromechanical device in which connections in one circuit are opened or closed by changes in another circuit.

REM: Rear Electronic Module.

Repetitive Spark: Multiple firings of individual spark plugs at engine speeds below 1000 RPM to improve idle quality and improve emissions.

RF: Radio Frequency.

RFI: Radio Frequency Interference.

RFS: Returnless Fuel System.

RM: Relay Module. A module containing two or more relays.

ROM: Read—Only Memory. Computer memory that can be accessed and utilized, but not altered.

RON: Research Octane Number.

Routine: A group of related tasks, such as a series of diagnostic tests.

RPM: Revolutions Per Minute.

RS: Reverse Switch.

RTN: Return. A dedicated sensor ground circuit.

RWD: Rear Wheel Drive.

SCP: Standard Corporate Protocol.

Self—Test: See Quick Test.

Sensor: A device that detects the value or change in a physical quantity, such as temperature, pressure or flow rate, and converts the data into an electrical signal.

SFI: Sequential Multiport Fuel Injection. A multiport fuel delivery system where each injector is individually energized and timed relative to its cylinder intake event.

Shield: A conducting sleeve that surrounds wires to be electronically isolated from electromagnetic interference (EMI).

Acronyms and Definitions (Continued)

Short Circuit: An undesirable condition in a circuit where it is terminated at a point other than that intended.

SHRT FT: Short—Term Fuel Trim. Fuel flow adjustment in response to the HO2S sensor(s) input during closed—loop operation.

SIG RTN: Signal Return. A dedicated sensor ground circuit that is common to two or more sensors.

SIL: Shift Indicator Lamp.

Smart Driver: A PCM or ECU output driver that can detect faults (open or shorts) on its output circuit.

SME: Society of Manufacturing Engineers.

SOF: Shift—On—the—Fly.

SOHC: Single Overhead Cam.

Solenoid: A device consisting of an electrical coil which produces a magnetic field in a plunger and pulled to a central position.

ST: Scan Tool. A device that interfaces with and communicates information on a data link.

Stoichiometry: An air/fuel mixture that is neither too rich nor too lean. Stoichiometric ratio is 14.7 parts of air for every 1 part of fuel.

Switch: A device for making, breaking, or changing the connections in an electrical circuit.

TA: Traction Assist.

TACH: Tachometer.

TC: 1. Traction Control. Combines anti—lock braking and axle torque reduction to control wheel slippage. 2. Turbocharger.

TDC: Top Dead Center.

Tear Tag: The two—piece adhesive label attached to the PCM to identify its calibration.

Thermistor: A temperature dependent resistor, like that used in CHT and ECT sensors.

Timing: Relationship between spark plug firing and piston position expressed in crankshaft degrees before (BTDC) or after (ATDC) top dead center of the compression stroke.

TMAP: Thermal Manifold Absolute Pressure Sensor. A MAP Sensor that includes a thermistor to measure intake air temperature.

TP: Throttle Position (sensor). A three—wire potentiometer that provides throttle angle and rate information for the PCM.

TP V: Throttle Position Sensor Voltage.

Transducer: A device that receives energy from one medium and transfers it to another. For example, thermal energy is converted to an electrical signal through a temperature probe.

TSB: Technical Service Bulletin. Notifies service personnel of any known vehicle concerns, procedures, or general service information.

Underspeed Mode: A control mode that prevents the engine from stalling in the event it stumbles while running. Also used during engine crank.

Vacuum: Manifold pressure that is reduced below the ambient atmospheric pressure.

Acronyms and Definitions (Continued)

Variable Reluctance: A process of passing a varying magnetic field through wire windings and inducing a voltage.

VCT: Variable Camshaft Timing.

VDF: Visctronic Drive Fan.

VECI: Vehicle Emission Control Information label.

VIN: Vehicle Identification Number. A unique identification number given to every vehicle produced. Includes information about the year, model, engine, and plant origin of the vehicle.

VMV: Vapor Management Valve. Controls the flow of fuel vapors out of the carbon canister.

VOM: Volt—Ohm Meter. Readings are indicated by sweep hand on a printed scale rather than a digital (DVOM) display.

VBPWR: Vehicle Buffered Power. A PCM supplied power source that supplies regulated voltage.

VPWR: Vehicle Power. A switched circuit that provides power to the EEC system. Compare "Battery Voltage (B+)."

VREF: Reference Voltage. A dedicated circuit that provides approximately a 5.0 volt signal used as a reference by certain sensors.

WAC: Wide Open Throttle A/C Cut—Off. Turns A/C system off during wide open throttle or certain other operating conditions.

Wastegate Control: A device that opens the wastegate in case of overboost from a turbocharger.

WOT: Wide Open Throttle. A condition of maximum airflow through the throttle body.

Zip Tube: Another name for "fresh air duct" or "air inlet duct".

4x4L: 4x4 Low.

A/T: Automatic Transmission.

CCS: Coast Clutch Solenoid.

CCSF: Coast Clutch Solenoid Fault. Displays a YES if fault exists.

EPC: Electronic Pressure Control.

EPCV: Electronic Pressure Control Volts.

ESS: Electronic Shift Scheduling.

M5OD: Manual 5—Speed transmission with overdrive (RWD).

M/T: Manual Transmission/Transaxle.

NPS: Neutral Pressure Switch or its signal input to the PCM.

OCS: Overdrive Cancel Switch.

OSS: Output Shaft Speed. Indicates rotational speed of the transmission output shaft.

Acronyms and Definitions (Continued)

PNP: Park/Neutral Position switch. Also known as Neutral Drive Switch (NDS), Neutral Gear Switch (NGS), and Transmission Switch Neutral (TSN).

REVERSE or REV: Transmission Reverse Switch Input.

SIL: Shift Indicator Lamp. A lamp that indicates the preferred shift points on select manual transmission/transaxle vehicles.

SS1/SS2/SS3: Shift solenoids. Devices that control the shifting in an automatic transmission.

TCC: Torque Converter Clutch. When energized, causes a mechanical engagement and disengagement of the Torque Converter Clutch.

TCIL: Transmission Control Indicator Lamp. Indicates that the TCS has been activated.

TCS: Transmission Control Switch. Modifies the operation of electronically controlled transmissions.

Torque converter: A device which by its design multiplies the torque in a fluid coupling between an engine and transmission/transaxle.

TFT: Transmission Fluid Temperature. Indicates temperature of transmission fluid.

Transmission: A device which selectively increases or decreases the ratio of relative rotation between its input and output shafts. Rear—wheel drive applications.

TR: Transmission Range. The range in which the transmission is operating.

TR Sensor: Formerly known as Manual Lever Position Sensor (MLPS). Provides information to the PCM on the transmission range selector position.

TR V: Transmission Range Voltage.

TSS: Turbine Shaft Speed. Indicates rotational speed of the transmission turbine shaft.

VSS: Vehicle Speed Sensor. A magnetic pickup device that generates an AC signal that is proportional to vehicle speed.